Yangians and Classical Lie Algebras

Mathematical
Surveys
and
Monographs

Volume 143

Yangians and Classical Lie Algebras

Alexander Molev

American Mathematical Society

2000 *Mathematics Subject Classification.* Primary 17B37, 81R50; Secondary 17B10, 17B20, 81R10, 05E10, 05E15.

For additional information and updates on this book, visit
www.ams.org/bookpages/SURV-143

Library of Congress Cataloging-in-Publication Data

Molev, Alexander, 1961–
Yangians and classical Lie algebras / Alexander Molev.
p. cm. — (Mathematical surveys and monographs, ISSN 0076-5376 ; v. 143)
Includes bibliographical references and index.
ISBN 978-0-8218-4374-1 (alk. paper)
1. Lie algebras. 2. Quantum groups. 3. Matrices. 4. Representations of algebras. 5. Symmetry (Mathematics). I. Title.

QA252.3.M65 2007
512′.482—dc22 2007060781

10 9 8 7 6 5 4 3 2 1 12 11 10 09 08 07

Посвящается Б.-Я.

Contents

Preface

The theory of classical matrix Lie algebras can be viewed from at least two related but different perspectives. On the one hand, the special linear, orthogonal and symplectic Lie algebras form four infinite series, A_n, B_n, C_n, D_n, which together with five exceptional Lie algebras, E_6, E_7, E_8, F_4, G_2, comprise a complete list of the simple Lie algebras over the field of complex numbers. The structure of these Lie algebras is uniformly described in terms of certain finite sets of vectors in a Euclidean space called the root systems. The symmetries of the root systems play a key role in the representation theory of all simple Lie algebras providing the dimension and character formulas for the representations. On the other hand, the matrix realizations of the classical Lie algebras allow some specific tools to be used for their study which are not always available for the exceptional Lie algebras. The theory of *Yangians* and *twisted Yangians* which we develop in this book is one of such tools bringing in new symmetries and shedding new light on this classical subject.

The Yangians and twisted Yangians are associative algebras whose defining relations are written in a specific matrix form. We describe the structure of these algebras and classify their finite-dimensional irreducible representations. The results exhibit many analogies with the representation theory of the classical Lie algebras themselves, including the triangular decompositions of the (twisted) Yangians and the parametrization of the representations by their highest weights. In the simplest cases explicit constructions of the irreducible representations are also given. Then we apply the Yangian symmetries to the classical Lie algebras. The applications include constructions of several families of Casimir elements, derivations of the characteristic identities and Capelli identities, and explicit constructions of all finite-dimensional irreducible representations of the classical Lie algebras via weight bases of Gelfand–Tsetlin type.

Let us discuss the relationship between the classical Lie algebras and the (twisted) Yangians in more detail. The term *Yangian* was introduced by V. G. Drinfeld (in honor of C. N. Yang) in his fundamental paper (1985). In that paper, Drinfeld also defined the *quantized Kac–Moody algebras*, which together with the work of M. Jimbo (1985), who introduced these algebras independently, marked the beginning of the era of *quantum groups*. The Yangians form a remarkable family of quantum groups related to rational solutions of the classical Yang–Baxter equation. For each simple finite-dimensional Lie algebra $\mathfrak{a}$ over the field $\mathbb{C}$ of complex numbers, the corresponding Yangian is defined as a canonical deformation of the universal enveloping algebra $\mathrm{U}(\mathfrak{a}[z])$ for the polynomial current Lie algebra $\mathfrak{a}[z]$. Importantly, the deformation is considered in the class of Hopf algebras, which guarantees its uniqueness under some natural homogeneity conditions. Another presentation of the Yangian for $\mathfrak{a}$ was given later by Drinfeld (1988).

A few years earlier, the algebra, which is now called the *Yangian for the general linear Lie algebra* $\mathfrak{gl}_N$ and denoted by $\mathrm{Y}(\mathfrak{gl}_N)$, was considered in the work of L. D. Faddeev and the St. Petersburg (Leningrad) school. The defining relations of the Yangian $\mathrm{Y}(\mathfrak{gl}_N)$ can be written in the form of a single *ternary* (or *RTT*) relation on the matrix of generators. This relation has a rich and extensive background. It originates from the *quantum inverse scattering method*; see for instance L. A. Takhtajan and Faddeev (1979), P. P. Kulish and E. K. Sklyanin (1982), and Faddeev (1984). The Yangians were primarily regarded as a vehicle for producing rational solutions of the Yang–Baxter equation; cf. Drinfeld (1985). Conversely, the ternary relation is a powerful tool for studying quantum groups themselves; see e.g. N. Yu. Reshetikhin, Takhtajan and Faddeev (1990). The Hopf algebra structure of $\mathrm{Y}(\mathfrak{gl}_N)$ can also be conveniently described in a matrix form.

From the algebraic point of view, the algebra $\mathrm{Y}(\mathfrak{gl}_N)$ and the closely related Yangian $\mathrm{Y}(\mathfrak{sl}_N)$ for the special linear Lie algebra $\mathfrak{sl}_N$ are exceptional in the following sense. For any simple Lie algebra $\mathfrak{a}$, the corresponding Yangian contains the universal enveloping algebra $\mathrm{U}(\mathfrak{a})$ as a subalgebra. However, only in the case $\mathfrak{a} = \mathfrak{sl}_N$ does there exist a homomorphism from the Yangian to $\mathrm{U}(\mathfrak{a})$ (the *evaluation homomorphism*) which is identical on the subalgebra $\mathrm{U}(\mathfrak{a})$ (Drinfeld, 1985). This property plays a key role in the applications of the Yangians to the conventional representation theory. In this book we concentrate on these distinguished algebras $\mathrm{Y}(\mathfrak{gl}_N)$ and $\mathrm{Y}(\mathfrak{sl}_N)$.

We will use the symbol $\mathfrak{g}_N$ to denote either the orthogonal Lie algebra $\mathfrak{o}_N$ or symplectic Lie algebra $\mathfrak{sp}_N$, assuming $N = 2n$ even for the latter. For each of these Lie algebras G. Olshanski (1992) introduced another algebra which he called the *twisted Yangian*. We will denote it by $\mathrm{Y}(\mathfrak{g}_N)$. When $\mathfrak{a} = \mathfrak{g}_N$, the twisted Yangian $\mathrm{Y}(\mathfrak{g}_N)$ should not be confused with the Yangian for $\mathfrak{g}_N$ defined by Drinfeld. The latter Yangian will not be considered in the main exposition of the present book; see, however, Examples 2.16.2 and 4.6.1.

The classical Lie algebra $\mathfrak{g}_N$ can be regarded as a fixed point subalgebra of an appropriate involution σ of the Lie algebra $\mathfrak{gl}_N$. Then the twisted Yangian $\mathrm{Y}(\mathfrak{g}_N)$ can be defined as a subalgebra of $\mathrm{Y}(\mathfrak{gl}_N)$. The algebra $\mathrm{Y}(\mathfrak{g}_N)$ is a deformation of the universal enveloping algebra for the twisted polynomial current Lie algebra

$$\mathfrak{gl}_N[z]^\sigma = \{\, A(z) \in \mathfrak{gl}_N[z] \mid \sigma\big(A(z)\big) = A(-z)\,\}.$$

This is *not* a Hopf algebra deformation. However, the twisted Yangian $\mathrm{Y}(\mathfrak{g}_N)$ contains the universal enveloping algebra $\mathrm{U}(\mathfrak{g}_N)$ as a subalgebra, and there exists a homomorphism $\mathrm{Y}(\mathfrak{g}_N) \to \mathrm{U}(\mathfrak{g}_N)$ identical on the subalgebra $\mathrm{U}(\mathfrak{g}_N)$. It is called the *evaluation homomorphism* by analogy with the $\mathfrak{gl}_N$ case. Moreover, the twisted Yangian turns out to be a (left) coideal of the Hopf algebra $\mathrm{Y}(\mathfrak{gl}_N)$.

Similar to the Yangian for $\mathfrak{gl}_N$, the twisted Yangians can be equivalently presented by generators and defining relations which can be written as a *quaternary* (or *reflection*) equation for the matrix of generators, together with a *symmetry* relation. Relations of this type appeared for the first time in the papers by I. V. Cherednik (1984) and Sklyanin (1988), where integrable systems with boundary conditions were studied.

This matrix form of the defining relations for the Yangian and the twisted Yangians allows special algebraic techniques (the so-called *R-matrix formalism*) to be used to describe the structure and to study representations of these algebras. On the other hand, the defining relations can also be observed inside the enveloping algebras. To be more precise, consider the general linear Lie algebra $\mathfrak{gl}_N$ with its standard basis E_{ij}, $i,j = 1,\dots,N$. The commutation relations are given by

$$[E_{ij}, E_{kl}] = \delta_{kj} E_{il} - \delta_{il} E_{kj}, \tag{0.1}$$

where δ_{ij} is the Kronecker delta. Introduce the $N \times N$ matrix E whose ij-th entry is E_{ij}. The matrix elements of the powers of the matrix E are known to satisfy the relations

$$[E_{ij}, (E^s)_{kl}] = \delta_{kj} (E^s)_{il} - \delta_{il} (E^s)_{kj}. \tag{0.2}$$

These, in particular, imply that the traces of powers of E are central elements of the universal enveloping algebra $\mathrm{U}(\mathfrak{gl}_N)$ known as the *Gelfand invariants*. The following generalization of (0.2) which can be verified by induction, appears to be less known:

$$[(E^{r+1})_{ij}, (E^s)_{kl}] - [(E^r)_{ij}, (E^{s+1})_{kl}] = (E^r)_{kj}(E^s)_{il} - (E^s)_{kj}(E^r)_{il}, \tag{0.3}$$

where $r, s \geqslant 0$ and $E^0 = 1$ is the identity matrix. The definition of the Yangian $\mathrm{Y}(\mathfrak{gl}_N)$ can be motivated by these relations: replacing $(E^r)_{ij}$ by an abstract generator $t_{ij}^{(r)}$ we obtain the Yangian defining relations; see (1.1) below. Introducing the generating series

$$e_{ij}(u) = \delta_{ij} + \sum_{r=1}^{\infty} (E^r)_{ij} u^{-r},$$

where u is a formal (complex) parameter, we can rewrite (0.3) in the form

$$(u - v)\,[e_{ij}(u), e_{kl}(v)] = e_{kj}(u)\, e_{il}(v) - e_{kj}(v)\, e_{il}(u) \tag{0.4}$$

which is equivalent to the *RTT* relation; see (1.19) below.

Alternatively, the generators of the Yangian can be realized as the *Capelli minors*. Keeping the notation E for the matrix of the basis elements of $\mathfrak{gl}_N$, we introduce the *Capelli determinant*

$$\det(1+Eu^{-1}) = \sum_{p\in\mathfrak{S}_N} \operatorname{sgn} p\cdot(1+Eu^{-1})_{p(1),1}\cdots(1+E(u-N+1)^{-1})_{p(N),N}. \tag{0.5}$$

When multiplied by $u(u-1)\cdots(u-N+1)$ this determinant becomes a polynomial in u whose coefficients (with the exception of the leading coefficient 1) constitute a family of algebraically independent generators of the center of $\mathrm{U}(\mathfrak{gl}_N)$. The value of this polynomial at $u = N-1$ is the distinguished central element whose image in a natural representation of $\mathfrak{gl}_N$ by differential operators is given by the celebrated *Capelli identity*. For a positive integer $M \leqslant N$ introduce the subsets of indices $\mathcal{B}_i = \{i, M+1, M+2, \dots, N\}$ and for any $1 \leqslant i, j \leqslant M$ consider the Capelli minor

$$\det(1 + Eu^{-1})_{\mathcal{B}_i\mathcal{B}_j} = \delta_{ij} + c_{ij}^{(1)} u^{-1} + c_{ij}^{(2)} u^{-2} + \cdots$$

defined as in (0.5), whose rows and columns are respectively enumerated by $\mathcal{B}_i$ and $\mathcal{B}_j$. These minors turn out to satisfy the Yangian defining relations; i.e., there is an algebra homomorphism

$$\mathrm{Y}(\mathfrak{gl}_M) \to \mathrm{U}(\mathfrak{gl}_N), \qquad t_{ij}^{(r)} \mapsto c_{ij}^{(r)}.$$

These two interpretations of the Yangian defining relations (which will reappear in Sections 1.4 and 1.12) indicate a close relationship between the representation theory of the algebra $\mathrm{Y}(\mathfrak{gl}_N)$ and the representation theory of the general linear Lie algebra.

Similar calculations applied to the orthogonal and symplectic Lie algebras lead to the defining relations for the corresponding *twisted Yangians*. For instance, consider the orthogonal Lie algebra $\mathfrak{o}_N$ as the subalgebra of $\mathfrak{gl}_N$ spanned by the skew-symmetric matrices. The elements $F_{ij} = E_{ij} - E_{ji}$ with $i < j$ form a basis of $\mathfrak{o}_N$. Introduce the $N \times N$ matrix F whose ij-th entry is F_{ij}. The matrix elements of the powers of the matrix F are known to satisfy the relations

$$(0.6) \qquad [F_{ij}, (F^s)_{kl}] = \delta_{kj}(F^s)_{il} - \delta_{il}(F^s)_{kj} - \delta_{ik}(F^s)_{jl} + \delta_{lj}(F^s)_{ki}.$$

A counterpart of (0.3) for the elements $(F^r)_{ij}$ can be explicitly written down. Introduce the generating series

$$f_{ij}(u) = \delta_{ij} + \sum_{r=1}^{\infty} (F^r)_{ij} \Big(u + \frac{N-1}{2}\Big)^{-r}.$$

Then we have the relations for these series analogous to (0.4):

$$(0.7) \qquad \begin{aligned}(u^2 - v^2)\,[f_{ij}(u), f_{kl}(v)] = (u+v)\,\big(f_{kj}(u)\,f_{il}(v) - f_{kj}(v)\,f_{il}(u)\big)\\ - (u-v)\,\big(f_{ik}(u)\,f_{jl}(v) - f_{ki}(v)\,f_{lj}(u)\big)\\ + f_{ki}(u)\,f_{jl}(v) - f_{ki}(v)\,f_{jl}(u).\end{aligned}$$

This motivates the defining relations of the twisted Yangians; see (2.6) below. They also include a symmetry relation which reflects the fact that the matrix F is skew-symmetric.

We now describe the contents of the book in more detail. Chapters 1 and 2 contain a detailed exposition of the algebraic properties of the Yangian $\mathrm{Y}(\mathfrak{gl}_N)$ and the twisted Yangian $\mathrm{Y}(\mathfrak{g}_N)$. We develop the R-matrix techniques as a powerful instrument to investigate the structure of these algebras. The key results there are the constructions of special formal power series called the *quantum determinant* and the *Sklyanin determinant* originating from the works of A. G. Izergin and V. E. Korepin (1981), Kulish and Sklyanin (1982), and Olshanski (1992) (more detailed discussions of the origins of these constructions and other results contained in this book can be found in the bibliographical notes at the end of each chapter). The coefficients of these power series generate the centers of the Yangian and twisted Yangian, respectively. The *quantum Liouville formula*, which is originally due to M. L. Nazarov (1991), explicit formulas for the quantum determinant and the Sklyanin determinant, as well as factorizations of these determinants will be important for the applications to the corresponding classical Lie algebras.

In Chapters 3 and 4 we prove classification theorems for the irreducible finite-dimensional representations of the algebras $\mathrm{Y}(\mathfrak{gl}_N)$ and $\mathrm{Y}(\mathfrak{g}_N)$, respectively. For the Yangian $\mathrm{Y}(\mathfrak{gl}_N)$ the classification results are a part of the Drinfeld theorem (1988). Our approach employs the *RTT* presentation of the Yangian, and it is based on the original work of V. O. Tarasov (1985, 1986). Note that an alternative exposition of the Yangian representation theory was given in the book by V. Chari and A. Pressley (1994, Chapter 12), whose methods rely on the *Drinfeld presentation* of the Yangians. In Chapter 3 we give a proof of the isomorphism theorem between the two presentations of $\mathrm{Y}(\mathfrak{gl}_N)$ following J. Brundan and A. Kleshchev (2005). For both the Yangian and twisted Yangians we give a complete description of the irreducible finite-dimensional representations in terms of their *highest weights* and *Drinfeld polynomials*. In the simplest case $N = 2$ explicit constructions of all such representations as tensor products of the evaluation modules are also given.

Chapters 5 and 6 are devoted to explicit constructions of finite-dimensional irreducible representations of the Yangian $\mathrm{Y}(\mathfrak{gl}_N)$. For a wide class of such representations it is possible to construct a basis and produce explicit formulas for the matrix elements of the generators of the Yangian in this basis (Chapter 5). The crucial observation here is the fact that the *lowering operators* for the reduction $\mathrm{Y}(\mathfrak{gl}_N) \downarrow \mathrm{Y}(\mathfrak{gl}_{N-1})$ can be written in terms of quantum minors. This basis may be regarded as a generalization of the original basis of I. M. Gelfand and M. L. Tsetlin (1950) for the representations of the Lie algebra $\mathfrak{gl}_N$. The techniques of lowering operators are developed further in Chapter 6, where we prove an irreducibility criterion for tensor products of the Yangian evaluation modules. An essential ingredient here is the *fusion procedure* for the symmetric group due to Cherednik (1986), which we also discuss in detail. It allows us to apply Weyl's tensor approach to the Yangian evaluation modules and realize them as submodules in the tensor products of the vector representations of the Lie algebra $\mathfrak{gl}_N$. This makes it possible to establish an important *binary property* for the tensor product modules due to Nazarov and Tarasov (2002).

In the remaining part of the book (Chapters 7–9) we consider various applications of the Yangian theory to the classical Lie algebras. Taking the images of the central elements of the (twisted) Yangian onto the corresponding classical enveloping algebra with respect to the evaluation homomorphism, we get Casimir elements for the corresponding Lie algebra. Several families of Casimir elements are discussed in Chapter 7. Many of them are well known; some have appeared quite recently. The Yangian perspective provides a unifying picture of all these families and relations between them. We also consider the images of the Casimir elements under the natural representations of the Lie algebras in the differential operators, thus providing various generalizations of the classical Capelli identity. These include the *higher Capelli identities* originally discovered by A. Okounkov (1996), which are related to a distinguished linear basis of the center of $\mathrm{U}(\mathfrak{gl}_N)$ formed by the *quantum immanants*.

Chapter 8 contains an account of the *centralizer construction*, which was the original motivation for Olshanski to discover the twisted Yangians. In order to explain the main ideas of the construction and its applications to the weight bases in Chapter 9, consider a complex reductive Lie algebra $\mathfrak{g}$ and let $\mathfrak{a} \subset \mathfrak{g}$ be a reductive subalgebra. Suppose that V is a finite-dimensional irreducible $\mathfrak{g}$-module and consider its restriction to the subalgebra $\mathfrak{a}$. This restriction is isomorphic to a direct sum of irreducible finite-dimensional $\mathfrak{a}$-modules W_μ which occur with certain multiplicities m_μ,

$$V|_{\mathfrak{a}} \cong \bigoplus_{\mu} m_\mu \, W_\mu.$$

If the decomposition is multiplicity-free (i.e., $m_\mu \leqslant 1$ for all μ) and each W_μ is provided with a basis, then it can be used to get a basis of V as the union of the bases of the spaces W_μ which occur in the decomposition. This was a key observation for the constructions of the bases for the representations of the general linear and orthogonal Lie algebras given by Gelfand and Tsetlin (1950). Alternatively, if the decomposition is not necessarily multiplicity-free, we can interpret it as the vector space isomorphism

$$V \cong \bigoplus_{\mu} U_\mu \otimes W_\mu, \tag{0.8}$$

where

$$U_\mu = \mathrm{Hom}_{\mathfrak{a}}(W_\mu, V), \qquad \dim U_\mu = m_\mu.$$

It is well known that the vector space U_μ is an irreducible module over the algebra $\mathrm{C}(\mathfrak{g}, \mathfrak{a}) = \mathrm{U}(\mathfrak{g})^{\mathfrak{a}}$, the centralizer of $\mathfrak{a}$ in the universal enveloping algebra $\mathrm{U}(\mathfrak{g})$. Now, if some bases of the spaces U_μ and W_μ are given, then the decomposition (0.8) yields the natural tensor product basis of V. The general difficulty of this approach is the complicated structure of the algebra $\mathrm{C}(\mathfrak{g}, \mathfrak{a})$. As was observed by Olshanski, for each pair of the classical Lie algebras

$$(\mathfrak{g}, \mathfrak{a}) \quad = \quad (\mathfrak{gl}_N, \mathfrak{gl}_M), \qquad (\mathfrak{g}_N, \mathfrak{g}_M),$$

there exist algebra homomorphisms

$$\mathrm{Y}(\mathfrak{gl}_{N-M}) \to \mathrm{C}(\mathfrak{gl}_N, \mathfrak{gl}_M), \qquad \mathrm{Y}(\mathfrak{g}_{N-M}) \to \mathrm{C}(\mathfrak{g}_N, \mathfrak{g}_M). \tag{0.9}$$

These homomorphisms turn out to be consistent for different values of M and N provided the difference $N - M$ is fixed, which allows one to embed the (twisted) Yangian into a projective limit of the centralizers. The structure of the "limit" algebras is much simpler than that of the corresponding centralizers $\mathrm{C}(\mathfrak{g}, \mathfrak{a})$: the (twisted) Yangian can be presented by quadratic and linear defining relations. Furthermore, the $\mathrm{C}(\mathfrak{g}, \mathfrak{a})$-module U_μ in (0.8) can be equipped with the structure of a representation of the Yangian or twisted Yangian, respectively, via the homomorphisms (0.9). Luckily, in the case $N - M = 2$ the $\mathrm{Y}(\mathfrak{g}_2)$-module U_μ admits an extension to a module over the Yangian $\mathrm{Y}(\mathfrak{gl}_2)$. This fact plays a key role in the construction: using the embedding

$$\mathrm{Y}(\mathfrak{gl}_1) \subset \mathrm{Y}(\mathfrak{gl}_2)$$

we can get a natural basis of Gelfand–Tsetlin type for the $\mathrm{Y}(\mathfrak{gl}_2)$-module U_μ and then, by induction, a weight basis of the representation V of the orthogonal and symplectic Lie algebra. Moreover, the matrix elements of the generators of the Lie algebras can be written down in an explicit form.

On the other hand, the (twisted) Yangian modules U_μ emerging from the centralizer construction give rise to a natural class of *skew representations*. These are described in Chapter 8 for the Yangian case, where we calculate their highest weights, Drinfeld polynomials and the Gelfand–Tsetlin characters. The skew representations of the twisted Yangians are considered in Chapter 9 in connection with the weight bases for representations of the orthogonal and symplectic Lie algebras.

The basis vectors of these representations are expressed explicitly in terms of the *lowering operators*. We describe these operators in the context of the *Mickelsson algebra theory* developed by D. P. Zhelobenko (1990, 1994). A brief account of this theory is given in the beginning of Chapter 9. The lowering operators were first used by J. G. Nagel and M. Moshinsky (1964) to construct the Gelfand–Tsetlin bases for the representations of $\mathfrak{gl}_N$. A similar construction of the bases for the orthogonal case was produced by S. C. Pang and K. T. Hecht (1967) and M. K. F. Wong (1967). J. Mickelsson (1972) gave some formulas for basis vectors of the representations of the symplectic Lie algebra as ordered products of the lowering operators. However, the action of the Lie algebra generators in such a basis does not seem to be computable. The reason is the fact that, unlike the cases of $\mathfrak{gl}_N$ and $\mathfrak{o}_N$, the lowering operators do not commute so that the basis depends on the chosen ordering. A "hidden symmetry" has been needed to make a natural choice of the appropriate combination of the lowering operators. That symmetry was provided by the action of the twisted Yangian $\mathrm{Y}(\mathfrak{sp}_2)$ on the homomorphism space U_μ in (0.8), and this action can be written in terms of the lowering operators. The basis of the representation of $\mathfrak{sp}_{2n}$ is then obtained by induction with the use of the chain of subalgebras

$$\mathfrak{sp}_2 \subset \mathfrak{sp}_4 \subset \cdots \subset \mathfrak{sp}_{2n}.$$

The same method can be applied to the pairs of orthogonal Lie algebras where we use the "two step reductions" $\mathfrak{o}_N \downarrow \mathfrak{o}_{N-2}$ instead of the restrictions of the representations of $\mathfrak{o}_N$ to the subalgebra $\mathfrak{o}_{N-1}$ used by Gelfand and Tsetlin. To compare the two constructions, note that the basis of Gelfand and Tsetlin lacks the *weight* property; i.e., the basis vectors are not eigenvectors for the Cartan subalgebra. The reason for that is the fact that the restrictions $\mathfrak{o}_N \downarrow \mathfrak{o}_{N-1}$ involve Lie algebras of different types (B and D) and the embeddings are not compatible with the root systems. In the new approach we use instead the chains

$$\mathfrak{o}_2 \subset \mathfrak{o}_4 \subset \cdots \subset \mathfrak{o}_{2n} \qquad \text{and} \qquad \mathfrak{o}_3 \subset \mathfrak{o}_5 \subset \cdots \subset \mathfrak{o}_{2n+1}.$$

The embeddings here "respect" the root systems so that the bases possess the weight property in both the symplectic and orthogonal cases. However, the new weight bases, in their turn, lack the *orthogonality* property of the Gelfand–Tsetlin bases: the latter are orthogonal with respect to the standard inner product in the representation space.

At the end of each chapter we give brief bibliographical comments pointing towards the original articles and give further references. This book was intended to be an introductory text on the Yangian theory and its applications, and so the list of topics covered here is by no means complete. In particular, we do not discuss in detail the *Bethe subalgebras* of the (twisted) Yangians (see Section 1.14 for the definition and some basic properties in the Yangian case). These are commutative subalgebras playing an important role in the theory of quantum integrable models in relation with the *Bethe ansatz*; see e.g. Takhtajan and Faddeev (1979), A. N. Kirillov and Reshetikhin (1986), Sklyanin (1992), and Nazarov and Olshanski (1996). The *Drinfeld functor* connecting the representation theory of the degenerate affine Hecke algebras with that of the Yangians is not considered either; see Drinfeld (1986), T. Arakawa and T. Suzuki (1998), Arakawa (1999), S. Khoroshkin and Nazarov (2006, 2007). An application of this functor leads to the character formulas for the finite-dimensional irreducible representations of $\mathrm{Y}(\mathfrak{gl}_N)$ expressing the characters in terms of the Kazhdan–Lusztig polynomials, as computed by Arakawa (1999); see also Brundan and Kleshchev (2007).

The Yangians, as well as their super and q-analogues, have found numerous applications in different areas of physics, including the theory of integrable models in statistical mechanics, conformal field theory, and quantum gravity. We do not discuss them in the book, although we give some references in the Bibliography to indicate at least a few directions for such applications. Some versions of the theorems proved in the book hold for other types of algebras, in particular, for the quantized enveloping algebra $\mathrm{U}_q(\mathfrak{gl}_N)$ and the quantum affine algebra $\mathrm{U}_q(\widehat{\mathfrak{gl}}_N)$. We briefly discuss these versions in the examples which follow each chapter, and we also formulate some open problems there.

We have tried to keep the exposition as self-contained as possible relying only on the basic facts of the Lie algebra representation theory; see e.g. J. E. Humphreys (1972), W. Fulton and J. Harris (1991), R. Goodman and N. R. Wallach (1998).

However, some applications in Chapters 6–9 use a few results from the representation theory of the symmetric group, theory of symmetric functions, and the Mickelsson algebra theory. Appropriate references are given in those chapters. The part of the exposition devoted to the Yangian $\mathrm{Y}(\mathfrak{gl}_N)$ can be read separately. The corresponding results are contained in Chapters 1, 3, 5, 6 and the respective parts of Chapters 7 and 8. Throughout the book we use the standard notation $\mathbb{Z}$ and $\mathbb{C}$ to indicate the sets of integers and complex numbers, respectively, while $\mathbb{Z}_+$ is used to denote the set of nonnegative integers. The vector spaces and algebras are considered over the field of complex numbers, except for a few examples, where some extensions of $\mathbb{C}$ are needed.

I am pleased to thank my colleagues and the participants of the Algebra Seminar at the University of Sydney for many encouraging discussions. My thanks also extend to Yuly Billig, Alexei Bondal, Tony Bracken, Ivan Cherednik, Vladimir Drinfeld, Boris Feigin, William Fulton, Vyacheslav Futorny, Mark Gould, Alexei Isaev, Minoru Itoh, Gordon James, Anthony Joseph, Sergei Khoroshkin, Anatol Kirillov, Tom Koornwinder, Alan Lascoux, Bernard Leclerc, Masatoshi Noumi, Andrei Okounkov, Sergei Ovsienko, Andrew Pressley, Pavel Pyatov, Eric Ragoucy, Arun Ram, Nicolai Reshetikhin, Vladimir Retakh, Bruce Sagan, Paul Sorba, Alexander Stolin, Vitaly Tarasov, Jean-Yves Thibon, Valeri Tolstoy and Tôru Umeda for discussions of various aspects of the theory during the last few years. I am especially grateful to Alexandre Alexandrovich Kirillov, my first representation theory teacher, and to Grigori Olshanski, who introduced me to the Yangians. My long-term collaboration with Maxim Nazarov and Grigori Olshanski on various projects related to the Yangians and classical Lie algebras has lead to several results and constructions which are discussed in this book. I express my deep gratitude to both of them.

Alexander Molev

Sydney, August 2007

CHAPTER 1

Yangian for $\mathfrak{gl}_N$

As we pointed out in the Preface, although the discovery of the Yangians was motivated by the quantum inverse scattering theory, the Yangian defining relations can be "observed" from a purely algebraic viewpoint. We regard (0.3) as an algebraic motivation for the definition of the Yangian for the general linear Lie algebra $\mathfrak{gl}_N$. We demonstrate that the defining relations can be written in a matrix form which provides a starting point for special algebraic techniques to study the Yangian structure. These techniques play an essential role in the construction of the quantum determinant and description of the center of the Yangian.

1.1. Defining relations

DEFINITION 1.1.1. The *Yangian* for $\mathfrak{gl}_N$ is a unital associative algebra over $\mathbb{C}$ with countably many generators $t_{ij}^{(1)}, t_{ij}^{(2)}, \dots$ where $i, j = 1, \dots, N$, and the defining relations

$$[t_{ij}^{(r+1)}, t_{kl}^{(s)}] - [t_{ij}^{(r)}, t_{kl}^{(s+1)}] = t_{kj}^{(r)} t_{il}^{(s)} - t_{kj}^{(s)} t_{il}^{(r)}, \tag{1.1}$$

where $r, s = 0, 1, \dots$ and $t_{ij}^{(0)} = \delta_{ij}$. This algebra is denoted by $\mathrm{Y}(\mathfrak{gl}_N)$. □

Introducing the formal generating series

$$t_{ij}(u) = \delta_{ij} + t_{ij}^{(1)} u^{-1} + t_{ij}^{(2)} u^{-2} + \dots \in \mathrm{Y}(\mathfrak{gl}_N)[[u^{-1}]], \tag{1.2}$$

we can write (1.1) in the form

$$(u - v)\,[t_{ij}(u), t_{kl}(v)] = t_{kj}(u)\, t_{il}(v) - t_{kj}(v)\, t_{il}(u); \tag{1.3}$$

the indeterminates u and v are considered to be commuting with each other and with the elements of the Yangian.

The following is an equivalent form of the defining relations of the algebra $\mathrm{Y}(\mathfrak{gl}_N)$.

PROPOSITION 1.1.2. *The system of relations* (1.1) *is equivalent to the system*

$$[t_{ij}^{(r)}, t_{kl}^{(s)}] = \sum_{a=1}^{\min\{r,s\}} \Big(t_{kj}^{(a-1)} t_{il}^{(r+s-a)} - t_{kj}^{(r+s-a)} t_{il}^{(a-1)} \Big). \tag{1.4}$$

PROOF. Observe that the multiplication of both sides of (1.3) by the formal series $\sum_{p=0}^{\infty} u^{-p-1} v^p$ yields an equivalent relation

$$[t_{ij}(u), t_{kl}(v)] = \Big(t_{kj}(u) t_{il}(v) - t_{kj}(v) t_{il}(u) \Big) \sum_{p=0}^{\infty} u^{-p-1} v^p.$$

Taking the coefficients of $u^{-r}v^{-s}$ on both sides gives

$$[t_{ij}^{(r)}, t_{kl}^{(s)}] = \sum_{a=1}^{r} \Big(t_{kj}^{(a-1)} t_{il}^{(r+s-a)} - t_{kj}^{(r+s-a)} t_{il}^{(a-1)} \Big).$$

This agrees with (1.4) in the case $r \leqslant s$. Finally, if $r > s$ observe that

$$\sum_{a=s+1}^{r} \Big(t_{kj}^{(a-1)} t_{il}^{(r+s-a)} - t_{kj}^{(r+s-a)} t_{il}^{(a-1)} \Big) = 0,$$

completing the proof. □

We will often use formal series to define or describe maps between various algebras. If $a(u)$ and $b(u)$ are formal power series in u^{-1} with coefficients in certain algebras, then assignments of the type $a(u) \mapsto b(u)$ are understood in the sense that every coefficient of $a(u)$ is mapped to the corresponding coefficient of $b(u)$.

Many applications of the Yangian are based on the following simple observation.

PROPOSITION 1.1.3. *The assignment*

$$\pi_N : t_{ij}(u) \mapsto \delta_{ij} + E_{ij}\, u^{-1} \tag{1.5}$$

defines a surjective homomorphism $\mathrm{Y}(\mathfrak{gl}_N) \to \mathrm{U}(\mathfrak{gl}_N)$. *Moreover, the assignment*

$$E_{ij} \mapsto t_{ij}^{(1)} \tag{1.6}$$

defines an embedding $\mathrm{U}(\mathfrak{gl}_N) \hookrightarrow \mathrm{Y}(\mathfrak{gl}_N)$.

PROOF. By Definition 1.1.1, we need to verify the equality

$$(u-v)[E_{ij}, E_{kl}]\, u^{-1}v^{-1} = (\delta_{kj} + E_{kj}u^{-1})(\delta_{il} + E_{il}v^{-1}) - (\delta_{kj} + E_{kj}v^{-1})(\delta_{il} + E_{il}u^{-1}).$$

But this clearly holds due to the commutation relations in $\mathfrak{gl}_N$, which proves the first part of the proposition. In order to prove the second part, put $r = s = 1$ in (1.4) which gives

$$[t_{ij}^{(1)}, t_{kl}^{(1)}] = \delta_{kj} t_{il}^{(1)} - \delta_{il} t_{kj}^{(1)}.$$

Thus, (1.6) is an algebra homomorphism. Its injectivity follows from the observation that by applying (1.6) and then (1.5), we get the identity map on $\mathrm{U}(\mathfrak{gl}_N)$. □

The homomorphism π_N is called the *evaluation homomorphism*. By its virtue, any representation of the Lie algebra $\mathfrak{gl}_N$ can be regarded as a representation of $\mathrm{Y}(\mathfrak{gl}_N)$. Any irreducible representation of $\mathfrak{gl}_N$ remains irreducible over $\mathrm{Y}(\mathfrak{gl}_N)$, by surjectivity of π_N.

Note that an alternative form of the defining relations (1.1) is also common in the literature. It is obtained by swapping the indices r and s on the right-hand side of (1.1); see also Remark 1.2.3 below.

1.2. Matrix form of the defining relations

Introduce the $N \times N$ matrix $T(u)$ whose ij-th entry is the series $t_{ij}(u)$. One can regard $T(u)$ as an element of the algebra $\mathrm{End}\,\mathbb{C}^N \otimes \mathrm{Y}(\mathfrak{gl}_N)[[u^{-1}]]$. Then

$$T(u) = \sum_{i,j=1}^{N} e_{ij} \otimes t_{ij}(u), \tag{1.7}$$

where $e_{ij} \in \operatorname{End}\mathbb{C}^N$ are the standard matrix units. If $e_1, \dots, e_N$ are the standard basis vectors of $\mathbb{C}^N$, then $T(u)\, e_j$ is interpreted as the linear combination

$$T(u)\, e_j = \sum_{i=1}^{N} e_i \otimes t_{ij}(u) \in \mathbb{C}^N \otimes \mathrm{Y}(\mathfrak{gl}_N)[[u^{-1}]].$$

Informally speaking, $T(u)$ may be considered as an "operator" on $\mathbb{C}^N$ with the "coefficients" in the algebra $\mathrm{Y}(\mathfrak{gl}_N)[[u^{-1}]]$.

For any positive integer m we will be using algebras of the form

$$(\operatorname{End}\mathbb{C}^N)^{\otimes m} \otimes \mathrm{Y}(\mathfrak{gl}_N). \tag{1.8}$$

For any $a \in \{1, \dots, m\}$ we denote by $T_a(u)$ the matrix $T(u)$ which corresponds to the a-th copy of the algebra $\operatorname{End}\mathbb{C}^N$ in the tensor product (1.8). That is, $T_a(u)$ is a formal power series in u^{-1} with the coefficients from the algebra (1.8),

$$T_a(u) = \sum_{i,j=1}^{N} 1^{\otimes(a-1)} \otimes e_{ij} \otimes 1^{\otimes(m-a)} \otimes t_{ij}(u), \tag{1.9}$$

where 1 is the identity matrix. If

$$C = \sum_{i,j,k,l=1}^{N} c_{ijkl}\, e_{ij} \otimes e_{kl} \in \operatorname{End}\mathbb{C}^N \otimes \operatorname{End}\mathbb{C}^N,$$

then for any two indices $a, b \in \{1, \dots, m\}$ such that $a < b$, we define the element C_{ab} of the algebra $(\operatorname{End}\mathbb{C}^N)^{\otimes m}$ by

$$C_{ab} = \sum_{i,j,k,l=1}^{N} c_{ijkl}\, 1^{\otimes(a-1)} \otimes e_{ij} \otimes 1^{\otimes(b-a-1)} \otimes e_{kl} \otimes 1^{\otimes(m-b)}. \tag{1.10}$$

Here the tensor factors e_{ij} and e_{kl} belong to the a-th and b-th copies of $\operatorname{End}\mathbb{C}^N$, respectively. The element C_{ab} can be identified with the element $C_{ab} \otimes 1$ of the algebra (1.8). If

$$t : \operatorname{End}\mathbb{C}^N \to \operatorname{End}\mathbb{C}^N, \qquad e_{ij} \mapsto e_{ji},$$

is the matrix transposition, then for any $a \in \{1, \dots, m\}$ we will denote by t_a the corresponding partial transposition on the algebra (1.8). It acts as t on the a-th copy of $\operatorname{End}\mathbb{C}^N$ and as the identity map on all the other tensor factors.

Consider now the permutation operator

$$P = \sum_{i,j=1}^{N} e_{ij} \otimes e_{ji} \in \operatorname{End}\mathbb{C}^N \otimes \operatorname{End}\mathbb{C}^N. \tag{1.11}$$

The rational function

$$R(u) = 1 - P u^{-1} \tag{1.12}$$

with values in $\operatorname{End}\mathbb{C}^N \otimes \operatorname{End}\mathbb{C}^N$ is called the *Yang R-matrix*. Here and below we write 1 instead of $1 \otimes 1$, for brevity. We will be frequently using the identity

$$R(u)\, R(-u) = 1 - u^{-2}. \tag{1.13}$$

We will also work with the rational function

$$R^t(u) = 1 - Q\, u^{-1} \tag{1.14}$$

where

$$Q = \sum_{i,j=1}^{N} e_{ij} \otimes e_{ij} = P^{t_1} = P^{t_2}. \tag{1.15}$$

We should write either $R^{t_1}(u)$ or $R^{t_2}(u)$ instead of $R^t(u)$, but we will *not* do so. Observe that Q is a one-dimensional operator on $\mathbb{C}^N \otimes \mathbb{C}^N$ satisfying $Q^2 = N\,Q$. Therefore

$$R^t(u)^{-1} = 1 + Q\,(u-N)^{-1}. \tag{1.16}$$

PROPOSITION 1.2.1. *In the algebra* $(\operatorname{End}\mathbb{C}^N)^{\otimes 3}(u,v)$ *we have the identity*

$$R_{12}(u)\,R_{13}(u+v)\,R_{23}(v) = R_{23}(v)\,R_{13}(u+v)\,R_{12}(u). \tag{1.17}$$

PROOF. Multiplying both sides of the relation (1.17) by $uv(u+v)$ we come to verify the identity

$$(u-P_{12})(u+v-P_{13})(v-P_{23}) = (v-P_{23})(u+v-P_{13})(u-P_{12}). \tag{1.18}$$

Each operator P_{ij} is the image of the corresponding transposition $(i\,j) \in \mathfrak{S}_3$ under the natural action of the symmetric group $\mathfrak{S}_3$ on $(\mathbb{C}^N)^{\otimes 3}$ by permutations of the tensor factors. Therefore, (1.18) is immediate from the relations in the group algebra $\mathbb{C}[\mathfrak{S}_3]$. □

The relation (1.17) is known as the *Yang–Baxter equation.* The Yang R-matrix is its simplest nontrivial solution. Below we regard $T_1(u)$ and $T_2(v)$ as formal power series with the coefficients from the algebra (1.8), where $m = 2$ copies of $\operatorname{End}\mathbb{C}^N$ are taken. We also identify $R(u-v)$ and the rational function $R(u-v) \otimes 1$ taking values in this algebra.

PROPOSITION 1.2.2. *The defining relations* (1.1) *of the algebra* $\mathrm{Y}(\mathfrak{gl}_N)$ *can be written in the equivalent form*

$$R(u-v)\,T_1(u)\,T_2(v) = T_2(v)\,T_1(u)\,R(u-v). \tag{1.19}$$

PROOF. Let us apply the "operators" on both sides of the relation (1.19) to an arbitrary basis vector $e_j \otimes e_l \in \mathbb{C}^N \otimes \mathbb{C}^N$. For the left-hand side we get

$$\sum_{i,k} t_{ij}(u)\,t_{kl}(v) \otimes e_i \otimes e_k - \frac{1}{u-v} \sum_{i,k} t_{ij}(u)\,t_{kl}(v) \otimes e_k \otimes e_i,$$

while the right-hand side gives

$$\sum_{i,k} t_{kl}(v)\,t_{ij}(u) \otimes e_i \otimes e_k - \frac{1}{u-v} \sum_{i,k} t_{kj}(v)\,t_{il}(u) \otimes e_i \otimes e_k.$$

Multiplying by $u-v$ and equating the coefficients of $e_i \otimes e_k$ we recover (1.3). □

The relation (1.19) will be referred to as the *ternary* (or the *RTT*) *relation.*

REMARK 1.2.3. There are two traditions in the literature to write the defining relations (1.19). The alternative way is to replace $R(u)$ with $\widetilde{R}(u) = 1 + Pu^{-1}$. The respective forms of (1.1) and (1.3) are then obtained by multiplying their right-hand sides by -1. The algebra defined in this way is isomorphic to $\mathrm{Y}(\mathfrak{gl}_N)$. An isomorphism, written in terms of the respective generating series of the two algebras, can be given by $t_{ij}(u) \mapsto \widetilde{t}_{ji}(u)$. □

1.3. Automorphisms and anti-automorphisms

In this section, we will use the $N \times N$ matrix $T(u)$ to define several distinguished automorphisms and anti-automorphisms of the associative unital algebra $\mathrm{Y}(\mathfrak{gl}_N)$. For each of them, we will describe the $N \times N$ matrix whose ij-entry is the formal power series in u^{-1} with the coefficients being the images of the corresponding coefficients of the series $t_{ij}(u)$. For example, the assignment (1.25) below means that $t_{ij}^{(r)} \mapsto (-1)^r \, t_{ij}^{(r)}$ for all indices $r = 1, 2, \dots$ and $i, j = 1, \dots, N$.

To begin, let $f(u)$ be any formal power series in u^{-1} with the leading term 1,

$$f(u) = 1 + f_1 \, u^{-1} + f_2 \, u^{-2} + \dots \in \mathbb{C}[[u^{-1}]].$$

Let c be any complex number and B be any nonsingular complex $N \times N$ matrix.

PROPOSITION 1.3.1. *Each of the mappings*

$$T(u) \mapsto f(u)\, T(u), \tag{1.20}$$

$$T(u) \mapsto T(u-c), \tag{1.21}$$

$$T(u) \mapsto B\, T(u)\, B^{-1} \tag{1.22}$$

defines an automorphism of $\mathrm{Y}(\mathfrak{gl}_N)$.

PROOF. The ternary relation (1.19) will be satisfied by the image of $T(u)$ with respect to each of the three mappings. This is immediate for (1.20) and (1.21), while for (1.22) we need to verify that

$$\begin{aligned} R(u-v)\, B_1\, T_1(u)\, B_1^{-1}\, B_2\, T_2(v)\, B_2^{-1} \\ = B_2\, T_2(v)\, B_2^{-1}\, B_1\, T_1(u)\, B_1^{-1}\, R(u-v). \end{aligned} \tag{1.23}$$

However, B_1 is permutable with each of B_2 and $T_2(v)$, and B_2 is permutable with $T_1(u)$. Moreover, since $PB_1 = B_2P$ and $PB_2 = B_1P$ we also have

$$R(u-v)\, B_1\, B_2 = B_2\, B_1\, R(u-v). \tag{1.24}$$

Now (1.23) clearly follows from (1.19). Finally, each of the three mappings is obviously invertible, which completes the proof. □

REMARK 1.3.2. The calculation of the images of the elements $t_{ij}^{(r)}$ under the shift automorphism (1.21) amounts to expanding expressions like $(u-c)^{-k}$ into power series in u^{-1}. Such images are well-defined elements of the Yangian $\mathrm{Y}(\mathfrak{gl}_N)$. We emphasize that the use of negative powers of u here is essential, as the shift automorphism is *not* defined on the algebra $\mathrm{Y}(\mathfrak{gl}_N)[[u]]$. □

We may regard the element $T(u)$ defined by (1.7) as a formal power series in u^{-1} whose coefficients are matrices with the entries from the algebra $\mathrm{Y}(\mathfrak{gl}_N)$. Since the leading term of this series is the identity matrix, the element $T(u)$ is invertible. We denote by $T^{-1}(u)$ the inverse element. Further, denote by $T^{\,t}(u)$ the transposed matrix for $T(u)$. Then

$$T^{\,t}(u) = \sum_{i,j=1}^{N} e_{ij} \otimes t_{ji}(u).$$

PROPOSITION 1.3.3. *Each of the mappings*

$$\sigma_N : T(u) \mapsto T(-u), \tag{1.25}$$

$$t : T(u) \mapsto T^t(u), \tag{1.26}$$

$$\mathrm{S} : T(u) \mapsto T^{-1}(u) \tag{1.27}$$

defines an anti-automorphism of $\mathrm{Y}(\mathfrak{gl}_N)$.

PROOF. The images $t^{\circ}_{ij}(u)$ of the series $t_{ij}(u)$ under any anti-automorphism of $\mathrm{Y}(\mathfrak{gl}_N)$ must satisfy the relations (1.3) with the opposite multiplication:

$$(u-v)\,[t^{\circ}_{ij}(u), t^{\circ}_{kl}(v)] = t^{\circ}_{il}(u)t^{\circ}_{kj}(v) - t^{\circ}_{il}(v)t^{\circ}_{kj}(u).$$

Exactly as in the proof of Proposition 1.2.2, one can show that these relations can be equivalently written in the following matrix form:

$$R(u-v)\,T^{\circ}_2(v)\,T^{\circ}_1(u) = T^{\circ}_1(u)\,T^{\circ}_2(v)\,R(u-v), \tag{1.28}$$

where $T^{\circ}(u)$ is the $N \times N$ matrix whose ij-th entry is $t^{\circ}_{ij}(u)$. The relation

$$R(u-v)\,T_2(-v)\,T_1(-u) = T_1(-u)\,T_2(-v)\,R(u-v) \tag{1.29}$$

follows from (1.19) if we conjugate both sides by P and then replace (u, v) by $(-v, -u)$. This shows that (1.25) defines an anti-homomorphism. Furthermore, the application of the partial transposition operator t_1 to both sides of the ternary relation (1.19) yields

$$T^t_1(u)\,R^t(u-v)\,T_2(v) = T_2(v)\,R^t(u-v)\,T^t_1(u). \tag{1.30}$$

Next, since $R(u-v)$ is stable under the composition $t_2 \circ t_1$, applying t_2 to (1.30) we get

$$T^t_1(u)\,T^t_2(v)\,R(u-v) = R(u-v)\,T^t_2(v)\,T^t_1(u).$$

This shows that (1.26) is an anti-homomorphism. Finally, for (1.27) observe that the relation

$$R(u-v)\,T^{-1}_2(v)\,T^{-1}_1(u) = T^{-1}_1(u)\,T^{-1}_2(v)\,R(u-v)$$

is equivalent to (1.19). Note now that the mappings (1.25) and (1.26) are involutive and so these anti-homomorphisms are bijective.

To complete the proof take the composition of the anti-homomorphisms (1.25) and (1.27), where (1.25) is applied first. Denote the composition by ω_N, so that

$$\omega_N : T(u) \mapsto T^{-1}(-u). \tag{1.31}$$

Then ω_N is an algebra homomorphism. Let us show that it is involutive. Indeed, apply ω_N to both sides of the identity

$$\omega_N(T(u)) \cdot T(-u) = 1.$$

This gives

$$\omega^2_N(T(u)) \cdot T^{-1}(u) = 1.$$

So the square ω^2_N of ω_N is the identity map. In particular, the homomorphism ω_N is bijective, and so is the anti-homomorphism (1.27). □

In the course of the proof of Proposition 1.3.3 we established the following fact.

COROLLARY 1.3.4. *The mapping* (1.31) *defines an involutive automorphism of the algebra* $\mathrm{Y}(\mathfrak{gl}_N)$.

The involutive anti-automorphisms (1.25) and (1.26) commute with each other. Their composition is an involutive automorphism of $\mathrm{Y}(\mathfrak{gl}_N)$ such that

$$\tau_N : T(u) \mapsto T^t(-u). \tag{1.32}$$

This automorphism of the algebra $\mathrm{Y}(\mathfrak{gl}_N)$ will play an important role in Chapter 2.

REMARK 1.3.5. The anti-automorphisms (1.25) and (1.27) do *not* commute with each other, and the anti-automorphism (1.27) is *not* involutive. This is the antipodal map of the Hopf algebra $\mathrm{Y}(\mathfrak{gl}_N)$; see Section 1.5. The square S^2 of this anti-automorphism will be calculated in Theorem 1.9.9. □

1.4. Poincaré–Birkhoff–Witt theorem

The following is a Yangian analogue of the classical Poincaré–Birkhoff–Witt theorem for universal enveloping algebras; see e.g. Dixmier [**89**, Section 2.1].

THEOREM 1.4.1. *Given an arbitrary linear order on the set of generators* $t_{ij}^{(r)}$, *any element of the algebra* $\mathrm{Y}(\mathfrak{gl}_N)$ *can be uniquely written as a linear combination of ordered monomials in these generators.*

PROOF. There are two natural ascending filtrations on the algebra $\mathrm{Y}(\mathfrak{gl}_N)$. Here we use the one defined by

$$\deg t_{ij}^{(r)} = r;$$

the other ascending filtration will be introduced in the next section. It is immediate from the defining relations (1.4) that the corresponding graded algebra $\mathrm{gr}\,\mathrm{Y}(\mathfrak{gl}_N)$ is commutative. Denote by $\bar{t}_{ij}^{(r)}$ the image of $t_{ij}^{(r)}$ in the r-th component of $\mathrm{gr}\,\mathrm{Y}(\mathfrak{gl}_N)$. It will be sufficient to show that the elements $\bar{t}_{ij}^{(r)}$ are algebraically independent.

By the defining relations (1.3), for any $M \geqslant 0$ there is a natural homomorphism

$$\iota_M : \mathrm{Y}(\mathfrak{gl}_N) \to \mathrm{Y}(\mathfrak{gl}_{N+M}), \tag{1.33}$$

such that $t_{ij}^{(r)} \mapsto t_{ij}^{(r)}$. Take the composition

$$\zeta_M = \pi_{N+M} \circ \omega_{N+M} \circ \iota_M,$$

where the automorphism ω_{N+M} of the algebra $\mathrm{Y}(\mathfrak{gl}_{N+M})$ is defined using (1.31), and the homomorphism $\pi_{N+M} : \mathrm{Y}(\mathfrak{gl}_{N+M}) \to \mathrm{U}(\mathfrak{gl}_{N+M})$ is defined via (1.5). Thus

$$\zeta_M : \mathrm{Y}(\mathfrak{gl}_N) \to \mathrm{U}(\mathfrak{gl}_{N+M}).$$

Due to (1.5) and (1.31),

$$\zeta_M : t_{ij}^{(r)} \mapsto (E^r)_{ij}$$

where E is the $(N+M)\times(N+M)$ matrix whose ij-th entry is E_{ij}. Observe that the homomorphism ζ_M is filtration-preserving so that it defines a homomorphism of the corresponding graded algebras

$$\bar{\zeta}_M : \mathrm{gr}\,\mathrm{Y}(\mathfrak{gl}_N) \to \mathrm{S}(\mathfrak{gl}_{N+M}),$$

where $\mathrm{S}(\mathfrak{gl}_{N+M})$ is the symmetric algebra of $\mathfrak{gl}_{N+M}$. We will consider the elements of $\mathrm{S}(\mathfrak{gl}_{N+M})$ as polynomial functions on $\mathfrak{gl}_{N+M}$. Thus the image of $\bar{t}_{ij}^{(r)}$ under $\bar{\zeta}_M$ is the polynomial $p_{ij}^{(r)}$ such that

$$p_{ij}^{(r)}(X) = (X^r)_{ij} \quad \text{for any} \quad X \in \mathfrak{gl}_{N+M}.$$

Let us consider an arbitrary finite family of elements $\bar{t}_{ij}^{(r)}$. Suppose that R is a positive integer such that for all the upper indices of this family we have $1 \leqslant r \leqslant R$. It will now be sufficient to demonstrate that the parameter M can be chosen in such a way that all the polynomials $p_{ij}^{(r)}$ with $1 \leqslant i, j \leqslant N$ and $1 \leqslant r \leqslant R$ are algebraically independent. For any such triple of indices (i, j, r) choose a subset

$$\mathcal{O}_{ij}^{(r)} \subset \{N+1, N+2, \dots\}$$

of cardinality $r-1$ in such a way that all these subsets are disjoint. Let M be so large that all of them are contained in $\{N+1, N+2, \dots, N+M\}$. Let $y_{ij}^{(r)}$ be complex parameters. Define a linear operator $x_{ij}^{(r)}$ in $\mathbb{C}^{N+M}$ depending on $y_{ij}^{(r)}$ as follows. Let $e_1, \dots, e_{N+M}$ be the standard basis in $\mathbb{C}^{N+M}$ and $a_1 < \dots < a_{r-1}$ be all the elements of the set $\mathcal{O}_{ij}^{(r)}$. Then put

$$\begin{aligned} x_{ij}^{(r)} : \quad & e_j \mapsto y_{ij}^{(r)} e_{a_{r-1}} , \ e_{a_{r-1}} \mapsto e_{a_{r-2}} , \ \dots , \ e_{a_1} \mapsto e_i , \\ & e_k \mapsto 0 \quad \text{for} \quad k \notin \{j\} \cup \mathcal{O}_{ij}^{(r)}. \end{aligned}$$

In the case $r = 1$, the first line above reads as $x_{ij}^{(1)} : e_j \mapsto y_{ij}^{(1)} e_i$. Now set

$$X = \sum_{i,j,r} x_{ij}^{(r)}. \tag{1.34}$$

Then for any matrix X of this form we have

$$p_{ij}^{(r)}(X) = y_{ij}^{(r)} + \psi$$

for a certain polynomial ψ in the $y_{kl}^{(s)}$ with $s < r$. Thus the polynomials $p_{ij}^{(r)}$ are algebraically independent even if they are restricted to the affine subspace of matrices of the form (1.34). □

COROLLARY 1.4.2. $\operatorname{gr} \mathrm{Y}(\mathfrak{gl}_N)$ *is the algebra of polynomials in the variables* $\bar{t}_{ij}^{(r)}$.

COROLLARY 1.4.3. *The homomorphism* (1.33) *is an embedding.*

REMARK 1.4.4. By Theorem 1.4.1, the Yangian $\mathrm{Y}(\mathfrak{gl}_N)$ can be viewed as a flat deformation of the algebra of polynomials in countably many variables. To see this, for each $h \in \mathbb{C}$ consider the algebra $\mathrm{Y}(\mathfrak{gl}_N, h)$ with the generators $t_{ij}^{(r)}$ and the relations obtained from (1.3) by multiplying the right-hand side by h:

$$[t_{ij}^{(r)}, t_{kl}^{(s)}] = h \cdot \sum_{a=1}^{\min\{r,s\}} \Big(t_{kj}^{(a-1)} t_{il}^{(r+s-a)} - t_{kj}^{(r+s-a)} t_{il}^{(a-1)} \Big).$$

The algebras $\mathrm{Y}(\mathfrak{gl}_N, h)$ with $h \neq 0$ are all isomorphic to each other; an isomorphism $\mathrm{Y}(\mathfrak{gl}_N, h) \to \mathrm{Y}(\mathfrak{gl}_N)$ can be defined by $t_{ij}^{(r)} \mapsto t_{ij}^{(r)} h^r$. On the other hand, $\mathrm{Y}(\mathfrak{gl}_N, 0)$ is the algebra of polynomials in the generators $t_{ij}^{(r)}$. Moreover, this algebra is equipped with a Poisson bracket given by

$$\{t_{ij}^{(r)}, t_{kl}^{(s)}\} = \sum_{a=1}^{\min\{r,s\}} \Big(t_{kj}^{(a-1)} t_{il}^{(r+s-a)} - t_{kj}^{(r+s-a)} t_{il}^{(a-1)} \Big).$$

□

1.5. Hopf algebra structure

A *coalgebra* (over the field $\mathbb{C}$) is a vector space A equipped with linear maps $\Delta : A \to A \otimes A$, the *comultiplication*, and $\varepsilon : A \to \mathbb{C}$, the *counit*, satisfying the axioms given by the following three commutative diagrams:

$$\begin{array}{ccc} A \otimes A \otimes A & \xleftarrow{\Delta \otimes \mathrm{id}} & A \otimes A \\ {\scriptstyle \mathrm{id}\otimes\Delta}\uparrow & & \uparrow{\scriptstyle \Delta} \\ A \otimes A & \xleftarrow[\Delta]{} & A \end{array}$$

(the *coassociativity* of Δ) and

$$\begin{array}{ccc} A \otimes \mathbb{C} & \xleftarrow{\mathrm{id}\otimes\varepsilon} & A \otimes A \\ {\scriptstyle \cong}\uparrow & & \uparrow{\scriptstyle \Delta} \\ A & \xleftarrow[\mathrm{id}]{} & A \end{array} \qquad \begin{array}{ccc} \mathbb{C} \otimes A & \xleftarrow{\varepsilon\otimes\mathrm{id}} & A \otimes A \\ {\scriptstyle \cong}\uparrow & & \uparrow{\scriptstyle \Delta} \\ A & \xleftarrow[\mathrm{id}]{} & A \end{array}$$

A *bialgebra* (over $\mathbb{C}$) is an associative unital algebra A equipped with a coalgebra structure such that Δ and ε are algebra homomorphisms. In particular, then we have $\Delta(1) = 1 \otimes 1$ and $\varepsilon(1) = 1$. A bialgebra A is called a *Hopf algebra* if it is also equipped with an anti-automorphism $\mathrm{S} : A \to A$, the *antipode*, such that the following two diagrams commute:

$$\begin{array}{ccc} A \otimes A & \xrightarrow{\mathrm{S}\otimes\mathrm{id}} & A \otimes A \\ {\scriptstyle \Delta}\uparrow & & \downarrow{\scriptstyle \mu} \\ A & \xrightarrow[\delta\circ\varepsilon]{} & A \end{array} \qquad \begin{array}{ccc} A \otimes A & \xrightarrow{\mathrm{id}\otimes\mathrm{S}} & A \otimes A \\ {\scriptstyle \Delta}\uparrow & & \downarrow{\scriptstyle \mu} \\ A & \xrightarrow[\delta\circ\varepsilon]{} & A \end{array}$$

Here $\mu : A \otimes A \to A$ is the algebra multiplication and $\delta : \mathbb{C} \to A$ is the unit map of the algebra A; that is $\delta(c) = c \cdot 1$ for any $c \in \mathbb{C}$.

THEOREM 1.5.1. *The Yangian* $\mathrm{Y}(\mathfrak{gl}_N)$ *is a Hopf algebra with comultiplication*

$$\Delta : t_{ij}(u) \mapsto \sum_{k=1}^{N} t_{ik}(u) \otimes t_{kj}(u), \tag{1.35}$$

the antipode (1.27) *and the counit* $\varepsilon : T(u) \mapsto 1$.

PROOF. We start by verifying the axiom that $\Delta : \mathrm{Y}(\mathfrak{gl}_N) \to \mathrm{Y}(\mathfrak{gl}_N) \otimes \mathrm{Y}(\mathfrak{gl}_N)$ is an algebra homomorphism. We will slightly generalize the notation used in Section 1.2. Let m and n be positive integers. Introduce the algebra

$$(\mathrm{End}\,\mathbb{C}^N)^{\otimes m} \otimes \mathrm{Y}(\mathfrak{gl}_N)^{\otimes n}. \tag{1.36}$$

For all $a \in \{1, \dots, m\}$ and $b \in \{1, \dots, n\}$ consider the formal power series in u^{-1} with the coefficients in this algebra,

$$T_{a[b]}(u) = \sum_{i,j=1}^{N} 1^{\otimes(a-1)} \otimes e_{ij} \otimes 1^{\otimes(m-a)} \otimes 1^{\otimes(b-1)} \otimes t_{ij}(u) \otimes 1^{\otimes(n-b)}.$$

The definition of Δ can now be written in a matrix form,

$$\Delta : T(u) \mapsto T_{[1]}(u)\, T_{[2]}(u), \tag{1.37}$$

where $T_{[b]}(u)$ is an abbreviation for the series $T_{1[b]}(u)$ with the coefficients from the algebra (1.36) where $m = 1$ and $n = 2$. We need to show that $\Delta(T(u))$ satisfies the ternary relation (1.19), i.e.

$$\begin{aligned} &R(u-v)\, T_{1[1]}(u)\, T_{1[2]}(u)\, T_{2[1]}(v)\, T_{2[2]}(v) \\ &= T_{2[1]}(v)\, T_{2[2]}(v)\, T_{1[1]}(u)\, T_{1[2]}(u)\, R(u-v). \end{aligned}$$

Here $m = n = 2$, and $R(u-v)$ is identified with $R(u-v) \otimes 1 \otimes 1$. But this relation is implied by the ternary relation (1.19) and by the observation that the elements $T_{1[2]}(u)$ and $T_{2[1]}(v)$ commute, as well as the elements $T_{1[1]}(u)$ and $T_{2[2]}(v)$.

Furthermore, S is an anti-automorphism of $\mathrm{Y}(\mathfrak{gl}_N)$ by Proposition 1.3.3. The axioms involving S are readily verified since

$$(\mathrm{S} \otimes \mathrm{id}) \circ \Delta : T(u) \mapsto T^{-1}_{[1]}(u)\, T_{[2]}(u)$$

and

$$(\mathrm{id} \otimes \mathrm{S}) \circ \Delta : T(u) \mapsto T_{[1]}(u)\, T^{-1}_{[2]}(u)$$

so that the subsequent application of μ yields the identity matrix in both cases. The remaining axioms are obviously satisfied. $\square$

Expanding the formal power series in u^{-1} in (1.35), we obtain a more explicit definition of the comultiplication Δ on $\mathrm{Y}(\mathfrak{gl}_N)$,

$$\Delta\big(t_{ij}^{(r)}\big) = \sum_{k=1}^{N} \sum_{s=0}^{r} t_{ik}^{(s)} \otimes t_{kj}^{(r-s)}. \tag{1.38}$$

Also $\varepsilon\big(t_{ij}^{(r)}\big) = 0$ for any $r \geqslant 1$.

Now introduce another filtration on $\mathrm{Y}(\mathfrak{gl}_N)$ by setting

$$\deg' t_{ij}^{(r)} = r - 1$$

for every $r \geqslant 1$; cf. Section 1.4. We denote by $\mathrm{gr}'\,\mathrm{Y}(\mathfrak{gl}_N)$ the corresponding graded algebra. Let $\bar{t}_{ij}^{(r)}$ be the image of $t_{ij}^{(r)}$ in the $(r-1)$-th component of $\mathrm{gr}'\,\mathrm{Y}(\mathfrak{gl}_N)$.

The graded algebra $\mathrm{gr}'\,\mathrm{Y}(\mathfrak{gl}_N)$ inherits from $\mathrm{Y}(\mathfrak{gl}_N)$ the Hopf algebra structure. Namely, using (1.38), for any $r \geqslant 1$ we get

$$\Delta\big(\bar{t}_{ij}^{(r)}\big) = \bar{t}_{ij}^{(r)} \otimes 1 + 1 \otimes \bar{t}_{ij}^{(r)}, \quad \varepsilon\big(\bar{t}_{ij}^{(r)}\big) = 0 \quad \text{and} \quad \mathrm{S}\big(\bar{t}_{ij}^{(r)}\big) = -\bar{t}_{ij}^{(r)}. \tag{1.39}$$

For any Lie algebra $\mathfrak{g}$ over the field $\mathbb{C}$ consider the universal enveloping algebra $\mathrm{U}(\mathfrak{g})$. There is a natural Hopf algebra structure on $\mathrm{U}(\mathfrak{g})$. The comultiplication Δ, the counit ε and the antipode S on $\mathrm{U}(\mathfrak{g})$ are defined by setting for $X \in \mathfrak{g}$

$$\Delta(X) = X \otimes 1 + 1 \otimes X, \quad \varepsilon(X) = 0 \quad \text{and} \quad \mathrm{S}(X) = -X. \tag{1.40}$$

In the next proposition $\mathfrak{g}$ is $\mathfrak{gl}_N[z] \cong \mathfrak{gl}_N \otimes \mathbb{C}[z]$, the polynomial current Lie algebra corresponding to the Lie algebra $\mathfrak{gl}_N$.

PROPOSITION 1.5.2. *The Hopf algebra* $\mathrm{gr}'\,\mathrm{Y}(\mathfrak{gl}_N)$ *is isomorphic to the universal enveloping algebra* $\mathrm{U}(\mathfrak{gl}_N[z])$.

PROOF. Using the defining relations (1.4) we get

$$[\bar{t}_{ij}^{(r)}, \bar{t}_{kl}^{(s)}] = \delta_{kj}\bar{t}_{il}^{(r+s-1)} - \delta_{il}\bar{t}_{kj}^{(r+s-1)}.$$

Hence the assignment $E_{ij}\, z^{r-1} \mapsto \bar{t}_{ij}^{(r)}$ for $r \geqslant 1$ defines a surjective homomorphism of associative algebras

$$\mathrm{U}(\mathfrak{gl}_N[z]) \to \mathrm{gr}'\, \mathrm{Y}(\mathfrak{gl}_N). \tag{1.41}$$

The kernel of this homomorphism is trivial due to Theorem 1.4.1. By comparing the definitions (1.39) and (1.40), we complete the proof. □

REMARK 1.5.3. The Yangian $\mathrm{Y}(\mathfrak{gl}_N)$ may be regarded as a flat deformation of the universal enveloping algebra $\mathrm{U}(\mathfrak{gl}_N[z])$ as a Hopf algebra. Indeed, renormalize the generators of $\mathrm{Y}(\mathfrak{gl}_N)$ by multiplying $t_{ij}^{(r)}$ by h^{r-1}. This results in the following modification of the defining relations (1.4):

$$[t_{ij}^{(r)}, t_{kl}^{(s)}] = \delta_{kj}\, t_{il}^{(r+s-1)} - \delta_{il}\, t_{kj}^{(r+s-1)} + h \sum_{a=2}^{\min\{r,s\}} \Big(t_{kj}^{(a-1)} t_{il}^{(r+s-a)} - t_{kj}^{(r+s-a)} t_{il}^{(a-1)}\Big).$$

Let $\mathrm{Y}'(\mathfrak{gl}_N, h)$ be the algebra defined by these modified relations. By (1.38),

$$\Delta(t_{ij}^{(r)}) = t_{ij}^{(r)} \otimes 1 + 1 \otimes t_{ij}^{(r)} + h \sum_{k=1}^{N} \sum_{s=1}^{r-1} t_{ik}^{(s)} \otimes t_{kj}^{(r-s)}.$$

Also $\varepsilon(t_{ij}^{(r)}) = 0$ for any $r \geqslant 1$ and

$$\mathrm{S}(t_{ij}^{(r)}) \in -t_{ij}^{(r)} + h\, \mathrm{Y}'(\mathfrak{gl}_N, h).$$

We have $\mathrm{Y}'(\mathfrak{gl}_N, 1) = \mathrm{Y}(\mathfrak{gl}_N)$, while $\mathrm{Y}'(\mathfrak{gl}_N, 0)$ is isomorphic to $\mathrm{U}(\mathfrak{gl}_N[z])$ as a Hopf algebra, due to Proposition 1.5.2. The flatness of deformation is guaranteed by Theorem 1.4.1; cf. Remark 1.4.4. □

Later on we will also use the comultiplication Δ' on $\mathrm{Y}(\mathfrak{gl}_N)$ *opposite* the comultiplication Δ defined by the assignment (1.35). By definition, the map

$$\Delta' : \mathrm{Y}(\mathfrak{gl}_N) \to \mathrm{Y}(\mathfrak{gl}_N) \otimes \mathrm{Y}(\mathfrak{gl}_N)$$

is the composition of the comultiplication Δ with the linear operator on the tensor product $\mathrm{Y}(\mathfrak{gl}_N) \otimes \mathrm{Y}(\mathfrak{gl}_N)$ exchanging the two tensor factors. Thus we have

$$\Delta' : t_{ij}(u) \mapsto \sum_{k=1}^{N} t_{kj}(u) \otimes t_{ik}(u). \tag{1.42}$$

Similarly to (1.37), the definition of Δ' can be written in a matrix form as

$$\Delta' : T(u) \mapsto T_{[2]}(u)\, T_{[1]}(u). \tag{1.43}$$

With the comultiplication Δ' the Yangian $\mathrm{Y}(\mathfrak{gl}_N)$ also becomes a Hopf algebra. The antipode and the counit for Δ' are the same as for the comultiplication Δ. By comparing the definitions (1.35) and (1.42), we immediately get

PROPOSITION 1.5.4. *The involutive automorphism* (1.32) *of associative algebra* $\mathrm{Y}(\mathfrak{gl}_N)$ *exchanges the comultiplications* Δ *and* Δ'*; that is*

$$\Delta \circ \tau_N = (\tau_N \otimes \tau_N) \circ \Delta'.$$

1.6. Quantum determinant and quantum minors

For any $m \geqslant 2$ introduce the rational function $R(u_1, \dots, u_m)$ with values in the tensor product algebra $(\operatorname{End} \mathbb{C}^N)^{\otimes m}$ by

$$R(u_1, \dots, u_m) = (R_{m-1,m})(R_{m-2,m} R_{m-2,m-1}) \dots (R_{1m} \dots R_{12}), \tag{1.44}$$

where $u_1, \dots, u_m$ are independent complex variables and we abbreviate $R_{ij} = R_{ij}(u_i - u_j)$. Applying the Yang–Baxter equation (1.17) and using the obvious fact that R_{ij} and R_{kl} commute if the indices i, j, k, l are distinct, we can write (1.44) in a different form. In particular, a simple induction argument shows that

$$R(u_1, \dots, u_m) = (R_{12} \dots R_{1m}) \dots (R_{m-2,m-1} R_{m-2,m})(R_{m-1,m}). \tag{1.45}$$

Below we will use the formal power series $T_1(u_1), \dots, T_m(u_m)$ with the coefficients from the algebra (1.8). We also identify $R(u_1, \dots, u_m)$ with the rational function $R(u_1, \dots, u_m) \otimes 1$ taking values in that algebra.

PROPOSITION 1.6.1. *We have the relation*

$$R(u_1, \dots, u_m)\, T_1(u_1) \dots T_m(u_m) = T_m(u_m) \dots T_1(u_1)\, R(u_1, \dots, u_m).$$

PROOF. To simplify the notation, set $T_a = T_a(u_a)$. First note that the identity

$$(R_{1m} \dots R_{12})\, T_1\, (T_2 \dots T_m) = (T_2 \dots T_m)\, T_1\, (R_{1m} \dots R_{12})$$

is verified by repeated application of (1.19), where we also use the fact that the matrices R_{ij} and T_k with disjoint indices are pairwise permutable. Finally, since

$$R(u_1, \dots, u_m) = R(u_2, \dots, u_m)\, (R_{1m} \dots R_{12}), \tag{1.46}$$

the induction on m completes the proof. □

Let $\mathfrak{S}_m$ denote the symmetric group acting on the set $\{1, \dots, m\}$. Consider the anti-symmetrizer in the group algebra of $\mathfrak{S}_m$,

$$\sum_{p \in \mathfrak{S}_m} \operatorname{sgn} p \cdot p \in \mathbb{C}[\mathfrak{S}_m].$$

Denote by A_m the image of this anti-symmetrizer under the natural action of $\mathfrak{S}_m$ on the tensor space $(\mathbb{C}^N)^{\otimes m}$. Keeping the notation $e_1, \dots, e_N$ for the standard basis vectors of $\mathbb{C}^N$ we thus have

$$A_m(e_{i_1} \otimes \dots \otimes e_{i_m}) = \sum_{p \in \mathfrak{S}_m} \operatorname{sgn} p \cdot e_{i_{p(1)}} \otimes \dots \otimes e_{i_{p(m)}}.$$

Note that this operator satisfies the relation $A_m^2 = m!\, A_m$.

PROPOSITION 1.6.2. *If* $u_i - u_{i+1} = 1$ *for all* $i = 1, \dots, m-1$, *then*

$$R(u_1, \dots, u_m) = A_m.$$

PROOF. We use induction on m. Case $m = 2$ is obvious for

$$A_2 = R_{12}(1) = 1 - P_{12}.$$

Due to (1.46), by the induction hypothesis we have for $m > 2$

$$R(u_1, \dots, u_m) = A'_{m-1}\, (R_{1m} \dots R_{12}),$$

where A'_{m-1} denotes the anti-symmetrizer corresponding to the subset of indices $\{2, \dots, m\}$. Observe that for any indices $m \geqslant i_1 > i_2 > \dots > i_k > 1$ we have

$$A'_{m-1}\, P_{1i_1}\, P_{1i_2} \dots P_{1i_k} = A'_{m-1}\, P_{i_1 i_2} \dots P_{i_1 i_k}\, P_{1i_1} = (-1)^{k-1}\, A'_{m-1}\, P_{1i_1}. \tag{1.47}$$

Therefore,

$$A'_{m-1}\,(R_{1m}\dots R_{12}) = A'_{m-1}\left(1-\frac{P_{1m}}{m-1}\right)\dots\left(1-\frac{P_{12}}{1}\right) \tag{1.48}$$
$$= A'_{m-1}\,(1-\alpha_2\,P_{12}-\dots-\alpha_m\,P_{1m})$$

with

$$\alpha_r = \frac{1}{r-1}\sum_{k\geqslant 1}\ \sum_{r>i_2>\dots>i_k>1}\frac{1}{(i_2-1)\dots(i_k-1)} = 1$$

for any $r = 2,\dots,m$. It remains to note that

$$A'_{m-1}\,(1-P_{12}-\dots-P_{1m}) = A_m. \qquad \square$$

Similarly, denote by H_m the image of the symmetrizer

$$\sum_{p\in\mathfrak{S}_m} p \in \mathbb{C}[\mathfrak{S}_m]$$

under the same action of $\mathfrak{S}_m$ on $(\mathbb{C}^N)^{\otimes m}$. We have

$$H_m(e_{i_1}\otimes\dots\otimes e_{i_m}) = \sum_{p\in\mathfrak{S}_m} e_{i_{p(1)}}\otimes\dots\otimes e_{i_{p(m)}}.$$

Note the relation $H_m^2 = m!\,H_m$. The following is a symmetric counterpart of Proposition 1.6.2 which is proved by the same argument.

PROPOSITION 1.6.3. *If $u_i - u_{i+1} = -1$ for all $i = 1,\dots,m-1$, then*

$$R(u_1,\dots,u_m) = H_m. \qquad \square$$

REMARK 1.6.4. Propositions 1.6.2 and 1.6.3 are particular cases of a more general result associated with the *fusion procedure*; see Theorem 6.4.7 below. This provides another proof of these propositions. □

By Propositions 1.6.1 and 1.6.2 we have the equality of the formal power series in u^{-1} with coefficients in the algebra (1.8),

$$A_m\,T_1(u)\dots T_m(u-m+1) = T_m(u-m+1)\dots T_1(u)\,A_m. \tag{1.49}$$

Here we identify A_m with $A_m\otimes 1$. Now take $m = N$. The image of the operator A_N on $(\mathbb{C}^N)^{\otimes N}$ is one-dimensional. Therefore, the element (1.49) with $m = N$ equals A_N times a scalar series with coefficients in $\mathrm{Y}(\mathfrak{gl}_N)$. This prompts the following definition.

DEFINITION 1.6.5. The *quantum determinant* of the matrix $T(u)$ with the coefficients in $\mathrm{Y}(\mathfrak{gl}_N)$ is the formal series

$$\operatorname{qdet} T(u) = 1 + d_1 u^{-1} + d_2 u^{-2} + \dots$$

such that the element (1.49) with $m = N$ equals $A_N \operatorname{qdet} T(u)$.

PROPOSITION 1.6.6. *For any permutation $q\in\mathfrak{S}_N$ we have*

$$\operatorname{qdet} T(u) = \operatorname{sgn} q \sum_{p\in\mathfrak{S}_N} \operatorname{sgn} p\cdot t_{p(1),q(1)}(u)\dots t_{p(N),q(N)}(u-N+1) \tag{1.50}$$

$$= \operatorname{sgn} q \sum_{p\in\mathfrak{S}_N} \operatorname{sgn} p\cdot t_{q(1),p(1)}(u-N+1)\dots t_{q(N),p(N)}(u). \tag{1.51}$$

In particular,

$$\operatorname{qdet} T(u) = \sum_{p\in\mathfrak{S}_N} \operatorname{sgn} p \cdot t_{p(1),1}(u)\dots t_{p(N),N}(u-N+1) \tag{1.52}$$

$$= \sum_{p\in\mathfrak{S}_N} \operatorname{sgn} p \cdot t_{1,p(1)}(u-N+1)\dots t_{N,p(N)}(u). \tag{1.53}$$

PROOF. Let us apply both sides of the identity

$$A_N\, T_1(u)\dots T_N(u-N+1) = A_N \operatorname{qdet} T(u)$$

to the vector $e_{q(1)}\otimes\dots\otimes e_{q(N)}$. By applying the left-hand side, we get

$$\sum_{i_1,\dots,i_N} A_N(e_{i_1}\otimes\dots\otimes e_{i_N})\otimes t_{i_1,q(1)}(u)\dots t_{i_N,q(N)}(u-N+1).$$

The vector $A_N(e_{i_1}\otimes\dots\otimes e_{i_N})$ is zero unless the sequence of indices $i_1,\dots,i_N$ is a permutation $p(1),\dots,p(N)$ of the sequence $1,\dots,N$ for some $p\in\mathfrak{S}_N$. Then

$$A_N\big(e_{p(1)}\otimes\dots\otimes e_{p(N)}\big) = \operatorname{sgn} p\cdot A_N(e_1\otimes\dots\otimes e_N).$$

The application of the right-hand side gives

$$\operatorname{sgn} q\cdot \operatorname{qdet} T(u)\cdot A_N(e_1\otimes\dots\otimes e_N),$$

which completes the proof of (1.50). For the proof of (1.51) we use the identity

$$T_N(u-N+1)\dots T_1(u)\, A_N = A_N \operatorname{qdet} T(u).$$

Apply both sides to the vector $e_N\otimes\dots\otimes e_1$ and compare the coefficients of the vector $e_{q(N)}\otimes\dots\otimes e_{q(1)}$. $\square$

EXAMPLE 1.6.7. In the case $N=2$ we have

$$\begin{aligned}\operatorname{qdet} T(u) &= t_{11}(u)\,t_{22}(u-1) - t_{21}(u)\,t_{12}(u-1)\\ &= t_{22}(u)\,t_{11}(u-1) - t_{12}(u)\,t_{21}(u-1)\\ &= t_{11}(u-1)\,t_{22}(u) - t_{12}(u-1)\,t_{21}(u)\\ &= t_{22}(u-1)\,t_{11}(u) - t_{21}(u-1)\,t_{12}(u).\end{aligned}$$ $\square$

More generally, assuming that $m\leqslant N$ is arbitrary, we can define $m\times m$ quantum minors as the matrix elements of the operator (1.49). Namely, the operator (1.49) can be written as

$$\sum e_{a_1b_1}\otimes\dots\otimes e_{a_mb_m}\otimes t^{a_1\dots a_m}_{b_1\dots b_m}(u),$$

summed over the indices $a_i,b_i\in\{1,\dots,N\}$, where $t^{a_1\dots a_m}_{b_1\dots b_m}(u)\in \mathrm{Y}(\mathfrak{gl}_N)[[u^{-1}]]$. We call these elements the *quantum minors* of the matrix $T(u)$. The following formulas are obvious generalizations of (1.50) and (1.51),

$$t^{a_1\dots a_m}_{b_1\dots b_m}(u) = \sum_{p\in\mathfrak{S}_m} \operatorname{sgn} p\cdot t_{a_{p(1)}b_1}(u)\dots t_{a_{p(m)}b_m}(u-m+1) \tag{1.54}$$

$$= \sum_{p\in\mathfrak{S}_m} \operatorname{sgn} p\cdot t_{a_1b_{p(1)}}(u-m+1)\dots t_{a_mb_{p(m)}}(u). \tag{1.55}$$

It is clear from the definition that the quantum minors are skew-symmetric with respect to permutations of the upper indices and of the lower indices:

$$t^{a_{p(1)}\dots a_{p(m)}}_{b_1\dots b_m}(u) = \operatorname{sgn} p\cdot t^{a_1\dots a_m}_{b_1\dots b_m}(u) \quad\text{and}\quad t^{a_1\dots a_m}_{b_{p(1)}\dots b_{p(m)}}(u) = \operatorname{sgn} p\cdot t^{a_1\dots a_m}_{b_1\dots b_m}(u)$$

for any $p \in \mathfrak{S}_m$. The following quantum analogues of the row and column expansion formulas take place.

PROPOSITION 1.6.8. *We have the relations*

$$\begin{aligned} t^{a_1\dots a_m}_{b_1\dots b_m}(u) &= \sum_{l=1}^{m}(-1)^{m-l}\, t^{a_1\dots\widehat{a}_l\dots a_m}_{b_1\dots b_{m-1}}(u)\, t_{a_l b_m}(u-m+1) \\ &= \sum_{l=1}^{m}(-1)^{m-l}\, t^{a_1\dots a_{m-1}}_{b_1\dots\widehat{b}_l\dots b_m}(u-1)\, t_{a_m b_l}(u) \\ &= \sum_{l=1}^{m}(-1)^{l-1}\, t_{a_l b_1}(u)\, t^{a_1\dots\widehat{a}_l\dots a_m}_{b_2\dots b_m}(u-1) \\ &= \sum_{l=1}^{m}(-1)^{l-1}\, t_{a_1 b_l}(u-m+1)\, t^{a_2\dots a_m}_{b_1\dots\widehat{b}_l\dots b_m}(u), \end{aligned}$$

where the hats indicate the indices to be omitted.

PROOF. All relations are immediate from the explicit formulas for the quantum minors. □

PROPOSITION 1.6.9. *The images of quantum minors under the comultiplication* (1.35) *are given by*

$$\Delta\big(t^{a_1\dots a_m}_{b_1\dots b_m}(u)\big) = \sum_{c_1<\dots<c_m} t^{a_1\dots a_m}_{c_1\dots c_m}(u)\otimes t^{c_1\dots c_m}_{b_1\dots b_m}(u), \tag{1.56}$$

summed over all subsets of indices $\{c_1,\dots,c_m\}$ *from* $\{1,\dots,N\}$.

PROOF. Using the notation of Section 1.5 we can write the image of the left-hand side of (1.49) under the comultiplication Δ as the expression

$$A_m\, T_{1[1]}(u)T_{1[2]}(u)\dots T_{m[1]}(u-m+1)T_{m[2]}(u-m+1).$$

Here the element $A_m \in (\operatorname{End}\mathbb{C}^N)^{\otimes m}$ is identified with the element $A_m \otimes 1^{\otimes n}$ of the algebra (1.36). Permuting the elements $T_{a[b]}(u)$ with distinct indices we write this expression as

$$A_m\, T_{1[1]}(u)\dots T_{m[1]}(u-m+1)\, T_{1[2]}(u)\dots T_{m[2]}(u-m+1).$$

Next, using $m!\,A_m = A_m^2$ and the relation (1.49) we bring this to the form

$$\frac{1}{m!}\, A_m\, T_{m[1]}(u-m+1)\dots T_{1[1]}(u)\, A_m\, T_{1[2]}(u)\dots T_{m[2]}(u-m+1)$$

which, by (1.49), coincides with

$$\frac{1}{m!}\, A_m\, T_{1[1]}(u)\dots T_{m[1]}(u-m+1)\, A_m\, T_{1[2]}(u)\dots T_{m[2]}(u-m+1).$$

Taking here the matrix elements and using the skew-symmetry of the quantum minors we come to (1.56). □

The following corollary means that under the comultiplication Δ on $\mathrm{Y}(\mathfrak{gl}_N)$, the quantum determinant $\operatorname{qdet} T(u)$ is *comultiplicative.*

COROLLARY 1.6.10. *We have*

$$\Delta : \operatorname{qdet} T(u) \mapsto \operatorname{qdet} T(u) \otimes \operatorname{qdet} T(u).$$

PROOF. This follows from Proposition 1.6.9 as $\operatorname{qdet} T(u) = t^{1\dots N}_{1\dots N}(u)$. □

1.7. Center of the algebra $\mathrm{Y}(\mathfrak{gl}_N)$

In the following k, l, a_i, b_j are arbitrary indices from the set $\{1, \dots, N\}$.

PROPOSITION 1.7.1. *We have the relations*

$$(u-v)\,[\,t_{kl}(u), t^{a_1\dots a_m}_{b_1\dots b_m}(v)] = \sum_{i=1}^{m} t_{a_i l}(u)\, t^{a_1\dots k\dots a_m}_{b_1\ \dots\ \ b_m}(v) - \sum_{i=1}^{m} t^{a_1\ \dots\ \ a_m}_{b_1\dots l\dots b_m}(v)\, t_{kb_i}(u)$$

and

$$(u-v+m-1)\,[\,t_{kl}(u), t^{a_1\dots a_m}_{b_1\dots b_m}(v)]$$
$$= \sum_{i=1}^{m} t^{a_1\dots k\dots a_m}_{b_1\ \dots\ \ b_m}(v)\, t_{a_i l}(u) - \sum_{i=1}^{m} t_{kb_i}(u)\, t^{a_1\ \dots\ \ a_m}_{b_1\dots l\dots b_m}(v),$$

where the indices k and l in the quantum minors replace a_i and b_i, respectively.

PROOF. By Proposition 1.6.1, we have the relation

$$\begin{aligned}(1.57)\quad & R(u, v, v-1, \dots, v-m+1)\, T_0(u)\, T_1(v) \dots T_m(v-m+1)\\ &= T_m(v-m+1) \dots T_1(v)\, T_0(u)\, R(u, v, v-1, \dots, v-m+1),\end{aligned}$$

where we have used an extra copy of the algebra $\operatorname{End}\mathbb{C}^N$ labelled by 0. Using (1.44) and Proposition 1.6.2 we get

$$R(u, v, v-1, \dots, v-m+1) = A_m\, R_{0m}(u-v+m-1) \dots R_{01}(u-v).$$

Then we proceed exactly as in the proof of Proposition 1.6.2 by using the relation (1.47) adjusted to our present notation. The calculation gives the following generalization of (1.48):

$$(1.58)\qquad R(u, v, v-1, \dots, v-m+1) = A_m \left(1 - \frac{1}{u-v}(P_{01} + \dots + P_{0m})\right).$$

Now apply both sides of (1.57) to the vector $e_l \otimes e_{b_1} \otimes \dots \otimes e_{b_m}$ and compare the coefficients of the vector $e_k \otimes e_{a_1} \otimes \dots \otimes e_{a_m}$. This yields the first relation. In order to prove the second, we use the relation

$$\begin{aligned}& R(v, v-1, \dots, v-m+1, u)\, T_1(v) \dots T_m(v-m+1)\, T_{m+1}(u)\\ &\qquad = T_{m+1}(u)\, T_m(v-m+1) \dots T_1(v)\, R(v, v-1, \dots, v-m+1, u)\end{aligned}$$

implied by Proposition 1.6.1. Similar to the above argument, we show with the use of (1.45) that

$$R(v, v-1, \dots, v-m+1, u) = A_m \left(1 + \frac{1}{u-v+m-1}(P_{1,m+1} + \dots + P_{m,m+1})\right).$$

The argument is now completed in the same way as for the first formula. ☐

COROLLARY 1.7.2. *For any indices i, j we have*

$$[t_{a_i b_j}(u), t^{a_1\dots a_m}_{b_1\dots b_m}(v)] = 0.$$

PROOF. The quantum minor is zero if it has two repeated upper or lower indices. Therefore, the right-hand side of either relation of Proposition 1.7.1 with $k = a_i$ and $l = b_j$ will be the commutator which occurs in the left-hand side. Hence the commutator is zero. ☐

We record the following relation derived in the proof of Proposition 1.7.1.

COROLLARY 1.7.3. *We have*

$$A_m\, R_{0m}(u-v+m-1)\dots R_{01}(u-v) = A_m\left(1-\frac{1}{u-v}(P_{01}+\dots+P_{0m})\right).$$

For the proof of the next theorem we need a general property of the universal enveloping algebras of the form $\mathrm{U}(\mathfrak{g}[z])$, where z is an indeterminate and $\mathfrak{g}[z]\cong\mathfrak{g}\otimes\mathbb{C}[z]$ is the polynomial current Lie algebra corresponding to a Lie algebra $\mathfrak{g}$.

LEMMA 1.7.4. *Let $\mathfrak{a}$ be a subalgebra of a finite-dimensional Lie algebra $\mathfrak{g}$ such that $\mathfrak{a}$ is reductive in $\mathfrak{g}$. Let $\mathfrak{b}$ be the centralizer of $\mathfrak{a}$ in $\mathfrak{g}$. Then the centralizer of $\mathrm{U}(\mathfrak{a}[z])$ in $\mathrm{U}(\mathfrak{g}[z])$ is equal to $\mathrm{U}(\mathfrak{b}[z])$. Moreover, if the center of $\mathfrak{g}$ is trivial, then the center of $\mathrm{U}(\mathfrak{g}[z])$ is also trivial.*

PROOF. For the first statement, it suffices to prove that the space of invariants in the symmetric algebra $\mathrm{S}(\mathfrak{g}[z])$ regarded as the adjoint $\mathfrak{a}[z]$-module is equal to $\mathrm{S}(\mathfrak{b}[z])$. Since $\mathfrak{g}$ is completely reducible as an adjoint $\mathfrak{a}$-module by our assumption, we can pick an $\mathfrak{a}$-invariant complement $\mathfrak{b}'$ to the trivial $\mathfrak{a}$-submodule $\mathfrak{b}$ in $\mathfrak{g}$. Let $\{Y_1,\dots,Y_p\}$ be a basis of $\mathfrak{b}'$ and let $\{Y_{p+1},\dots,Y_n\}$ be a basis of $\mathfrak{b}$. Suppose that $B\in\mathrm{S}(\mathfrak{g}[z])$ is an $\mathfrak{a}[z]$-invariant element. Define m to be the minimum nonnegative integer such that B has the form

$$B=\sum_l B_l\,(Y_1\,z^m)^{l_1}\dots(Y_p\,z^m)^{l_p},$$

where $l=(l_1,\dots,l_p)$ ranges over p-tuples of nonnegative integers and B_l is a polynomial in the variables $Y_i\,z^r$ for $1\leqslant i\leqslant p$ and $r<m$ together with the variables $Y_j\,z^r$ for $p<j\leqslant n$ and $r\geqslant 0$. Pick a basis $\{X_1,\dots,X_q\}$ for $\mathfrak{a}$ and let

$$[X_i,\,Y_j]=\sum_{k=1}^{p} c_{ij}^k\,Y_k,$$

where $c_{ij}^k\in\mathbb{C}$. By the definition of B we have

$$\mathrm{ad}(X_i\,z)(B)=0\quad\text{for}\quad i=1,\dots,q.$$

The component of the left-hand side that contains the elements of the form $Y_k\,z^{m+1}$ must be zero; that is,

$$\sum_l B_l\sum_{j=1}^p l_j\,(Y_1\,z^m)^{l_1}\dots(Y_j\,z^m)^{l_j-1}\dots(Y_p\,z^m)^{l_p}\sum_{k=1}^p c_{ij}^k\,Y_k z^{m+1}=0. \tag{1.59}$$

Taking here the coefficient of $Y_k\,z^{m+1}$ we obtain that

$$\sum_l B_l\sum_{j=1}^p l_j\,c_{ij}^k\,(Y_1\,z^m)^{l_1}\dots(Y_j\,z^m)^{l_j-1}\dots(Y_p\,z^m)^{l_p}=0.$$

Thus, for any p-tuple of nonnegative integers $l=(l_1,\dots,l_p)$ we have

$$\sum_{j=1}^n B_{l+\delta_j}(l_j+1)\,c_{ij}^k=0\qquad\text{for}\quad i=1,\dots,q,\quad k=1,\dots,p, \tag{1.60}$$

where $l+\delta_j$ denotes the tuple $(l_1,\dots,l_j+1,\dots,l_p)$. Fix l and observe that the elements

$$Y_j'=(l_j+1)\,Y_j\quad\text{where}\quad j=1,\dots,p$$

also form a basis of $\mathfrak{b}'$. Since $\mathfrak{a}$ has no nontrivial invariants in $\mathfrak{b}'$, the system of linear equations for the variables $b_1, \dots, b_p$,

$$[X_i\,, \sum_{j=1}^{p} b_j\, Y_j'] = 0 \quad \text{where} \quad i = 1, \dots, q,$$

has only trivial solution. This system can be rewritten as the system of pq equations

$$\sum_{j=1}^{p} b_j (l_j + 1)\, c_{ij}^k = 0 \quad \text{where} \quad i = 1, \dots, q, \quad k = 1, \dots, p.$$

Comparing the latter system with (1.60) we see that $B_{l+\delta_j} = 0$. Thus we obtain that $B_l = 0$ for all $l \neq 0$, and so by the minimality of m we must have $m = 0$ and $B \in \mathrm{S}(\mathfrak{b}\,[z])$.

For the second statement of the lemma, we observe that in the case $\mathfrak{a} = \mathfrak{g}$ and $\mathfrak{b} = 0$ the above argument applies without the assumption that $\mathfrak{g}$ is reductive. □

Consider the elements $d_i \in \mathrm{Y}(\mathfrak{gl}_N)$ introduced in Definition 1.6.5.

THEOREM 1.7.5. *The coefficients* $d_1, d_2, \dots$ *of the series* $\operatorname{qdet} T(u)$ *belong to the center of the algebra* $\mathrm{Y}(\mathfrak{gl}_N)$. *Moreover, these elements are algebraically independent and generate the center of* $\mathrm{Y}(\mathfrak{gl}_N)$.

PROOF. The first claim of the theorem follows from Corollary 1.7.2. To prove the second claim, let us consider the filtration on the algebra $\mathrm{Y}(\mathfrak{gl}_N)$ introduced in Section 1.5. The corresponding graded algebra $\mathrm{gr}'\,\mathrm{Y}(\mathfrak{gl}_N)$ is isomorphic to the universal enveloping algebra $\mathrm{U}(\mathfrak{gl}_N[z])$ due to Proposition 1.5.2.

Now we derive from (1.52) that the coefficient d_r of $\operatorname{qdet} T(u)$ has the form

$$d_r = t_{11}^{(r)} + \dots + t_{NN}^{(r)} \quad \text{plus terms of degree less than} \quad r - 1.$$

Therefore, the image of d_r in the $(r-1)$-th component of $\mathrm{gr}'\,\mathrm{Y}(\mathfrak{gl}_N)$ coincides with $I\,z^{r-1}$ where $I = E_{11} + \dots + E_{NN}$. The elements $I\,z^{r-1}$ with $r \geqslant 1$ are algebraically independent, hence so are the elements d_r. By Lemma 1.7.4, the center of $\mathrm{U}(\mathfrak{sl}_N[z])$ is trivial so that the elements $I\,z^{r-1}$ generate the center of $\mathrm{U}(\mathfrak{gl}_N[z])$. This implies that $d_1, d_2, \dots$ generate the center of the algebra $\mathrm{Y}(\mathfrak{gl}_N)$. □

1.8. Yangian for $\mathfrak{sl}_N$

In this section we regard the special linear Lie algebra $\mathfrak{sl}_N$ as a subalgebra in $\mathfrak{gl}_N$. For any series $f(u) \in 1 + u^{-1}\mathbb{C}[[u^{-1}]]$ consider the automorphism (1.20) of the associative algebra $\mathrm{Y}(\mathfrak{gl}_N)$.

DEFINITION 1.8.1. The *Yangian* for $\mathfrak{sl}_N$ is the subalgebra $\mathrm{Y}(\mathfrak{sl}_N)$ of $\mathrm{Y}(\mathfrak{gl}_N)$ which consists of the elements stable under all automorphisms (1.20). □

Let us denote by $\mathrm{ZY}(\mathfrak{gl}_N)$ the center of the algebra $\mathrm{Y}(\mathfrak{gl}_N)$.

THEOREM 1.8.2. *The algebra* $\mathrm{Y}(\mathfrak{gl}_N)$ *is isomorphic to the tensor product of its subalgebras*

$$\mathrm{Y}(\mathfrak{gl}_N) = \mathrm{ZY}(\mathfrak{gl}_N) \otimes \mathrm{Y}(\mathfrak{sl}_N).$$

In particular, the center of $\mathrm{Y}(\mathfrak{sl}_N)$ *is trivial.*

PROOF. It is straightforward to verify that there exists a unique formal power series

$$\widetilde{d}(u) = 1 + \widetilde{d}_1\, u^{-1} + \widetilde{d}_2\, u^{-2} + \cdots \in \mathrm{ZY}(\mathfrak{gl}_N)[[u^{-1}]]$$

which satisfies

$$\widetilde{d}(u)\, \widetilde{d}(u-1) \dots \widetilde{d}(u-N+1) = \operatorname{qdet} T(u).$$

Then by Proposition 1.6.6, the automorphism (1.20) maps

$$\widetilde{d}(u) \mapsto f(u)\, \widetilde{d}(u). \tag{1.61}$$

This implies that all coefficients of the series

$$\widetilde{t}_{ij}(u) = \widetilde{d}(u)^{-1}\, t_{ij}(u) \tag{1.62}$$

belong to the subalgebra $\mathrm{Y}(\mathfrak{sl}_N)$. By multiplying the series $\widetilde{t}_{ij}(u)$ by $\widetilde{d}(u)$, we get the series $t_{ij}(u)$ back. Hence every element of the Yangian $\mathrm{Y}(\mathfrak{gl}_N)$ can be presented as a polynomial in $\widetilde{d}_1, \widetilde{d}_2, \dots$ with coefficients in $\mathrm{Y}(\mathfrak{sl}_N)$. In order to show that such presentation is unique, suppose on the contrary that for some positive integer n there exists a nonzero polynomial B in n variables with the coefficients from the algebra $\mathrm{Y}(\mathfrak{sl}_N)$ such that

$$B(\widetilde{d}_1, \dots, \widetilde{d}_n) = 0. \tag{1.63}$$

Now consider the minimal n with this property. All coefficients of the polynomial B are stable under any automorphism (1.20). Hence by applying to the equality (1.63) the automorphism (1.20) where $f(u) = 1 + c u^{-n}$ and $c \in \mathbb{C}$, we get

$$B(\widetilde{d}_1, \dots, \widetilde{d}_n + c) = 0$$

for every $c \in \mathbb{C}$. This means that the polynomial B does not depend on its n-th variable, which contradicts the choice of n. □

COROLLARY 1.8.3. *The algebra* $\mathrm{Y}(\mathfrak{sl}_N)$ *is isomorphic to the quotient of* $\mathrm{Y}(\mathfrak{gl}_N)$ *by the ideal generated by the elements* $d_1, d_2, \dots$; *i.e.,*

$$\mathrm{Y}(\mathfrak{sl}_N) \cong \mathrm{Y}(\mathfrak{gl}_N)/(\operatorname{qdet} T(u) = 1).$$

PROOF. Let I be the ideal of $\mathrm{Y}(\mathfrak{gl}_N)$ generated by the coefficients $d_1, d_2, \dots$ of $\operatorname{qdet} T(u)$. Then Theorem 1.8.2 implies that $\mathrm{Y}(\mathfrak{gl}_N) = \mathrm{I} \oplus \mathrm{Y}(\mathfrak{sl}_N)$. □

PROPOSITION 1.8.4. *The subalgebra* $\mathrm{Y}(\mathfrak{sl}_N)$ *of* $\mathrm{Y}(\mathfrak{gl}_N)$ *is a Hopf algebra whose comultiplication, antipode and counit are obtained by restricting those from* $\mathrm{Y}(\mathfrak{gl}_N)$.

PROOF. The coefficients of all the series (1.62) clearly generate the subalgebra $\mathrm{Y}(\mathfrak{sl}_N)$. On the other hand, Corollary 1.6.10 implies that

$$\Delta : \widetilde{d}(u) \mapsto \widetilde{d}(u) \otimes \widetilde{d}(u). \tag{1.64}$$

Therefore the image of $\mathrm{Y}(\mathfrak{sl}_N)$ under the comultiplication on $\mathrm{Y}(\mathfrak{gl}_N)$ is contained in $\mathrm{Y}(\mathfrak{sl}_N) \otimes \mathrm{Y}(\mathfrak{sl}_N)$. Using Definition 1.6.5, we find that the image of $\operatorname{qdet} T(u)$ under the antipode S is $(\operatorname{qdet} T(u))^{-1}$, and so

$$\mathrm{S} : \widetilde{d}(u)^{-1}\, T(u) \mapsto \widetilde{d}(u)\, T^{-1}(u).$$

The automorphism (1.20) maps $\widetilde{d}(u) \mapsto f(u)\, \widetilde{d}(u)$, leaving the product $\widetilde{d}(u)\, T^{-1}(u)$ invariant. Therefore the subalgebra $\mathrm{Y}(\mathfrak{sl}_N)$ of $\mathrm{Y}(\mathfrak{gl}_N)$ is stable under S. □

1.9. Quantum Liouville formula

DEFINITION 1.9.1. The *quantum comatrix* $\widehat{T}(u)$ is defined by

$$\widehat{T}(u)\, T(u-N+1) = \operatorname{qdet} T(u). \tag{1.65}$$ □

PROPOSITION 1.9.2. *The entries $\widehat{t}_{ij}(u)$ of the matrix $\widehat{T}(u)$ are given by*

$$\widehat{t}_{ij}(u) = (-1)^{i+j}\, t^{\,1\dots\widehat{j}\dots N}_{\,1\dots\widehat{i}\dots N}(u), \tag{1.66}$$

where the hats on the right-hand side indicate the indices to be omitted. Moreover, we have the relation

$$\widehat{T}^{\,t}(u-1)\, T^{\,t}(u) = \operatorname{qdet} T(u). \tag{1.67}$$

PROOF. Using Definition 1.6.5 we derive from (1.65) that

$$A_N\, T_1(u) \dots T_{N-1}(u-N+2) = A_N\, \widehat{T}_N(u).$$

Taking here the matrix elements we obtain the equality (1.66). Using Proposition 1.6.6 we find that under the automorphism (1.32),

$$\widehat{T}(u) \mapsto \widehat{T}^{\,t}(-u+N-2) \quad \text{and} \quad d(u) \mapsto d(-u+N-1)$$

where $d(u) = \operatorname{qdet} T(u)$. Now applying (1.32) to the equality (1.65) and replacing $-u+N-1$ by u, we get (1.67). □

PROPOSITION 1.9.3. *The mapping*

$$T(u) \mapsto \widehat{T}(u)$$

defines an anti-automorphism of $\mathrm{Y}(\mathfrak{gl}_N)$.

PROOF. By definition of the quantum comatrix, we have

$$\widehat{T}(u) = \operatorname{qdet} T(u)\, T^{-1}(u-N+1).$$

Since the coefficients of $\operatorname{qdet} T(u)$ are central in $\mathrm{Y}(\mathfrak{gl}_N)$, we conclude from Propositions 1.3.1 and 1.3.3 that relation (1.28) is satisfied by the matrix $\widehat{T}(u)$. We will now verify that the mapping is invertible. It suffices to show that the map

$$T(u) \mapsto \operatorname{qdet} T(u)\, T(u)$$

is invertible. However, the inverse map of the latter is given by

$$T(u) \mapsto h(u)\, T(u),$$

where $h(u)$ is the series with coefficients in the center of the Yangian which are uniquely determined by the relation

$$h(u)^2\, h(u-1) \dots h(u-N+1) \operatorname{qdet} T(u) = 1.$$ □

COROLLARY 1.9.4. *The mapping*

$$t_{ij}(u) \mapsto (-1)^{i+j}\, t^{\,1\dots\widehat{j}\dots N}_{\,1\dots\widehat{i}\dots N}(-u)$$

defines an automorphism of $\mathrm{Y}(\mathfrak{gl}_N)$.

PROOF. This automorphism is the composition of the anti-automorphisms given by (1.25) and Proposition 1.9.3. □

We will now construct another family of generators of the center of $\mathrm{Y}(\mathfrak{gl}_N)$. Consider the series $z(u)$ with coefficients from $\mathrm{Y}(\mathfrak{gl}_N)$ given by the formula

$$z(u)^{-1} = \frac{1}{N}\,\mathrm{tr}\,\Big(T(u)\,T^{-1}(u-N)\Big), \tag{1.68}$$

so that

$$z(u) = 1 + z_2u^{-2} + z_3u^{-3} + \dots \quad \text{where} \quad z_i \in \mathrm{Y}(\mathfrak{gl}_N).$$

THEOREM 1.9.5. *We have the relation*

$$z(u) = \frac{\mathrm{qdet}\, T(u-1)}{\mathrm{qdet}\, T(u)}. \tag{1.69}$$

PROOF. From (1.68) and (1.65) we find

$$z(u)^{-1} = \frac{1}{N}\,\mathrm{tr}\,\big(T(u)\,\widehat{T}(u-1)\,(\mathrm{qdet}\, T(u-1))^{-1}\big).$$

Using the centrality of $\mathrm{qdet}\, T(u)$ and (1.67) we get

$$T^t(u)\,\widehat{T}^t(u-1) = \mathrm{qdet}\, T(u)$$

and so

$$\mathrm{tr}\,\big(T(u)\,\widehat{T}(u-1)\big) = N\,\mathrm{qdet}\, T(u),$$

implying (1.69). □

REMARK 1.9.6. Relation (1.69) may be regarded as a "quantum analogue" of the classical Liouville formula for the derivative of the determinant of a matrix-valued function. To see this, for each nonzero $h \in \mathbb{C}$ consider the algebra $\mathrm{Y}(\mathfrak{gl}_N, h)$ introduced in Remark 1.4.4. Define the corresponding generating series $t_{ij}(u)$ for the elements $t_{ij}^{(1)}, t_{ij}^{(2)}, \dots \in \mathrm{Y}(\mathfrak{gl}_N, h)$ and form the matrix $T(u)$. The quantum determinant and the series $z(u)$ for the new algebra are respectively given by

$$\mathrm{qdet}\, T(u) = \sum_{p\in\mathfrak{S}_N} \mathrm{sgn}\, p \cdot t_{p(1),1}(u)\, t_{p(2),2}(u-h) \dots t_{p(N),N}(u - Nh + h)$$

and

$$z(u)^{-1} = \frac{1}{N}\,\mathrm{tr}\,\Big(T(u)\,T^{-1}(u-Nh)\Big).$$

Then the coefficients of the series $\mathrm{qdet}\, T(u)$ and $z(u)$ are central in $\mathrm{Y}(\mathfrak{gl}_N, h)$. The equality (1.69) is generalized to

$$z(u) = \frac{\mathrm{qdet}\, T(u-h)}{\mathrm{qdet}\, T(u)},$$

which can be rewritten as

$$\begin{aligned} &\mathrm{tr}\,\Big(T^{-1}(u-Nh)\cdot\frac{T(u)-T(u-Nh)}{Nh}\Big)\\ &= \frac{1}{\mathrm{qdet}\, T(u-h)}\cdot\frac{\mathrm{qdet}\, T(u) - \mathrm{qdet}\, T(u-h)}{h}. \end{aligned}$$

In the limit $h \to 0$ the entries of the matrix $T(u)$ become commutative while the quantum determinant tends to the usual $\det T(u)$. In this limit we obtain from the last displayed equality that

$$\mathrm{tr}\,\Big(T^{-1}(u)\frac{d}{du}T(u)\Big) = \frac{1}{\det T(u)}\cdot\frac{d}{du}\det T(u)$$

which is the Liouville formula. For this reason we refer to (1.69) as the *quantum Liouville formula* for the T-matrix. □

COROLLARY 1.9.7. *The coefficients* $z_2, z_3, \dots$ *of the formal series* $z(u)$ *are algebraically independent generators of the center of* $\mathrm{Y}(\mathfrak{gl}_N)$.

PROOF. This follows from Theorem 1.9.5 by writing the relation

$$z(u)\,\mathrm{qdet}\,T(u) = \mathrm{qdet}\,T(u-1)$$

in terms of the coefficients z_i and d_i. □

The series $z(u)$ can be defined equivalently in a way similar to the quantum determinant; cf. Definition 1.6.5. Let Q be the operator on the vector space $\mathbb{C}^N \otimes \mathbb{C}^N$, defined by (1.15).

PROPOSITION 1.9.8. *We have the relations*

$$Q\,T_1(u)\,T_2^{-1}(u-N)^t = T_2^{-1}(u-N)^t\,T_1(u)\,Q = Q\,z(u)^{-1}$$

and

$$Q\,T_1^t(u)^{-1}T_2(u-N) = T_2(u-N)\,T_1^t(u)^{-1}\,Q = Q\,z(u). \tag{1.70}$$

PROOF. Let us multiply both sides of the ternary relation (1.19) by $T_2^{-1}(v)$ from the left and take the transposition with respect to the second copy of $\mathrm{End}\,\mathbb{C}^N$. Using the definition (1.14), we obtain the relation

$$R^t(u-v)\,\widetilde{T}_2(v)\,T_1(u) = T_1(u)\,\widetilde{T}_2(v)\,R^t(u-v),$$

where $\widetilde{T}(u) = \big(T^{-1}(u)\big)^t$. Multiplying this relation by $R^t(u-v)^{-1}$ from both sides and taking residue at $u = v+N$, due to (1.16) we get

$$Q\,T_1(v+N)\,\widetilde{T}_2(v) = \widetilde{T}_2(v)\,T_1(v+N)\,Q. \tag{1.71}$$

Since the image of the operator Q is one-dimensional, each side of the equality (1.71) must be equal to Q times a certain series in v^{-1} with the coefficients in $\mathrm{Y}(\mathfrak{gl}_N)$. Denote this series by $\xi(v)$. Applying the left-hand side of (1.71) to the basis vector $e_i \otimes e_j$ we obtain the equality

$$\sum_{a=1}^{N} t_{ai}(v+N)\,t'_{ja}(v) = \delta_{ij}\,\xi(v), \tag{1.72}$$

where the $t'_{ij}(u)$ denote the entries of the inverse matrix $T^{-1}(u)$. Putting $i=j$ and taking the sum over $i = 1,\dots,N$ we derive that $\xi(v) = z(v+N)^{-1}$. This proves the first relation. In order to prove the second, apply the transposition with respect to the first copy of $\mathrm{End}\,\mathbb{C}^N$ to the ternary relation (1.19) to get

$$T_1^t(u)\,R^t(u-v)\,T_2(v) = T_2(v)\,R^t(u-v)\,T_1^t(u).$$

Then multiply the relation by the inverses to $T_1^t(u)$ and $R^t(u-v)$ from both sides and take the residue at $v = u-N$. This gives the first equality in (1.70), which can also be rewritten in the equivalent form

$$Q\,T_2^{-1}(u-N)\,T_1^t(u) = T_1^t(u)\,T_2^{-1}(u-N)\,Q. \tag{1.73}$$

This operator must have the form $Q\,\eta(u)$ for a series $\eta(u)$ with the coefficients in $\mathrm{Y}(\mathfrak{gl}_N)$. Applying the right-hand side of (1.73) to the vector $e_i \otimes e_i$ and taking the coefficient of $e_i \otimes e_i$ we come to

$$\eta(u) = \sum_{a=1}^{N} t_{ai}(u)\, t'_{ia}(u-N).$$

The summation over i then gives $\eta(u) = z(u)^{-1}$, completing the proof. □

THEOREM 1.9.9. *The square of the antipode* S *is the automorphism of* $\mathrm{Y}(\mathfrak{gl}_N)$ *given by*

$$\mathrm{S}^2 : T(u) \mapsto z(u+N)\, T(u+N).$$

In particular, $\operatorname{qdet} T(u)$ *is stable under* S^2.

PROOF. Applying the anti-automorphism S to both sides of the identity

$$\sum_{a=1}^{N} t_{ja}(v)\, t'_{ai}(v) = \delta_{ij},$$

we get

$$\sum_{a=1}^{N} t''_{ai}(v)\, t'_{ja}(v) = \delta_{ij}, \tag{1.74}$$

where $t''_{ai}(v)$ is the image of $t_{ai}(v)$ under the automorphism S^2. Comparing (1.72) and (1.74) we conclude that $t''_{ai}(v) = t_{ai}(v+N)\, z(v+N)$. The second claim follows from Theorem 1.9.5. □

1.10. Factorization of the quantum determinant

Let $A = [a_{ij}]$ be an $N \times N$ matrix over a ring with 1. Denote by A^{ij} the matrix obtained from A by deleting the i-th row and j-th column. Suppose that the matrix A^{ij} is invertible.

DEFINITION 1.10.1. The *ij-th quasideterminant of A* is defined by the formula

$$|A|_{ij} = a_{ij} - r_i^j\, (A^{ij})^{-1}\, c_j^i,$$

where r_i^j is the row matrix obtained from the i-th row of A by deleting the element a_{ij}, and c_j^i is the column matrix obtained from the j-th column of A by deleting the element a_{ij}. □

EXAMPLE 1.10.2. For a 2×2 matrix A the four quasideterminants are

$$|A|_{11} = a_{11} - a_{12}\, a_{22}^{-1}\, a_{21}, \qquad |A|_{12} = a_{12} - a_{11}\, a_{21}^{-1}\, a_{22},$$
$$|A|_{21} = a_{21} - a_{22}\, a_{12}^{-1}\, a_{11}, \qquad |A|_{22} = a_{22} - a_{21}\, a_{11}^{-1}\, a_{12}.$$ □

In a more graphic fashion, the quasideterminant $|A|_{ij}$ is denoted by boxing the entry a_{ij},

$$|A|_{ij} = \begin{vmatrix} a_{11} & \cdots & a_{1j} & \cdots & a_{1N} \\ & \cdots & & \cdots & \\ a_{i1} & \cdots & \boxed{a_{ij}} & \cdots & a_{iN} \\ & \cdots & & \cdots & \\ a_{N1} & \cdots & a_{Nj} & \cdots & a_{NN} \end{vmatrix}.$$

LEMMA 1.10.3. *Suppose that the submatrix A^{11} of the matrix A is invertible. Then the system*

$$\begin{cases} a_{11}x_1 + \dots + a_{1N}x_N = y_1 \\ a_{21}x_1 + \dots + a_{2N}x_N = 0 \\ \dots \\ a_{N1}x_1 + \dots + a_{NN}x_N = 0 \end{cases}$$

implies that $y_1 = |A|_{11}\, x_1$.

PROOF. The column vector with coordinates $x_2, \dots, x_N$ can be expressed from the last $N-1$ equations as

$$\begin{bmatrix} x_2 \\ \vdots \\ x_N \end{bmatrix} = -\left(A^{11}\right)^{-1} \begin{bmatrix} a_{21} \\ \vdots \\ a_{N1} \end{bmatrix} x_1.$$

Substituting this into the first equation we get

$$\left(a_{11} - r_1^1 (A^{11})^{-1} c_1^1\right) x_1 = y_1.$$

By Definition 1.10.1, this is the desired relation. □

PROPOSITION 1.10.4. *Suppose that there exists the inverse matrix A^{-1} and its ji-th entry $(A^{-1})_{ji}$ is an invertible element of the ring. Then the ij-th quasideterminant of A is defined and given by*

$$|A|_{ij} = \left((A^{-1})_{ji}\right)^{-1}.$$

PROOF. First consider the case $i = j = 1$. Let $B = A^{-1}$ so that the element b_{11} is invertible. Using block multiplication of matrices we derive from $AB = 1$ that

$$A^{11}\left(B^{11} - \begin{bmatrix} b_{21} \\ \vdots \\ b_{N1} \end{bmatrix} b_{11}^{-1}\,[b_{12} \dots b_{1N}]\right) = 1,$$

proving that A^{11} is invertible so that $|A|_{11}$ is defined. Now let b be the first column of B. Then $A\,b = e_1$, where e_1 is the column vector which has 1 on the first position and zeros elsewhere. Lemma 1.10.3 gives $|A|_{11}\, b_{11} = 1$, proving the claim. In the case of arbitrary i and j we rearrange the matrix A as follows: move the i-th row up to the top position and then move the j-th column to the leftmost position. Rearranging the matrix B accordingly, we reduce the argument to the previous case, which gives $|A|_{ij}\, b_{ji} = 1$. □

If the entries of the matrix A belong to a commutative ring, then the ij-th quasideterminant of A can be given by

$$|A|_{ij} = (-1)^{i+j} \frac{\det A}{\det A^{ij}}. \tag{1.75}$$

For $m = 1, \dots, N$ denote by $T^{(m)}(u)$ the submatrix of $T(u)$ corresponding to the first m rows and columns. The coefficients of the series $t_{ij}(u)$ with $1 \leqslant i, j \leqslant m$ can be regarded as generators of the Yangian $\mathrm{Y}(\mathfrak{gl}_m)$; see Corollary 1.4.3. Hence the quantum determinant $\operatorname{qdet} T^{(m)}(u)$ is well-defined.

THEOREM 1.10.5. *The quantum determinant* $\operatorname{qdet} T(u)$ *admits the factorization in the algebra* $\mathrm{Y}(\mathfrak{gl}_N)[[u^{-1}]]$

$$\operatorname{qdet} T(u) = t_{11}(u)\, \big|T^{(2)}(u-1)\big|_{22} \cdots \big|T^{(N)}(u-N+1)\big|_{NN}.$$

Moreover, the N *factors on the right-hand side of this equality pairwise commute.*

PROOF. By Definition 1.9.1 we have

$$\widehat{T}(u) = \operatorname{qdet} T(u)\, T^{-1}(u-N+1). \tag{1.76}$$

Taking the NN-th entry we come to

$$\operatorname{qdet} T(u)\, \big(T^{-1}(u-N+1)\big)_{NN} = \widehat{t}_{NN}(u).$$

Propositions 1.9.2 and 1.10.4 give

$$\operatorname{qdet} T(u) = \operatorname{qdet} T^{(N-1)}(u)\, \big|T^{(N)}(u-N+1)\big|_{NN}.$$

Note that the factors here commute by the centrality of the quantum determinant. An obvious induction completes the proof. □

REMARK 1.10.6. Another decomposition of $\operatorname{qdet} T(u)$ can be obtained by starting the induction argument from the $(1,1)$ entry in (1.76). To write down the resulting formula, we will use the subscript (k) of a matrix to indicate its submatrix obtained by removing the first $k-1$ rows and columns. Then the alternative decomposition reads

$$\operatorname{qdet} T(u) = \big|T_{(1)}(u-N+1)\big|_{11} \cdots \big|T_{(N-1)}(u-1)\big|_{N-1,N-1}\, t_{NN}(u). \qquad \square$$

Fix an integer m such that $0 \leqslant m \leqslant N$. For any two subsets

$$\mathcal{P} = \{i_1, \dots, i_m\} \quad \text{and} \quad \mathcal{Q} = \{j_1, \dots, j_m\}$$

of $\{1, \dots, N\}$ with cardinality m let

$$\overline{\mathcal{P}} = \{i_{m+1}, \dots, i_N\} \quad \text{and} \quad \overline{\mathcal{Q}} = \{j_{m+1}, \dots, j_N\}$$

be their set complements in $\{1, \dots, N\}$. We assume that

$$i_1 < \dots < i_m \quad \text{and} \quad j_1 < \dots < j_m,$$
$$i_{m+1} < \dots < i_N \quad \text{and} \quad j_{m+1} < \dots < j_N.$$

For any $N \times N$ matrix X, we will denote by $X_{\mathcal{P}\mathcal{Q}}$ the submatrix whose rows and columns are enumerated by the elements of the sets $\mathcal{P}$ and $\mathcal{Q}$ respectively.

Each of the sequences $i_1, \dots, i_N$ and $j_1, \dots, j_N$ above is a permutation of the sequence $1, \dots, N$. Denote these two permutations by p and q respectively. If A is a $N \times N$ matrix with complex entries and B is the inverse matrix, then the following identity for complementary minors of A and B holds:

$$\det A \cdot b^{\, j_{m+1} \cdots j_N}_{\, i_{m+1} \cdots i_N} = \operatorname{sgn} p \cdot \operatorname{sgn} q \cdot a^{\, i_1 \cdots i_m}_{\, j_1 \cdots j_m}. \tag{1.77}$$

Here $a^{\, i_1 \cdots i_m}_{\, j_1 \cdots j_m}$ is the minor of A corresponding to the submatrix $A_{\mathcal{P}\mathcal{Q}}$, and $b^{\, j_{m+1} \cdots j_N}_{\, i_{m+1} \cdots i_N}$ is the minor of B corresponding to the submatrix $B_{\overline{\mathcal{Q}}\,\overline{\mathcal{P}}}$. Let us give an analogue of the identity (1.77) for the matrix $T(u)$. We will employ the quantum minors of the matrix $T(u)$, introduced in Section 1.6. Recall that by Corollary 1.3.4, the assignment (1.31) determines an automorphism of the algebra $\mathrm{Y}(\mathfrak{gl}_N)$.

THEOREM 1.10.7. *We have the identity*

$$\operatorname{qdet} T(u) \cdot \omega_N\big(t^{\, j_{m+1} \cdots j_N}_{\, i_{m+1} \cdots i_N}(-u+N-1)\big) = \operatorname{sgn} p \cdot \operatorname{sgn} q \cdot t^{\, i_1 \cdots i_m}_{\, j_1 \cdots j_m}(u).$$

PROOF. By Definition 1.6.5,

$$\operatorname{qdet} T(u)\, A_N = A_N\, T_1 \dots T_N, \tag{1.78}$$

where $T_i = T_i(u - i + 1)$ for $i = 1, \dots, N$. Let us multiply both sides of (1.78) by $T_N^{-1} \dots T_{m+1}^{-1}$ from the right. Then (1.78) takes the form

$$\operatorname{qdet} T(u)\, A_N\, T_N^{-1} \dots T_{m+1}^{-1} = A_N\, T_1 \dots T_m. \tag{1.79}$$

Now we apply both sides of (1.79) to the basis vector

$$e_{j_1} \otimes \dots \otimes e_{j_m} \otimes e_{i_{m+1}} \otimes \dots \otimes e_{i_N} \in (\mathbb{C}^N)^{\otimes N}. \tag{1.80}$$

By applying the right-hand side of (1.79) to this vector, we get

$$\sum_{a_1,\dots,a_m} A_N(e_{a_1} \otimes \dots \otimes e_{a_m} \otimes e_{i_{m+1}} \otimes \dots \otimes e_{i_N}) \otimes t_{a_1 j_1}(u) \dots t_{a_m j_m}(u-m+1). \tag{1.81}$$

The summation here can obviously be restricted to the sequences $a_1, \dots, a_m$ which are permutations of the sequence $i_1, \dots, i_m$. Consider the vector $A_N(e_1 \otimes \dots \otimes e_N)$. The sum (1.81) is proportional to this vector, with the coefficient

$$\operatorname{sgn} p \cdot t^{i_1 \dots i_m}_{j_1 \dots j_m}(u).$$

We will keep the notation $t'_{ij}(u)$ for the entries of the matrix $T^{-1}(u)$; this notation was introduced in the proof of Proposition 1.9.8. Then, by applying the left-hand side of (1.79) to the basis vector (1.80), we obtain

$$\operatorname{qdet} T(u) \sum_{b_{m+1},\dots,b_N} A_N(e_{j_1} \otimes \dots \otimes e_{j_m} \otimes e_{b_{m+1}} \otimes \dots \otimes e_{b_N})$$
$$\otimes\, t'_{b_N i_N}(u - N + 1) \dots t'_{b_{m+1} i_{m+1}}(u - m).$$

By (1.54), here the coefficient of $A_N(e_1 \otimes \dots \otimes e_N)$ equals

$$\operatorname{sgn} q \cdot \operatorname{qdet} T(u) \cdot \omega_N\big(t^{j_{m+1} \dots j_N}_{i_{m+1} \dots i_N}(-u + N - 1)\big).$$

Thus Theorem 1.10.7 follows from the equality (1.79). □

Recall that by definition of the anti-automorphisms (1.25) and (1.27) we have $\mathrm{S} = \omega_N \circ \sigma_N$. Using the notation of Theorem 1.10.7 we get the following corollary.

COROLLARY 1.10.8. *We have the identity*

$$\operatorname{qdet} T(u) \cdot \mathrm{S}\big(t^{j_{m+1} \dots j_N}_{i_{m+1} \dots i_N}(u - m)\big) = \operatorname{sgn} p \cdot \operatorname{sgn} q \cdot t^{i_1 \dots i_m}_{j_1 \dots j_m}(u).$$

PROOF. It is enough to observe that due to (1.54), the anti-automorphism σ_N acts on the quantum minors by

$$\sigma_N : t^{a_1 \dots a_m}_{b_1 \dots b_m}(u) \mapsto t^{a_1 \dots a_m}_{b_1 \dots b_m}(-u + m - 1).$$ □

Applying Corollary 1.10.8 twice, we derive

$$\mathrm{S}^2 : t_{ij}(u) \mapsto \frac{\operatorname{qdet} T(u + N - 1)}{\operatorname{qdet} T(u + N)}\, t_{ij}(u + N),$$

which together with Theorem 1.9.5 gives one more proof of Theorem 1.9.9.

The following is another corollary to Theorem 1.10.7, where we use the notation $d(u) = \operatorname{qdet} T(u)$.

COROLLARY 1.10.9. *We have the identity*

$$d(u) \cdot \omega_N(d(-u + N - 1)) = 1.$$ □

PROPOSITION 1.10.10. *We have the relations*

$$(u-v)\,[t_{ij}(u),t'_{kl}(v)] = \delta_{kj}\sum_{a=1}^{N} t_{ia}(u)\,t'_{al}(v) - \delta_{il}\sum_{a=1}^{N} t'_{ka}(v)\,t_{aj}(u).$$

In particular, the entries of the matrices $T(u)_{\mathcal{P}\mathcal{Q}}$ *and* $T^{-1}(v)_{\overline{\mathcal{Q}}\,\overline{\mathcal{P}}}$ *commute with each other.*

PROOF. Multiply the ternary relation (1.19) by $T_2^{-1}(v)$ from both sides to get

$$T_2(v)^{-1}\,R(u-v)\,T_1(u) = T_1(u)\,R(u-v)\,T_2(v)^{-1}.$$

Now apply the operators on both sides to the basis vector $e_j \otimes e_l$ multiplied by $u-v$. On the left-hand side we get

$$\sum_{i,k=1}^{N}\Big((u-v)\,e_i\otimes e_k\otimes t'_{kl}(v)\,t_{ij}(u) - e_l\otimes e_k\otimes t'_{ki}(v)\,t_{ij}(u)\Big)$$

while the right-hand side gives

$$\sum_{i,k=1}^{N}\Big((u-v)\,e_i\otimes e_k\otimes t_{ij}(u)\,t'_{kl}(v) - e_i\otimes e_j\otimes t_{ik}(u)\,t'_{kl}(v)\Big).$$

Comparing the coefficients of $e_i\otimes e_k$ we obtain the desired relation. □

1.11. Gauss decomposition

The following lemma applies to matrices over an arbitrary ring with 1. Take any matrix of size $(M+N)\times(M+N)$ over the ring, and write it as the block matrix

$$\begin{bmatrix} A & B\\ C & D\end{bmatrix} \tag{1.82}$$

where A, B, C, D are matrices of sizes $M\times M$, $M\times N$, $N\times M$, $N\times N$ respectively.

LEMMA 1.11.1. *Suppose the matrix* (1.82) *is invertible. Suppose the matrices* A *and* D *are also invertible. Then the matrices* $A-B\,D^{-1}C$ *and* $D-C\,A^{-1}B$ *are invertible too, and*

$$\begin{bmatrix} A & B\\ C & D\end{bmatrix}^{-1} = \begin{bmatrix} (A-B\,D^{-1}C)^{-1} & -A^{-1}B\,(D-C\,A^{-1}B)^{-1}\\ -D^{-1}C\,(A-B\,D^{-1}C)^{-1} & (D-C\,A^{-1}B)^{-1}\end{bmatrix}.$$

PROOF. Let

$$\begin{bmatrix} A & B\\ C & D\end{bmatrix}^{-1} = \begin{bmatrix} A' & B'\\ C' & D'\end{bmatrix}$$

be the corresponding block partition for the inverse matrix. By block multiplication,

$$AB'+BD'=0 \qquad\text{and}\qquad CB'+DD'=1.$$

This gives the desired expressions for B' and D'. Similarly, the expressions for A' and C' follow from the relations

$$AA'+BC'=1 \qquad\text{and}\qquad CA'+DC'=0.$$ □

For any $M \geqslant 0$ introduce the homomorphism φ_M

$$\varphi_M : \mathrm{Y}(\mathfrak{gl}_N) \to \mathrm{Y}(\mathfrak{gl}_{M+N}), \tag{1.83}$$

which takes $t_{ij}(u)$ to $t_{M+i,M+j}(u)$. By Corollary 1.4.3 this homomorphism is injective. Consider the composition

$$\psi_M = \omega_{M+N} \circ \varphi_M \circ \omega_N, \tag{1.84}$$

where ω_N is the involutive automorphism of $\mathrm{Y}(\mathfrak{gl}_N)$ defined in (1.31); see also Corollary 1.3.4. Then ψ_M is an injective algebra homomorphism

$$\psi_M : \mathrm{Y}(\mathfrak{gl}_N) \to \mathrm{Y}(\mathfrak{gl}_{M+N}).$$

Its action on the generators of $\mathrm{Y}(\mathfrak{gl}_N)$ can be expressed in terms of quasideterminants (see Definition 1.10.1) as follows.

LEMMA 1.11.2. *For any $1 \leqslant i, j \leqslant N$, we have*

$$\psi_M : t_{ij}(u) \mapsto \begin{vmatrix} t_{11}(u) & \dots & t_{1M}(u) & t_{1,M+j}(u) \\ \vdots & \ddots & \vdots & \vdots \\ t_{M1}(u) & \dots & t_{MM}(u) & t_{M,M+j}(u) \\ t_{M+i,1}(u) & \dots & t_{M+i,M}(u) & \boxed{t_{M+i,M+j}(u)} \end{vmatrix}.$$

PROOF. Introduce the sets of indices

$$\mathcal{P} = \{1, \dots, M\} \quad \text{and} \quad \mathcal{Q} = \{M+1, \dots, M+N\}.$$

Consider the block partitioned matrix

$$\begin{bmatrix} A(u) & B(u) \\ C(u) & D(u) \end{bmatrix}$$

whose entries are the series $t_{ij}(u)$ so that

$$A(u) = T(u)_{\mathcal{P}\mathcal{P}}, \quad B(u) = T(u)_{\mathcal{P}\mathcal{Q}}, \quad C(u) = T(u)_{\mathcal{Q}\mathcal{P}} \quad \text{and} \quad D(u) = T(u)_{\mathcal{Q}\mathcal{Q}}.$$

Let

$$\begin{bmatrix} A(u) & B(u) \\ C(u) & D(u) \end{bmatrix}^{-1} = \begin{bmatrix} A'(u) & B'(u) \\ C'(u) & D'(u) \end{bmatrix}.$$

By its definition, the homomorphism ψ_M maps $\omega_N(T(u)) = T^{-1}(-u)$ to $D'(-u)$. Hence, by Lemma 1.11.1,

$$\psi_M : T(u) \mapsto D(u) - C(u)A(u)^{-1}B(u).$$

Taking the ij-th entry and using Definition 1.10.1 we get the desired formula for the image of $t_{ij}(u)$. □

By Lemma 1.11.2, the description of $\psi_M(t_{ij}(u))$ does not depend on N. Therefore, the maps ψ_M are compatible with the standard embeddings ι_M; see (1.33). In other words, the following diagram commutes:

$$\begin{array}{ccccccc} \mathrm{Y}(\mathfrak{gl}_1) & \xrightarrow{\iota_1} & \mathrm{Y}(\mathfrak{gl}_2) & \xrightarrow{\iota_2} & \mathrm{Y}(\mathfrak{gl}_3) & \xrightarrow{\iota_3} & \dots \\ {\scriptstyle\psi_M}\downarrow & & {\scriptstyle\psi_M}\downarrow & & {\scriptstyle\psi_M}\downarrow & & \\ \mathrm{Y}(\mathfrak{gl}_{M+1}) & \xrightarrow[\iota_{M+1}]{} & \mathrm{Y}(\mathfrak{gl}_{M+2}) & \xrightarrow[\iota_{M+2}]{} & \mathrm{Y}(\mathfrak{gl}_{M+3}) & \xrightarrow[\iota_{M+3}]{} & \dots . \end{array}$$

Note also the property

$$\psi_L \circ \psi_M = \psi_{L+M} \tag{1.85}$$

implied by the definition of ψ_M. The action of ψ_M on the quantum minors (see Section 1.6) is described in the next lemma.

LEMMA 1.11.3. *We have*

$$\psi_M : t^{a_1\dots a_m}_{b_1\dots b_m}(u) \mapsto \big(t^{1\dots M}_{1\dots M}(u+M)\big)^{-1} \cdot t^{1\dots M, M+a_1\dots M+a_m}_{1\dots M, M+b_1\dots M+b_m}(u+M),$$

where $a_i, b_i \in \{1, \dots, N\}$.

PROOF. The images of the quantum minors under the automorphism ω_N are provided by Theorem 1.10.7. Therefore, the statement follows from the definition of ψ_M and the use of Theorem 1.10.7 twice. □

In particular, we obtain a description of ψ_M alternative to Lemma 1.11.2.

COROLLARY 1.11.4. *For any* $1 \leqslant i, j \leqslant N$, *we have*

$$\psi_M : t_{ij}(u) \mapsto \big(t^{1\dots M}_{1\dots M}(u+M)\big)^{-1} \cdot t^{1\dots M, M+i}_{1\dots M, M+j}(u+M). \qquad \square$$

In the next theorem we give a Gauss decomposition of the matrix $T(u)$. We will employ the following general result concerning the Gauss decomposition for a matrix over an arbitrary ring with 1.

LEMMA 1.11.5. *Let T be an $N \times N$ matrix over a ring with 1 such that for all $m = 1, \dots, N$ the submatrices of T determined by the first m rows and columns of T are invertible. Then there exist unique matrices $H = \operatorname{diag}\,[h_1, \dots, h_N]$ and*

$$E = \begin{bmatrix} 1 & e_{12} & \dots & e_{1N} \\ 0 & 1 & \dots & e_{2N} \\ \vdots & \vdots & \ddots & \vdots \\ 0 & 0 & \dots & 1 \end{bmatrix}, \qquad F = \begin{bmatrix} 1 & 0 & \dots & 0 \\ f_{21} & 1 & \dots & 0 \\ \vdots & \vdots & \ddots & \vdots \\ f_{N1} & f_{N2} & \dots & 1 \end{bmatrix}$$

with entries in the ring such that

$$T = F\,H\,E. \tag{1.86}$$

Moreover, for any $1 \leqslant i \leqslant N$

$$h_i = \begin{vmatrix} t_{11} & \dots & t_{1,i-1} & t_{1i} \\ \vdots & \ddots & \vdots & \vdots \\ t_{i-1,1} & \dots & t_{i-1,i-1} & t_{i-1,i} \\ t_{i1} & \dots & t_{i,i-1} & \boxed{t_{ii}} \end{vmatrix}, \tag{1.87}$$

while for $1 \leqslant i < j \leqslant N$

$$f_{ji} = \begin{vmatrix} t_{11} & \dots & t_{1,i-1} & t_{1i} \\ \vdots & \ddots & \vdots & \vdots \\ t_{i-1,1} & \dots & t_{i-1,i-1} & t_{i-1,i} \\ t_{j1} & \dots & t_{j,i-1} & \boxed{t_{ji}} \end{vmatrix} \cdot h_i^{-1} \tag{1.88}$$

and

$$e_{ij} = h_i^{-1} \cdot \begin{vmatrix} t_{11} & \dots & t_{1,i-1} & t_{1j} \\ \vdots & \ddots & \vdots & \vdots \\ t_{i-1,1} & \dots & t_{i-1,i-1} & t_{i-1,j} \\ t_{i1} & \dots & t_{i,i-1} & \boxed{t_{ij}} \end{vmatrix}. \tag{1.89}$$

PROOF. We will use the superscript i to indicate the principle submatrices determined by the first i rows and columns. Arguing by induction on i, suppose that for a fixed value of $i \geqslant 1$ the matrices $F^{(i)}$, $H^{(i)}$ and $E^{(i)}$ are uniquely determined by the relation

$$T^{(i)} = F^{(i)} H^{(i)} E^{(i)}. \tag{1.90}$$

For any $j > i$ denote by t_j and t^j the row and column vectors with the entries $t_{j1}, \dots, t_{ji}$ and $t_{1j}, \dots, t_{ij}$, respectively. Define also f_j and e^j in a similar way. Considering the principal submatrices of size $i+1$ in (1.86) we obtain

$$f_{i+1} H^{(i)} E^{(i)} = t_{i+1}, \qquad F^{(i)} H^{(i)} e^{i+1} = t^{i+1}$$

and

$$f_{i+1} H^{(i)} e^{i+1} + h_{i+1} = t_{i+1,i+1}.$$

These relations uniquely determine f_{i+1}, e^{i+1} and h_{i+1}. Express f_{i+1} and e^{i+1} from the first two of the relations and substitute into the third. Together with (1.90) this gives

$$h_{i+1} = t_{i+1,i+1} - t_{i+1} (T^{(i)})^{-1} t^{i+1}.$$

Using Definition 1.10.1, we can write this as

$$h_{i+1} = \left|T^{(i+1)}\right|_{i+1,i+1},$$

thus proving (1.87) with i replaced by $i+1$. Furthermore, by considering the principal submatrices of size j in (1.86) we obtain

$$f_j H^{(i)} E^{(i)} = t_j, \qquad F^{(i)} H^{(i)} e^j = t^j.$$

Applying (1.90), we can bring these relations to the form

$$f_j (F^{(i)})^{-1} = t_j (T^{(i)})^{-1}, \qquad (E^{(i)})^{-1} e^j = (T^{(i)})^{-1} t^j.$$

Equating the last entries in these rows and columns, respectively, we come to

$$f_{ji} = \sum_{a=1}^{i} t_{ja} \left((T^{(i)})^{-1}\right)_{ai}, \qquad e_{ij} = \sum_{a=1}^{i} \left((T^{(i)})^{-1}\right)_{ia} t_{aj}.$$

The first of these relations can be presented in the form

$$f_{ji} = \Big(t_{ji} - \sum_{a,b=1}^{i-1} t_{ja} \left((T^{(i-1)})^{-1}\right)_{ab} t_{bi}\Big) \left((T^{(i)})^{-1}\right)_{ii}.$$

Indeed, this is immediate from the identity

$$\left((T^{(i)})^{-1}\right)_{ai} + \sum_{b=1}^{i-1} \left((T^{(i-1)})^{-1}\right)_{ab} t_{bi} \left((T^{(i)})^{-1}\right)_{ii} = 0$$

which holds for any $1 \leqslant a \leqslant i-1$. In order to verify the latter, equate the first $i-1$ entries of the last columns of the matrix identity $T^{(i)} (T^{(i)})^{-1} = 1$. This gives

$$\sum_{b=1}^{i-1} t_{ab} \left((T^{(i)})^{-1}\right)_{bi} + t_{ai} \left((T^{(i)})^{-1}\right)_{ii} = 0$$

for any $1 \leqslant a \leqslant i-1$ implying the required identity. Due to Definition 1.10.1 and Proposition 1.10.4, this proves (1.88). The formula (1.89) is derived in a similar fashion by transforming the corresponding expression for e_{ij}. □

The application of Lemma 1.11.5 to the matrix $T(u)$ yields the corresponding *Gauss decomposition*

$$T(u) = F(u)\, H(u)\, E(u) \tag{1.91}$$

for unique matrices $F(u)$, $H(u)$ and $E(u)$ of the required form with the entries in $\mathrm{Y}(\mathfrak{gl}_N)[[u^{-1}]]$.

THEOREM 1.11.6. *The entries of the matrices $F(u)$, $H(u)$ and $E(u)$ can be given by the formulas*

$$h_i(u) = t^{1\dots i}_{1\dots i}(u+i-1) \cdot \big(t^{1\dots i-1}_{1\dots i-1}(u+i-1)\big)^{-1}$$

for $1 \leqslant i \leqslant N$, and

$$f_{ji}(u) = t^{1\dots i-1,\, j}_{1\dots i}(u+i-1) \cdot \big(t^{1\dots i}_{1\dots i}(u+i-1)\big)^{-1},$$
$$e_{ij}(u) = \big(t^{1\dots i}_{1\dots i}(u+i-1)\big)^{-1} \cdot t^{1\dots i}_{1\dots i-1,\, j}(u+i-1)$$

for $1 \leqslant i < j \leqslant N$.

PROOF. Applying the quasideterminant formulas of Lemma 1.11.5 to the matrix $T(u)$ instead of T and using Lemma 1.11.2 we obtain

$$\begin{aligned} h_i(u) &= \psi_{i-1}(t_{11}(u)), \\ f_{ji}(u) &= \psi_{i-1}\big(t_{j-i+1,1}(u)\, t_{11}(u)^{-1}\big), \\ e_{ij}(u) &= \psi_{i-1}\big(t_{11}(u)^{-1}\, t_{1,j-i+1}(u)\big). \end{aligned} \tag{1.92}$$

The proof is completed by the application of Corollaries 1.7.2 and 1.11.4. □

The following corollary provides formulas for the action of the transposition t of the matrix $T(u)$ on the series $h_i(u)$, $e_{ij}(u)$ and $f_{ji}(u)$. Recall that t defines an anti-automorphism of $\mathrm{Y}(\mathfrak{gl}_N)$; see Proposition 1.3.3.

COROLLARY 1.11.7. *Under the action of the anti-automorphism t,*

$$h_i(u) \mapsto h_i(u), \qquad e_{ij}(u) \mapsto f_{ji}(u), \qquad f_{ji}(u) \mapsto e_{ij}(u).$$

PROOF. This is immediate from Theorem 1.11.6, since t acts on the quantum minors by

$$t^{a_1\dots a_m}_{b_1\dots b_m}(u) \mapsto t^{b_1\dots b_m}_{a_1\dots a_m}(u);$$

see (1.54) and (1.55). □

The following is immediate either from Theorem 1.10.5 or Theorem 1.11.6.

COROLLARY 1.11.8. *We have the decomposition of the quantum determinant,*

$$\operatorname{qdet} T(u) = h_1(u)\, h_2(u-1) \dots h_N(u-N+1).$$ □

1.12. Quantum Sylvester theorem

Suppose that $A = [a_{ij}]$ is a numerical $(M+N)\times(M+N)$ matrix. For any indices $i, j = 1, \dots, N$ introduce the minors c_{ij} of A corresponding to the rows $1, \dots, M, M+i$ and columns $1, \dots, M, M+j$ so that

$$c_{ij} = a^{1\dots M, M+i}_{1\dots M, M+j}.$$

Let $A^{(M)}$ be the submatrix of A determined by the first M rows and columns. The classical Sylvester theorem provides a formula for the determinant of the matrix $C = [c_{ij}]$:

$$\det C = \det A \cdot \big(\det A^{(M)}\big)^{N-1}.$$

We now give a Yangian analogue of this theorem where minors of A are replaced by quantum minors of the matrix $T(u)$. With the same assumptions on the indices i, j as above, introduce the following series with coefficients in $\mathrm{Y}(\mathfrak{gl}_{M+N})$

$$t^{\sharp}_{ij}(u) = t^{\,1\dots\, M, M+i}_{\,1\dots\, M, M+j}(u)$$

and combine them into the matrix $T^{\sharp}(u) = \big[t^{\sharp}_{ij}(u)\big]$. Let $T^{(M)}(u)$ be the submatrix of $T(u)$ determined by the first M rows and columns.

THEOREM 1.12.1. *The mapping*

$$t_{ij}(u) \mapsto t^{\sharp}_{ij}(u), \qquad 1 \leqslant i, j \leqslant N, \tag{1.93}$$

defines a homomorphism $\mathrm{Y}(\mathfrak{gl}_N) \to \mathrm{Y}(\mathfrak{gl}_{M+N})$. *Moreover, we have the identity*

$$\operatorname{qdet} T^{\sharp}(u) = \operatorname{qdet} T(u) \cdot \operatorname{qdet} T^{(M)}(u-1) \dots \operatorname{qdet} T^{(M)}(u-N+1).$$

PROOF. By Corollary 1.11.4, we have

$$\psi_M : t_{ij}(u-M) \mapsto \big(\operatorname{qdet} T^{(M)}(u)\big)^{-1} \cdot t^{\sharp}_{ij}(u). \tag{1.94}$$

Corollary 1.7.2 implies that the coefficients of the series $\operatorname{qdet} T^{(M)}(u)$ commute with those of the series $t^{\sharp}_{ij}(v)$. Hence, since ψ_M is a homomorphism, we can conclude by the application of the shift automorphism (1.21) that the mapping (1.93) defines a homomorphism. Furthermore, by Lemma 1.11.3,

$$\psi_M : t^{\,1\dots\, N}_{\,1\dots\, N}(u-M) \mapsto \big(\operatorname{qdet} T^{(M)}(u)\big)^{-1} \cdot \operatorname{qdet} T(u).$$

On the other hand, expanding the quantum minor, we obtain from (1.94) that

$$\psi_M : t^{\,1\dots\, N}_{\,1\dots\, N}(u-M) \mapsto \big(\operatorname{qdet} T^{(M)}(u) \dots \operatorname{qdet} T^{(M)}(u-N+1)\big)^{-1} \cdot \operatorname{qdet} T^{\sharp}(u),$$

completing the proof of the desired identity for $\operatorname{qdet} T^{\sharp}(u)$. □

We conclude this section with a "dual" version of Theorem 1.12.1. Let us fix an integer m such that $1 \leqslant m \leqslant N$. For any indices $i, j = 1, \dots, m$ introduce the following series with coefficients in $\mathrm{Y}(\mathfrak{gl}_N)$,

$$t^{\flat}_{ij}(u) = t^{\,i, m+1\dots\, N}_{\,j, m+1\dots\, N}(u)$$

and combine them into the matrix $T^{\flat}(u) = \big[t^{\flat}_{ij}(u)\big]$. According to the notation of Remark 1.10.6, let $T_{(m+1)}(u)$ be the submatrix of $T(u)$ obtained by removing the first m rows and columns. The quantum determinant $\operatorname{qdet} T_{(m+1)}(u)$ is well-defined since the assignment

$$t_{kl}(u) \mapsto t_{k+m, l+m}(u) \quad \text{for} \quad k, l = 1, \dots, N-m$$

determines an embedding $\mathrm{Y}(\mathfrak{gl}_{N-m}) \to \mathrm{Y}(\mathfrak{gl}_N)$; see (1.83).

THEOREM 1.12.2. *The mapping*

$$t_{ij}(u) \mapsto t^{\flat}_{ij}(u), \qquad 1 \leqslant i, j \leqslant m,$$

defines a homomorphism $\mathrm{Y}(\mathfrak{gl}_m) \to \mathrm{Y}(\mathfrak{gl}_N)$. *Moreover, we have the identity*

$$\operatorname{qdet} T^{\flat}(u) = \operatorname{qdet} T(u) \cdot \operatorname{qdet} T_{(m+1)}(u-1) \dots \operatorname{qdet} T_{(m+1)}(u-m+1).$$

PROOF. This is immediate from Theorem 1.12.1 by the replacement of $M+N$ by N and application of the automorphism of $\mathrm{Y}(\mathfrak{gl}_N)$ given by

$$\upsilon_N:\ t_{ij}(u)\mapsto t_{N-i+1,N-j+1}(u) \tag{1.95}$$

which is a particular case of the automorphism (1.22) with the anti-diagonal matrix $B=[\delta_{i,N-j+1}]$. □

REMARK 1.12.3. It is easy to prove that the homomorphisms of Theorems 1.12.1 and 1.12.2 are injective; cf. the proof of Proposition 1.9.3. □

1.13. Gelfand–Tsetlin subalgebra

For any $M\geqslant 1$ we identify the Yangian $\mathrm{Y}(\mathfrak{gl}_M)$ with a subalgebra of $\mathrm{Y}(\mathfrak{gl}_{M+N})$ via the embedding ι_M; see Corollary 1.4.3.

PROPOSITION 1.13.1. *The centralizer of the subalgebra* $\mathrm{Y}(\mathfrak{gl}_M)$ *in* $\mathrm{Y}(\mathfrak{gl}_{M+N})$ *is equal to* $\mathrm{ZY}(\mathfrak{gl}_M)\,\psi_M(\mathrm{Y}(\mathfrak{gl}_N))$.

PROOF. By Corollaries 1.7.2 and 1.11.4, the subalgebra $\mathrm{ZY}(\mathfrak{gl}_M)\,\psi_M(\mathrm{Y}(\mathfrak{gl}_N))$ centralizes $\mathrm{Y}(\mathfrak{gl}_M)$, so we need to show only that the centralizer is no larger. The associated graded algebra $\mathrm{gr}'\,\mathrm{Y}(\mathfrak{gl}_{M+N})$ is isomorphic to the universal enveloping algebra $\mathrm{U}(\mathfrak{gl}_{M+N}[z])$; see Proposition 1.5.2. Furthermore, using Lemma 1.11.2, we find that the image of $\psi_M(t_{ij}^{(r)})$ in the $(r-1)$-th component of $\mathrm{gr}'\,\mathrm{Y}(\mathfrak{gl}_{M+N})$ coincides with $E_{M+i,M+j}\,z^{r-1}$. Hence, the graded algebra $\mathrm{gr}'\,\psi_M(\mathrm{Y}(\mathfrak{gl}_N))$ is isomorphic to $\mathrm{U}(\mathfrak{gl}_N[z])$, where $\mathfrak{gl}_N$ is identified with a subalgebra of $\mathfrak{gl}_{M+N}$ via the embedding $E_{ij}\mapsto E_{M+i,M+j}$. Now we apply Lemma 1.7.4. Since the centralizer of $\mathfrak{gl}_M$ in $\mathfrak{gl}_{M+N}$ is spanned by $\mathfrak{gl}_N$ and the element $I=E_{11}+\dots+E_{MM}$, the centralizer of $\mathrm{U}(\mathfrak{gl}_M[z])$ in $\mathrm{U}(\mathfrak{gl}_{M+N}[z])$ is equal to $\mathrm{Z}(\mathfrak{gl}_M[z])\,\mathrm{U}(\mathfrak{gl}_N[z])$, where $\mathrm{Z}(\mathfrak{gl}_M[z])$ denotes the center of $\mathrm{U}(\mathfrak{gl}_M[z])$. As we have seen in the proof of Theorem 1.7.5, the associated graded algebra $\mathrm{gr}'\,\mathrm{ZY}(\mathfrak{gl}_M)$ is isomorphic to $\mathrm{Z}(\mathfrak{gl}_M[z])$, completing the proof. □

Consider now the chain of subalgebras

$$\mathrm{Y}(\mathfrak{gl}_1)\subset\mathrm{Y}(\mathfrak{gl}_2)\subset\dots\subset\mathrm{Y}(\mathfrak{gl}_N). \tag{1.96}$$

DEFINITION 1.13.2. The *Gelfand–Tsetlin subalgebra of* $\mathrm{Y}(\mathfrak{gl}_N)$ is the (commutative) subalgebra H_N generated by the centers $\mathrm{ZY}(\mathfrak{gl}_1)$, $\mathrm{ZY}(\mathfrak{gl}_2),\dots,\mathrm{ZY}(\mathfrak{gl}_N)$ of the subalgebras of the chain (1.96). □

By Theorem 1.7.5, the algebra $\mathrm{ZY}(\mathfrak{gl}_N)$ is generated by the coefficients of the quantum determinant $\mathrm{qdet}\,T(u)=t^{1\cdots N}_{1\cdots N}(u)$. Therefore, the subalgebra H_N is generated by the coefficients of all series $t^{1\cdots m}_{1\cdots m}(u)$ with $m=1,\dots,N$. Equivalently, H_N is generated by the coefficients of the series $h_1(u),\dots,h_N(u)$, defined in Theorem 1.11.6.

THEOREM 1.13.3. *The Gelfand–Tsetlin subalgebra* H_N *of* $\mathrm{Y}(\mathfrak{gl}_N)$ *is maximal commutative.*

PROOF. We will demonstrate by induction on N that H_N coincides with its own centralizer in $\mathrm{Y}(\mathfrak{gl}_N)$. By (1.85) and (1.92) we have $H_N=\mathrm{Y}(\mathfrak{gl}_1)\,\psi_1(H_{N-1})$. Proposition 1.13.1 implies that the centralizer of $\mathrm{Y}(\mathfrak{gl}_1)$ in $\mathrm{Y}(\mathfrak{gl}_N)$ coincides with $\mathrm{ZY}(\mathfrak{gl}_1)\,\psi_1(\mathrm{Y}(\mathfrak{gl}_{N-1}))$. By the induction hypothesis, the subalgebra H_{N-1} coincides with its own centralizer in $\mathrm{Y}(\mathfrak{gl}_{N-1})$. Hence, since $\mathrm{ZY}(\mathfrak{gl}_1)=\mathrm{Y}(\mathfrak{gl}_1)$, the centralizer of H_N in $\mathrm{Y}(\mathfrak{gl}_N)$ is $\mathrm{ZY}(\mathfrak{gl}_1)\,\psi_1(H_{N-1})=H_N$. □

Note that an alternative proof of Theorem 1.13.3 can be obtained with the use of Lemma 1.7.4 and the fact that the diagonal subalgebra of $\mathfrak{gl}_N$ coincides with its own centralizer.

1.14. Bethe subalgebras

Using the quantum minors defined in Section 1.6, for each $k = 1, \dots, N$ set

$$\tau_k(u) = \sum_{a_1 < \dots < a_k} t^{a_1 \dots a_k}_{a_1 \dots a_k}(u),$$

summed over all subsets of indices $\{a_1, \dots, a_k\}$ from $\{1, \dots, N\}$. In particular, $\tau_1(u) = \operatorname{tr} T(u)$ and $\tau_N(u)$ coincides with the quantum determinant $\operatorname{qdet} T(u)$.

PROPOSITION 1.14.1. *All coefficients of the series $\tau_1(u), \dots, \tau_N(u)$ commute.*

PROOF. Recalling that the quantum minors are the matrix elements of the operator which occurs in (1.49), we can write the series $\tau_k(u)$ in an equivalent form as

$$\tau_k(u) = \frac{1}{k!} \operatorname{tr} A_k\, T_1(u) \dots T_k(u - k + 1).$$

Here the trace is the linear map

$$\operatorname{tr} : (\operatorname{End} \mathbb{C}^N)^{\otimes k} \to \mathbb{C} \tag{1.97}$$

acting on the basis elements by

$$\operatorname{tr} (e_{i_1 j_1} \otimes \dots \otimes e_{i_k j_k}) = \delta_{i_1 j_1} \dots \delta_{i_k j_k}.$$

In order to verify that $\tau_k(u)$ commutes with $\tau_l(v)$ we use Proposition 1.6.1, where we take $m = k + l$ and specialize the variables u_i as follows:

$$u_i = u - i + 1, \quad i = 1, \dots, k \qquad \text{and} \qquad u_{k+j} = v - j + 1, \quad j = 1, \dots, l.$$

Then by Proposition 1.6.2 we have

$$R(u_1, \dots, u_k) = A_k, \qquad R(u_{k+1}, \dots, u_{k+l}) = A'_l, \tag{1.98}$$

where by A'_l we denote the anti-symmetrizer corresponding to the subset of indices $\{k + 1, \dots, k + l\}$. Applying (1.17) and using the notation of Section 1.6, we get

$$\begin{aligned} R(u_1, \dots, u_{k+l}) &= \prod^{\longrightarrow}_{j=1,\dots,l} \Big(\prod^{\longleftarrow}_{i=1,\dots,k} R_{i,k+j} \Big) A_k\, A'_l \\ &= A_k\, A'_l \prod^{\longleftarrow}_{j=1,\dots,l} \Big(\prod^{\longrightarrow}_{i=1,\dots,k} R_{i,k+j} \Big), \end{aligned} \tag{1.99}$$

where the left and right arrows indicate that the products are taken in the decreasing and increasing orders of the indices, respectively. Our next step is to show that (1.99) takes the form

$$R(u_1, \dots, u_{k+l}) = \widetilde{R}(u, v)\, A_k\, A'_l = A_k\, A'_l\, \widetilde{R}(u, v), \tag{1.100}$$

for the operator

$$\widetilde{R}(u,v) = \sum_{p=0}^{\min\{k,l\}} \frac{(-1)^p\, p!}{(u-v-k+1)\dots(u-v-k+p)} \times \sum_{\substack{1\leqslant i_1<\dots<i_p\leqslant k\\ 1\leqslant j_1<\dots<j_p\leqslant l}} P_{i_1,k+j_1}\dots P_{i_p,k+j_p}.$$

Let us begin with the first equality in (1.99). If indices $i_1,\dots,i_s \in \{1,\dots,k\}$ are distinct, then

$$P_{i_1,k+j}\dots P_{i_s,k+j}\, A_k = (-1)^{s-1} P_{i_s,k+j}\, A_k.$$

This implies that for each j,

$$R_{k,k+j}\dots R_{1,k+j}\, A_k = \Big(1 - \frac{P_{1,k+j}+\dots+P_{k,k+j}}{u-v-k+j}\Big) A_k;$$

cf. Corollary 1.7.3. Similarly, if indices $j_1,\dots,j_s \in \{1,\dots,l\}$ are distinct, then

$$P_{i,k+j_1}\dots P_{i,k+j_s}\, A'_l = (-1)^{s-1} P_{i,k+j_s}\, A'_l.$$

Moreover, if $i_1 \neq i_2$ and $j_1 \neq j_2$, then

$$P_{i_1,k+j_1} P_{i_2,k+j_2}\, A_k\, A'_l = P_{i_1,k+j_2} P_{i_2,k+j_1}\, A_k\, A'_l.$$

It is now straightforward to bring the expression to the desired form, thus completing the proof of the first relation in (1.100). The second is verified by a similar calculation. Hence Proposition 1.6.1 gives

$$\begin{aligned}(1.101)\qquad &\widetilde{R}(u,v)\, A_k\, T_1(u)\dots T_k(u-k+1)\, A'_l\, T_{k+1}(v)\dots T_{k+l}(v-l+1)\\ &= A'_l\, T_{k+1}(v)\dots T_{k+l}(v-l+1)\, A_k\, T_1(u)\dots T_k(u-k+1)\, \widetilde{R}(u,v).\end{aligned}$$

By (1.100) the operator $\widetilde{R}(u,v)$ preserves the subspace $A_k\, A'_l\, (\mathbb{C}^N)^{\otimes(k+l)}$, and its restriction to this subspace is invertible. The latter follows from (1.99), as all factors $R_{i,k+j}$ are invertible due to (1.13). Using the trace map (1.97) taken over all $k+l$ copies of $\operatorname{End}\mathbb{C}^N$, we find that the product $k!\, l!\, \tau_k(u)\, \tau_l(v)$ equals

$$\begin{aligned}&\operatorname{tr} A_k\, T_1(u)\dots T_k(u-k+1)\, A'_l\, T_{k+1}(v)\dots T_{k+l}(v-l+1)\\ &= \operatorname{tr} \widetilde{R}(u,v)^{-1} A'_l\, T_{k+1}(v)\dots T_{k+l}(v-l+1)\, A_k\, T_1(u)\dots T_k(u-k+1)\, \widetilde{R}(u,v),\end{aligned}$$

which coincides with $k!\, l!\, \tau_l(v)\, \tau_k(u)$ by the cyclic property of trace. □

Now fix an arbitrary $N\times N$ matrix C whose entries are complex numbers. For any $k=1,\dots,N$ set

$$\tau_k(u,C) = \frac{1}{k!} \operatorname{tr} A_k\, C_1\dots C_k\, T_1(u)\dots T_k(u-k+1),$$

where the trace map is taken over the k copies of $\operatorname{End}\mathbb{C}^N$.

PROPOSITION 1.14.2. *The coefficients of the series $\tau_1(u,C),\dots,\tau_N(u,C)$ generate a commutative subalgebra of* $\mathrm{Y}(\mathfrak{gl}_N)$.

PROOF. The ternary relation (1.19) will hold if $T(u)$ is replaced with the matrix $C\,T(u)$. Therefore the arguments used in the proof of Proposition 1.14.1 also apply to this matrix instead of $T(u)$. □

Using another $N \times N$ matrix D over $\mathbb{C}$, for any $k = 1, \dots, N$ set

$$\sigma_k(u, D) = \frac{1}{N!} \operatorname{tr} A_N \, T_1(u) \dots T_k(u - k + 1) \, D_{k+1} \dots D_N,$$

where the trace map is taken over the N copies of $\operatorname{End} \mathbb{C}^N$. In particular, the series $\sigma_N(u, D)$ does not depend on the matrix D and coincides with the quantum determinant $\operatorname{qdet} T(u)$.

PROPOSITION 1.14.3. *The coefficients of the series* $\sigma_1(u, D), \dots, \sigma_N(u, D)$ *generate a commutative subalgebra of* $\mathrm{Y}(\mathfrak{gl}_N)$.

PROOF. As $\sigma_k(u, D)$ depends polynomially on the entries of D, it will be sufficient to verify the claim for invertible matrices D. Since

$$A_N \, D_1 \dots D_N = A_N \, \det D,$$

we have

$$\sigma_k(u, D) = \frac{\det D}{N!} \operatorname{tr} A_N \, D_1^{-1} \, T_1(u) \dots D_k^{-1} \, T_k(u - k + 1).$$

Therefore, up to a nonzero factor, $\sigma_k(u, D)$ coincides with $\tau_k(u, C)$ for $C = D^{-1}$. The statement now follows from Proposition 1.14.2. □

The commutative subalgebras of the Yangian $\mathrm{Y}(\mathfrak{gl}_N)$, which are defined in Propositions 1.14.2 and 1.14.3 are called the *Bethe subalgebras*.

REMARK 1.14.4. It can be proved that if the matrix D has simple spectrum, then the coefficients of $u^{-1}, u^{-2}, \dots$ of the series $\sigma_1(u, D), \dots, \sigma_N(u, D)$ are algebraically independent and generate a maximal commutative subalgebra of $\mathrm{Y}(\mathfrak{gl}_N)$. Also, in this case the images of the coefficients under the evaluation homomorphism (1.5) generate a maximal commutative subalgebra of $\mathrm{U}(\mathfrak{gl}_N)$. □

1.15. Examples

1. A q-analogue of the universal enveloping algebra $\mathrm{U}(\mathfrak{gl}_N)$ can be introduced by using ternary relations of type (1.19). Let q be a nonzero complex parameter. Consider the R-matrix

$$R = q \sum_i e_{ii} \otimes e_{ii} + \sum_{i \neq j} e_{ii} \otimes e_{jj} + (q - q^{-1}) \sum_{i<j} e_{ij} \otimes e_{ji},$$

where the indices run over the set $\{1, \dots, N\}$. In the notation of Section 1.2, this element of $(\operatorname{End} \mathbb{C}^N)^{\otimes 2}$ satisfies the equation in $(\operatorname{End} \mathbb{C}^N)^{\otimes 3}$,

$$R_{12} \, R_{13} \, R_{23} = R_{23} \, R_{13} \, R_{12} \, .$$

The *quantized enveloping algebra* $\mathrm{U}_q(\mathfrak{gl}_N)$ is generated by elements t_{ij} and $\bar{t}_{ij}$ with $1 \leqslant i, j \leqslant N$ subject to the relations

$$t_{ij} = \bar{t}_{ji} = 0, \qquad 1 \leqslant i < j \leqslant N,$$
$$t_{ii} \, \bar{t}_{ii} = \bar{t}_{ii} \, t_{ii} = 1, \qquad 1 \leqslant i \leqslant N,$$
$$R T_1 T_2 = T_2 T_1 R, \qquad R \overline{T}_1 \overline{T}_2 = \overline{T}_2 \overline{T}_1 R, \qquad R \overline{T}_1 T_2 = T_2 \overline{T}_1 R,$$

where T and $\overline{T}$ are the matrices

$$T = \sum_{i,j} e_{ij} \otimes t_{ij}, \qquad \overline{T} = \sum_{i,j} e_{ij} \otimes \bar{t}_{ij}.$$

As $q \to 1$ the algebra $\mathrm{U}_q(\mathfrak{gl}_N)$ specializes to $\mathrm{U}(\mathfrak{gl}_N)$ in the following precise sense. For the discussion of this point, regard q as a formal variable and $\mathrm{U}_q(\mathfrak{gl}_N)$ as an algebra over $\mathbb{C}(q)$. Then set $\mathcal{A} = \mathbb{C}[q, q^{-1}]$ and consider the $\mathcal{A}$-subalgebra $\mathrm{U}_{\mathcal{A}}$ of $\mathrm{U}_q(\mathfrak{gl}_N)$ generated by the elements

$$\tau_{ij} = \frac{t_{ij}}{q - q^{-1}} \quad \text{for} \quad i > j, \qquad \bar{\tau}_{ij} = \frac{\bar{t}_{ij}}{q - q^{-1}} \quad \text{for} \quad i < j,$$

and

$$\tau_{ii} = \frac{t_{ii} - 1}{q - 1}, \qquad \bar{\tau}_{ii} = \frac{\bar{t}_{ii} - 1}{q - 1},$$

for $i = 1, \dots, N$. Then we have an isomorphism

$$\mathrm{U}_{\mathcal{A}} \otimes_{\mathcal{A}} \mathbb{C} \cong \mathrm{U}(\mathfrak{gl}_N)$$

with the action of $\mathcal{A}$ on $\mathbb{C}$ defined via the evaluation $q = 1$. The elements τ_{ij} and $\bar{\tau}_{ij}$ respectively specialize to E_{ij} and $-E_{ij}$.

The quantized enveloping algebra $\mathrm{U}_q(\mathfrak{gl}_N)$ is a Hopf algebra with the comultiplication

$$\Delta : t_{ij} \mapsto \sum_{k=1}^{N} t_{ik} \otimes t_{kj}, \qquad \bar{t}_{ij} \mapsto \sum_{k=1}^{N} \bar{t}_{ik} \otimes \bar{t}_{kj},$$

the antipode

$$\mathrm{S} : T \mapsto T^{-1}, \qquad \overline{T} \mapsto \overline{T}^{-1},$$

and the counit

$$\varepsilon : T \mapsto 1, \qquad \overline{T} \mapsto 1.$$

2. The *braid group* $\mathfrak{B}_N$ is generated by elements $\beta_1, \dots, \beta_{N-1}$ subject to the defining relations

$$\beta_i \beta_{i+1} \beta_i = \beta_{i+1} \beta_i \beta_{i+1}, \qquad i = 1, \dots, N-2$$

and

$$\beta_i \beta_j = \beta_j \beta_i, \qquad |i - j| > 1.$$

The group $\mathfrak{B}_N$ acts on the algebra $\mathrm{U}_q(\mathfrak{gl}_N)$ by automorphisms. For the action of β_i with $i \in \{1, \dots, N-1\}$ we have

$$\beta_i : t_{ii} \mapsto t_{i+1,i+1}, \qquad t_{i+1,i+1} \mapsto t_{ii}, \qquad t_{kk} \mapsto t_{kk} \qquad \text{if} \quad k \neq i, i+1,$$

$$\begin{aligned} \beta_i : t_{i+1,i} &\mapsto q^{-1}\, \bar{t}_{i,i+1}\, t_{ii}^2 & & \\ t_{ik} &\mapsto q\, t_{ik}\, t_{i+1,i}\, \bar{t}_{ii} - t_{i+1,k}, & t_{i+1,k} &\mapsto q^{-1}\, t_{ik}, & &\text{if} \quad k \leqslant i-1 \\ t_{li} &\mapsto q^{-1}\, \bar{t}_{i,i+1}\, t_{li}\, t_{ii} - t_{l,i+1}, & t_{l,i+1} &\mapsto q\, t_{li}, & &\text{if} \quad l \geqslant i+2 \\ t_{kl} &\mapsto t_{kl} & &\text{in all remaining cases,} & & \end{aligned}$$

and

$$\begin{aligned} \beta_i : \bar{t}_{i,i+1} &\mapsto q\, \bar{t}_{ii}^{\,2}\, t_{i+1,i} & & \\ \bar{t}_{ki} &\mapsto q^{-1}\, t_{ii}\, \bar{t}_{i,i+1} \bar{t}_{ki} - \bar{t}_{k,i+1}, & \bar{t}_{k,i+1} &\mapsto q\, \bar{t}_{ki}, & &\text{if} \quad k \leqslant i-1 \\ \bar{t}_{il} &\mapsto q\, \bar{t}_{ii}\, \bar{t}_{il}\, t_{i+1,i} - \bar{t}_{i+1,l}, & \bar{t}_{i+1,l} &\mapsto q^{-1}\, \bar{t}_{il}, & &\text{if} \quad l \geqslant i+2 \\ \bar{t}_{kl} &\mapsto \bar{t}_{kl} & &\text{in all remaining cases.} & & \end{aligned}$$

3. Consider the R-matrix depending on parameters u and v given by

$$R(u,v) = (u-v)\sum_{i\neq j} e_{ii}\otimes e_{jj} + (q^{-1}u - qv)\sum_{i} e_{ii}\otimes e_{ii}$$
$$+ (q^{-1}-q)\,u\sum_{i>j} e_{ij}\otimes e_{ji} + (q^{-1}-q)\,v\sum_{i<j} e_{ij}\otimes e_{ji}.$$

It satisfies the Yang–Baxter equation

$$R_{12}(u,v)\,R_{13}(u,w)\,R_{23}(v,w) = R_{23}(v,w)\,R_{13}(u,w)\,R_{12}(u,v)\,.$$

The *quantum affine algebra* $\mathrm{U}_q(\widehat{\mathfrak{gl}}_N)$ (with the trivial central charge) is generated by elements $t_{ij}^{(r)}$ and $\bar{t}_{ij}^{(r)}$, where $1\leqslant i,j\leqslant N$ and r runs over nonnegative integers, subject to the relations

$$\begin{aligned}
t_{ij}^{(0)} &= \bar{t}_{ji}^{(0)} = 0, \qquad 1\leqslant i<j\leqslant N,\\
t_{ii}^{(0)}\,\bar{t}_{ii}^{(0)} &= \bar{t}_{ii}^{(0)}\,t_{ii}^{(0)} = 1, \qquad 1\leqslant i\leqslant N,\\
R(u,v)\,T_1(u)T_2(v) &= T_2(v)T_1(u)R(u,v),\\
R(u,v)\,\overline{T}_1(u)\overline{T}_2(v) &= \overline{T}_2(v)\overline{T}_1(u)R(u,v),\\
R(u,v)\,\overline{T}_1(u)T_2(v) &= T_2(v)\overline{T}_1(u)R(u,v),
\end{aligned}$$

where

$$T(u) = \sum_{i,j=1}^{N} e_{ij}\otimes t_{ij}(u), \qquad \overline{T}(u) = \sum_{i,j=1}^{N} e_{ij}\otimes \bar{t}_{ij}(u),$$

and

$$t_{ij}(u) = \sum_{r=0}^{\infty} t_{ij}^{(r)}\,u^{-r}, \qquad \bar{t}_{ij}(u) = \sum_{r=0}^{\infty} \bar{t}_{ij}^{(r)}\,u^{r}.$$

Similar to Example 1, as $q\to 1$ the algebra $\mathrm{U}_q(\widehat{\mathfrak{gl}}_N)$ specializes to $\mathrm{U}\big(\mathfrak{gl}_N[z,z^{-1}]\big)$, where z is a variable. For this reason, the algebra $\mathrm{U}_q(\widehat{\mathfrak{gl}}_N)$ (as well as the algebra $\mathrm{U}_q(\widehat{\mathfrak{sl}}_N)$ introduced in Example 4 below) is often called the *quantum loop algebra.*

The quantum affine algebra $\mathrm{U}_q(\widehat{\mathfrak{gl}}_N)$ is a Hopf algebra with the comultiplication

$$\Delta: t_{ij}(u)\mapsto \sum_{k=1}^{N} t_{ik}(u)\otimes t_{kj}(u), \qquad \bar{t}_{ij}(u)\mapsto \sum_{k=1}^{N} \bar{t}_{ik}(u)\otimes \bar{t}_{kj}(u),$$

the antipode

$$\mathrm{S}: T(u)\mapsto T(u)^{-1}, \qquad \overline{T}(u)\mapsto \overline{T}(u)^{-1},$$

and the counit

$$\varepsilon: T(u)\mapsto 1, \qquad \overline{T}(u)\mapsto 1.$$

The *q-Yangian* $\mathrm{Y}_q(\mathfrak{gl}_N)$ is the standard Borel subalgebra of $\mathrm{U}_q(\widehat{\mathfrak{gl}}_N)$ generated by the elements $t_{ij}^{(r)}$ with $i,j\in\{1,\dots,N\}$ and $r\geqslant 0$.

4. Let $f(u)$ and $\bar{f}(u)$ be formal power series in u^{-1} and u, respectively,

$$\begin{aligned}
f(u) &= f_0 + f_1\,u^{-1} + f_2\,u^{-2} + \dots,\\
\bar{f}(u) &= \bar{f}_0 + \bar{f}_1\,u + \bar{f}_2\,u^{2} + \dots,
\end{aligned}$$

such that $f_0\,\bar{f}_0 = 1$. The mapping

$$T(u)\mapsto f(u)\,T(u), \qquad \overline{T}(u)\mapsto \bar{f}(u)\,\overline{T}(u) \tag{1.102}$$

defines an automorphism of the algebra $\mathrm{U}_q(\widehat{\mathfrak{gl}}_N)$. The *quantum affine algebra* $\mathrm{U}_q(\widehat{\mathfrak{sl}}_N)$ (with the trivial central charge) can be defined as the subalgebra of $\mathrm{U}_q(\widehat{\mathfrak{gl}}_N)$ which consists of the elements stable under all automorphisms (1.102); cf. Definition 1.8.1.

5. The quantized enveloping algebra $\mathrm{U}_q(\mathfrak{gl}_N)$ is a natural subalgebra of $\mathrm{U}_q(\widehat{\mathfrak{gl}}_N)$ defined via the embedding

$$t_{ij} \mapsto t_{ij}^{(0)}, \qquad \bar{t}_{ij} \mapsto \bar{t}_{ij}^{(0)}.$$

Moreover, there exists a surjective homomorphism $\mathrm{U}_q(\widehat{\mathfrak{gl}}_N) \to \mathrm{U}_q(\mathfrak{gl}_N)$ given by

$$T(u) \mapsto T - \overline{T}\, u^{-1}, \qquad \overline{T}(u) \mapsto \overline{T} - T\, u.$$

Its restriction to the q-Yangian provides a homomorphism $\mathrm{Y}_q(\mathfrak{gl}_N) \to \mathrm{U}_q(\mathfrak{gl}_N)$, which is also surjective. These are known as the *evaluation homomorphisms*; cf. Proposition 1.1.3. A more general homomorphism $\mathrm{U}_q(\widehat{\mathfrak{gl}}_N) \to \mathrm{U}_q(\mathfrak{gl}_N)$ can be obtained by twisting the evaluation homomorphism by the automorphism of $\mathrm{U}_q(\widehat{\mathfrak{gl}}_N)$ defined by

$$T(u) \mapsto T(au), \qquad \overline{T}(u) \mapsto \overline{T}(au)$$

for any nonzero complex number a.

6. Analogues of the quantum determinant for the q-Yangian and the quantum affine algebra can be constructed by analogy with $\operatorname{qdet} T(u)$; see Section 1.6. Here we assume that the complex parameter q is nonzero and not a root of unity. Let us define $R(u_1, \dots, u_m)$ by the formula (1.44), where R_{ij} now equals $R_{ij}(u_i, u_j)$. Then Proposition 1.6.1 holds in the same form for each of the matrices $T(u)$ and $\overline{T}(u)$. The action of the symmetric group $\mathfrak{S}_m$ on the space $(\mathbb{C}^N)^{\otimes m}$ can be defined by setting $s_i \mapsto P^q_{s_i} := P^q_{i,i+1}$, where s_i denotes the transposition $(i, i+1)$ and P^q is the q-permutation operator

$$P^q = \sum_i e_{ii} \otimes e_{ii} + q \sum_{i>j} e_{ij} \otimes e_{ji} + q^{-1} \sum_{i<j} e_{ij} \otimes e_{ji}.$$

This operator is an involution. If $s = s_{i_1} \dots s_{i_l}$ is a reduced decomposition of an element $s \in \mathfrak{S}_m$, we set $P^q_s = P^q_{s_{i_1}} \dots P^q_{s_{i_l}}$. We denote by A^q_m the q-antisymmetrizer

$$A^q_m = \sum_{s \in \mathfrak{S}_m} \operatorname{sgn} s \cdot P^q_s.$$

The following analogue of Proposition 1.6.2 holds:

$$R(1, q^{-2}, \dots, q^{-2m+2}) = \prod_{0 \leqslant i < j \leqslant m-1} (q^{-2i} - q^{-2j}) \cdot A^q_m.$$

Hence we have the identity

$$A^q_m\, T_1(u) \dots T_m(q^{-2m+2}u) = T_m(q^{-2m+2}u) \dots T_1(u)\, A^q_m.$$

The quantum determinant $\operatorname{qdet} T(u)$ is defined as the power series in u^{-1} with coefficients in $\mathrm{U}_q(\widehat{\mathfrak{gl}}_N)$ such that this operator with $m = N$ equals $A^q_N \operatorname{qdet} T(u)$. Explicitly,

$$\operatorname{qdet} T(u) = \sum_{s \in \mathfrak{S}_N} (-q)^{-l(s)} \cdot t_{s(1)1}(u) \dots t_{s(N)N}(q^{-2N+2}u),$$

where $l(s)$ is the length of a reduced decomposition of s. The quantum determinant $\operatorname{qdet} \overline{T}(u)$ is given by the same formula, where the $t_{ij}(u)$ are respectively replaced with $\bar{t}_{ij}(u)$.

The quantum determinants $\operatorname{qdet} T(u)$ and $\operatorname{qdet} \overline{T}(u)$ are comultiplicative: with respect to the comultiplication Δ on $\mathrm{U}_q(\widehat{\mathfrak{gl}}_N)$,

$$\operatorname{qdet} T(u) \mapsto \operatorname{qdet} T(u) \otimes \operatorname{qdet} T(u),$$
$$\operatorname{qdet} \overline{T}(u) \mapsto \operatorname{qdet} \overline{T}(u) \otimes \operatorname{qdet} \overline{T}(u).$$

7. The coefficients d_k and $\bar{d}_k$ of the quantum determinants defined by

$$\operatorname{qdet} T(u) = \sum_{k=0}^{\infty} d_k \, u^{-k}, \qquad \operatorname{qdet} \overline{T}(u) = \sum_{k=0}^{\infty} \bar{d}_k \, u^{k}$$

belong to the center of the algebra $\mathrm{U}_q(\widehat{\mathfrak{gl}}_N)$. The image of $\operatorname{qdet} T(u)$ under the evaluation homomorphism $\mathrm{Y}_q(\mathfrak{gl}_N) \to \mathrm{U}_q(\mathfrak{gl}_N)$ from Example 5 is a series whose coefficients are central elements of $\mathrm{U}_q(\mathfrak{gl}_N)$; see Example 7.7.1.

8. The following generalization of Proposition 1.7.1 holds:

$$[t^{\,a_1 \dots a_m}_{\,b_1 \dots b_m}(u), t^{\,c_1 \dots c_l}_{\,d_1 \dots d_l}(v)] = \sum_{p=1}^{\min\{m,l\}} \frac{(-1)^{p-1}\, p!}{(u-v-m+1)\dots(u-v-m+p)}$$

$$\times \sum_{\substack{i_1<\dots<i_p \\ j_1<\dots<j_p}} \Big(t^{\,a_1 \dots c_{j_1} \dots c_{j_p} \dots a_m}_{\,b_1 \;\; \dots \;\; b_m}(u)\, t^{\,c_1 \dots a_{i_1} \dots a_{i_p} \dots c_l}_{\,d_1 \;\; \dots \;\; d_l}(v)$$

$$- t^{\,c_1 \;\; \dots \;\; c_l}_{\,d_1 \dots b_{i_1} \dots b_{i_p} \dots d_l}(v)\, t^{\,a_1 \;\; \dots \;\; a_m}_{\,b_1 \dots d_{j_1} \dots d_{j_p} \dots b_m}(u) \Big).$$

Here the p-tuples of upper indices $(a_{i_1}, \dots, a_{i_p})$ and $(c_{j_1}, \dots, c_{j_p})$ are respectively interchanged in the first summand on the right-hand side while the p-tuples of lower indices $(b_{i_1}, \dots, b_{i_p})$ and $(d_{j_1}, \dots, d_{j_p})$ are interchanged in the second summand. The relation follows from (1.101).

9. Consider the $\mathbb{Z}_2$-graded vector space $\mathbb{C}^{M|N}$. For any $i = 1, \dots, M+N$ let $\bar{\imath}$ be the element of $\mathbb{Z}_2$ which is 0 if $i \leqslant M$ and 1 if $i > M$. Then the $\mathbb{Z}_2$-degree of the standard basis vector e_i of $\mathbb{C}^{M|N}$ is $\bar{\imath}$. The graded permutation operator is defined by

$$P = \sum_{i,j=1}^{M+N} (-1)^{\bar{\jmath}}\, e_{ij} \otimes e_{ji} \in \operatorname{End} \mathbb{C}^{M|N} \otimes \operatorname{End} \mathbb{C}^{M|N}.$$

The R-matrix $R(u)$ is given by the same formula (1.12) with the new permutation operator P. The Yangian $\mathrm{Y}(\mathfrak{gl}_{M|N})$ for the general linear Lie superalgebra $\mathfrak{gl}_{M|N}$ is a $\mathbb{Z}_2$-graded unital associative algebra generated by elements $t_{ij}^{(1)}, t_{ij}^{(2)}, \dots$ where $i, j = 1, \dots, M+N$ and the $\mathbb{Z}_2$-degree of $t_{ij}^{(r)}$ is $\bar{\imath} + \bar{\jmath}$. The defining relations have the form of the ternary relation (1.19), where the element

$$T(u) \in \operatorname{End} \mathbb{C}^{M|N} \otimes \mathrm{Y}(\mathfrak{gl}_{M|N})[[u^{-1}]] \tag{1.103}$$

is defined by (1.2) and (1.7). Here $e_{ij} \in \operatorname{End} \mathbb{C}^{M|N}$ is a standard matrix unit; its $\mathbb{Z}_2$-degree is $\bar{\imath} + \bar{\jmath}$. Note that in the new defining relation (1.19) the tensor product

$$\operatorname{End} \mathbb{C}^{M|N} \otimes \operatorname{End} \mathbb{C}^{M|N} \otimes \mathrm{Y}(\mathfrak{gl}_{M|N})$$

is that of $\mathbb{Z}_2$-graded algebras. For any two $\mathbb{Z}_2$-graded associative algebras $\mathcal{A}$ and $\mathcal{B}$, their tensor product $\mathcal{A} \otimes \mathcal{B}$ is a $\mathbb{Z}_2$-graded algebra such that for any homogeneous elements $x, y \in \mathcal{A}$ and $z, w \in \mathcal{B}$

$$(x \otimes z)\,(y \otimes w) = (-1)^{\deg y \deg z}\,(x\,y) \otimes (z\,w) \tag{1.104}$$

and

$$\deg\,(x \otimes z) = \deg x + \deg z.$$

Here deg stands for the $\mathbb{Z}_2$-degree. In terms of the matrix elements, the relations in $\mathrm{Y}(\mathfrak{gl}_{M|N})$ take the form

$$(u - v)\,[t_{ij}(u), t_{kl}(v)] = (-1)^{\bar{\imath}\bar{k}+\bar{\imath}\bar{l}+\bar{k}\bar{l}}\big(t_{kj}(u)t_{il}(v) - t_{kj}(v)t_{il}(u)\big),$$

where the square brackets denote the *supercommutator*. In general, for any homogeneous elements x, y of a $\mathbb{Z}_2$-graded associative algebra we have

$$[x, y] = xy - (-1)^{\deg x \deg y}\, yx.$$

10. Let E_{ij} with $i, j = 1, \dots, M + N$ denote the standard basis of the general linear Lie superalgebra $\mathfrak{gl}_{M|N}$. We have the commutation relations

$$[E_{ij}, E_{kl}] = \delta_{kj} E_{il} - \delta_{il} E_{kj} (-1)^{(\bar{\imath}+\bar{\jmath})(\bar{k}+\bar{l})},$$

where the square brackets stand for the supercommutator. The mapping

$$t_{ij}(u) \mapsto \delta_{ij} + E_{ij}(-1)^{\bar{\imath}\bar{\jmath}}\, u^{-1}$$

defines the evaluation homomorphism $\mathrm{Y}(\mathfrak{gl}_{M|N}) \to \mathrm{U}(\mathfrak{gl}_{M|N})$.

11. Consider the Yangian $\mathrm{Y}(\mathfrak{gl}_{M|N})$ defined in Example 9 and let $T^{-1}(u)$ be the element inverse to (1.103). Determine the series $t'_{ij}(u)$ with the coefficients in $\mathrm{Y}(\mathfrak{gl}_{M|N})$ by the decomposition

$$T^{-1}(u) = \sum_{i,j=1}^{M+N} e_{ij} \otimes t'_{ij}(u).$$

Note that here we employ the convention (1.104) so that the ij-entry of the $N \times N$ matrix inverse to $[\,t_{ij}(u)]$, is equal to $(-1)^{\bar{\imath}+\bar{\jmath}}\, t'_{ij}(u)$ rather than to $t'_{ij}(u)$. The series

$$\begin{aligned} b(u) = \sum_{p \in \mathfrak{S}_M} &\operatorname{sgn} p \cdot t_{p(1),1}(u) \dots t_{p(M),M}(u - M + 1) \\ &\times \sum_{q \in \mathfrak{S}_N} \operatorname{sgn} q \cdot t'_{M+1,M+q(1)}(u - M + 1) \dots t'_{M+N,M+q(N)}(u - M + N) \end{aligned}$$

is called the *quantum Berezinian* of the element $T(u)$. All the coefficients of the series $b(u)$ belong to the center of the Yangian $\mathrm{Y}(\mathfrak{gl}_{M|N})$ and generate the center. The coefficients of the image of $b(u)$ under the evaluation homomorphism of Example 10 generate the center of the universal enveloping algebra $\mathrm{U}(\mathfrak{gl}_{M|N})$.

Bibliographical notes

1.2. For the origins of the ternary (RTT) relation and the quantum inverse scattering method, see for instance the papers Kulish and Sklyanin [**251**], Reshetikhin, Takhtajan and Faddeev [**403**], Takhtajan and Faddeev [**435**], Faddeev [**110, 111**], and Sklyanin [**419**]. The statistical mechanics background of quantum groups was briefly explained in the book by Chari and Pressley [**65**, Chapter 7]. An introduction to the quantum inverse scattering method and its physical applications can be found in the book by Korepin, Bogoliubov and Izergin [**241**]; see also the lectures by Semenov-Tian-Shansky [**414**] for more remarks on the history of the method.

1.4. The Poincaré–Birkhoff–Witt theorem for general Yangians was known to Drinfeld, although he did not publish a proof. The proof of this theorem for the Yangian $\mathrm{Y}(\mathfrak{gl}_N)$ given here is due to Olshanski [**374**]. For three more proofs see Brundan and Kleshchev [**51**], Hopkins and Molev [**174**] and Nazarov [**342**]. A proof of the Poincaré–Birkhoff–Witt theorem for all Drinfeld Yangians was given by Levendorskiĭ [**268**].

1.6. The definition of the quantum determinant $\mathrm{qdet}\, T(u)$ for the Yangian $\mathrm{Y}(\mathfrak{gl}_N)$ originally appeared in Izergin and Korepin [**193**], in the case $N = 2$. The basic ideas and formulas associated with the quantum determinant for an arbitrary N are contained in the survey paper by Kulish and Sklyanin [**251**]. Detailed proofs can be found in Molev, Nazarov and Olshanski [**316**]. Proposition 1.6.9 is amongst several facts "well known to the specialists" which are collected in Iohara [**180**]; it is also proved in Nazarov and Tarasov [**351**]. Lemma 1.7.4 can be found in Brundan and Kleshchev [**51**]. It generalizes an earlier result by Chari and Ilangovan [**60**]; see also [**316**].

1.8. Following the general approach of Drinfeld [**100**], the Yangian for $\mathfrak{sl}_N$ should be defined as a quotient algebra of $\mathrm{Y}(\mathfrak{gl}_N)$. The fact that it can also be realized as a (Hopf) subalgebra of $\mathrm{Y}(\mathfrak{gl}_N)$ was observed by Olshanski, while the characterization of $\mathrm{Y}(\mathfrak{sl}_N)$ in terms of the automorphisms (1.20) is due to Drinfeld.

1.9. The quantum Liouville formula is due to Nazarov [**338**]; the series $z(u)$ was also introduced in that paper. The argument given in Section 1.9 is a simplified version of his proof in [**338**] and follows [**316**].

1.10. A general theory of quasideterminants of matrices over noncommutative rings was developed by Gelfand and Retakh [**132, 133**]. In particular, these papers contain quasideterminant factorizations of quantum determinants for the quantized algebra of functions on the group GL_N. In a more general context of Hopf algebras such factorizations were constructed by Etingof and Retakh [**109**]. For further properties of quasideterminants see Gelfand, Krob, Lascoux, Leclerc, Retakh and Thibon [**131**], Gelfand, Gelfand, Retakh and Wilson [**129**], and Krob and Leclerc [**246**].

1.11. The Gauss decomposition for a general noncommutative matrix in terms of quasideterminants is due to Gelfand and Retakh [**133**]. The Gauss decomposition of the matrix $T(u)$ was given by Iohara [**180**] following the approach of Ding and Frenkel [**86**]. Our exposition follows the paper by Brundan and Kleshchev [**51**], where the embedding ψ_M was used to construct analogues of Levi subalgebras in the Yangian and produce its new presentations. This embedding was also employed in an earlier work by Nazarov and Tarasov [**351**]. In particular, Lemma 1.11.3 was proved in that paper.

1.12. A general noncommutative analogue of the Sylvester theorem was given by Gelfand and Retakh [**132**]. A quantum version of this theorem for the quantized algebra of functions on GL_N is due to Krob and Leclerc [**246**]. Our proof of Theorem 1.12.1 essentially follows their approach with some modifications implied by the use of the embedding ψ_M, as in Brundan and Kleshchev [**51**]. A different proof was given by the author [**307**]. Some noncommutative versions of the Sylvester identity were given by Konvalinka [**239**]. Other quantum analogues of the classical minor relations are collected in Iohara [**180**].

1.13. Proposition 1.13.1 and Theorem 1.13.3 are due to Cherednik [**73, 74**]; see also Olshanski [**374**]. Our proofs follow Brundan and Kleshchev [**51**].

1.14. The commutative subalgebras in the Yangian $\mathrm{Y}(\mathfrak{gl}_N)$ were studied in the work of the St. Petersburg school in relation with the Bethe ansatz; see e.g. Kulish and Sklyanin [**251**], Kirillov and Reshetikhin [**227**], and Korepin, Bogoliubov and Izergin [**241**]. The name 'Bethe subalgebras' was used in the paper by Nazarov and Olshanski [**348**], where proofs of the results stated in Remark 1.14.4 were given. Moreover, the paper contains analogues of these results for the twisted Yangians and corresponding classical enveloping algebras; see also Mishchenko and Fomenko [**293**], Vinberg [**462**], and Rybnikov [**409**]. In the $\mathfrak{gl}_n$ case the Bethe subalgebras were used by Talalaev [**436**] to construct higher Hamiltonians for the Gaudin model; see also Chervov and Talalaev [**76**]. By a uniqueness theorem of Rybnikov [**410**], the Hamiltonians give rise to the same family of commuting operators, as provided by a construction of Feigin, Frenkel and Reshetikhin [**113**].

1.15. In Examples 1–7 we sketched the R-matrix approach to the study of quantized enveloping algebras and quantum affine algebras in the spirit of Jimbo [**198**] and Reshetikhin, Takhtajan and Faddeev [**403**]. Other presentations of these algebras and isomorphisms between them can be found in Ding and Frenkel [**86**], and Frenkel and Mukhin [**115**]. An analogue of Theorem 1.8.2 holds for certain extensions of the quantum affine algebras; see e.g. [**115**]. The action of the braid group on the quantized enveloping algebras was found by Lusztig [**276**]. Explicit formulas of Example 2 are obtained from [**276**] by re-writing the action in terms of the RTT presentation of $\mathrm{U}_q(\mathfrak{gl}_N)$. A relationship between the quantum affine algebras and Yangians was noted by Drinfeld [**100**]. On the other hand, in a more explicit form it can be obtained via a class of algebras called *Drinfeldians* introduced by Tolstoy [**445, 446**]. A proof of the relation of Example 8 is given by the author [**307**]. Some other quantum minor relations, including Laplace expansions generalizing Proposition 1.6.8, were proved by Lauve [**260**]. Together with Example 8 they were used to construct noncommutative flags associated with the Yangian $\mathrm{Y}(\mathfrak{gl}_N)$. Examples 9–11 are due to Nazarov [**338**], where it was conjectured that the coefficients of $b(u)$ generate the center of the Yangian $\mathrm{Y}(\mathfrak{gl}_{M|N})$. This conjecture was proved by Gow [**158**]. A new proof of the fact that the coefficients of the quantum Berezinian belong to the center of the Yangian $\mathrm{Y}(\mathfrak{gl}_{M|N})$ was given in another paper of Gow [**157**]. This paper also contains an analogue of Theorem 1.10.5 for the quantum Berezinian.

A class of 'deformed Yangians' was introduced by Khoroshkin, Stolin and Tolstoy [**214, 215**]. These algebras are quantizations of some rational solutions of the classical Yang–Baxter equation which were classified by Stolin [**425, 426**]; see also Kulish and Stolin [**254, 427**]. A q-analogue of the super-Yangian was introduced by Zhang [**476**]. A relationship between the (super)-Yangians and finite

$\mathcal{W}$-algebras was discovered by Ragoucy and Sorba [**397, 398**]; see also Briot and Ragoucy [**47, 49**]. This led Brundan and Kleshchev [**51, 52, 53**] to the discovery of the so-called 'shifted Yangians' which they used to obtain a deep generalization of this relationship. The representation theory of the shifted Yangians was developed in [**53**]. A natural analogue of the Yangian $\mathrm{Y}(\mathfrak{gl}_N)$ for the queer Lie superalgebra was introduced and studied by Nazarov [**339, 342**]. This analogue shares many properties of $\mathrm{Y}(\mathfrak{gl}_N)$; see also Nazarov and Sergeev [**349**]. The R-matrix approach to the theory of quantum (super)-groups together with its applications in theoretical physics are discussed in the review paper by Isaev [**182**].

CHAPTER 2

Twisted Yangians

In this chapter we introduce the twisted Yangians associated with the orthogonal and symplectic Lie algebras $\mathfrak{o}_N$ and $\mathfrak{sp}_N$ and describe their algebraic structure. In particular, we extend the R-matrix techniques developed in Chapter 1 to construct an analogue of the quantum determinant for the twisted Yangians and establish its relationship with the quantum determinant for the Yangian $\mathrm{Y}(\mathfrak{gl}_N)$.

2.1. Defining relations

We equip $\mathbb{C}^N$ with a nonsingular bilinear form which may be either symmetric or alternating. The alternating case can occur only if N is even. Let $G = [g_{ij}]$ be the corresponding matrix so that G is nonsingular with $G^t = \pm G$. Throughout this chapter we will use the following convention. Whenever the double sign $\pm$ or $\mp$ occurs, the upper sign corresponds to the symmetric case and the lower sign to the alternating case. Introduce the elements F_{ij} of the Lie algebra $\mathfrak{gl}_N$ by the formulas

$$F_{ij} = \sum_{k=1}^{N} (E_{ik}\, g_{kj} \mp E_{jk}\, g_{ki}). \tag{2.1}$$

Equivalently, denote by F the $N \times N$ matrix whose ij-th entry is F_{ij}. This matrix is then related to the matrix $E = [E_{ij}]$ by

$$F = EG \mp (EG)^t = EG - GE^t. \tag{2.2}$$

Obviously,

$$F_{ji} = \mp F_{ij} \tag{2.3}$$

and the elements F_{ij} satisfy the commutation relations

$$[F_{ij}, F_{kl}] = g_{kj}\, F_{il} - g_{il}\, F_{kj} - g_{ik}\, F_{jl} + g_{lj}\, F_{ki}. \tag{2.4}$$

The Lie subalgebra of $\mathfrak{gl}_N$ spanned by the elements F_{ij} is isomorphic to the orthogonal Lie algebra $\mathfrak{o}_N$ in the symmetric case and to the symplectic Lie algebra $\mathfrak{sp}_N$ in the alternating case. This Lie algebra will be denoted by $\mathfrak{g}_N$. Thus,

$$\mathfrak{g}_N = \mathfrak{o}_N \quad \text{or} \quad \mathfrak{g}_N = \mathfrak{sp}_N,$$

where in the latter case N is supposed to be even.

DEFINITION 2.1.1. The *twisted Yangian* $\mathrm{Y}_G(\mathfrak{g}_N)$ corresponding to the Lie algebra $\mathfrak{g}_N$ and the matrix G is a unital associative algebra with generators $s_{ij}^{(1)}, s_{ij}^{(2)}, \dots$ where $1 \leqslant i, j \leqslant N$, and the defining relations written in terms of the generating series

$$s_{ij}(u) = g_{ij} + s_{ij}^{(1)} u^{-1} + s_{ij}^{(2)} u^{-2} + \dots \tag{2.5}$$

as follows

$$(u^2 - v^2)\,[s_{ij}(u), s_{kl}(v)] = (u+v)\,\big(s_{kj}(u)s_{il}(v) - s_{kj}(v)s_{il}(u)\big) \\ - (u-v)\,\big(s_{ik}(u)s_{jl}(v) - s_{ki}(v)s_{lj}(u)\big) \\ + s_{ki}(u)s_{jl}(v) - s_{ki}(v)s_{jl}(u) \tag{2.6}$$

and

$$s_{ji}(-u) = \pm\, s_{ij}(u) + \frac{s_{ij}(u) - s_{ij}(-u)}{2u}. \tag{2.7}$$ □

It will be shown below (see Corollary 2.3.2) that the algebras $\mathrm{Y}_G(\mathfrak{g}_N)$ corresponding to different matrices G are isomorphic to each other so that $\mathrm{Y}_G(\mathfrak{g}_N)$ depends only on $\mathfrak{g}_N$. Therefore, when the matrix G is fixed, we will usually omit the subscript G and denote the twisted Yangian simply by $\mathrm{Y}(\mathfrak{g}_N)$. This twisted Yangian should not be confused with the Yangian of the Lie algebra $\mathfrak{g}_N$ as defined by Drinfeld in [**97**]; see Example 2.16.2 at the end of this chapter.

The following relation between the twisted Yangians and the corresponding classical Lie algebras plays a key role in many applications; cf. Proposition 1.1.3.

PROPOSITION 2.1.2. *The assignment*

$$s_{ij}(u) \mapsto g_{ij} + F_{ij}\Big(u \pm \frac{1}{2}\Big)^{-1} \tag{2.8}$$

defines an algebra epimorphism $\varrho_N : \mathrm{Y}(\mathfrak{g}_N) \to \mathrm{U}(\mathfrak{g}_N)$. *Moreover, the assignment*

$$F_{ij} \mapsto s_{ij}^{(1)} \tag{2.9}$$

defines an embedding $\mathrm{U}(\mathfrak{g}_N) \hookrightarrow \mathrm{Y}(\mathfrak{g}_N)$.

PROOF. We need to verify that the relations (2.6) and (2.7) are satisfied when $s_{ij}(u)$ is replaced with $\varphi_{ij}(u) = g_{ij} + F_{ij}\big(u \pm 1/2\big)^{-1}$. By (2.4) we have

$$(u \pm 1/2)\,(v \pm 1/2)\,[\varphi_{ij}(u), \varphi_{kl}(v)] = g_{kj}\,F_{il} - g_{il}\,F_{kj} - g_{ik}\,F_{jl} + g_{lj}\,F_{ki},$$

while

$$(u \pm 1/2)\,(v \pm 1/2)\,\big(\varphi_{kj}(u)\varphi_{il}(v) - \varphi_{kj}(v)\varphi_{il}(u)\big) = (u-v)(g_{kj}\,F_{il} - g_{il}\,F_{kj}),$$

$$(u \pm 1/2)\,(v \pm 1/2)\,\big(\varphi_{ik}(u)\varphi_{jl}(v) - \varphi_{ki}(v)\varphi_{lj}(u)\big) = (u+v\pm 1)(g_{ik}\,F_{jl} - g_{lj}\,F_{ki})$$

and

$$(u \pm 1/2)\,(v \pm 1/2)\,\big(\varphi_{ki}(u)\varphi_{jl}(v) - \varphi_{ki}(v)\varphi_{jl}(u)\big) = \pm(u-v)(g_{ik}\,F_{jl} - g_{lj}\,F_{ki})$$

which verifies (2.6). Next,

$$\varphi_{ji}(-u) \mp \varphi_{ij}(u) = F_{ij}(u^2 - 1/4)^{-1} = \frac{\varphi_{ij}(u) - \varphi_{ij}(-u)}{2u}$$

verifying (2.7).

Now divide both sides of (2.6) by $u^2 - v^2$ and use the formal expansion

$$(u-v)^{-1} = u^{-1} + u^{-2}v + u^{-3}v^2 + \dots \tag{2.10}$$

to represent the right-hand side of (2.6) as a Laurent series in u and v. Comparing the coefficients of $u^{-1}v^{-1}$ on both sides we get

$$[s_{ij}^{(1)}, s_{kl}^{(1)}] = g_{kj}\,s_{il}^{(1)} - g_{il}\,s_{kj}^{(1)} - g_{ik}\,s_{jl}^{(1)} + g_{lj}\,s_{ki}^{(1)}.$$

Furthermore, taking the coefficients of u^{-1} on both sides of (2.7) we get

$$-s_{ji}^{(1)} = \pm\, s_{ij}^{(1)}.$$

Due to (2.3) and (2.4), the above relations imply that the map (2.9) is a homomorphism $\mathrm{U}(\mathfrak{g}_N) \to \mathrm{Y}(\mathfrak{g}_N)$. Since the composition of ϱ_N with (2.9) is the identity map, the kernel of (2.9) is trivial. □

By analogy with (1.5), we will call ϱ_N the *evaluation homomorphism.*

2.2. Matrix form of the defining relations

Introduce the $N \times N$ matrix $S(u)$ whose ij-th entry is the series $s_{ij}(u)$. Following the notation of Section 1.2, we regard it as an element of the algebra $\mathrm{End}\,\mathbb{C}^N \otimes \mathrm{Y}(\mathfrak{g}_N)[[u^{-1}]]$ given by

$$S(u) = \sum_{i,j=1}^{N} e_{ij} \otimes s_{ij}(u).$$

Consider the R-matrix $R(u)$ defined in (1.12) and its transpose $R^t(u)$ defined in (1.14). In accordance with Section 1.2, we denote by $S_1(u)$ and $S_2(u)$ the elements of the algebra $\mathrm{End}\,\mathbb{C}^N \otimes \mathrm{End}\,\mathbb{C}^N \otimes \mathrm{Y}(\mathfrak{g}_N)[[u^{-1}]]$ defined as in (1.9).

PROPOSITION 2.2.1. *The defining relations* (2.6) *and* (2.7) *of* $\mathrm{Y}(\mathfrak{g}_N)$ *can be written, respectively, in the equivalent matrix form*

$$R(u-v)\, S_1(u)\, R^t(-u-v)\, S_2(v) = S_2(v)\, R^t(-u-v)\, S_1(u)\, R(u-v) \tag{2.11}$$

and

$$S^t(-u) = \pm\, S(u) + \frac{S(u) - S(-u)}{2u}. \tag{2.12}$$

PROOF. Let us apply the operators on both sides of (2.11) to an arbitrary basis vector $e_j \otimes e_l \in \mathbb{C}^N \otimes \mathbb{C}^N$. For the left-hand side we get

$$\sum_{i,k} e_i \otimes e_k \otimes s_{ij}(u)\, s_{kl}(v) + \frac{1}{u+v} \sum_{i,k} e_i \otimes e_k \otimes s_{ik}(u)\, s_{jl}(v)$$
$$- \frac{1}{u-v} \sum_{i,k} e_k \otimes e_i \otimes s_{ij}(u)\, s_{kl}(v) - \frac{1}{u^2-v^2} \sum_{i,k} e_k \otimes e_i \otimes s_{ik}(u)\, s_{jl}(v)$$

while the right-hand side gives

$$\sum_{i,k} e_i \otimes e_k \otimes s_{kl}(v)\, s_{ij}(u) - \frac{1}{u-v} \sum_{i,k} e_i \otimes e_k \otimes s_{kj}(v)\, s_{il}(u)$$
$$+ \frac{1}{u+v} \sum_{i,k} e_i \otimes e_k \otimes s_{ki}(v)\, s_{lj}(u) - \frac{1}{u^2-v^2} \sum_{i,k} e_i \otimes e_k \otimes s_{ki}(v)\, s_{jl}(u).$$

Equating the coefficients of $e_i \otimes e_k$ we get (2.6). Obviously, (2.12) is equivalent to (2.7). □

We will refer to (2.11) and (2.12) as the *quaternary* and *symmetry* relations, respectively.

2.3. Automorphisms and anti-automorphisms

Let B be an arbitrary nonsingular complex $N \times N$ matrix.

PROPOSITION 2.3.1. *The matrix $B\,S(u)\,B^t$ satisfies both the quaternary and symmetry relations.*

PROOF. Since the transpose of $B\,S(u)\,B^t$ is $B\,S^t(u)\,B^t$, the symmetry relation is satisfied by the matrix $B\,S(u)\,B^t$. Further, we need to verify that

$$R(u-v)\,B_1\,S_1(u)\,B_1^t\,R^t(-u-v)\,B_2\,S_2(v)\,B_2^t = B_2\,S_2(v)\,B_2^t\,R^t(-u-v)\,B_1\,S_1(u)\,B_1^t\,R(u-v). \tag{2.13}$$

Applying the transposition t_1 to both sides of (1.24) we obtain the relation

$$B_1^t\,R^t(-u-v)\,B_2 = B_2\,R^t(-u-v)\,B_1^t. \tag{2.14}$$

Therefore, the left-hand side of (2.13) can be written as

$$R(u-v)\,B_1\,B_2\,S_1(u)\,R^t(-u-v)\,S_2(v)\,B_1^t\,B_2^t.$$

Next, apply (1.24), then the quaternary relation (2.11) to bring this expression to the form

$$B_2\,B_1\,S_2(v)\,R^t(-u-v)\,S_1(u)\,R(u-v)\,B_1^t\,B_2^t. \tag{2.15}$$

Finally, applying (1.24) and (2.14) to the matrix B^t we arrive at the right-hand side of (2.13). □

Suppose that G and G' are two nonsingular symmetric (respectively, skew-symmetric) $N\times N$-matrices. Consider the corresponding twisted Yangians $\mathrm{Y}_G(\mathfrak{g}_N)$ and $\mathrm{Y}_{G'}(\mathfrak{g}_N)$ associated with the Lie algebra $\mathfrak{g}_N$; see Definition 2.1.1.

COROLLARY 2.3.2. *The algebras $\mathrm{Y}_G(\mathfrak{g}_N)$ and $\mathrm{Y}_{G'}(\mathfrak{g}_N)$ are isomorphic to each other.*

PROOF. Let $S(u)$ and $S'(u)$ be the generator matrices for the algebras $\mathrm{Y}_G(\mathfrak{g}_N)$ and $\mathrm{Y}_{G'}(\mathfrak{g}_N)$, respectively. Choose a matrix B such that $BGB^t = G'$. Then B is nonsingular. By Proposition 2.3.1, the mapping

$$S'(u) \mapsto BS(u)B^t \tag{2.16}$$

defines an algebra isomorphism $\mathrm{Y}_{G'}(\mathfrak{g}_N) \to \mathrm{Y}_G(\mathfrak{g}_N)$. □

Suppose now that B is a matrix which preserves the bilinear form defined by the matrix G, that is, $BGB^t = G$. Also, consider an arbitrary even formal series which begins with 1,

$$g(u) = 1 + g_1\,u^{-2} + g_2\,u^{-4} + \cdots \in \mathbb{C}[[u^{-2}]].$$

PROPOSITION 2.3.3. *Each of the mappings*

$$\nu_g : S(u) \mapsto g(u)\,S(u), \tag{2.17}$$

$$S(u) \mapsto B\,S(u)\,B^t \tag{2.18}$$

defines an automorphism of $\mathrm{Y}(\mathfrak{g}_N)$.

PROOF. The assertion is obvious for the map (2.17). By Proposition 2.3.1, the quaternary and symmetry relations will be satisfied if the matrix $S(u)$ is replaced by its image under the map (2.18). Moreover, the map is obviously invertible. □

PROPOSITION 2.3.4. *The mapping*

$$S(u) \mapsto \pm\, S^t(u),$$

defines an anti-automorphism of $\mathrm{Y}(\mathfrak{g}_N)$.

PROOF. It is clear from the defining relations (2.6) that the images of the series $s_{ij}(u)$ under the mapping will satisfy (2.6) with the opposite multiplication. Moreover, the matrix $\pm\, S^t(u)$ satisfies the symmetry relation (2.7) and its constant term component is G. □

PROPOSITION 2.3.5. *The matrix* $S^{-1}\big(-u-N/2\big)$ *satisfies the quaternary relation* (2.11).

PROOF. By (1.13) and (1.16) we have

$$R(u)^{-1} = \frac{u^2}{u^2-1}\, R(-u), \qquad R^t(u)^{-1} = R^t(-u+N). \tag{2.19}$$

Inverting both sides of (2.11) we get

$$R(-u+v)\, S_1^{-1}(u)\, R^t(u+v+N)\, S_2^{-1}(v)$$
$$= S_2^{-1}(v)\, R^t(u+v+N)\, S_1^{-1}(u)\, R(-u+v).$$

It remains to replace the parameters u and v in the last relation by $-u-N/2$ and $-v-N/2$, respectively. □

Note that the matrix $S^{-1}\big(-u-N/2\big)$ does not satisfy the symmetry relation (2.12); cf. Sections 2.13 and 2.15 below.

2.4. Embedding into the Yangian

We will demonstrate that the twisted Yangian $\mathrm{Y}(\mathfrak{g}_N)$ can be regarded as a subalgebra of the Yangian $\mathrm{Y}(\mathfrak{gl}_N)$.

LEMMA 2.4.1. *We have the identity*

$$R(u-v)\, G_1\, R^t(-u-v)\, G_2 = G_2\, R^t(-u-v)\, G_1\, R(u-v). \tag{2.20}$$

PROOF. By Proposition 2.2.1, the claim is immediate from the observation that (2.6) will hold if the $s_{ij}(u)$ are respectively replaced with g_{ij}. □

COROLLARY 2.4.2. *The mapping* $S(u) \mapsto G$ *defines a (one-dimensional) representation of the algebra* $\mathrm{Y}(\mathfrak{g}_N)$.

PROOF. Due to Lemma 2.4.1, we need to verify only that the symmetry relation (2.12) is satisfied if $S(u)$ is replaced by G, which is obvious. □

THEOREM 2.4.3. *The mapping*

$$S(u) \mapsto T(u)\, G\, T^t(-u) \tag{2.21}$$

defines an embedding $\mathrm{Y}(\mathfrak{g}_N) \hookrightarrow \mathrm{Y}(\mathfrak{gl}_N)$.

PROOF. Let us show first that the mapping defines an algebra homomorphism. Obviously, the constant term component of the matrix $T(u)\, G\, T^t(-u)$ is G. Now verify that this matrix satisfies the quaternary relation, that is,

$$R(u-v)\, T_1(u)\, G_1\, T_1^t(-u)\, R^t(-u-v)\, T_2(v)\, G_2\, T_2^t(-v) \tag{2.22}$$
$$= T_2(v)\, G_2\, T_2^t(-v)\, R^t(-u-v)\, T_1(u)\, G_1\, T_1^t(-u)\, R(u-v).$$

Indeed, arguing as in the proof of Proposition 2.3.1, transform the left-hand side by first applying the relation (1.30), then the ternary relation (1.19). This brings it to the form

$$T_2(v)\, T_1(u)\, R(u-v)\, G_1\, R^t(-u-v)\, G_2\, T_1^t(-u)\, T_2^t(-v), \tag{2.23}$$

where we have also used the fact that arbitrary elements A_i and B_j of the algebra (1.8) commute if $i \neq j$; see Section 1.2. Since (1.32) is an automorphism of the algebra $\mathrm{Y}(\mathfrak{gl}_N)$, we have

$$R(u-v)\, T_1^t(-u)\, T_2^t(-v) = T_2^t(-v)\, T_1^t(-u)\, R(u-v). \tag{2.24}$$

The application of the partial transposition t_1 to both sides gives

$$T_1(-u)\, R^t(u-v)\, T_2^t(-v) = T_2^t(-v)\, R^t(u-v)\, T_1(-u). \tag{2.25}$$

Now, transform further (2.23) by first applying Lemma 2.4.1, then (2.24) and (2.25). This yields the right-hand side of (2.22).

The ji-th entry of the matrix $T(u)\, G\, T^t(-u)$ is given by

$$\sum_{k,l} g_{kl}\, t_{jk}(u)\, t_{il}(-u).$$

However, by the defining relations (1.3) we have

$$t_{jk}(u)\, t_{il}(-u) = t_{il}(-u)\, t_{jk}(u) + \frac{t_{ik}(u)\, t_{jl}(-u) - t_{ik}(-u)\, t_{jl}(u)}{2u}.$$

Since $g_{lk} = \pm\, g_{kl}$, multiplying both sides by g_{kl} and taking sum over k and l we see that the matrix $T(u)\, G\, T^t(-u)$ satisfies the symmetry relation (2.12).

It remains to show that the kernel of the homomorphism (2.21) is trivial. Define an ascending filtration on $\mathrm{Y}(\mathfrak{g}_N)$ by setting $\deg s_{ij}^{(r)} = r$ (another filtration will be considered in Section 2.8). Also consider the filtration on the Yangian $\mathrm{Y}(\mathfrak{gl}_N)$ introduced in Section 1.4. The homomorphism (2.21) is filtration-preserving and hence it defines a homomorphism of the corresponding graded algebras

$$\mathrm{gr}\, \mathrm{Y}(\mathfrak{g}_N) \to \mathrm{gr}\, \mathrm{Y}(\mathfrak{gl}_N). \tag{2.26}$$

It will be sufficient to show that the kernel of this homomorphism is trivial. Denote by $\bar{s}_{ij}^{(r)}$ the image of $s_{ij}^{(r)}$ in the r-th component of $\mathrm{gr}\, \mathrm{Y}(\mathfrak{g}_N)$. Then the image of $\bar{s}_{ij}^{(r)}$ under the homomorphism (2.26) is

$$\sum_k g_{kj}\, \bar{t}_{ik}^{(r)} + (-1)^r \sum_l g_{il}\, \bar{t}_{jl}^{(r)} + \sum_{a+b=r} \sum_{k,l} (-1)^b g_{kl}\, \bar{t}_{ik}^{(a)} \bar{t}_{jl}^{(b)}, \tag{2.27}$$

where a and b run over positive integers. Now observe that the algebra $\mathrm{gr}\, \mathrm{Y}(\mathfrak{g}_N)$ is commutative. Indeed, divide both sides of (2.6) by $u^2 - v^2$, then use the formal expansion (2.10) and compare the coefficients of $u^{-r} v^{-p}$ on both sides to get $[\bar{s}_{ij}^{(r)}, \bar{s}_{kl}^{(p)}] = 0$. The symmetry relation (2.7) gives $(-1)^r\, \bar{s}_{ji}^{(r)} = \pm\, \bar{s}_{ij}^{(r)}$. So, it is enough to show that the elements

$$\bar{s}_{ij}^{(2p)} \quad \text{with} \quad i \geqslant j \qquad \text{and} \qquad \bar{s}_{ij}^{(2p-1)} \quad \text{with} \quad i > j \qquad \text{for} \quad p = 1, 2, \dots,$$

in the orthogonal case, and the elements

$$\bar{s}_{ij}^{(2p)} \quad \text{with} \quad i > j \qquad \text{and} \qquad \bar{s}_{ij}^{(2p-1)} \quad \text{with} \quad i \geqslant j \qquad \text{for} \quad p = 1, 2, \dots,$$

in the symplectic case, are algebraically independent. However, by Corollary 1.4.2, the elements $\bar{t}_{ij}^{(r)}$ of $\mathrm{gr}\, \mathrm{Y}(\mathfrak{gl}_N)$ are algebraically independent. Note that the sum over

a and b in (2.27) only involves generators $\bar{t}_{ij}^{(p)}$ with $p < r$. On the other hand, the linear part of (2.27) is an element of $\operatorname{gr} \mathrm{Y}(\mathfrak{gl}_N)$ depending on the indices i, j, r. Since the matrix G is nonsingular, these elements parameterized by the triples (i, j, r) with the respective restrictions in the orthogonal and symplectic case are algebraically independent. This implies the desired property of the elements $\bar{s}_{ij}^{(r)}$. □

Using Theorem 2.4.3 we will often identify $\mathrm{Y}(\mathfrak{g}_N)$ with its image under the embedding (2.21) and write

$$s_{ij}(u) = \sum_{k,l} g_{kl}\, t_{ik}(u)\, t_{jl}(-u). \tag{2.28}$$

The following is an analogue of the Poincaré–Birkhoff–Witt theorem for the algebra $\mathrm{Y}(\mathfrak{g}_N)$.

COROLLARY 2.4.4. *Given an arbitrary linear order on the set of generators*

$$s_{ij}^{(2p)} \quad \textit{with} \quad i \geqslant j \qquad \textit{and} \qquad s_{ij}^{(2p-1)} \quad \textit{with} \quad i > j \qquad \textit{for} \quad p = 1, 2, \dots,$$

in the orthogonal case, and

$$s_{ij}^{(2p)} \quad \textit{with} \quad i > j \qquad \textit{and} \qquad s_{ij}^{(2p-1)} \quad \textit{with} \quad i \geqslant j \qquad \textit{for} \quad p = 1, 2, \dots,$$

in the symplectic case, any element of the algebra $\mathrm{Y}(\mathfrak{g}_N)$ is uniquely written as a linear combination of ordered monomials in the generators.

PROOF. Due to the symmetry relation (2.7), the chosen elements generate the algebra $\mathrm{Y}(\mathfrak{g}_N)$. As we have seen in the proof of Theorem 2.4.3, their images in the graded algebra $\operatorname{gr} \mathrm{Y}(\mathfrak{g}_N)$ are algebraically independent, which implies the statement. □

REMARK 2.4.5. By Corollary 2.4.4, the twisted Yangian $\mathrm{Y}(\mathfrak{g}_N)$ can be viewed as a flat deformation of the algebra of polynomials in countably many variables. Indeed, for each $h \in \mathbb{C}$ consider the algebra $\mathrm{Y}(\mathfrak{g}_N, h)$ with the generators $s_{ij}^{(r)}$ and the relations

$$\begin{aligned}(u^2 - v^2)\,[s_{ij}(u), s_{kl}(v)] = {} & h\,(u+v)\,\big(s_{kj}(u)s_{il}(v) - s_{kj}(v)s_{il}(u)\big)\\ & - h\,(u-v)\,\big(s_{ik}(u)s_{jl}(v) - s_{ki}(v)s_{lj}(u)\big)\\ & + h^2\,\big(s_{ki}(u)s_{jl}(v) - s_{ki}(v)s_{jl}(u)\big)\end{aligned}$$

and

$$s_{ji}(-u) = \pm\, s_{ij}(u) + h\, \frac{s_{ij}(u) - s_{ij}(-u)}{2u},$$

where $s_{ij}(u)$ is given by (2.5). The algebras $\mathrm{Y}(\mathfrak{g}_N, h)$ with $h \neq 0$ are all isomorphic to each other; an isomorphism $\mathrm{Y}(\mathfrak{g}_N, h) \to \mathrm{Y}(\mathfrak{g}_N)$ can be given by $s_{ij}^{(r)} \mapsto s_{ij}^{(r)} h^r$. On the other hand, $\mathrm{Y}(\mathfrak{g}_N, 0)$ is an algebra of polynomials with an inherited Poisson structure; cf. Remark 1.4.4. □

REMARK 2.4.6. Let us regard the twisted Yangian $\mathrm{Y}(\mathfrak{g}_N)$ as a subalgebra of $\mathrm{Y}(\mathfrak{gl}_N)$ using Theorem 2.4.3. Consider the restriction of the evaluation homomorphism π_N defined by (1.5) to the subalgebra $\mathrm{Y}(\mathfrak{g}_N) \subset \mathrm{Y}(\mathfrak{gl}_N)$. This restriction does not coincide with the homomorphism ϱ_N defined by (2.8). Due to (2.2), under the homomorphism ϱ_N

$$S(u) \mapsto G + (EG - GE^t)\Big(u \pm \frac{1}{2}\Big)^{-1},$$

while under the homomorphism π_N

$$S(u) \mapsto (1 + Eu^{-1})\, G\, (1 - E^t u^{-1}) = G + (EG - GE^t)u^{-1} - EGE^t u^{-2}.$$

In fact, the entries of the matrix EGE^t do not belong to the subalgebra $\mathrm{U}(\mathfrak{g}_N)$ of $\mathrm{U}(\mathfrak{gl}_N)$. □

2.5. Sklyanin determinant

Here we use the notation of Section 1.6 for elements of the tensor product algebra $(\operatorname{End}\mathbb{C}^N)^{\otimes m} \otimes \mathrm{Y}(\mathfrak{g}_N)$. Let $u_1, \dots, u_m$ be independent complex variables. Consider the rational function $R(u_1, \dots, u_m)$ defined by (1.44) and set

$$S_i = S_i(u_i), \quad 1 \leqslant i \leqslant m$$

and

$$R^t_{ij} = R^t_{ji} = R^t_{ij}(-u_i - u_j), \quad 1 \leqslant i < j \leqslant m,$$

where $S_a(u)$ and $R^t_{ab}(u)$ are defined as in (1.9) and (1.10). For an arbitrary permutation $(p_1, \dots, p_m)$ of the indices $1, \dots, m$, we abbreviate

$$\langle S_{p_1}, \dots, S_{p_m} \rangle = S_{p_1} (R^t_{p_1 p_2} \dots R^t_{p_1 p_m}) S_{p_2} (R^t_{p_2 p_3} \dots R^t_{p_2 p_m}) \dots S_{p_m}. \tag{2.29}$$

PROPOSITION 2.5.1. *We have the identity*

$$R(u_1, \dots, u_m) \langle S_1, \dots, S_m \rangle = \langle S_m, \dots, S_1 \rangle R(u_1, \dots, u_m). \tag{2.30}$$

PROOF. Let i, j, k be distinct indices from the set $\{1, \dots, m\}$. Note the following relations

$$R_{ij}\, S_i\, R^t_{ij}\, S_j = S_j\, R^t_{ji}\, S_i\, R_{ij} \tag{2.31}$$

and

$$R_{ij}\, R^t_{ik}\, R^t_{jk} = R^t_{jk}\, R^t_{ik}\, R_{ij}. \tag{2.32}$$

Indeed, (2.31) is just the quaternary relation (2.11) applied to the i-th and j-th copies of $\operatorname{End}\mathbb{C}^N$ in the tensor product $(\operatorname{End}\mathbb{C}^N)^{\otimes m} \otimes \mathrm{Y}(\mathfrak{g}_N)$. To derive (2.32), write the Yang–Baxter equation (1.17) in the form

$$R_{ij}\, R_{ik}\, R_{jk} = R_{jk}\, R_{ik}\, R_{ij}.$$

Now apply the partial transposition t_k to both sides to get

$$R_{ij}\, R^t_{jk}\, R^t_{ik} = R^t_{ik}\, R^t_{jk}\, R_{ij}. \tag{2.33}$$

Further, since $P_{ij}\, R^t_{jk}\, P_{ij} = R^t_{ik}$, conjugating both sides of (2.33) by P_{ij}, we get (2.32). We will also need the following generalization of (2.33),

$$R_{ij}(R^t_{ik_1} \dots R^t_{ik_r})(R^t_{jk_1} \dots R^t_{jk_r}) = (R^t_{jk_1} \dots R^t_{jk_r})(R^t_{ik_1} \dots R^t_{ik_r}) R_{ij}, \tag{2.34}$$

provided $i, j, k_1, \dots, k_r$ are pairwise distinct. To verify (2.34) we observe that $R^t_{ik_a}$ and $R^t_{jk_b}$ commute when $a \neq b$, so that (2.34) can be rewritten as

$$R_{ij}\, (R^t_{ik_1} R^t_{jk_1}) \dots (R^t_{ik_r} R^t_{jk_r}) = (R^t_{ik_r} R^t_{jk_r}) \dots (R^t_{ik_1} R^t_{jk_1})\, R_{ij}$$

which is an immediate consequence of (2.33).

Let us prove that for any $i = 1, \dots, m-1$ and any permutation $(p_1, \dots, p_m)$ of the indices $1, \dots, m$ we have

$$R_{p_i p_{i+1}} \langle S_{p_1}, \dots, S_{p_m} \rangle = \langle S_{p_1}, \dots, S_{p_{i-1}}, S_{p_{i+1}}, S_{p_i}, S_{p_{i+2}}, \dots, S_{p_m} \rangle R_{p_i p_{i+1}}. \tag{2.35}$$

First, we examine the segment of the product $\langle S_{p_1}, \dots, S_{p_m} \rangle$ which precedes S_{p_i}. All the factors of this segment commute with $R_{p_i p_{i+1}}$ except for $R^t_{p_k p_i}$ and $R^t_{p_k p_{i+1}}$, where $k = 1, \dots, i-1$. To permute $R_{p_i p_{i+1}}$ with these factors we use the rule

$$R_{p_i p_{i+1}} R^t_{p_k p_i} R^t_{p_k p_{i+1}} = R^t_{p_k p_{i+1}} R^t_{p_k p_i} R_{p_i p_{i+1}};$$

see (2.32). After these transformations the segment under consideration takes the same form as the corresponding segment in the right-hand side of (2.35).

Next, we examine the segment

$$S_{p_i} R^t_{p_i p_{i+1}} \Big(\prod_{k>i+1} R^t_{p_i p_k} \Big) S_{p_{i+1}} \Big(\prod_{k>i+1} R^t_{p_{i+1} p_k} \Big).$$

Since $R^t_{p_i p_k}$ and $S_{p_{i+1}}$ commute for any $k > i+1$, we may rewrite this segment as

$$S_{p_i} R^t_{p_i p_{i+1}} S_{p_{i+1}} \Big(\prod_{k>i+1} R^t_{p_i p_k} \Big) \Big(\prod_{k>i+1} R^t_{p_{i+1} p_k} \Big). \tag{2.36}$$

To permute $R_{p_i p_{i+1}}$ with (2.36), we first use (2.31) then (2.34). Then we rewrite $R^t_{p_i p_{i+1}}$ as $R^t_{p_{i+1} p_i}$ and permute S_{p_i} with the product

$$\prod_{k>i+1} R^t_{p_{i+1} p_k}.$$

Again, after these transformations our segment takes the same form as the corresponding segment in the right-hand side of (2.35). Finally, the remaining segment is just $\langle S_{p_{i+2}}, \dots, S_{p_m} \rangle$. All the factors of this segment commute with $R_{p_i p_{i+1}}$.

It remains to observe that (2.30) is a consequence of (2.35). In fact, using (2.35) repeatedly, we permute R_{12} with $\langle S_1, \dots, S_m \rangle$, then we permute R_{13} with $\langle S_2, S_1, S_3, \dots, S_m \rangle$, etc. The total effect of the permutation with all the factors R_{ij} occurring in $R(u_1, \dots, u_m)$ clearly amounts to rearranging the factors S_i into reverse order, just as they appear in the right-hand side of (2.30). □

Now take $m = N$ and specialize the variables u_i as

$$u_i = u - i + 1, \qquad i = 1, \dots, N.$$

By Propositions 1.6.2 and 2.5.1 we have

$$A_N \langle S_1, \dots, S_N \rangle = \langle S_N, \dots, S_1 \rangle A_N. \tag{2.37}$$

By analogy with Definition 1.6.5 we come to the following.

DEFINITION 2.5.2. The *Sklyanin determinant* of the matrix $S(u)$ with coefficients in $\mathrm{Y}(\mathfrak{g}_N)$ is the formal series

$$\operatorname{sdet} S(u) = c_0 + c_1 u^{-1} + c_2 u^{-2} + \dots$$

such that the element (2.37) equals $A_N \operatorname{sdet} S(u)$. □

In the next theorem we regard $\mathrm{Y}(\mathfrak{g}_N)$ as a subalgebra of $\mathrm{Y}(\mathfrak{gl}_N)$.

THEOREM 2.5.3. *We have*

$$\operatorname{sdet} S(u) = \gamma_{n,G}(u) \operatorname{qdet} T(u) \operatorname{qdet} T(-u + N - 1),$$

where

$$\gamma_{n,G}(u) = \begin{cases} \det G & \text{if } \mathfrak{g}_N = \mathfrak{o}_N, \\[1em] \dfrac{2u+1}{2u-2n+1} \det G & \text{if } \mathfrak{g}_N = \mathfrak{sp}_{2n}. \end{cases}$$

In particular, $c_0 = \det G$.

PROOF. Recall the automorphism τ_N of $\mathrm{Y}(\mathfrak{gl}_N)$ given by (1.32) and observe that $S_i = T_i\, G_i\, T_i^{\sigma}$, where $T_i = T_i(u-i+1)$ and $T_i^{\sigma} = T_i^t(-u+i-1)$. Therefore, the left-hand side of (2.37) takes the form

$$A_N\, T_1\, G_1\, T_1^{\sigma}\, R_{12}^t \dots R_{1N}^t\, T_2\, G_2\, T_2^{\sigma}\, R_{23}^t \times \dots R_{2N}^t\, T_3\, G_3\, T_3^{\sigma} \dots T_{N-1}\, G_{N-1}\, T_{N-1}^{\sigma}\, R_{N-1,N}^t\, T_N\, G_N\, T_N^{\sigma}, \tag{2.38}$$

where $R_{ij}^t = R_{ij}^t(-2u+i+j-2)$. By (1.30),

$$T_i^{\sigma}\, R_{ij}^t\, T_j = T_j\, R_{ji}^t\, T_i^{\sigma}. \tag{2.39}$$

Since the elements T_i and T_i^{σ} commute with R_{jk}^t for $i \neq j, k$, we can rewrite (2.38) in the following way:

$$A_N\, T_1\, G_1 (T_1^{\sigma} R_{12}^t T_2) R_{13}^t \dots R_{1N}^t\, G_2 (T_2^{\sigma} R_{23}^t T_3) \times \dots\, G_{N-1} (T_{N-1}^{\sigma} R_{N-1,N}^t T_N)\, G_N\, T_N^{\sigma}.$$

Applying (2.39) to the products enclosed in brackets, we obtain the expression

$$A_N\, T_1 T_2\, G_1 R_{12}^t (T_1^{\sigma} R_{13}^t T_3) R_{14}^t \dots R_{1N}^t\, G_2\, R_{23}^t (T_2^{\sigma} R_{24}^t T_4) \times \dots (T_{N-2}^{\sigma} R_{N-2,N}^t T_N)\, G_{N-1}\, R_{N-1,N}^t\, G_N\, T_{N-1}^{\sigma} T_N^{\sigma}.$$

Further applying (2.39) repeatedly, we bring (2.38) to the form

$$A_N T_1 \dots T_N\, G_1\, R_{12}^t \dots R_{1N}^t\, G_2\, R_{23}^t \dots R_{2N}^t \dots R_{N-1,N}^t\, G_N\, T_1^{\sigma} \dots T_N^{\sigma}.$$

Using Definition 1.6.5 and notation (2.29) (with $S_i(u_i)$ replaced by G_i) we write this as

$$\operatorname{qdet} T(u)\, A_N \langle G_1, \dots, G_N \rangle\, T_1^{\sigma} \dots T_N^{\sigma}.$$

By Corollary 2.4.2, the mapping $S(u) \mapsto G$ defines a representation of the algebra $\mathrm{Y}(\mathfrak{g}_N)$. Therefore, (2.37) gives

$$A_N \langle G_1, \dots, G_N \rangle = \langle G_N, \dots, G_1 \rangle\, A_N = A_N\, \gamma_{n,G}(u) \tag{2.40}$$

for a scalar function $\gamma_{n,G}(u)$. By Definition 1.6.5,

$$A_N\, T_1^{\sigma} \dots T_N^{\sigma} = \tau_N\big(\operatorname{qdet} T(u)\big).$$

On the other hand, applying τ_N to both sides of (1.52) and using (1.53) we derive that

$$\tau_N\big(\operatorname{qdet} T(u)\big) = \operatorname{qdet} T(-u+N-1).$$

In order to complete the proof we need to calculate the scalar $\gamma_{n,G}(u)$. Recall that any constant matrix B satisfies the ternary relation; see (1.24). Moreover, we have the relations

$$A_N B_1 \dots B_N = A_N B_1^t \dots B_N^t = A_N \det B. \tag{2.41}$$

Therefore, repeating the previous argument with the product $T(u)\, G\, T^t(-u)$ replaced by BGB^t we come to the following conclusion. If the matrix G is replaced by BGB^t, then the scalar $\gamma_{n,G}(u)$ is replaced by $\gamma_{n,G}(u)(\det B)^2$. This allows us to consider, without loss of generality, a particular matrix G for the calculation of $\gamma_{n,G}(u)$. Consider the orthogonal case first and take $G = 1$, the identity matrix. By definition (1.15) of the operator Q we have $PQ = Q$ and so

$$(1-P)\, R^t(u) = 1 - P.$$

Hence, for any $i < j$ we have

$$A_N\, R^t_{ij} = \frac{1}{2} A_N(1 - P_{ij}) R^t_{ij} = \frac{1}{2} A_N(1 - P_{ij}) = A_N.$$

This yields $\gamma_{n,1}(u) = 1$. In the symplectic case we take the $2n \times 2n$ matrix G with

$$g_{2k-1,2k} = -g_{2k,2k-1} = 1, \qquad k = 1, \dots, n$$

and zeros elsewhere. Apply the left-hand side of (2.40) to the basis vector

$$v = e_{2n-1} \otimes e_{2n-3} \otimes \dots \otimes e_1 \otimes e_2 \otimes e_4 \otimes \dots \otimes e_{2n}.$$

Since $G\, e_{2k} = e_{2k-1}$ the resulting element can be written as

$$A_{2n} G_1 R^t_{12} \dots R^t_{1,2n} G_2 \dots G_n R^t_{n,n+1} \dots R^t_{n,2n}\, w, \tag{2.42}$$

where

$$w = e_{2n-1} \otimes e_{2n-3} \otimes \dots \otimes e_1 \otimes e_1 \otimes e_3 \otimes \dots \otimes e_{2n-1}.$$

Let $A^{(i)}_{2n}$ denote the anti-symmetrizer over the copies of $\operatorname{End}\mathbb{C}^N$ numbered by the indices $\{i+1, i+2, \dots, 2n\}$. Then $A^{(0)}_{2n} = A_{2n}$ and $A^{(1)}_{2n} = A'_{2n}$ in the notation of Section 1.6. We claim that (2.42) can also be written as

$$c \cdot A_{2n} G_1 R^t_{1,2n} \dots R^t_{12}\, A^{(1)}_{2n} G_2 R^t_{2,2n} \dots R^t_{23}\, A^{(3)}_{2n} G_3 \\ \times \dots A^{(n-1)}_{2n} G_n R^t_{n,n+1} \dots R^t_{n,2n}\, w,$$

where c is a nonzero constant. Indeed, this follows by induction from the relations $(2n-1)!\, A_{2n} = A_{2n}\, A'_{2n}$ and

$$A'_{2n} R^t_{12} \dots R^t_{1,2n} = R^t_{1,2n} \dots R^t_{12}\, A'_{2n}. \tag{2.43}$$

In order to verify the latter, take $m = 2n$ and set $u_i = u - i + 1$ for $i = 2, \dots, 2n$ and $u_1 = -u$ in (1.44). Then apply Proposition 1.6.2 and equate the right-hand sides of (1.44) and (1.45). The application of the partial transposition t_1 gives (2.43).

Now, if $j > n+1$, then $R^t_{n,j}\, w = w$. Furthermore, in the tensor product of the n-th and $(n+1)$-th copies of $\mathbb{C}^N$ we have

$$R^t_{n,n+1}\, (e_1 \otimes e_1) = e_1 \otimes e_1 + \frac{1}{2u - 2n + 1} \sum_{k=1}^{2n} e_k \otimes e_k.$$

Next apply the operator G_n and note that due to the subsequent application of the anti-symmetrizer $A^{(n-1)}_{2n}$ it suffices to keep the linear combination of the tensor products containing e_1 or e_2 on the n-th and $(n+1)$-th places. Hence, we may represent (2.42) as

$$\frac{2u - 2n + 3}{2u - 2n + 1} A_{2n} G_1 R^t_{12} \dots R^t_{1,2n} G_2 \dots G_{n-1} R^t_{n-1,n} \dots R^t_{n-1,2n}\, w_{n-1}$$

with

$$w_{n-1} = e_{2n-1} \otimes e_{2n-3} \otimes \dots \otimes e_3 \otimes e_1 \otimes e_2 \otimes e_3 \otimes \dots \otimes e_{2n-1}.$$

Continuing the calculation in a similar manner, we prove by induction that for any $a = 1, \dots, n$ we have

$$A_{2n} G_1 R^t_{12} \dots R^t_{1,2n} G_2 \dots G_a R^t_{a,a+1} \dots R^t_{a,2n}\, w_a = \frac{2u + 1}{2u - 2a + 1} A_{2n}\, v,$$

where

$$w_a = e_{2n-1} \otimes \dots \otimes e_1 \otimes e_2 \otimes \dots \otimes e_{2n-2a} \otimes e_{2n-2a+1} \otimes \dots \otimes e_{2n-1}.$$

Since $\det G = 1$ taking $a = n$ gives the required value for the scalar $\gamma_{n,G}(u)$. □

The following relation is immediate from Theorem 2.5.3.

COROLLARY 2.5.4. *The Sklyanin determinant* $\operatorname{sdet} S(u)$ *has the symmetry property*

$$\gamma_{n,G}(-u+N-1)\cdot \operatorname{sdet} S(u) = \gamma_{n,G}(u)\cdot \operatorname{sdet} S(-u+N-1). \qquad \square$$

Suppose that G and G' are two nonsingular symmetric (respectively, skew-symmetric) $N\times N$ matrices. As we saw in the proof of Corollary 2.3.2, an isomorphism $\mathrm{Y}_{G'}(\mathfrak{g}_N)\to \mathrm{Y}_G(\mathfrak{g}_N)$ can be given by the mapping (2.16). We record the following observation made in the proof of Theorem 2.5.3.

COROLLARY 2.5.5. *The image of the Sklyanin determinant* $\operatorname{sdet} S'(u)$ *under the isomorphism* (2.16) *coincides with* $(\det B)^2\cdot \operatorname{sdet} S(u)$. □

2.6. Sklyanin minors

We will define analogues of the quantum minors (see Section 1.6) for the twisted Yangians by extending Definition 2.5.2. We also establish some analogues of the row and column expansion formulas; cf. Proposition 1.6.8.

Apply Propositions 1.6.2 and 2.5.1 and specialize the variables u_i as

$$u_i = u - i + 1, \qquad i = 1,\dots,m.$$

We get the relation

$$A_m\,\langle S_1,\dots,S_m\rangle = \langle S_m,\dots,S_1\rangle\, A_m. \tag{2.44}$$

This element of the tensor product $(\operatorname{End}\mathbb{C}^N)^{\otimes m}\otimes \mathrm{Y}(\mathfrak{g}_N)[[u^{-1}]]$ can be written as

$$\sum e_{a_1b_1}\otimes\dots\otimes e_{a_mb_m}\otimes s^{a_1\dots a_m}_{b_1\dots b_m}(u),$$

summed over the indices $a_i, b_i\in\{1,\dots,N\}$, where $s^{a_1\dots a_m}_{b_1\dots b_m}(u)\in \mathrm{Y}(\mathfrak{g}_N)[[u^{-1}]]$. We call these elements the *Sklyanin minors* of the matrix $S(u)$. Clearly, the Sklyanin minors are skew-symmetric with respect to permutations of the upper indices and of the lower indices:

$$s^{a_{p(1)}\dots a_{p(m)}}_{b_1\dots b_m}(u) = \operatorname{sgn} p\cdot s^{a_1\dots a_m}_{b_1\dots b_m}(u) \quad\text{and}\quad s^{a_1\dots a_m}_{b_{p(1)}\dots b_{p(m)}}(u) = \operatorname{sgn} p\cdot s^{a_1\dots a_m}_{b_1\dots b_m}(u)$$

for any $p\in\mathfrak{S}_m$. We have $s^a_b(u) = s_{ab}(u)$ and

$$s^{1\dots N}_{1\dots N}(u) = \operatorname{sdet} S(u) \tag{2.45}$$

by Definition 2.5.2.

We will also use auxiliary minors $\check{s}^{a_1\dots a_m}_{b_1\dots b_{m-1},c}(u)\in \mathrm{Y}(\mathfrak{g}_N)[[u^{-1}]]$ defined by

$$\begin{aligned} A_m\,\langle S_1,\dots,S_{m-1}\rangle\, R^t_{1m}\cdots R^t_{m-1,m} \\ = \sum e_{a_1b_1}\otimes\dots\otimes e_{a_{m-1}b_{m-1}}\otimes e_{a_mc}\otimes \check{s}^{a_1\dots a_m}_{b_1\dots b_{m-1},c}(u),\end{aligned} \tag{2.46}$$

summed over $a_i, b_i, c\in\{1,\dots,N\}$. We obviously have

$$\check{s}^{a_{p(1)}\dots a_{p(m)}}_{b_1\dots b_{m-1},c}(u) = \operatorname{sgn} p\cdot \check{s}^{a_1\dots a_m}_{b_1\dots b_{m-1},c}(u) \tag{2.47}$$

for any $p\in\mathfrak{S}_m$. Furthermore, we have $(m-1)!\,A_m = A_m\,A_{m-1}$. Hence using (2.44) and the relation

$$A_{m-1}\,R^t_{1m}\cdots R^t_{m-1,m} = R^t_{m-1,m}\cdots R^t_{1m}\,A_{m-1}$$

verified in the same way as (2.43), we can represent the left-hand side of (2.46) in the form

$$\frac{1}{(m-1)!}\, A_m \,\langle S_{m-1},\dots,S_1\rangle\, R^t_{m-1,m}\cdots R^t_{1m}\, A_{m-1}.$$

This implies another property of the auxiliary minors

$$\check{s}^{\,a_1\dots a_m}_{\,b_{p(1)}\dots b_{p(m-1)},c}(u) = \operatorname{sgn} p\cdot \check{s}^{\,a_1\dots a_m}_{\,b_1\dots b_{m-1},c}(u) \tag{2.48}$$

for any $p\in\mathfrak{S}_{m-1}$.

PROPOSITION 2.6.1. *We have the relation*

$$s^{\,a_1\dots a_m}_{\,b_1\dots b_m}(u) = \sum_{c=1}^{N} \check{s}^{\,a_1\dots a_m}_{\,b_1\dots b_{m-1},c}(u)\, s_{c b_m}(u-m+1).$$

PROOF. This is immediate from the formula

$$\langle S_1,\dots,S_{m-1}\rangle\, R^t_{1m}\cdots R^t_{m-1,m}\, S_m = \langle S_1,\dots,S_m\rangle$$

which is an obvious consequence of (2.29). □

PROPOSITION 2.6.2. *Suppose that* $b_1\in\{a_1,\dots,a_m\}$ *and* $c\notin\{b_2,\dots,b_{m-1}\}$. *Then*

$$\check{s}^{\,a_1\dots a_m}_{\,b_1\dots b_{m-1},c}(u)=0 \qquad \textit{if}\quad c\notin\{a_1,\dots,a_m\},\qquad \textit{and}$$

$$\check{s}^{\,a_1\dots a_m}_{\,b_1\dots b_{m-1},c}(u)=\frac{1\mp 2u}{1-2u}\sum_{r=1}^{m-1}(-1)^{r-1}\, s^t_{a_r b_1}(-u)\, s^{\,a_1\dots\widehat{a}_r\dots a_{m-1}}_{\,b_2\dots b_{m-1}}(u-1)$$

if $c=a_m$, *where the* $s^t_{ij}(u)$ *denote the entries of the matrix* $S^t(u)$.

PROOF. Let us apply the operator on the left-hand side of (2.46) to the basis vector $e_{b_1}\otimes\dots\otimes e_{b_{m-1}}\otimes e_c$. Due to the assumption $c\notin\{b_2,\dots,b_{m-1}\}$ we have

$$R^t_{2m}\cdots R^t_{m-1,m}(e_{b_1}\otimes\dots\otimes e_{b_{m-1}}\otimes e_c) = e_{b_1}\otimes\dots\otimes e_{b_{m-1}}\otimes e_c.$$

Therefore the image of the vector can be written as

$$A_m S_1\, R^t_{12}\cdots R^t_{1m}\langle S_2,\dots,S_{m-1}\rangle(e_{b_1}\otimes\dots\otimes e_{b_{m-1}}\otimes e_c). \tag{2.49}$$

As in Section 1.6, let A'_m denote the anti-symmetrizer over the copies of $\operatorname{End}\mathbb{C}^N$ numbered by the indices $\{2,\dots,m\}$. Then $(m-1)!\,A_m = A_m\, A'_m$, and by (2.43) we have

$$A'_m R^t_{12}\cdots R^t_{1m} = R^t_{1m}\cdots R^t_{12}\, A'_m.$$

We claim that this operator can also be written as

$$A'_m\left(1+\frac{1}{2u-1}(Q_{12}+\dots+Q_{1m})\right).$$

Indeed, by Corollary 1.7.3 we have

$$A'_m\, R_{1m}(-2u+m-1)\dots R_{12}(-2u+1) = A'_m\left(1+\frac{1}{2u-1}(P_{12}+\dots+P_{1m})\right)$$

so that the application of the partial transposition t_1 gives the result. Since A'_m commutes with S_1, the expression (2.49) can be brought to the form

$$\frac{1}{(m-1)!}\, A_m S_1\left(1+\frac{1}{2u-1}(Q_{12}+\dots+Q_{1m})\right) \times A'_m\langle S_2,\dots,S_{m-1}\rangle(e_{b_1}\otimes\dots\otimes e_{b_{m-1}}\otimes e_c). \tag{2.50}$$

Now write $(m-2)!\, A'_m = A'_m A'_{m-1}$ and use the expansion

$$A'_{m-1}\langle S_2,\dots,S_{m-1}\rangle(e_{b_1}\otimes\dots\otimes e_{b_{m-1}}\otimes e_c)\\ = \sum_{d_2,\dots,d_{m-1}} e_{b_1}\otimes e_{d_2}\otimes\dots\otimes e_{d_{m-1}}\otimes e_c\otimes s^{\,d_2\dots\, d_{m-1}}_{\,b_2\dots\, b_{m-1}}(u-1)$$

to represent (2.50) as

$$\frac{1}{(m-2)!\,(m-1)!}\, A_m S_1\left(1+\frac{1}{2u-1}(Q_{12}+\dots+Q_{1m})\right)\\ \times A'_m \sum_{d_2,\dots,d_{m-1}} e_{b_1}\otimes e_{d_2}\otimes\dots\otimes e_{d_{m-1}}\otimes e_c\otimes s^{\,d_2\dots\, d_{m-1}}_{\,b_2\dots\, b_{m-1}}(u-1).$$

This equals

$$\frac{1}{(m-2)!}\, A_m S_1\left(1+\frac{1}{2u-1}(Q_{12}+\dots+Q_{1m})\right)\\ \times \sum_{d_2,\dots,d_{m-1}} e_{b_1}\otimes e_{d_2}\otimes\dots\otimes e_{d_{m-1}}\otimes e_c\otimes s^{\,d_2\dots\, d_{m-1}}_{\,b_2\dots\, b_{m-1}}(u-1). \tag{2.51}$$

Now, if c does not occur among the indices $a_1,\dots,a_m$, then, in particular, $c\neq b_1$ and hence

$$Q_{1m}(e_{b_1}\otimes e_{d_2}\otimes\dots\otimes e_{d_{m-1}}\otimes e_c)=0. \tag{2.52}$$

Therefore, the expansion of (2.51) as a linear combination of the basis vectors $e_{c_1}\otimes\dots\otimes e_{c_m}$ will contain only those with the property $c\in\{c_1,\dots,c_m\}$. This proves the first assertion of the proposition.

Now suppose that $c=a_m$ and assume first that $a_m\neq b_1$. Then the equality (2.52) still holds. Expanding (2.51) we obtain

$$\frac{1}{(m-2)!}\, A_m \sum_{d_1,\dots,d_{m-1}} e_{d_1}\otimes\dots\otimes e_{d_{m-1}}\otimes e_{a_m}\\ \otimes\left(s_{d_1 b_1}(u)\, s^{\,d_2\dots\, d_{m-1}}_{\,b_2\dots\, b_{m-1}}(u-1)+\frac{1}{2u-1}\sum_{k=2}^{m-1} s_{d_1 d_k}(u)\, s^{\,d_2\dots b_1\dots\, d_{m-1}}_{\,b_2\dots\, b_{m-1}}(u-1)\right),$$

where the index b_1 replaces d_k in the second Sklyanin minor. By the skew symmetry property (2.47) of the auxiliary minors, in order to calculate the coefficient of $e_{a_1}\otimes\dots\otimes e_{a_m}$ we may assume without loss of generality that $b_1=a_1$. Then the coefficient in question is

$$\frac{1}{(m-2)!}\sum_{p\in\mathfrak{S}_{m-1}} \operatorname{sgn} p\cdot\Big(s_{a_{p(1)}a_1}(u)\, s^{\,a_{p(2)}\dots\, a_{p(m-1)}}_{\,b_2\dots\, b_{m-1}}(u-1)\\ +\frac{1}{2u-1}\sum_{k=2}^{m-1} s_{a_{p(1)}a_{p(k)}}(u)\, s^{\,a_{p(2)}\dots a_1\dots\, a_{p(m-1)}}_{\,b_2\dots\, b_{m-1}}(u-1)\Big),$$

where a_1 replaces $a_{p(k)}$ in the second Sklyanin minor. By the skew symmetry of the Sklyanin minors, the expression simplifies to

$$s_{a_1a_1}(u)\, s^{a_2\dots a_{m-1}}_{b_2\dots b_{m-1}}(u-1)$$
$$+\sum_{r=2}^{m-1}(-1)^{r-1}\Big(\frac{2u}{2u-1}\, s_{a_ra_1}(u)-\frac{1}{2u-1}\, s_{a_1a_r}(u)\Big)\, s^{a_1\dots\widehat{a}_r\dots a_{m-1}}_{b_2\dots b_{m-1}}(u-1).$$

However, the symmetry relation (2.7) gives

$$\frac{2u}{2u-1}\, s_{ij}(u)-\frac{1}{2u-1}\, s_{ji}(u)=\frac{1\mp 2u}{1-2u}\, s_{ji}(-u)=\frac{1\mp 2u}{1-2u}\, s^{t}_{ij}(-u).$$

This completes the proof in the case under consideration. It remains to consider the case where $c=a_m=b_1$. The summation in (2.51) may be taken only over the indices $d_i\neq b_1$ for $i=2,\dots,m-1$. Indeed, if $d_i=b_1=c$ for some index i, then the basis vector $e_{b_1}\otimes e_{d_2}\otimes\dots\otimes e_{d_{m-1}}\otimes e_c$ is annihilated by the anti-symmetrizer A'_m. Therefore, the coefficient of $e_{a_1}\otimes\dots\otimes e_{a_m}$ in the expansion of (2.51) equals

$$\sum_{r=1}^{m-1}(-1)^{r-1}\Big(\frac{2u}{2u-1}\, s_{a_rb_1}(u)-\frac{1}{2u-1}\, s_{b_1a_r}(u)\Big)\, s^{a_1\dots\widehat{a}_r\dots a_{m-1}}_{b_2\dots b_{m-1}}(u-1).$$

The proof is completed by another application of the symmetry relation. □

2.7. Explicit formula for the Sklyanin determinant

There is an explicit expression for the Sklyanin determinant in terms of the generators $s_{ij}(u)$. It uses a special map

$$\varphi_N:\mathfrak{S}_N\to\mathfrak{S}_N,\qquad p\mapsto p' \tag{2.53}$$

from the symmetric group $\mathfrak{S}_N$ into itself which is defined by the following inductive procedure. Given a set of positive integers $a_1<\dots<a_N$ we regard $\mathfrak{S}_N$ as the group of their permutations. If $N=2$ we define φ_2 as the map $\mathfrak{S}_2\to\mathfrak{S}_2$ whose image is the identity permutation. For $N>2$ define a map from the set of ordered pairs (a_k,a_l) with $k\neq l$ into itself by the rule

$$\begin{aligned}(a_k,a_l)&\mapsto(a_l,a_k), && k,l<N,\\ (a_k,a_N)&\mapsto(a_{N-1},a_k), && k<N-1,\\ (a_N,a_k)&\mapsto(a_k,a_{N-1}), && k<N-1,\\ (a_{N-1},a_N)&\mapsto(a_{N-1},a_{N-2}), &&\\ (a_N,a_{N-1})&\mapsto(a_{N-1},a_{N-2}). &&\end{aligned}\tag{2.54}$$

Let $p=(p_1,\dots,p_N)$ be a permutation of the indices $a_1,\dots,a_N$. Its image under the map φ_N is the permutation $p'=(p'_1,\dots,p'_{N-1},a_N)$, where the pair (p'_1,p'_{N-1}) is the image of the ordered pair (p_1,p_N) under the map (2.54). Then the pair (p'_2,p'_{N-2}) is found as the image of (p_2,p_{N-1}) under the map (2.54), which is defined on the set of ordered pairs of elements obtained from $(a_1,\dots,a_N)$ by deleting p_1 and p_N. The procedure is completed in the same manner by determining consecutively the pairs (p'_i,p'_{N-i}).

The map φ_N has curious combinatorial properties, which are outlined in Example 2.16.7. In particular, each fiber of this map is an interval in $\mathfrak{S}_N$ with respect to the Bruhat order, and this interval is isomorphic to a Boolean poset.

EXAMPLE 2.7.1. The diagrams below show the Hasse diagram for the Bruhat order on $\mathfrak{S}_3$ and the fibers of the map φ_3. □

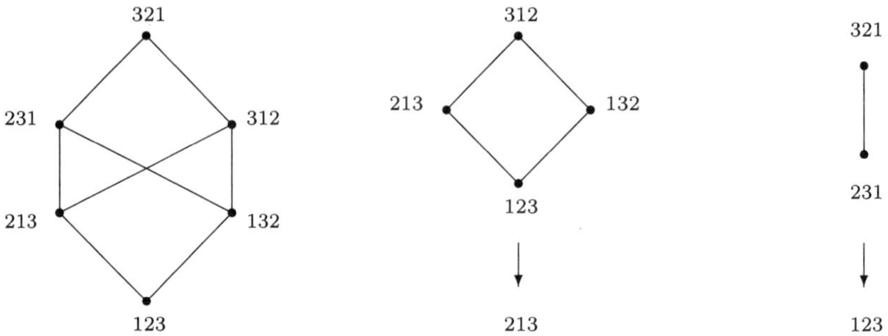

As before, we denote the matrix elements of the transposed matrix $S^t(u)$ by $s^t_{ij}(u)$. For any permutation $p \in \mathfrak{S}_N$ we denote by p' its image under the map φ_N. We will suppose that $N = 2n$ or $N = 2n+1$, where n is a positive integer. Introduce the scalar $\gamma_n(u)$ by

$$\gamma_n(u) = \begin{cases} 1 & \text{if} \quad \mathfrak{g}_N = \mathfrak{o}_N, \\ (-1)^n \dfrac{2u+1}{2u-2n+1} & \text{if} \quad \mathfrak{g}_N = \mathfrak{sp}_{2n}. \end{cases} \tag{2.55}$$

THEOREM 2.7.2. *We have*

$$\operatorname{sdet} S(u) = \gamma_n(u) \sum_{p \in \mathfrak{S}_N} \operatorname{sgn} pp' \cdot s^t_{p(1),p'(1)}(-u) \dots s^t_{p(n),p'(n)}(-u+n-1)$$
$$\times\, s_{p(n+1),p'(n+1)}(u-n) \dots s_{p(N),p'(N)}(u-N+1)$$

and also

$$\operatorname{sdet} S(u) = \gamma_n(u) \sum_{p \in \mathfrak{S}_N} \operatorname{sgn} pp' \cdot s_{p'(1),p(1)}(u-N+1) \dots s_{p'(n),p(n)}(u-N+n)$$
$$\times\, s^t_{p'(n+1),p(n+1)}(-u+N-n-1) \dots s^t_{p'(N),p(N)}(-u).$$

EXAMPLE 2.7.3. For $N = 2$ we have

$$\operatorname{sdet} S(u) = \frac{1 \mp 2u}{1-2u} \big(s^t_{11}(-u)\, s_{22}(u-1) - s^t_{21}(-u)\, s_{12}(u-1)\big)$$
$$= \frac{1 \mp 2u}{1-2u} \big(s_{11}(u-1)\, s^t_{22}(-u) - s_{12}(u-1)\, s^t_{21}(-u)\big).$$

If $N = 3$ then

$$\operatorname{sdet} S(u) = s^t_{22}(-u)\, s_{11}(u-1)\, s_{33}(u-2) + s^t_{12}(-u)\, s_{31}(u-1)\, s_{23}(u-2)$$
$$+ s^t_{21}(-u)\, s_{32}(u-1)\, s_{13}(u-2) - s^t_{12}(-u)\, s_{21}(u-1)\, s_{33}(u-2)$$
$$- s^t_{32}(-u)\, s_{11}(u-1)\, s_{23}(u-2) - s^t_{31}(-u)\, s_{22}(u-1)\, s_{13}(u-2).$$

□

PROOF OF THEOREM 2.7.2. It is sufficient to prove the first of the two formulas since the second is implied by the first and Corollary 2.5.4. Consider the case $\mathfrak{g}_N = \mathfrak{o}_N$ first. We will combine Propositions 2.6.1 and 2.6.2 to get a recurrence relation for the Sklyanin minors of the form $s^{a_1\dots a_m}_{a_1\dots a_{m-1},b_m}(u)$. Proposition 2.6.1 gives

$$s^{a_1\dots a_m}_{a_1\dots a_{m-1},b_m}(u)=\sum_{c=1}^{N}\check{s}^{a_1\dots a_m}_{a_1\dots a_{m-1},c}(u)\,s_{cb_m}(u-m+1).$$

Due to the first part of Proposition 2.6.2, the summation in this formula may be restricted to the set of indices $c\in\{a_1,\dots,a_m\}$ and so,

$$s^{a_1\dots a_m}_{a_1\dots a_{m-1},b_m}(u)=\sum_{k=1}^{m}\check{s}^{a_1\dots a_m}_{a_1\dots a_{m-1},a_k}(u)\,s_{a_k b_m}(u-m+1). \tag{2.56}$$

In order to satisfy the conditions of the second part of Proposition 2.6.2, we use the skew symmetry properties (2.47) and (2.48) to write

$$\check{s}^{a_1\dots a_m}_{a_1\dots a_{m-1},a_k}(u)=(-1)^{m-1}\,\check{s}^{a_1\dots\widehat{a}_k\dots a_m,a_k}_{a_k,a_1\dots\widehat{a}_k\dots a_{m-1},a_k}(u)$$

for $k=1,\dots,m-1$. Applying the second relation of Proposition 2.6.2 we will bring (2.56) to the form

$$\begin{aligned}
s^{a_1\dots a_m}_{a_1\dots a_{m-1},b_m}(u)&=s^{\,t}_{a_{m-1}a_{m-1}}(-u)\,s^{a_1\dots a_{m-2}}_{a_1\dots a_{m-2}}(u-1)\,s_{a_m b_m}(u-m+1)\\
&-\sum_{l=1}^{m-2}s^{\,t}_{a_l a_{m-1}}(-u)\,s^{a_1\dots\widehat{a}_l\dots a_{m-1}}_{a_1\dots\widehat{a}_l\dots a_{m-2},a_l}(u-1)\,s_{a_m b_m}(u-m+1)\\
&-\sum_{k=1}^{m-1}\Big\{s^{\,t}_{a_m a_k}(-u)\,s^{a_1\dots\widehat{a}_k\dots a_{m-1}}_{a_1\dots\widehat{a}_k\dots a_{m-1}}(u-1)\,s_{a_k b_m}(u-m+1)\\
&-\sum_{l=1}^{k-1}s^{\,t}_{a_l a_k}(-u)\,s^{a_1\dots\widehat{a}_l\dots\widehat{a}_k\dots a_m}_{a_1\dots\widehat{a}_l\dots\widehat{a}_k\dots a_{m-1},a_l}(u-1)\,s_{a_k b_m}(u-m+1)\\
&-\sum_{l=k+1}^{m-1}s^{\,t}_{a_l a_k}(-u)\,s^{a_1\dots\widehat{a}_k\dots\widehat{a}_l\dots a_m}_{a_1\dots\widehat{a}_k\dots\widehat{a}_l\dots a_{m-1},a_l}(u-1)\,s_{a_k b_m}(u-m+1)\Big\}.
\end{aligned}$$

Using (2.45) and starting with $s^{1\dots N}_{1\dots N}(u)$ we apply this recurrence relation repeatedly to get the following expression for the series $\operatorname{sdet} S(u)$ in terms of the generators $s_{ij}(u)$:

$$\begin{aligned}
\operatorname{sdet} S(u)=\sum_{p\in\mathfrak{S}_N}\alpha(p)\cdot s^{\,t}_{p(1),p'(1)}(-u)\dots s^{\,t}_{p(n),p'(n)}(-u+n-1)\\
\times s_{p(n+1),p'(n+1)}(u-n)\dots s_{p(N),p'(N)}(u-N+1),
\end{aligned}$$

where the coefficient $\alpha(p)$ equals 1 or -1. It remains to verify that $\alpha(p)=\operatorname{sgn} pp'$. To do this, we note that given a permutation $p\in\mathfrak{S}_N$, the corresponding summand in the expansion of $s^{1\dots N}_{1\dots N}(u)$ by the above recurrence relation is

$$\varepsilon(p)\cdot s^{\,t}_{p(1),p'(1)}(-u)\,s^{1\dots\widehat{p(1)}\dots\widehat{p(N)}\dots N}_{1\dots\widehat{p(1)}\dots\widehat{p(N)}\dots p'(N-1)}(u-1)\,s_{p(N),p'(N)}(u-N+1),$$

where

$$\varepsilon(p)=\begin{cases}1, & \text{if either}\quad p(1),\ p(N)<N\quad\text{or}\quad p(1)=N-1,\ p(N)=N\\ -1, & \text{if either}\quad p(1)<N-1,\ p(N)=N\quad\text{or}\quad p(1)=N,\ p(N)<N.\end{cases}$$

For any sequence of distinct positive integers $a=(a_1,\dots,a_m)$ denote by $\operatorname{inv} a$ the number of inversions in a. For a permutation $p\in\mathfrak{S}_N$ we let $\widetilde{p}$ and $\widetilde{p}'$ denote the sequences $(p(2), \dots, p(N-1))$ and $(p'(2), \dots, p'(N-2))$, respectively. It suffices to verify the following identity:

$$(-1)^{\operatorname{inv} p} \cdot (-1)^{\operatorname{inv} p'} = \varepsilon(p)\, (-1)^{\operatorname{inv} \widetilde{p}} \cdot (-1)^{\operatorname{inv} \widetilde{p}'}. \tag{2.57}$$

Then the required relation $\alpha(p) = \operatorname{sgn} pp'$ will follow from (2.57) by induction.

Suppose that $p(1) = k$, $p(N) = l$. If $k, l < N$, then clearly $\varepsilon(p) = 1$ while

$$\operatorname{inv} p = (k-1) + (N-l) + \operatorname{inv} \widetilde{p},$$
$$\operatorname{inv} p' = (l-1) + (N-k-2) + \operatorname{inv} \widetilde{p}' \quad \text{if} \quad k < l;$$

and

$$\operatorname{inv} p = (k-1) + (N-l-1) + \operatorname{inv} \widetilde{p},$$
$$\operatorname{inv} p' = (l-1) + (N-k-1) + \operatorname{inv} \widetilde{p}' \quad \text{if} \quad k > l,$$

so that in both cases (2.57) holds. Similarly, if $k < N-1$, $l = N$, then $\varepsilon(p) = -1$ and

$$\operatorname{inv} p = (k-1) + \operatorname{inv} \widetilde{p},$$
$$\operatorname{inv} p' = (N-2) + (N-k-2) + \operatorname{inv} \widetilde{p}';$$

if $k = N$, $l < N-1$, then $\varepsilon(p) = -1$ and

$$\operatorname{inv} p = (N-1) + (N-l-1) + \operatorname{inv} \widetilde{p},$$
$$\operatorname{inv} p' = (l-1) + \operatorname{inv} \widetilde{p}';$$

if $k = N-1$, $l = N$, then $\varepsilon(p) = 1$ and

$$\operatorname{inv} p = (N-2) + \operatorname{inv} \widetilde{p}, \qquad \operatorname{inv} p' = (N-2) + \operatorname{inv} \widetilde{p}';$$

if $k = N$, $l = N-1$, then $\varepsilon(p) = -1$ and

$$\operatorname{inv} p = (N-1) + \operatorname{inv} \widetilde{p}, \qquad \operatorname{inv} p' = (N-2) + \operatorname{inv} \widetilde{p}'.$$

This proves (2.57) and the theorem in the case under consideration.

In the symplectic case the same argument is applied. The only difference comes from the factor on the right-hand side of the second formula of Proposition 2.6.2. The resulting expression will differ from the orthogonal case by the factor

$$\frac{1+2u}{1-2u} \cdot \frac{1+2(u-1)}{1-2(u-1)} \cdots \frac{1+2(u-n+1)}{1-2(u-n+1)} \tag{2.58}$$

which equals $\gamma_n(u)$. □

REMARK 2.7.4. Since for any permutation $(a_1, \dots, a_N)$ of the indices $(1, \dots, N)$ we have

$$\operatorname{sdet} S(u) = s^{\,a_1 \dots a_N}_{\,a_1 \dots a_N}(u)$$

we can begin the induction argument in the proof of Theorem 2.7.2 with $s^{\,a_1 \dots a_N}_{\,a_1 \dots a_N}(u)$ instead of $s^{\,1 \dots N}_{\,1 \dots N}(u)$. Repeating the argument word by word, we derive the following more general expressions for the Sklyanin determinant:

$$\operatorname{sdet} S(u) = \gamma_n(u) \sum_{p \in \mathfrak{S}_N} \operatorname{sgn} pp' \cdot s^{\,t}_{a_{p(1)}, a_{p'(1)}}(-u) \dots s^{\,t}_{a_{p(n)}, a_{p'(n)}}(-u+n-1)$$
$$\times\, s_{a_{p(n+1)}, a_{p'(n+1)}}(u-n) \dots s_{a_{p(N)}, a_{p'(N)}}(u-N+1)$$

and also

$$\operatorname{sdet} S(u) = \gamma_n(u) \sum_{p\in\mathfrak{S}_N} \operatorname{sgn} pp' \cdot s_{a_{p'(1)},a_{p(1)}}(u-N+1)\dots s_{a_{p'(n)},a_{p(n)}}(u-N+n)$$
$$\times\, s^t_{a_{p'(n+1)},a_{p(n+1)}}(-u+N-n-1)\dots s^t_{a_{p'(N)},a_{p(N)}}(-u).$$

□

The next lemma implies that the highest degree component of the Sklyanin determinant coincides with the usual determinant; see Section 8.9 below.

LEMMA 2.7.5. *The map $\mathfrak{S}_N \to \mathfrak{S}_N$ defined by $p \mapsto p\,(p')^{-1}$is bijective.*

PROOF. Suppose that p and q are two elements of $\mathfrak{S}_N$ such that $p\,(p')^{-1} = q\,(q')^{-1}$. It suffices to show that $p = q$. By definition of the map φ_N we have $p'_N = q'_N = N$ which implies that $p_N = q_N$. Then, due to the formulas (2.54), we have $p'_1 = q'_1$. Hence, $p_1 = q_1$. Now, since the pairs (p_1, p_N) and (q_1, q_N) coincide, so do their images under the map (2.54). In particular, $p'_{N-1} = q'_{N-1}$. This implies that $p_{N-1} = q_{N-1}$, and the proof is completed by repeating this argument for the pairs (p_{i+1}, p_{N-i}) and (q_{i+1}, q_{N-i}) with $i = 1, 2, \dots$. □

We will need the following analogue of the quantum comatrix; cf. Definition 1.9.1.

DEFINITION 2.7.6. The *Sklyanin comatrix* $\widehat{S}(u)$ is defined by

$$\widehat{S}(u)\, S(u-N+1) = \operatorname{sdet} S(u).$$

PROPOSITION 2.7.7. *The matrix elements $\widehat{s}_{ij}(u)$ of the matrix $\widehat{S}(u)$ are given by*

$$\widehat{s}_{ij}(u) = (-1)^{N-i}\, \check{s}^{\,1\dots N}_{\,1\dots\widehat{i}\dots N,\, j}(u).$$

Moreover,

$$\widehat{s}_{ii}(u) = s^{\,1\dots\widehat{i}\dots N}_{\,1\dots\widehat{i}\dots N}(u).$$

PROOF. By Definition 2.5.2, we have

$$A_N \langle S_1, \dots, S_N\rangle = A_N \operatorname{sdet} S(u).$$

Multiplying by S_N^{-1} from the right we get

$$A_N \langle S_1, \dots, S_{N-1}\rangle\, R^t_{1N} \dots R^t_{N-1,N} = A_N\, \widehat{S}_N(u). \tag{2.59}$$

Due to (2.46), the desired formula for $\widehat{s}_{ij}(u)$ follows by applying the operators on both sides of (2.59) to the basis vector $v_{ij} = e_1 \otimes \dots \widehat{e}_i \dots \otimes e_N \otimes e_j$ and comparing the coefficients of $e_1 \otimes \dots \otimes e_N$. Similarly, the formula for $\widehat{s}_{ii}(u)$ is obtained by applying the operators on both sides of (2.59) to the basis vector v_{ii} and using the observation $R^t_{jN} v_{ii} = v_{ii}$ for any $j = 1, \dots, N-1$. □

An explicit formula for the entries $\widehat{s}_{ij}(u)$ of the Sklyanin comatrix can be given with the use of the map $p \mapsto p'$ defined in (2.53).

PROPOSITION 2.7.8. *For any permutation $(a_1, \dots, a_N)$ of the indices $(1, \dots, N)$ and any index $k \in \{1, \dots, N\}$ we have*

$$\widehat{s}_{a_N a_k}(u) = \gamma_n(u) \sum_{p\in\mathfrak{S}_N,\ p(N)=k} \operatorname{sgn} pp' \cdot s^t_{a_{p(1)},a_{p'(1)}}(-u)\dots s^t_{a_{p(n)},a_{p'(n)}}(-u+n-1)$$
$$\times\, s_{a_{p(n+1)},a_{p'(n+1)}}(u-n)\dots s_{a_{p(N-1)},a_{p'(N-1)}}(u-N+2).$$

PROOF. The starting point is the following expression for $\widehat{s}_{a_N a_k}(u)$ given by Proposition 2.7.7:

$$\widehat{s}_{a_N a_k}(u) = \check{s}\,^{a_1 \dots a_N}_{a_1 \dots a_{N-1}, a_k}(u).$$

Then we use the recurrence relations for the Sklyanin minors provided by Propositions 2.6.1 and 2.6.2 and proceed with the same argument as for the proof of Theorem 2.7.2; cf. Remark 2.7.4. □

2.8. The center of the twisted Yangian

The following analogue of Lemma 1.7.4 will be used in the proof of Theorem 2.8.2. Let $\mathfrak{a}$ be a Lie algebra and σ an involutive automorphism of $\mathfrak{a}$. Denote by $\mathfrak{a}_0$ (respectively $\mathfrak{a}_1$) the set of elements $a \in \mathfrak{a}$ such that $\sigma(a) = a$ (respectively $\sigma(a) = -a$). Let $\mathfrak{a}[z]^\sigma$ denote the corresponding twisted polynomial current Lie algebra:

$$\mathfrak{a}[z]^\sigma = \mathfrak{a}_0 \oplus \mathfrak{a}_1 z \oplus \mathfrak{a}_0 z^2 \oplus \mathfrak{a}_1 z^3 \oplus \dots.$$

LEMMA 2.8.1. *Suppose that the Lie algebra $\mathfrak{a}$ does not have any nonzero elements invariant under the adjoint action of the subalgebra $\mathfrak{a}_0$. Then the center of the universal enveloping algebra $\mathrm{U}(\mathfrak{a}[z]^\sigma)$ is trivial.*

PROOF. Let $\{X_1, \dots, X_r\}$ and $\{Y_1, \dots, Y_n\}$ be bases of $\mathfrak{a}_0$ and $\mathfrak{a}_1$ respectively. Then for $1 \leqslant i \leqslant r$ and $1 \leqslant j \leqslant n$

$$[X_i, Y_j] = \sum_{k=1}^{n} c_{ij}^k Y_k,$$

where c_{ij}^k are the structure constants. As in the proof of Lemma 1.7.4, it suffices to verify that if an element $A \in \mathrm{S}(\mathfrak{a}[z]^\sigma)$ is invariant under the adjoint action of $\mathfrak{a}[z]^\sigma$, then $A = 0$. Let m be the minimum nonnegative integer such that A can be written in the form

$$A = \sum_{l} A_l (Y_1 z^m)^{l_1} \dots (Y_n z^m)^{l_n},$$

where $l = (l_1, \dots, l_n)$, $l_1 \geqslant 0, \dots, l_n \geqslant 0$ and A_l is a polynomial in the variables $Y_i z^s$, $s < m$, with coefficients from the subalgebra $\mathrm{S}(\mathfrak{a}_0[z^2])$. Just as in the proof of Lemma 1.7.4, we deduce the equalities of the form (1.59) and (1.60) from the relations $\mathrm{ad}(X_i z)(A) = 0$ for $i = 1, \dots, r$. Repeating again the arguments of the proof of Lemma 1.7.4, we conclude that $A_l = 0$ for all $l \neq 0$; i.e. A belongs to the subalgebra $\mathrm{S}(\mathfrak{a}_0[z^2])$. To complete the proof, it remains to apply Lemma 1.7.4 to the Lie algebra $\mathfrak{a}_0$ instead of $\mathfrak{a}$. □

Due to Corollary 2.5.4, the odd coefficients of the series $\operatorname{sdet} S(u)$ can be expressed in terms of the even ones; see Definition 2.5.2.

THEOREM 2.8.2. *All coefficients of the series $\operatorname{sdet} S(u)$ belong to the center of the algebra $\mathrm{Y}(\mathfrak{g}_N)$. Moreover, the even coefficients $c_2, c_4, \dots$ are algebraically independent and generate the center of $\mathrm{Y}(\mathfrak{g}_N)$.*

PROOF. The first assertion is immediate from Theorems 1.7.5 and 2.5.3. To prove the second introduce a filtration on $\mathrm{Y}(\mathfrak{g}_N)$ by setting

$$\deg' s_{ij}^{(r)} = r - 1.$$

This filtration coincides with the restriction of the filtration on the Yangian $\mathrm{Y}(\mathfrak{gl}_N)$ introduced in the proof of Theorem 1.7.5. To describe the corresponding graded algebra $\mathrm{gr}'\,\mathrm{Y}(\mathfrak{g}_N)$ consider the involutive automorphism σ of the Lie algebra $\mathfrak{gl}_N$ defined by

$$\sigma : A \mapsto -G^{-1}A^tG. \tag{2.60}$$

The elements F_{ij} defined in (2.1) are stable under σ, and the fixed point subalgebra of σ coincides with the Lie subalgebra $\mathfrak{g}_N$. Consider the twisted polynomial current Lie algebra

$$\mathfrak{gl}_N[z]^\sigma = \{\, A(z) \in \mathfrak{gl}_N[z] \,|\, \sigma\big(A(z)\big) = A(-z)\,\}.$$

The elements of $\mathfrak{gl}_N[z]^\sigma$ are polynomials of the form

$$A(z) = A_0 + A_1 z + \cdots + A_k z^k,$$

where the even coefficients A_{2i} belong to the subalgebra $\mathfrak{g}_N$ while the odd coefficients A_{2i+1} belong to the (-1)-eigenspace of σ. Thus, the Lie algebra $\mathfrak{gl}_N[z]^\sigma$ is spanned by the elements

$$\Big(\sum_{k=1}^{N}(E_{ik}\, g_{kj} \pm (-1)^r\, E_{jk}\, g_{ki})\Big)\, z^{r-1}, \qquad r = 1, 2, \dots. \tag{2.61}$$

Our next step is to show that there is an algebra isomorphism

$$\mathrm{gr}'\,\mathrm{Y}(\mathfrak{g}_N) \cong \mathrm{U}(\mathfrak{gl}_N[z]^\sigma).$$

Consider the isomorphism $\mathrm{U}(\mathfrak{gl}_N[z]) \to \mathrm{gr}'\,\mathrm{Y}(\mathfrak{gl}_N)$ constructed in the proof of Proposition 1.5.2. The images of the elements (2.61) have the form

$$\sum_{k=1}^{N}\big(\bar{t}_{ik}^{(r)}\, g_{kj} \pm (-1)^r\, \bar{t}_{jk}^{(r)}\, g_{ki}\big).$$

However, by formula (2.28), these are precisely the images of the generators $s_{ij}^{(r)}$ in the $(r-1)$-th component of the graded algebra $\mathrm{gr}'\,\mathrm{Y}(\mathfrak{g}_N)$, as desired.

Now apply Theorem 2.5.3. As we established in the proof of Theorem 1.7.5, the image of the coefficient d_r of the quantum determinant $\mathrm{qdet}\, T(u)$ in the $(r-1)$-th component of $\mathrm{gr}'\,\mathrm{Y}(\mathfrak{g}_N)$ is $I\, z^{r-1}$ where $I = E_{11} + \cdots + E_{NN}$. Therefore, the image of the coefficient c_{2m} in the $(2m-1)$-th component of $\mathrm{gr}'\,\mathrm{Y}(\mathfrak{g}_N)$ coincides with $2 \det G \cdot I\, z^{2m-1}$. To complete the proof of the theorem, it remains to show that the center of the algebra $\mathrm{U}(\mathfrak{gl}_N[z]^\sigma)$ is generated by $Iz, Iz^3, Iz^5, \dots$. Observe that

$$\mathrm{U}(\mathfrak{gl}_N[z]^\sigma) = \mathbb{C}[Iz, Iz^3, \dots] \otimes \mathrm{U}(\mathfrak{sl}_N[z]^\sigma).$$

On the other hand, the center of $\mathrm{U}(\mathfrak{sl}_N[z]^\sigma)$ is trivial. With the exception of the case $N = 2$ and G is symmetric, this follows from Lemma 2.8.1 applied to the Lie algebra $\mathfrak{a} = \mathfrak{sl}_N$ and the involution σ defined in (2.60). In the remaining case $\mathfrak{a}_0 = \mathfrak{o}_2$ is the one-dimensional abelian Lie algebra so that Lemma 2.8.1 does not apply. The triviality of the center of $\mathrm{U}(\mathfrak{sl}_2[z]^\sigma)$ can be established directly using arguments similar to those used in the proof of Lemma 2.8.1. □

Note that the centrality of the Sklyanin determinant can also be proved by using the matrix form of the defining relations of $\mathrm{Y}(\mathfrak{g}_N)$; see Section 2.13.

REMARK 2.8.3. The algebra $\mathrm{Y}(\mathfrak{g}_N)$ may be considered as a flat deformation of the algebra $\mathrm{U}(\mathfrak{gl}_N[z]^\sigma)$; cf. Remark 2.4.5. To see this, introduce new generators of $\mathrm{Y}(\mathfrak{g}_N)$ by

$$\widetilde{s}_{ij}^{(r)} = s_{ij}^{(r)} h^{r-1},$$

where $h \in \mathbb{C} \setminus \{0\}$ is the deformation parameter and write the defining relations for the twisted Yangian in terms of the new generators. Denote by $\mathrm{Y}'(\mathfrak{g}_N, h)$ the (abstract) algebra with generators $\widetilde{s}_{ij}^{(r)}$ and those defining relations. Then the algebras $\mathrm{Y}'(\mathfrak{g}_N, h)$ with $h \neq 0$ are easily seen to be isomorphic to $\mathrm{Y}(\mathfrak{g}_N) = \mathrm{Y}'(\mathfrak{g}_N, 1)$, while setting $h = 0$ we get the defining relations of the algebra $\mathrm{U}(\mathfrak{gl}_N[z]^\sigma)$. The flatness of the deformation is guaranteed by the Poincaré–Birkhoff–Witt theorem for $\mathrm{Y}(\mathfrak{g}_N)$; see Corollary 2.4.4. □

2.9. The special twisted Yangian

In the next definition we regard $\mathrm{Y}(\mathfrak{g}_N)$ as a subalgebra of the Yangian $\mathrm{Y}(\mathfrak{gl}_N)$ using the embedding (2.21). Recall the definition of the Yangian $\mathrm{Y}(\mathfrak{sl}_N)$ from Section 1.8.

DEFINITION 2.9.1. The *special twisted Yangian* $\mathrm{SY}(\mathfrak{g}_N)$ is the subalgebra of $\mathrm{Y}(\mathfrak{g}_N)$ defined by

$$\mathrm{SY}(\mathfrak{g}_N) = \mathrm{Y}(\mathfrak{sl}_N) \cap \mathrm{Y}(\mathfrak{g}_N).$$ □

Equivalently, $\mathrm{SY}(\mathfrak{g}_N)$ is the subalgebra of $\mathrm{Y}(\mathfrak{g}_N)$ which consists of the elements stable under all automorphisms of the form (2.17).

THEOREM 2.9.2. *The algebra $\mathrm{Y}(\mathfrak{g}_N)$ is isomorphic to the tensor product of its center $\mathrm{ZY}(\mathfrak{g}_N)$ and the subalgebra $\mathrm{SY}(\mathfrak{g}_N)$,*

$$\mathrm{Y}(\mathfrak{g}_N) = \mathrm{ZY}(\mathfrak{g}_N) \otimes \mathrm{SY}(\mathfrak{g}_N).$$

In particular, the center of $\mathrm{SY}(\mathfrak{g}_N)$ is trivial.

PROOF. We will derive the statement from Theorem 1.8.2. First, prove the decomposition

$$\mathrm{Y}(\mathfrak{g}_N) = \mathrm{ZY}(\mathfrak{g}_N)\, \mathrm{SY}(\mathfrak{g}_N). \tag{2.62}$$

We will use the notation of Section 1.8. Introduce the series

$$\widetilde{s}_{ij}(u) = \big(\widetilde{d}(u)\, \widetilde{d}(-u)\big)^{-1} s_{ij}(u) \tag{2.63}$$

which coincides with

$$\sum_{k,l} g_{kl}\, \widetilde{t}_{ik}(u)\, \widetilde{t}_{jl}(-u)$$

by (1.62) and (2.28). Let us verify that all coefficients of the series $\widetilde{d}(u)\, \widetilde{d}(-u)$ belong to $\mathrm{ZY}(\mathfrak{g}_N)$. By Theorem 2.5.3,

$$\begin{aligned}\gamma_{n,G}(u)^{-1} \mathrm{sdet}\, S(u) &= \mathrm{qdet}\, T(u)\, \mathrm{qdet}\, T(-u+N-1)\\ &= \big(\widetilde{d}(u)\, \widetilde{d}(-u)\big)\big(\widetilde{d}(u-1)\, \widetilde{d}(-u+1)\big) \dots \big(\widetilde{d}(u-N+1)\, \widetilde{d}(-u+N-1)\big).\end{aligned}$$

Therefore, all coefficients of the series $\widetilde{d}(u)\, \widetilde{d}(-u)$ can be expressed as polynomials in the coefficients of the series $\mathrm{sdet}\, S(u)$. Due to Theorem 2.8.2, this proves that $\widetilde{d}(u)\, \widetilde{d}(-u) \in \mathrm{ZY}(\mathfrak{g}_N)[[u^{-1}]]$. Note that $\widetilde{s}_{ij}(u) \in \mathrm{SY}(\mathfrak{g}_N)[[u^{-1}]]$, since these series are stable under all automorphisms of the form (2.17); see (1.61). Now (2.62) follows from the decomposition

$$s_{ij}(u) = \widetilde{d}(u)\, \widetilde{d}(-u)\, \widetilde{s}_{ij}(u).$$

Finally, the desired tensor product decomposition is implied by Theorem 1.8.2 and the inclusions $\mathrm{ZY}(\mathfrak{g}_N) \subset \mathrm{ZY}(\mathfrak{gl}_N)$ and $\mathrm{SY}(\mathfrak{g}_N) \subset \mathrm{Y}(\mathfrak{sl}_N)$. □

Due to the symmetry property of sdet $S(u)$ provided by Corollary 2.5.4, we have the expansion

$$\gamma_{n,G}\big(u+(N-1)/2\big)^{-1}\text{sdet}\, S\big(u+(N-1)/2\big) = 1 + c_1' u^{-2} + c_2' u^{-4} + \dots$$

which involves only even powers of u. Theorem 2.8.2 implies that the coefficients $c_1', c_2', \dots$ are algebraically independent and generate the center of $\mathrm{Y}(\mathfrak{g}_N)$.

COROLLARY 2.9.3. *The algebra* $\mathrm{SY}(\mathfrak{g}_N)$ *is isomorphic to the quotient of* $\mathrm{Y}(\mathfrak{g}_N)$ *by the ideal generated by the elements* $c_1', c_2', \dots$, *i.e.*,

$$\mathrm{SY}(\mathfrak{g}_N) \cong \mathrm{Y}(\mathfrak{g}_N)/\big(\text{sdet}\, S(u) = \gamma_{n,G}(u)\big).$$

PROOF. Theorems 2.8.2 and 2.9.2 imply that $\mathrm{Y}(\mathfrak{g}_N) = \mathrm{I} \oplus \mathrm{SY}(\mathfrak{g}_N)$, where I is the ideal of $\mathrm{Y}(\mathfrak{g}_N)$ generated by the elements $c_1', c_2', \dots$. □

2.10. Coideal property

Unlike the Yangian $\mathrm{Y}(\mathfrak{gl}_N)$, and unlike the Yangian for the simple Lie algebra $\mathfrak{g}_N$ defined by Drinfeld in [**97**], the twisted Yangian $\mathrm{Y}(\mathfrak{g}_N)$ seems not to possess any natural Hopf algebra structure. Nevertheless, the following coideal property with respect to the coproduct in the Yangian $\mathrm{Y}(\mathfrak{gl}_N)$ takes place.

THEOREM 2.10.1. *The subalgebra* $\mathrm{Y}(\mathfrak{g}_N)$ *is a left coideal of the Hopf algebra* $\mathrm{Y}(\mathfrak{gl}_N)$, *i.e.*,

$$\Delta\big(\mathrm{Y}(\mathfrak{g}_N)\big) \subset \mathrm{Y}(\mathfrak{gl}_N) \otimes \mathrm{Y}(\mathfrak{g}_N).$$

Moreover,

$$\Delta : s_{ij}(u) \mapsto \sum_{a,b=1}^{N} t_{ia}(u)\, t_{jb}(-u) \otimes s_{ab}(u). \tag{2.64}$$

PROOF. It is enough to prove (2.64). Using (1.35) and (2.28) we get

$$\begin{aligned}\Delta : s_{ij}(u) \mapsto &\sum_{k,l=1}^{N} \sum_{a,b=1}^{N} g_{kl}\, \big(t_{ia}(u) \otimes t_{ak}(u)\big)\, \big(t_{jb}(-u) \otimes t_{bl}(-u)\big)\\ &= \sum_{a,b=1}^{N} t_{ia}(u)\, t_{jb}(-u) \otimes s_{ab}(u).\end{aligned}$$

□

COROLLARY 2.10.2. *The subalgebra* $\mathrm{SY}(\mathfrak{g}_N)$ *is a left coideal of the Hopf algebra* $\mathrm{Y}(\mathfrak{sl}_N)$.

PROOF. We have to verify that

$$\Delta\big(\mathrm{SY}(\mathfrak{g}_N)\big) \subset \mathrm{Y}(\mathfrak{sl}_N) \otimes \mathrm{SY}(\mathfrak{g}_N).$$

Due to the tensor product decomposition stated in Theorem 2.9.2, the coefficients of the series $\widetilde{s}_{ij}(u)$ introduced in the proof of that theorem generate the subalgebra $\mathrm{SY}(\mathfrak{g}_N)$. By (1.64) and (2.64) we have

$$\begin{aligned}\Delta : \widetilde{s}_{ij}(u) \mapsto \big(\widetilde{d}(u)\, \widetilde{d}(-u)\big)^{-1} \otimes \big(\widetilde{d}(u)\, \widetilde{d}(-u)\big)^{-1} &\sum_{a,b=1}^{N} t_{ia}(u)\, t_{jb}(-u) \otimes s_{ab}(u)\\ = &\sum_{a,b=1}^{N} \widetilde{t}_{ia}(u)\, \widetilde{t}_{jb}(-u) \otimes \widetilde{s}_{ab}(u)\end{aligned}$$

which proves the claim. □

2.11. Quantum Liouville formula

Here we introduce an analogue of the series $z(u)$ (see Section 1.9) for the twisted Yangian $\mathrm{Y}(\mathfrak{g}_N)$.

Let us multiply the quaternary relation (2.11) from both sides consecutively by the inverses to $R(u-v)$, $S_1(u)$ and $R^t(-u-v)$. Then applying (2.19) we get

$$R^t(u+v+N)\, S_1^{-1}(u)\, R(-u+v)\, S_2(v)$$
$$= S_2(v)\, R(-u+v)\, S_1^{-1}(u)\, R^t(u+v+N).$$

Taking the residue at $u=-v-N$ we arrive at

$$Q\, S_1^{-1}(-v-N)\, R(2v+N)\, S_2(v) = S_2(v)\, R(2v+N)\, S_1^{-1}(-v-N)\, Q. \tag{2.65}$$

Recalling that Q is a one-dimensional operator satisfying $Q^2 = N\,Q$, we conclude that (2.65) must be equal to Q times a series in v^{-1} with coefficients in $\mathrm{Y}(\mathfrak{g}_N)$. Therefore, each of the following relations uniquely determines the series $y(u)$, where the scalar factor is chosen for convenience to simplify the quantum Liouville formula (Theorem 2.11.2 below), and v in the above formula is replaced with $u-N$:

$$\frac{(1\mp 2u)(2u-N-1)}{(1-2u)(2u-N)}\, y(u)\, Q = Q\, S_1^{-1}(-u)\, R(2u-N)\, S_2(u-N) \tag{2.66}$$

and

$$\frac{(1\mp 2u)(2u-N-1)}{(1-2u)(2u-N)}\, y(u)\, Q = S_2(u-N)\, R(2u-N)\, S_1^{-1}(-u)\, Q. \tag{2.67}$$

PROPOSITION 2.11.1. *We have the following formulas for the series $y(u)$ and its inverse:*

$$y(u) = \frac{1-2u}{N(1\mp 2u)}\, \mathrm{tr}\Big\{\Big(\frac{2u-N}{2u-N-1} S^{-1}(-u)^t - \frac{1}{2u-N-1} S^{-1}(-u)\Big) S(u-N)\Big\}$$

and

$$y(u)^{-1} = \frac{1\mp 2u}{N(1-2u)}\, \mathrm{tr}\Big\{\Big(\frac{2u-N}{2u-N+1} S^t(-u) + \frac{1}{2u-N+1} S(-u)\Big) S^{-1}(u-N)\Big\}.$$

PROOF. In order to prove the first formula it is sufficient to apply the operators on both sides of (2.66) to the basis vector $e_i\otimes e_i$ and take the summation over i. For the proof of the second formula, we use (2.19) to rewrite (2.67) in the equivalent form

$$\frac{(1-2u)(2u-N+1)}{(1\mp 2u)(2u-N)}\, y(u)^{-1}\, Q = S_1(-u)\, R(-2u+N)\, S_2^{-1}(u-N)\, Q.$$

Since $P\,Q = Q\,P = Q$, conjugating both sides by P we get

$$\frac{(1-2u)(2u-N+1)}{(1\mp 2u)(2u-N)}\, y(u)^{-1}\, Q = S_2(-u)\, R(-2u+N)\, S_1^{-1}(u-N)\, Q.$$

Now apply the operators on both sides to the basis vector $e_i\otimes e_i$, compare the coefficients of $e_k\otimes e_k$, and sum over $k=1,\dots,N$. □

The following is a counterpart of the quantum Liouville formula for the twisted Yangian $\mathrm{Y}(\mathfrak{g}_N)$; cf. Theorem 1.9.5.

THEOREM 2.11.2. *We have the relation*

$$y(u) = \frac{\operatorname{sdet} S(u-1)}{\operatorname{sdet} S(u)}.$$

PROOF. Regarding $\mathrm{Y}(\mathfrak{g}_N)$ as a subalgebra of the Yangian $\mathrm{Y}(\mathfrak{gl}_N)$ write $S(u) = T(u)\, G\, T^t(-u)$ so that the right-hand side of (2.66) will take the form

$$Q\, T_1^t(u)^{-1} G_1^{-1} T_1^{-1}(-u)\, R(2u-N)\, T_2(u-N)\, G_2 T_2^t(-u+N). \tag{2.68}$$

By relation (1.29) we have

$$T_1^{-1}(u)\, R(2u-N)\, T_2(u-N) = T_2(u-N)\, R(2u-N)\, T_1^{-1}(u)$$

so that (2.68) can be written as

$$\begin{aligned}&Q\, T_1^t(u)^{-1} G_1^{-1} T_2(u-N)\, R(2u-N)\, T_1^{-1}(-u)\, G_2\, T_2^t(-u+N)\\ &\qquad = Q\, T_1^t(u)^{-1} T_2(u-N)\, G_1^{-1}\, R(2u-N)\, G_2\, T_1^{-1}(-u)\, T_2^t(-u+N).\end{aligned}$$

By Proposition 1.9.8, $Q\, T_1^t(u)^{-1} T_2(u-N) = Q\, z(u)$, where the series $z(u)$ is defined by (1.68). Furthermore, we have

$$\begin{aligned}Q\, G_1^{-1}\, R(2u-N)\, G_2 &= Q\, G_1^{-1}\Big(1 - \frac{P}{2u-N}\Big)\, G_2\\ &= Q\, G_1^{-1}\, G_2 - \frac{1}{2u-N}\, Q\, G_1^{-1} P\, G_2 = Q\,\Big(\pm 1 - \frac{1}{2u-N}\Big),\end{aligned}$$

where we have used the equalities $G_1^{-1}P = P\, G_2^{-1}$, $QP = Q$ and

$$Q\, G_1^{-1} = Q\,\big(G_2^{-1}\big)^t = \pm Q\, G_2^{-1}.$$

To complete the calculation, observe that

$$Q\, T_1^{-1}(-u)\, T_2^t(-u+N) = Q\, z(-u+N)^{-1}.$$

Indeed, since the operator (1.73) equals $z(u)^{-1}$ it suffices to conjugate (1.73) by P and replace u by $-u+N$.

Thus, we arrive at the relation

$$\frac{(1 \mp 2u)(2u-N-1)}{(1-2u)(2u-N)}\, y(u) = \Big(\pm 1 - \frac{1}{2u-N}\Big)\, z(u)\, z(-u+N)^{-1}.$$

Now, the quantum Liouville formula for the Yangian $\mathrm{Y}(\mathfrak{gl}_N)$ (Theorem 1.9.5) gives

$$z(u)\, z(-u+N)^{-1} = \frac{\operatorname{qdet} T(u-1)}{\operatorname{qdet} T(u)}\; \frac{\operatorname{qdet} T(-u+N)}{\operatorname{qdet} T(-u+N-1)}.$$

On the other hand, by Theorem 2.5.3,

$$\operatorname{qdet} T(u)\, \operatorname{qdet} T(-u+N-1) = \gamma_{n,G}(u)^{-1} \operatorname{sdet} S(u),$$

so that a simple calculation with the scalar factors completes the proof. □

Theorems 2.8.2 and 2.11.2 imply the following.

COROLLARY 2.11.3. *The coefficients of the series $y(u)$ generate the center of the twisted Yangian $\mathrm{Y}(\mathfrak{g}_N)$.* □

2.12. Factorization of the Sklyanin determinant

Here we give a quasideterminant factorization of the Sklyanin determinant analogous to Theorem 1.10.5. We need an additional assumption on the matrix G. We will assume that for each $1 \leqslant M \leqslant N-1$ the submatrix of G corresponding to the first M rows and columns is nonsingular. Hereby the symplectic case $\mathfrak{g}_N = \mathfrak{sp}_N$ is excluded. Below we prove the factorization theorem only for the orthogonal case where such matrices G do exist, e.g., $G = 1$. An analogous factorization of the Sklyanin determinant for both the orthogonal and symplectic cases can be obtained for a different presentation of the twisted Yangian $\mathrm{Y}(\mathfrak{g}_N)$; see Theorem 4.1.7 below.

For $1 \leqslant M \leqslant N$ denote by $S^{(M)}(u)$ the submatrix of $S(u)$ corresponding to the first M rows and columns. The coefficients of the series $s_{ij}(u)$ with $1 \leqslant i, j \leqslant M$ can be regarded as generators of the twisted Yangian $\mathrm{Y}(\mathfrak{g}_M)$; see Corollary 2.4.4. In particular, the Sklyanin determinant $\operatorname{sdet} S^{(M)}(u)$ is well-defined.

We use the quasideterminants introduced in Definition 1.10.1.

THEOREM 2.12.1. *Under the above assumption on the matrix G, the Sklyanin determinant* $\operatorname{sdet} S(u)$ *for the twisted Yangian* $\mathrm{Y}(\mathfrak{o}_N)$ *admits the factorization in the algebra* $\mathrm{Y}(\mathfrak{o}_N)[[u^{-1}]]$

$$\operatorname{sdet} S(u) = s_{11}(u)\,\big|S^{(2)}(u-1)\big|_{22} \cdots \big|S^{(N)}(u-N+1)\big|_{NN}.$$

Moreover, the N factors on the right-hand side of this equality pairwise commute.

PROOF. By Definition 2.7.6,

$$\widehat{S}(u) = \operatorname{sdet} S(u)\, S^{-1}(u-N+1). \tag{2.69}$$

Taking the NN-th entry gives

$$\widehat{s}_{NN}(u) = \operatorname{sdet} S(u)\,\big(S^{-1}(u-N+1)\big)_{NN}$$

and hence, by Proposition 1.10.4,

$$\operatorname{sdet} S(u) = \widehat{s}_{NN}(u)\,\big|S(u-N+1)\big|_{NN}. \tag{2.70}$$

Now, Proposition 2.7.7 gives

$$\widehat{s}_{NN}(u) = s^{\,1\dots N-1}_{\,1\dots N-1}(u).$$

However, the Sklyanin minor $s^{\,1\dots N-1}_{\,1\dots N-1}(u)$ coincides with $\operatorname{sdet} S^{(N-1)}(u)$. Indeed, this is implied by the recurrence relation for the Sklyanin minors obtained in the proof of Theorem 2.7.2. Thus, we can write

$$\operatorname{sdet} S(u) = \operatorname{sdet} S^{(N-1)}(u)\,\big|S^{(N)}(u-N+1)\big|_{NN}.$$

Note that the factors here commute by the centrality of the Sklyanin determinant. The proof is completed by an obvious induction. □

We will need the following analogue of Proposition 1.10.10. Now we consider both cases $\mathfrak{g}_N = \mathfrak{o}_N$ and $\mathfrak{sp}_N$ simultaneously. Denote by $s'_{ij}(u)$ the matrix elements of the matrix $S^{-1}(u)$.

PROPOSITION 2.12.2. *We have the relations*

$$(u^2-v^2)\,[s_{ij}(u),s'_{kl}(v)] = (u+v)\Big(\delta_{kj}\sum_{a=1}^{N} s_{ia}(u)\,s'_{al}(v) - \delta_{il}\sum_{a=1}^{N} s'_{ka}(v)\,s_{aj}(u)\Big)$$
$$-(u-v)\Big(\delta_{ik}\sum_{a=1}^{N} s_{aj}(u)\,s'_{al}(v) - \delta_{jl}\sum_{a=1}^{N} s'_{ka}(v)\,s_{ia}(u)\Big)$$
$$+\delta_{ik}\sum_{a=1}^{N} s_{ja}(u)\,s'_{al}(v) - \delta_{jl}\sum_{a=1}^{N} s'_{ka}(v)\,s_{ai}(u).$$

PROOF. Multiply the quaternary relation (2.11) by $S_2^{-1}(v)$ from both sides to get

$$S_2^{-1}(v)\,R(u-v)\,S_1(u)\,R^t(-u-v) = R^t(-u-v)\,S_1(u)\,R(u-v)\,S_2^{-1}(v).$$

Now apply the operators on both sides to the basis vector $e_j\otimes e_l$ and compare the coefficients of $e_i\otimes e_k$. A calculation analogous to the proof of Proposition 1.10.10 yields the desired relation. □

An important property of the Sklyanin comatrix is provided by the following proposition. Recall the scalar $\gamma_{n,G}(u)$ defined in Theorem 2.5.3.

PROPOSITION 2.12.3. *The mapping*

$$S(u)\mapsto \gamma_{n,G^{-1}}(u)\cdot G\,\widehat{S}(-u+N/2-1)\,G$$

defines an automorphism of the algebra $\mathrm{Y}(\mathfrak{g}_N)$.

PROOF. Regarding $\mathrm{Y}(\mathfrak{g}_N)$ as a subalgebra of $\mathrm{Y}(\mathfrak{gl}_N)$, we can write $S(u) = T(u)\,G\,T^t(-u)$. Note that the mapping

$$\psi: T(u)\mapsto G\,\widehat{T}^t(u+N/2-1)\,G^{-1}$$

defines an automorphism of the algebra $\mathrm{Y}(\mathfrak{gl}_N)$. Indeed, the mapping is the composition of the anti-automorphism of Proposition 1.9.3, the transposition $T(u)\mapsto T^t(u)$, and the shift and conjugation automorphisms; see Propositions 1.3.1 and 1.3.3. Since $G^t=\pm G$, for the image of the matrix $S(u)$ we have

$$\psi: S(u)\mapsto G\,\widehat{T}^t(u+N/2-1)\,G^{-1}\widehat{T}(-u+N/2-1)\,G.$$

Using Definition 1.9.1 and Proposition 1.9.2, we obtain

$$\begin{aligned}\psi\big(S(-u+N/2-1)\big)\,G^{-1}\,S(u-N+1)&\\ &= G\,\widehat{T}^t(-u+N-2)\,G^{-1}\,\widehat{T}(u)\,T(u-N+1)\,G\,T^t(-u+N-1)\\ &= \mathrm{qdet}\,T(u)\,G\,\widehat{T}^t(-u+N-2)\,T^t(-u+N-1)\\ &= \mathrm{qdet}\,T(u)\,\mathrm{qdet}\,T(-u+N-1)\,G\end{aligned}$$

which coincides with $\gamma_{n,G}(u)^{-1}\,\mathrm{sdet}\,S(u)\,G$ by Theorem 2.5.3. Hence,

$$\psi: S(-u+N/2-1)\mapsto \gamma_{n,G}(u)^{-1}\,G\,\widehat{S}(u)\,G.$$

Replacing u by $-u+N/2-1$ and observing that

$$\gamma_{n,G^{-1}}(u)\,\gamma_{n,G}(-u+N/2-1)=1,$$

we find that the image of $S(u)$ under ψ is found by the formula given in the statement of the proposition. This means that the subalgebra $\mathrm{Y}(\mathfrak{g}_N)$ of $\mathrm{Y}(\mathfrak{gl}_N)$ is stable under the automorphism ψ, and thus the restriction of ψ to $\mathrm{Y}(\mathfrak{g}_N)$ defines an automorphism of the latter. □

2.13. Extended twisted Yangian

Here we show that the twisted Yangian $\mathrm{Y}(\mathfrak{g}_N)$ may be regarded as a quotient of a 'covering' algebra by an ideal generated by central elements. As before, we fix a nonsingular symmetric or skew-symmetric matrix G.

DEFINITION 2.13.1. The *extended twisted Yangian* $\mathrm{X}_G(\mathfrak{g}_N)$ corresponding to the Lie algebra $\mathfrak{g}_N$ and the matrix G is a unital associative algebra with generators $s_{ij}^{(1)}, s_{ij}^{(2)}, \dots$, where $1 \leqslant i, j \leqslant N$, and the defining relations (2.6) with the series $s_{ij}(u)$ defined in (2.5). □

The use of the same symbols as for the generators of $\mathrm{Y}_G(\mathfrak{g}_N)$ should not cause confusion, as we always specify which algebra is considered at any moment.

The proof of Proposition 2.2.1 shows that the defining relations of the algebra $\mathrm{X}_G(\mathfrak{g}_N)$ can be written in matrix form as the quaternary relation (2.11). Moreover, Proposition 2.3.1 implies that the algebras $\mathrm{X}_G(\mathfrak{g}_N)$ corresponding to different symmetric (respectively, skew-symmetric) matrices G are isomorphic to each other. When the matrix G is fixed, we will simply write $\mathrm{X}(\mathfrak{g}_N)$ instead of $\mathrm{X}_G(\mathfrak{g}_N)$.

PROPOSITION 2.13.2. *There exists a formal series*

$$x(u) = 1 + x_1\, u^{-1} + x_2\, u^{-2} + \cdots \in \mathrm{X}(\mathfrak{g}_N)[[u^{-1}]]$$

such that

$$Q\, S_1(u)\, R(2u)\, S_2^{-1}(-u) = S_2^{-1}(-u)\, R(2u)\, S_1(u)\, Q = \Big(\pm 1 - \frac{1}{2u} \Big)\, x(u)\, Q. \tag{2.71}$$

Moreover, this series satisfies $x(u)\, x(-u) = 1$.

PROOF. By multiplying the quaternary relation (2.11) by $S_2^{-1}(v)$ from both sides we obtain the relation

$$S_2^{-1}(v)\, R(u-v)\, S_1(u) R^t(-u-v) = R^t(-u-v)\, S_1(u)\, R(u-v)\, S_2^{-1}(v). \tag{2.72}$$

Taking the residue at $v = -u$, we get the first equality in (2.71). Since Q is a one-dimensional operator on $\mathbb{C}^N \otimes \mathbb{C}^N$ satisfying $Q^2 = N\, Q$, there exists a formal power series $x(u)$ with coefficients in $\mathrm{X}(\mathfrak{g}_N)$ satisfying (2.71). The constant term of $x(u)$ is 1 since $Q\, G_1\, G_2^{-1} = Q\, G_2^t\, G_2^{-1} = \pm\, Q$.

Furthermore, (2.71) implies

$$Q\, S_2(-u)\, R(2u)^{-1}\, S_1^{-1}(u) = \Big(\pm 1 - \frac{1}{2u} \Big)^{-1} x(u)^{-1}\, Q.$$

Now apply (2.19), replace u by $-u$ and conjugate both sides by the permutation operator P. This gives $x(-u)^{-1} = x(u)$. □

THEOREM 2.13.3. *All coefficients of the series $x(u)$ belong to the center of the algebra* $\mathrm{X}(\mathfrak{g}_N)$.

PROOF. Consider the algebra $\operatorname{End}\mathbb{C}^N \otimes \operatorname{End}\mathbb{C}^N \otimes \operatorname{End}\mathbb{C}^N \otimes \mathrm{X}(\mathfrak{g}_N)$ and enumerate the copies of $\operatorname{End}\mathbb{C}^N$ by the indices $0, 1, 2$. Maintaining the matrix notation of Section 1.2, set

$$S_i = S_i(u_i), \qquad i = 0, 1, 2$$

and

$$R_{ij} = R_{ij}(u_i - u_j), \quad R^t_{ij} = R^t_{ij}(-u_i - u_j); \qquad 0 \leqslant i < j \leqslant 2,$$

where u_0, u_1, u_2 are formal variables. We will be proving the relation

$$S_0\, x(u_1)\, Q_{12} = x(u_1)\, S_0\, Q_{12}, \tag{2.73}$$

which implies the statement of the theorem. We start by verifying the following auxiliary identities which hold provided that $u_1 + u_2 = 0$:

$$Q_{12} R^t_{01} R_{02} = R_{02} R^t_{01} Q_{12} = Q_{12}\big(1 - (u_0 + u_1)^{-2}\big), \tag{2.74}$$

$$Q_{12} R^t_{02} R_{01} = R_{01} R^t_{02} Q_{12} = Q_{12}\big(1 - (u_0 - u_1)^{-2}\big). \tag{2.75}$$

Indeed, by (2.32),

$$R^t_{12} R^t_{01} R_{02} = R_{02} R^t_{01} R^t_{12}.$$

Taking the residue at $u_2 = -u_1$, we obtain the first equality in (2.74). The second equality in (2.74) is now verified by applying the operator $Q_{12}\, R^t_{01}\, R_{02}$ to the basis vectors $e_i \otimes e_1 \otimes e_1$ with $i = 1, \dots, N$. Applying the conjugation by P_{12} to (2.74) we get (2.75).

Our next step is to prove that for arbitrary formal variables u_0, u_1, u_2 the following identity holds:

$$R_{01} R^t_{02}\, S_0 R_{02} R^t_{01}\, S_2^{-1} R_{12}\, S_1 R^t_{12} = S_2^{-1} R_{12}\, S_1 R^t_{12} R^t_{01} R_{02}\, S_0 R^t_{02} R_{01}. \tag{2.76}$$

This is obtained by the following sequence of transformations:

$$\begin{aligned}
&R_{01}(R^t_{02}\, S_0 R_{02}\, S_2^{-1}) R^t_{01} R_{12}\, S_1 R^t_{12} \\
&\qquad = R_{01}\, S_2^{-1} R_{02}\, S_0 (R^t_{02} R^t_{01} R_{12})\, S_1 R^t_{12} && \text{by (2.72)} \\
&\qquad = R_{01}\, S_2^{-1} R_{02}\, S_0 R_{12} R^t_{01} R^t_{02}\, S_1 R^t_{12} && \text{by (2.32)} \\
&\qquad = S_2^{-1} (R_{01} R_{02} R_{12})\, S_0 R^t_{01}\, S_1 R^t_{02} R^t_{12} \\
&\qquad = S_2^{-1} R_{12} R_{02} (R_{01}\, S_0 R^t_{01}\, S_1) R^t_{02} R^t_{12} && \text{by (1.17)} \\
&\qquad = S_2^{-1} R_{12} R_{02}\, S_1 R^t_{01}\, S_0 (R_{01} R^t_{02} R^t_{12}) && \text{by (2.11)} \\
&\qquad = S_2^{-1} R_{12} R_{02}\, S_1 R^t_{01}\, S_0 R^t_{12} R^t_{02} R_{01} && \text{by (2.32)} \\
&\qquad = S_2^{-1} R_{12}\, S_1 (R_{02} R^t_{01} R^t_{12})\, S_0 R^t_{02} R_{01},
\end{aligned}$$

which coincides with the right-hand side of (2.76) by (2.32).

Finally, take the residues of both sides of (2.76) at $u_2 = -u_1$. We obtain the equality

$$R_{01} R^t_{02}\, S_0 R_{02} R^t_{01} (\, S_2^{-1} R_{12}\, S_1 Q_{12}) = (\, S_2^{-1} R_{12}\, S_1 Q_{12}) R^t_{01} R_{02}\, S_0 R^t_{02} R_{01}.$$

By Proposition 2.13.2, this may be rewritten as

$$R_{01} R^t_{02}\, S_0 R_{02} R^t_{01} Q_{12}\, x(u_1) = x(u_1)\, Q_{12} R^t_{01} R_{02}\, S_0 R^t_{02} R_{01}. \tag{2.77}$$

By (2.74) and (2.75), the left-hand side equals

$$\big(1 - (u_0 + u_1)^{-2}\big)\big(1 - (u_0 - u_1)^{-2}\big)\, S_0\, x(u_1)\, Q_{12},$$

while the right-hand side is

$$\big(1 - (u_0 + u_1)^{-2}\big)\big(1 - (u_0 - u_1)^{-2}\big)\, x(u_1)\, S_0\, Q_{12}.$$

This proves (2.73) and the theorem. □

THEOREM 2.13.4. *The relation $x(u) = 1$ in the algebra $\mathrm{X}(\mathfrak{g}_N)$ is equivalent to the symmetry relation* (2.12).

PROOF. Apply the operator $Q\, S_1(u)\, R(2u)\, S_2^{-1}(-u)$ to the basis vector $e_j \otimes e_l$. This gives the expression

$$\sum_{i=1}^{N} \Big(s_{ij}(u)\, s'_{il}(-u) - \frac{1}{2u}\, s_{ji}(u)\, s'_{il}(-u) \Big)\, \eta, \qquad \eta = \sum_{k=1}^{N} e_k \otimes e_k.$$

By Proposition 2.13.2, this expression must coincide with

$$\delta_{jl} \Big(\pm 1 - \frac{1}{2u} \Big)\, x(u)\, \eta.$$

We thus obtain the matrix relation

$$\Big(S^t(u) - \frac{1}{2u}\, S(u) \Big)\, S^{-1}(-u) = \Big(\pm 1 - \frac{1}{2u} \Big)\, x(u)$$

and hence

$$S^t(u) - \frac{1}{2u}\, S(u) = \Big(\pm 1 - \frac{1}{2u} \Big)\, x(u)\, S(-u).$$

Clearly, $x(u) = 1$ is equivalent to (2.12). □

COROLLARY 2.13.5. *The twisted Yangian* $\mathrm{Y}(\mathfrak{g}_N)$ *is isomorphic to the quotient of the extended twisted Yangian* $\mathrm{X}(\mathfrak{g}_N)$ *by the ideal generated by all the coefficients* $x_1, x_2, \ldots$ *of the series* $x(u)$. □

We introduce the Sklyanin minors $s^{a_1 \ldots a_m}_{b_1 \ldots b_m}(u)$ of the matrix $S(u)$ and auxiliary minors $\check{s}^{a_1 \ldots a_m}_{b_1 \ldots b_{m-1}, c}(u)$ for the extended twisted Yangian $\mathrm{X}(\mathfrak{g}_N)$ in exactly the same way as in Section 2.6. These are series in u^{-1} with coefficients in $\mathrm{X}(\mathfrak{g}_N)$ which are respectively defined by the expansions of (2.44) and (2.46). Clearly, they possess the same skew-symmetry properties as their counterparts in the twisted Yangian $\mathrm{Y}(\mathfrak{g}_N)$.

PROPOSITION 2.13.6. *We have the relations in* $\mathrm{X}(\mathfrak{g}_N)$

$$\begin{aligned}
&(u^2 - v^2)\, [s_{kl}(u), s^{a_1 \ldots a_m}_{b_1 \ldots b_m}(v)] \\
&\quad = (u+v) \sum_{i=1}^{m} \Big(s_{a_i l}(u)\, s^{a_1 \ldots k \ldots a_m}_{b_1 \ldots \ldots b_m}(v) - s^{a_1 \ldots \ldots a_m}_{b_1 \ldots l \ldots b_m}(v)\, s_{k b_i}(u) \Big) \\
&\quad - (u-v) \sum_{i=1}^{m} \Big(s_{k a_i}(u)\, s^{a_1 \ldots l \ldots a_m}_{b_1 \ldots \ldots b_m}(v) - s^{a_1 \ldots \ldots a_m}_{b_1 \ldots k \ldots b_m}(v)\, s_{b_i l}(u) \Big) \\
&\quad + \sum_{i=1}^{m} \Big(s_{a_i k}(u)\, s^{a_1 \ldots l \ldots a_m}_{b_1 \ldots \ldots b_m}(v) - s^{a_1 \ldots \ldots a_m}_{b_1 \ldots k \ldots b_m}(v)\, s_{l b_i}(u) \Big) \\
&\quad + \sum_{i \neq j} \Big(s_{a_i a_j}(u)\, s^{a_1 \ldots k \ldots l \ldots a_m}_{b_1 \ldots \ldots b_m}(v) - s^{a_1 \ldots \ldots a_m}_{b_1 \ldots k \ldots l \ldots b_m}(v)\, s_{b_i b_j}(u) \Big),
\end{aligned}$$

where in the Sklyanin minors the indices k *and* l *replace* a_i *and* b_i*, respectively, in the first sum; the indices* l *and* k *replace* a_i *and* b_i*, respectively, in the second and third sums; and* k *and* l *replace* a_i *and* a_j*, and* b_i *and* b_j*, respectively, in the fourth sum.*

PROOF. By Proposition 2.5.1, we have the relation

$$\begin{aligned}
(2.78) \quad & R(u, v, v-1, \ldots, v-m+1)\, \langle S_0, \ldots, S_m \rangle \\
&= \langle S_m, \ldots, S_0 \rangle\, R(u, v, v-1, \ldots, v-m+1),
\end{aligned}$$

where we have used an extra copy of the algebra $\operatorname{End}\mathbb{C}^N$ labelled by 0, and the parameters are specialized by

$$u_0 = u \quad \text{and} \quad u_i = v - i + 1 \quad \text{for} \quad i = 1, \dots, m.$$

Now use (1.58) and its consequence

$$A_m\, R^t_{01} \dots R^t_{0m} = A_m \Big(1 + \frac{1}{u+v}(Q_{01} + \dots + Q_{0m})\Big) \tag{2.79}$$

obtained by the application of the transposition over the zeroth copy of $\operatorname{End}\mathbb{C}^N$ and by replacing u by $-u$. Hence (2.78) takes the form

$$\Big(1 - \frac{1}{u-v} \sum_{k=1}^{m} P_{0k}\Big)\, S_0(u) \Big(1 + \frac{1}{u+v} \sum_{k=1}^{m} Q_{0k}\Big)\, A_m\, \langle S_1, \dots, S_m \rangle$$
$$= \langle S_m, \dots, S_1 \rangle\, A_m \Big(1 + \frac{1}{u+v} \sum_{k=1}^{m} Q_{0k}\Big)\, S_0(u) \Big(1 - \frac{1}{u-v} \sum_{k=1}^{m} P_{0k}\Big).$$

It remains to apply both sides to the vector $e_l \otimes e_{b_1} \otimes \dots \otimes e_{b_m}$ and compare the coefficients of the vector $e_k \otimes e_{a_1} \otimes \dots \otimes e_{a_m}$. □

COROLLARY 2.13.7. *Suppose that for some indices $i, j, l, r \in \{1, \dots, m\}$ we have $a_i = b_l$ and $b_j = a_r$. Then*

$$[s_{a_i b_j}(u), s^{\,a_1 \dots a_m}_{\,b_1 \dots b_m}(v)] = 0.$$

PROOF. Using the skew-symmetry of Sklyanin minors, we derive from Proposition 2.13.6 that

$$(u - v - 1)(u + v + 1)\, [s_{a_i b_j}(u), s^{\,a_1 \dots a_m}_{\,b_1 \dots b_m}(v)] = 0,$$

which gives the desired relation. □

The Sklyanin determinant $\operatorname{sdet} S(u) \in \mathrm{X}(\mathfrak{g}_N)[[u^{-1}]]$ is defined in the same way as for the twisted Yangian; see Definition 2.5.2. Equivalently, we can set

$$\operatorname{sdet} S(u) = s^{\,1 \dots N}_{\,1 \dots N}(u).$$

Corollary 2.13.7 implies that all coefficients of this series belong to the center of the algebra $\mathrm{X}(\mathfrak{g}_N)$. In particular, this gives an alternative proof of the centrality of the Sklyanin determinant in the twisted Yangian; cf. Theorem 2.8.2.

We will now prove some analogs of the complementary minor identity (1.77) for the algebra $\mathrm{X}(\mathfrak{g}_N)$; cf. Theorem 1.10.7. It will be convenient to use the following transformation ϖ_N on the Sklyanin minors and auxiliary minors of the matrix $S(u)$. Given an expansion of a minor in terms of the series $s_{ij}(u)$, the image of the minor under ϖ_N is obtained by the replacement of each series $s_{ij}(u)$ with $s'_{ij}(-u - N/2)$, where, as before, the $s'_{ij}(u)$ denote the entries of the inverse matrix $S^{-1}(u)$. Since, by Proposition 2.3.5, the matrix $S^{-1}(-u - N/2)$ satisfies the quaternary relation (2.11), the result of this replacement does not depend on the expansion of the minor. So, the transformation ϖ_N is well-defined. We will denote this by writing

$$\varpi_N : S(u) \rightsquigarrow S^{-1}(-u - N/2). \tag{2.80}$$

Note that the constant term matrix of $S^{-1}(-u - N/2)$ is G^{-1} so that the transformation (2.80) should not be confused with an automorphism of $\mathrm{X}(\mathfrak{g}_N)$. Similar to the map (1.31), the transformation ϖ_N is involutive, $\varpi_N \circ \varpi_N = 1$.

REMARK 2.13.8. Alternatively, ϖ_N can be interpreted as the isomorphism

$$\varpi_N : \mathrm{X}_G(\mathfrak{g}_N) \to \mathrm{X}_{G^{-1}}(\mathfrak{g}_N), \qquad S(u) \mapsto \widetilde{S}^{-1}(-u - N/2), \tag{2.81}$$

where $\widetilde{S}(u)$ denotes the generator matrix for $\mathrm{X}_{G^{-1}}(\mathfrak{g}_N)$. The results of this section can be reformulated in terms of the isomorphism (2.81); see also Section 2.14. □

Fix an integer M such that $0 \leqslant M \leqslant N$ and let $a, b \in \{M+1, \dots, N\}$.

THEOREM 2.13.9. *We have the identities*

$$\operatorname{sdet} S(u) \cdot \varpi_N\big(s\,{}^{M+1\dots N}_{M+1\dots N}(-u + N/2 - 1)\big) = s\,{}^{1\dots M}_{1\dots M}(u) \tag{2.82}$$

and

$$\operatorname{sdet} S(u) \cdot \varpi_N\big(\check{s}\,{}^{M+1\dots N}_{M+1\dots \widehat{a}\dots N,\,b}(-u + N/2 - 1)\big) = (-1)^{N-a}\, s\,{}^{1\dots M,\,a}_{1\dots M,\,b}(u). \tag{2.83}$$

In particular,

$$\operatorname{sdet} S(u) \cdot \varpi_N\big(\operatorname{sdet} S(-u + N/2 - 1)\big) = 1. \tag{2.84}$$

PROOF. By definition of the Sklyanin determinant,

$$A_N \langle S_1, \dots, S_N \rangle = A_N \operatorname{sdet} S(u). \tag{2.85}$$

This implies the relation

$$A_N \langle S_1, \dots, S_M \rangle \prod_{i=1,\dots,M}^{\longrightarrow} (R^t_{i,M+1} \cdots R^t_{iN}) = A_N \operatorname{sdet} S(u) \tag{2.86}$$

$$\times\, S_N^{-1} (R^t_{N-1,N})^{-1} S_{N-1}^{-1} \cdots S_{M+2}^{-1} (R^t_{M+1,N})^{-1} \cdots (R^t_{M+1,M+2})^{-1} S_{M+1}^{-1}.$$

Since $R^t(u)^{-1} = R^t(-u + N)$, the right-hand side can be written as

$$A_N \operatorname{sdet} S(u)\, S_N^\circ R_{N-1,N}^\circ\, S_{N-1}^\circ \cdots S_{M+2}^\circ\, R_{M+1,N}^\circ \cdots R_{M+1,M+2}^\circ\, S_{M+1}^\circ,$$

where we have used the notation $S^\circ(u) = \varpi_N(S(u))$ and

$$S_i^\circ = S_i^\circ(u_i^\circ), \qquad R_{ij}^\circ = R_{ij}^t(-u_i^\circ - u_j^\circ), \qquad u_i^\circ = -u_i - N/2 \tag{2.87}$$

with $u_i = u - i + 1$ for $i = 1, \dots, N$. Now apply both sides of (2.86) to the basis vector $v = e_1 \otimes \dots \otimes e_N$ and compare the coefficients of $A_N\, v$. Using the skew-symmetry of the Sklyanin minors we get (2.82).

The proof of (2.83) is similar with the use of the relation

$$A_N \langle S_1, \dots, S_{M+1} \rangle \prod_{i=1,\dots,M}^{\longrightarrow} (R^t_{i,M+2} \cdots R^t_{iN}) \tag{2.88}$$

$$= A_N \operatorname{sdet} S(u)\, S_N^{-1} (R^t_{N-1,N})^{-1} S_{N-1}^{-1} \cdots S_{M+2}^{-1} (R^t_{M+1,N})^{-1} \cdots (R^t_{M+1,M+2})^{-1}$$

implied by (2.85). Using (2.87) we can rewrite the right-hand side as

$$A_N \operatorname{sdet} S(u)\, S_N^\circ R_{N-1,N}^\circ\, S_{N-1}^\circ \cdots S_{M+2}^\circ R_{M+1,N}^\circ \cdots R_{M+1,M+2}^\circ.$$

The proof is completed by applying both sides of (2.88) to the vector

$$e_1 \otimes \dots \otimes e_M \otimes e_b \otimes e_{M+1} \otimes \dots \otimes \widehat{e}_a \otimes \dots \otimes e_N$$

and comparing the coefficients of $A_N(e_1 \otimes \dots \otimes e_N)$. □

PROPOSITION 2.13.10. *For any $1 \leqslant M \leqslant N$ there exist expansions of the minors*

$$s^{1\dots M}_{1\dots M}(u) \qquad \text{and} \qquad \check{s}^{\,1\dots M}_{1\dots\widehat{k}\dots M,\,l}(u), \qquad k,l\in\{1,\dots,M\}$$

in terms of the generator series $s_{ij}(u)$ of the algebra $\mathrm{X}(\mathfrak{g}_N)$ which only involve the series with the indices satisfying $1 \leqslant i,j \leqslant M$.

PROOF. By definition, the Sklyanin minor $s^{1\dots M}_{1\dots M}(u)$ in $\mathrm{X}(\mathfrak{g}_N)$ is the coefficient of the basis vector $e_1\otimes\dots\otimes e_M$ in the expansion of

$$A_M\,\langle S_1,\dots,S_M\rangle(e_1\otimes\dots\otimes e_M); \tag{2.89}$$

see (2.44). We note the following identity, analogous to (2.79), which is verified in the same way:

$$A_{m-1}\,R^t_{1m}\dots R^t_{m-1,m} = A_{m-1}\Big(1+\frac{Q_{1m}+\dots+Q_{m-1,m}}{2u-m+1}\Big).$$

Therefore, we can write the operator which occurs in (2.89) in the form

$$A_M\,S_1\Big(1+\frac{Q_{12}}{2u-1}\Big)\,S_2\Big(1+\frac{Q_{13}+Q_{23}}{2u-2}\Big)\times\dots S_{M-1}\Big(1+\frac{Q_{1M}+\dots+Q_{M-1,M}}{2u-M+1}\Big)\,S_M;$$

cf. the proof of Proposition 2.6.2. We have

$$\Big(1+\frac{Q_{1M}+\dots+Q_{M-1,M}}{2u-M+1}\Big)\,S_M(e_1\otimes\dots\otimes e_M) = \sum_{a=1}^{N}\Big(e_1\otimes\dots\otimes e_{M-1}\otimes e_a\otimes s_{aM}(u-M+1) + \frac{1}{2u-M+1}\sum_{i=1}^{M-1} e_1\otimes\dots\otimes e_a\otimes\dots\otimes e_{M-1}\otimes e_a\otimes s_{iM}(u-M+1)\Big),$$

where the first e_a takes the i-th position in the product of the second sum. This implies that the terms corresponding to the indices $a>M$ do not contribute to the coefficient of the vector $e_1\otimes\dots\otimes e_M$ in the expansion of (2.89). Arguing by induction, we can conclude that the coefficient in question will be represented as a combination of products of the series of type $s_{ij}(u-k+1)$ with $i,j\in\{1,\dots,M\}$. This proves the first statement of the proposition. The second is verified by the same argument with the use of the definition of the auxiliary minors (2.46). □

2.14. Quantum Sylvester theorem

Consider the extended twisted Yangian $\mathrm{X}(\mathfrak{g}_{M+N})$ corresponding to a nonsingular symmetric or skew-symmetric matrix G. We will be assuming throughout this section that the submatrix $G^{(M)}$ of G determined by the first M rows and columns is nonsingular. This implies that both parameters M and N are even in the symplectic case.

Similar to (2.80), introduce the transformation υ_M which takes the generator series of $\mathrm{X}(\mathfrak{g}_N)$ to the generator series of $\mathrm{X}(\mathfrak{g}_{M+N})$ by the rule

$$\upsilon_M: s_{ij}(u)\rightsquigarrow s_{M+i,M+j}(u), \qquad i,j\in\{1,\dots,N\}. \tag{2.90}$$

The defining relations of $\mathrm{X}(\mathfrak{g}_N)$ are preserved by υ_M so that by Proposition 2.13.10, the extension of this transformation to the Sklyanin minors and auxiliary minors is well-defined. It takes the minors in $\mathrm{X}(\mathfrak{g}_N)$ to minors in $\mathrm{X}(\mathfrak{g}_{M+N})$.

Consider the composition of the transformations

$$\vartheta_M = \varpi_{M+N} \circ \upsilon_M \circ \varpi_N, \tag{2.91}$$

where ϖ_N is defined by the formula (2.80). Repeating the argument of the proof of Lemma 1.11.2, we deduce the following.

LEMMA 2.14.1. *For any $1 \leqslant i, j \leqslant N$, we have*

$$\vartheta_M : s_{ij}(u + M/2) \rightsquigarrow \begin{vmatrix} s_{11}(u) & \dots & s_{1M}(u) & s_{1,M+j}(u) \\ \vdots & \ddots & \vdots & \vdots \\ s_{M1}(u) & \dots & s_{MM}(u) & s_{M,M+j}(u) \\ s_{M+i,1}(u) & \dots & s_{M+i,M}(u) & \boxed{s_{M+i,M+j}(u)} \end{vmatrix}.$$

Note that due to the assumption that the matrix $G^{(M)}$ is nonsingular, the quasideterminant is well-defined; see Definition 1.10.1. We can also express the images of $s_{ij}(u)$ under ϑ_M in terms of Sklyanin minors; cf. Corollary 1.11.4.

LEMMA 2.14.2. *For any $1 \leqslant i, j \leqslant N$, we have*

$$\vartheta_M : s_{ij}(u) \rightsquigarrow \big[s^{\,1 \dots M}_{\,1 \dots M}(u + M/2)\big]^{-1} \cdot s^{\,1 \dots M, M+i}_{\,1 \dots M, M+j}(u + M/2).$$

PROOF. Applying ϖ_N to both sides of (2.83) with $M = 0$ and using (2.84) we obtain

$$\varpi_N : s_{ij}(u) \rightsquigarrow (-1)^{N-i}\big[s^{\,1 \dots N}_{\,1 \dots N}(-u + N/2 - 1)\big]^{-1} \cdot \check{s}^{\,1 \dots N}_{\,1 \dots \hat{i} \dots N,\, j}(-u + N/2 - 1).$$

Now apply υ_M and then ϖ_{M+N} with the use of (2.82) and (2.83) to complete the argument. □

For any $i, j \in \{1 \dots, N\}$ introduce the series with coefficients in the extended twisted Yangian $\mathrm{X}_G(\mathfrak{g}_{M+N})$ by

$$s^{\sharp}_{ij}(u) = s^{\,1 \dots M, M+i}_{\,1 \dots M, M+j}(u + M/2)$$

and combine these series into the matrix $S^{\sharp}(u) = [s^{\sharp}_{ij}(u)]$. Let $G^{\sharp}$ denote the constant term component of this matrix. Equating the constant terms in the identity

$$s^{\sharp}_{ij}(u + M/2) = s^{\,1 \dots M}_{\,1 \dots M}(u + M) \cdot \vartheta_M\big(s_{ij}(u + M/2)\big),$$

provided by Lemma 2.14.2, we derive from Lemma 2.14.1 that $G^{\sharp}$ is symmetric (respectively, skew-symmetric) as soon as G is. Using our assumptions on G we could also deduce from the classical Sylvester theorem (see Section 1.12) that $G^{\sharp}$ is nonsingular. However, this will also be implied by the following analogue of the Sylvester theorem for the extended twisted Yangian.

Let $S^{(M)}(u)$ be the submatrix of $S(u)$ determined by the first M rows and columns.

THEOREM 2.14.3. *The assignment*

$$s_{ij}(u) \mapsto s^{\sharp}_{ij}(u)$$

defines an algebra homomorphism $X_{G^\sharp}(\mathfrak{g}_N) \to X_G(\mathfrak{g}_{M+N})$. *Moreover, we have the identity*

$$\operatorname{sdet} S^\sharp(u) = \operatorname{sdet} S(u+M/2) \\ \times \operatorname{sdet} S^{(M)}(u+M/2-1)\dots \operatorname{sdet} S^{(M)}(u+M/2-N+1).$$

PROOF. By Lemma 2.14.2, we have

$$\vartheta_M : s_{ij}(u) \rightsquigarrow \big[\operatorname{sdet} S^{(M)}(u+M/2)\big]^{-1} \cdot s^\sharp_{ij}(u). \tag{2.92}$$

Proposition 2.13.10 provides an expansion of $\operatorname{sdet} S^{(M)}(u)$ which contains only the series $s_{kl}(u)$ with $k,l \in \{1,\dots,M\}$ (with some shifts in u). Hence, by Corollary 2.13.7 the series $\operatorname{sdet} S^{(M)}(u)$ commutes with $s^\sharp_{ij}(v)$ for any $i,j \in \{1,\dots,N\}$. Since ϑ_M preserves the extended twisted Yangian defining relations, we can conclude that the assignment $s_{ij}(u) \mapsto s^\sharp_{ij}(u)$ defines a homomorphism.

Furthermore, applying the first relation of Theorem 2.13.9 twice, we deduce

$$\vartheta_M : s^{1\dots N}_{1\dots N}(u) \rightsquigarrow \big[\operatorname{sdet} S^{(M)}(u+M/2)\big]^{-1} \cdot \operatorname{sdet} S(u+M/2).$$

The expansion of the Sklyanin minor $s^{1\dots N}_{1\dots N}(u)$ in terms of the matrix elements has the form of a linear combination of the products

$$s_{a_1b_1}(u)\, s_{a_2b_2}(u-1)\dots s_{a_Nb_N}(u-N+1), \qquad a_i, b_i \in \{1,\dots,N\},$$

where the coefficients are rational functions in u. Therefore, using the relation (2.92) we derive the desired formula. □

The Sklyanin comatrix $\widehat{S}(u)$ for the extended twisted Yangian can be defined in the same way as for the twisted Yangian; see Definition 2.7.6. Formulas for its entries $\widehat{s}_{ij}(u)$ are given by Proposition 2.7.7.

Denote by $\widehat{\sigma}_{ij}(u)$ the entries of the Sklyanin comatrix corresponding to the matrix $S^\sharp(u)$.

PROPOSITION 2.14.4. *For any* $i,j \in \{1,\dots,N\}$ *we have the relation*

$$\widehat{\sigma}_{ij}(u) = \widehat{s}_{ij}(u+M/2) \\ \times \operatorname{sdet} S^{(M)}(u+M/2-1)\dots \operatorname{sdet} S^{(M)}(u+M/2-N+2).$$

PROOF. We use the same argument as in the proof of Theorem 2.14.3. Let us calculate the image of the auxiliary minor $\check{s}^{1\dots N}_{1\dots\widehat{i}\dots N,j}(u)$ under the map ϑ_M in two different ways; see (2.91). Using (2.83) we get

$$\varpi_N : \check{s}^{1\dots N}_{1\dots\widehat{i}\dots N,j}(u) \rightsquigarrow (-1)^{N-i}\big[s^{1\dots N}_{1\dots N}(-u+N/2-1)\big]^{-1} \cdot s_{ij}(-u+N/2-1).$$

Now apply υ_M and ϖ_{M+N} using (2.82) and (2.83) to conclude that

$$\vartheta_M : \check{s}^{1\dots N}_{1\dots\widehat{i}\dots N,j}(u) \rightsquigarrow \big[\operatorname{sdet} S^{(M)}(u+M/2)\big]^{-1} \cdot \widehat{s}_{ij}(u+M/2),$$

where we have taken into account Proposition 2.7.7. On the other hand, due to Proposition 2.13.10, the expansion of the auxiliary minor $\check{s}^{1\dots N}_{1\dots\widehat{i}\dots N,j}(u)$ in terms of the matrix elements has the form of a linear combination of products

$$s_{a_1b_1}(u)\, s_{a_2b_2}(u-1)\dots s_{a_{N-1}b_{N-1}}(u-N+2), \qquad a_i, b_i \in \{1,\dots,N\},$$

the coefficients being rational functions in u. The proof is completed by the application of (2.92). □

As we will see below, the homomorphism of Theorem 2.14.3 respects the symmetry relation (2.12) in the orthogonal case, while in the symplectic case a minor correction is needed to obtain the corresponding homomorphism of the twisted Yangians. In order to treat both cases simultaneously, introduce the following notation

$$\alpha_p(u) = \begin{cases} 1 & \text{in the orthogonal case} \\ \dfrac{2u+1}{2u-p+1} & \text{in the symplectic case.} \end{cases} \tag{2.93}$$

We can now prove a quantum Sylvester theorem for the twisted Yangian. We maintain the notation used in Theorem 2.14.3.

THEOREM 2.14.5. *The mapping*

$$s_{ij}(u) \mapsto \alpha_{-M}(u)\, s^{\sharp}_{ij}(u)$$

defines an algebra homomorphism $\mathrm{Y}_{G^{\sharp}}(\mathfrak{g}_N) \to \mathrm{Y}_G(\mathfrak{g}_{M+N})$. *Moreover,*

$$\begin{aligned} \operatorname{sdet}\big[\alpha_{-M}(u)\, S^{\sharp}(u)\big] &= \alpha(u)\cdot \operatorname{sdet} S(u+M/2) \\ &\times \operatorname{sdet} S^{(M)}(u+M/2-1)\dots \operatorname{sdet} S^{(M)}(u+M/2-N+1), \end{aligned}$$

where

$$\alpha(u) = \alpha_{-M}(u)\,\alpha_{-M}(u-1)\dots\alpha_{-M}(u-N+1).$$

PROOF. Using Propositions 2.3.1 and 2.12.3 we find that both the quaternary and symmetry relations (2.11) and (2.12) in $\mathrm{Y}(\mathfrak{g}_N)$ are preserved by the transformation

$$\varpi^*_N : S(u) \rightsquigarrow \alpha_N(u)\, \widehat{S}(-u+N/2-1); \tag{2.94}$$

cf. (2.80). Then by Corollary 2.5.4 and (2.69), we have

$$\varpi^*_N\big(s_{ij}(u)\big) = \gamma_N(u)\, \varpi_N\big(s_{ij}(u)\big), \qquad \gamma_N(u) = \frac{\operatorname{sdet} S(u+N/2)}{\alpha_N(u+N/2)}. \tag{2.95}$$

By analogy with (2.91), consider the composition

$$\vartheta^*_M = \varpi^*_{M+N} \circ \upsilon_M \circ \varpi^*_N,$$

which obviously preserves the twisted Yangian defining relations. By Proposition 2.7.7,

$$\varpi^*_N : s_{ij}(u) \rightsquigarrow (-1)^{N-i}\, \alpha_N(u)\, \check{s}^{\,1\dots N}_{\,1\dots\widehat{i}\dots N,\,j}(-u+N/2-1).$$

Next, apply υ_M followed by ϖ^*_{M+N}, where for the latter we use (2.83) and (2.95). This gives

$$\begin{aligned} \vartheta^*_M : s_{ij}(u) \rightsquigarrow \alpha_N(u)\cdot \big[s^{\,1\dots M+N}_{\,1\dots M+N}(u+M/2)\big]^{-1} &\cdot s^{\,1\dots M,M+i}_{\,1\dots M,M+j}(u+M/2) \\ &\times \prod_{i=1}^{N-1} \gamma_{M+N}(-u+N/2-i). \end{aligned}$$

Using Corollary 2.5.4, we can rewrite this in the form

$$\vartheta^*_M : s_{ij}(u) \rightsquigarrow \alpha_{-M}(u)\,\gamma(u)\, s^{\sharp}_{ij}(u),$$

where

$$\gamma(u) = \prod_{i=2}^{N-1} \gamma_{M+N}(-u+N/2-i).$$

Due to Corollary 2.5.4, we have $\gamma_{M+N}(-u) = \gamma_{M+N}(u-1)$ which implies $\gamma(u) = \gamma(-u)$. Hence, the multiplication of the generator series $s_{ij}(u)$ by $\gamma(u)^{-1}$ preserves the twisted Yangian defining relations, thus proving the first part of the theorem.

The formula for the Sklyanin determinant of the matrix $\alpha_{-M}(u)S^{\sharp}(u)$ is immediate from Theorem 2.14.3 and the definition of sdet $S^{\sharp}(u)$. □

We complete this section by an application of the arguments used in the proofs of Theorems 2.13.9 and 2.14.5 to the calculation of the entries of the Sklyanin comatrix.

LEMMA 2.14.6. *In the twisted Yangian* $\mathrm{Y}(\mathfrak{g}_N)$ *we have*

$$A_N\, S_1(u)\, R^t_{12}(-2u+1)\dots R^t_{1N}(-2u+N-1) = \frac{1\mp 2u}{1-2u}\, A_N\, S^t_1(-u). \tag{2.96}$$

PROOF. The argument is essentially contained in the proof of Proposition 2.6.2. We have $(N-1)!\, A_N = A_N\, A'_{N-1}$, where A'_{N-1} denotes the anti-symmetrizer corresponding to the subset of indices $\{2,\dots,N\}$. By (2.79),

$$A'_{N-1}\, R^t_{12}(-2u+1)\dots R^t_{1N}(-2u+N-1) = A'_{N-1}\Big(1+\frac{Q_{12}+\dots+Q_{1N}}{2u-1}\Big).$$

Therefore, the left-hand side of (2.96) takes the form

$$A_N\, S_1(u)\Big(1+\frac{Q_{12}+\dots+Q_{1N}}{2u-1}\Big).$$

Apply this operator to a basis vector

$$v_{ij} = e_j\otimes e_1\otimes\dots\otimes\widehat{e}_i\otimes\dots\otimes e_N, \qquad i,j\in\{1,\dots,N\}.$$

The coefficient of $A_N\, v_{ii}$ will be equal to

$$\frac{2u}{2u-1}s_{ij}(u) - \frac{1}{2u-1}s_{ji}(u) = \frac{1\mp 2u}{1-2u}\, s_{ji}(-u) = \frac{1\mp 2u}{1-2u}\, s^t_{ij}(-u)$$

due to the symmetry relation (2.7). This coincides with the coefficient of $A_N\, v_{ii}$ obtained by the application of the right-hand side of (2.96) to v_{ij}. □

We can now derive a formula for the entries of the Sklyanin comatrix $\widehat{S}(u)$ in terms of the Sklyanin minors; cf. Proposition 1.9.2.

PROPOSITION 2.14.7. *For any* $i,j\in\{1,\dots,N\}$ *we have the relation in* $\mathrm{Y}(\mathfrak{g}_N)$

$$\widehat{s}_{ij}(u) = \pm\,\alpha_{2N-2}(u)\cdot(-1)^{i+j}\cdot s^{\,1\dots\widehat{i}\dots N}_{\,1\dots\widehat{j}\dots N}(-u+N-2).$$

PROOF. By Definition 2.7.6 and relation (2.88) with $M=N-2$ we have

$$A_N\,\langle S_1,\dots,S_{N-1}\rangle = A_N\,\widehat{S}_N(u)\,(R^t_{N-1,N})^{-1}\cdots(R^t_{1,N})^{-1}. \tag{2.97}$$

Using the notation (2.87) and setting $S^*(u) = \varpi^*_N\big(S(u)\big)$ we can write

$$\widehat{S}(u) = \alpha_N(u)\, S^*(u^{\circ}_N); \tag{2.98}$$

see (2.94). Hence, (2.97) becomes

$$A_N\,\langle S_1,\dots,S_{N-1}\rangle = \alpha_N(u)\, A_N\, S^*_N(u^{\circ}_N)\, R^{\circ}_{N-1,N}\cdots R^{\circ}_{1N}. \tag{2.99}$$

However, the transformation $S(u) \rightsquigarrow S^*(u)$ preserves both the quaternary and symmetry relations (2.11) and (2.12). Therefore, applying Lemma 2.14.6 we can simplify the right-hand side of (2.99) as

$$\alpha_N(u)\, A_N\, S^*_N(u^\circ_N)\, R^\circ_{N-1,N} \cdots R^\circ_{1N} = \alpha_N(u)\, \frac{1 \mp 2u^\circ_N}{1 - 2u^\circ_N}\, A_N\, S^{*t}_N(-u^\circ_N).$$

Hence, we come to the identity

$$A_N\, \langle S_1, \dots, S_{N-1}\rangle = \pm\, \alpha_{N-2}(u)\, A_N\, S^{*t}_N(-u^\circ_N).$$

Applying both sides to the basis vector $e_1 \otimes \cdots \otimes \widehat{e}_j \otimes \cdots \otimes e_N \otimes e_i$ we get

$$\pm\, \alpha_{N-2}(u)\, s^*_{ij}(u - N/2 + 1) = (-1)^{i+j}\, s^{\,1 \dots \widehat{i} \dots N}_{\,1 \dots \widehat{j} \dots N}(u),$$

where the $s^*_{ij}(u)$ denote the entries of the matrix $S^*(u)$. The argument is completed by using (2.98). □

2.15. An equivalent presentation of $\mathrm{Y}(\mathfrak{g}_N)$

The twisted Yangian $\mathrm{Y}(\mathfrak{g}_N)$ admits a presentation with modified quaternary and symmetry relations. This presentation will be more convenient for the study of representations of $\mathrm{Y}(\mathfrak{g}_N)$. As before, we fix a symmetric or alternating nonsingular bilinear form on $\mathbb{C}^N$ with the matrix G. For any $N \times N$ matrix A we will denote by A' its transpose with respect to the form, i.e.,

$$A' = G\, A^t G^{-1}. \tag{2.100}$$

Consider the R-matrix $R(u)$ defined in (1.12) and denote by $R'(u)$ its partial transpose, so that

$$R'(u) = 1 - Q' u^{-1}, \qquad Q' = G_1\, Q\, G_1^{-1} = G_2\, Q\, G_2^{-1}, \tag{2.101}$$

where Q is defined in (1.15) and the second equality for Q' follows from the relations $G_1\, Q = G_2^t\, Q$ and $Q\, G_1 = Q\, G_2^t$.

Now consider a new family of generators $\mathsf{s}_{ij}^{(r)}$ where $i, j = 1, \dots, N$ and $r = 1, 2, \dots$. Arrange them into the formal series

$$\mathsf{s}_{ij}(u) = \delta_{ij} + \mathsf{s}_{ij}^{(1)} u^{-1} + \mathsf{s}_{ij}^{(2)} u^{-2} + \dots.$$

Then combine the series into the $N \times N$ matrix $\mathcal{S}(u)$ (to be distinguished from the matrix $S(u)$) whose ij-th entry is $\mathsf{s}_{ij}(u)$.

PROPOSITION 2.15.1. *The twisted Yangian* $\mathrm{Y}(\mathfrak{g}_N)$ *is isomorphic to the algebra generated by the elements* $\mathsf{s}_{ij}^{(r)}$ *with the defining relations given by*

$$R(u-v)\, \mathcal{S}_1(u)\, R'(-u-v)\, \mathcal{S}_2(v) = \mathcal{S}_2(v)\, R'(-u-v)\, \mathcal{S}_1(u)\, R(u-v) \tag{2.102}$$

and

$$\mathcal{S}'(-u) = \mathcal{S}(u) \pm \frac{\mathcal{S}(u) - \mathcal{S}(-u)}{2u}. \tag{2.103}$$

PROOF. We claim that the map

$$S(u) \mapsto \mathcal{S}(u)\, G \tag{2.104}$$

provides the desired isomorphism. Indeed, since the map is obviously invertible, we only need to verify that the image of the matrix $S(u)$ satisfies both the quaternary relation (2.11) and symmetry relation (2.12). For the left-hand side of (2.11) we have

$$R(u-v)\,\mathcal{S}_1(u)\,G_1\,R^t(-u-v)\,\mathcal{S}_2(v)\,G_2 = R(u-v)\,\mathcal{S}_1(u)\,R'(-u-v)\,\mathcal{S}_2(v)\,G_1\,G_2. \tag{2.105}$$

Now apply (2.102) and the relation

$$R(u-v)\,G_1\,G_2 = G_2\,G_1\,R(u-v)$$

to bring (2.105) to the form

$$\mathcal{S}_2(v)\,R'(-u-v)\,\mathcal{S}_1(u)\,G_1\,G_2\,R(u-v) = \mathcal{S}_2(v)\,G_2\,R^t(-u-v)\,\mathcal{S}_1(u)\,G_1\,R(u-v),$$

which is the right-hand side of (2.11).

The symmetry relation (2.12) is also satisfied by the matrix $\mathcal{S}(u)\,G$; this is immediate from the equalities

$$\big(\mathcal{S}(u)\,G\big)^t = G^t\,\mathcal{S}^t(u) = \pm\,G\,\mathcal{S}^t(u) = \pm\,\mathcal{S}'(u)\,G$$

and the relation (2.103). □

All the results and formulas of this chapter can be accordingly modified and reformulated in terms of the realization of the twisted Yangian $\mathrm{Y}(\mathfrak{g}_N)$ provided by Proposition 2.15.1. In the rest of this section we give some of these reformulations to be used later on.

The realization of Lie algebra $\mathfrak{g}_N$ can be modified accordingly so that $\mathfrak{g}_N$ is now regarded as the subalgebra of $\mathfrak{gl}_N$ spanned by the matrix elements F_{ij} of the matrix

$$F = E - E' = E - G\,E^t\,G^{-1}$$

instead of (2.2) (we use the same notation for the generators of $\mathfrak{g}_N$ since at any moment only one realization of $\mathfrak{g}_N$ will be considered). Then the evaluation homomorphism $\mathrm{Y}(\mathfrak{g}_N) \to \mathrm{U}(\mathfrak{g}_N)$ of Proposition 2.1.2 reads as

$$\varrho_N : \mathrm{s}_{ij}(u) \mapsto \delta_{ij} + F_{ij}\Big(u \pm \frac{1}{2}\Big)^{-1}, \tag{2.106}$$

while for the embedding $\mathrm{U}(\mathfrak{g}_N) \hookrightarrow \mathrm{Y}(\mathfrak{g}_N)$ we have

$$F_{ij} \mapsto \mathrm{s}_{ij}^{(1)}. \tag{2.107}$$

The embedding $\mathrm{Y}(\mathfrak{g}_N) \hookrightarrow \mathrm{Y}(\mathfrak{gl}_N)$ is now given by

$$\mathcal{S}(u) \mapsto T(u)\,T'(-u). \tag{2.108}$$

Note that, due to Propositions 1.3.1 and 1.3.3, the transposition $T(u) \mapsto T'(u)$ is an anti-automorphism of the Yangian $\mathrm{Y}(\mathfrak{gl}_N)$.

The Sklyanin minors and auxiliary minors are defined in the same way as in Section 2.6. The defining relations of Proposition 2.15.1 imply that Proposition 2.5.1 holds in the same form, with the standard transposition t replaced by the transposition (2.100) everywhere. Then set $u_i = u - i + 1$ for $i = 1, \dots, m$. Together with Proposition 1.6.2 this implies the following counterpart of (2.44),

$$A_m\,\langle \mathcal{S}_1, \dots, \mathcal{S}_m \rangle' = \langle \mathcal{S}_m, \dots, \mathcal{S}_1 \rangle'\,A_m, \tag{2.109}$$

where the prime indicates that the transposition (2.100) is used instead of t as compared with (2.44). The Sklyanin minors are the coefficients of the expansion of this element of $(\operatorname{End}\mathbb{C}^N)^{\otimes m}\otimes \mathrm{Y}(\mathfrak{g}_N)[[u^{-1}]]$,

$$\sum e_{a_1b_1}\otimes\dots\otimes e_{a_mb_m}\otimes \mathbf{s}^{a_1\dots a_m}_{b_1\dots b_m}(u),$$

summed over the indices $a_i, b_i \in \{-n,\dots,n\}$. The Sklyanin minors $\mathbf{s}^{a_1\dots a_m}_{b_1\dots b_m}(u)$ retain the skew-symmetry property with respect to permutations of the upper or lower indices. As in Section 2.6, we define the auxiliary minors $\check{\mathbf{s}}^{a_1\dots a_m}_{b_1\dots b_{m-1},c}(u)$ by

$$A_m\,\langle \mathcal{S}_1,\dots,\mathcal{S}_{m-1}\rangle'\,R'_{1m}\cdots R'_{m-1,m} = \sum e_{a_1b_1}\otimes\dots\otimes e_{a_{m-1}b_{m-1}}\otimes e_{a_mc}\otimes \check{\mathbf{s}}^{a_1\dots a_m}_{b_1\dots b_{m-1},c}(u), \tag{2.110}$$

summed over $a_i, b_i, c\in\{1,\dots,N\}$.

The Sklyanin determinant is defined by $\operatorname{sdet}\mathcal{S}(u) = \mathbf{s}^{a_1\dots a_N}_{a_1\dots a_N}(u)$ for any permutation $(a_1,\dots,a_N)$ of the indices $(1,\dots,N)$. Equivalently, taking $m=N$ in (2.109), we get

$$A_N\,\langle \mathcal{S}_1,\dots,\mathcal{S}_N\rangle' = \langle \mathcal{S}_N,\dots,\mathcal{S}_1\rangle'\,A_N, \tag{2.111}$$

and the Sklyanin determinant $\operatorname{sdet}\mathcal{S}(u)$ is uniquely determined by the condition that the operator (2.111) equals $A_N \operatorname{sdet}\mathcal{S}(u)$.

PROPOSITION 2.15.2. *The image of the series* $\operatorname{sdet} S(u)$ *under the isomorphism* (2.104) *is given by*

$$\operatorname{sdet} S(u)\mapsto \det G\cdot \operatorname{sdet}\mathcal{S}(u).$$

PROOF. Let us consider the left-hand side of (2.37) and rewrite it by using the equalities $S_i = \mathcal{S}_i\, G_i$ and $G_i\, R^t_{ij} = R'_{ij}\, G_i$. Then for any $i=1,\dots,N-1$ we obviously have

$$S_i\, R^t_{i,i+1}\cdots R^t_{iN} = \mathcal{S}_i\, R'_{i,i+1}\cdots R'_{iN}\, G_i.$$

We thus come to the identity

$$A_N\,\langle S_1,\dots,S_N\rangle = A_N\,\langle \mathcal{S}_1,\dots,\mathcal{S}_N\rangle'\, G_1\dots G_N.$$

It remains to apply (2.111) and (2.41). □

Proposition 2.15.2 and Theorem 2.5.3 immediately imply the following.

COROLLARY 2.15.3. *We have*

$$\operatorname{sdet}\mathcal{S}(u) = \alpha_N(u)\operatorname{qdet} T(u)\operatorname{qdet} T(-u+N-1),$$

where $\alpha_N(u)$ *is defined in* (2.93). □

We define the Sklyanin comatrix $\widehat{\mathcal{S}}(u)$ corresponding to $\mathcal{S}(u)$ by the formula

$$\widehat{\mathcal{S}}(u)\,\mathcal{S}(u-N+1) = \operatorname{sdet}\mathcal{S}(u); \tag{2.112}$$

cf. Definition 2.7.6. The following corollary is implied by Proposition 2.15.2.

COROLLARY 2.15.4. *The image of the Sklyanin comatrix under the isomorphism* (2.104) *is given by*

$$\widehat{S}(u)\mapsto \widehat{G}\,\widehat{\mathcal{S}}(u),$$

where $\widehat{G}$ *is the comatrix of* G *defined by* $\widehat{G}\,G=\det G$. □

We have the following consequence of Proposition 2.12.3.

COROLLARY 2.15.5. *The mapping*

$$\mathcal{S}(u) \mapsto \alpha_N(u)\,\widehat{\mathcal{S}}(-u+N/2-1)$$

defines an automorphism of the algebra $\mathrm{Y}(\mathfrak{g}_N)$.

PROOF. It suffices to use the isomorphism (2.104) and apply Proposition 2.15.2. □

We will denote by $\mathbf{y}(u)$ the image of the series $y(u)$ defined in Section 2.11 under the isomorphism (2.104). The following formula for the inverse series $\mathbf{y}(u)^{-1}$ follows from the second relation of Proposition 2.11.1,

$$\mathbf{y}(u)^{-1} = \frac{2u+1}{N(2u\pm 1)} \times \operatorname{tr}\Big\{\Big(\frac{2u-N}{2u-N+1}\mathcal{S}'(-u) \pm \frac{1}{2u-N+1}\mathcal{S}(-u)\Big)\mathcal{S}^{-1}(u-N)\Big\}. \tag{2.113}$$

Using Theorem 2.11.2 and Proposition 2.15.2, we obtain the following version of the quantum Liouville formula.

COROLLARY 2.15.6. *We have the relation*

$$\mathbf{y}(u) = \frac{\operatorname{sdet}\mathcal{S}(u-1)}{\operatorname{sdet}\mathcal{S}(u)}.$$

Dropping the symmetry relation (2.103) in Proposition 2.15.1 we obtain an equivalent presentation of the extended twisted Yangian $\mathrm{X}(\mathfrak{g}_N)$. The results of Section 2.13 can be easily carried over to this presentation by using the isomorphism (2.104). Let $\mathcal{S}(u)$ be the generator matrix of $\mathrm{X}(\mathfrak{g}_N)$ so that the defining relations for $\mathrm{X}(\mathfrak{g}_N)$ are given by the quaternary relation (2.102). The following proposition is verified in the same way as the corresponding property of the map (1.31); cf. (2.80).

PROPOSITION 2.15.7. *The mapping*

$$\varpi_N : \mathcal{S}(u) \mapsto \mathcal{S}^{-1}\big(-u-N/2\big)$$

defines an involutive automorphism of $\mathrm{X}(\mathfrak{g}_N)$. □

2.16. Examples

1. The converse of Lemma 2.4.1 is also true: if G is an arbitrary complex $N\times N$ matrix satisfying the relation (2.20), then G is either symmetric or skew-symmetric.

2. Given a symmetric or skew-symmetric nonsingular matrix G, for the corresponding Lie algebra $\mathfrak{g}_N$ introduce the parameter $\kappa = N/2 \mp 1$. Consider the R-matrix

$$R(u) = 1 - \frac{P}{u} + \frac{Q'}{u-\kappa},$$

where Q' is defined in (2.101). This R-matrix is a solution of the Yang–Baxter equation (1.17). The *extended Drinfeld Yangian* $\mathrm{X}^D(\mathfrak{g}_N)$ is generated by elements $t_{ij}^{(r)}$, where $1 \leqslant i,j \leqslant N$ and r runs over positive integers, subject to the relations

$$R(u-v)\,T_1(u)\,T_2(v) = T_2(v)\,T_1(u)\,R(u-v), \tag{2.114}$$

where

$$T(u)=\sum_{i,j=1}^{N} e_{ij}\otimes t_{ij}(u)$$

and

$$t_{ij}(u)=\delta_{ij}+\sum_{r=1}^{\infty} t_{ij}^{(r)}\,u^{-r}.$$

As with the twisted Yangians, the algebras $\mathrm{X}^D(\mathfrak{g}_N)$ corresponding to different symmetric (respectively, skew-symmetric) matrices G are isomorphic to each other; cf. Corollary 2.3.2.

The extended Drinfeld Yangian $\mathrm{X}^D(\mathfrak{g}_N)$ is a Hopf algebra with the comultiplication

$$\Delta: t_{ij}(u)\mapsto\sum_{k=1}^{N} t_{ik}(u)\otimes t_{kj}(u),$$

the antipode $\mathrm{S}: T(u)\mapsto T(u)^{-1}$, and the counit $\varepsilon: T(u)\mapsto 1$.

Taking the residue of both sides of (2.114) at $u-v=\kappa$ we obtain

$$Q'\,T_1(u+\kappa)\,T_2(u)=T_2(u)\,T_1(u+\kappa)\,Q'.$$

Since Q'/N is a projection operator in $\mathbb{C}^N\otimes\mathbb{C}^N$ with a one-dimensional image, the expression on each side of this relation equals Q' times a series $z(u)$ with coefficients in $\mathrm{X}^D(\mathfrak{g}_N)$. All coefficients of $z(u)$ belong to the center of $\mathrm{X}^D(\mathfrak{g}_N)$. Moreover, they are algebraically independent and generate the center. The *Drinfeld Yangian* $\mathrm{Y}^D(\mathfrak{g}_N)$ is the quotient of the extended Yangian $\mathrm{X}^D(\mathfrak{g}_N)$ by the ideal generated by the coefficients of the series $z(u)$, i.e., by the relations $z(u)=1$. Also, $\mathrm{Y}^D(\mathfrak{g}_N)$ is isomorphic to a subalgebra of $\mathrm{X}^D(\mathfrak{g}_N)$.

3. We have algebra isomorphisms

$$\mathrm{X}^D(\mathfrak{sp}_2)\cong\mathrm{Y}(\mathfrak{gl}_2),\qquad \mathrm{X}^D(\mathfrak{o}_3)\cong\mathrm{Y}(\mathfrak{gl}_2),\qquad \mathrm{X}^D(\mathfrak{o}_4)\cong\mathrm{Y}(\mathfrak{sl}_2)\otimes\mathrm{Y}(\mathfrak{gl}_2).$$

4. Fix a partition of the number N into a sum of two nonnegative integers, $N=k+l$. Denote by D the diagonal matrix $D=\mathrm{diag}(\varepsilon_1,\dots,\varepsilon_N)$ where $\varepsilon_i=1$ for $i\leqslant k$, and $\varepsilon_i=-1$ for $i>k$. The *reflection equation algebra* $\mathcal{B}(N,l)$ is generated by elements $b_{ij}^{(r)}$ where r runs over positive integers and i and j satisfy $1\leqslant i,j\leqslant N$. The defining relations are given by the *reflection equation*

$$R(u-v)\,B_1(u)\,R(u+v)\,B_2(v)=B_2(v)\,R(u+v)\,B_1(u)\,R(u-v)$$

together with the *unitary condition* $B(u)B(-u)=1$, where

$$B(u)=\sum_{i,j=1}^{N} e_{ij}\otimes b_{ij}(u)$$

and

$$b_{ij}(u)=\sum_{r=0}^{\infty} b_{ij}^{(r)}u^{-r},\qquad b_{ij}^{(0)}=\delta_{ij}\,\varepsilon_i.$$

5. The mapping $B(u)\mapsto T(u)\,D\,T^{-1}(-u)$ defines an embedding of the reflection equation algebra $\mathcal{B}(N,l)$ into the Yangian $\mathrm{Y}(\mathfrak{gl}_N)$. Similar to the twisted Yangians, the subalgebra $\mathcal{B}(N,l)$ of $\mathrm{Y}(\mathfrak{gl}_N)$ is a left coideal with respect to the coproduct. An analogue of the Sklyanin determinant $\mathrm{sdet}\,B(u)$ for the algebra

$\mathcal{B}(N,l)$ can be defined along the lines of Section 2.5. It coincides with the product of quantum determinants

$$\operatorname{qdet} T(u) \operatorname{qdet} T(-u+N-1)^{-1}$$

up to a scalar factor. The coefficients of $\operatorname{sdet} B(u)$ generate the center of $\mathcal{B}(N,l)$.

Open problem: Find an explicit formula for $\operatorname{sdet} B(u)$ in terms of the $b_{ij}(u)$; cf. Theorem 2.7.2.

6. The *extended reflection equation algebra* $\widetilde{\mathcal{B}}(N,l)$ is defined in the same way as the algebra $\mathcal{B}(N,l)$ (Example 4), but with the unitary condition dropped. The following algebras are isomorphic: $\mathrm{X}(\mathfrak{sp}_2) \cong \widetilde{\mathcal{B}}(2,0)$ and $\mathrm{X}(\mathfrak{o}_2) \cong \widetilde{\mathcal{B}}(2,1)$; see Definition 2.13.1.

7. Let N and k be positive integers. The *signless Stirling number of the first kind* $c\,(N,k)$ is defined as the number of elements of the symmetric group $\mathfrak{S}_N$ with exactly k cycles. Equivalently, the numbers $c\,(N,k)$ can be defined by

$$\sum_{k=1}^{N} c\,(N,k)\, x^k = x(x+1)\dots(x+N-1).$$

The *Boolean poset* B_n consists of the 2^n subsets of the set $\{1,2,\dots,n\}$. One defines $S \leqslant T$ if $S \subset T$ as sets. The *Bruhat order* on the symmetric group $\mathfrak{S}_N$ is defined as follows: if $q=(q_1,\dots,q_N)\in\mathfrak{S}_N$, then a reduction of q is a permutation obtained from q by interchanging two elements q_i and q_j, where $i<j$ and $q_i>q_j$. Then $p \leqslant q$ with respect to the Bruhat order if p can be obtained from q by a sequence of reductions. We may regard (2.53) as the map φ_N from $\mathfrak{S}_N$ to $\mathfrak{S}_{N-1}$. Then each fiber of the map is an interval in $\mathfrak{S}_N$ with respect to the Bruhat order. Moreover, each of these intervals is isomorphic to B_k as a poset. For any k the number of intervals isomorphic to B_k is $c\,(N-1,k)$.

8. Twisted (or nonstandard) deformations of the classical universal enveloping algebras $\mathrm{U}(\mathfrak{g}_N)$ can be introduced as subalgebras of $\mathrm{U}_q(\mathfrak{gl}_N)$; see Example 1.15.1. Let $G=1$ be the identity matrix in the orthogonal case and let

$$G = q\sum_{k=1}^{n} e_{2k-1,2k} - \sum_{k=1}^{n} e_{2k,2k-1}$$

in the symplectic case. The *twisted quantized enveloping algebra* $\mathrm{U}'_q(\mathfrak{o}_N)$ is the subalgebra of $\mathrm{U}_q(\mathfrak{gl}_N)$ generated by the matrix elements of the matrix $S=T\,G\,\overline{T}^{\,t}$. Equivalently, $\mathrm{U}'_q(\mathfrak{o}_N)$ is the algebra generated by the matrix elements s_{ij} of the matrix S subject only to the relations written in the form of a quaternary-type relation

$$R\,S_1\,R^{t_1}S_2 = S_2\,R^{t_1}S_1\,R, \tag{2.115}$$

where S is considered to be lower triangular with ones on the diagonal.

Similarly, the *twisted quantized enveloping algebra* $\mathrm{U}'_q(\mathfrak{sp}_{2n})$ is the subalgebra of $\mathrm{U}_q(\mathfrak{gl}_{2n})$ generated by the matrix elements s_{ij} of the matrix $S=T\,G\,\overline{T}^{\,t}$ and by the elements $s^{-1}_{i,i+1}=q^{-1}\bar{t}_{ii}\,t_{i+1,i+1}$ for each odd $i=1,3,\dots,2n-1$.

Equivalently, $\mathrm{U}'_q(\mathfrak{sp}_{2n})$ is the algebra generated by the matrix elements s_{ij} of the matrix S having a block lower triangular form with n diagonal blocks of size

2×2, and additional elements $s_{i,i+1}^{-1}$ with odd $i = 1, 3, \ldots, 2n-1$ subject only to the following relations. These are given by (2.115) together with

$$s_{i,i+1}\, s_{i,i+1}^{-1} = s_{i,i+1}^{-1}\, s_{i,i+1} = 1$$

and

$$s_{i+1,i+1}\, s_{ii} - q^2\, s_{i+1,i} s_{i,i+1} = q^3$$

for each $i = 1, 3, \ldots, 2n-1$.

In both cases the entries of another matrix $\overline{S} = \overline{T} G T^t$ also belong to the subalgebra $\mathrm{U}_q'(\mathfrak{g}_N)$. This subalgebra possesses a coideal property with respect to the coproduct in $\mathrm{U}_q(\mathfrak{gl}_N)$.

9. The defining relations of the algebra $\mathrm{U}_q'(\mathfrak{o}_N)$ can be written in the equivalent form

$$\begin{aligned}
s_{ij}\, s_{kl} - s_{kl}\, s_{ij} &= 0 && \text{if} \quad i > j > k > l\\
s_{ij}\, s_{kl} - s_{kl}\, s_{ij} &= 0 && \text{if} \quad i > k > l > j\\
s_{ij}\, s_{kl} - s_{kl}\, s_{ij} &= (q - q^{-1})\,(s_{kj} s_{il} - s_{ik} s_{jl}) && \text{if} \quad i > k > j > l\\
q\, s_{ij}\, s_{jl} - s_{jl}\, s_{ij} &= (q - q^{-1})\, s_{il} && \text{if} \quad i > j > l\\
q\, s_{ij}\, s_{il} - s_{il}\, s_{ij} &= (q - q^{-1})\, s_{lj} && \text{if} \quad i > l > j\\
q\, s_{ij}\, s_{kj} - s_{kj}\, s_{ij} &= (q - q^{-1})\, s_{ki} && \text{if} \quad k > i > j.
\end{aligned}$$

Suppose now that the nonzero parameter q satisfies $q^2 \neq 1$. One more presentation of $\mathrm{U}_q'(\mathfrak{o}_N)$ can be obtained by using the elements

$$I_i = \frac{s_{i+1,i}}{q - q^{-1}}, \qquad i = 1, \ldots, N-1.$$

The algebra $\mathrm{U}_q'(\mathfrak{o}_N)$ is generated by the elements $I_1, \ldots, I_{N-1}$ subject only to the relations

$$\begin{aligned}
I_i\, I_{i+1}^2 - (q + q^{-1})\, I_{i+1}\, I_i\, I_{i+1} + I_{i+1}^2\, I_i &= -q^{-1}\, I_i,\\
I_i^2\, I_{i+1} - (q + q^{-1})\, I_i\, I_{i+1}\, I_i + I_{i+1}\, I_i^2 &= -q^{-1}\, I_{i+1},
\end{aligned}$$

for $i = 1, \ldots, N-2$ (the *Serre type relations*), and

$$I_i\, I_j = I_j\, I_i, \qquad |i - j| > 1.$$

10. The braid group $\mathfrak{B}_N$ (see Example 1.15.2) acts on the algebra $\mathrm{U}_q'(\mathfrak{o}_N)$ by automorphisms. For the action of β_i with $i \in \{1, \ldots, N-1\}$ we have

$$\begin{aligned}
\beta_i : I_{i+1} &\mapsto q\, I_{i+1}\, I_i - I_i\, I_{i+1}\\
I_{i-1} &\mapsto I_i\, I_{i-1} - q\, I_{i-1}\, I_i\\
I_i &\mapsto -I_i\\
I_k &\mapsto I_k \qquad \text{if} \quad k \neq i-1, i, i+1.
\end{aligned}$$

In terms of the generators s_{kl}, this action can be written in the form

$$\begin{aligned}
&\beta_i : s_{i+1,i} \mapsto -s_{i+1,i}\\
&\quad s_{ik} \mapsto q\, s_{i+1,k} - q\, s_{i+1,i}\, s_{ik}, && s_{i+1,k} \mapsto s_{ik}, && \text{if} \quad k \leqslant i-1\\
&\quad s_{li} \mapsto q^{-1}\, s_{l,i+1} - s_{li}\, s_{i+1,i}, && s_{l,i+1} \mapsto s_{li}, && \text{if} \quad l \geqslant i+2\\
&\quad s_{kl} \mapsto s_{kl} && \text{in all remaining cases.}
\end{aligned}$$

11. Here we regard q as a formal variable and consider $\mathrm{U}'_q(\mathfrak{o}_N)$ as an algebra over the field of rational functions $\mathbb{C}(q)$. Set $\mathcal{A} = \mathbb{C}[q, q^{-1}]$ and introduce the $\mathcal{A}$-subalgebra $\mathrm{U}'_{\mathcal{A}}$ of $\mathrm{U}'_q(\mathfrak{o}_N)$ generated by the elements s_{ij}. Then we have an isomorphism

$$\mathrm{U}'_{\mathcal{A}} \otimes_{\mathcal{A}} \mathbb{C} \cong \mathcal{P}_N, \tag{2.116}$$

where the action of $\mathcal{A}$ on $\mathbb{C}$ is defined via the evaluation $q = 1$ and $\mathcal{P}_N$ denotes the algebra of polynomials in independent variables a_{ij} with $1 \leqslant j < i \leqslant N$. The elements a_{ij} are respective images of the s_{ij} under the isomorphism (2.116). Furthermore, the algebra $\mathcal{P}_N$ is equipped with the Poisson bracket $\{\cdot,\cdot\}$ defined by

$$\{f, h\} = \frac{\widetilde{f}\,\widetilde{h} - \widetilde{h}\,\widetilde{f}}{1-q}\Big|_{q=1},$$

where $f, h \in \mathcal{P}_N$ and $\widetilde{f}$ and $\widetilde{h}$ are elements of $\mathrm{U}'_{\mathcal{A}}$ whose images in $\mathcal{P}_N$ under the specialization $q = 1$ coincide with f and h, respectively. Using the defining relations of Example 9 we get

$$\begin{aligned}
\{a_{ij}, a_{kl}\} &= 0 && \text{if} \quad i > j > k > l\\
\{a_{ij}, a_{kl}\} &= 0 && \text{if} \quad i > k > l > j\\
\{a_{ij}, a_{kl}\} &= 2\,(a_{ik}a_{jl} - a_{kj}a_{il}) && \text{if} \quad i > k > j > l\\
\{a_{ij}, a_{jl}\} &= a_{ij}a_{jl} - 2\,a_{il} && \text{if} \quad i > j > l\\
\{a_{ij}, a_{il}\} &= a_{ij}a_{il} - 2\,a_{lj} && \text{if} \quad i > l > j\\
\{a_{ij}, a_{kj}\} &= a_{ij}a_{kj} - 2\,a_{ki} && \text{if} \quad k > i > j.
\end{aligned}$$

The braid group $\mathfrak{B}_N$ acts on the algebra $\mathcal{P}_N$ by

$$\begin{aligned}
\beta_i :\ & a_{i+1,i} \mapsto -a_{i+1,i}\\
& a_{ik} \mapsto a_{i+1,k} - a_{i+1,i}\,a_{ik}, && a_{i+1,k} \mapsto a_{ik}, && \text{if} \quad k \leqslant i-1\\
& a_{li} \mapsto a_{l,i+1} - a_{li}\,a_{i+1,i}, && a_{l,i+1} \mapsto a_{li}, && \text{if} \quad l \geqslant i+2\\
& a_{kl} \mapsto a_{kl} && \text{in all remaining cases,}
\end{aligned}$$

where $i = 1, \dots, N-1$. The Poisson bracket on $\mathcal{P}_N$ is invariant under this action.

12. Similar to Example 11, consider the extended twisted quantized enveloping algebra $\hat{\mathrm{U}}'_q(\mathfrak{sp}_{2n})$ which is an associative algebra over $\mathbb{C}(q)$ generated by elements s_{ij}, $i, j \in \{1, \dots, 2n\}$. The defining relations are given by (2.115) together with the conditions that $s_{ij} = 0$ for $j = i+1$ with even i, and for $j \geqslant i+2$ and all values of i. As $q \to 1$ the algebra $\hat{\mathrm{U}}'_q(\mathfrak{sp}_{2n})$ specializes to the algebra $\mathcal{P}_{2n}$ of polynomials in $2n^2 + 2n$ variables, which we denote by a_{ij}. The algebra $\mathcal{P}_{2n}$ possesses a Poisson bracket defined by

$$\begin{aligned}
\{a_{ij}, a_{kl}\} &= \big(\delta_{ik} + \delta_{jk} - \delta_{il} - \delta_{jl}\big)\, a_{ij}\, a_{kl}\\
&\quad - 2\,\big(\delta_{l<j} - \delta_{i<k}\big)\, a_{kj}\, a_{il} - 2\,\delta_{l<i}\, a_{ki}\, a_{lj} + 2\,\delta_{j<k}\, a_{ik}\, a_{jl},
\end{aligned}$$

where $\delta_{i<j}$ equals 1 if $i < j$, and 0 otherwise.

Open problem: Find an analogue of the braid group action of Example 11 on the (extended) twisted quantized enveloping algebra or the Poisson algebra $\mathcal{P}_{2n}$.

13. Using the same matrix G as in Example 8, define the *twisted q-Yangian* $\mathrm{Y}'_q(\mathfrak{g}_N)$ as the subalgebra of $\mathrm{U}_q(\widehat{\mathfrak{gl}}_N)$ (see Example 1.15.3) generated by the coefficients $s_{ij}^{(r)}$ of the matrix elements of the matrix $S(u) = T(u)\, G\, \overline{T}(u^{-1})^t$, together

with the inverse elements $s_{i,i+1}^{(0)\,-1}$, $i = 1, 3, \ldots, 2n-1$ in the symplectic case. The coefficients of the matrix elements of the matrix $\overline{S}(u) = \overline{T}(u)\, G\, T(u^{-1})^t$ belong to the same subalgebra.

Equivalently, $\mathrm{Y}_q'(\mathfrak{g}_N)$ is the algebra generated by the coefficients of the matrix elements of the matrix $S(u)$, together with additional elements $s_{i,i+1}^{(0)\,-1}$, $i = 1, 3, \ldots, 2n-1$ in the symplectic case, subject to the defining relations

$$R(u,v)\, S_1(u)\, R^{t_1}(u^{-1},v)\, S_2(v) = S_2(v)\, R^{t_1}(u^{-1},v)\, S_1(u)\, R(u,v)$$

with the assumption that the constant term $S^{(0)}$ of the series $S(u)$ satisfies the same conditions as the matrix S in Example 8.

14. The mappings

$$S(u) \mapsto S + q^{-1} u^{-1} \overline{S} \qquad \text{and} \qquad S(u) \mapsto S + q\, u^{-1} \overline{S}$$

define surjective homomorphisms $\mathrm{Y}_q'(\mathfrak{o}_N) \to \mathrm{U}_q'(\mathfrak{o}_N)$ and $\mathrm{Y}_q'(\mathfrak{sp}_{2n}) \to \mathrm{U}_q'(\mathfrak{sp}_{2n})$, respectively.

15. The Sklyanin determinant $\operatorname{sdet} S(u)$ for the algebra $\mathrm{Y}_q'(\mathfrak{g}_N)$ can be constructed by analogy with the case of the twisted Yangian by using the q-antisymmetrizer of Example 1.15.6. The series $\operatorname{sdet} S(u)$ coincides with the product of quantum determinants

$$c(u) = \operatorname{qdet} T(u)\, \operatorname{qdet} \overline{T}(q^{2N-2} u^{-1}),$$

up to a factor which is a rational function in u. Moreover, there exists an analogue of Theorem 2.7.2 for the twisted q-Yangian $\mathrm{Y}_q'(\mathfrak{o}_N)$. Assuming $N = 2n$ or $2n+1$, the formula reads

$$c(u) = \sum_{p \in \mathfrak{S}_N} (-q)^{-l(p)+l(p')}\, \bar{s}^{\,t}_{p(1),p'(1)}(u^{-1}) \ldots \bar{s}^{\,t}_{p(n),p'(n)}(q^{2n-2} u^{-1})$$
$$\times\, s_{p(n+1),p'(n+1)}(q^{-2n} u) \ldots s_{p(N),p'(N)}(q^{-2N+2} u),$$

where the $\bar{s}^{\,t}_{ij}(u)$ denote the matrix elements of the transposed matrix $\overline{S}^{\,t}(u)$ and $p \mapsto p'$ is the map (2.53). The coefficients of $c(u)$ belong to the center of $\mathrm{Y}_q'(\mathfrak{o}_N)$.

Open problem: Find an analogous expression for the series $c(u)$ in the symplectic case.

16. Let G be a nonsingular block-diagonal matrix such that the upper-left block is a symmetric $m \times m$ matrix, while the lower-right block is a skew-symmetric $2n \times 2n$ matrix. The *twisted Yangian* $\mathrm{Y}(\mathfrak{osp}_{m|2n})$ for the orthosymplectic Lie superalgebra $\mathfrak{osp}_{m|2n}$ can be defined by using the R-matrix defined in Example 1.15.9 where n is replaced by $2n$. The super-analogues of the quaternary and symmetry relations take the form

$$R(u-v)\, S_1(u)\, R^{t_1}(-u-v)\, S_2(v) = S_2(v)\, R^{t_2}(-u-v)\, S_1(u)\, R(u-v)$$

and

$$J\, S^t(-u) = S(u) + \frac{S(u) - S(-u)}{2u},$$

where J is the diagonal matrix which has the first m diagonal entries equal to 1 and the remaining $2n$ entries equal to -1; t is the matrix super-transposition such that $(A^t)_{ij} = A_{ji}(-1)^{\bar{\imath}(\bar{\jmath}+1)}$. There exists an evaluation homomorphism $\mathrm{Y}(\mathfrak{osp}_{m|2n}) \to \mathrm{U}(\mathfrak{osp}_{m|2n})$.

Open problem: Construct an analogue of the quantum Berezinian for the twisted Yangian $Y(\mathfrak{osp}_{m|2n})$.

Bibliographical notes

2.1–2.5. The twisted Yangians were introduced by Olshanski [**376**]. In a comment in [**317**] he acknowledges an invaluable contribution by M. Nazarov which led to the discovery of the fundamental relation (2.11). The paper [**376**] also contains a sketch of the basic properties of the twisted Yangians, including the construction of an analogue of the quantum determinant. It was named the 'Sklyanin determinant' in the subsequent paper by Molev, Nazarov and Olshanski [**316**] in recognition of the pioneering work of E. K. Sklyanin [**418**]. Our exposition basically follows Olshanski [**376**] and Molev, Nazarov and Olshanski [**316**], although those papers deal with a particular matrix G and the realization of the twisted Yangian given in Section 2.15.

2.6–2.7. The 'short' formula for the Sklyanin determinant (Theorem 2.7.2) was proved in the author's paper [**296**] where a particular matrix G was used.

2.8–2.10. These results were outlined in Olshanski [**376**] and detailed proofs were given in [**316**].

2.11. Theorem 2.11.2 is due to Nazarov; its proof was given in [**316**].

2.12. The quasideterminant factorization of the Sklyanin determinant is due to the author [**297, 300**].

2.13. Theorems 2.13.3 and 2.13.4 are due to Olshanski [**376**], [**316**].

2.14. The quantum Sylvester theorem for the twisted Yangian was proved in the author's paper [**311**].

2.16. The R-matrix presentation of the Drinfeld Yangians of series B, C and D introduced in Example 2 was studied in detail in the papers by Arnaudon *et al.* [**10, 14**]. In particular, the explicit isomorphisms of Example 3 were constructed in [**14**]. The R-matrices associated with the classical Lie algebras appear in A. Zamolodchikov and Al. Zamolodchikov [**472**], and Ogievetsky, Reshetikhin and Wiegmann [**364**]. The reflection equation algebras discussed in Example 4 were first introduced by Sklyanin [**418**]. They were also studied in the physics literature in connection with the integrable models with boundary conditions and the nonlinear Schrödinger equation; see e.g. Kulish and Sklyanin [**252**], Kulish, Sasaki and Schwiebert [**253**], Kuznetsov, Jørgensen and Christiansen [**257**], Liguori, Mintchev and Zhao [**271**], and Mintchev, Ragoucy, Sorba and Zaugg [**291, 292**]. The action of such algebras on spaces of hypergeometric functions was studied by Koornwinder and Kuznetsov [**240**]. Coideal subalgebras in the Drinfeld Yangians were introduced by Delius, MacKay and Short [**85**] in connection with the principal chiral models with boundaries. These subalgebras are defined in terms of the presentation of the Yangian with a finite set of generators as in Drinfeld [**97**]. A different R-matrix presentation of coideal subalgebras in the (super)-Yangian is given by Arnaudon, Avan, Crampé, Frappat and Ragoucy [**10**]. Field theoretical applications of the coideal subalgebras in the quantum affine algebras have been studied by Delius and MacKay [**84**]. A general universal R-matrix approach to the reflection equation and twisted Yangians was developed by Mudrov [**326**]. Examples 5 and 6 are due to Molev and Ragoucy [**319**]. Ragoucy [**396**] established a relationship between the twisted Yangians and folded $\mathcal{W}$-algebras. The combinatorial properties of the map (2.53) described in Example 7 were observed by A. Lascoux. Proofs can be found in [**300**]. The twisted (or nonstandard) deformations of the classical enveloping algebras were introduced independently in the orthogonal case by Gavrilik and Klimyk [**127**] (in terms of the I_i as in Example 9) and by Nelson and Regge [**354**] (in terms of the s_{ij} as in Example 9). The R-matrix presentation of Example 8 in both the orthogonal and symplectic case is due to Noumi [**357**]. In the latter work these deformations were considered in the context of the quantum analogues of symmetric spaces; see also Noumi and Sugitani [**358**]. Automorphisms of both the algebra $\mathrm{U}'_q(\mathfrak{o}_N)$ and the Poisson bracket on $\mathcal{P}_N$ (Example 10) were given by Nelson and Regge [**355, 356**], although the explicit group relations between them were discussed only in some low-dimensional cases. The formulas of Example 10 were produced independently by Chekhov [**69**] and Molev and Ragoucy [**320**]. The Poisson bracket on $\mathcal{P}_N$ (Example 11) was known to Nelson and Regge [**354**]. It was re-discovered by Dubrovin [**106**] and Ugaglia [**450**] and by Bondal [**39, 40**]; see also Boalch [**38**] and Xu [**471**]. The action of the braid group on $\mathcal{P}_N$ was given by Dubrovin [**106**] and Bondal [**39**]. A quantization of this Poisson algebra leading to the algebra $\mathrm{U}'_q(\mathfrak{o}_N)$ was constructed by Ciccoli and Gavarini [**78**] in the context of the general "quantum quality principle"; see also Gavarini [**124**]. The algebra $\mathrm{U}'_q(\mathfrak{o}_N)$ appears as the symmetry algebra for the q-oscillator representation of the quantized enveloping algebra $\mathrm{U}_q(\mathfrak{sp}_{2n})$; see Noumi, Umeda and Wakayama [**360, 361**], and Umeda and Wakayama [**457**]. Different families of Casimir elements for these algebras were constructed by Noumi, Umeda and Wakayama [**361**], Gavrilik and Iorgov [**126**], and Molev, Ragoucy and Sorba [**321**]. A general description of the coideal subalgebras of $\mathrm{U}_q(\mathfrak{g})$ associated with an arbitrary irreducible symmetric pair $(\mathfrak{g}, \mathfrak{k})$ was given by Letzter [**267**]. Example 12 is due to Molev and Ragoucy [**320**], where a solution to the open problem was given for $n = 2$. Examples 13–15 are due to Molev, Ragoucy and Sorba [**321**]. The twisted Yangians of Example 16, their representations and connections to $\mathcal{W}$-algebras were studied by Briot and Ragoucy [**48, 49**].

CHAPTER 3

Irreducible representations of $\mathrm{Y}(\mathfrak{gl}_N)$

In this chapter we prove the classification theorems of Drinfeld and Tarasov for finite-dimensional irreducible representations of the Yangian $\mathrm{Y}(\mathfrak{gl}_N)$. As in the representation theory of reductive Lie algebras, these representations are parameterized by their "highest weights" with respect to a commutative Cartan-type subalgebra in $\mathrm{Y}(\mathfrak{gl}_N)$ whose role is played by the Gelfand–Tsetlin subalgebra H_N; see Section 1.13. We begin with constructing a presentation of the Yangian where the generators of H_N comprise a subset of the total set of generators of $\mathrm{Y}(\mathfrak{gl}_N)$.

3.1. Drinfeld presentation of the Yangian

Consider the series $h_i(u)$, $e_{ij}(u)$ and $f_{ji}(u)$ which occur in the Gauss decomposition of the matrix $T(u)$; see Theorem 1.11.6. Let us set

$$e_i(u) = e_{i,i+1}(u) \qquad \text{and} \qquad f_i(u) = f_{i+1,i}(u).$$

Then relations (1.92) can be written as

$$\begin{aligned} h_i(u) &= \psi_{i-1}(h_1(u)) = \psi_{i-1}(t_{11}(u)), \\ f_i(u) &= \psi_{i-1}(f_1(u)) = \psi_{i-1}\big(t_{21}(u)\, t_{11}(u)^{-1}\big), \\ e_i(u) &= \psi_{i-1}(e_1(u)) = \psi_{i-1}\big(t_{11}(u)^{-1}\, t_{12}(u)\big), \end{aligned} \tag{3.1}$$

where the homomorphism ψ_M is defined in (1.84). Explicitly, we have

$$\begin{aligned} h_i(u) &= t\,{}^{1\dots i}_{1\dots i}(u+i-1)\cdot\big(t\,{}^{1\dots i-1}_{1\dots i-1}(u+i-1)\big)^{-1}, \\ f_i(u) &= t\,{}^{1\dots i-1,\, i+1}_{1\dots i}(u+i-1)\cdot\big(t\,{}^{1\dots i}_{1\dots i}(u+i-1)\big)^{-1}, \\ e_i(u) &= \big(t\,{}^{1\dots i}_{1\dots i}(u+i-1)\big)^{-1}\cdot t\,{}^{1\dots i}_{1\dots i-1,\, i+1}(u+i-1). \end{aligned} \tag{3.2}$$

Recall that the coefficients of the series $h_i(u)$ for $i = 1,\dots,N$ generate a commutative subalgebra of $\mathrm{Y}(\mathfrak{gl}_N)$ which we call the Gelfand–Tsetlin subalgebra; see Section 1.13. We also record the following relations which are implied by Corollary 1.7.2:

$$[h_i(u), e_j(v)] = 0 \quad \text{and} \quad [h_i(u), f_j(v)] = 0 \quad \text{for} \quad i \neq j, j+1, \tag{3.3}$$

$$[e_i(u), e_j(v)] = 0 \quad \text{for} \quad |i-j| > 1, \tag{3.4}$$

$$[e_i(u), f_j(v)] = 0 \quad \text{for} \quad i \neq j. \tag{3.5}$$

Let us introduce the coefficients of the series $h_i(u)$, $e_i(u)$ and $f_i(u)$ by

$$h_i(u) = 1 + \sum_{r=1}^{\infty} h_i^{(r)}\, u^{-r}, \quad e_i(u) = \sum_{r=1}^{\infty} e_i^{(r)}\, u^{-r}, \quad f_i(u) = \sum_{r=1}^{\infty} f_i^{(r)}\, u^{-r}. \tag{3.6}$$

The relations (3.2) yield

$$e_i^{(1)} = t_{i,i+1}^{(1)} \qquad \text{and} \qquad f_i^{(1)} = t_{i+1,i}^{(1)}.$$

Moreover, using Proposition 1.7.1 and Theorem 1.11.6 we deduce that for any $i < j$

$$e_{i,j+1}(u) = [e_{ij}(u), e_j^{(1)}] \qquad \text{and} \qquad f_{j+1,i}(u) = [f_j^{(1)}, f_{ji}(u)]. \tag{3.7}$$

In this section we aim to produce a presentation of the Yangian $\mathrm{Y}(\mathfrak{gl}_N)$ where the elements $h_i^{(r)}$, $e_i^{(r)}$ and $f_i^{(r)}$ play the role of generators; see Theorem 3.1.5 below. The formulas (3.1) indicate that the defining relations in this presentation are largely determined by those in the Yangian $\mathrm{Y}(\mathfrak{gl}_N)$ with $N \leqslant 3$. We start by proving a sequence of lemmas describing relations between $h_i(u)$, $e_i(u)$ and $f_i(u)$ for small values of the index i. We will use the notation $h_i'(u) = h_i(u)^{-1}$.

LEMMA 3.1.1. *The following relations hold in* $\mathrm{Y}(\mathfrak{gl}_2)$:

$$(u-v)\,[h_1(u), e_1(v)] = h_1(u)\,(e_1(v) - e_1(u)), \tag{3.8}$$

$$(u-v)\,[e_1(u), h_2'(v)] = (e_1(u) - e_1(v))\,h_2'(v), \tag{3.9}$$

$$(u-v)\,[e_1(u), f_1(v)] = h_1'(u)\,h_2(u) - h_1'(v)\,h_2(v), \tag{3.10}$$

$$(u-v)\,[e_1(u), e_1(v)] = (e_1(v) - e_1(u))^2. \tag{3.11}$$

PROOF. The argument is based on the relations between the entries of the matrix $T(u)$ and its inverse, provided by Proposition 1.10.10. We have

$$(u-v)\,[t_{11}(u), t_{12}'(v)] = t_{11}(u)\,t_{12}'(v) + t_{12}(u)\,t_{22}'(v). \tag{3.12}$$

By the Gauss decomposition for $T(u)$ (Theorem 1.11.6), we have

$$\begin{bmatrix} t_{11}(u) & t_{12}(u) \\ t_{21}(u) & t_{22}(u) \end{bmatrix} = \begin{bmatrix} h_1(u) & h_1(u)\,e_1(u) \\ f_1(u)\,h_1(u) & h_2(u) + f_1(u)\,h_1(u)\,e_1(u) \end{bmatrix},$$

and inverting both sides of (1.91) we get

$$\begin{bmatrix} t_{11}'(u) & t_{12}'(u) \\ t_{21}'(u) & t_{22}'(u) \end{bmatrix} = \begin{bmatrix} h_1'(u) + e_1(u)\,h_2'(u)\,f_1(u) & -\,e_1(u)\,h_2'(u) \\ -\,h_2'(u)\,f_1(u) & h_2'(u) \end{bmatrix}.$$

Substituting from these relations into (3.12) and using the fact that $h_1(u)$ commutes with $h_2'(v)$ gives the identity

$$(u-v)\,[h_1(u), e_1(v)]\,h_2'(v) = h_1(u)\,(e_1(v) - e_1(u))\,h_2'(v),$$

proving (3.8). The proof of (3.9) is the same with the use of the relation

$$(u-v)\,[t_{12}(u), t_{22}'(v)] = t_{11}(u)\,t_{12}'(v) + t_{12}(u)\,t_{22}'(v)$$

instead of (3.12). Furthermore, another particular case of Proposition 1.10.10 gives

$$(u-v)\,[t_{12}(u), t_{21}'(v)] = t_{11}(u)\,t_{11}'(v) + t_{12}(u)\,t_{21}'(v) - t_{21}'(v)\,t_{12}(u) - t_{22}'(v)\,t_{22}(u).$$

Rearranging this in the same manner as above, we get

$$\begin{aligned} h_1(u)\,h_1'(v) &+ h_1(u)\big(e_1(v)\,h_2'(v) + (u-v-1)\,e_1(u)\,h_2'(v)\big)\,f_1(v) \\ &= h_2'(v)\,h_2(u) + h_2'(v)\big(f_1(u)\,h_1(u) + (u-v-1)\,f_1(v)\,h_1(u)\big)\,e_1(u). \end{aligned}$$

On the other hand, (3.9) is equivalent to

$$e_1(v)\,h_2'(v) + (u-v-1)\,e_1(u)\,h_2'(v) = (u-v)\,h_2'(v)\,e_1(u) \tag{3.13}$$

while applying Corollary 1.11.7, we deduce from (3.8) that

$$f_1(u)\,h_1(u) + (u-v-1)\,f_1(v)\,h_1(u) = (u-v)\,h_1(u)\,f_1(v).$$

Substituting these respectively for the expressions in the brackets in the previous formula, we come to (3.10). Finally, using Proposition 1.10.10 once more, we get

$$[t_{12}(u), t'_{12}(v)] = 0$$

which implies

$$h_1(u)\, e_1(u)\, e_1(v)\, h'_2(v) = e_1(v)\, h_1(u)\, h'_2(v)\, e_1(u).$$

Multiply this by $(u-v)^2$ and transform the right-hand side by using (3.13) and the following form of (3.8):

$$(u-v)\, e_1(v) h_1(u) = h_1(u) e_1(u) + (u-v-1)\, h_1(u) e_1(v).$$

We obtain

$$(u-v)^2\, e_1(u)\, e_1(v) = \big(e_1(u) + (u-v-1)\, e_1(v)\big)\big(e_1(v) + (u-v-1)\, e_1(u)\big),$$

which is equivalent to (3.11). □

Observe that by (3.1), all relations of Lemma 3.1.1 will be valid in the Yangian $\mathrm{Y}(\mathfrak{gl}_N)$ if the subscripts 1 and 2 in these relations are replaced by i and $i+1$ respectively, for any $1 \leqslant i < N$.

LEMMA 3.1.2. *The following relations hold in the algebra* $\mathrm{Y}(\mathfrak{gl}_3)$:

$$(u-v)\,[e_1(u), e_2(v)] = e_1(u)\, e_2(v) - e_1(v)\, e_2(v) - e_{13}(u) + e_{13}(v), \tag{3.14}$$

$$[e_{13}(u), e_2(v)] = e_2(v)\,[e_1(u), e_2(v)], \tag{3.15}$$

$$[e_1(u), e_{13}(v) - e_1(v)\, e_2(v)] = -[e_1(u), e_2(v)]\, e_1(u). \tag{3.16}$$

PROOF. As in the proof of the previous lemma, our argument relies on Proposition 1.10.10 and Theorem 1.11.6 applied to the Yangian $\mathrm{Y}(\mathfrak{gl}_3)$. Starting with the relation

$$(u-v)\,[t_{12}(u), t'_{23}(v)] = t_{11}(u)\, t'_{13}(v) + t_{12}(u)\, t'_{23}(v) + t_{13}(u)\, t'_{33}(v)$$

and calculating $t_{12}(u)$ and $t'_{23}(v)$ from the decomposition (1.91), we get

$$\begin{aligned}(u-v)\,&[h_1(u)\, e_1(u),\ e_2(v)\, h'_3(v)]\\ &= h_1(u)\big(e_1(u)\, e_2(v) - e_1(v)\, e_2(v) - e_{13}(u) + e_{13}(v)\big) h'_3(v),\end{aligned}$$

which together with (3.3) gives (3.14). Similarly, the relation

$$[t_{13}(u), t'_{23}(v)] = 0$$

implies

$$[h_1(u)\, e_{13}(u),\ e_2(v)\, h'_3(v)] = 0,$$

which is equivalent to

$$[e_{13}(u),\ e_2(v)\, h'_3(v)] = 0. \tag{3.17}$$

By (3.9), we can write

$$(u-v)\,[h'_3(v), e_2(u)] = (e_2(v) - e_2(u))\, h'_3(v).$$

Taking the coefficient of u^0 we obtain

$$[h'_3(v), e_2^{(1)}] = e_2(v)\, h'_3(v).$$

Hence, taking into account (3.7), we get

$$[e_{13}(u), h_3'(v)] = [[e_1(u), e_2^{(1)}], h_3'(v)] = [e_1(u), [e_2^{(1)}, h_3'(v)]]$$
$$= -[e_1(u),\, e_2(v)\, h_3'(v)] = -[e_1(u),\, e_2(v)]\, h_3'(v),$$

where we have also used (3.3). Now, transform (3.17) to obtain

$$0 = [e_{13}(u),\, e_2(v)\, h_3'(v)] = [e_{13}(u),\, e_2(v)]\, h_3'(v) + e_2(v)\, [e_{13}(u), h_3'(v)]$$
$$= \big([e_{13}(u),\, e_2(v)] - e_2(v)\, [e_1(u),\, e_2(v)]\big)\, h_3'(v),$$

proving (3.15). Finally, using the relation

$$[t_{12}(u), t_{13}'(v)] = 0$$

we get

$$[h_1(u)\, e_1(u),\, \big(e_{13}(v) - e_1(v)\, e_2(v)\big)\, h_3'(v)] = 0$$

which implies

$$[h_1(u)\, e_1(u),\, e_{13}(v) - e_1(v)\, e_2(v)] = 0. \tag{3.18}$$

Taking the coefficient of u^0 in (3.14) gives

$$[e_1^{(1)}, e_2(v)] = e_{13}(v) - e_1(v)\, e_2(v).$$

Similarly, we find from (3.8) that $[h_1(u), e_1^{(1)}] = h_1(u)\, e_1(u)$. Therefore,

$$[h_1(u),\, e_{13}(v) - e_1(v)\, e_2(v)] = [h_1(u),\, [e_1^{(1)}, e_2(v)]] = [[h_1(u),\, e_1^{(1)}],\, e_2(v)]$$
$$= [h_1(u)\, e_1(u),\, e_2(v)] = h_1(u)\, [e_1(u),\, e_2(v)].$$

Thus, we can transform (3.18) as

$$0 = [h_1(u)\, e_1(u),\, e_{13}(v) - e_1(v)\, e_2(v)] = h_1(u)\, [e_1(u),\, e_2(v)]\, e_1(u)$$
$$+ h_1(u)\, [e_1(u),\, e_{13}(v) - e_1(v)\, e_2(v)],$$

proving (3.16). □

LEMMA 3.1.3. *The following relations hold in the algebra* Y($\mathfrak{gl}_3$):

$$[[e_1(u),\, e_2(v)],\, e_2(v)] = 0, \tag{3.19}$$
$$[e_1(u),\, [e_1(u),\, e_2(v)]] = 0. \tag{3.20}$$

PROOF. Using (3.14) and (3.15), we obtain

$$(u-v)\, [[e_1(u),\, e_2(v)],\, e_2(v)] = [e_1(u)\, e_2(v) - e_1(v)\, e_2(v) - e_{13}(u) + e_{13}(v),\, e_2(v)]$$
$$= [e_1(u),\, e_2(v)]\, e_2(v) - [e_1(v),\, e_2(v)]\, e_2(v)$$
$$+ e_2(v)\, [e_1(v),\, e_2(v)] - e_2(v)\, [e_1(u),\, e_2(v)]$$
$$= [[e_1(u),\, e_2(v)],\, e_2(v)] - [[e_1(v),\, e_2(v)],\, e_2(v)].$$

Hence,

$$(u-v-1)\, [[e_1(u),\, e_2(v)],\, e_2(v)] = -[[e_1(v),\, e_2(v)],\, e_2(v)].$$

Note that the right-hand side is zero, which is seen by setting $u = v+1$. This proves (3.19). A similar calculation with the use of (3.16) instead of (3.15) proves (3.20). □

LEMMA 3.1.4. *The following relations hold in the algebra* $\mathrm{Y}(\mathfrak{gl}_3)$:

$$[[e_1(u), e_2(v)], e_2(w)] + [[e_1(u), e_2(w)], e_2(v)] = 0, \tag{3.21}$$

$$[e_1(u), [e_1(v), e_2(w)]] + [e_1(v), [e_1(u), e_2(w)]] = 0. \tag{3.22}$$

PROOF. Transforming the expression

$$(u-v)(u-w)(v-w)\,[[e_1(u), e_2(v)], e_2(w)] \tag{3.23}$$

by using (3.14), we can write it as

$$(u-w)(v-w)\,[e_1(u)\,e_2(v) - e_1(v)\,e_2(v) + e_{13}(v) - e_{13}(u), e_2(w)].$$

Applying (3.15) and (3.19) we bring this to the form

$$\begin{aligned}
&(u-w)(v-w)\,[e_1(u), e_2(w)]\,e_2(v) + (u-w)(v-w)\,e_1(u)\,[e_2(v), e_2(w)]\\
&-(u-w)(v-w)\,[e_1(v), e_2(w)]\,e_2(v) - (u-w)(v-w)\,e_1(v)\,[e_2(v), e_2(w)]\\
&+(u-w)(v-w)\,[e_1(v), e_2(w)]\,e_2(w) - (u-w)(v-w)\,[e_1(u), e_2(w)]\,e_2(w).
\end{aligned}$$

Finally, expanding the commutators with the use of (3.11) and (3.14), we get

$$\begin{aligned}
&(v-w)\,\big(e_1(u)\,e_2(w)\,e_2(v) - e_1(w)\,e_2(w)\,e_2(v) + e_{13}(w)\,e_2(v) - e_{13}(u)\,e_2(v)\big)\\
&+(u-w)\,\big(e_1(u)\,e_2(v)^2 - e_1(u)\,e_2(v)\,e_2(w) - e_1(u)\,e_2(w)\,e_2(v) + e_1(u)\,e_2(w)^2\big)\\
&-(u-w)\,\big(e_1(v)\,e_2(w)\,e_2(v) - e_1(w)\,e_2(w)\,e_2(v) + e_{13}(w)\,e_2(v) - e_{13}(v)\,e_2(v)\big)\\
&-(u-w)\,\big(e_1(v)\,e_2(v)^2 - e_1(v)\,e_2(v)\,e_2(w) - e_1(v)\,e_2(w)\,e_2(v) + e_1(v)\,e_2(w)^2\big)\\
&+(u-w)\,\big(e_1(v)\,e_2(w)\,e_2(w) - e_1(w)\,e_2(w)\,e_2(w) + e_{13}(w)\,e_2(w) - e_{13}(v)\,e_2(w)\big)\\
&-(v-w)\,\big(e_1(u)\,e_2(w)\,e_2(w) - e_1(w)\,e_2(w)\,e_2(w) + e_{13}(w)\,e_2(w) - e_{13}(u)\,e_2(w)\big).
\end{aligned}$$

Observe that this expression is symmetric in v and w; hence so is (3.23). This proves (3.21). A similar argument with the use of (3.16) and (3.20) shows that the expression

$$(u-v)(u-w)(v-w)\,[e_1(u), [e_1(v), e_2(w)]]$$

is symmetric in u and v, proving (3.22). □

In addition to the coefficients of the series $h_i(u)$, $e_i(u)$ and $f_i(u)$ introduced in (3.6), let us define the $h_i^{\prime(r)}$ by

$$h_i'(u) = 1 + \sum_{r=1}^{\infty} h_i^{\prime(r)}\, u^{-r}.$$

THEOREM 3.1.5. *The algebra* $\mathrm{Y}(\mathfrak{gl}_N)$ *is generated by the elements* $h_i^{(r)}$, $h_i^{\prime(r)}$ *with* $i = 1, \dots, N$ *and* $r \geqslant 0$, *and by* $e_i^{(r)}$, $f_i^{(r)}$ *with* $i = 1, \dots, N-1$ *and* $r \geqslant 1$, *subject only to the following relations:*

$$h_i^{(0)} = 1, \qquad \sum_{t=0}^{r} h_i^{(t)}\, h_i^{\prime(r-t)} = \delta_{r0}, \qquad [h_i^{(r)}, h_j^{(s)}] = 0,$$

$$[e_i^{(r)}, f_j^{(s)}] = -\,\delta_{ij} \sum_{t=0}^{r+s-1} h_i'^{(t)}\, h_{i+1}^{(r+s-t-1)}, \tag{3.24}$$

$$[h_i^{(r)}, e_j^{(s)}] = (\delta_{ij} - \delta_{i,j+1}) \sum_{t=0}^{r-1} h_i^{(t)}\, e_j^{(r+s-t-1)}, \tag{3.25}$$

$$[h_i^{(r)}, f_j^{(s)}] = (\delta_{i,j+1} - \delta_{ij}) \sum_{t=0}^{r-1} f_j^{(r+s-t-1)}\, h_i^{(t)}, \tag{3.26}$$

$$[e_i^{(r)}, e_i^{(s+1)}] - [e_i^{(r+1)}, e_i^{(s)}] = e_i^{(r)} e_i^{(s)} + e_i^{(s)} e_i^{(r)}, \tag{3.27}$$

$$[f_i^{(r+1)}, f_i^{(s)}] - [f_i^{(r)}, f_i^{(s+1)}] = f_i^{(r)} f_i^{(s)} + f_i^{(s)} f_i^{(r)}, \tag{3.28}$$

$$[e_i^{(r)}, e_{i+1}^{(s+1)}] - [e_i^{(r+1)}, e_{i+1}^{(s)}] = -e_i^{(r)} e_{i+1}^{(s)}, \tag{3.29}$$

$$[f_i^{(r+1)}, f_{i+1}^{(s)}] - [f_i^{(r)}, f_{i+1}^{(s+1)}] = -f_{i+1}^{(s)} f_i^{(r)}, \tag{3.30}$$

$$[e_i^{(r)}, e_j^{(s)}] = 0 \qquad \text{if} \quad |i-j| > 1, \tag{3.31}$$

$$[f_i^{(r)}, f_j^{(s)}] = 0 \qquad \text{if} \quad |i-j| > 1, \tag{3.32}$$

$$[e_i^{(r)}, [e_i^{(s)}, e_j^{(t)}]] + [e_i^{(s)}, [e_i^{(r)}, e_j^{(t)}]] = 0 \qquad \text{if} \quad |i-j| = 1, \tag{3.33}$$

$$[f_i^{(r)}, [f_i^{(s)}, f_j^{(t)}]] + [f_i^{(s)}, [f_i^{(r)}, f_j^{(t)}]] = 0 \qquad \text{if} \quad |i-j| = 1, \tag{3.34}$$

for all admissible i, j, r, s, t.

PROOF. We show first that all these relations are satisfied in $\mathrm{Y}(\mathfrak{gl}_N)$. This is clear for the first three. Due to (3.5), we only need to verify (3.24) for $i = j$. However, this follows from (3.1) and (3.10), with the use of the following observation: for any formal series

$$g(u) = \sum_{r \geqslant 1} g^{(r)} u^{-r}$$

we have the identity

$$\frac{g(u) - g(v)}{u - v} = -\sum_{r,s \geqslant 1} g^{(r+s-1)} u^{-r} v^{-s}. \tag{3.35}$$

Similarly, by (3.1) and (3.3), it suffices to verify (3.25) for two cases $i = j = 1$ and $i = 2, j = 1$. But these respectively follow from (3.8) and (3.9). Next, (3.26) follows from (3.25) by the application of the anti-automorphism t; see Corollary 1.11.7. In the same way, (3.28), (3.30), (3.32) and (3.34) follow from their respective counterparts involving the generators $e_i^{(r)}$. Finally, (3.31) is a restatement of (3.4), while (3.27), (3.29) and (3.33) respectively follow from (3.11), (3.14) and Lemma 3.1.4, again taking (3.1) into account. Note that the terms on the right-hand side of (3.14) which only depend on u or v do not contribute to the right-hand side of (3.29) because of the restrictions $r, s \geqslant 1$ in that relation.

To complete the proof, let us consider the algebra $\widehat{\mathrm{Y}}(\mathfrak{gl}_N)$ with generators and relations as in the statement of the theorem. By the arguments above, there is a surjective algebra homomorphism

$$\widehat{\mathrm{Y}}(\mathfrak{gl}_N) \to \mathrm{Y}(\mathfrak{gl}_N) \tag{3.36}$$

which takes the generators $h_i^{(r)}$, $h_i'^{(r)}$, $e_i^{(r)}$ and $f_i^{(r)}$ of $\widehat{\mathrm{Y}}(\mathfrak{gl}_N)$ to the elements of $\mathrm{Y}(\mathfrak{gl}_N)$ denoted by the same symbols. We need to show that this homomorphism is injective. We start by showing that the set of monomials in

$$h_i^{(r)} \quad \text{with} \quad i = 1, \dots, N, \quad r \geqslant 1$$

and

$$e_{ij}^{(r)}, \quad f_{ji}^{(r)} \quad \text{with} \quad 1 \leqslant i < j \leqslant N, \quad r \geqslant 1,$$

taken in some fixed order is linearly independent in the Yangian $\mathrm{Y}(\mathfrak{gl}_N)$. We employ Proposition 1.5.2. Applying (1.87)–(1.89) to the matrix $T(u)$ we deduce that the images of the elements $h_i^{(r)}$, $e_{ij}^{(r)}$ and $f_{ji}^{(r)}$ in the $(r-1)$-th component of the graded algebra $\mathrm{gr}' \, \mathrm{Y}(\mathfrak{gl}_N)$ respectively correspond to the elements

$$E_{ii} \, z^{r-1}, \qquad E_{ij} \, z^{r-1} \qquad \text{and} \qquad E_{ji} \, z^{r-1}$$

of the universal enveloping algebra $\mathrm{U}(\mathfrak{gl}_N[z])$. Hence the claim follows from the Poincaré–Birkhoff–Witt theorem for $\mathrm{U}(\mathfrak{gl}_N[z])$.

For any $1 \leqslant i < j \leqslant N$ define elements $e_{ij}^{(r)}$ and $f_{ji}^{(r)}$ of $\widehat{\mathrm{Y}}(\mathfrak{gl}_N)$ inductively by the relations $e_{i,i+1}^{(r)} = e_i^{(r)}$, $f_{i+1,i}^{(r)} = f_i^{(r)}$ and

$$e_{i,j+1}^{(r)} = [e_{ij}^{(r)}, e_j^{(1)}], \qquad f_{j+1,i}^{(r)} = [f_j^{(1)}, f_{ji}^{(r)}], \qquad \text{for} \quad j > i,$$

which are obviously consistent with (3.7). Theorem 3.1.5 will follow if we prove that the algebra $\widehat{\mathrm{Y}}(\mathfrak{gl}_N)$ is spanned by the monomials in $h_i^{(r)}$, $e_{ij}^{(r)}$ and $f_{ji}^{(r)}$ taken in some fixed order. In order to see this, denote by $\widehat{E}_N$, $\widehat{F}_N$ and $\widehat{H}_N$ the subalgebras of $\widehat{\mathrm{Y}}(\mathfrak{gl}_N)$ respectively generated by all elements of the form $e_i^{(r)}$, $f_i^{(r)}$ and $h_i^{(r)}$. Define an ascending filtration on $\widehat{E}_N$ by setting $\deg' e_i^{(r)} = r-1$. Denote by $\mathrm{gr}' \, \widehat{E}_N$ the corresponding graded algebra. Let $\bar{e}_{ij}^{(r)}$ be the image of $e_{ij}^{(r)}$ in the $(r-1)$-th component of the graded algebra $\mathrm{gr}' \, \widehat{E}_N$. We claim that these images satisfy

$$[\bar{e}_{ij}^{(r)}, \bar{e}_{kl}^{(s)}] = \delta_{kj} \, \bar{e}_{il}^{(r+s-1)} - \delta_{il} \, \bar{e}_{kj}^{(r+s-1)}. \tag{3.37}$$

In order to verify this, we start by recording the relations between the elements $\bar{e}_{ij}^{(r)}$ which are immediate from the defining relations. First of all, (3.31) implies that for $|i-k| > 1$ we have

$$[\bar{e}_{i,i+1}^{(r)}, \bar{e}_{k,k+1}^{(s)}] = 0. \tag{3.38}$$

Since (3.27) is equivalent to

$$(u-v) \, [e_i(u), e_i(v)] = (e_i(v) - e_i(u))^2,$$

we deduce, using (3.35), the following alternative form of (3.27):

$$[e_i^{(r)}, e_i^{(s)}] = \sum_{t=1}^{s-1} e_i^{(t)} \, e_i^{(r+s-t-1)} - \sum_{t=1}^{r-1} e_i^{(t)} \, e_i^{(r+s-t-1)}.$$

This shows that (3.38) holds for $i = k$ as well. Further, if $|i-k| = 1$, then using (3.29) and (3.33) we get

$$[\bar{e}_{i,i+1}^{(r+1)}, \bar{e}_{k,k+1}^{(s)}] = [\bar{e}_{i,i+1}^{(r)}, \bar{e}_{k,k+1}^{(s+1)}], \tag{3.39}$$

$$[\bar{e}_{i,i+1}^{(r)}, [\bar{e}_{i,i+1}^{(s)}, \bar{e}_{k,k+1}^{(t)}]] = -[\bar{e}_{i,i+1}^{(s)}, [\bar{e}_{i,i+1}^{(r)}, \bar{e}_{k,k+1}^{(t)}]]. \tag{3.40}$$

By (3.39) and the definition of the elements $e_{ij}^{(r)}$, we also have

$$\bar{e}_{ij}^{(r)} = [\bar{e}_{i,j-1}^{(r)}, \bar{e}_{j-1,j}^{(1)}] = [\bar{e}_{i,i+1}^{(1)}, \bar{e}_{i+1,j}^{(r)}] \qquad \text{for} \quad j-i>1, \tag{3.41}$$

where the second equality easily follows by induction on $j-i$. We will verify (3.37) case by case, considering all relations between the pairs of indices $i<j$ and $k<l$. If $j<k$, then (3.37) follows immediately from (3.38) and (3.41). If $j=k$, then using (3.39) and (3.41) we deduce

$$[\bar{e}_{j-1,j}^{(r)}, \bar{e}_{j,j+1}^{(s)}] = \bar{e}_{j-1,j+1}^{(r+s-1)}.$$

Now take the bracket of both sides with $\bar{e}_{j+1,j+2}^{(1)}, \dots, \bar{e}_{l-1,l}^{(1)}$ to obtain

$$[\bar{e}_{j-1,j}^{(r)}, \bar{e}_{jl}^{(s)}] = \bar{e}_{j-1,l}^{(r+s-1)}.$$

Then take the bracket with $\bar{e}_{j-2,j-1}^{(1)}, \dots, \bar{e}_{i,i+1}^{(1)}$ to get (3.37) for $j=k$. Next we show that

$$[\bar{e}_{i,i+2}^{(r)}, \bar{e}_{i+1,i+3}^{(s)}] = 0.$$

The argument is the same for all i, so take $i=1$. The commutator $[\bar{e}_{13}^{(r)}, \bar{e}_{24}^{(s)}]$ equals

$$\begin{aligned}[][[\bar{e}_{12}^{(r)}, \bar{e}_{23}^{(1)}], [\bar{e}_{23}^{(1)}, \bar{e}_{34}^{(s)}]] &= [\bar{e}_{23}^{(1)}, [[\bar{e}_{12}^{(r)}, \bar{e}_{23}^{(1)}], \bar{e}_{34}^{(s)}]] = [\bar{e}_{23}^{(1)}, [\bar{e}_{12}^{(r)}, [\bar{e}_{23}^{(1)}, \bar{e}_{34}^{(s)}]]] \\ &= [[\bar{e}_{23}^{(1)}, \bar{e}_{12}^{(r)}], [\bar{e}_{23}^{(1)}, \bar{e}_{34}^{(s)}]] = -[[\bar{e}_{12}^{(r)}, \bar{e}_{23}^{(1)}], [\bar{e}_{23}^{(1)}, \bar{e}_{34}^{(s)}]],\end{aligned}$$

where we have used the relations

$$[[\bar{e}_{12}^{(r)}, \bar{e}_{23}^{(1)}], \bar{e}_{23}^{(1)}] = 0 \quad \text{and} \quad [\bar{e}_{23}^{(1)}, [\bar{e}_{23}^{(1)}, \bar{e}_{34}^{(s)}]] = 0$$

implied by (3.40). Hence the commutator is zero. Now we prove (3.37) with $i<k$ and $j=l$. First consider a particular case. Using (3.40) we obtain

$$[\bar{e}_{13}^{(r)}, \bar{e}_{23}^{(s)}] = [[\bar{e}_{12}^{(r)}, \bar{e}_{23}^{(1)}], \bar{e}_{23}^{(s)}] = -[[\bar{e}_{12}^{(r)}, \bar{e}_{23}^{(s)}], \bar{e}_{23}^{(1)}].$$

By (3.39) this equals

$$-[[\bar{e}_{12}^{(r+s-1)}, \bar{e}_{23}^{(1)}], \bar{e}_{23}^{(1)}],$$

which is zero by (3.40). This also implies that

$$[\bar{e}_{14}^{(r)}, \bar{e}_{23}^{(s)}] = [[\bar{e}_{13}^{(r)}, \bar{e}_{34}^{(1)}], e_{23}^{(s)}] = -[e_{13}^{(r)}, e_{24}^{(s)}] = 0.$$

Applying (3.41), we then deduce that

$$[\bar{e}_{ij}^{(r)}, \bar{e}_{k,k+1}^{(s)}] = 0 \tag{3.42}$$

for any $i<k$ and $k+1<j$. Using (3.42) along with (3.41) we can complete the proof of (3.37) in the case under consideration by reducing it to the particular case

$$[\bar{e}_{j-2,j}^{(r)}, \bar{e}_{j-1,j}^{(s)}] = 0.$$

The calculation is the same for all j, and for $j=3$ it was performed above. If $i=k$ and $j<l$, then the application of (3.41) and (3.42) reduces the calculation to showing that

$$[\bar{e}_{i,i+1}^{(r)}, \bar{e}_{i,i+2}^{(s)}] = 0,$$

which is similar to the previous case. If $i=k$ and $j=l$ with $j=i+1$, then the claim follows by (3.38) with $i=k$. For $j>i+1$ write

$$[\bar{e}_{ij}^{(r)}, \bar{e}_{ij}^{(s)}] = [[\bar{e}_{i,j-1}^{(r)}, \bar{e}_{j-1,j}^{(1)}], \bar{e}_{ij}^{(s)}] = [[\bar{e}_{i,j-1}^{(r)}, \bar{e}_{ij}^{(s)}], \bar{e}_{j-1,j}^{(1)}] + [\bar{e}_{i,j-1}^{(r)}, [\bar{e}_{j-1,j}^{(1)}, \bar{e}_{ij}^{(s)}]],$$

which is zero by the two previous cases. If $i < k < j < l$, then due to (3.41) and (3.42) it suffices to verify that

$$[\bar{e}^{(r)}_{i,i+2}, \bar{e}^{(s)}_{i+1,i+3}] = 0.$$

However, this has been done above for $i = 1$ and the argument for general i is the same. Similarly, the case $i < k < l < j$ is reduced to (3.42). This completes the verification of the relations (3.37). They imply that the graded algebra $\mathrm{gr}'\,\widehat{E}_N$ is spanned by the set of monomials in the elements $\bar{e}^{(r)}_{ij}$ taken in some fixed order. So the algebra $\widehat{E}_N$ is spanned by the corresponding monomials in the elements $e^{(r)}_{ij}$.

Furthermore, it is immediate from the defining relation of $\widehat{\mathrm{Y}}(\mathfrak{gl}_N)$ that the mapping

$$h^{(r)}_i \mapsto h^{(r)}_i, \qquad e^{(r)}_i \mapsto f^{(r)}_i, \qquad f^{(r)}_i \mapsto e^{(r)}_i$$

defines an anti-automorphism of $\widehat{\mathrm{Y}}(\mathfrak{gl}_N)$. By applying this anti-automorphism, we deduce that the ordered monomials in the elements $f^{(r)}_{ji}$ span the subalgebra $\widehat{F}_N$. Note also that the ordered monomials in $h^{(r)}_i$ span $\widehat{H}_N$. By the defining relations of $\widehat{\mathrm{Y}}(\mathfrak{gl}_N)$, the multiplication map

$$\widehat{F}_N \otimes \widehat{H}_N \otimes \widehat{E}_N \to \widehat{\mathrm{Y}}(\mathfrak{gl}_N)$$

is surjective. Thus, ordering the elements $h^{(r)}_i$, $e^{(r)}_{ij}$ and $f^{(r)}_{ji}$ in such a way that the elements of $\widehat{F}_N$ precede the elements of $\widehat{H}_N$, and the latter precede the elements of $\widehat{E}_N$, we can conclude that the ordered monomials in these elements span $\widehat{\mathrm{Y}}(\mathfrak{gl}_N)$. This proves that (3.36) is an isomorphism. □

In the course of the proof we have obtained another version of the Poincaré–Birkhoff–Witt theorem for $\mathrm{Y}(\mathfrak{gl}_N)$; cf. Theorem 1.4.1. Recall the definition of the Gelfand–Tsetlin subalgebra H_N (Section 1.13). Denote by E_N and F_N the subalgebras of $\mathrm{Y}(\mathfrak{gl}_N)$ generated by all elements of the form $e^{(r)}_i$ and $f^{(r)}_i$, respectively.

COROLLARY 3.1.6. (i) *The set of all monomials in the elements $h^{(r)}_i$ with $i = 1, \dots, N$ and $r \geqslant 1$, taken in some fixed order, forms a basis of H_N.*

(ii) *The set of all monomials in the elements $e^{(r)}_{ij}$ with $1 \leqslant i < j \leqslant N$ and $r \geqslant 1$, taken in some fixed order, forms a basis of E_N.*

(iii) *The set of all monomials in the elements $f^{(r)}_{ji}$ with $1 \leqslant i < j \leqslant N$ and $r \geqslant 1$, taken in some fixed order, forms a basis of F_N.*

(iv) *The set of all monomials in the elements $h^{(r)}_i$, $e^{(r)}_{ij}$ and $f^{(r)}_{ji}$ with the above conditions on the indices, taken in some fixed order, forms a basis of $\mathrm{Y}(\mathfrak{gl}_N)$.*

We now give a presentation of the Yangian $\mathrm{Y}(\mathfrak{sl}_N)$ (see Section 1.8) implied by Theorem 3.1.5. Set

$$\begin{aligned} \widetilde{h}_i(u) &= v_N\Big(h_{N-i+1}\Big(u - N + \frac{i+1}{2}\Big)\Big), \\ \xi^+_i(u) &= v_N\Big(f_{N-i}\Big(u - N + \frac{i+1}{2}\Big)\Big), \\ \xi^-_i(u) &= v_N\Big(e_{N-i}\Big(u - N + \frac{i+1}{2}\Big)\Big), \end{aligned} \tag{3.43}$$

where v_N is the automorphism of $\mathrm{Y}(\mathfrak{gl}_N)$ which sends $t_{ij}(u)$ to $t_{N-i+1,N-j+1}(u)$; see (1.95). Using Theorem 1.11.6, we obtain explicit formulas for these series in terms of the quantum minors:

$$\widetilde{h}_i(u) = t^{\,i\dots N}_{\,i\dots N}\Big(u - \frac{i-1}{2}\Big) \cdot \Big(t^{\,i+1\dots N}_{\,i+1\dots N}\Big(u - \frac{i-1}{2}\Big)\Big)^{-1} \tag{3.44}$$

for $1 \leqslant i \leqslant N$, and

$$\xi^+_i(u) = t^{\,i,i+2\dots N}_{\,i+1\dots N}\Big(u - \frac{i+1}{2}\Big) \cdot \Big(t^{\,i+1\dots N}_{\,i+1\dots N}\Big(u - \frac{i+1}{2}\Big)\Big)^{-1},$$

$$\xi^-_i(u) = \Big(t^{\,i+1\dots N}_{\,i+1\dots N}\Big(u - \frac{i+1}{2}\Big)\Big)^{-1} \cdot t^{\,i+1\dots N}_{\,i,i+2\dots N}\Big(u - \frac{i+1}{2}\Big)$$

for $1 \leqslant i < N$. Define the elements κ_{ik} and $\xi^\pm_{ik}$ with $i = 1,\dots,N-1$ and $k \geqslant 0$ by the relations

$$\kappa_i(u) = 1 + \sum_{k\geqslant 0} \kappa_{ik}\, u^{-k-1} = \widetilde{h}_i(u)\, \widetilde{h}_{i+1}(u-1/2)^{-1}, \tag{3.45}$$

$$\xi^+_i(u) = \sum_{k\geqslant 0} \xi^+_{ik}\, u^{-k-1}, \qquad \xi^-_i(u) = \sum_{k\geqslant 0} \xi^-_{ik}\, u^{-k-1}.$$

By Definition 1.8.1, all these elements belong to the subalgebra $\mathrm{Y}(\mathfrak{sl}_N)$, which is implied by the explicit formulas (1.54) for the quantum minors.

Let $[a_{ij}]$ denote the Cartan matrix of type A_{N-1} so that $a_{ii} = 2$, $a_{ij} = -1$ if $|i-j| = 1$, and $a_{ij} = 0$ if $|i-j| > 1$. In any of the relations (3.48) to (3.51) below, where the double signs $\pm$ occur, only the upper or only the lower signs are to be taken simultaneously.

COROLLARY 3.1.7. *The algebra* $\mathrm{Y}(\mathfrak{sl}_N)$ *is generated by the elements* κ_{ik} *and* $\xi^\pm_{ik}$ *with* $i = 1,\dots,N-1$ *and* $k \geqslant 0$, *subject only to the following relations:*

$$[\kappa_{ik}, \kappa_{jl}] = 0, \tag{3.46}$$

$$[\xi^+_{ik}, \xi^-_{jl}] = \delta_{ij}\, \kappa_{i,k+l}, \tag{3.47}$$

$$[\kappa_{i0}, \xi^\pm_{jl}] = \pm\, a_{ij}\, \xi^\pm_{jl}, \tag{3.48}$$

$$[\kappa_{i,k+1}, \xi^\pm_{jl}] - [\kappa_{ik}, \xi^\pm_{j,l+1}] = \pm\, \frac{a_{ij}}{2}\, (\kappa_{ik}\, \xi^\pm_{jl} + \xi^\pm_{jl}\, \kappa_{ik}), \tag{3.49}$$

$$[\xi^\pm_{i,k+1}, \xi^\pm_{jl}] - [\xi^\pm_{ik}, \xi^\pm_{j,l+1}] = \pm\, \frac{a_{ij}}{2}\, (\xi^\pm_{ik}\, \xi^\pm_{jl} + \xi^\pm_{jl}\, \xi^\pm_{ik}), \tag{3.50}$$

$$\sum_{p\in\mathfrak{S}_n} [\xi^\pm_{ik_{p(1)}}, [\xi^\pm_{ik_{p(2)}}, \dots [\xi^\pm_{ik_{p(n)}}, \xi^\pm_{jl}] \dots]] = 0, \tag{3.51}$$

where the last relation holds for all $i \neq j$, *and we denoted* $n = 1 - a_{ij}$.

PROOF. We verify first that the elements κ_{ik} and $\xi^\pm_{ik}$ generate the subalgebra $\mathrm{Y}(\mathfrak{sl}_N)$ of $\mathrm{Y}(\mathfrak{gl}_N)$. In view of Theorem 1.8.2, it suffices to check that these elements together with the coefficients d_k of the quantum determinant $\operatorname{qdet} T(u)$ generate the entire algebra $\mathrm{Y}(\mathfrak{gl}_N)$. By (3.44) we have

$$\operatorname{qdet} T(u) = \widetilde{h}_1(u)\, \widetilde{h}_2(u+1/2) \dots \widetilde{h}_N\big(u + (N-1)/2\big),$$

which implies the identity

$$\widetilde{h}_1(u)\,\widetilde{h}_1(u+1)\dots\widetilde{h}_1(u+N-1)$$
$$= \operatorname{qdet} T(u) \prod_{i=1}^{N-1} \kappa_i\Big(u+\frac{i+1}{2}\Big)\kappa_i\Big(u+\frac{i+3}{2}\Big)\dots\kappa_i\Big(u+N-\frac{i+1}{2}\Big).$$

Hence the coefficients of the series $\widetilde{h}_1(u)$ are polynomials in the elements d_k and κ_{ik}. The same property for the coefficients of the series $\widetilde{h}_i(u)$ for $i = 2,\dots,N$ follows by induction from the relations

$$\widetilde{h}_{i+1}(u) = \widetilde{h}_i(u+1/2)\,\kappa_i(u+1/2)^{-1}, \qquad i = 1,\dots,N-1.$$

So, since $\mathrm{Y}(\mathfrak{gl}_N)$ is generated by the coefficients of the series $\widetilde{h}_i(u)$ and $\xi_i^{\pm}(u)$, the claim follows.

Now we verify that the relations hold in $\mathrm{Y}(\mathfrak{gl}_N)$. All the coefficients of the series $h_i(u)$ are contained in the commutative subalgebra H_N of $\mathrm{Y}(\mathfrak{gl}_N)$. Therefore, their images under the automorphism υ_N commute with each other, which gives (3.46). Further, using (3.1) and (3.10) we obtain

$$(u-v)\,[e_i(u), f_i(v)] = h_i'(u)\,h_{i+1}(u) - h_i'(v)\,h_{i+1}(v).$$

Now replace i by $N-i$, apply the automorphism υ_N and then replace u and v respectively by $v - N + (i+1)/2$ and $u - N + (i+1)/2$ to get

$$(u-v)\,[\xi_i^+(u), \xi_i^-(v)] = \kappa_i(v) - \kappa_i(u).$$

Applying (3.35) and (3.5) gives (3.47).

Furthermore, using (3.1), we deduce from (3.8) and (3.9) that

$$(u-v)\,[h_i'(u)\,h_{i+1}(u),\, e_i(v)]$$
$$= (e_i(u) - e_i(v))\,h_i'(u)\,h_{i+1}(u) + h_i'(u)\,h_{i+1}(u)\,(e_i(u) - e_i(v)).$$

In the same way as above, we obtain

$$(u-v)\,[\kappa_i(u), \xi_i^-(v)] = \kappa_i(u)\,(\xi_i^-(u) - \xi_i^-(v)) + (\xi_i^-(u) - \xi_i^-(v))\,\kappa_i(u).$$

Now equate the coefficients of $u^{-k-1}v^{-l-1}$ to get the corresponding relations (3.48) and (3.49) with $i = j$. Next, using (3.3) and (3.8) we get

$$(u-v)\,[h_i'(u)\,h_{i+1}(u),\, e_{i+1}(v)] = h_i'(u)\,h_{i+1}(u)\,(e_{i+1}(v) - e_{i+1}(u)).$$

Replace i by $N-i$, apply the automorphism υ_N and then replace u and v respectively by $u - N + (i+1)/2$ and $v - N + i/2$ to get

$$(u - v + 1/2)\,[\kappa_i(u), \xi_{i-1}^-(v)] = \kappa_i(u)\,\big(\xi_{i-1}^-(v) - \xi_{i-1}^-(u+1/2)\big),$$

which implies

$$(u-v)\,[\kappa_i(u), \xi_{i-1}^-(v)] = \frac{1}{2}\Big(\kappa_i(u)\,\xi_{i-1}^-(v) + \xi_{i-1}^-(v)\,\kappa_i(u)\Big) - \kappa_i(u)\,\xi_{i-1}^-(u+1/2).$$

Equating the coefficients of $u^{-k-1}v^{-l-1}$ we deduce the corresponding case of (3.49) with $j = i-1$. The remaining cases of (3.48) and (3.49) are verified in the same way, using (3.3), (3.8), (3.9) and observing that the anti-automorphism t acts by

$$\text{(3.52)} \qquad t: \quad \kappa_{ik} \mapsto \kappa_{ik}, \qquad \xi_{ik}^+ \mapsto \xi_{ik}^-, \qquad \xi_{ik}^- \mapsto \xi_{ik}^+;$$

see Corollary 1.11.7. A similar argument with the use of (3.4), (3.11) and (3.14) proves (3.50). Finally, for $|i-j| > 1$ the relation (3.51) follows from (3.31), while for $|i-j|=1$ it follows from (3.1) and Lemma 3.1.4 by the application of the automorphism υ_N and using (3.52).

Thus, if we denote by $\widehat{\mathrm{Y}}(\mathfrak{sl}_N)$ the algebra with generators and relations as in Corollary 3.1.7, then there is a surjective homomorphism $\widehat{\mathrm{Y}}(\mathfrak{sl}_N) \to \mathrm{Y}(\mathfrak{sl}_N)$ which sends the generators κ_{ik} and $\xi^{\pm}_{ik}$ of $\widehat{\mathrm{Y}}(\mathfrak{sl}_N)$ to the elements of $\mathrm{Y}(\mathfrak{sl}_N)$ denoted by the same symbols. Corollary 3.1.6 implies the vector space decomposition

$$\mathrm{Y}(\mathfrak{sl}_N) = \upsilon_N\big(E_N\big) \otimes \big(\upsilon_N(H_N) \cap \mathrm{Y}(\mathfrak{sl}_N)\big) \otimes \upsilon_N\big(F_N\big). \tag{3.53}$$

Note that the image $\upsilon_N(H_N)$ of the Gelfand–Tsetlin subalgebra H_N under the automorphism υ_N is the commutative subalgebra of $\mathrm{Y}(\mathfrak{gl}_N)$ generated by all coefficients of the series $\widetilde{h}_1(u), \dots, \widetilde{h}_N(u)$. The subalgebras $\upsilon_N(E_N)$ and $\upsilon_N(F_N)$ are respectively generated by the elements ξ^-_{ik} and ξ^+_{ik} with $i = 1, \dots, N-1$ and $k \geqslant 0$. Now the injectivity of the homomorphism $\widehat{\mathrm{Y}}(\mathfrak{sl}_N) \to \mathrm{Y}(\mathfrak{sl}_N)$ follows from the decomposition (3.53) and the corresponding argument of the proof of Theorem 3.1.5 applied to the subalgebras of $\widehat{\mathrm{Y}}(\mathfrak{sl}_N)$ generated by the elements ξ^+_{ik} and ξ^-_{ik}, respectively. □

REMARK 3.1.8. The mapping

$$\begin{aligned} \kappa_i(u) &\mapsto h_i\Big(u - \frac{i-1}{2}\Big)^{-1} h_{i+1}\Big(u - \frac{i-1}{2}\Big), \\ \xi^+_i(u) &\mapsto f_i\Big(u - \frac{i-1}{2}\Big), \\ \xi^-_i(u) &\mapsto e_i\Big(u - \frac{i-1}{2}\Big) \end{aligned}$$

defines another isomorphism between the presentations of the Yangian $\mathrm{Y}(\mathfrak{sl}_N)$; cf. (3.43). We have used the composition with the automorphism υ_N in order to obtain a more natural correspondence between the parameters of the representations of $\mathrm{Y}(\mathfrak{gl}_N)$ and $\mathrm{Y}(\mathfrak{sl}_N)$; see Corollary 3.4.9 below. □

3.2. Highest weight representations

We will often be using formal series to describe representations of various algebras. Suppose that $a(u)$ and $b(u)$ are formal power series in u^{-1} with coefficients in a certain algebra. If ζ is an element of a module over this algebra, then the relation $a(u)\,\zeta = b(u)\,\zeta$ is understood in the sense that the actions on ζ of the corresponding coefficients of $a(u)$ and $b(u)$ coincide.

The following definition employs the presentation of the Yangian $\mathrm{Y}(\mathfrak{gl}_N)$ provided by Theorem 3.1.5.

DEFINITION 3.2.1. A representation L of the Yangian $\mathrm{Y}(\mathfrak{gl}_N)$ is called a *highest weight representation* if there exists a nonzero vector $\zeta \in L$ such that L is generated by ζ and the following relations hold:

$$e_i(u)\,\zeta = 0 \qquad \text{for} \quad 1 \leqslant i \leqslant N-1, \qquad \text{and} \tag{3.54}$$

$$h_i(u)\,\zeta = \lambda_i(u)\,\zeta \qquad \text{for} \quad 1 \leqslant i \leqslant N, \tag{3.55}$$

for some formal series

$$\lambda_i(u) = 1 + \lambda_i^{(1)} u^{-1} + \lambda_i^{(2)} u^{-2} + \dots, \qquad \lambda_i^{(r)} \in \mathbb{C}. \tag{3.56}$$

The vector ζ is called the *highest vector* of L, and the N-tuple of formal series $\lambda(u) = \big(\lambda_1(u), \dots, \lambda_N(u)\big)$ is the *highest weight* of L. □

Alternatively, highest weight representations can be described in terms of the original presentation of $\mathrm{Y}(\mathfrak{gl}_N)$; see Definition 1.1.1.

PROPOSITION 3.2.2. *A representation L of $\mathrm{Y}(\mathfrak{gl}_N)$ is a highest weight representation with the highest weight $\lambda(u) = \big(\lambda_1(u), \dots, \lambda_N(u)\big)$ if and only if there exists a nonzero vector $\zeta \in L$ such that L is generated by ζ and the following relations hold:*

$$t_{ij}(u)\,\zeta = 0 \qquad \text{for} \quad 1 \leqslant i < j \leqslant N, \qquad \text{and} \tag{3.57}$$

$$t_{ii}(u)\,\zeta = \lambda_i(u)\,\zeta \qquad \text{for} \quad 1 \leqslant i \leqslant N. \tag{3.58}$$

PROOF. Suppose that L is a highest weight representation with the highest weight $\lambda(u)$ and the highest vector $\zeta \in L$. Then using (3.7) we deduce that $e_{ij}(u)\,\zeta = 0$ for all $1 \leqslant i < j \leqslant N$. Using the expression for $e_{ij}(u)$ provided by Theorem 1.11.6, we obtain

$$t^{\,1\dots i}_{\,1\dots i-1,\,j}(u+i-1)\,\zeta = 0. \tag{3.59}$$

Now proceed by induction on i. If $i = 1$, then (3.57) coincides with (3.59). For $i > 1$ apply the relation (1.54) to rewrite (3.59) as

$$\sum_{p \in \mathfrak{S}_i} \operatorname{sgn} p \cdot t_{p(1),1}(u+i-1) \dots t_{p(i),j}(u)\,\zeta = 0.$$

By the induction hypothesis, this implies

$$t^{\,1\dots i-1}_{\,1\dots i-1}(u+i-1)\,t_{ij}(u)\,\zeta = 0,$$

which proves (3.57). Furthermore, applying again (1.54) and using (3.57) we get

$$t^{\,1\dots i}_{\,1\dots i}(u+i-1)\,\zeta = t^{\,1\dots i-1}_{\,1\dots i-1}(u+i-1)\,t_{ii}(u)\,\zeta.$$

Hence, due to Corollary 1.7.2 and (3.2), the relation (3.55) implies (3.58).

The converse statement is now immediate. □

We now introduce universal highest weight representations.

DEFINITION 3.2.3. Let $\lambda(u) = \big(\lambda_1(u), \dots, \lambda_N(u)\big)$ be an arbitrary tuple of formal series of the form (3.56). The *Verma module* $M(\lambda(u))$ is the quotient of $\mathrm{Y}(\mathfrak{gl}_N)$ by the left ideal generated by the elements $e_i^{(r)}$ with $1 \leqslant i \leqslant N-1$ and $r \geqslant 1$, and by $h_i^{(r)} - \lambda_i^{(r)}$ with $1 \leqslant i \leqslant N$ and $r \geqslant 1$.

Equivalently, $M(\lambda(u))$ is the quotient of $\mathrm{Y}(\mathfrak{gl}_N)$ by the left ideal generated by all the coefficients of the series $t_{ij}(u)$ for $1 \leqslant i < j \leqslant N$ and $t_{ii}(u) - \lambda_i(u)$ for $1 \leqslant i \leqslant N$. □

The equivalence of the two definitions of $M(\lambda(u))$ follows from the proof of Proposition 3.2.2. Clearly, $M(\lambda(u))$ is a highest weight representation of $\mathrm{Y}(\mathfrak{gl}_N)$ with the highest weight $\lambda(u)$ and the highest vector $1_{\lambda(u)}$ which is the image of the element $1 \in \mathrm{Y}(\mathfrak{gl}_N)$ in the quotient. Moreover, if L is a highest weight representation of $\mathrm{Y}(\mathfrak{gl}_N)$ with the highest weight $\lambda(u)$ and the highest vector ζ, then the mapping $1_{\lambda(u)} \mapsto \zeta$ defines a surjective $\mathrm{Y}(\mathfrak{gl}_N)$-module homomorphism $M(\lambda(u)) \to L$. Hence, L is isomorphic to the quotient of $M(\lambda(u))$ by the kernel of this homomorphism.

The next proposition is immediate from the two versions of the Poincaré–Birkhoff–Witt theorem; see Theorem 1.4.1 and Corollary 3.1.6.

PROPOSITION 3.2.4. *For any given order on the set of generators $f_{ji}^{(r)}$ with $1 \leqslant i < j \leqslant N$ and $r \geqslant 1$, the elements*

$$f_{j_1 i_1}^{(r_1)} \dots f_{j_m i_m}^{(r_m)} 1_{\lambda(u)}, \qquad m \geqslant 0,$$

with ordered products of the generators, form a basis of the Verma module $M(\lambda(u))$. Moreover, given any order on the set of generators $t_{ji}^{(r)}$ with $1 \leqslant i < j \leqslant N$ and $r \geqslant 1$, the elements

$$t_{j_1 i_1}^{(r_1)} \dots t_{j_m i_m}^{(r_m)} 1_{\lambda(u)}, \qquad m \geqslant 0,$$

with ordered products of the generators, form a basis of $M(\lambda(u))$. □

Recall that the quantum determinant $\operatorname{qdet} T(u)$ is a series in u^{-1} with coefficients in $\mathrm{Y}(\mathfrak{gl}_N)$; see Definition 1.6.5.

PROPOSITION 3.2.5. *Suppose that L is a highest weight representation of $\mathrm{Y}(\mathfrak{gl}_N)$ with the highest weight $\lambda(u) = (\lambda_1(u), \dots, \lambda_N(u))$. Then each coefficient of the quantum determinant $\operatorname{qdet} T(u)$ acts on L as multiplication by a scalar determined by*

$$\operatorname{qdet} T(u)|_L = \lambda_1(u) \dots \lambda_N(u - N + 1).$$

PROOF. By Theorem 1.7.5, all coefficients of the quantum determinant belong to the center of the Yangian $\mathrm{Y}(\mathfrak{gl}_N)$. Therefore, Proposition 3.2.4 implies that the action of $\operatorname{qdet} T(u)$ on L is determined by the application of $\operatorname{qdet} T(u)$ to the highest vector ζ of L. Now we use the expression for $\operatorname{qdet} T(u)$ provided by (1.52). Proposition 3.2.2 gives

$$\operatorname{qdet} T(u)\, \zeta = \lambda_1(u) \dots \lambda_N(u - N + 1)\, \zeta,$$

thus proving the claim. □

Identifying the elements $E_{ij} \in \mathfrak{gl}_N$ with their images $t_{ij}^{(1)}$ in $\mathrm{Y}(\mathfrak{gl}_N)$ under the embedding (1.6) and taking the coefficient of v^0 in (1.3) we deduce that

$$[E_{ij}, t_{kl}(u)] = \delta_{kj}\, t_{il}(u) - \delta_{il}\, t_{kj}(u), \tag{3.60}$$

and, hence, for any $r \geqslant 1$ we have

$$[E_{ij}, t_{kl}^{(r)}] = \delta_{kj}\, t_{il}^{(r)} - \delta_{il}\, t_{kj}^{(r)}. \tag{3.61}$$

Deploying the embedding (1.6), we may regard $M(\lambda(u))$ as a $\mathfrak{gl}_N$-module. For any N-tuple $\mu = (\mu_1, \dots, \mu_N)$ of complex numbers, set

$$M(\lambda(u))_\mu = \{\eta \in M(\lambda(u)) \mid E_{ii}\, \eta = \mu_i\, \eta, \quad i = 1, \dots, N\}.$$

Using the standard Lie algebra representation theory terminology, we call μ a *weight* of $M(\lambda(u))$ if $M(\lambda(u))_\mu \neq 0$; then $M(\lambda(u))_\mu$ is the corresponding weight space. Let us also introduce the standard partial ordering on the set of weights. For the diagonal Cartan subalgebra $\mathfrak{h}$ of $\mathfrak{gl}_N$, we denote by $\mathfrak{h}^*$ the dual vector space of $\mathfrak{h}$. We let $\varepsilon_1, \dots, \varepsilon_N$ denote the basis vectors of $\mathfrak{h}^*$ dual to the basis elements $E_{11}, \dots, E_{NN}$ of $\mathfrak{h}$, respectively. The weight μ can then be identified with the element $\mu_1\varepsilon_1 + \dots + \mu_N\varepsilon_N \in \mathfrak{h}^*$, where $\mu_i = \mu(E_{ii})$. Now, if α and β are two weights, then α *precedes* β if $\beta - \alpha$ is a $\mathbb{Z}_+$-linear combination of the N-tuples $\varepsilon_i - \varepsilon_j$ with $i < j$. By (3.58) and (3.61), the set of weights of $M(\lambda(u))$ coincides with that of the $\mathfrak{gl}_N$-Verma module with the highest weight $\lambda^{(1)} = (\lambda_1^{(1)}, \dots, \lambda_N^{(1)})$. In particular, this set has a unique maximal weight $\lambda^{(1)}$. The corresponding weight space $M(\lambda(u))_{\lambda^{(1)}}$ is one-dimensional and spanned by the highest vector $1_{\lambda(u)}$. By Proposition 3.2.4, we have the weight space decomposition

$$M(\lambda(u)) = \bigoplus_{\mu} M(\lambda(u))_{\mu}, \tag{3.62}$$

summed over all weights μ of $M(\lambda(u))$. Let K be a submodule of $M(\lambda(u))$. Acting on a vector $\eta \in K$ by the elements E_{ii} we easily deduce that all weight components of η must also belong to K. In other words, we have the weight space decomposition

$$K = \bigoplus_{\mu} K_{\mu}, \qquad K_{\mu} = K \cap M(\lambda(u))_{\mu}.$$

Since the Verma module $M(\lambda(u))$ is generated by the vector $1_{\lambda(u)}$, any proper submodule of $M(\lambda(u))$ has zero intersection with the weight space $M(\lambda(u))_{\lambda^{(1)}}$. Hence, the sum of all proper submodules is the unique maximal proper submodule of $M(\lambda(u))$. This motivates our next definition.

DEFINITION 3.2.6. The *irreducible highest weight representation* $L(\lambda(u))$ of $\mathrm{Y}(\mathfrak{gl}_N)$ with the highest weight $\lambda(u)$ is defined as the quotient of the Verma module $M(\lambda(u))$ by the unique maximal proper submodule. □

It is clear from the previous discussion that $L(\lambda(u))$ is isomorphic to the unique irreducible quotient of an arbitrary highest weight representation L with the highest weight $\lambda(u)$. Moreover, two irreducible highest weight representations are isomorphic if and only if they have the same highest weight.

THEOREM 3.2.7. *Every finite-dimensional irreducible representation L of the Yangian $\mathrm{Y}(\mathfrak{gl}_N)$ is a highest weight representation. Moreover, L contains a unique, up to a constant factor, highest vector.*

PROOF. Introduce the following subspace of L,

$$L^0 = \{\xi \in L \mid t_{ij}(u)\, \xi = 0, \qquad 1 \leqslant i < j \leqslant N\}. \tag{3.63}$$

We show first that L^0 is nonzero. Consider the set of weights of L, where L is regarded as a $\mathfrak{gl}_N$-module defined via the embedding (1.6). This set is finite and hence contains a maximal weight μ with respect to the partial ordering introduced above. The corresponding weight vector ξ belongs to L^0. Indeed, by (3.60) the weight of $t_{ij}(u)\, \xi$ is $\mu + \varepsilon_i - \varepsilon_j$. So, if $i < j$, then by the maximality of μ we must have $t_{ij}(u)\, \xi = 0$.

Next, we show that the subspace L^0 is invariant with respect to the action of all elements $t_{kk}^{(r)}$. Indeed, by (1.3) we have

$$(u - v)\, [t_{ij}(u), t_{kk}(v)] = t_{kj}(u)\, t_{ik}(v) - t_{kj}(v)\, t_{ik}(u) = t_{ik}(v)\, t_{kj}(u) - t_{ik}(u)\, t_{kj}(v).$$

Suppose that $i < j$. Using the first equality for $i < k$ and the second for $i \geqslant k$ we conclude that if $\xi \in L^0$, then $t_{kk}(v)\, \xi \in L^0$; that is, $t_{kk}^{(r)}\, \xi \in L^0$ for any r and k.

Furthermore, (1.3) implies that

$$[t_{ii}(u), t_{ii}(v)] = 0,$$

while for any $i < j$ we have

$$(u - v)\, [t_{ii}(u), t_{jj}(v)] = t_{ji}(u)\, t_{ij}(v) - t_{ji}(v)\, t_{ij}(u).$$

This shows that the elements $t_{kk}^{(r)}$ with $k = 1, \dots, N$ and $r \geqslant 1$ act on L^0 as pairwise commuting operators. Hence, any simultaneous eigenvector $\zeta \in L^0$ for these operators will satisfy the conditions of Proposition 3.2.2, because ζ generates

L due to the assumption that L is irreducible. In particular, ζ is a weight vector with a certain weight μ.

Finally, by Proposition 3.2.4 the vector space L is spanned by the elements

$$t_{j_1 i_1}^{(r_1)} \dots t_{j_m i_m}^{(r_m)} \zeta, \qquad m \geqslant 0,$$

with ordered products of the generators. Then (3.61) implies that the weight space L_μ is one-dimensional and spanned by the vector ζ. Moreover, if ν is a weight of L and $\nu \neq \mu$, then ν strictly precedes μ. This proves that the highest vector ζ of L is determined uniquely, up to a constant factor. □

Theorem 3.2.7 implies that every finite-dimensional irreducible representation of the Yangian $\mathrm{Y}(\mathfrak{gl}_N)$ is isomorphic to a unique irreducible highest weight representation $L(\lambda(u))$. Therefore, in order to describe all finite-dimensional irreducible representations, we need to find necessary and sufficient conditions on the highest weight $\lambda(u)$ which would determine whether $L(\lambda(u))$ is finite-dimensional. We will do this in the next sections, considering the key case of $\mathrm{Y}(\mathfrak{gl}_2)$ first.

The following property of highest weight representations (not necessarily finite-dimensional) follows by an obvious modification of the argument used in the proof of Theorem 3.2.7.

COROLLARY 3.2.8. *Suppose that L is an irreducible highest weight representation of $\mathrm{Y}(\mathfrak{gl}_N)$. Then the subspace L^0 defined in (3.63) is one-dimensional and spanned by the highest vector of L.* □

Given an N-tuple of complex numbers $\lambda = (\lambda_1, \dots, \lambda_N)$ we will denote by $L(\lambda)$ the irreducible representation of the Lie algebra $\mathfrak{gl}_N$ with the highest weight λ. So, $L(\lambda)$ is generated by a nonzero vector ζ such that

$$\begin{aligned} E_{ij}\,\zeta &= 0 \qquad \text{for} \quad 1 \leqslant i < j \leqslant N, \qquad \text{and} \\ E_{ii}\,\zeta &= \lambda_i\,\zeta \qquad \text{for} \quad 1 \leqslant i \leqslant N. \end{aligned} \tag{3.64}$$

The representation $L(\lambda)$ is finite-dimensional if and only if $\lambda_i - \lambda_{i+1} \in \mathbb{Z}_+$ for all $i = 1, \dots, N-1$. The evaluation homomorphism (1.5) allows us to equip any $L(\lambda)$ with a structure of $\mathrm{Y}(\mathfrak{gl}_N)$-module. Namely, for any indices $i, j \in \{1, \dots, N\}$ the generator $t_{ij}^{(1)}$ acts on $L(\lambda)$ as E_{ij}, while any generator $t_{ij}^{(r)}$ with $r \geqslant 2$ acts as the zero operator. We will keep the same notation $L(\lambda)$ for this $\mathrm{Y}(\mathfrak{gl}_N)$-module and call it the *evaluation module*. It should always be clear from the context whether $L(\lambda)$ is regarded as a representation of the Lie algebra $\mathfrak{gl}_N$ or the Yangian $\mathrm{Y}(\mathfrak{gl}_N)$. Note that $L(\lambda)$ is obviously a highest weight representation of the Yangian with the highest vector ζ, and the components of the highest weight are given by

$$\lambda_i(u) = 1 + \lambda_i u^{-1}, \qquad i = 1, \dots, N. \tag{3.65}$$

If L and M are any two $\mathrm{Y}(\mathfrak{gl}_N)$-modules, then the tensor product space $L \otimes M$ can be equipped with a $\mathrm{Y}(\mathfrak{gl}_N)$-action with the use of the comultiplication Δ on $\mathrm{Y}(\mathfrak{gl}_N)$ (see Section 1.5) by the rule

$$y \cdot (\xi \otimes \eta) = \Delta(y)(\xi \otimes \eta), \qquad y \in \mathrm{Y}(\mathfrak{gl}_N), \quad \xi \in L, \quad \eta \in M.$$

By the coassociativity of Δ, we may unambiguously define multiple tensor product modules of the form

$$L(\lambda^{(1)}) \otimes L(\lambda^{(2)}) \otimes \dots \otimes L(\lambda^{(k)}), \tag{3.66}$$

where each $L(\lambda^{(m)})$ is an evaluation module with $\lambda^{(m)} = (\lambda_1^{(m)}, \dots, \lambda_N^{(m)}) \in \mathbb{C}^N$. We let ζ_m denote the highest vector of $L(\lambda^{(m)})$ and set

$$\zeta = \zeta_1 \otimes \dots \otimes \zeta_k .$$

PROPOSITION 3.2.9. *The submodule* $\mathrm{Y}(\mathfrak{gl}_N)\zeta$ *of the* $\mathrm{Y}(\mathfrak{gl}_N)$*-module* (3.66) *is a highest weight representation with the highest vector* ζ *and the highest weight* $(\lambda_1(u), \dots, \lambda_N(u))$, *where*

$$\lambda_i(u) = (1 + \lambda_i^{(1)} u^{-1})(1 + \lambda_i^{(2)} u^{-1}) \dots (1 + \lambda_i^{(k)} u^{-1}). \tag{3.67}$$

PROOF. By definition, the submodule is generated by the vector ζ, so we only need to verify the conditions (3.57) and (3.58) for this vector. Using the definition (1.35) of Δ, for any $\eta_m \in L(\lambda^{(m)})$ with $m = 1, \dots, k$ we can write

$$\begin{aligned} &t_{ij}(u)(\eta_1 \otimes \dots \otimes \eta_k) \\ &\qquad = \sum_{a_1, \dots, a_{k-1}} t_{i a_1}(u)\, \eta_1 \otimes t_{a_1 a_2}(u)\, \eta_2 \otimes \dots \otimes t_{a_{k-1} j}(u)\, \eta_k , \end{aligned} \tag{3.68}$$

summed over $a_1, \dots, a_{k-1} \in \{1, \dots, N\}$. If $i < j$ and for every $m = 1, \dots, k$ we have $\eta_m = \zeta_m$, then each summand in the sum (3.68) is zero because it contains a factor of the form $t_{kl}(u)\, \zeta_m$ with $k < l$, which is zero. Similarly, if $i = j$, then the only nonzero summand corresponds to the case where each index a_m equals i. Using (3.65) we get the desired formula for $\lambda_i(u)$. □

PROPOSITION 3.2.10. *Suppose that the representation* (3.66) *of* $\mathrm{Y}(\mathfrak{gl}_N)$ *is irreducible and all tensor factors* $L(\lambda^{(m)})$ *are finite-dimensional. Then any permutation of the tensor factors in* (3.66) *gives an isomorphic representation of* $\mathrm{Y}(\mathfrak{gl}_N)$.

PROOF. Let us denote the $\mathrm{Y}(\mathfrak{gl}_N)$-module (3.66) by L and let L' be the representation obtained from L by some permutation of the tensor factors. By Proposition 3.2.9, the submodule of L' generated by the tensor product of the highest vectors of the $L(\lambda^{(m)})$ is a highest weight representation with the highest weight given by (3.67). Therefore, L is isomorphic to a subquotient of L'. Since L and L' have the same dimension, they have to be isomorphic. □

PROPOSITION 3.2.11. *Every generator* $t_{ij}^{(r)}$ *of* $\mathrm{Y}(\mathfrak{gl}_N)$ *with* $r \geqslant k+1$ *acts in the module* (3.66) *as the zero operator.*

PROOF. By the definition of the evaluation modules, if $\eta_m \in L(\lambda^{(m)})$, then $t_{kl}(u)\, \eta_m$ is a polynomial in u^{-1} of degree not more than 1. Therefore, the left-hand side of (3.68) is a polynomial in u^{-1} of degree not more than k. □

Given any finite-dimensional vector space L over $\mathbb{C}$ we will denote by L^* the dual vector space whose elements are lineal maps $\omega : L \to \mathbb{C}$. Suppose that an infinite-dimensional vector space L is a module over $\mathfrak{gl}_N$ with finite-dimensional weight subspaces,

$$L = \bigoplus_{\mu} L_{\mu}, \qquad \dim L_{\mu} < \infty. \tag{3.69}$$

Then we define the (restricted) dual vector space to L by

$$L^* = \bigoplus_{\mu} L^*_{\mu}. \tag{3.70}$$

In other words, elements of L^* are finite linear combinations of the vectors dual to the basis vectors of any weight basis of L. The space L^* can be equipped with a $\mathfrak{gl}_N$-module structure by

$$(E_{ij}\,\omega)(\eta) = \omega(-E_{N-i+1,N-j+1}\,\eta), \qquad \omega \in L^*, \ \eta \in L. \tag{3.71}$$

Denote by $\varkappa_N$ the anti-automorphism of the algebra $\mathrm{Y}(\mathfrak{gl}_N)$, defined by

$$\varkappa_N : \ t_{ij}(u) \mapsto t_{N-i+1,N-j+1}(-u). \tag{3.72}$$

This is the composition of the (commuting) anti-automorphism (1.25) and automorphism (1.95) of $\mathrm{Y}(\mathfrak{gl}_N)$. Suppose now that the $\mathfrak{gl}_N$-action on L is obtained by restriction of a $\mathrm{Y}(\mathfrak{gl}_N)$-action. Then the $\mathfrak{gl}_N$-module structure on L^* can be regarded as the restriction of the $\mathrm{Y}(\mathfrak{gl}_N)$-module structure defined by

$$(y\,\omega)(\eta) = \omega(\varkappa_N(y)\,\eta) \quad \text{for} \quad y \in \mathrm{Y}(\mathfrak{gl}_N) \quad \text{and} \quad \omega \in L^*, \ \eta \in L. \tag{3.73}$$

Observe that if K is a submodule of the $\mathrm{Y}(\mathfrak{gl}_N)$-module L, then its annihilator

$$\operatorname{Ann} K = \{\omega \in L^* \mid \omega(\eta) = 0 \quad \text{for all} \quad \eta \in K\} \tag{3.74}$$

is a submodule of the $\mathrm{Y}(\mathfrak{gl}_N)$-module L^*. Similarly, since the anti-automorphism $\varkappa_N$ is involutive, given a submodule M of L^*, the subspace

$$\operatorname{Ker} M = \{\eta \in L \mid \omega(\eta) = 0 \quad \text{for all} \quad \omega \in M\}$$

is a submodule of L.

We can now describe the dual Yangian module L^*, where L is the $\mathrm{Y}(\mathfrak{gl}_N)$-module (3.66). Note that the weight subspaces of L are finite-dimensional so that (3.69) holds. For any $\lambda = (\lambda_1, \dots, \lambda_N)$ we set $\widetilde{\lambda} = (-\lambda_N, \dots, -\lambda_1)$.

PROPOSITION 3.2.12. *The* $\mathrm{Y}(\mathfrak{gl}_N)$*-module* L^* *is isomorphic to the tensor product module*

$$L(\widetilde{\lambda}^{(1)}) \otimes L(\widetilde{\lambda}^{(2)}) \otimes \dots \otimes L(\widetilde{\lambda}^{(k)}).$$

PROOF. Verify first that the dual $\mathrm{Y}(\mathfrak{gl}_N)$-module $L(\lambda)^*$ to the evaluation module $L(\lambda)$ is isomorphic to $L(\widetilde{\lambda})$. Indeed, due to the above observations, the module $L(\lambda)^*$ is irreducible. Moreover, the highest vector of $L(\lambda)^*$ is the vector ζ^* defined by $\zeta^*(\zeta) = 1$ and $\zeta^*(\eta) = 0$ for all $\eta \in L(\lambda)_\mu$ with $\mu \neq \lambda$. The claim now follows by calculating the highest weight of $L(\lambda)^*$.

Furthermore, the vector space L^* can be naturally identified with the tensor product space

$$L(\lambda^{(1)})^* \otimes L(\lambda^{(2)})^* \otimes \dots \otimes L(\lambda^{(k)})^*.$$

Using the $\mathrm{Y}(\mathfrak{gl}_N)$-module isomorphisms $L(\lambda^{(i)})^* \cong L(\widetilde{\lambda}^{(i)})$, the claim is now deduced from the fact that the anti-automorphism $\varkappa_N$ commutes with the comultiplication Δ in the sense that

$$\Delta \circ \varkappa_N = (\varkappa_N \otimes \varkappa_N) \circ \Delta;$$

cf. Proposition 1.5.4. □

3.3. Representations of $\mathrm{Y}(\mathfrak{gl}_2)$

Consider the irreducible highest weight representation $L(\lambda(u))$ of $\mathrm{Y}(\mathfrak{gl}_2)$ with an arbitrary highest weight $\lambda(u) = \big(\lambda_1(u), \lambda_2(u)\big)$.

PROPOSITION 3.3.1. *If* $\dim L(\lambda(u)) < \infty$, *then there exists a formal series*

$$f(u) = 1 + f_1 u^{-1} + f_2 u^{-2} + \dots, \qquad f_r \in \mathbb{C},$$

such that $f(u)\lambda_1(u)$ *and* $f(u)\lambda_2(u)$ *are polynomials in* u^{-1}.

PROOF. By twisting the action of $\mathrm{Y}(\mathfrak{gl}_2)$ on $L(\lambda(u))$ by the automorphism (1.20) with $f(u) = \lambda_2(u)^{-1}$, we obtain a module over $\mathrm{Y}(\mathfrak{gl}_2)$ which is isomorphic to the irreducible highest weight representation $L(\nu(u), 1)$ with $\nu(u) = \lambda_1(u)/\lambda_2(u)$. So, we may assume without loss of generality that the highest weight of $L(\lambda(u))$ has the form $\lambda(u) = (\nu(u), 1)$. Let ζ denote the highest vector of $L(\nu(u), 1)$. Since this representation is finite-dimensional, the vectors $t_{21}^{(i)}\zeta \in L(\nu(u), 1)$ with $i \geqslant 1$ are linearly dependent. Therefore, the corresponding Verma module $M(\nu(u), 1)$ contains a nonzero vector ξ of the form

$$\xi = \sum_{i=1}^{m} c_i \, t_{21}^{(i)} \, 1_{\lambda(u)}, \qquad c_i \in \mathbb{C},$$

which belongs to the maximal proper submodule K of $M(\nu(u), 1)$. Here m is a positive integer, and we may assume that $c_m \neq 0$. Then we have $t_{12}^{(r)}\xi = 0$ for all $r \geqslant 1$ because otherwise the highest vector $1_{\lambda(u)}$ would belong to K. Write

$$\nu(u) = 1 + \nu^{(1)} u^{-1} + \nu^{(2)} u^{-2} + \dots, \qquad \nu^{(i)} \in \mathbb{C}.$$

By the defining relations (1.4), in $M(\nu(u), 1)$ we have

$$t_{12}^{(r)} \, t_{21}^{(i)} \, 1_{\lambda(u)} = \sum_{a=1}^{\min\{r,i\}} \Big(t_{22}^{(a-1)} t_{11}^{(r+i-a)} - t_{22}^{(r+i-a)} t_{11}^{(a-1)} \Big) 1_{\lambda(u)} = \nu^{(r+i-1)} \, 1_{\lambda(u)}.$$

Hence, for all $r \geqslant 1$ we have the relations

$$\sum_{i=1}^{m} c_i \, \nu^{(r+i-1)} = 0.$$

They imply

$$\nu(u) \, (c_1 + c_2 \, u + \dots + c_m \, u^{m-1}) = b_1 + b_2 \, u + \dots + b_m \, u^{m-1}$$

for some coefficients $b_i \in \mathbb{C}$ with $b_m = c_m$. Thus, taking now

$$f(u) = c_m^{-1} \sum_{i=1}^{m} c_i \, u^{-m+i}$$

we conclude that both $f(u)\,\nu(u)$ and $f(u)\,1$ are polynomials in u^{-1}. □

Proposition 3.3.1 implies that taking the composition of the representation of $\mathrm{Y}(\mathfrak{gl}_2)$ on $L(\lambda(u))$ with an appropriate automorphism of the form (1.20), we can get another highest weight representation of $\mathrm{Y}(\mathfrak{gl}_2)$ where both components of the highest weight are polynomials in u^{-1}. We now aim to investigate such representations of $\mathrm{Y}(\mathfrak{gl}_2)$.

For any $\alpha, \beta \in \mathbb{C}$ consider the irreducible highest weight representation $L(\alpha, \beta)$ of the Lie algebra $\mathfrak{gl}_2$ and equip it with a $\mathrm{Y}(\mathfrak{gl}_2)$-module structure as defined in Section 3.2. Let ζ denote the highest vector of $L(\alpha, \beta)$. Then

$$E_{11}\,\zeta = \alpha\,\zeta, \qquad E_{22}\,\zeta = \beta\,\zeta, \qquad E_{12}\,\zeta = 0.$$

Moreover, if $\alpha - \beta \in \mathbb{Z}_+$, then the vectors $(E_{21})^r\zeta$ with $r = 0, 1, \dots, \alpha - \beta$ form a basis of $L(\alpha, \beta)$ so that $\dim L(\alpha, \beta) = \alpha - \beta + 1$. If $\alpha - \beta \notin \mathbb{Z}_+$, then a basis of $L(\alpha, \beta)$ is formed by the vectors $(E_{21})^r\zeta$, where r runs over all nonnegative integers.

Now let $\lambda_1(u)$ and $\lambda_2(u)$ be polynomials in u^{-1} of degree not more than k. Write the decompositions

$$\begin{aligned} \lambda_1(u) &= (1 + \alpha_1 u^{-1}) \dots (1 + \alpha_k u^{-1}), \\ \lambda_2(u) &= (1 + \beta_1 u^{-1}) \dots (1 + \beta_k u^{-1}), \end{aligned} \tag{3.75}$$

where the constants α_i and β_i are complex numbers (some of them are zero if the degree of the corresponding polynomial is strictly less than k).

PROPOSITION 3.3.2. *Suppose that for every $i = 1, \dots, k-1$ the following condition holds: if the multiset $\{\alpha_p - \beta_q \mid i \leqslant p, q \leqslant k\}$ contains nonnegative integers, then $\alpha_i - \beta_i$ is minimal amongst them. Then the representation $L(\lambda_1(u), \lambda_2(u))$ of $\mathrm{Y}(\mathfrak{gl}_2)$ is isomorphic to the tensor product module*

$$L(\alpha_1, \beta_1) \otimes L(\alpha_2, \beta_2) \otimes \dots \otimes L(\alpha_k, \beta_k). \tag{3.76}$$

PROOF. Let us denote the module (3.76) by L and let ζ_i be the highest vector of $L(\alpha_i, \beta_i)$ for $i = 1, \dots, k$. By Proposition 3.2.9, the cyclic span $\mathrm{Y}(\mathfrak{gl}_2)\zeta$ of the vector $\zeta = \zeta_1 \otimes \dots \otimes \zeta_k$ is a highest weight module with the highest weight $(\lambda_1(u), \lambda_2(u))$. Therefore, the proposition will follow if we prove that the module L is irreducible.

We claim that any vector $\xi \in L$ satisfying $t_{12}(u)\xi = 0$ is proportional to ζ. We will prove this claim by induction on k. This is obvious for $k = 1$ so suppose that $k \geqslant 2$. Write any such vector ξ, which is assumed to be nonzero, in the form

$$\xi = \sum_{r=0}^{p} (E_{21})^r \zeta_1 \otimes \xi_r, \qquad \text{where} \quad \xi_r \in L(\alpha_2, \beta_2) \otimes \dots \otimes L(\alpha_k, \beta_k)$$

and p is some nonnegative integer. Moreover, if $\alpha_1 - \beta_1 \in \mathbb{Z}_+$, then we may and will assume that $p \leqslant \alpha_1 - \beta_1$. We also assume that $\xi_p \neq 0$. Applying $t_{12}(u)$ to ξ, with the use of (1.35) we get

$$\sum_{r=0}^{p} \Big(t_{11}(u)(E_{21})^r\zeta_1 \otimes t_{12}(u)\xi_r + t_{12}(u)(E_{21})^r\zeta_1 \otimes t_{22}(u)\xi_r \Big) = 0. \tag{3.77}$$

Using the definition of the Yangian action on $L(\alpha_1, \beta_1)$ and commutation relations in $\mathfrak{gl}_2$, we obtain

$$t_{11}(u)(E_{21})^r\zeta_1 = (1 + E_{11}u^{-1})(E_{21})^r\zeta_1 = (1 + (\alpha_1 - r)\,u^{-1})(E_{21})^r\zeta_1,$$

and

$$t_{12}(u)(E_{21})^r\zeta_1 = u^{-1}E_{12}\,(E_{21})^r\zeta_1 = u^{-1}\,r(\alpha_1 - \beta_1 - r + 1)(E_{21})^{r-1}\zeta_1.$$

Hence, taking the coefficient of $(E_{21})^p\zeta_1$ in (3.77) gives

$$(1 + (\alpha_1 - p)\,u^{-1})\,t_{12}(u)\xi_p = 0,$$

implying the relation $t_{12}(u)\,\xi_p = 0$. By the induction hypothesis, applied to the $Y(\mathfrak{gl}_2)$-module $L(\alpha_2,\beta_2)\otimes\ldots\otimes L(\alpha_k,\beta_k)$, the vector ξ_p must be proportional to $\zeta_2\otimes\ldots\otimes\zeta_k$. Therefore, using (3.68) we get

$$t_{22}(u)\,\xi_p = (1+\beta_2 u^{-1})\ldots(1+\beta_k u^{-1})\,\xi_p. \tag{3.78}$$

In order to complete the proof of the claim it now suffices to show that p must be equal to zero. Suppose by way of contradiction that $p\geqslant 1$. Then taking the coefficient of $(E_{21})^{p-1}\zeta_1$ in (3.77) we derive

$$(1+(\alpha_1-p+1)u^{-1})\,t_{12}(u)\,\xi_{p-1}+u^{-1}p(\alpha_1-\beta_1-p+1)\,t_{22}(u)\,\xi_p=0.$$

Hence, multiplying by u^k and taking into account (3.78) we get

$$(u+\alpha_1-p+1)u^{k-1}t_{12}(u)\,\xi_{p-1}+p(\alpha_1-\beta_1-p+1)(u+\beta_2)\ldots(u+\beta_k)\,\xi_p=0.$$

By Proposition 3.2.11, the vector $u^{k-1}t_{12}(u)\,\xi_{p-1}$ depends on u polynomially, so that by taking the value $u=-\alpha_1+p-1$ we obtain the relation

$$p(\alpha_1-\beta_1-p+1)(\alpha_1-\beta_2-p+1)\ldots(\alpha_1-\beta_k-p+1)=0.$$

But this is impossible due to the conditions on the parameters α_i and β_i. Thus, p must be zero and the claim follows.

Suppose now that M is a nonzero submodule of L. Then M must contain a nonzero vector ξ such that $t_{12}(u)\,\xi=0$. Indeed, this is immediate from (3.60) and the fact that the set of $\mathfrak{gl}_2$-weights of L has a maximal element. The above argument thus shows that M contains the vector ζ. It remains to prove that the cyclic span $K=Y(\mathfrak{gl}_2)\,\zeta$ coincides with L. By Proposition 3.2.12, the dual $Y(\mathfrak{gl}_2)$-module L^* is isomorphic to the tensor product

$$L(-\beta_1,-\alpha_1)\otimes\ldots\otimes L(-\beta_k,-\alpha_k).$$

Moreover, the highest vector ζ_i^* of the module $L(-\beta_i,-\alpha_i)\cong L(\alpha_i,\beta_i)^*$ can be identified with the element of $L(\alpha_i,\beta_i)^*$ such that $\zeta_i^*(\zeta_i)=1$ and $\zeta_i^*(\eta_i)=0$ for all weight vectors $\eta_i\in L(\alpha_i,\beta_i)$ whose weights are different from (α_i,β_i). Suppose on the contrary that the submodule K of L is proper. Then its annihilator $\operatorname{Ann}K$ defined as in (3.74) is a nonzero submodule of L^*, which does not contain the vector $\zeta_1^*\otimes\ldots\otimes\zeta_k^*$. However, this contradicts the claim verified in the first part of the proof, because the condition on the parameters α_i and β_i remain satisfied after we replace each α_i by $-\beta_i$ and each β_i by $-\alpha_i$. □

Note that the condition of Proposition 3.3.2 is not restrictive. Indeed, given arbitrary decompositions (3.75), choose a minimal nonnegative integer difference amongst all differences $\alpha_p-\beta_q$, if it exists. Re-enumerating the parameters if necessary, we may assume that this difference is $\alpha_1-\beta_1$. Then proceed by induction, considering the differences $\alpha_p-\beta_q$ with $p,q\geqslant 2$, etc.

THEOREM 3.3.3. *The irreducible highest weight representation $L(\lambda_1(u),\lambda_2(u))$ of $Y(\mathfrak{gl}_2)$ is finite-dimensional if and only if there exists a monic polynomial $P(u)$ in u such that*

$$\frac{\lambda_1(u)}{\lambda_2(u)}=\frac{P(u+1)}{P(u)}. \tag{3.79}$$

In this case $P(u)$ is unique.

PROOF. Suppose that the representation $L(\lambda_1(u), \lambda_2(u))$ is finite-dimensional. Then by Proposition 3.3.1 we can find a formal series $f(u)$ such that

$$f(u)\,\lambda_1(u) = (1 + \alpha_1 u^{-1}) \dots (1 + \alpha_k u^{-1}),$$
$$f(u)\,\lambda_2(u) = (1 + \beta_1 u^{-1}) \dots (1 + \beta_k u^{-1}),$$

for some $k \geqslant 0$ and some complex parameters α_i, β_i. Re-enumerate these parameters, if necessary, to satisfy the condition of Proposition 3.3.2. By this proposition, all differences $\alpha_i - \beta_i$ must be nonnegative integers because the representation $L(\lambda_1(u), \lambda_2(u))$ is finite-dimensional. Then the polynomial

$$P(u) = \prod_{i=1}^{k} (u + \beta_i)(u + \beta_i + 1) \dots (u + \alpha_i - 1) \tag{3.80}$$

obviously satisfies (3.79).

Conversely, suppose (3.79) holds for a polynomial $P(u) = (u + \gamma_1) \dots (u + \gamma_p)$. Set

$$\mu_1(u) - (1 + (\gamma_1 + 1)u^{-1}) \dots (1 + (\gamma_p + 1)u^{-1}),$$
$$\mu_2(u) = (1 + \gamma_1 u^{-1}) \dots (1 + \gamma_p u^{-1}),$$

and consider the tensor product module

$$L = L(\gamma_1 + 1, \gamma_1) \otimes L(\gamma_2 + 1, \gamma_2) \otimes \dots \otimes L(\gamma_p + 1, \gamma_p)$$

of $Y(\mathfrak{gl}_2)$. Obviously, this module is finite-dimensional. By Proposition 3.2.9, the cyclic $Y(\mathfrak{gl}_2)$-span of the tensor product of the highest vectors of $L(\gamma_i + 1, \gamma_i)$ is a highest weight module with the highest weight $(\mu_1(u), \mu_2(u))$. Since this submodule is finite-dimensional, then so is its irreducible quotient $L(\mu_1(u), \mu_2(u))$. Since

$$\frac{\mu_1(u)}{\mu_2(u)} = \frac{\lambda_1(u)}{\lambda_2(u)},$$

there exists an automorphism of $Y(\mathfrak{gl}_2)$ of the form (1.20) such that its composition with the representation $L(\mu_1(u), \mu_2(u))$ is isomorphic to $L(\lambda_1(u), \lambda_2(u))$. Thus, the latter is also finite-dimensional.

Finally, suppose that $Q(u)$ is another monic polynomial in u and

$$\frac{P(u+1)}{P(u)} = \frac{Q(u+1)}{Q(u)}.$$

This means that the ratio $P(u)/Q(u)$ is periodic in u, which is only possible for $P(u) = Q(u)$. □

The polynomial $P(u)$ defined in Theorem 3.3.3 is called the *Drinfeld polynomial* of the finite-dimensional representation $L(\lambda_1(u), \lambda_2(u))$.

Relation (3.79) motivates the notation which will be used in the following. For two formal series $\lambda_1(u)$ and $\lambda_2(u)$ in u^{-1} we will write

$$\lambda_1(u) \longrightarrow \lambda_2(u) \tag{3.81}$$

if there exists a monic polynomial $P(u)$ in u such that (3.79) holds. If $\lambda_1(u)$ and $\lambda_2(u)$ are polynomials in u^{-1} as in (3.75), then (3.81) is equivalent to the conditions

$$\alpha_i - \beta_i \in \mathbb{Z}_+, \qquad i = 1, \dots, k, \tag{3.82}$$

for an appropriate renumbering of the α_i.

Recall that the Yangian $Y(\mathfrak{sl}_2)$ is the subalgebra of $Y(\mathfrak{gl}_2)$ which consists of the elements stable under all automorphisms of the form (1.20); see Section 1.8.

COROLLARY 3.3.4. *The isomorphism classes of finite-dimensional irreducible representations of the Yangian* $\mathrm{Y}(\mathfrak{sl}_2)$ *are parameterized by monic polynomials in* u. *Every such representation is isomorphic to the restriction of a* $\mathrm{Y}(\mathfrak{gl}_2)$*-module of the form*

$$L(\alpha_1, \beta_1) \otimes L(\alpha_2, \beta_2) \otimes \ldots \otimes L(\alpha_k, \beta_k), \tag{3.83}$$

where each difference $\alpha_i - \beta_i$ *is a positive integer.*

PROOF. The elements of the center $\mathrm{ZY}(\mathfrak{gl}_2)$ of the Yangian $\mathrm{Y}(\mathfrak{gl}_2)$ act on any finite-dimensional irreducible representation of $\mathrm{Y}(\mathfrak{gl}_2)$ as multiplications by scalars. Hence, by Theorem 1.8.2, the restriction of any such representation to the subalgebra $\mathrm{Y}(\mathfrak{sl}_2)$ remains irreducible. Moreover, every finite-dimensional irreducible representation L of $\mathrm{Y}(\mathfrak{sl}_2)$ can be obtained by such a restriction. Indeed, due to Theorem 1.8.2, L can be extended to a representation of $\mathrm{Y}(\mathfrak{gl}_2)$ by letting the elements of the center $\mathrm{ZY}(\mathfrak{gl}_2)$ act on L as multiplications by certain scalars. Therefore, by Theorem 3.3.3, we only need to describe possible restrictions of the finite-dimensional representations $L(\lambda_1(u), \lambda_2(u))$ to $\mathrm{Y}(\mathfrak{sl}_2)$. By the definition of $\mathrm{Y}(\mathfrak{sl}_2)$, this restriction depends only on the ratio $\lambda_1(u)/\lambda_2(u)$. Conversely, consider the series $\widetilde{t}_{ij}(u)$ with coefficients in $\mathrm{Y}(\mathfrak{sl}_2)$, defined by (1.62). The eigenvalues for the action of $\widetilde{t}_{11}(u)$ and $\widetilde{t}_{22}(u)$ on the highest vector ζ of $L(\lambda_1(u), \lambda_2(u))$ uniquely determine the ratio $\lambda_1(u)/\lambda_2(u)$. This proves the first part of the corollary.

Given a monic polynomial $P(u)$ we may find polynomials $\lambda_1(u)$ and $\lambda_2(u)$ in u^{-1} such that relation (3.79) holds; see the proof of Theorem 3.3.3. Then by Proposition 3.3.2 the representation $L(\lambda_1(u), \lambda_2(u))$ of $\mathrm{Y}(\mathfrak{gl}_2)$ is isomorphic to a tensor product (3.83), where all differences $\alpha_i - \beta_i$ are nonnegative integers, and its restriction to $\mathrm{Y}(\mathfrak{sl}_2)$ yields the finite-dimensional irreducible representation corresponding to $P(u)$. Possible zero differences can be excluded, since the restrictions of $L(\lambda_1(u), \lambda_2(u))$ and $L(\gamma(u)\lambda_1(u), \gamma(u)\lambda_2(u))$ with $\gamma(u) = 1 + \gamma u^{-1}$ to the subalgebra $\mathrm{Y}(\mathfrak{sl}_2)$ are isomorphic. □

We will now establish a criterion of irreducibility of the representations of the form (3.83). Define the *string* corresponding to a pair of complex numbers (α, β) with $\alpha - \beta \in \mathbb{Z}_+$ as the set

$$S(\alpha, \beta) = \{\beta, \beta + 1, \ldots, \alpha - 1\}.$$

If $\alpha = \beta$, then the set $S(\alpha, \beta)$ is regarded to be empty.

DEFINITION 3.3.5. Two strings S_1 and S_2 are *in general position* if either

(i) $S_1 \cup S_2$ is not a string, or
(ii) $S_1 \subset S_2$, or $S_2 \subset S_1$. □

Denote by L the tensor product (3.83), where all differences $\alpha_i - \beta_i$ are nonnegative integers.

COROLLARY 3.3.6. *The representation* L *of* $\mathrm{Y}(\mathfrak{gl}_2)$ *(or* $\mathrm{Y}(\mathfrak{sl}_2)$*) is irreducible if and only if the strings* $S(\alpha_1, \beta_1), \ldots, S(\alpha_k, \beta_k)$ *are pairwise in general position.*

PROOF. As every irreducible finite-dimensional representation of $\mathrm{Y}(\mathfrak{gl}_2)$ remains irreducible when restricted to $\mathrm{Y}(\mathfrak{sl}_2)$, it suffices to consider L as a $\mathrm{Y}(\mathfrak{gl}_2)$-module.

Suppose that the strings are pairwise in general position and assume first that $\alpha_1 - \beta_1 \leqslant \dots \leqslant \alpha_k - \beta_k$. Then one easily verifies that the condition of Proposition 3.3.2 is satisfied for the parameters α_i and β_i, and so L is irreducible. Then any permutation of the tensor factors yields an isomorphic (and hence, irreducible) representation by Proposition 3.2.10.

Conversely, let $k = 2$ and let $L(\alpha_1, \beta_1) \otimes L(\alpha_2, \beta_2)$ be irreducible. Suppose that the strings $S(\alpha_1, \beta_1)$ and $S(\alpha_2, \beta_2)$ are not in general position. Then $\alpha_2 - \beta_1$, $\alpha_1 - \beta_2 \in \mathbb{Z}_+$ and the strings $S(\alpha_1, \beta_2)$ and $S(\alpha_2, \beta_1)$ are in general position. Hence, the representation $L(\alpha_1, \beta_2) \otimes L(\alpha_2, \beta_1)$ of $\mathrm{Y}(\mathfrak{gl}_2)$ is irreducible due to the first part of the proof. Since it has the same highest weight as the irreducible representation $L(\alpha_1, \beta_1) \otimes L(\alpha_2, \beta_2)$, these two representations have to be isomorphic and, in particular, have the same dimension:

$$(\alpha_1 - \beta_1 + 1)(\alpha_2 - \beta_2 + 1) = (\alpha_1 - \beta_2 + 1)(\alpha_2 - \beta_1 + 1).$$

This implies that $(\alpha_1 - \alpha_2)(\beta_1 - \beta_2) = 0$, and hence that $S(\alpha_1, \beta_1)$ and $S(\alpha_2, \beta_2)$ are in general position. This is a contradiction.

Now consider the general case. Suppose that L is irreducible, but a pair of strings $S(\alpha_i, \beta_i)$ and $S(\alpha_j, \beta_j)$ is not in general position. Applying Proposition 3.2.10 and permuting the tensor factors in (3.83) if necessary, we may assume that i and j are adjacent. However, the representation $L(\alpha_i, \beta_i) \otimes L(\alpha_j, \beta_j)$ of $\mathrm{Y}(\mathfrak{gl}_2)$ is reducible as shown above. This implies that L is reducible, a contradiction. □

Since the Drinfeld polynomial for the representation (3.83) is given by the formula (3.80), Corollaries 3.3.4 and 3.3.6 imply that any monic polynomial $P(u)$ can be written in the form

$$P(u) = \prod_{i=1}^{k} \prod_{\gamma \in S_i} (u + \gamma), \tag{3.84}$$

where $S_1, \dots, S_k$ are nonempty strings which are pairwise in general position. By writing each string S_i in the form $S(\alpha_i, \beta_i)$, for some parameters α_i, β_i, we thus recover the corresponding irreducible Yangian module (3.83).

PROPOSITION 3.3.7. *For any monic polynomial $P(u)$, the multiset of strings $S(\alpha_1, \beta_1), \dots, S(\alpha_k, \beta_k)$ determined by the expansion (3.84) is unique. Moreover, the $\mathrm{Y}(\mathfrak{sl}_2)$-module of the form (3.83) corresponding to the Drinfeld polynomial $P(u)$ is determined uniquely, up to a permutation of the tensor factors.*

PROOF. Suppose that

$$P(u) = \prod_{i=1}^{l} \prod_{\gamma \in S_i'} (u + \gamma)$$

is another expansion, where $S_1', \dots, S_l'$ are nonempty strings which are pairwise in general position. Then for each i the string S_i' has the form $S(\alpha_i', \beta_i')$ for some parameters α_i', β_i' such that $\alpha_i' - \beta_i'$ is a positive integer. Then the $\mathrm{Y}(\mathfrak{sl}_2)$-module (3.83) is isomorphic to the tensor product

$$L(\alpha_1', \beta_1') \otimes L(\alpha_2', \beta_2') \otimes \dots \otimes L(\alpha_l', \beta_l'). \tag{3.85}$$

Considering the $\mathrm{Y}(\mathfrak{gl}_2)$-highest weights of the modules (3.83) and (3.85), we deduce from (3.79) that

$$\frac{(u+\alpha_1)\dots(u+\alpha_k)}{(u+\beta_1)\dots(u+\beta_k)} = \frac{(u+\alpha'_1)\dots(u+\alpha'_l)}{(u+\beta'_1)\dots(u+\beta'_l)}.$$

However, both fractions are in a reduced form, as the equalities $\alpha_i = \beta_j$ or $\alpha'_i = \beta'_j$ are impossible because each multiset of strings is pairwise in general position. This implies that $l = k$ and

$$\{\alpha_1, \dots, \alpha_k\} = \{\alpha'_1, \dots, \alpha'_k\}, \qquad \{\beta_1, \dots, \beta_k\} = \{\beta'_1, \dots, \beta'_k\}.$$

Permuting the tensor factors in (3.85), if necessary, we may assume that $\alpha'_i = \alpha_i$ for each i. Then $\beta'_i = \beta_{p(i)}$ for a permutation $p \in \mathfrak{S}_k$. Both claims of the proposition will follow if we show that $\beta_{p(i)} = \beta_i$ for all i. Considering for any fixed i the multisets of strings corresponding to the subset of indices of the form $\{i, p(i), p^2(i), \dots\}$, we can reduce the task to the particular case where p is the cycle $(1\,2\,\dots\,k)$. Thus we need to show that if the strings $S(\alpha_1, \beta_2), S(\alpha_2, \beta_3), \dots, S(\alpha_k, \beta_1)$ are pairwise in general position, then $\beta_1 = \beta_2 = \dots = \beta_k$ or $\alpha_1 = \alpha_2 = \dots = \alpha_k$. Suppose this is not the case. Observe that if $\beta_i = \beta_{i+1}$ for some $i \in \{1, \dots, k\}$ (assuming $\beta_{k+1} = \beta_1$), then the strings $S(\alpha_i, \beta_i)$ and $S(\alpha_i, \beta_{i+1})$ coincide. Excluding such strings, we may assume that $\beta_i \neq \beta_{i+1}$ and, similarly, $\alpha_i \neq \alpha_{i+1}$ for all $i \in \{1, \dots, k\}$. Now, $\beta_2 - \beta_1$ is an integer because both $\alpha_1 - \beta_1$ and $\alpha_1 - \beta_2$ are positive integers. Suppose first that $\beta_2 - \beta_1$ is positive. Then $\beta_2 \in S(\alpha_1, \beta_1)$. Since $S(\alpha_1, \beta_1)$ and $S(\alpha_2, \beta_2)$ are in general position, we must also have $\alpha_2 \in S(\alpha_1, \beta_2)$. Then, since $S(\alpha_1, \beta_2)$ and $S(\alpha_2, \beta_3)$ are in general position, we must have $\beta_3 \in S(\alpha_2, \beta_2)$. Continuing by induction, we see that $\alpha_k \in S(\alpha_{k-1}, \beta_k)$. However, $\beta_k - \beta_1$ is a positive integer, which implies that $S(\alpha_{k-1}, \beta_k)$ and $S(\alpha_k, \beta_1)$ are not in general position, a contradiction. The same argument applies to the case where $\beta_1 - \beta_2$ is a positive integer. □

We complete this section by producing bases of Gelfand–Tsetlin type for a class of representations of the Yangian $\mathrm{Y}(\mathfrak{gl}_2)$. By Proposition 3.2.11, the generators $t_{ij}^{(r)}$ with $r \geqslant k+1$ act as zero operators in the representation L of $\mathrm{Y}(\mathfrak{gl}_2)$ given in (3.83). Therefore, the operators in L defined by $T_{ij}(u) = u^k\, t_{ij}(u)$ are polynomials in u:

$$T_{ij}(u) = \delta_{ij}\, u^k + t_{ij}^{(1)} u^{k-1} + \dots + t_{ij}^{(k)}. \tag{3.86}$$

Suppose that each difference $\alpha_i - \beta_i$ is a nonnegative integer and let ζ_i be the highest vector of the $\mathfrak{gl}_2$-module $L(\alpha_i, \beta_i)$ for $i = 1, \dots, k$. By Proposition 3.2.9, for the vector $\zeta = \zeta_1 \otimes \dots \otimes \zeta_k$ we have $T_{12}(u)\,\zeta = 0$ and

$$\begin{aligned} T_{11}(u)\,\zeta &= (u+\alpha_1)\dots(u+\alpha_k)\,\zeta, \\ T_{22}(u)\,\zeta &= (u+\beta_1)\dots(u+\beta_k)\,\zeta. \end{aligned} \tag{3.87}$$

Let a k-tuple $\gamma = (\gamma_1, \dots, \gamma_k)$ of complex numbers satisfy the conditions

$$\alpha_i - \gamma_i \in \mathbb{Z}_+, \qquad \gamma_i - \beta_i \in \mathbb{Z}_+, \qquad i = 1, \dots, k. \tag{3.88}$$

Define the corresponding vector $\eta_\gamma \in L$ by

$$\eta_\gamma = \prod_{i=1}^{k} T_{21}(-\gamma_i + 1)\, T_{21}(-\gamma_i + 2) \dots T_{21}(-\beta_i)\,\zeta. \tag{3.89}$$

In particular, $\eta_\beta = \zeta$ for $\beta = (\beta_1, \dots, \beta_k)$. Note that the ordering of the factors in the product is irrelevant as the operators $T_{21}(u)$ and $T_{21}(v)$ in L commute due to the defining relations in $\mathrm{Y}(\mathfrak{gl}_2)$.

THEOREM 3.3.8. *Suppose that the* $\mathrm{Y}(\mathfrak{gl}_2)$*-module* L *is irreducible and the strings* $S(\alpha_i, \beta_i)$, $i = 1, \dots, k$, *are pairwise disjoint. Then the vectors* η_γ *with* γ *satisfying* (3.88) *form a basis of* L. *Moreover, the generators of* $\mathrm{Y}(\mathfrak{gl}_2)$ *act in this basis by the rules*

$$T_{21}(-\gamma_i)\, \eta_\gamma = \eta_{\gamma+\delta_i}, \tag{3.90}$$

$$T_{12}(-\gamma_i)\, \eta_\gamma = -\prod_{j=1}^{k} (\alpha_j - \gamma_i + 1)(\beta_j - \gamma_i)\, \eta_{\gamma-\delta_i}, \tag{3.91}$$

for $i = 1, \dots, k$, *and*

$$T_{11}(u)\, \eta_\gamma = \prod_{i=1}^{k} \frac{(u + \alpha_i)(u + \beta_i - 1)}{u + \gamma_i - 1}\, \eta_\gamma + \prod_{i=1}^{k} \frac{1}{u + \gamma_i - 1}\, T_{21}(u)\, T_{12}(u-1)\, \eta_\gamma, \tag{3.92}$$

$$T_{22}(u)\, \eta_\gamma = (u + \gamma_1) \dots (u + \gamma_k)\, \eta_\gamma. \tag{3.93}$$

Here the k*-tuple* $\gamma \pm \delta_i$ *is obtained from* γ *by replacing* γ_i *with* $\gamma_i \pm 1$, *and* $\eta_{\gamma\pm\delta_i}$ *is considered to be zero if the* k*-tuple* $\gamma \pm \delta_i$ *does not satisfy* (3.88).

PROOF. Note that the polynomials $T_{12}(u)$ and $T_{21}(u)$ have degree $\leqslant k - 1$. Therefore, the actions of $T_{12}(u)$ and $T_{21}(u)$ on η_γ are determined by the Lagrange interpolation formula with the interpolation points $-\gamma_1, \dots, -\gamma_k$ which are distinct by the condition on the strings $S(\alpha_i, \beta_i)$. For instance,

$$T_{21}(u) = \sum_{i=1}^{k} \frac{(u + \gamma_1) \cdots \wedge_i \dots (u + \gamma_k)}{(\gamma_1 - \gamma_i) \cdots \wedge_i \dots (\gamma_k - \gamma_i)}\, T_{21}(-\gamma_i),$$

where $\wedge_i$ indicates that the i-th factor in a product is omitted. A similar formula holds for $T_{12}(u)$.

By the defining relations (1.3), we have

$$T_{22}(u)\, T_{21}(v) = \frac{u - v + 1}{u - v}\, T_{21}(v)\, T_{22}(u) - \frac{1}{u - v}\, T_{21}(u)\, T_{22}(v).$$

So, if $\gamma_i - \beta_i \geqslant 1$ for some i, then

$$\begin{aligned} T_{22}(u)\, \eta_\gamma &= T_{22}(u)\, T_{21}(-\gamma_i + 1)\, \eta_{\gamma-\delta_i} \\ &= \frac{u + \gamma_i}{u + \gamma_i - 1}\, T_{21}(-\gamma_i + 1)\, T_{22}(u)\, \eta_{\gamma-\delta_i} \\ &- \frac{1}{u + \gamma_i - 1}\, T_{21}(u)\, T_{22}(-\gamma_i + 1)\, \eta_{\gamma-\delta_i}. \end{aligned}$$

Now (3.93) follows by induction with the use of (3.87).

Furthermore, recall that the quantum determinant $\operatorname{qdet} T(u)$ for the Yangian $\mathrm{Y}(\mathfrak{gl}_2)$ is given in Example 1.6.7. The series $D(u) = u^k (u-1)^k \operatorname{qdet} T(u)$ acts in L

as a polynomial in u of the form

$$D(u) = T_{11}(u)\,T_{22}(u-1) - T_{21}(u)\,T_{12}(u-1) \tag{3.94}$$

$$= T_{11}(u-1)\,T_{22}(u) - T_{12}(u-1)\,T_{21}(u). \tag{3.95}$$

By Theorem 1.7.5, $D(u)$ acts in L as multiplication by a polynomial in u. This polynomial is found by the application of $D(u)$ to the highest vector ζ of L. Since $T_{12}(u)\,\zeta = 0$ we find from (3.87) and (3.94) that

$$D(u)|_L = \prod_{j=1}^{k} (u + \alpha_j)(u + \beta_j - 1). \tag{3.96}$$

If $\gamma_i - \beta_i \geqslant 1$, then by (3.95) we have

$$\begin{aligned} T_{12}(-\gamma_i)\,\eta_\gamma &= T_{12}(-\gamma_i)\,T_{21}(-\gamma_i+1)\,\eta_{\gamma-\delta_i} \\ &= \Big(T_{11}(-\gamma_i)\,T_{22}(-\gamma_i+1) - D(-\gamma_i+1)\Big)\,\eta_{\gamma-\delta_i}. \end{aligned}$$

So, the application of (3.93) and (3.96) proves (3.91) in the case under consideration. If $\gamma_i = \beta_i$, then (3.91) is verified by the same induction argument as (3.93), with the use of the relation

$$[T_{12}(u), T_{21}(v)] = \frac{1}{u-v}\Big(T_{11}(v)\,T_{22}(u) - T_{11}(u)\,T_{22}(v)\Big)$$

implied by (1.3).

We can now prove that the vectors η_γ with γ satisfying (3.88) form a basis of L. Indeed, $\eta_\gamma \neq 0$ since the application of appropriate operators $T_{12}(v)$ with the use of (3.91) yields the highest vector ζ of L with a nonzero coefficient. Furthermore, the vectors η_γ are linearly independent as they are eigenvectors for $T_{22}(u)$ with distinct eigenvalues. The number of these vectors is

$$\prod_{i=1}^{k} (\alpha_i - \beta_i + 1),$$

which coincides with the dimension of L, thus proving the claim.

Relation (3.92) is immediate from (3.93) and (3.96). Finally, if both γ and $\gamma+\delta_i$ satisfy the conditions (3.88) for some $i \in \{1, \dots, k\}$, then (3.90) is immediate from the definition (3.89) of the vectors η_γ. It remains to verify that $T_{21}(-\gamma_i)\,\eta_\gamma = 0$ if $\gamma_i = \alpha_i$. The argument used for the proof of (3.93) shows that if the vector $\eta' = T_{21}(-\gamma_i)\,\eta_\gamma$ is nonzero, then η' is an eigenvector for $T_{22}(u)$ with the eigenvalue

$$(u+\gamma_1)\dots(u+\gamma_{i-1})\,(u+\alpha_i+1)\,(u+\gamma_{i+1})\dots(u+\gamma_k).$$

However, as we have seen above, the module L admits a basis which consists of eigenvectors of $T_{22}(u)$ with distinct eigenvalues. We come to a contradiction since none of these eigenvalues coincides with the eigenvalue of η'. So, $\eta' = 0$. □

Note that the action of the generator $t_{ij}^{(r)}$ on the basis vector η_γ can be found by taking the coefficient of u^{k-r} in the polynomial $T_{ij}(u)$. In particular,

$$t_{21}^{(1)}\,\eta_\gamma = \sum_{i=1}^{k} \frac{1}{(\gamma_1-\gamma_i)\cdots\wedge_i\dots(\gamma_k-\gamma_i)}\,\eta_{\gamma+\delta_i}.$$

This gives an alternative way to calculate the action of $T_{11}(u)$.

COROLLARY 3.3.9. *Under the assumptions of Theorem 3.3.8, we have*

$$T_{11}(u)\,\eta_\gamma = T_{22}(u)\,\eta_\gamma + \big[T_{12}(u), t_{21}^{(1)}\big]\,\eta_\gamma.$$

PROOF. This is immediate from the relation

$$\big[t_{12}(u), t_{21}^{(1)}\big] = t_{11}(u) - t_{22}(u)$$

implied by (1.3). □

3.4. Representations of $\mathrm{Y}(\mathfrak{gl}_N)$

We are now in a position to prove the classification theorem for the finite-dimensional irreducible representations of the Yangian $\mathrm{Y}(\mathfrak{gl}_N)$, where N is an arbitrary positive integer. As we proved in Section 3.2, any finite-dimensional irreducible representation of $\mathrm{Y}(\mathfrak{gl}_N)$ is isomorphic to a highest weight representation $L(\lambda(u))$, where $\lambda(u)$ is an N-tuple of formal series in u^{-1},

$$\lambda(u) = (\lambda_1(u), \dots, \lambda_N(u)).$$

We now give necessary and sufficient conditions on an arbitrary N-tuple $\lambda(u)$ which ensure that $L(\lambda(u))$ is finite-dimensional. We will be using the notation (3.81).

THEOREM 3.4.1. *The irreducible highest weight representation $L(\lambda(u))$ of the Yangian $\mathrm{Y}(\mathfrak{gl}_N)$ is finite-dimensional if and only if the following relation holds:*

$$\lambda_1(u) \longrightarrow \lambda_2(u) \longrightarrow \dots \longrightarrow \lambda_N(u). \tag{3.97}$$

PROOF. Suppose that $\dim L(\lambda(u)) < \infty$. For any $k \in \{0, \dots, N-2\}$ consider the homomorphism $\mathrm{Y}(\mathfrak{gl}_2) \to \mathrm{Y}(\mathfrak{gl}_N)$ which sends $t_{ij}(u)$ to $t_{k+i,k+j}(u)$ for any $i, j \in \{1, 2\}$. Let $\mathrm{Y}(\mathfrak{gl}_2)$ act on $L(\lambda(u))$ via this homomorphism. The cyclic span $\mathrm{Y}(\mathfrak{gl}_2)\zeta$ of the highest vector ζ of $L(\lambda(u))$ is a highest weight representation of $\mathrm{Y}(\mathfrak{gl}_2)$ with the highest weight $(\lambda_{k+1}(u), \lambda_{k+2}(u))$. Its irreducible quotient is finite-dimensional, and so Theorem 3.3.3 implies $\lambda_{k+1}(u) \longrightarrow \lambda_{k+2}(u)$, proving (3.97).

Conversely, let (3.97) hold. Then for any $i = 1, \dots, N-1$ we have

$$\frac{\lambda_i(u)}{\lambda_{i+1}(u)} = \frac{P_i(u+1)}{P_i(u)} \tag{3.98}$$

for certain monic polynomials $P_i(u)$. Suppose that $k_i = \deg P_i(u)$ and write

$$P_i(u) = (u + \gamma_i^{(1)}) \dots (u + \gamma_i^{(k_i)}), \qquad \gamma_i^{(l)} \in \mathbb{C}.$$

Consider the irreducible highest weight representation $L(\mu(u))$ of $\mathrm{Y}(\mathfrak{gl}_N)$ with the highest weight $\mu(u) = (\mu_1(u), \dots, \mu_N(u))$, given by

$$\mu_i(u) = u^{-k} P_1(u) \dots P_{i-1}(u)\, P_i(u+1) \dots P_{N-1}(u+1),$$

where $k = k_1 + \dots + k_{N-1}$. Equivalently,

$$\mu_i(u) = (1 + \mu_i^{(1)} u^{-1}) \dots (1 + \mu_i^{(k)} u^{-1}),$$

and for any $1 \leqslant j < N$ and $1 \leqslant l \leqslant k_j$ we have

$$\mu_i^{(k_1 + \dots + k_{j-1} + l)} = \begin{cases} \gamma_j^{(l)} & \text{for} \quad j < i, \\ \gamma_j^{(l)} + 1 & \text{for} \quad j \geqslant i. \end{cases}$$

Let us set $\mu^{(j)} = (\mu_1^{(j)}, \dots, \mu_N^{(j)})$. Then each module $L(\mu^{(j)})$ is finite-dimensional and, hence, so is the $\mathrm{Y}(\mathfrak{gl}_N)$-module

$$L(\mu^{(1)}) \otimes \dots \otimes L(\mu^{(k)}).$$

By Proposition 3.2.9, the tensor product of the highest vectors of the modules $L(\mu^{(j)})$ generates a highest weight representation of $\mathrm{Y}(\mathfrak{gl}_N)$ with the highest weight $\mu(u)$. Since the irreducible quotient of this representation is isomorphic to $L(\mu(u))$, we may conclude that $L(\mu(u))$ is finite-dimensional.

Finally, by definition,

$$\frac{\mu_i(u)}{\mu_{i+1}(u)} = \frac{P_i(u+1)}{P_i(u)}$$

for any $i = 1, \dots, N-1$. Therefore, (3.98) implies that there exists a formal series $f(u)$ such that $\lambda_i(u) = f(u)\,\mu_i(u)$ for all i. Hence, the composition of the representation of $\mathrm{Y}(\mathfrak{gl}_N)$ on $L(\mu(u))$ and the automorphism (1.20) is isomorphic to $L(\lambda(u))$, proving that the latter is also finite-dimensional. □

COROLLARY 3.4.2. *Finite-dimensional irreducible representations of the Yangian* $\mathrm{Y}(\mathfrak{gl}_N)$ *are parameterized by the tuples* $\big(f(u), P_1(u), \dots, P_{N-1}(u)\big)$, *where* $f(u)$ *is a formal power series in* u^{-1} *with constant term* 1, *and the* $P_i(u)$ *are monic polynomials in* u.

PROOF. Given an irreducible highest weight representation $L(\lambda(u))$ with $\lambda(u)$ satisfying (3.97), define the polynomials $P_i(u)$ by (3.98) and set $f(u) = \lambda_N(u)$. This gives the desired correspondence. □

DEFINITION 3.4.3. The polynomials $P_i(u)$ with $i = 1, \dots, N-1$ are called the *Drinfeld polynomials* of the corresponding representation of $\mathrm{Y}(\mathfrak{gl}_N)$. □

EXAMPLE 3.4.4. Suppose that the representation $L(\lambda)$ of the Lie algebra $\mathfrak{gl}_N$ is finite-dimensional. Then the Drinfeld polynomials of the evaluation representation $L(\lambda)$ of $\mathrm{Y}(\mathfrak{gl}_N)$ defined in Section 3.2 can be easily found from (3.65) and are given by

$$P_i(u) = (u + \lambda_{i+1})(u + \lambda_{i+1} + 1) \dots (u + \lambda_i - 1)$$

for $i = 1, \dots, N-1$. □

PROPOSITION 3.4.5. *Suppose that* L *and* M *are finite-dimensional irreducible representations of* $\mathrm{Y}(\mathfrak{gl}_N)$ *and suppose that*

$$\big(f(u), P_1(u), \dots, P_{N-1}(u)\big) \qquad \text{and} \qquad \big(g(u), Q_1(u), \dots, Q_{N-1}(u)\big)$$

are the corresponding tuples of series and Drinfeld polynomials, respectively. Then the irreducible quotient of the cyclic $\mathrm{Y}(\mathfrak{gl}_N)$*-span of the tensor product of the highest vectors of* L *and* M *corresponds to the tuple*

$$\big(f(u)\,g(u), P_1(u)\,Q_1(u), \dots, P_{N-1}(u)\,Q_{N-1}(u)\big).$$

PROOF. Let $\zeta \in L$ and $\xi \in M$ be the highest vectors and let

$$\lambda(u) = \big(\lambda_1(u), \dots, \lambda_N(u)\big) \qquad \text{and} \qquad \mu(u) = \big(\mu_1(u), \dots, \mu_N(u)\big)$$

be the corresponding highest weights of L and M, respectively. Arguing as in the proof of Proposition 3.2.9, we deduce that the submodule $\mathrm{Y}(\mathfrak{gl}_N)(\zeta \otimes \xi)$ is a highest weight representation with the highest weight

$$\big(\lambda_1(u)\,\mu_1(u), \dots, \lambda_N(u)\,\mu_N(u)\big).$$

The statement now follows by calculating the corresponding series and Drinfeld polynomials as in the proof of Corollary 3.4.2. □

A finite-dimensional irreducible representation of $\mathrm{Y}(\mathfrak{gl}_N)$ is called *fundamental* if for some $i \in \{1, \dots, N-1\}$ and $\alpha \in \mathbb{C}$ its Drinfeld polynomials are given by

$$P_i(u) = u + \alpha \qquad \text{and} \qquad P_j(u) = 1 \quad \text{if} \quad j \neq i. \tag{3.99}$$

The evaluation $\mathrm{Y}(\mathfrak{gl}_N)$-module $L(\alpha+1, \dots, \alpha+1, \alpha, \dots, \alpha)$ with i copies of $\alpha+1$ is fundamental, and its Drinfeld polynomials are given by (3.99); see Example 3.4.4. Moreover, any fundamental representation with these Drinfeld polynomials is obtained as the composition of the action of $\mathrm{Y}(\mathfrak{gl}_N)$ on $L(\alpha+1, \dots, \alpha+1, \alpha, \dots, \alpha)$ with an automorphism of the form (1.20).

The following corollary is immediate either from Proposition 3.4.5 or from the proof of Theorem 3.4.1, as each evaluation module $L(\mu^{(j)})$ used in that proof is fundamental.

COROLLARY 3.4.6. *Every finite-dimensional irreducible representation of the Yangian $\mathrm{Y}(\mathfrak{gl}_N)$ is isomorphic to a subquotient of a tensor product of fundamental representations.* □

Now consider the Yangian $\mathrm{Y}(\mathfrak{sl}_N)$, as defined in Section 1.8.

PROPOSITION 3.4.7. *Every finite-dimensional irreducible representation of the Yangian $\mathrm{Y}(\mathfrak{gl}_N)$ remains irreducible when restricted to the subalgebra $\mathrm{Y}(\mathfrak{sl}_N)$. Moreover, such restrictions exhaust all finite-dimensional irreducible representations of $\mathrm{Y}(\mathfrak{sl}_N)$.*

PROOF. Both statements follow from Theorem 1.8.2 because the elements of the center $\mathrm{ZY}(\mathfrak{gl}_N)$ of $\mathrm{Y}(\mathfrak{gl}_N)$ act on any finite-dimensional irreducible representation of $\mathrm{Y}(\mathfrak{gl}_N)$ as multiplications by scalars; cf. the proof of Corollary 3.3.4. □

We will call a representation of $\mathrm{Y}(\mathfrak{sl}_N)$ *fundamental* if it can be obtained as the restriction of a fundamental representation of $\mathrm{Y}(\mathfrak{gl}_N)$.

COROLLARY 3.4.8. *Finite-dimensional irreducible representations of the Yangian $\mathrm{Y}(\mathfrak{sl}_N)$ are parameterized by the tuples $\big(P_1(u), \dots, P_{N-1}(u)\big)$ of monic polynomials in u. Moreover, every such representation is isomorphic to a subquotient of a tensor product of fundamental representations.*

PROOF. The statements follow from Corollaries 3.4.2 and 3.4.6 respectively, by the same argument as in the proof of Corollary 3.3.4. □

We conclude this section with a description of the finite-dimensional irreducible representations of the Yangian $\mathrm{Y}(\mathfrak{sl}_N)$ in terms of the presentation provided by Corollary 3.1.7.

COROLLARY 3.4.9. *Every finite-dimensional irreducible representation of the Yangian $\mathrm{Y}(\mathfrak{sl}_N)$ contains a unique, up to a constant factor, vector $\zeta \neq 0$ such that*

$$\xi_i^+(u)\,\zeta = 0 \quad \textit{for} \quad i = 1, \dots, N-1. \tag{3.100}$$

Moreover, this vector satisfies

$$\kappa_i(u)\,\zeta = \frac{Q_i(u+1)}{Q_i(u)}\,\zeta \quad \textit{for} \quad i = 1, \dots, N-1, \tag{3.101}$$

where each $Q_i(u)$ is a monic polynomial in u and the ratio $Q_i(u+1)/Q_i(u)$ has to be expanded as a power series in u^{-1}. The tuple of polynomials $\big(Q_1(u),\dots,Q_{N-1}(u)\big)$ determines the representation up to an isomorphism.

PROOF. By the Yangian defining relation (1.3) we have

$$t_{11}(u+1)\,t_{21}(u) = t_{21}(u+1)\,t_{11}(u).$$

Hence we can rewrite the definition (3.1) of the series $f_i(u)$ in the equivalent form

$$f_i(u) = \psi_{i-1}\big(t_{11}(u+1)^{-1}\,t_{21}(u+1)\big).$$

Therefore, the definition (3.43) of $\xi_i^+(u)$ can be written in terms of the quantum minors as follows:

$$\xi_i^+(u) = \Big(t^{\,i+1\dots N}_{\,i+1\dots N}\Big(u-\frac{i-1}{2}\Big)\Big)^{-1}\cdot t^{\,i,i+2\dots N}_{\,i+1\dots N}\Big(u-\frac{i-1}{2}\Big).$$

Suppose now that L is a finite-dimensional irreducible representation of $\mathrm{Y}(\mathfrak{sl}_N)$. By Proposition 3.4.7, L can be extended to a representation of $\mathrm{Y}(\mathfrak{gl}_N)$. Now, using the same argument as in the proof of Proposition 3.2.2, we deduce that the highest vector ζ of the $\mathrm{Y}(\mathfrak{gl}_N)$-module L satisfies (3.100). Similarly, the conditions (3.100) imply that $t_{ij}(u)\,\zeta = 0$ for all $i<j$. Then by Corollary 3.2.8, ζ is the highest vector of the $\mathrm{Y}(\mathfrak{gl}_N)$-module L, which proves the first statement. Furthermore, we have $t_{ii}(u)\,\zeta = \lambda_i(u)\,\zeta$ for certain formal series $\lambda_i(u)$, $i=1,\dots,N$. Then, due to (1.54),

$$t^{\,i\dots N}_{\,i\dots N}(u)\,\zeta = \lambda_i(u)\,\lambda_{i+1}(u-1)\dots\lambda_N(u-N+i)\,\zeta.$$

Now using (3.45) we deduce that

$$\kappa_i(u)\,\zeta = \frac{\lambda_i(u-(i-1)/2)}{\lambda_{i+1}(u-(i-1)/2)}\,\zeta.$$

So, by (3.98), the relation (3.101) holds for the polynomials $Q_i(u) = P_i(u-(i-1)/2)$, where the $P_i(u)$ are the Drinfeld polynomials of the representation L. This also proves the last statement of the corollary, since the tuple of Drinfeld polynomials determines the corresponding $\mathrm{Y}(\mathfrak{sl}_N)$-module up to an isomorphism; see Corollary 3.4.8. □

Note that Corollary 3.4.9 can also be proved without the use of the relationship between the algebras $\mathrm{Y}(\mathfrak{sl}_N)$ and $\mathrm{Y}(\mathfrak{gl}_N)$. That proof is based on the presentation of $\mathrm{Y}(\mathfrak{sl}_N)$ given in Corollary 3.1.7 and extends to the Drinfeld Yangians associated with arbitrary simple Lie algebras.

REMARK 3.4.10. Contrary to the case $N=2$, it is not true for $N\geqslant 3$ that every finite-dimensional irreducible representation of $\mathrm{Y}(\mathfrak{sl}_N)$ is isomorphic to a tensor product of evaluation modules; cf. Corollary 3.3.4. For example, the $\mathrm{Y}(\mathfrak{sl}_3)$-module $L(\lambda(u))$ with

$$\lambda_1(u) = (1+3u^{-1})(1+u^{-1}),\qquad \lambda_2(u) = 1+3u^{-1},\qquad \lambda_3(u) = 1+2u^{-1}$$

is 8-dimensional, which can be verified with the use of Theorem 8.5.4 below (in fact, the $\mathrm{Y}(\mathfrak{sl}_3)$-module $L(\lambda(u))$ is isomorphic to the skew representation $L(\lambda)^+_\mu$ with $\lambda=(3,3,1,1)$ and $\mu=(2)$). On the other hand, the possible dimensions of the evaluation modules are $1,3,6,8,\dots$ so that $L(\lambda(u))$ cannot be isomorphic to a tensor product of such modules. □

3.5. Examples

1. The Yangian $\mathrm{Y}(\mathfrak{sl}_2)$ is isomorphic to the Hopf algebra with six generators $e, f, h, J(e), J(f), J(h)$ subject to the defining relations

$$[e,f]=h, \qquad [h,e]=2e, \qquad [h,f]=-2f,$$
$$[x,J(y)]=J([x,y]), \qquad J(ax)=a\,J(x),$$

where $x,y\in\{e,f,h\}$, $a\in\mathbb{C}$, and

$$\big[[J(e),J(f)],J(h)\big]=\big(J(e)f-e\,J(f)\big)\,h.$$

The Hopf algebra structure is defined by

$$\begin{aligned}
\Delta:&\quad x\mapsto x\otimes 1+1\otimes x, && J(x)\mapsto J(x)\otimes 1+1\otimes J(x)+\frac{1}{2}[x\otimes 1,t],\\
\mathrm{S}:&\quad x\mapsto -x, && J(x)\mapsto -J(x)+x,\\
\varepsilon:&\quad x\mapsto 0, && J(x)\mapsto 0,
\end{aligned}$$

where $t=e\otimes f+f\otimes e+\frac{1}{2}h\otimes h$. The isomorphism is given by

$$e\mapsto t_{12}^{(1)}, \qquad f\mapsto t_{21}^{(1)}, \qquad h\mapsto t_{11}^{(1)}-t_{22}^{(1)},$$

$$\begin{aligned}
J(e)&\mapsto t_{12}^{(2)}-\frac{1}{2}(t_{11}^{(1)}+t_{22}^{(1)}-1)\,t_{12}^{(1)},\\
J(f)&\mapsto t_{21}^{(2)}-\frac{1}{2}(t_{11}^{(1)}+t_{22}^{(1)}-1)\,t_{21}^{(1)},\\
J(h)&\mapsto t_{11}^{(2)}-t_{22}^{(2)}-\frac{1}{2}(t_{11}^{(1)}+t_{22}^{(1)}-1)(t_{11}^{(1)}-t_{22}^{(1)}).
\end{aligned}$$

2. In terms of the realization of $\mathrm{Y}(\mathfrak{sl}_2)$ provided by Corollary 3.1.7, the comultiplication map Δ has the form (we omit the subscript 1 of the generating series):

$$\Delta:\xi^+(u)\mapsto \xi^+(u)\otimes 1+\sum_{k=0}^{\infty}(-1)^k\xi^-(u+1)^k\,\kappa(u)\,\otimes\xi^+(u)^{k+1},$$

$$\Delta:\xi^-(u)\mapsto 1\otimes\xi^-(u)+\sum_{k=0}^{\infty}(-1)^k\xi^-(u)^{k+1}\otimes\kappa(u)\,\xi^+(u+1)^k,$$

$$\Delta:\quad \kappa(u)\mapsto\sum_{k=0}^{\infty}(-1)^k(k+1)\,\xi^-(u+1)^k\,\kappa(u)\otimes\kappa(u)\,\xi^+(u+1)^k.$$

3. The Verma module $M(\lambda(u))$ over the Yangian $\mathrm{Y}(\mathfrak{gl}_N)$ is reducible if and only if for some index $i\in\{1,\dots,N-1\}$ the series $\lambda_i(u)/\lambda_{i+1}(u)$ is the Laurent expansion at $u=\infty$ of a rational function in u, i.e.,

$$\frac{\lambda_i(u)}{\lambda_{i+1}(u)}=\frac{P(u)}{Q(u)},$$

where $P(u)$ and $Q(u)$ are monic polynomials in u of the same degree.

4. The irreducible quotient $L(\lambda(u))$ inherits the weight space decomposition of the Verma module (3.62), so that

$$L(\lambda(u))=\bigoplus_{\mu}L(\lambda(u))_{\mu}.$$

All weight subspaces $L(\lambda(u))_\mu$ of the module $L(\lambda(u))$ are finite-dimensional if and only if for each index $i \in \{1, \dots, N-1\}$ the series $\lambda_i(u)/\lambda_{i+1}(u)$ is the Laurent expansion at $u = \infty$ of a rational function in u.

5. Consider the quantized enveloping algebra $\mathrm{U}_q(\mathfrak{gl}_N)$; see Example 1.15.1. Suppose that the complex parameter q is nonzero and not a root of unity. Let $\lambda = (\lambda_1, \dots, \lambda_N)$ be an N-tuple of integers such that $\lambda_1 \geqslant \dots \geqslant \lambda_N$. The corresponding irreducible highest weight representation $L(\lambda)$ of $\mathrm{U}_q(\mathfrak{gl}_N)$ is generated by a nonzero vector ζ such that

$$\begin{aligned} \bar{t}_{ij}\,\zeta &= 0 \qquad \text{for} \quad 1 \leqslant i < j \leqslant N, \\ t_{ii}\,\zeta &= q^{\lambda_i}\,\zeta \quad \text{for} \quad 1 \leqslant i \leqslant N. \end{aligned}$$

This representation is a q-analogue of the irreducible $\mathfrak{gl}_N$-module with the highest weight λ. In particular, these modules have the same dimension.

For any nonzero complex number d and any N-tuple $(\varepsilon_1, \dots, \varepsilon_N)$ with each ε_i equal to 1 or -1, the assignment

$$t_{ij} \mapsto \varepsilon_i\, d\, t_{ij}, \qquad \bar{t}_{ij} \mapsto \varepsilon_i\, d^{-1}\, \bar{t}_{ij}$$

defines an automorphism of $\mathrm{U}_q(\mathfrak{gl}_N)$. Every finite-dimensional irreducible representation of $\mathrm{U}_q(\mathfrak{gl}_N)$ can be obtained from a representation of the form $L(\lambda)$ by twisting with such an automorphism.

6. Consider the quantum affine algebra $\mathrm{U}_q(\widehat{\mathfrak{gl}}_N)$; see Example 1.15.3. Suppose that the complex parameter q is nonzero and not a root of unity. The irreducible *pseudo-highest weight representation* $L(\nu(u), \bar{\nu}(u))$ of $\mathrm{U}_q(\widehat{\mathfrak{gl}}_N)$ with the *pseudo-highest weight* $(\nu(u), \bar{\nu}(u))$ is generated by a nonzero vector ζ such that

$$\begin{aligned} t_{ij}(u)\,\zeta &= 0, & \bar{t}_{ij}(u)\,\zeta &= 0 & \text{for} \quad 1 \leqslant i < j \leqslant N, \\ t_{ii}(u)\,\zeta &= \nu_i(u)\,\zeta, & \bar{t}_{ii}(u)\,\zeta &= \bar{\nu}_i(u)\,\zeta & \text{for} \quad 1 \leqslant i \leqslant N, \end{aligned}$$

where $\nu(u) = (\nu_1(u), \dots, \nu_N(u))$ and $\nu(u) = (\bar{\nu}_1(u), \dots, \bar{\nu}_N(u))$ are certain N-tuples of formal power series in u^{-1} and u, respectively.

Suppose that there exist polynomials $P_1(u), \dots, P_{N-1}(u)$ in u, all with constant term 1, such that

$$\frac{\nu_i(u)}{\nu_{i+1}(u)} = q^{-\deg P_i} \cdot \frac{P_i(uq^2)}{P_i(u)} = \frac{\bar{\nu}_i(u)}{\bar{\nu}_{i+1}(u)} \tag{3.102}$$

for any $i = 1, \dots, N-1$. The first equality in (3.102) is understood in the sense that the ratio of polynomials has to be expanded as a power series in u^{-1}, while for the second equality the same ratio has to be expanded as a power series in u. Then the corresponding representation $L(\nu(u), \bar{\nu}(u))$ is finite-dimensional. It is called a *type* **1** *representation*. The $P_i(u)$ are called its *Drinfeld polynomials*; cf. Definition 3.4.3.

Every finite-dimensional irreducible representation of $\mathrm{U}_q(\widehat{\mathfrak{gl}}_N)$ is isomorphic to the composition of a type **1** representation and an automorphism of the form

$$t_{ij}(u) \mapsto \varepsilon_i\, t_{ij}(u), \qquad \bar{t}_{ij}(u) \mapsto \varepsilon_i\, \bar{t}_{ij}(u), \tag{3.103}$$

where each ε_i equals 1 or -1. Moreover, finite-dimensional irreducible type **1** representations of the quantum affine algebra $\mathrm{U}_q(\widehat{\mathfrak{sl}}_N)$ (see Example 1.15.4) are parameterized by the tuples $\big(P_1(u), \dots, P_{N-1}(u)\big)$ of polynomials in u with constant term 1; cf. Corollary 3.4.8.

These classification theorems can be proved by appropriate modifications of the arguments of Sections 3.2–3.4. In particular, the counterpart of Theorem 3.2.7 can be proved with the use of the embedding $\mathrm{U}_q(\mathfrak{gl}_N) \hookrightarrow \mathrm{U}_q(\widehat{\mathfrak{gl}}_N)$; see Example 1.15.5. The reduction to the case $N = 2$ is similar to the proof of Theorem 3.4.1. In Examples 7–9 below we sketch the arguments for the proof of the counterpart of the key theorem (Theorem 3.3.3).

7. Every finite-dimensional irreducible representation of $\mathrm{U}_q(\widehat{\mathfrak{gl}}_2)$ is isomorphic to a pseudo-highest weight representation $L\big(\nu_1(u), \nu_2(u); \bar{\nu}_1(u), \bar{\nu}_2(u)\big)$. By twisting this representation with an appropriate automorphism of the form (1.102), we may assume without loss of generality that $\nu_2(u) = \bar{\nu}_2(u) = 1$. Write

$$\nu_1(u) = \nu_1^{(0)} + \nu_1^{(1)}\, u^{-1} + \nu_1^{(2)}\, u^{-2} + \dots,$$
$$\bar{\nu}_1(u) = \bar{\nu}_1^{(0)} + \bar{\nu}_1^{(1)}\, u + \bar{\nu}_1^{(2)}\, u^2 + \dots.$$

Note that $\nu_1^{(0)}\,\bar{\nu}_1^{(0)} = 1$ by the relation $t_{11}^{(0)}\,\bar{t}_{11}^{(0)} = 1$.

Since the representation $L\big(\nu_1(u), 1; \bar{\nu}_1(u), 1\big)$ is finite-dimensional, there exist integers $n \geqslant 0$, $m \geqslant 1$ and complex numbers c_i, d_j such that

$$\sum_{i=0}^{n} c_i\, t_{21}^{(i)} \zeta + \sum_{j=1}^{m} d_j\, \bar{t}_{21}^{(j)} \zeta = 0,$$

and $c_n, d_m \neq 0$. Denote the linear combination which occurs on the left-hand side by ξ. Then $t_{12}^{(r)}\,\xi = 0$ for all $r \geqslant 1$. On the other hand, by the defining relations of $\mathrm{U}_q(\widehat{\mathfrak{gl}}_2)$, we have

$$t_{12}^{(r)}\, t_{21}^{(s)} - t_{21}^{(s)}\, t_{12}^{(r)} = (q - q^{-1}) \sum_{p=1}^{r} \big(t_{22}^{(r-p)}\, t_{11}^{(s+p)} - t_{22}^{(s+p)}\, t_{11}^{(r-p)}\big)$$

and

$$t_{12}^{(r)}\, \bar{t}_{21}^{(s)} - \bar{t}_{21}^{(s)}\, t_{12}^{(r)} = (q - q^{-1}) \sum_{p=1}^{\min\{r,s\}} \big(t_{22}^{(r-p)}\, \bar{t}_{11}^{(s-p)} - \bar{t}_{22}^{(s-p)}\, t_{11}^{(r-p)}\big).$$

Hence, taking the coefficient of ζ in $t_{12}^{(r)}\,\xi = 0$ we get

$$\sum_{i=0}^{n} c_i \nu_1^{(r+i)} + \sum_{j=1}^{m} d_j \big(\bar{\nu}_1^{(j-r)} - \nu_1^{(r-j)}\big) = 0$$

for all $r \geqslant 1$, where we assume $\nu_1^{(s)} = \bar{\nu}_1^{(s)} = 0$ for $s < 0$. This is equivalent to the relation

$$\nu_1(u) \left(\sum_{i=0}^{n} c_i u^i - \sum_{j=1}^{m} d_j u^{-j} \right) = \sum_{i=0}^{n} c_i u^i \big(\nu_1^{(0)} + \dots + \nu_1^{(i)} u^{-i}\big)$$
$$- \sum_{j=1}^{m} d_j u^{-j} \big(\bar{\nu}_1^{(0)} + \dots + \bar{\nu}_1^{(j-1)} u^{j-1}\big).$$

Similarly, the relations $\bar{t}_{12}^{(r)}\,\xi = 0$, $r \geqslant 0$, yield

$$\bar{\nu}_1(u)\Bigg(\sum_{i=0}^{n} c_i u^i - \sum_{j=1}^{m} d_j u^{-j}\Bigg) = \sum_{i=0}^{n} c_i u^i \,\big(\nu_1^{(0)} + \dots + \nu_1^{(i)} u^{-i}\big)$$
$$- \sum_{j=1}^{m} d_j u^{-j}\,\big(\bar{\nu}_1^{(0)} + \dots + \bar{\nu}_1^{(j-1)} u^{j-1}\big).$$

This implies that both series $\nu_1(u)$ and $\bar{\nu}_1(u)$ are expansions of the same rational function in u,

$$\nu_1(u) = \frac{Q(u)}{R(u)} = \bar{\nu}_1(u).$$

This establishes an analogue of Proposition 3.3.1 for the algebra $\mathrm{U}_q(\widehat{\mathfrak{gl}}_2)$. From now on we may work with the q-Yangian $\mathrm{Y}_q(\mathfrak{gl}_2)$ instead of the quantum affine algebra (see Example 1.15.3).

8. By Example 7, applying an automorphism of the form (1.102) if necessary, we may now assume that both series $\nu_1(u)$ and $\nu_2(u)$ are polynomials in u^{-1} so that for some positive integer k

$$\nu_1(u) = \nu_1^{(0)} + \dots + \nu_1^{(k)} u^{-k},$$
$$\nu_2(u) = \nu_2^{(0)} + \dots + \nu_2^{(k)} u^{-k},$$

with $\nu_1^{(0)}\nu_1^{(k)} = \nu_2^{(0)}\nu_2^{(k)} = (-1)^k$. The cyclic span $\mathrm{Y}_q(\mathfrak{gl}_2)\,\zeta$ is a highest weight representation of the q-Yangian $\mathrm{Y}_q(\mathfrak{gl}_2)$ with the highest weight $\nu(u) = (\nu_1(u), \nu_2(u))$, which is defined in the same way as for the Yangian $\mathrm{Y}(\mathfrak{gl}_2)$; see Proposition 3.2.2. Now we establish an analogue of Proposition 3.3.2 for irreducible highest weight representations $L(\nu_1(u), \nu_2(u))$ of $\mathrm{Y}_q(\mathfrak{gl}_2)$. Write the decompositions

$$\nu_1(u) = (\alpha_1 - \alpha_1^{-1}u^{-1}) \dots (\alpha_k - \alpha_k^{-1}u^{-1}), \tag{3.104}$$
$$\nu_2(u) = (\beta_1 - \beta_1^{-1}u^{-1}) \dots (\beta_k - \beta_k^{-1}u^{-1}).$$

Similar to Example 5, for any pair of nonzero complex numbers α and β consider the corresponding irreducible highest weight representation $L(\alpha, \beta)$ of $\mathrm{U}_q(\mathfrak{gl}_2)$. That is, $L(\alpha, \beta)$ is generated by a nonzero vector ζ such that

$$\bar{t}_{12}\,\zeta = 0, \qquad t_{11}\,\zeta = \alpha\zeta, \qquad t_{22}\,\zeta = \beta\zeta.$$

The representation $L(\alpha, \beta)$ is finite-dimensional if and only if $\alpha/\beta = \pm q^m$ for some nonnegative integer m. We make $L(\alpha, \beta)$ into a module over the q-Yangian $\mathrm{Y}_q(\mathfrak{gl}_2)$ via the evaluation homomorphism $\mathrm{Y}_q(\mathfrak{gl}_2) \to \mathrm{U}_q(\mathfrak{gl}_2)$; see Example 1.15.5. The highest weight of this evaluation module is the pair of polynomials of degree 1 in u^{-1} given by

$$\big(\alpha - \alpha^{-1}u^{-1},\ \beta - \beta^{-1}u^{-1}\big).$$

Re-enumerating the parameters in the decompositions (3.104) if necessary, we may assume that for every $i = 1, \dots, k-1$ the following condition holds: if the multiset $\{\alpha_p/\beta_q \mid i \leqslant p, q \leqslant k\}$ contains numbers of the form $\pm q^m$ with nonnegative integers m, then $\alpha_i/\beta_i = \pm q^{m_0}$ and m_0 is minimal amongst these integers. Then the representation $L(\nu_1(u), \nu_2(u))$ of $\mathrm{Y}_q(\mathfrak{gl}_2)$ is isomorphic to the tensor product module

$$L(\alpha_1, \beta_1) \otimes L(\alpha_2, \beta_2) \otimes \dots \otimes L(\alpha_k, \beta_k). \tag{3.105}$$

This is verified by a straightforward modification of the proof of Proposition 3.3.2.

9. The final step in the proof of the counterpart of Theorem 3.3.3 is to observe that the condition $\dim L(\alpha,\beta)<\infty$ can be written in terms of a polynomial $P(u)$. Namely, if $\alpha/\beta = q^m$ for a nonnegative integer m, then $L(\alpha,\beta)$ is a type **1** representation with

$$\frac{\alpha-\alpha^{-1}u^{-1}}{\beta-\beta^{-1}u^{-1}} = q^{-\deg P}\cdot\frac{P(uq^2)}{P(u)},$$

where

$$P(u) = (1-\beta^2 u)(1-\beta^2 q^2 u)\dots(1-\beta^2 q^{2m-2} u).$$

The representation $L(\alpha,\beta)$ with $\alpha/\beta = -q^m$ is obtained from a type **1** representation by twisting with an appropriate automorphism of the form (3.103).

Note that the above arguments also imply that, up to twisting with an automorphism of the form (1.102), every finite-dimensional irreducible representation of $\mathrm{Y}_q(\mathfrak{gl}_2)$ (or $\mathrm{U}_q(\widehat{\mathfrak{gl}}_2)$) is isomorphic to a tensor product of the form (3.105); cf. Corollary 3.3.4. Moreover, an irreducibility criterion of the tensor product (3.105) of finite-dimensional evaluation representations $L(\alpha_i,\beta_i)$ with $\alpha_i/\beta_i = q^{m_i}$ can be formulated in terms of the associated *q-strings*

$$S_q(\alpha_i,\beta_i) = \{\beta_i, \beta_i q, \dots, \beta_i q^{m_i-1}\};$$

cf. Definition 3.3.5 and Corollary 3.3.6.

10. The representation $L(\lambda)$ of $\mathrm{U}_q(\mathfrak{gl}_N)$ defined in Example 5 can be extended to a representation of $\mathrm{U}_q(\widehat{\mathfrak{gl}}_N)$ via the evaluation homomorphism; see Example 1.15.5. Such a $\mathrm{U}_q(\widehat{\mathfrak{gl}}_N)$-module is called an *evaluation module.* Its Drinfeld polynomials are given by

$$P_i(u) = (1-q^{2\lambda_{i+1}}u)(1-q^{2\lambda_{i+1}+2}u)\dots(1-q^{2\lambda_i-2}u),$$

for $i=1,\dots,N-1$; cf. Example 3.4.4.

11. If the complex parameter q is nonzero and not a root of unity, then the restriction of any finite-dimensional irreducible representation of the quantum affine algebra $\mathrm{U}_q(\widehat{\mathfrak{gl}}_N)$ to the q-Yangian $\mathrm{Y}_q(\mathfrak{gl}_N)$ is irreducible; see Example 1.15.3.

Bibliographical notes

3.1. The presentation of $\mathrm{Y}(\mathfrak{sl}_N)$ provided by Corollary 3.1.7 is due to Drinfeld [**99**]. We used the same notation for the generators as in [**99**]. Following the title of that paper, some authors call this presentation the 'new realization' of $\mathrm{Y}(\mathfrak{sl}_N)$. Such a presentation is given in [**99**] for the Yangian associated with an arbitrary complex simple Lie algebra. In the case of quantum affine algebras, a proof of the equivalence of the new realization and Drinfeld's original presentation of [**97**] was given by Beck [**24**]. Theorem 3.1.5 is due to Brundan and Kleshchev [**51**]. Our proof of Corollary 3.1.7 also follows [**51**]. However, the isomorphism between the presentations of the Yangian $\mathrm{Y}(\mathfrak{sl}_N)$ used in that paper is different from ours; see Remark 3.1.8.

3.2. Theorem 3.2.7 is due to Drinfeld [**99**, Theorem 2]. His theorem applies to any Yangian associated with a complex simple Lie algebra.

3.3. Classification of the finite-dimensional irreducible representations of $\mathrm{Y}(\mathfrak{gl}_2)$ is due to Tarasov [**437, 438**]. These papers need to be translated from the language of the quantum inverse scattering theory and integrable models; the Yangian terminology had not been introduced before Drinfeld's paper [**97**]. Definition 3.3.5 is due to Chari and Pressley [**62**]; Corollary 3.3.6 and Proposition 3.3.7 are also contained in [**62**]. These results together with Theorem 3.3.8 go back to Tarasov [**438**].

3.4. Results of this section are due to Drinfeld [**99**]. Corollary 3.4.9 is a particular case of his general classification theorem for the finite-dimensional irreducible representations of the Yangians associated with complex simple Lie algebras, which was stated in [**99**, Theorem 2].

3.5. For an arbitrary N there exists a presentation of the Yangian $\mathrm{Y}(\mathfrak{sl}_N)$ with a finite set of generators analogous to the one given in Example 1 for $N = 2$. It was originally given by Drinfeld in [**97**], where the Yangian associated with an arbitrary complex simple Lie algebra was introduced. The formulas for the comultiplication map in Example 2 are due to the author; see e.g. Khoroshkin and Tolstoy [**217**]. These formulas are easily derived from the definition of the comultiplication on $\mathrm{Y}(\mathfrak{gl}_2)$. They were generalized to the Yangian $\mathrm{Y}(\mathfrak{sl}_3)$ by Solov'ev [**423**], and to $\mathrm{Y}(\mathfrak{sl}_N)$ with arbitrary N by Iohara [**180**] and Crampé [**81**]. Example 3 is contained in Billig, Futorny and Molev [**35**] and Brundan and Kleshchev [**53**]. Example 4 is taken from the paper [**35**], which also contains corresponding results for all Drinfeld Yangians. A different class of infinite-dimensional representations of the Yangians was constructed by Gerasimov, Kharchev, Lebedev and Oblezin [**139, 140**] establishing a new connection between the quantum inverse scattering method and representation theory. For the results presented in Example 5 and more general representation theory of the quantized enveloping algebras, see Lusztig [**274, 278**], Rosso [**406**], Joseph [**201**], and Chari and Pressley [**65**]. Example 6 is a particular case of the classification theorem for representations of the quantum affine algebras proved by Chari and Pressley [**64, 65**]. We stated their results in terms of the RTT-presentation of the quantum affine algebra $\mathrm{U}_q(\widehat{\mathfrak{gl}}_N)$; see Reshetikhin and Semenov-Tian-Shansky [**402**], Reshetikhin, Takhtajan and Faddeev [**403**], and Frenkel and Mukhin [**115**]. The arguments outlined in Examples 7–9 go back to Tarasov's work [**438**] and provide an alternative proof of the classification theorem. The term "pseudo-highest weight" was used in [**65**]; in the classical limit it does not correspond to the usual notion of highest weight for representations of the affine Lie algebras; see Kac [**205**]. Example 11 is a "folklore" theorem. An outline of the proof can be found in Molev, Tolstoy and Zhang [**324**]. A generalization of this theorem to the quantum affine algebras of arbitrary types was given by Bowman [**44**] relying on the earlier work of Benkart and Terwillinger [**26**]. Finite-dimensional irreducible representations of the Yangian $\mathrm{Y}(\mathfrak{gl}_{M|N})$ for the general linear Lie superalgebra $\mathfrak{gl}_{M|N}$ were classified by Zhang [**475, 477**]; see Example 1.15.9. Analogues of Theorem 3.1.5 and Corollary 3.1.7 for $\mathrm{Y}(\mathfrak{gl}_{M|N})$ were proved by Gow [**158**]. This provides an isomorphism between the RTT presentation of the Yangian $\mathrm{Y}(\mathfrak{gl}_{M|N})$ and the Drinfeld type presentation given by Stukopin [**428, 429**].

CHAPTER 4

Irreducible representations of $Y(\mathfrak{g}_N)$

Our aim in this chapter is to describe finite-dimensional irreducible representations of the twisted Yangians $Y(\mathfrak{g}_N)$. The results turn out to be parallel to those of the previous chapter. We prove that the representations of $Y(\mathfrak{g}_N)$ are parameterized by their highest weights and give necessary and sufficient conditions for a highest weight to correspond to a finite-dimensional representation. In the simplest nontrivial case $N = 2$ we produce an explicit construction of every finite-dimensional irreducible representation. In the first section we introduce a particular presentation of the twisted Yangian and reformulate some of the results of Chapter 2 for this presentation.

4.1. Split presentation of the twisted Yangian

In Chapter 2 we used an arbitrary symmetric or alternating form on $\mathbb{C}^N$ associated with a nonsingular matrix G to describe the structure of the twisted Yangian $Y(\mathfrak{g}_N)$. It will be convenient to choose certain particular matrices G for the purposes of this chapter. In the orthogonal case we take G to be the matrix which has 1's on the second diagonal and 0's elsewhere. This choice of G corresponds to the so-called *split realization* of $\mathfrak{o}_N$. In the symplectic case the split realization corresponds to a skew-symmetric matrix G which has ± 1's on the second diagonal and 0's elsewhere. This realization of $\mathfrak{g}_N$ in both cases is convenient for the parametrization of its finite-dimensional irreducible representations, as the diagonal elements of $\mathfrak{g}_N$ form a Cartan subalgebra. Given a positive integer N, we will number the rows and columns of $N \times N$ matrices by the indices $\{-n, \dots, -1, 0, 1, \dots, n\}$ if $N = 2n + 1$, and by $\{-n, \dots, -1, 1, \dots, n\}$ if $N = 2n$. Then for the matrix elements of $G = [g_{ij}]$ we have

$$g_{ij} = \begin{cases} \delta_{i,-j} & \text{in the orthogonal case,} \\ \delta_{i,-j} \cdot \operatorname{sgn} i & \text{in the symplectic case.} \end{cases} \tag{4.1}$$

We will be working with the presentation of the twisted Yangian $Y(\mathfrak{g}_N)$ described in Section 2.15. It will be convenient to use the symbol θ_{ij} defined by

$$\theta_{ij} = \begin{cases} 1 & \text{in the orthogonal case,} \\ \operatorname{sgn} i \cdot \operatorname{sgn} j & \text{in the symplectic case.} \end{cases}$$

The matrix transposition (2.100) with respect to the form G takes the form of the symmetry with respect to the second diagonal with an additional sign change in the symplectic case: the ij-th entry a'_{ij} of the transposed matrix A' of the matrix $A = [a_{ij}]$ is found by

$$a'_{ij} = \theta_{ij}\, a_{-j,-i}. \tag{4.2}$$

The defining relations of Proposition 2.15.1 take the following form (we will simply write $s_{ij}(u)$ instead of $\mathbf{s}_{ij}(u)$, and denote the generator matrix by $S(u)$ because only one presentation of the twisted Yangian will be used here):

$$(4.3)\qquad (u^2-v^2)\,[s_{ij}(u),s_{kl}(v)] = (u+v)\big(s_{kj}(u)s_{il}(v)-s_{kj}(v)s_{il}(u)\big) \\ -(u-v)\big(\theta_{k,-j}s_{i,-k}(u)s_{-j,l}(v)-\theta_{i,-l}s_{k,-i}(v)s_{-l,j}(u)\big) \\ +\theta_{i,-j}\big(s_{k,-i}(u)s_{-j,l}(v)-s_{k,-i}(v)s_{-j,l}(u)\big)$$

and

$$(4.4)\qquad \theta_{ij}\,s_{-j,-i}(-u) = s_{ij}(u) \pm \frac{s_{ij}(u)-s_{ij}(-u)}{2u},$$

where, as before, the generators $s_{ij}^{(r)}$ of $\mathrm{Y}(\mathfrak{g}_N)$ with $-n \leqslant i,j \leqslant n$ and $r \geqslant 1$ are arranged into the formal series

$$s_{ij}(u) = \delta_{ij} + s_{ij}^{(1)}\,u^{-1} + s_{ij}^{(2)}\,u^{-2} + \dots.$$

The relations (4.3) and (4.4) are easily derived from (2.6) and (2.7), respectively, by using the isomorphism (2.104). We keep to the convention on the double signs $\pm$ and $\mp$ adopted in Chapter 2: the upper and lower signs correspond to the orthogonal and symplectic case, respectively.

For use in this and subsequent chapters, we will produce the corresponding versions of the formulas for the Sklyanin determinant, Sklyanin minors and the entries of the quantum comatrix in the split realization of the twisted Yangians; see the definitions in Section 2.15. The following are respective counterparts of Propositions 2.6.1 and 2.6.2 which are proved in the same way with straightforward modifications.

PROPOSITION 4.1.1. *We have the relation*

$$s^{\,a_1\dots\, a_m}_{\,b_1\dots\, b_m}(u) = \sum_{c=-n}^{n} \check{s}^{\,a_1\dots\, a_m}_{\,b_1\dots\, b_{m-1},c}(u)\, s_{cb_m}(u-m+1).$$ □

PROPOSITION 4.1.2. *Suppose* $-b_1 \in \{a_1,\dots,a_m\}$ *and* $c \notin \{-b_2,\dots,-b_{m-1}\}$. *Then*

$$\check{s}^{\,a_1\dots\, a_m}_{\,b_1\dots\, b_{m-1},c}(u) = 0 \qquad \text{if}\quad c \notin \{a_1,\dots,a_m\}, \quad \text{and}$$

$$\check{s}^{\,a_1\dots\, a_m}_{\,b_1\dots\, b_{m-1},c}(u) = \frac{2u+1}{2u\pm 1}\sum_{r=1}^{m-1}(-1)^{r-1}\, s'_{a_r b_1}(-u)\, s^{\,a_1\dots\widehat{a}_r\dots\, a_{m-1}}_{\,b_2\dots\, b_{m-1}}(u-1)$$

if $c=a_m$, *where the* $s'_{ij}(u)=\theta_{ij}\,s_{-j,-i}(u)$ *are the entries of the transposed matrix* $S'(u)$. □

We will need a reformulation of Theorem 2.7.2; see also Remark 2.7.4. As in Section 2.15, we let the scalar $\alpha_p(u)$ be defined by

$$(4.5)\qquad \alpha_p(u) = \begin{cases} 1 & \text{in the orthogonal case} \\ \dfrac{2u+1}{2u-p+1} & \text{in the symplectic case.} \end{cases}$$

Let $(a_1,\dots,a_N)$ be a permutation of the indices $(-n,-n+1,\dots,n)$. Also recall the map $p\mapsto p'$ of the symmetric group $\mathfrak{S}_N$ into itself given by (2.53).

PROPOSITION 4.1.3. *The following formula for the Sklyanin determinant holds:*

$$\begin{aligned}\operatorname{sdet} S(u) &= (-1)^n\, \alpha_N(u)\\ &\times \sum_{p\in\mathfrak{S}_N} \operatorname{sgn} pp' \cdot s'_{-a_{p(1)},a_{p'(1)}}(-u)\dots s'_{-a_{p(n)},a_{p'(n)}}(-u+n-1)\\ &\qquad\times s_{-a_{p(n+1)},a_{p'(n+1)}}(u-n)\dots s_{-a_{p(N)},a_{p'(N)}}(u-N+1).\end{aligned}$$

PROOF. Let us apply the first formula of Remark 2.7.4 to the permutation $(-a_1,\dots,-a_N)$ of the indices $(-n,-n+1,\dots,n)$ to write down an expression for the Sklyanin determinant. The claim follows by the application of the isomorphism (2.104) to this expression and use of Proposition 2.15.2. □

Recall that the Sklyanin comatrix $\widehat{S}(u)$ is defined by the formula

$$\widehat{S}(u)\, S(u-N+1) = \operatorname{sdet} S(u); \tag{4.6}$$

see Section 2.15. The following is a reformulation of Proposition 2.7.8 for the split presentation of the twisted Yangian.

PROPOSITION 4.1.4. *Let $(a_1,\dots,a_N)$ be an arbitrary permutation of the indices $(-n,\dots,n)$ and let $k\in\{1,\dots,N\}$. We have*

$$\begin{aligned}\widehat{s}_{a_N,-a_k}(u) &= (-1)^n\, \alpha_N(u)\\ &\times \sum_{p\in\mathfrak{S}_N,\ p(N)=k} \operatorname{sgn} pp' \cdot s'_{-a_{p(1)},a_{p'(1)}}(-u)\dots s'_{-a_{p(n)},a_{p'(n)}}(-u+n-1)\\ &\times s_{-a_{p(n+1)},a_{p'(n+1)}}(u-n)\dots s_{-a_{p(N-1)},a_{p'(N-1)}}(u-N+2).\end{aligned}$$

PROOF. The formula follows by the application of the isomorphism (2.104) to the expression for the entries of the Sklyanin comatrix provided by Proposition 2.7.8 and the use of Proposition 2.15.2 and Corollary 2.15.4. □

The following is the counterpart of Proposition 2.14.7, where we use the notation $(a_1,\dots,a_N)=(-n,\dots,n)$.

PROPOSITION 4.1.5. *For any $i,j\in\{1,\dots,N\}$ we have the relation*

$$\widehat{s}^{\,\prime}_{a_ia_j}(u) = (-1)^{i+j}\cdot \alpha_{2N-2}(u)\cdot s^{\,a_1\dots\widehat{a}_j\dots a_N}_{\,a_1\dots\widehat{a}_i\dots a_N}(-u+N-2),$$

where the $\widehat{s}^{\,\prime}_{ij}(u)=\theta_{ij}\,\widehat{s}_{-j,-i}(u)$ are the entries of the transposed matrix $\widehat{S}^{\,\prime}(u)$.

PROOF. The formula is verified by a slight modification of the argument used in the proof of Proposition 2.14.7. □

A different expression for some entries of the Sklyanin comatrix is given by the next lemma. For fixed indices $a,b\in\{-n,n\}$ denote by $\widetilde{S}(u)$ the matrix with the rows numbered by $-n+1,\dots,n-1,a$ and columns numbered by $-n+1,\dots,n-1,b$ such that its (i,j) entry is $s_{ij}(u)$. Also, let $S^{(n-1)}(u)$ denote the submatrix of $S(u)$ corresponding to the rows and columns numbered by $-n+1,-n+2,\dots,n-1$. We use the quasideterminants introduced in Definition 1.10.1.

LEMMA 4.1.6. *If $a,b\in\{-n,n\}$, then the entry $\widehat{s}_{ab}(u)$ of the Sklyanin comatrix $\widehat{S}(u)$ can be given by*

$$\widehat{s}_{ab}(u) = \frac{2u+1}{2u-1}\,\big|\widetilde{S}(-u)\big|_{ab}\cdot \operatorname{sdet} S^{(n-1)}(u-1)$$

in the symplectic case, and

$$\widehat{s}_{ab}(u) = \operatorname{sgn} ab \cdot \big|\widetilde{S}(-u)\big|_{ab} \cdot \operatorname{sdet} S^{(n-1)}(u-1)$$

in the orthogonal case.

PROOF. Recalling that (2.111) coincides with $A_N \operatorname{sdet} S(u)$, we derive the following relation from (4.6):

$$A_N S_1(u) R'_{12} \dots R'_{1N} S_2(u-1) R'_{23} \dots R'_{2N} S_3(u-2)$$
$$\times \dots S_{N-1}(u-N+2) R'_{N-1,N} = A_N \widehat{S}_N(u),$$

where $R'_{ij} = R'_{ij}(-2u+i+j-2)$. Applying the isomorphism (2.104) to the relation of Lemma 2.14.6, we obtain the following counterpart of that relation for the split presentation of the twisted Yangian,

$$A_N S_1(u) R'_{12} \dots R'_{1N} = \frac{2u+1}{2u \pm 1} A_N S'_1(-u).$$

This implies

$$\frac{2u+1}{2u\pm 1} A_N S'_1(-u)\, S_2(u-1) R'_{23} \dots R'_{2,N-1} S_3(u-2)$$
$$\times \dots S_{N-1}(u-N+2) R'_{2N} \dots R'_{N-1,N} = A_N \widehat{S}_N(u).$$

Apply the operators in both sides of this formula to the basis vectors

$$v_{ib} = e_{-i} \otimes e_{-n+1} \otimes e_{-n+2} \otimes \dots \otimes e_{n-1} \otimes e_b, \qquad i \in \{-n+1, \dots, n-1, a\}.$$

For the right-hand side we clearly have

$$A_N \widehat{S}_N(u) v_{ib} = \delta_{ia}\, \widehat{s}_{ab}(u)\, A_N\, v_{aa}.$$

To calculate the left-hand side we note that

$$R'_{2N} \dots R'_{N-1,N}\, v_{ib} = v_{ib}.$$

Furthermore, using the definition of the Sklyanin minors in Section 2.15, for the left-hand side we can write

$$\frac{2u+1}{2u\pm 1} \sum_{k=1}^{N-1} (-1)^{k-1}\, s'_{a_k,-i}(-u)\, s^{\,a_1 \dots \widehat{a}_k \dots a_{N-1}}_{\,-n+1 \dots n-1}(u-1)\, A_N\, v_{bb},$$

where $(a_1, \dots, a_{N-1}) = (-b, -n+1, \dots, n-1)$. Arguing as in the proof of Proposition 2.13.10, we find that $s^{\,-n+1 \dots n-1}_{\,-n+1 \dots n-1}(u-1) = \operatorname{sdet} S^{(n-1)}(u-1)$. Note also that $A_N\, v_{aa} = \operatorname{sgn} ab \cdot A_N\, v_{bb}$ and $s'_{a_k,-i}(-u) = \theta_{a_k,-i}\, s_{i,-a_k}(-u)$. Now apply Lemma 1.10.3 to the system of equations

$$\frac{2u+1}{2u\pm 1} \sum_{k=1}^{N-1} (-1)^{k-1}\, \theta_{a_k,-i}\, s_{i,-a_k}(-u)\, s^{\,a_1 \dots \widehat{a}_k \dots a_{N-1}}_{\,-n+1 \dots n-1}(u-1) = \delta_{ia} \operatorname{sgn} ab \cdot \widehat{s}_{ab}(u),$$

where $i = -n+1, \dots, n-1, a$. This gives

$$\operatorname{sgn} ab \cdot \widehat{s}_{ab}(u) = \frac{2u+1}{2u\pm 1}\, \theta_{ab}\, \big|\widetilde{S}(-u)\big|_{ab} \cdot \operatorname{sdet} S^{(n-1)}(u-1),$$

completing the proof. □

We will now prove a factorization formula for the Sklyanin determinant; cf. Theorem 2.12.1. It will be convenient to introduce the series

$$\mathbf{c}(u) = \alpha_N\big(u + (N-1)/2\big)^{-1} \cdot \operatorname{sdet} S\big(u + (N-1)/2\big).$$

Note that $\mathbf{c}(-u) = \mathbf{c}(u)$ by Corollary 2.15.3.

For any $1 \leqslant m \leqslant n$ denote by $S^{(m)}(u)$ the submatrix of $S(u)$ corresponding to the rows and columns numbered by $-m, -m+1, \dots, m$, and by $\widetilde{S}^{(m)}(u)$ the submatrix of $S^{(m)}(u)$ obtained by removing the row and column enumerated by the index $-m$.

THEOREM 4.1.7. *If $N = 2n$, then*

$$\mathbf{c}(u) = \big|\widetilde{S}^{(1)}(-u-1/2)\big|_{11} \cdot \big|S^{(1)}(u-1/2)\big|_{11} \\ \times \dots \big|\widetilde{S}^{(n)}(-u-n+1/2)\big|_{nn} \cdot \big|S^{(n)}(u-n+1/2)\big|_{nn}.$$

If $N = 2n+1$, then

$$\mathbf{c}(u) = s_{00}(u) \cdot \big|\widetilde{S}^{(1)}(-u-1)\big|_{11} \cdot \big|S^{(1)}(u-1)\big|_{11} \\ \times \dots \big|\widetilde{S}^{(n)}(-u-n)\big|_{nn} \cdot \big|S^{(n)}(u-n)\big|_{nn}.$$

Moreover, all the factors on the right-hand side of each expression commute.

PROOF. By (4.6) we have

$$\widehat{S}(u) = \operatorname{sdet} S(u)\, S^{-1}(u-N+1).$$

Taking the nn-th entry gives

$$\widehat{s}_{nn}(u) = \operatorname{sdet} S(u)\, \big(S^{-1}(u-N+1)\big)_{nn},$$

and hence, by Proposition 1.10.4,

$$\operatorname{sdet} S(u) = \widehat{s}_{nn}(u)\, \big|S(u-N+1)\big|_{nn}.$$

Now using Lemma 4.1.6 with $a = b = n$, we obtain the recurrence relation

$$\operatorname{sdet} S(u) = \frac{2u+1}{2u\pm 1}\, \big|\widetilde{S}^{(n)}(-u)\big|_{nn} \cdot \operatorname{sdet} S^{(n-1)}(u-1) \cdot \big|S(u-N+1)\big|_{nn}. \tag{4.7}$$

Applying the isomorphism (2.104) to the relation of Proposition 2.12.2, we find that the series $\big|S(u-N+1)\big|_{nn}$ commutes with the matrix elements of the matrix $S^{(n-1)}(v)$ and hence with the series $\operatorname{sdet} S^{(n-1)}(u-1)$. Since the coefficients of $\operatorname{sdet} S(u)$ belong to the center of the algebra $\mathrm{Y}(\mathfrak{g}_N)$, all the factors on the right-hand side of (4.7) are mutually permutable. So, (4.7) can be rewritten as

$$\operatorname{sdet} S(u) = \frac{2u+1}{2u\pm 1} \operatorname{sdet} S^{(n-1)}(u-1) \cdot \big|\widetilde{S}^{(n)}(-u)\big|_{nn} \cdot \big|S(u-N+1)\big|_{nn}.$$

The proof is completed by expanding $\operatorname{sdet} S^{(n-1)}(u-1)$ further by induction and replacing u by $u + (N-1)/2$ in the resulting expression. □

Now we reformulate Proposition 2.13.6 for the split presentation of the twisted Yangian, which is proved by a straightforward modification of the corresponding arguments of Section 2.13. The proposition actually holds for the corresponding extended twisted Yangian. However, we need it only for the quotient algebra $\mathrm{Y}(\mathfrak{g}_N)$.

PROPOSITION 4.1.8. *We have the following relations in the twisted Yangian* $Y(\mathfrak{g}_N)$:

$$
\begin{aligned}
&(u^2-v^2)\,[s_{kl}(u), s^{\,a_1\dots a_m}_{\,b_1\dots b_m}(v)]\\
&= (u+v)\sum_{i=1}^{m}\Big(s_{a_i l}(u)\, s^{\,a_1\dots k\dots a_m}_{\,b_1\,\dots\quad b_m}(v) - s^{\,a_1\,\dots\quad a_m}_{\,b_1\dots l\dots b_m}(v)\, s_{k b_i}(u)\Big)\\
&- (u-v)\sum_{i=1}^{m}\Big(\theta_{a_i,-l}\, s_{k,-a_i}(u)\, s^{\,a_1\dots -l\dots a_m}_{\,b_1\,\dots\quad b_m}(v) - \theta_{k,-b_i}\, s^{\,a_1\,\dots\quad a_m}_{\,b_1\dots -k\dots b_m}(v)\, s_{-b_i,l}(u)\Big)\\
&+ \theta_{k,-l}\sum_{i=1}^{m}\Big(s_{a_i,-k}(u)\, s^{\,a_1\dots -l\dots a_m}_{\,b_1\,\dots\quad b_m}(v) - s^{\,a_1\,\dots\quad a_m}_{\,b_1\dots -k\dots b_m}(v)\, s_{-l,b_i}(u)\Big)\\
&+ \sum_{i\neq j}\Big(\theta_{a_j,-l}\, s_{a_i,-a_j}(u)\, s^{\,a_1\dots k\dots -l\dots a_m}_{\,b_1\,\dots\quad b_m}(v) - \theta_{k,-b_i}\, s^{\,a_1\,\dots\quad a_m}_{\,b_1\dots -k\dots l\dots b_m}(v)\, s_{-b_i,b_j}(u)\Big),
\end{aligned}
$$

where in the Sklyanin minors the indices k and l replace a_i and b_i, respectively, in the first sum; the indices $-l$ and $-k$ replace a_i and b_i, respectively, in the second and third sums; in the fourth sum k and $-l$ replace a_i and a_j, respectively, and $-k$ and l replace b_i and b_j, respectively. □

COROLLARY 4.1.9. *Suppose that for some indices $i, j, l, r \in \{1, \dots, m\}$ we have $a_i = -b_l$ and $b_j = -a_r$. Then*

$$[s_{a_i b_j}(u), s^{\,a_1\dots a_m}_{\,b_1\dots b_m}(v)] = 0.$$

PROOF. By the skew-symmetry property, the Sklyanin minor is zero if it has two repeated upper or lower indices. Hence we may assume that $i = r$ if and only if $j = l$. Suppose first that $i \neq r$. Then using the skew-symmetry of Sklyanin minors, we derive from Proposition 4.1.8 that

$$(u-v-1)(u+v+1)\,[s_{a_i b_j}(u), s^{\,a_1\dots a_m}_{\,b_1\dots b_m}(v)] = \theta_{a_i,-b_j}\,[s_{-b_j,-a_i}(u), s^{\,a_1\dots a_m}_{\,b_1\dots b_m}(v)].$$

The same relation holds with i and j replaced by r and l, respectively, which proves the claim in the case under consideration. If $a_i = -b_j$, then Proposition 4.1.8 immediately gives

$$(u-v-1)(u+v+1)\,[s_{a_i b_j}(u), s^{\,a_1\dots a_m}_{\,b_1\dots b_m}(v)] = 0,$$

completing the proof. □

We will also use two versions of the first part of Theorem 2.14.5 for the split presentation of the twisted Yangian. Both of them are verified by adjusting the corresponding arguments of Section 2.14 to the split presentation. Take a nonnegative integer M such that $N - M$ is even so that if $N = 2n$ or $N = 2n+1$, then $M = 2m$ or $M = 2m+1$, respectively, for some $m \leqslant n$. We use the notation (4.5).

PROPOSITION 4.1.10. *The mapping*

$$s_{ij}(u) \mapsto \alpha_{M-N}(u)\cdot s^{\,-n\,\dots\,-m-1,\,i,\,m+1\dots\,n}_{\,-n\,\dots\,-m-1,\,j,\,m+1\dots\,n}(u+n-m), \qquad -m \leqslant i, j \leqslant m$$

defines an algebra homomorphism $Y(\mathfrak{g}_M) \to Y(\mathfrak{g}_N)$. □

PROPOSITION 4.1.11. *The mapping*

$$s_{ab}(u) \mapsto \alpha_{-M}(u)\cdot s^{\,-m\dots\,m,\,a}_{\,-m\dots\,m,\,b}(u+M/2), \qquad m+1 \leqslant |a|, |b| \leqslant n \tag{4.8}$$

defines an algebra homomorphism $Y(\mathfrak{g}_{N-M}) \to Y(\mathfrak{g}_N)$. □

Now consider the particular case $M = N - 2$. The homomorphism of Proposition 4.1.11 will play an important role in Chapter 9. This homomorphism turns out to be closely related to the one obtained by restriction of the automorphism of Corollary 2.15.5. More precisely, identify the subalgebra of $\mathrm{Y}(\mathfrak{g}_N)$ generated by the elements $s_{ab}^{(r)}$ for $a, b \in \{-n, n\}$ with the twisted Yangian $\mathrm{Y}(\mathfrak{g}_2)$. The restriction of the automorphism of Corollary 2.15.5 to the subalgebra $\mathrm{Y}(\mathfrak{g}_2)$ defines the homomorphism

$$\mathrm{Y}(\mathfrak{g}_2) \to \mathrm{Y}(\mathfrak{g}_N), \qquad s_{ab}(u) \mapsto \alpha_N(u)\, \widehat{s}_{ab}(-u + N/2 - 1). \tag{4.9}$$

The defining relations of the twisted Yangian imply that the map

$$s_{ab}(u) \mapsto \operatorname{sgn} a \cdot \operatorname{sgn} b \cdot s_{ab}(u), \qquad a, b \in \{-n, n\} \tag{4.10}$$

defines an automorphism $\mathrm{Y}(\mathfrak{g}_2) \to \mathrm{Y}(\mathfrak{g}_2)$.

COROLLARY 4.1.12. *The homomorphism* (4.9) *coincides with the homomorphism* (4.8) *in the symplectic case, while in the orthogonal case the homomorphism* (4.9) *coincides with the composition of* (4.8) *with the automorphism* (4.10).

PROOF. This is immediate from Proposition 4.1.5 and the skew-symmetry of the Sklyanin minors. □

4.2. Highest weight representations

DEFINITION 4.2.1. A representation V of the twisted Yangian $\mathrm{Y}(\mathfrak{g}_N)$ is called a *highest weight representation* if there exists a nonzero vector $\xi \in V$ such that V is generated by ξ and the following relations hold:

$$s_{ij}(u)\, \xi = 0 \qquad \text{for} \quad -n \leqslant i < j \leqslant n, \qquad \text{and} \tag{4.11}$$

$$s_{ii}(u)\, \xi = \mu_i(u)\, \xi \tag{4.12}$$

for some formal series

$$\mu_i(u) = 1 + \mu_i^{(1)} u^{-1} + \mu_i^{(2)} u^{-2} + \dots, \qquad \mu_i^{(r)} \in \mathbb{C}, \tag{4.13}$$

where the index i in (4.12) takes the values $1, \dots, n$ if $N = 2n$, and $0, 1, \dots, n$ if $N = 2n + 1$. The vector ξ is called the *highest vector* of V, and the tuple of formal series $\mu(u) = (\mu_1(u), \dots, \mu_n(u))$ or $\mu(u) = (\mu_0(u), \dots, \mu_n(u))$, respectively, is the *highest weight* of V. □

Note that ξ is also an eigenvector for the action of $s_{ii}(u)$ with negative i. Indeed, the symmetry relation (4.4) gives

$$s_{-i,-i}(u) = s_{ii}(-u) \pm \frac{s_{ii}(u) - s_{ii}(-u)}{2u}. \tag{4.14}$$

Therefore, if $i > 0$, then

$$s_{-i,-i}(u)\, \xi = \Big(\mu_i(-u) \pm \frac{\mu_i(u) - \mu_i(-u)}{2u} \Big)\, \xi. \tag{4.15}$$

Moreover, (4.14) implies that $s_{00}(u) = s_{00}(-u)$ in the case where N is odd. Hence, the series $\mu_0(u)$ has to be even, $\mu_0(u) = \mu_0(-u)$.

DEFINITION 4.2.2. Let $\mu(u) = (\mu_1(u), \dots, \mu_n(u))$ or $\mu(u) = (\mu_0(u), \dots, \mu_n(u))$ be an arbitrary tuple of formal series of the form (4.13), where the series $\mu_0(u)$ is even. The *Verma module* $M(\mu(u))$ over the twisted Yangian $\mathrm{Y}(\mathfrak{g}_N)$ with $N = 2n$ or $N = 2n + 1$, respectively, is the quotient of $\mathrm{Y}(\mathfrak{g}_N)$ by the left ideal generated by all coefficients of the series $s_{ij}(u)$ for $-n \leqslant i < j \leqslant n$ and $s_{ii}(u) - \mu_i(u)$ for $i = 1, \dots, n$ or $i = 0, 1, \dots, n$. □

Similar to the case of the Yangian $\mathrm{Y}(\mathfrak{gl}_N)$ (see Section 3.2), the Verma module $M(\mu(u))$ is a highest weight representation of $\mathrm{Y}(\mathfrak{g}_N)$ with the highest weight $\mu(u)$ and the highest vector $1_{\mu(u)}$ which is the image of the element $1 \in \mathrm{Y}(\mathfrak{g}_N)$ in the quotient. Moreover, any highest weight representation of $\mathrm{Y}(\mathfrak{g}_N)$ with the highest weight $\mu(u)$ is isomorphic to a quotient of $M(\mu(u))$.

Using the isomorphism (2.104), we deduce the following version of the Poincaré–Birkhoff–Witt theorem for the algebra $\mathrm{Y}(\mathfrak{g}_N)$ from Corollary 2.4.4.

COROLLARY 4.2.3. *Given an arbitrary linear order on the set of generators*

$$s_{ij}^{(2p)} \quad \text{with} \quad i+j \geqslant 0 \quad \text{and} \quad s_{ij}^{(2p-1)} \quad \text{with} \quad i+j > 0 \qquad \text{for} \quad p = 1, 2, \dots,$$

in the orthogonal case, and

$$s_{ij}^{(2p)} \quad \text{with} \quad i+j > 0 \quad \text{and} \quad s_{ij}^{(2p-1)} \quad \text{with} \quad i+j \geqslant 0 \qquad \text{for} \quad p = 1, 2, \dots,$$

in the symplectic case, any element of the algebra $\mathrm{Y}(\mathfrak{g}_N)$ *can be uniquely written as a linear combination of ordered monomials in the generators.* □

Consider the subset of the generators introduced in Corollary 4.2.3 which satisfy the additional condition $i > j$. Then, given any order on this subset, the corresponding ordered monomials

$$s_{i_1 j_1}^{(r_1)} \dots s_{i_m j_m}^{(r_m)} 1_{\mu(u)}, \qquad m \geqslant 0, \tag{4.16}$$

form a basis of $M(\mu(u))$; cf. Proposition 3.2.4.

PROPOSITION 4.2.4. *Suppose that* V *is a highest weight representation of* $\mathrm{Y}(\mathfrak{g}_N)$ *with the highest weight* $\mu(u)$. *Then each coefficient of the Sklyanin determinant* $\operatorname{sdet} S(u)$ *acts on* V *as multiplication by a scalar determined by*

$$\operatorname{sdet} S(u)|_V = \alpha_N(u)\, \mu_1(-u+n-1)\, \mu_1(u-n) \dots \mu_n(-u)\, \mu_n(u-N+1),$$

if $N = 2n$, *and*

$$\operatorname{sdet} S(u)|_V = \mu_0(u-n)\, \mu_1(-u+n-1)\, \mu_1(u-n-1) \dots \mu_n(-u)\, \mu_n(u-N+1),$$

if $N = 2n+1$, *where* $\alpha_N(u)$ *is defined in* (4.5).

PROOF. By Theorem 2.8.2, all coefficients of the Sklyanin determinant belong to the center of the twisted Yangian $\mathrm{Y}(\mathfrak{g}_N)$. Since V is a homomorphic image of the Verma module $M(\mu(u))$, the action of $\operatorname{sdet} S(u)$ on V is determined by the application of $\operatorname{sdet} S(u)$ to the highest vector ξ of V. Now we use the expression for $\operatorname{sdet} S(u)$ provided by Proposition 4.1.3 with $(a_1, \dots, a_N) = (-n, -n+1, \dots, n)$ to calculate $\operatorname{sdet} S(u)\, \xi$. By the definition of the map (2.53), we have $p'(N) = N$. Now, since $a_N = n$ and ξ is the highest vector, we have

$$s_{-a_{p(N)}, a_{p'(N)}}(u-N+1)\, \xi = 0$$

unless $a_{p(N)} = -n$; that is, $p(N) = 1$. Using the definition of the map (2.53) again, we obtain $p'(1) = 1$ so that $a_{p'(1)} = -n$. Hence, the application of the summand corresponding to a permutation p to ξ gives a nonzero contribution only if $a_{p(1)} = n$, that is, $p(1) = N$. This implies $p'(N-1) = N-1$, and we can proceed by induction to deduce that there is only one summand in the formula for $\operatorname{sdet} S(u)$ which gives a nonzero contribution for $\operatorname{sdet} S(u)\, \xi$. This summand corresponds to the permutation $p = (N, N-1, \ldots, 1)$. Its image under the map (2.53) is the identity permutation $p' = (1, 2, \ldots, N)$. Now the value of $\operatorname{sdet} S(u)\,\xi$ is easily calculated from Definition 4.2.1, completing the proof. □

In accordance with Section 2.15, the Lie algebra $\mathfrak{g}_N$ is spanned by the elements $F_{ij} \in \mathfrak{gl}_N$ given by

$$F_{ij} = E_{ij} - \theta_{ij}\, E_{-j,-i}, \qquad -n \leqslant i, j \leqslant n.$$

In particular, for any i, j we have

$$F_{-j,-i} = -\theta_{ij}\, F_{ij}.$$

Let us identify the elements F_{ij} with their images $s_{ij}^{(1)}$ in $\mathrm{Y}(\mathfrak{g}_N)$ under the embedding (2.107). Taking the coefficient of v in (4.3) we deduce that

$$(4.17)\quad [F_{ij}, s_{kl}(u)] = \delta_{kj} s_{il}(u) - \delta_{il} s_{kj}(u) - \theta_{k,-j}\delta_{i,-k} s_{-j,l}(u) + \theta_{i,-l}\delta_{-l,j} s_{k,-i}(u).$$

Regarding $M(\mu(u))$ as a $\mathfrak{g}_N$-module and arguing as in Section 3.2, we obtain the weight space decomposition

$$M(\mu(u)) = \bigoplus_{\nu} M(\mu(u))_{\nu},$$

summed over all $\mathfrak{g}_N$-weights $\nu = (\nu_1, \ldots, \nu_n)$ of $M(\mu(u))$, where

$$M(\mu(u))_{\nu} = \{\eta \in M(\mu(u)) \mid F_{ii}\,\eta = \nu_i\,\eta, \quad i = 1, \ldots, n\}.$$

By (4.17) the set of weights of $M(\mu(u))$ coincides with that of the $\mathfrak{g}_N$-Verma module with the highest weight $\mu^{(1)} = (\mu_1^{(1)}, \ldots, \mu_n^{(1)})$. More precisely, this set consists of all weights of the form $\mu^{(1)} - \omega$, where ω is a $\mathbb{Z}_+$-linear combination of the positive roots. We take the linear span of the elements $F_{11}, \ldots, F_{nn}$ as the Cartan subalgebra $\mathfrak{h}$ of $\mathfrak{g}_N$ and let $\varepsilon_1, \ldots, \varepsilon_n$ denote the basis vectors of the dual space $\mathfrak{h}^*$ such that $\varepsilon_i(F_{jj}) = \delta_{ij}$. We identify the elements $\nu = \nu_1\,\varepsilon_1 + \cdots + \nu_n\,\varepsilon_n$ of $\mathfrak{h}^*$ with the n-tuples $(\nu_1, \ldots, \nu_n)$. We consider the nonzero elements F_{ij} with $i < j$ as the positive root vectors. The corresponding positive roots are

$$-\varepsilon_i - \varepsilon_j, \quad \varepsilon_i - \varepsilon_j \quad \text{with} \quad 1 \leqslant i < j \leqslant n$$

for $\mathfrak{g}_N = \mathfrak{o}_{2n}$,

$$-2\,\varepsilon_i \quad \text{with} \quad 1 \leqslant i \leqslant n \quad \text{and} \quad -\varepsilon_i - \varepsilon_j, \quad \varepsilon_i - \varepsilon_j \quad \text{with} \quad 1 \leqslant i < j \leqslant n$$

for $\mathfrak{g}_N = \mathfrak{sp}_{2n}$, and

$$-\varepsilon_i \quad \text{with} \quad 1 \leqslant i \leqslant n \quad \text{and} \quad -\varepsilon_i - \varepsilon_j, \quad \varepsilon_i - \varepsilon_j \quad \text{with} \quad 1 \leqslant i < j \leqslant n$$

for $\mathfrak{g}_N = \mathfrak{o}_{2n+1}$. The standard partial ordering on the set of weights of any $\mathfrak{g}_N$-module is defined as follows; cf. Section 3.2. If α and β are two weights, then α *precedes* β if $\beta - \alpha$ is a $\mathbb{Z}_+$-linear combination of the positive roots.

Exactly as in Section 3.2, we deduce that any submodule K of $M(\mu(u))$ admits the weight space decomposition

$$K = \bigoplus_{\nu} K_{\nu}, \qquad K_{\nu} = K \cap M(\mu(u))_{\nu}.$$

This implies that the sum of all proper submodules is the unique maximal proper submodule of $M(\mu(u))$.

DEFINITION 4.2.5. The *irreducible highest weight representation* $V(\mu(u))$ of $\mathrm{Y}(\mathfrak{g}_N)$ with the highest weight $\mu(u)$ is defined as the quotient of the Verma module $M(\mu(u))$ by the unique maximal proper submodule. □

Clearly, $V(\mu(u))$ is isomorphic to the unique irreducible quotient of an arbitrary highest weight representation V with the highest weight $\mu(u)$. Moreover, two irreducible highest weight representations are isomorphic if and only if they have the same highest weight.

THEOREM 4.2.6. *Every finite-dimensional irreducible representation V of the twisted Yangian $\mathrm{Y}(\mathfrak{g}_N)$ is a highest weight representation. Moreover, V contains a unique, up to a constant factor, highest vector.*

PROOF. We argue as in the proof of Theorem 3.2.7. Set

$$V^0 = \{\eta \in V \mid s_{ij}(u)\,\eta = 0, \qquad -n \leqslant i < j \leqslant n\}. \tag{4.18}$$

Let us show that V^0 is nonzero. Consider the set of weights of V, regarded as a $\mathfrak{g}_N$-module defined via the embedding (2.107). This set is finite and hence contains a maximal weight ν with respect to the partial ordering on the set of weights of V. The corresponding weight vector η belongs to V^0. Indeed, if $i < j$, then by (4.17) the weight of $s_{ij}(u)\,\eta$ has the form $\nu + \alpha$ for a positive root α. By the maximality of ν, we have $s_{ij}(u)\,\eta = 0$.

Now, let us show that the subspace V^0 is invariant with respect to the action of all elements $s_{kk}^{(r)}$. It is sufficient to demonstrate that if $i < j$, then the commutator $[s_{ij}(u), s_{kk}(v)]$ acts on V^0 as the zero operator. Applying the symmetry relation (4.4), if necessary, we may assume that $k \geqslant 0$ and $j \geqslant 1$. Then the claim follows directly from (4.3) if $i < k$. If $i \geqslant k$, then $k < j$ and the claim is verified by the application of the formula (4.3) to the commutator $[s_{kk}(v), s_{ij}(u)]$.

Similarly, for any i and j the commutator $[s_{ii}(u), s_{jj}(v)]$ acts on V^0 as the zero operator. Indeed, by the symmetry relation (4.4), we may assume that $0 \leqslant i \leqslant j$. Now the claim follows easily from (4.3). This shows that the elements $s_{kk}^{(r)}$ act on V^0 as pairwise commuting operators. Let $\xi \in V^0$ be a simultaneous eigenvector for all these operators. Then $V = \mathrm{Y}(\mathfrak{g}_N)\xi$ since V is irreducible so that ξ is a highest vector of V. In particular, ξ is a $\mathfrak{g}_N$-weight vector with a certain weight ν.

Finally, since V is a homomorphic image of the Verma module $M(\mu(u))$, the vector space V is spanned by the vectors of the form

$$s_{i_1 j_1}^{(r_1)} \dots s_{i_m j_m}^{(r_m)}\,\xi, \qquad m \geqslant 0,$$

with ordered products of the generators and the same conditions on the indices as in (4.16). Hence, by (4.17) the weight space V_ν is one-dimensional and spanned by the vector ξ. Moreover, if ρ is a weight of V and $\rho \neq \nu$, then ρ strictly precedes ν. This proves that the highest vector ξ of V is determined uniquely, up to a constant factor. □

The following analogue of Corollary 3.2.8 is immediate from the proof of Theorem 4.2.6.

COROLLARY 4.2.7. *Suppose that V is an irreducible highest weight representation of $\mathrm{Y}(\mathfrak{g}_N)$. Then the subspace V^0 defined in (4.18) is one-dimensional and spanned by the highest vector of V.* □

Due to Theorem 4.2.6, all finite-dimensional irreducible representations of the twisted Yangian $\mathrm{Y}(\mathfrak{g}_N)$ have the form $V(\mu(u))$ for a certain highest weight $\mu(u)$. So, it remains to describe the set of highest weights $\mu(u)$ such that $V(\mu(u))$ is finite-dimensional. The following proposition provides a necessary condition on such highest weights. We use the notation (3.81).

PROPOSITION 4.2.8. *If the irreducible highest weight representation $V(\mu(u))$ of $\mathrm{Y}(\mathfrak{g}_N)$ is finite-dimensional, then the following condition holds:*

$$\mu_1(u) \longrightarrow \mu_2(u) \longrightarrow \dots \longrightarrow \mu_n(u).$$

PROOF. Let J be the left ideal of $\mathrm{Y}(\mathfrak{g}_N)$ generated by all the coefficients of the series $s_{-i,j}(u)$ for $i,j=1,\dots,n$. Consider the subspace V^J of $V(\mu(u))$ defined by

$$V^J = \{\eta \in V(\mu(u)) \mid s_{-i,j}(u)\,\eta = 0 \quad \text{for all} \quad i,j=1,\dots,n\}.$$

Note that the highest vector ξ of $V(\mu(u))$ belongs to V^J. By the defining relations (4.3), for any positive indices i,j,k,l we have

$$[s_{-i,j}(u), s_{kl}(v)] \equiv 0 \mod J.$$

This implies that the subspace V^J is stable under the action of $s_{ij}(u)$ for any positive indices i,j. Moreover, for any positive i,j,k,l the defining relations (4.3) give

$$(u-v)\,[s_{ij}(u), s_{kl}(v)] \equiv s_{kj}(u)s_{il}(v) - s_{kj}(v)s_{il}(u) \mod J.$$

Therefore, comparing this with the defining relations (1.3), we conclude that V^J can be equipped with the action of the Yangian $\mathrm{Y}(\mathfrak{gl}_n)$, defined by $t_{ij}(u) \mapsto s_{ij}(u)$. Furthermore, the cyclic span $\mathrm{Y}(\mathfrak{gl}_n)\,\xi$ is a finite-dimensional highest weight representation of $\mathrm{Y}(\mathfrak{gl}_n)$ with the highest weight $(\mu_1(u),\dots,\mu_n(u))$. Hence, by Theorem 3.4.1, this highest weight satisfies the required conditions. □

For any n-tuple of complex numbers $\mu = (\mu_1,\dots,\mu_n)$ we will denote by $V(\mu)$ the irreducible representation of the Lie algebra $\mathfrak{g}_N$ with the highest weight μ. That is, $V(\mu)$ is generated by a nonzero vector ξ such that

$$F_{ij}\,\xi = 0 \qquad \text{for} \quad -n \leqslant i < j \leqslant n, \qquad \text{and}$$
$$F_{ii}\,\xi = \mu_i\,\xi \qquad \text{for} \quad 1 \leqslant i \leqslant n. \tag{4.19}$$

The representation $V(\mu)$ is finite-dimensional if and only if

$$\mu_i - \mu_{i+1} \in \mathbb{Z}_+ \qquad \text{for} \quad i=1,\dots,n-1$$

and

$$\begin{aligned} -\mu_1 - \mu_2 &\in \mathbb{Z}_+ \qquad \text{if} \quad \mathfrak{g}_N = \mathfrak{o}_{2n}, \\ -\mu_1 &\in \mathbb{Z}_+ \qquad \text{if} \quad \mathfrak{g}_N = \mathfrak{sp}_{2n}, \\ -2\,\mu_1 &\in \mathbb{Z}_+ \qquad \text{if} \quad \mathfrak{g}_N = \mathfrak{o}_{2n+1}. \end{aligned}$$

Here we have assumed that $n \geqslant 2$ in the even orthogonal case $\mathfrak{g}_N = \mathfrak{o}_{2n}$. The abelian Lie algebra $\mathfrak{o}_2$ has a family of one-dimensional representations $V(\gamma)$, where γ runs over the set of complex numbers. The generator F_{11} acts on the basis vector ξ by $F_{11}\,\xi = \gamma\,\xi$.

Using the evaluation homomorphism (2.106), we can extend each representation $V(\mu)$ to a module over $\mathrm{Y}(\mathfrak{g}_N)$. This is obviously a highest weight representation of $\mathrm{Y}(\mathfrak{g}_N)$ with the highest weight given by

$$\mu_i(u) = \frac{1 + (\mu_i \pm 1/2)\,u^{-1}}{1 \pm 1/2\,u^{-1}}, \qquad i=1,\dots,n, \tag{4.20}$$

and $\mu_0(u) = 1$ if $N = 2n+1$. In the case of $\mathfrak{g}_N = \mathfrak{o}_2$ the highest weight of the $\mathrm{Y}(\mathfrak{o}_2)$-module $V(\gamma)$ is the single series given by

$$\mu(u) = \frac{1 + (\gamma + 1/2)\, u^{-1}}{1 + 1/2\, u^{-1}}. \tag{4.21}$$

We will usually identify the twisted Yangian $\mathrm{Y}(\mathfrak{g}_N)$ with a subalgebra of $\mathrm{Y}(\mathfrak{gl}_N)$ via the embedding (2.108). Then we will write

$$S(u) = T(u)\, T'(-u),$$

where the ij-th entry of the matrix $T'(u) = G T^t(u) G^{-1}$ is $\theta_{ij}\, t_{-j,-i}(u)$. Hence, the entries of $S(u)$ are given by

$$s_{ij}(u) = \sum_a \theta_{aj}\, t_{ia}(u)\, t_{-j,-a}(-u). \tag{4.22}$$

Using the isomorphism (2.104) or repeating the argument of the proof of Theorem 2.10.1 for the presentation of $\mathrm{Y}(\mathfrak{g}_N)$ under consideration, we can easily get the corresponding version of the formula (2.64) which reads

$$\Delta : s_{ij}(u) \mapsto \sum_{a,b} \theta_{bj}\, t_{iu}(u)\, t_{-j,-b}(-u) \otimes s_{ab}(u). \tag{4.23}$$

If L is a $\mathrm{Y}(\mathfrak{gl}_N)$-module and V is a $\mathrm{Y}(\mathfrak{g}_N)$-module, then using this formula, we can equip $L \otimes V$ with a structure of a $\mathrm{Y}(\mathfrak{g}_N)$-module by

$$y \cdot (\xi \otimes \eta) = \Delta(y)(\xi \otimes \eta), \qquad y \in \mathrm{Y}(\mathfrak{g}_N), \quad \xi \in L, \quad \eta \in V.$$

Let ζ denote the highest vector of the $\mathrm{Y}(\mathfrak{gl}_N)$-module $L(\lambda(u))$ with an arbitrary highest weight

$$\lambda(u) = (\lambda_{-n}(u), \dots, \lambda_n(u))$$

and let ξ denote the highest vector of the $\mathrm{Y}(\mathfrak{g}_N)$-module $V(\mu(u))$.

PROPOSITION 4.2.9. *The submodule $\mathrm{Y}(\mathfrak{g}_N)(\zeta \otimes \xi)$ of the tensor product module $L(\lambda(u)) \otimes V(\mu(u))$ over $\mathrm{Y}(\mathfrak{g}_N)$ is a highest weight representation with the highest vector $\zeta \otimes \xi$. Moreover, the i-th component of the highest weight is given by $\lambda_i(u)\, \lambda_{-i}(-u)\, \mu_i(u)$.*

PROOF. Set $\eta = \zeta \otimes \xi$. By (4.23) we can write

$$s_{ij}(u)\, \eta = \sum_{a,b} \theta_{bj}\, t_{ia}(u)\, t_{-j,-b}(-u)\, \zeta \otimes s_{ab}(u)\, \xi. \tag{4.24}$$

Suppose that $i < j$. If $a < b$, then $s_{ab}(u)\, \xi = 0$ so that we may restrict the summation to the set of indices with $a \geqslant b$. If $-j < -b$, then $t_{-j,-b}(-u)\, \zeta = 0$. Otherwise, $j \leqslant b$ and hence $i < a$. Then $t_{ia}(u)\, \zeta = 0$, and we can deduce that $s_{ij}(u)\, \zeta = 0$ by using the relations

$$\begin{aligned} 2\, u\, [t_{ia}(u), t_{-j,-b}(-u)] &= t_{-j,a}(u)\, t_{i,-b}(-u) - t_{-j,a}(-u)\, t_{i,-b}(u) \\ &= t_{i,-b}(-u)\, t_{-j,a}(u) - t_{i,-b}(u)\, t_{-j,a}(-u) \end{aligned}$$

implied by (1.3). Indeed, we either have the inequality $i < -b$ or $-j < a$, because otherwise $i \geqslant -b \geqslant -a \geqslant j$, contradicting $i < j$.

Similar argument shows that if $i = j$ is nonnegative (the zero value may occur only in the case $N = 2n+1$), then the only nonzero term in (4.24) corresponds to the values $a = b = i$, thus completing the proof. □

If $\mu_i(u) = 1$ for all values of i, then the representation $V(\mu(u))$ of $\mathrm{Y}(\mathfrak{g}_N)$ is one-dimensional. In this case the $\mathrm{Y}(\mathfrak{g}_N)$-module $L(\lambda(u)) \otimes V(\mu(u))$ is isomorphic to the restriction of the $\mathrm{Y}(\mathfrak{gl}_N)$-module $L(\lambda(u))$. This is immediate by the comparison of the formulas (4.22) and (4.23), since $s_{ab}(u)$ acts on $V(\mu(u))$ as multiplication by δ_{ab}. We have thus proved the following corollary.

COROLLARY 4.2.10. *Consider $L(\lambda(u))$ as a $\mathrm{Y}(\mathfrak{g}_N)$-module obtained by restriction of the $\mathrm{Y}(\mathfrak{gl}_N)$-module. Then its submodule $\mathrm{Y}(\mathfrak{g}_N)\zeta$ is a highest weight representation with the highest vector ζ. Moreover, the i-th component of the highest weight is given by $\lambda_i(u)\,\lambda_{-i}(-u)$.* □

For any n-tuple $\mu = (\mu_1, \dots, \mu_n)$ of complex numbers we can consider the tensor product

$$L(\lambda^{(1)}) \otimes L(\lambda^{(2)}) \otimes \dots \otimes L(\lambda^{(k)}) \otimes V(\mu) \tag{4.25}$$

of the $\mathrm{Y}(\mathfrak{gl}_N)$-module (3.66) with the evaluation $\mathrm{Y}(\mathfrak{g}_N)$-module $V(\mu)$. We write $\lambda^{(a)} = (\lambda^{(a)}_{-n}, \dots, \lambda^{(a)}_n)$ for the highest weights of the evaluation $\mathrm{Y}(\mathfrak{gl}_N)$-modules. Denote by ζ_a the highest vector of $L(\lambda^{(a)})$ and set $\zeta = \zeta_1 \otimes \dots \otimes \zeta_k \otimes \xi$.

PROPOSITION 4.2.11. *The submodule $\mathrm{Y}(\mathfrak{g}_N)\zeta$ of the $\mathrm{Y}(\mathfrak{g}_N)$-module (4.25) is a highest weight representation with the highest vector ζ. Moreover, the i-th component of the highest weight is given by $\lambda_i(u)\,\lambda_{-i}(-u)\,\mu_i(u)$, where the $\lambda_i(u)$ are defined by (3.67) and the $\mu_i(u)$ are defined by (4.20) or (4.21), respectively.*

PROOF. We argue by induction on k. For any $k \geqslant 1$ we can regard the module $\mathrm{Y}(\mathfrak{g}_N)\zeta$ as a submodule of $L(\lambda^{(1)}) \otimes V$, where V is the cyclic span over $\mathrm{Y}(\mathfrak{g}_N)$ of the vector $\zeta_2 \otimes \dots \otimes \zeta_k \otimes \xi$. It remains to apply the argument used in the proof of Proposition 4.2.9. □

In the above notation set $\zeta' = \zeta_1 \otimes \dots \otimes \zeta_k$. The following corollary is immediate from Proposition 4.2.11.

COROLLARY 4.2.12. *Consider the tensor product*

$$L(\lambda^{(1)}) \otimes L(\lambda^{(2)}) \otimes \dots \otimes L(\lambda^{(k)}) \tag{4.26}$$

as a $\mathrm{Y}(\mathfrak{g}_N)$-module obtained by restriction of the $\mathrm{Y}(\mathfrak{gl}_N)$-module. Then its submodule $\mathrm{Y}(\mathfrak{g}_N)\zeta'$ is a highest weight representation with the highest vector ζ'. Moreover, the i-th component of the highest weight is given by $\lambda_i(u)\,\lambda_{-i}(-u)$, where the $\lambda_i(u)$ are defined by (3.67). □

PROPOSITION 4.2.13. (i) *For any element η of the $\mathrm{Y}(\mathfrak{g}_N)$-module (4.25) and any indices i, j the expression $(1 \pm 1/2\, u^{-1})\, s_{ij}(u)\, \eta$ is a polynomial in u^{-1} of degree $\leqslant 2k+1$.*

(ii) *For any element η of the $\mathrm{Y}(\mathfrak{g}_N)$-module (4.26) and any indices i, j the expression $s_{ij}(u)\,\eta$ is a polynomial in u^{-1} of degree $\leqslant 2k$.*

PROOF. Using the definition of the evaluation module $V(\mu)$, we deduce from (2.106) that $(1 \pm 1/2\, u^{-1})\, s_{ab}(u)$ acts in $V(\mu)$ as the operator $\delta_{ab} + (F_{ab} \pm 1/2\, \delta_{ab})\, u^{-1}$. The first part now follows from (4.23) and Proposition 3.2.11. The second part is immediate from (4.22) and Proposition 3.2.11. □

With our convention on the indices, the anti-automorphism (3.72) of $\mathrm{Y}(\mathfrak{gl}_N)$ now takes the form

$$\varkappa_N : t_{ij}(u) \mapsto t_{-i,-j}(-u).$$

Then by (4.22),

$$\varkappa_N : s_{ij}(u) \mapsto \sum_a \theta_{aj}\, t_{ja}(u)\, t_{-i,-a}(-u) = \theta_{ij}\, s_{ji}(u).$$

This means that the subalgebra $\mathrm{Y}(\mathfrak{g}_N)$ of $\mathrm{Y}(\mathfrak{gl}_N)$ is stable under $\varkappa_N$ and the restriction of $\varkappa_N$ to $\mathrm{Y}(\mathfrak{g}_N)$ yields an anti-automorphism of the latter.

Given a $\mathfrak{g}_N$-module V with finite-dimensional weight subspaces,

$$V = \bigoplus_\mu V_\mu, \qquad \dim V_\mu < \infty, \tag{4.27}$$

we define the (restricted) dual vector space to V by

$$V^* = \bigoplus_\mu V_\mu^*.$$

If the $\mathfrak{g}_N$-action on V is obtained by restriction of a $\mathrm{Y}(\mathfrak{g}_N)$-action, we will equip the dual vector space V^* with a structure of a $\mathrm{Y}(\mathfrak{g}_N)$-module by

$$(y\,\omega)(\eta) = \omega(\varkappa_N(y)\eta) \quad \text{for} \quad y \in \mathrm{Y}(\mathfrak{g}_N) \quad \text{and} \quad \omega \in V^*,\ \eta \in V;$$

cf. Section 3.2. It is straightforward to verify that the dual module $V(\mu)^*$ to the evaluation module $V(\mu)$ is isomorphic to $V(\mu)$.

Denote the $\mathrm{Y}(\mathfrak{g}_N)$-module (4.25) by V. Its restriction to the subalgebra $\mathrm{U}(\mathfrak{g}_N)$ satisfies the condition (4.27). Indeed, this is implied by the fact that the restriction of any irreducible highest weight representation $L(\lambda)$ of $\mathfrak{gl}_N$ to the subalgebra $\mathfrak{g}_N$ has finite-dimensional weight subspaces. The dual $\mathrm{Y}(\mathfrak{g}_N)$-module V^* is described in the following proposition, where we use the notation of Proposition 3.2.12.

PROPOSITION 4.2.14. *The* $\mathrm{Y}(\mathfrak{g}_N)$*-module* V^* *is isomorphic to the tensor product module*

$$L(\widetilde{\lambda}^{(1)}) \otimes L(\widetilde{\lambda}^{(2)}) \otimes \ldots \otimes L(\widetilde{\lambda}^{(k)}) \otimes V(\mu).$$

PROOF. Let L denote the tensor product of the evaluation modules $L(\lambda^{(i)})$, as in Proposition 3.2.12. The vector space V^* can be naturally identified with the tensor product space $L^* \otimes V(\mu)^*$, where the dual space L^* can be understood in the sense of (3.70). Let $f \in L^*$ and $g \in V(\mu)^*$. Then by (4.23) and the definition of V^*, for any elements $\eta \in L$ and $\zeta \in V(\mu)$ we have

$$\begin{aligned}\big(s_{ij}(u)\,(f \otimes g)\big)(\eta \otimes \zeta) &= (f \otimes g)\big(\theta_{ij}\, s_{ji}(u)(\eta \otimes \zeta)\big)\\ &= (f \otimes g)\Big(\theta_{ij} \sum_{a,b} \theta_{bi}\, t_{ja}(u)\, t_{-i,-b}(-u)\,\eta \otimes s_{ab}(u)\,\zeta\Big)\\ &= \sum_{a,b} \theta_{bj}\, f(t_{ja}(u)\, t_{-i,-b}(-u)\,\eta) \otimes g(s_{ab}(u)\,\zeta).\end{aligned}$$

By the definition of L^* and $V(\mu)^*$ this coincides with

$$\Big(\sum_{a,b} \theta_{aj} t_{ib}(u)\, t_{-j,-a}(-u)\, f \otimes s_{ba}(u)\, g\Big)(\eta \otimes \zeta).$$

Hence, the proof is completed by the application of Proposition 3.2.12. □

Recall that the special twisted Yangian $\mathrm{SY}(\mathfrak{g}_N)$ is the subalgebra of $\mathrm{Y}(\mathfrak{g}_N)$ which consists of the elements stable under all automorphisms of the form (2.17); see Section 2.9.

PROPOSITION 4.2.15. *Every finite-dimensional irreducible representation of the twisted Yangian* $Y(\mathfrak{g}_N)$ *remains irreducible when restricted to the special twisted Yangian* $SY(\mathfrak{g}_N)$*. Moreover, such restrictions exhaust all finite-dimensional irreducible representation of* $SY(\mathfrak{g}_N)$.

PROOF. Both statements follow from Theorem 2.9.2 because the elements of the center $ZY(\mathfrak{g}_N)$ of $Y(\mathfrak{g}_N)$ act on any finite-dimensional irreducible representation of $Y(\mathfrak{g}_N)$ as multiplications by scalars. □

4.3. Representations of $Y(\mathfrak{sp}_{2n})$

As with the representations of the Yangian for $\mathfrak{gl}_N$, the case of $Y(\mathfrak{sp}_2)$ plays a key role in the description of the finite-dimensional irreducible representations of the twisted Yangian $Y(\mathfrak{sp}_{2n})$.

By Definition 4.2.5, the irreducible highest weight representations $V(\mu(u))$ of $Y(\mathfrak{sp}_2)$ are parameterized by formal series of the form

$$\mu(u) = 1 + \mu^{(1)}u^{-1} + \mu^{(2)}u^{-2} + \dots, \qquad \mu^{(r)} \in \mathbb{C}. \tag{4.28}$$

PROPOSITION 4.3.1. *If* $\dim V(\mu(u)) < \infty$, *then there exists an even formal series*

$$g(u) = 1 + g_1\, u^{-2} + g_2\, u^{-4} + \dots, \qquad g_r \in \mathbb{C},$$

such that $g(u)\mu(u)$ *is a polynomial in* u^{-1}.

PROOF. The symmetry relation (4.4) implies that for $i = -1, 1$ we have

$$(2u - 1)\, s_{i,-i}(u) = (-2u - 1)\, s_{i,-i}(-u). \tag{4.29}$$

Hence, for any $r \geqslant 1$ we have $s_{i,-i}^{(2r-1)} = 2\, s_{i,-i}^{(2r)}$ and

$$\frac{s_{i,-i}(u)}{2u + 1} = \frac{1}{2} \sum_{r=1}^{\infty} s_{i,-i}^{(2r-1)}\, u^{-2r}. \tag{4.30}$$

By twisting the action of $Y(\mathfrak{sp}_2)$ on $V(\mu(u))$ by the automorphism (2.17) with

$$g(u) = \frac{2}{\mu(u) + \mu(-u)}$$

we obtain a module over $Y(\mathfrak{sp}_2)$ which is isomorphic to the irreducible highest weight representation $V(g(u)\mu(u))$. So, we may assume without loss of generality that the highest weight of $V(\mu(u))$ satisfies $\mu(u) + \mu(-u) = 2$; that is, $\mu^{(2r)} = 0$ for all $r \geqslant 1$.

Let ξ denote the highest vector of $V(\mu(u))$. Since this representation is finite-dimensional, the vectors $s_{1,-1}^{(2i-1)}\xi \in V(\mu(u))$ with $i \geqslant 1$ are linearly dependent. Therefore, the corresponding Verma module $M(\mu(u))$ contains a nonzero vector η of the form

$$\eta = \sum_{i=1}^{m} c_i\, s_{1,-1}^{(2i-1)}\, 1_{\mu(u)}, \qquad c_i \in \mathbb{C},$$

which belongs to the maximal proper submodule K of $M(\mu(u))$. Here m is a positive integer, and we may assume that $c_m \neq 0$. Then we have $s_{-1,1}^{(2r-1)}\eta = 0$ for all $r \geqslant 1$ because otherwise the highest vector $1_{\mu(u)}$ would belong to K. By the defining relations (4.3), we have

$$(u^2 - v^2)\,[s_{-1,1}(u), s_{1,-1}(v)] = (u+v+1)\,\big(s_{11}(u)s_{-1,-1}(v) - s_{11}(v)s_{-1,-1}(u)\big) \\ + (u-v)\,\big(s_{-1,-1}(u)s_{-1,-1}(v) - s_{11}(v)s_{11}(u)\big).$$

Furthermore, by (4.15),

$$s_{-1,-1}(u)\,1_{\mu(u)} = \Big(\mu(-u) - \frac{\mu(u)-\mu(-u)}{2u}\Big)\,1_{\mu(u)}.$$

An easy calculation gives

$$(u^2-v^2)\,s_{-1,1}(u)\,s_{1,-1}(v)\,1_{\mu(u)} = (2u+1)(2v+1) \\ \times \Big(\frac{\mu(u)-\mu(-u)}{2u}\cdot\frac{\mu(v)+\mu(-v)}{2} - \frac{\mu(v)-\mu(-v)}{2v}\cdot\frac{\mu(u)+\mu(-u)}{2}\Big)\,1_{\mu(u)}.$$

Using the assumption $\mu(u)+\mu(-u)=2$, we can write this as

$$\frac{s_{-1,1}(u)}{2u+1}\,\frac{s_{1,-1}(v)}{2v+1}\,1_{\mu(u)} = \frac{1}{u^2-v^2}\Big(\frac{\mu(u)-1}{u} - \frac{\mu(v)-1}{v}\Big)\,1_{\mu(u)}.$$

Taking the coefficients of $u^{-2r}v^{-2i}$ on both sides and using (3.35) and (4.30), we come to

$$s_{-1,1}^{(2r-1)}\,s_{1,-1}^{(2i-1)}\,1_{\mu(u)} = -4\,\mu^{(2r+2i-3)}\,1_{\mu(u)}.$$

So, the relations $s_{-1,1}^{(2r-1)}\eta = 0$ imply

$$\sum_{i=1}^{m} c_i\,\mu^{(2r+2i-3)} = 0$$

for all $r \geqslant 1$. Hence

$$\frac{\mu(u)-1}{u}\cdot(c_1 + c_2\,u^2 + \dots + c_m\,u^{2m-2}) = b_1 + b_2\,u^2 + \dots + b_{m-1}\,u^{2m-4}$$

for some coefficients $b_i \in \mathbb{C}$. We have

$$\mu(u) = \frac{c_1 + b_1\,u + c_2\,u^2 + b_2\,u^3 + \dots + c_m\,u^{2m-2}}{c_1 + c_2\,u^2 + \dots + c_m\,u^{2m-2}}.$$

Thus, taking now

$$g(u) = c_m^{-1}\sum_{i=1}^{m} c_i\,u^{-2m+2i}$$

we conclude that $g(u)\,\mu(u)$ is a polynomial in u^{-1}. □

Proposition 4.3.1 implies that if $\dim V(\mu(u)) < \infty$, then the composition of the representation of $\mathrm{Y}(\mathfrak{sp}_2)$ in $V(\mu(u))$ with an automorphism (2.17) yields an irreducible highest weight representation where the highest weight is a polynomial in u^{-1}. Our aim now is to investigate such representations of $\mathrm{Y}(\mathfrak{sp}_2)$.

Suppose that $\mu(u)$ is a polynomial of degree $\leqslant 2k$ in u^{-1}. Write

$$\mu(u) = (1-\gamma_1 u^{-1})\dots(1-\gamma_{2k}u^{-1}), \tag{4.31}$$

where the γ_i are complex numbers (some of them are zero if the degree of $\mu(u)$ is strictly less than $2k$).

PROPOSITION 4.3.2. *Suppose that for every $i = 1, \dots, k$ the following condition holds: if the multiset $\{\gamma_p + \gamma_q \mid 2i-1 \leqslant p < q \leqslant 2k\}$ contains nonnegative integers, then $\gamma_{2i-1} + \gamma_{2i}$ is minimal amongst them. Then the representation $V(\mu(u))$ of $\mathrm{Y}(\mathfrak{sp}_2)$ is isomorphic to the tensor product*

$$L(\gamma_1, -\gamma_2) \otimes L(\gamma_3, -\gamma_4) \otimes \dots \otimes L(\gamma_{2k-1}, -\gamma_{2k}), \tag{4.32}$$

regarded as a $\mathrm{Y}(\mathfrak{sp}_2)$-module obtained by restriction of the $\mathrm{Y}(\mathfrak{gl}_2)$-module.

PROOF. Let us denote the $\mathrm{Y}(\mathfrak{sp}_2)$-module (4.32) by V and let ζ_i be the highest vector of $L(\gamma_{2i-1}, -\gamma_{2i})$ for $i = 1, \dots, k$. By Corollary 4.2.12, the cyclic span $\mathrm{Y}(\mathfrak{sp}_2)\zeta$ of the vector $\zeta = \zeta_1 \otimes \dots \otimes \zeta_k$ is a highest weight module with the highest weight $\mu(u)$. Therefore, the proposition will follow if we prove that the module V is irreducible.

We claim that any vector $\eta \in V$ satisfying $s_{-1,1}(u)\,\eta = 0$ is proportional to ζ. We will prove this claim by induction on k. This is obvious for $k = 1$, so suppose that $k \geqslant 2$. Write any such vector η, which may certainly be assumed to be nonzero, in the form

$$\eta = \sum_{r=0}^{p} (E_{1,-1})^r \zeta_1 \otimes \eta_r,$$

where $\eta_r \in L(\gamma_3, -\gamma_4) \otimes \dots \otimes L(\gamma_{2k-1}, -\gamma_{2k})$ and p is a certain nonnegative integer. Moreover, if $\gamma_1 + \gamma_2 \in \mathbb{Z}_+$, then $p \leqslant \gamma_1 + \gamma_2$. We may obviously assume that $\eta_p \neq 0$. Applying $s_{-1,1}(u)$ to η with the use of (4.23) we get

$$\begin{aligned} s_{-1,1}(u)\big((E_{1,-1})^r \zeta_1 \otimes \eta_r\big) = &- t_{-1,-1}(u)\, t_{-1,1}(-u)(E_{1,-1})^r \zeta_1 \otimes s_{-1,-1}(u)\,\eta_r \\ &+ t_{-1,-1}(u)\, t_{-1,-1}(-u)(E_{1,-1})^r \zeta_1 \otimes s_{-1,1}(u)\,\eta_r \\ &- t_{-1,1}(u)\, t_{-1,1}(-u)(E_{1,-1})^r \zeta_1 \otimes s_{1,-1}(u)\,\eta_r \\ &+ t_{-1,1}(u)\, t_{-1,-1}(-u)(E_{1,-1})^r \zeta_1 \otimes s_{1,1}(u)\,\eta_r. \end{aligned} \tag{4.33}$$

Using the definition of the Yangian action on $L(\gamma_1, -\gamma_2)$ and the commutation relations in $\mathfrak{gl}_2$, we obtain

$$t_{-1,-1}(u)(E_{1,-1})^r \zeta_1 = (1 + (\gamma_1 - r)\,u^{-1})(E_{1,-1})^r \zeta_1 \tag{4.34}$$

and

$$t_{-1,1}(u)(E_{1,-1})^r \zeta_1 = u^{-1}\, r(\gamma_1 + \gamma_2 - r + 1)(E_{1,-1})^{r-1} \zeta_1. \tag{4.35}$$

Taking the coefficient of $(E_{1,-1})^p \zeta_1$ in (4.33) gives

$$(1 - (\gamma_1 - p)\,u^{-2})\, s_{-1,1}(u)\,\eta_p = 0,$$

implying $s_{-1,1}(u)\,\eta_p = 0$. By the induction hypothesis, the vector η_p must be proportional to $\zeta_2 \otimes \dots \otimes \zeta_k$. Therefore, Corollary 4.2.12 implies

$$s_{11}(u)\,\eta_p = (1 - \gamma_3 u^{-1}) \dots (1 - \gamma_{2k} u^{-1})\,\eta_p. \tag{4.36}$$

To complete the proof of the claim we need to show that p is zero. Suppose by way of contradiction that $p \geqslant 1$. Then taking the coefficient of $(E_{1,-1})^{p-1}\zeta_1$ in the relation $s_{-1,1}(u)\,\eta = 0$ we deduce by using (4.33) that

$$\begin{aligned} &(1 - (\gamma_1 - p + 1)^2 u^{-2})\, s_{-1,1}(u)\,\eta_{p-1} \\ &\quad + u^{-1} p(\gamma_1 + \gamma_2 - p + 1)(1 + (\gamma_1 - p + 1)u^{-1})\, s_{-1,-1}(u)\,\eta_p \\ &\quad + u^{-1} p(\gamma_1 + \gamma_2 - p + 1)(1 - (\gamma_1 - p)u^{-1})\, s_{11}(u)\,\eta_p = 0. \end{aligned}$$

Multiplying both sides by $u^{2k}/(2u+1)$ and using (4.14) we get

(4.37)

$$\big(u^2-(\gamma_1-p+1)^2\big)\,u^{2k-2}\,\frac{s_{-1,1}(u)}{2u+1}\,\eta_{p-1}+p(\gamma_1+\gamma_2-p+1)$$
$$\times\frac{1}{2u}\Big((u+\gamma_1-p+1)u^{2k-2}s_{11}(-u)+(u-\gamma_1+p-1)u^{2k-2}s_{11}(u)\Big)\,\eta_p=0.$$

By Proposition 4.2.13(ii), the expression $u^{2k-2}\,s_{-1,1}(u)\,\eta_{p-1}$ is a polynomial in u. Hence, so is $u^{2k-2}\,(2u+1)^{-1}s_{-1,1}(u)\,\eta_{p-1}$ by (4.29). Now apply (4.36) and put $u=\gamma_1-p+1$ in (4.37) to get

$$p\,(\gamma_1+\gamma_2-p+1)(\gamma_1+\gamma_3-p+1)\dots(\gamma_1+\gamma_{2k}-p+1)=0.$$

However, this is impossible due to the conditions on the parameters γ_i. Thus, p must be zero and the claim is verified.

Suppose now that M is a nonzero submodule of V. Then M must contain a nonzero vector η such that $s_{-1,1}(u)\eta=0$. Indeed, this is immediate from (4.17) and the fact that the set of $\mathfrak{sp}_2$-weights of V has a maximal element. The above argument thus shows that M contains the vector ζ. It remains to prove that the cyclic span $K=Y(\mathfrak{sp}_2)\,\zeta$ coincides with V. By Proposition 4.2.14 (taking the trivial module $V(\mu)$), the dual $Y(\mathfrak{sp}_2)$-module V^* is isomorphic to the restriction of the $Y(\mathfrak{gl}_2)$-module

$$L(\gamma_2,-\gamma_1)\otimes\dots\otimes L(\gamma_{2k},-\gamma_{2k-1}).$$

Moreover, the highest vector ζ_i^* of the module $L(\gamma_{2i},-\gamma_{2i-1})\cong L(\gamma_{2i-1},-\gamma_{2i})^*$ can be identified with the element of $L(\gamma_{2i-1},-\gamma_{2i})^*$ such that $\zeta_i^*(\zeta_i)=1$ and $\zeta_i^*(\eta)=0$ for all F_{11}-weight vectors $\eta\in L(\gamma_{2i-1},-\gamma_{2i})$ whose weights are different from $-\gamma_{2i-1}-\gamma_{2i}$. Now, if the submodule K of V is proper, then its annihilator

$$\operatorname{Ann}K=\{\omega\in V^*\mid\omega(\eta)=0\quad\text{for all}\quad\eta\in K\}$$

is a nonzero submodule of V^* which does not contain the vector $\zeta_1^*\otimes\dots\otimes\zeta_k^*$. However, this contradicts the claim verified in the first part of the proof, because the condition on the parameters γ_i remains satisfied after we replace each γ_{2i-1} by γ_{2i} and each γ_{2i} by γ_{2i-1}. □

Note that given any polynomial (4.31), the condition of Proposition 4.3.2 on the parameters γ_i can be achieved by a re-enumeration of the indices; cf. the remark following the proof of Proposition 3.3.2.

Suppose now that $\mu(u)$ is an arbitrary series of the form (4.28).

THEOREM 4.3.3. *The irreducible highest weight representation $V(\mu(u))$ of the twisted Yangian $Y(\mathfrak{sp}_2)$ is finite-dimensional if and only if there exists a monic polynomial $P(u)$ in u such that $P(u)=P(-u+1)$ and*

$$\frac{\mu(-u)}{\mu(u)}=\frac{P(u+1)}{P(u)}. \tag{4.38}$$

In this case $P(u)$ is unique.

PROOF. Suppose that the representation $V(\mu(u))$ is finite-dimensional. Then by Proposition 4.3.1 we can find an even formal series $g(u)$ such that

$$g(u)\,\mu(u)=(1-\gamma_1u^{-1})\dots(1-\gamma_{2k}u^{-1}),$$

for some $k \geqslant 0$ and some complex parameters γ_i. Re-enumerate these parameters, if necessary, to satisfy the condition of Proposition 4.3.2. By this proposition, all numbers $\gamma_{2i-1} + \gamma_{2i}$ must be nonnegative integers because $V(\mu(u))$ is finite-dimensional. Set

$$\lambda_1(u) = (1 + \gamma_1 u^{-1})(1 + \gamma_3 u^{-1}) \dots (1 + \gamma_{2k-1} u^{-1}),$$
$$\lambda_2(u) = (1 - \gamma_2 u^{-1})(1 - \gamma_4 u^{-1}) \dots (1 - \gamma_{2k} u^{-1}).$$

As we showed in the proof of Theorem 3.3.3, there exists a monic polynomial $Q(u)$ in u such that

$$\frac{\lambda_1(u)}{\lambda_2(u)} = \frac{Q(u+1)}{Q(u)}.$$

Define $P(u)$ by

$$P(u) = Q(u)\, Q(-u+1)(-1)^{\deg Q}.$$

Note that $P(u)$ is monic and $P(u) = P(-u+1)$. Since

$$\mu(u) = \lambda_1(-u)\, \lambda_2(u)\, g(u)^{-1},$$

and $g(u)$ is even, (4.38) follows.

Conversely, let (4.38) hold for a monic polynomial $P(u)$ satisfying the condition $P(u) = P(-u+1)$. Then this polynomial has an even degree and the multiset of its roots is invariant with respect to the transformation $u \mapsto -u+1$. Therefore, we may write the roots in the form $\{-\delta_1, \dots, -\delta_p, \delta_1 + 1, \dots, \delta_p + 1\}$ for a certain $p \geqslant 0$. Set

$$\nu(u) = (1 - (\delta_1 + 1)u^{-1}) \dots (1 - (\delta_p + 1)u^{-1})(1 + \delta_1 u^{-1}) \dots (1 + \delta_p u^{-1})$$

and consider the $Y(\mathfrak{sp}_2)$-module

$$L(\delta_1 + 1, \delta_1) \otimes L(\delta_2 + 1, \delta_2) \otimes \dots \otimes L(\delta_p + 1, \delta_p).$$

By Corollary 4.2.12, the cyclic $Y(\mathfrak{sp}_2)$-span of the tensor product of the highest vectors of $L(\delta_i + 1, \delta_i)$ is a highest weight module with the highest weight $\nu(u)$. This submodule is finite-dimensional and so is its irreducible quotient $V(\nu(u))$. Since

$$\frac{\mu(-u)}{\mu(u)} = \frac{\nu(-u)}{\nu(u)} = \frac{P(u+1)}{P(u)},$$

the series $g(u) = \mu(u)/\nu(u)$ is even. The composition of the corresponding automorphism (2.17) of $Y(\mathfrak{sp}_2)$ with the representation $V(\nu(u))$ is isomorphic to $V(\mu(u))$. Thus, the latter is also finite-dimensional.

The uniqueness of $P(u)$ was established in the proof of Theorem 3.3.3. □

We call the polynomial $P(u)$ defined in Theorem 4.3.3 the *Drinfeld polynomial* of the finite-dimensional representation $V(\mu(u))$.

The following notation is motivated by Theorem 4.3.3; cf. (3.81). For two formal series $\mu(u)$ and $\nu(u)$ in u^{-1} we will write

$$\mu(u) \Longrightarrow \nu(u) \tag{4.39}$$

if there exists a monic polynomial $P(u)$ in u such that $P(u) = P(-u+1)$ and

$$\frac{\mu(u)}{\nu(u)} = \frac{P(u+1)}{P(u)}.$$

Obviously, this is only possible if $\mu(u)$ and $\nu(u)$ satisfy $\mu(u)\mu(-u) = \nu(u)\nu(-u)$.

Recall that the special twisted Yangian $\mathrm{SY}(\mathfrak{sp}_2)$ is the subalgebra of $\mathrm{Y}(\mathfrak{sp}_2)$ which consists of the elements stable under all automorphisms of the form (2.17); see Definition 2.9.1.

COROLLARY 4.3.4. *The isomorphism classes of finite-dimensional irreducible representations of the special twisted Yangian* $\mathrm{SY}(\mathfrak{sp}_2)$ *are parameterized by monic polynomials* $P(u)$ *in* u *satisfying* $P(u) = P(-u+1)$. *Every such representation is isomorphic to the restriction of a* $\mathrm{Y}(\mathfrak{gl}_2)$*-module of the form*

$$L(\alpha_1, \beta_1) \otimes L(\alpha_2, \beta_2) \otimes \ldots \otimes L(\alpha_k, \beta_k), \tag{4.40}$$

where each difference $\alpha_i - \beta_i$ *is a positive integer.*

PROOF. The argument is similar to the proof of Corollary 3.3.4 and is based on Theorem 2.9.2. This theorem implies that any finite-dimensional irreducible representation of the twisted Yangian $\mathrm{Y}(\mathfrak{sp}_2)$ remains irreducible when restricted to the subalgebra $\mathrm{SY}(\mathfrak{sp}_2)$, and every finite-dimensional irreducible representation of $\mathrm{SY}(\mathfrak{sp}_2)$ can be obtained by such a restriction. Therefore, by Theorem 4.3.3, we only need to describe possible restrictions of the finite-dimensional representations $V(\mu(u))$ to $\mathrm{SY}(\mathfrak{sp}_2)$. By the definition of $\mathrm{SY}(\mathfrak{sp}_2)$, this restriction is unchanged if $\mu(u)$ is replaced by $g(u)\,\mu(u)$ for any even formal series $g(u)$ in u^{-1}. Conversely, the action of the series $\widetilde{s}_{11}(u)$ with coefficients in $\mathrm{SY}(\mathfrak{sp}_2)$ (see (2.63)) on the highest vector ξ of $V(\mu(u))$ determines the series $\mu(u)$ up to a multiplication by an even formal series. Due to Theorem 4.3.3, this establishes a one-to-one correspondence between the finite-dimensional irreducible representation of $\mathrm{SY}(\mathfrak{sp}_2)$ and the monic polynomials $P(u)$ satisfying $P(u) = P(-u+1)$.

Given such a polynomial $P(u)$ we may find a polynomial $\mu(u)$ in u^{-1} such that relation (4.38) holds; see the proof of Theorem 4.3.3. Then by Proposition 4.3.2 the representation $V(\mu(u))$ of $\mathrm{Y}(\mathfrak{sp}_2)$ is isomorphic to a tensor product (4.40), and its restriction to $\mathrm{SY}(\mathfrak{sp}_2)$ yields the finite-dimensional irreducible representation corresponding to $P(u)$.

Finally, since $\dim V(\mu(u)) < \infty$, all differences $\alpha_i - \beta_i$ in (4.40) are nonnegative integers. Possible zero differences can be excluded because the highest weight of the $\mathrm{Y}(\mathfrak{sp}_2)$-module $L(\gamma, \gamma) \otimes V(\mu(u))$ is $g(u)\,\mu(u)$ with $g(u) = 1 - \gamma u^{-2}$, so that the restrictions of the representations $V(\mu(u))$ and $V(g(u)\,\mu(u))$ to the subalgebra $\mathrm{SY}(\mathfrak{sp}_2)$ are isomorphic. □

We will now establish a criterion of irreducibility of the representations of $\mathrm{Y}(\mathfrak{sp}_2)$ of the form (4.40). We use Definition 3.3.5. Denote by V the tensor product (4.40), where all differences $\alpha_i - \beta_i$ are nonnegative integers.

COROLLARY 4.3.5. *The representation* V *of* $\mathrm{Y}(\mathfrak{sp}_2)$ *is irreducible if and only if each pair of strings*

$$S(\alpha_i, \beta_i),\ S(\alpha_j, \beta_j) \qquad \textit{and} \qquad S(\alpha_i, \beta_i),\ S(-\beta_j, -\alpha_j)$$

is in general position for all $1 \leqslant i < j \leqslant k$. *Moreover, if* V *is irreducible, then each of the following operations yields an isomorphic representation of the algebra* $\mathrm{Y}(\mathfrak{sp}_2)$:

(i) *any permutation of tensor factors in* (4.40);

(ii) *replacement of the factor* $L(\alpha_j, \beta_j)$ *by* $L(-\beta_j, -\alpha_j)$ *for any* $j \in \{1, \ldots, k\}$.

PROOF. Suppose that the condition on the strings holds. Then the representation (4.40) of the Yangian $\mathrm{Y}(\mathfrak{gl}_2)$ is irreducible by Corollary 3.3.6. Furthermore, any permutation of the tensor factors gives an isomorphic representation. Hence, we may assume without loss of generality that $\alpha_1 - \beta_1 \leqslant \cdots \leqslant \alpha_k - \beta_k$. Then one easily verifies that the condition of Proposition 4.3.2 is satisfied for the parameters $\alpha_1, -\beta_1, \alpha_2, -\beta_2, \ldots, \alpha_k, -\beta_k$ and so, V is irreducible as a $Y(\mathfrak{sp}_2)$-module.

Conversely, let V be irreducible as a $Y(\mathfrak{sp}_2)$-module. Then V is certainly irreducible as a $Y(\mathfrak{gl}_2)$-module. By Corollary 3.3.6, the strings $S(\alpha_i, \beta_i)$ and $S(\alpha_j, \beta_j)$ are in general position for any $i < j$. Now fix an index $j \in \{1, \ldots, k\}$ and consider the $Y(\mathfrak{sp}_2)$-module V' obtained by replacement of the tensor factor $L(\alpha_j, \beta_j)$ by $L(-\beta_j, -\alpha_j)$. Then V' is isomorphic to V. Indeed, $\dim V = \dim V'$ and by Corollary 4.2.12, the module V is isomorphic to the irreducible quotient of the $Y(\mathfrak{sp}_2)$-submodule of V' generated by the tensor product of the highest weight vectors of the factors. So, V' is irreducible as a $Y(\mathfrak{sp}_2)$-module and, hence, as a $Y(\mathfrak{gl}_2)$-module. By Corollary 3.3.6, the string $S(-\beta_j, -\alpha_j)$ is in general position with any string $S(\alpha_i, \beta_i)$ for $i \neq j$.

The second part of the corollary was established in the above argument. □

Observe that the strings $S(\alpha, \beta)$ and $S(-\beta, -\alpha)$ are symmetric with respect to the line $\operatorname{Re} z = -1/2$ in the complex plane. For any string S we will write S' for the image of S under the mirror symmetry with respect to this line. Corollaries 4.3.4 and 4.3.5 imply that any monic polynomial $P(u)$ satisfying $P(u) = P(-u+1)$ can be written in the form

$$P(u) = \prod_{i=1}^{k} \prod_{\gamma \in S_i} (u + \gamma) \prod_{\gamma' \in S_i'} (u + \gamma'), \tag{4.41}$$

where $S_1, \ldots, S_k$ are nonempty strings such that for all pairs of indices $1 \leqslant i < j \leqslant k$ each pair of strings (S_i, S_j) and (S_i, S_j') is in general position. By writing each string S_i in the form $S(\alpha_i, \beta_i)$, for some parameters α_i, β_i, we get the corresponding irreducible module (4.40) over $Y(\mathfrak{sp}_2)$.

PROPOSITION 4.3.6. *Given any monic polynomial $P(u)$ satisfying the condition $P(u) = P(-u+1)$, the multiset of strings $S_1, \ldots, S_k, S_1', \ldots, S_k'$ determined by the expansion* (4.41) *is unique. Moreover, the irreducible* $SY(\mathfrak{sp}_2)$*-module of the form* (4.40) *corresponding to $P(u)$ is determined uniquely up to a permutation of the tensor factors and replacements of factors $L(\alpha_j, \beta_j)$ by $L(-\beta_j, -\alpha_j)$, respectively, for any $j \in \{1, \ldots, k\}$.*

PROOF. Given a multiset of strings $\mathcal{S} = \{S_1, \ldots, S_k, S_1', \ldots, S_k'\}$ determined by the expansion (4.41), we construct another multiset $\mathcal{S}^\circ$ as follows. For all indices $i \in \{1, \ldots, k\}$, if the strings S_i and S_i' are not in general position, then we replace them by the strings $S_i \cup S_i'$ and $S_i \cap S_i'$. After these replacements all strings in $\mathcal{S}^\circ$ will be pairwise in general position. However, by Proposition 3.3.7, $\mathcal{S}^\circ$ is uniquely determined by the polynomial $P(u)$. In particular, the multiset $\mathcal{S}^\circ$ is invariant under the symmetry with respect to the line $\operatorname{Re} z = -1/2$. So, it suffices to show that the map $\mathcal{S} \mapsto \mathcal{S}^\circ$ defines a bijection between the set of multisets of strings determined by the expansion (4.41) and the set of multisets of strings which are pairwise in general position and determined by the polynomial $P(u)$. In order to define the converse map, we may clearly reduce the task to the particular case where $\mathcal{S}^\circ = \{A_1, \ldots, A_m\}$, with all strings A_i symmetric with respect to the line $\operatorname{Re} z = -1/2$, and $A_1 \subset \cdots \subset A_m$. Suppose first that the real parts of the elements of the strings A_i are half-integers. Then m has to be even; $m = 2l$ due to the relation $P(u) = P(-u+1)$. For any $i \in \{1, \dots, l\}$ there exists a unique pair of strings (S_i, S_i') such that $A_{2i} = S_i \cup S_i'$ and $A_{2i-1} = S_i \cap S_i'$. This defines a multiset $\mathcal{S} = \{S_1, \dots, S_l, S_1', \dots, S_l'\}$. By the construction, for all pairs of indices $1 \leqslant i < j \leqslant l$ each pair of strings (S_i, S_j) and (S_i, S_j') is in general position and we have $\mathcal{S} \mapsto \mathcal{S}^\circ$. The construction of $\mathcal{S}$ in the case where the real parts of the elements of the strings A_i are integers is quite similar. This proves the first part of the proposition. The second part follows from the first. □

We conclude the discussion of representations of the twisted Yangian $\mathrm{Y}(\mathfrak{sp}_2)$ by providing bases for a class of its irreducible representations.

The defining relations (1.3) of the Yangian $\mathrm{Y}(\mathfrak{gl}_2)$ allow us to rewrite the formula (4.22) for $s_{1,-1}(u)$ in the form

$$s_{1,-1}(u) = \frac{u+1/2}{u}\Big(t_{1,-1}(u)\, t_{11}(-u) - t_{1,-1}(-u)\, t_{11}(u)\Big).$$

Consider the tensor product L given in (4.40), where all differences $\alpha_i - \beta_i$ are nonnegative integers and recall the operators $T_{ij}(u)$ in L defined in (3.86). The operator in L defined by

$$\begin{aligned} S_{1,-1}(u) &= \frac{u^{2k}}{u+1/2}\, s_{1,-1}(u) \\ &= \frac{(-1)^k}{u}\Big(T_{1,-1}(u)\, T_{11}(-u) - T_{1,-1}(-u)\, T_{11}(u)\Big) \end{aligned} \tag{4.42}$$

is an even polynomial in u of degree $\leqslant 2k-2$. Let ζ_i be the highest vector of the $\mathfrak{gl}_2$-module $L(\alpha_i, \beta_i)$ for $i = 1, \dots, k$, and let $\zeta = \zeta_1 \otimes \dots \otimes \zeta_k$. Suppose that a k-tuple $\gamma = (\gamma_1, \dots, \gamma_k)$ of complex numbers satisfies the conditions

$$\alpha_i - \gamma_i \in \mathbb{Z}_+, \qquad \gamma_i - \beta_i \in \mathbb{Z}_+, \qquad i = 1, \dots, k. \tag{4.43}$$

Define the corresponding vector $\zeta_\gamma \in L$ by

$$\zeta_\gamma = \prod_{i=1}^{k} S_{1,-1}(\gamma_i - 1)\, S_{1,-1}(\gamma_i - 2) \dots S_{1,-1}(\beta_i)\, \zeta. \tag{4.44}$$

Note that the ordering of the factors in the product is irrelevant, as the operators $S_{1,-1}(u)$ and $S_{1,-1}(v)$ in L commute due to the defining relations in $\mathrm{Y}(\mathfrak{sp}_2)$.

THEOREM 4.3.7. *Suppose that the $\mathrm{Y}(\mathfrak{sp}_2)$-module L is irreducible and the strings $S(\alpha_i, \beta_i)$, $i = 1, \dots, k$, satisfy the conditions*

$$S(\alpha_i, \beta_i) \cap S(\alpha_j, \beta_j) = \emptyset \qquad \text{and} \qquad S(\alpha_i, \beta_i) \cap S(-\beta_j, -\alpha_j) = \emptyset$$

for all $i < j$. Then the vectors ζ_γ with γ satisfying (4.43) form a basis of L.

PROOF. Consider the basis $\{\eta_\gamma\}$ of L constructed in Theorem 3.3.8. Using (4.42) and the formulas for the action of the operators $T_{1,-1}(u)$ and $T_{11}(u)$ in this basis which are provided by Theorem 3.3.8, we obtain

$$S_{1,-1}(\gamma_i)\, \eta_\gamma = 2 \prod_{a=1,\, a\neq i}^{k} (-\gamma_i - \gamma_a)\, \eta_{\gamma+\delta_i}, \qquad i = 1, \dots, k.$$

The second condition on the strings implies that $\gamma_i + \gamma_a \neq 0$ for all $a \neq i$. Therefore, each vector ζ_γ coincides with η_γ up to a nonzero constant factor. The statement now follows from Theorem 3.3.8. □

We can now prove the classification theorem for finite-dimensional irreducible representations of the twisted Yangian $\mathrm{Y}(\mathfrak{sp}_{2n})$, where n is an arbitrary positive integer. By the results of Section 4.2, any finite-dimensional irreducible representation of $\mathrm{Y}(\mathfrak{sp}_{2n})$ is isomorphic to a highest weight representation $V(\mu(u))$, where $\mu(u)$ is an n-tuple of formal series in u^{-1},

$$\mu(u) = (\mu_1(u), \dots, \mu_n(u)).$$

We give necessary and sufficient conditions on an arbitrary n-tuple $\mu(u)$ which ensure that $V(\mu(u))$ is finite-dimensional. We will be using the notation (3.81) and (4.39).

THEOREM 4.3.8. *The irreducible highest weight representation $V(\mu(u))$ of the twisted Yangian $\mathrm{Y}(\mathfrak{sp}_{2n})$ is finite-dimensional if and only if the following relation holds:*

$$\mu_1(-u) \Longrightarrow \mu_1(u) \longrightarrow \mu_2(u) \longrightarrow \dots \longrightarrow \mu_n(u). \tag{4.45}$$

PROOF. Suppose first that $\dim V(\mu(u)) < \infty$. Consider the homomorphism $\mathrm{Y}(\mathfrak{sp}_2) \to \mathrm{Y}(\mathfrak{sp}_{2n})$ which sends $s_{ij}(u)$ to the series with the same name in $\mathrm{Y}(\mathfrak{sp}_{2n})$ for any $i, j \in \{-1, 1\}$. Let $\mathrm{Y}(\mathfrak{sp}_2)$ act on $V(\mu(u))$ via this homomorphism. The cyclic span $\mathrm{Y}(\mathfrak{sp}_2)\xi$ of the highest vector ξ of $V(\mu(u))$ is a highest weight representation of $\mathrm{Y}(\mathfrak{sp}_2)$ with the highest weight $(\mu_1(u))$. Its irreducible quotient is finite-dimensional, and so Theorem 4.3.3 implies $\mu_1(-u) \Longrightarrow \mu_1(u)$. Together with Proposition 4.2.8 this gives (4.45).

Conversely, let (4.45) hold. Then

$$\frac{\mu_1(-u)}{\mu_1(u)} = \frac{P_1(u+1)}{P_1(u)} \tag{4.46}$$

and

$$\frac{\mu_{i-1}(u)}{\mu_i(u)} = \frac{P_i(u+1)}{P_i(u)}, \qquad i = 2, \dots, n, \tag{4.47}$$

for certain monic polynomials $P_i(u)$ with $P_1(u) = P_1(-u+1)$. Then we can write

$$P_1(u) = Q(u)\,Q(-u+1)(-1)^{\deg Q}$$

for a monic polynomial $Q(u)$ in u.

Consider the irreducible highest weight representation $L(\lambda(u))$ of the Yangian $\mathrm{Y}(\mathfrak{gl}_{2n})$ with the highest weight $\lambda(u) = (\lambda_{-n}(u), \dots, \lambda_n(u))$ given by

$$\lambda_i(u) = u^{-k} Q(u)\, P_2(u) \dots P_i(u)\, P_{i+1}(u+1) \dots P_n(u+1)$$

and

$$\lambda_{-i}(u) = u^{-k} Q(u+1)\, P_2(u+1) \dots P_n(u+1)$$

for $i = 1, \dots, n$, where k denotes the sum of the degrees of the polynomials $Q(u)$, $P_2(u), \dots, P_n(u)$. By Theorem 3.4.1, the representation $L(\lambda(u))$ of $\mathrm{Y}(\mathfrak{gl}_{2n})$ is finite-dimensional. Due to Corollary 4.2.10, the cyclic $\mathrm{Y}(\mathfrak{sp}_{2n})$-span of the highest vector of $L(\lambda(u))$ is a highest weight representation of $\mathrm{Y}(\mathfrak{sp}_{2n})$ with the highest weight $\mu'(u) = (\mu'_1(u), \dots, \mu'_n(u))$ where $\mu'_i(u) = \lambda_i(u)\lambda_{-i}(-u)$. So, $\dim V(\mu'(u)) < \infty$. However, the components $\mu'_i(u)$ satisfy (4.46) and (4.47). Therefore, there exists an even formal series $g(u)$ such that $\mu_i(u) = g(u)\,\mu'_i(u)$ for all i. Hence, the composition of the representation of $\mathrm{Y}(\mathfrak{sp}_{2n})$ on $V(\mu'(u))$ and the automorphism (2.17) is isomorphic to $V(\mu(u))$, proving that the latter is also finite-dimensional. □

DEFINITION 4.3.9. The polynomials $P_i(u)$ with $i = 1, \dots, n$ introduced in the proof of Theorem 4.3.8 are called the *Drinfeld polynomials* of the corresponding representation of $\mathrm{Y}(\mathfrak{sp}_{2n})$. □

EXAMPLE 4.3.10. Suppose that the representation $V(\mu)$ of the Lie algebra $\mathfrak{sp}_{2n}$ is finite-dimensional. Then the Drinfeld polynomials of the evaluation representation $V(\mu)$ of $\mathrm{Y}(\mathfrak{sp}_{2n})$ defined in Section 4.2 can be easily found from (4.20) and are given by

$$P_1(u) = (u + \mu_1 - 1/2)(u + \mu_1 + 1/2)\dots(u - 3/2) \\ \times (u + 1/2)(u + 3/2)\dots(u - \mu_1 - 1/2)$$

and

$$P_i(u) = (u + \mu_i - 1/2)(u + \mu_i + 1/2)\dots(u + \mu_{i-1} - 3/2)$$

for $i = 2, \dots, n$. □

COROLLARY 4.3.11. *Finite-dimensional irreducible representations of the special twisted Yangian* $\mathrm{SY}(\mathfrak{sp}_{2n})$ *are parameterized by the tuples* $\big(P_1(u), \dots, P_n(u)\big)$ *of monic polynomials in* u *such that* $P_1(u) = P_1(-u + 1)$.

PROOF. This follows from Theorem 4.3.8 and Proposition 4.2.15. □

4.4. Representations of $\mathrm{Y}(\mathfrak{o}_{2n})$

We start by considering the particular case $n = 1$. By Definition 4.2.5, the irreducible highest weight representations $V(\mu(u))$ of $\mathrm{Y}(\mathfrak{o}_2)$ are parameterized by formal series of the form

$$\mu(u) = 1 + \mu^{(1)} u^{-1} + \mu^{(2)} u^{-2} + \dots, \qquad \mu^{(r)} \in \mathbb{C}. \tag{4.48}$$

PROPOSITION 4.4.1. *If* $\dim V(\mu(u)) < \infty$, *then there exists an even formal series*

$$g(u) = 1 + g_1\, u^{-2} + g_2\, u^{-4} + \dots, \qquad g_r \in \mathbb{C},$$

such that $(1 + 1/2\, u^{-1})\, g(u)\mu(u)$ *is a polynomial in* u^{-1}.

PROOF. The symmetry relation (4.4) implies that for $i = -1, 1$ we have

$$s_{i,-i}(u) = s_{i,-i}(-u). \tag{4.49}$$

Hence, for any $r \geqslant 1$ we have $s_{i,-i}^{(2r-1)} = 0$. By twisting the action of $\mathrm{Y}(\mathfrak{o}_2)$ on $V(\mu(u))$ by the automorphism (2.17) with

$$g(u) = \frac{4u}{(2u+1)\,\mu(u) + (2u-1)\,\mu(-u)}$$

we obtain a module over $\mathrm{Y}(\mathfrak{o}_2)$ which is isomorphic to the irreducible highest weight representation $V(g(u)\mu(u))$. So, we may assume without loss of generality that the highest weight of $V(\mu(u))$ satisfies $(2u+1)\,\mu(u) + (2u-1)\,\mu(-u) = 4u$. Equivalently, all even coefficients $\mu'^{(2r)}$ of the series

$$\mu'(u) = (1 + 1/2\, u^{-1})\,\mu(u) \tag{4.50}$$

are zero for $r \geqslant 1$.

Let ξ denote the highest vector of $V(\mu(u))$. Since this representation is finite-dimensional, the vectors $s_{1,-1}^{(2i)}\xi \in V(\mu(u))$ with $i \geqslant 1$ are linearly dependent. Therefore, the corresponding Verma module $M(\mu(u))$ contains a nonzero vector η of the form

$$\eta = \sum_{i=1}^{m} c_i \, s_{1,-1}^{(2i)} \, 1_{\mu(u)}, \qquad c_i \in \mathbb{C},$$

which belongs to the maximal proper submodule K of $M(\mu(u))$. Here m is a positive integer, and we may assume that $c_m \neq 0$. Then we have $s_{-1,1}^{(2r)}\eta = 0$ for all $r \geqslant 1$ because otherwise the highest vector $1_{\mu(u)}$ would belong to K. By the defining relations (4.3), we have

$$\begin{aligned}(u^2 - v^2)\,[s_{-1,1}(u), s_{1,-1}(v)] = (u+v+1)\,\big(s_{11}(u)s_{-1,-1}(v) - s_{11}(v)s_{-1,-1}(u)\big)\\ - (u-v)\,\big(s_{-1,-1}(u)s_{-1,-1}(v) - s_{11}(v)s_{11}(u)\big).\end{aligned}$$

Furthermore, by (4.15),

$$s_{-1,-1}(u)\, 1_{\mu(u)} = \Big(\mu(-u) + \frac{\mu(u) - \mu(-u)}{2u}\Big)\, 1_{\mu(u)}.$$

An easy calculation gives

$$\begin{aligned}&(u^2 - v^2)\, s_{-1,1}(u)\, s_{1,-1}(v)\, 1_{\mu(u)} =\\ &\Big(u\,\big(\mu'(u) - \mu'(-u)\big)\big(\mu'(v) + \mu'(-v)\big) - v\,\big(\mu'(v) - \mu'(-v)\big)\big(\mu'(u) + \mu'(-u)\big)\Big)\, 1_{\mu(u)}.\end{aligned}$$

By the assumption, $\mu'(u) + \mu'(-u) = 2$, so we can write this relation as

$$s_{-1,1}(u)\, s_{1,-1}(v)\, 1_{\mu(u)} = \frac{2}{u^2 - v^2}\Big(u\,\big(\mu'(u) - 1\big) - v\,\big(\mu'(v) - 1\big)\Big)\, 1_{\mu(u)}.$$

Taking the coefficients of $u^{-2r}v^{-2i}$ on both sides and using (3.35), we come to

$$s_{-1,1}^{(2r)}\, s_{1,-1}^{(2i)}\, 1_{\mu(u)} = -2\, \mu'^{(2r+2i-3)}\, 1_{\mu(u)}.$$

So, the relations $s_{-1,1}^{(2r)}\eta = 0$ imply

$$\sum_{i=1}^{m} c_i\, \mu'^{(2r+2i-3)} = 0$$

for all $r \geqslant 1$. Hence

$$u\,\big(\mu'(u) - 1\big) \cdot (c_1 + c_2 u^2 + \dots + c_m u^{2m-2}) = b_2 u^2 + b_3 u^4 + \dots + b_m u^{2m-2}$$

for some coefficients $b_i \in \mathbb{C}$. We have

$$\mu'(u) = \frac{c_1 + b_2 u + c_2 u^2 + b_3 u^3 + \dots + c_m u^{2m-2}}{c_1 + c_2 u^2 + \dots + c_m u^{2m-2}}.$$

Thus, taking now

$$g(u) = c_m^{-1} \sum_{i=1}^{m} c_i\, u^{-2m+2i}$$

we conclude that $g(u)\,\mu'(u)$ is a polynomial in u^{-1}. □

Due to Proposition 4.4.1, we can concentrate on the representations $V(\mu(u))$ of $\mathrm{Y}(\mathfrak{o}_2)$ such that the series (4.50) is a polynomial in u^{-1}. Suppose that the degree of this polynomial does not exceed $2k+1$ for some nonnegative integer k. Write

$$\mu'(u) = (1-\gamma_1 u^{-1})\dots(1-\gamma_{2k+1}u^{-1}), \tag{4.51}$$

where the γ_i are complex numbers (some of them are zero if the degree of $\mu'(u)$ is strictly less than $2k+1$).

Recall that for any $\gamma \in \mathbb{C}$ the twisted Yangian $\mathrm{Y}(\mathfrak{o}_2)$ has a one-dimensional representation $V(\gamma)$ whose highest weight is given by (4.21).

PROPOSITION 4.4.2. *Suppose that for every $i = 1,\dots,k$ the following condition holds: if the multiset $\{\gamma_p+\gamma_q \mid 2i-1 \leqslant p < q \leqslant 2k+1\}$ contains nonnegative integers, then $\gamma_{2i-1}+\gamma_{2i}$ is minimal amongst them. Then the representation $V(\mu(u))$ of $\mathrm{Y}(\mathfrak{o}_2)$ is isomorphic to the tensor product*

$$L(\gamma_1,-\gamma_2)\otimes L(\gamma_3,-\gamma_4)\otimes\dots\otimes L(\gamma_{2k-1},-\gamma_{2k})\otimes V(-\gamma_{2k+1}-1/2). \tag{4.52}$$

PROOF. Denote the module (4.52) by V. Let ζ_i be the highest vector of $L(\gamma_{2i-1},-\gamma_{2i})$ for $i=1,\dots,k$ and let ξ bc the basis vector of the one-dimensional representation $V(-\gamma_{2k+1}-1/2)$. By Proposition 4.2.11, the cyclic span $\mathrm{Y}(\mathfrak{o}_2)\zeta$ of the vector $\zeta = \zeta_1\otimes\dots\otimes\zeta_k\otimes\xi$ is a highest weight module with the highest weight $\mu(u)$. Therefore, the proposition will follow if we prove that the module V is irreducible.

We claim that any vector $\eta\in V$ satisfying $s_{-1,1}(u)\eta = 0$ is proportional to ζ. We will prove this claim by induction on k. This is obvious for $k=0$, so suppose that $k\geqslant 1$. Write any such vector η, which may be assumed to be nonzero, in the form

$$\eta = \sum_{r=0}^{p}(E_{1,-1})^r\zeta_1\otimes\eta_r,$$

where $\eta_r \in L(\gamma_3,-\gamma_4)\otimes\dots\otimes L(\gamma_{2k-1},-\gamma_{2k})\otimes V(-\gamma_{2k+1}-1/2)$ and p is a certain nonnegative integer. Moreover, if $\gamma_1+\gamma_2\in\mathbb{Z}_+$, then $p\leqslant\gamma_1+\gamma_2$. We may obviously assume that $\eta_p\neq 0$. Applying $s_{-1,1}(u)$ to η with the use of (4.23) we get

$$\begin{aligned} s_{-1,1}(u)\big((E_{1,-1})^r\zeta_1\otimes\eta_r\big) &= t_{-1,-1}(u)\,t_{-1,1}(-u)(E_{1,-1})^r\zeta_1\otimes s_{-1,-1}(u)\,\eta_r\\ &+t_{-1,-1}(u)\,t_{-1,-1}(-u)(E_{1,-1})^r\zeta_1\otimes s_{-1,1}(u)\,\eta_r\\ &+t_{-1,1}(u)\,t_{-1,1}(-u)(E_{1,-1})^r\zeta_1\otimes s_{1,-1}(u)\,\eta_r\\ &+t_{-1,1}(u)\,t_{-1,-1}(-u)(E_{1,-1})^r\zeta_1\otimes s_{1,1}(u)\,\eta_r. \end{aligned} \tag{4.53}$$

Using (4.34), (4.35) and taking the coefficient of $(E_{1,-1})^p\zeta_1$ in (4.53) gives

$$(1-(\gamma_1-p)\,u^{-2})\,s_{-1,1}(u)\,\eta_p = 0,$$

implying $s_{-1,1}(u)\eta_p = 0$. By the induction hypothesis, the vector η_p must be proportional to $\zeta_2\otimes\dots\otimes\zeta_k\otimes\xi$. Therefore, Proposition 4.2.11 implies

$$(1+1/2\,u^{-1})\,s_{11}(u)\,\eta_p = (1-\gamma_3u^{-1})\dots(1-\gamma_{2k+1}u^{-1})\,\eta_p. \tag{4.54}$$

To complete the proof of the claim we need to show that p is zero. Suppose by way of contradiction that $p\geqslant 1$. Then taking the coefficient of $(E_{1,-1})^{p-1}\zeta_1$ in the relation $s_{-1,1}(u)\,\eta = 0$ we deduce by using (4.53) that

$$\begin{aligned}(1-(\gamma_1-p+1)^2u^{-2})\,s_{-1,1}(u)\,\eta_{p-1}\\ -u^{-1}\,p(\gamma_1+\gamma_2-p+1)(1+(\gamma_1-p+1)u^{-1})\,s_{-1,-1}(u)\,\eta_p\\ +u^{-1}\,p(\gamma_1+\gamma_2-p+1)(1-(\gamma_1-p)u^{-1})\,s_{11}(u)\,\eta_p=0.\end{aligned}$$

Multiplying both sides by u^{2k+1} and using (4.14) we get

$$\begin{aligned}(u^2-(\gamma_1-p+1)^2)\,u^{2k-1}\,s_{-1,1}(u)\,\eta_{p-1}+p(\gamma_1+\gamma_2-p+1)\\ \times\Big((1+1/2\,u^{-1})(u-\gamma_1+p-1)\,u^{2k-1}s_{11}(u)\\ -(1-1/2\,u^{-1})(u+\gamma_1-p+1)\,u^{2k-1}s_{11}(-u)\Big)\,\eta_p=0.\end{aligned}\tag{4.55}$$

Observe that the expression $u^{2k-1}\,s_{-1,1}(u)\,\eta_{p-1}$ is a polynomial in u. Indeed, the elements $s^{(r)}_{-1,1}$ and $s^{(r)}_{1,-1}$ act trivially in $V(-\gamma_{2k+1}-1/2)$. Hence, by (4.23) for any vector $\vartheta\in L(\gamma_3,-\gamma_4)\otimes\ldots\otimes L(\gamma_{2k-1},-\gamma_{2k})$ we have

$$\begin{aligned}s_{-1,1}(u)(\vartheta\otimes\xi)=t_{-1,1}(u)\,t_{-1,-1}(-u)\,\vartheta\otimes s_{11}(u)\,\xi\\ +t_{-1,-1}(u)\,t_{-1,1}(-u)\,\vartheta\otimes s_{-1,-1}(u)\,\xi.\end{aligned}$$

Using the defining relations (1.3) and the symmetry relation (4.4), we may rewrite this as

$$\begin{aligned}s_{-1,1}(u)(\vartheta\otimes\xi)=t_{-1,-1}(-u)\,t_{-1,1}(u)\,\vartheta\otimes(1+1/2\,u^{-1})\,s_{11}(u)\,\xi\\ +t_{-1,-1}(u)\,t_{-1,1}(-u)\,\vartheta\otimes(1-1/2\,u^{-1})\,s_{11}(-u)\,\xi\end{aligned}$$

so that multiplying by u^{2k-1} we get a polynomial in u by the definition of the $\mathrm{Y}(\mathfrak{o}_2)$-modules $V(\gamma)$ and Proposition 3.2.11.

Now apply (4.54) and put $u=\gamma_1-p+1$ in (4.55) (after dividing both sides by u in the case $\gamma_1-p+1=0$) to get

$$p\,(\gamma_1+\gamma_2-p+1)(\gamma_1+\gamma_3-p+1)\ldots(\gamma_1+\gamma_{2k+1}-p+1)=0.$$

However, this is impossible due to the conditions on the parameters γ_i. Thus, p must be zero and the claim is verified.

Suppose now that M is a nonzero submodule of V. Then M must contain a nonzero vector η such that $s_{-1,1}(u)\eta=0$. Indeed, this follows from (4.17) by considering the set of $\mathfrak{o}_2$-weights of V. The above argument thus shows that M contains the vector ζ. It remains to prove that the cyclic span $K=\mathrm{Y}(\mathfrak{o}_2)\zeta$ coincides with V. By Proposition 4.2.14, the dual $\mathrm{Y}(\mathfrak{o}_2)$-module V^* is isomorphic to the tensor product

$$L(\gamma_2,-\gamma_1)\otimes\ldots\otimes L(\gamma_{2k},-\gamma_{2k-1})\otimes V(-\gamma_{2k+1}-1/2).$$

Moreover, the highest vector ζ_i^* of the module $L(\gamma_{2i},-\gamma_{2i-1})\cong L(\gamma_{2i-1},-\gamma_{2i})^*$ can be identified with the element of $L(\gamma_{2i-1},-\gamma_{2i})^*$ such that $\zeta_i^*(\zeta_i)=1$ and $\zeta_i^*(\eta)=0$ for all F_{11}-weight vectors $\eta\in L(\gamma_{2i-1},-\gamma_{2i})$ whose weights are different from $-\gamma_{2i-1}-\gamma_{2i}$. Now, if the submodule K of V is proper, then its annihilator

$$\operatorname{Ann}K=\{\omega\in V^*\mid\omega(\eta)=0\quad\text{for all}\quad\eta\in K\}$$

is a nonzero submodule of V^* which does not contain the vector $\zeta_1^*\otimes\ldots\otimes\zeta_k^*\otimes\xi$. However, this contradicts the claim verified in the first part of the proof, because the condition on the parameters γ_i remains satisfied after we replace each γ_{2i-1} by γ_{2i} and each γ_{2i} by γ_{2i-1} for $i=1,\ldots,k$. □

Note that given any polynomial (4.51), the condition of Proposition 4.4.2 on the parameters γ_i can be achieved by a re-enumeration of the indices.

THEOREM 4.4.3. *The irreducible highest weight representation $V(\mu(u))$ of the twisted Yangian $\mathrm{Y}(\mathfrak{o}_2)$ is finite-dimensional if and only if there is a pair $(P(u),\gamma)$, where $P(u)$ is a monic polynomial in u with $P(u)=P(-u+1)$ and $\gamma\in\mathbb{C}$ with $P(\gamma)\neq 0$ such that*

$$\frac{\mu'(-u)}{\mu'(u)}=\frac{P(u+1)}{P(u)}\cdot\frac{u-\gamma}{u+\gamma}, \tag{4.56}$$

where $\mu'(u)=(1+1/2\,u^{-1})\,\mu(u)$. In this case the pair $(P(u),\gamma)$ is unique.

PROOF. Suppose that the representation $V(\mu(u))$ is finite-dimensional. Then by Proposition 4.4.1 we can find an even formal series $g(u)$ such that

$$g(u)\,\mu'(u)=(1-\gamma_1u^{-1})\dots(1-\gamma_{2k+1}u^{-1}),$$

for some $k\geqslant 0$ and some complex parameters γ_i. Re-enumerate these parameters, if necessary, to satisfy the condition of Proposition 4.4.2. By this proposition, all numbers $\gamma_{2i-1}+\gamma_{2i}$ with $i=1,\dots,k$ must be nonnegative integers because $V(\mu(u))$ is finite-dimensional. Set

$$\lambda_1(u)=(1+\gamma_1u^{-1})(1+\gamma_3u^{-1})\dots(1+\gamma_{2k-1}u^{-1}),$$
$$\lambda_2(u)=(1-\gamma_2u^{-1})(1-\gamma_4u^{-1})\dots(1-\gamma_{2k}u^{-1}).$$

As we showed in the proof of Theorem 3.3.3, there exists a monic polynomial $Q(u)$ in u such that

$$\frac{\lambda_1(u)}{\lambda_2(u)}=\frac{Q(u+1)}{Q(u)}.$$

Define $P(u)$ by

$$P(u)=Q(u)\,Q(-u+1)(-1)^{\deg Q}.$$

The polynomial $P(u)$ is monic and $P(u)=P(-u+1)$. Since

$$\mu'(u)=\lambda_1(-u)\,\lambda_2(u)\,(1-\gamma_{2k+1}u^{-1})\,g(u)^{-1},$$

and $g(u)$ is even, we derive (4.56), where $\gamma=-\gamma_{2k+1}$. Note that $P(\gamma)\neq 0$ by the condition of Proposition 4.4.2.

Conversely, let (4.56) hold for a monic polynomial $P(u)$ satisfying the condition $P(u)=P(-u+1)$, and some $\gamma\in\mathbb{C}$ with $P(\gamma)\neq 0$. The multiset of roots of $P(u)$ can be written as $\{-\delta_1,\dots,-\delta_p,\delta_1+1,\dots,\delta_p+1\}$ for some $p\geqslant 0$. Set

$$\begin{aligned}\nu(u)=(1-(\delta_1+1)u^{-1})\dots(1-(\delta_p+1)u^{-1})\\ \times(1+\delta_1u^{-1})\dots(1+\delta_pu^{-1})\,(1+\gamma u^{-1})(1+1/2\,u^{-1})^{-1}\end{aligned} \tag{4.57}$$

and consider the $\mathrm{Y}(\mathfrak{o}_2)$-module

$$L(\delta_1+1,\delta_1)\otimes L(\delta_2+1,\delta_2)\otimes\dots\otimes L(\delta_p+1,\delta_p)\otimes V(\gamma-1/2).$$

By Proposition 4.2.11, the cyclic $\mathrm{Y}(\mathfrak{o}_2)$-span of the tensor product of the highest vectors of $L(\delta_i+1,\delta_i)$ for $i=1,\dots,p$ and the basis vector of $V(\gamma-1/2)$ is a highest weight module with the highest weight $\nu(u)$. This submodule is finite-dimensional and so is its irreducible quotient $V(\nu(u))$. We have

$$\frac{\mu'(-u)}{\mu'(u)}=\frac{\nu'(-u)}{\nu'(u)}=\frac{P(u+1)}{P(u)}\cdot\frac{u-\gamma}{u+\gamma},$$

where $\nu'(u) = (1 + 1/2\,u^{-1})\,\nu(u)$. Consider the automorphism (2.17) of $Y(\mathfrak{o}_2)$ with the even series $g(u) = \mu'(u)/\nu'(u)$. The composition of this automorphism with the representation $V(\nu(u))$ is isomorphic to $V(\mu(u))$. Thus, the latter is also finite-dimensional.

To prove the uniqueness of the pair $(P(u), \gamma)$ suppose that $(R(u), \delta)$ is another pair, where $R(u)$ is a monic polynomial in u such that $R(u) = R(-u+1)$, $R(\delta) \neq 0$ and

$$\frac{P(u+1)}{P(u)} \cdot \frac{u-\gamma}{u+\gamma} = \frac{R(u+1)}{R(u)} \cdot \frac{u-\delta}{u+\delta}. \tag{4.58}$$

If $\gamma = \delta$, then we get $P(u) = R(u)$; see the proof of Theorem 3.3.3. Let $\gamma \neq \delta$. We prove by induction on the sum of the degrees of the polynomials $P(u)$ and $R(u)$ that (4.58) is impossible. If $P(u) = R(u) = 1$, then this is obvious. Now suppose that $\deg P(u) \geqslant 2$. Take a root u_0 of $P(u)$ such that $u_0 + 1$ is not a root. Rewriting (4.58) in the form

$$P(u)R(u+1)(u-\delta)(u+\gamma) = P(u+1)R(u)(u+\delta)(u-\gamma), \tag{4.59}$$

we find that either $R(u_0) = 0$ or $u_0 = -\delta$. If $R(u_0) = 0$, then we may write

$$P(u) = P'(u)(u-u_0)(u+u_0-1), \qquad R(u) = R'(u)(u-u_0)(u+u_0-1)$$

and use the induction hypothesis for the polynomials $P'(u)$ and $R'(u)$. If $u_0 = -\delta$ but $R(u_0) \neq 0$, then we write $P(u) = P'(u)(u-u_0)(u+u_0-1)$ and obtain from (4.59) that

$$P'(u)R(u+1)(u+u_0-1)(u+\gamma) = P'(u+1)R(u)(u-u_0+1)(u-\gamma).$$

Set $\delta' = -u_0 + 1$. Then $\delta' \neq \gamma$ because $P(\delta') = 0$. Moreover, $R(\delta') = R(u_0) \neq 0$, and we may apply the induction hypothesis to the polynomials $P'(u)$ and $R(u)$. □

We call the polynomial $P(u)$ defined in Theorem 4.4.3 the *Drinfeld polynomial* for the finite-dimensional representation $V(\mu(u))$.

The following corollary is a consequence of the arguments of the proof of Theorem 4.4.3. We record it here for use in the next section.

COROLLARY 4.4.4. *Suppose that $V(\mu(u))$ is an irreducible highest weight representation of the twisted Yangian $Y(\mathfrak{o}_2)$, where $\mu'(u) = (1+1/2\,u^{-1})\,\mu(u)$ is given by (4.51) for some complex parameters γ_i. Then $V(\mu(u))$ is finite-dimensional if and only if the γ_i can be renumbered in such a way that the sums $\gamma_{2i-1} + \gamma_{2i}$ are nonnegative integers for all $i = 1, \dots, k$.* □

Recall that the special twisted Yangian $SY(\mathfrak{o}_2)$ is the subalgebra of $Y(\mathfrak{o}_2)$ which consists of the elements stable under all automorphisms of the form (2.17); see Definition 2.9.1.

COROLLARY 4.4.5. *The isomorphism classes of finite-dimensional irreducible representations of the special twisted Yangian $SY(\mathfrak{o}_2)$ are parameterized by pairs $(P(u), \gamma)$, where $P(u)$ is a monic polynomial in u with $P(u) = P(-u+1)$ and $\gamma \in \mathbb{C}$ with $P(\gamma) \neq 0$. Every such representation is isomorphic to the restriction of a $Y(\mathfrak{o}_2)$-module of the form*

$$L(\alpha_1, \beta_1) \otimes L(\alpha_2, \beta_2) \otimes \ldots \otimes L(\alpha_k, \beta_k) \otimes V(\delta), \tag{4.60}$$

where each difference $\alpha_i - \beta_i$ is a positive integer.

PROOF. Theorem 2.9.2 implies that any finite-dimensional irreducible representation of the twisted Yangian $\mathrm{Y}(\mathfrak{o}_2)$ remains irreducible when restricted to the subalgebra $\mathrm{SY}(\mathfrak{o}_2)$, and every finite-dimensional irreducible representation of $\mathrm{SY}(\mathfrak{o}_2)$ can be obtained by such a restriction. Therefore, by Theorem 4.4.3, we only need to describe possible restrictions of the finite-dimensional representations $V(\mu(u))$ to $\mathrm{SY}(\mathfrak{o}_2)$. By the definition of $\mathrm{SY}(\mathfrak{o}_2)$, this restriction is unchanged if $\mu(u)$ is replaced by $g(u)\,\mu(u)$ for any even formal series $g(u)$ in u^{-1}. Conversely, the action of the series $\widetilde{s}_{11}(u)$ with coefficients in $\mathrm{SY}(\mathfrak{o}_2)$ (see (2.63)) on the highest vector ξ of $V(\mu(u))$ determines the series $\mu(u)$ up to multiplication by an even formal series. Due to Theorem 4.4.3, this establishes a one-to-one correspondence between the finite-dimensional irreducible representation of $\mathrm{SY}(\mathfrak{o}_2)$ and the pairs $(P(u),\gamma)$.

Given a pair $(P(u),\gamma)$ we may find a series $\mu(u)$ in u^{-1} such that relation (4.56) holds and $(1+1/2\,u^{-1})\,\mu(u)$ is a polynomial in u^{-1}; see the proof of Theorem 4.4.3. Then by Proposition 4.4.2 the representation $V(\mu(u))$ of $\mathrm{Y}(\mathfrak{o}_2)$ is isomorphic to a tensor product (4.60), and its restriction to $\mathrm{SY}(\mathfrak{o}_2)$ yields the finite-dimensional irreducible representation corresponding to the pair $(P(u),\gamma)$.

Finally, since $\dim V(\mu(u))<\infty$, all differences $\alpha_i-\beta_i$ in (4.60) are nonnegative integers. Possible zero differences can be excluded because the highest weight of the $\mathrm{Y}(\mathfrak{o}_2)$-module $L(\gamma,\gamma)\otimes V(\mu(u))$ is $g(u)\,\mu(u)$ with $g(u)=1-\gamma u^{-2}$, so that the restrictions of the representations $V(\mu(u))$ and $V(g(u)\,\mu(u))$ to the subalgebra $\mathrm{SY}(\mathfrak{o}_2)$ are isomorphic. □

Denote by V the tensor product (4.60), where all differences $\alpha_i-\beta_i$ are nonnegative integers.

COROLLARY 4.4.6. *The representation V of $\mathrm{Y}(\mathfrak{o}_2)$ is irreducible if and only if each pair of strings*

$$S(\alpha_i,\beta_i),\ S(\alpha_j,\beta_j) \qquad \text{and} \qquad S(\alpha_i,\beta_i),\ S(-\beta_j,-\alpha_j)$$

is in general position for all $1\leqslant i<j\leqslant k$, and

$$\delta-1/2\notin S(\alpha_i,\beta_i), \qquad \delta-1/2\notin S(-\beta_i,-\alpha_i) \tag{4.61}$$

for all $i=1,\dots,k$. Moreover, if V is irreducible, then each of the following operations yields an isomorphic representation of the algebra $\mathrm{Y}(\mathfrak{o}_2)$:

(i) *any permutation of the tensor factors $L(\alpha_j,\beta_j)$ in (4.60);*

(ii) *replacement of the factor $L(\alpha_j,\beta_j)$ by $L(-\beta_j,-\alpha_j)$ for any $j\in\{1,\dots,k\}$.*

PROOF. Suppose that the condition on the strings holds. Then the representation

$$L=L(\alpha_1,\beta_1)\otimes L(\alpha_2,\beta_2)\otimes\dots\otimes L(\alpha_k,\beta_k) \tag{4.62}$$

of the Yangian $\mathrm{Y}(\mathfrak{gl}_2)$ is irreducible by Corollary 3.3.6. Furthermore, any permutation of the tensor factors gives an isomorphic representation. Hence, we may assume without loss of generality that $\alpha_1-\beta_1\leqslant\dots\leqslant\alpha_k-\beta_k$. Then one easily verifies that the condition of Proposition 4.4.2 is satisfied for the parameters $\alpha_1,-\beta_1,\alpha_2,-\beta_2,\dots,\alpha_k,-\beta_k,-\delta-1/2$, and so V is irreducible as a $\mathrm{Y}(\mathfrak{o}_2)$-module.

Conversely, let V be irreducible as a $\mathrm{Y}(\mathfrak{o}_2)$-module. Then the $\mathrm{Y}(\mathfrak{gl}_2)$-module (4.62) is also irreducible. By Corollary 3.3.6, the strings $S(\alpha_i,\beta_i)$ and $S(\alpha_j,\beta_j)$ are in general position for any $i<j$. Now fix an index $j\in\{1,\dots,k\}$ and consider the $\mathrm{Y}(\mathfrak{gl}_2)$-module L' obtained from L by the replacement of the tensor factor

$L(\alpha_j,\beta_j)$ by $L(-\beta_j,-\alpha_j)$. Then the $\mathrm{Y}(\mathfrak{o}_2)$-module $V' = L'\otimes V(\delta)$ is isomorphic to V. Indeed, $\dim V = \dim V'$, and by Proposition 4.2.11, the module V is isomorphic to the irreducible quotient of the $\mathrm{Y}(\mathfrak{o}_2)$-submodule of V' generated by the tensor product of the highest weight vectors of the factors. So, V' is irreducible and, hence, L' is irreducible as a $\mathrm{Y}(\mathfrak{gl}_2)$-module. By Corollary 3.3.6, the string $S(-\beta_j,-\alpha_j)$ is in general position with any string $S(\alpha_i,\beta_i)$ for $i\neq j$. Furthermore, replacing the tensor factor $L(\alpha_i,\beta_i)$ by $L(-\beta_i,-\alpha_i)$ as above, if necessary, we find that it is sufficient to derive only one of the two conditions (4.61). Suppose on the contrary that the second condition in (4.61) is violated for some $i\in\{1,\dots,k\}$. Permuting the factors in (4.62), if necessary, we may assume without loss of generality that $i=k$. As the module V over $\mathrm{Y}(\mathfrak{o}_2)$ is irreducible, then so is the $\mathrm{Y}(\mathfrak{o}_2)$-module $L(\alpha_k,\beta_k)\otimes V(\delta)$. However, since $\delta-1/2\in S(-\beta_k,-\alpha_k)$, the tensor product module $L(-\delta-1/2,\beta_k)\otimes V(-\alpha_k-1/2)$ over $\mathrm{Y}(\mathfrak{o}_2)$ is also irreducible by Proposition 4.4.2. The modules $L(\alpha_k,\beta_k)\otimes V(\delta)$ and $L(-\delta-1/2,\beta_k)\otimes V(-\alpha_k-1/2)$ have the same highest weight and therefore must be isomorphic. But this is impossible, as they have different dimensions. This contradiction completes the proof of the first part of the corollary. The second part was established in the above argument. □

REMARK 4.4.7. (i) Note that $V(0)$ is the trivial representation of $\mathrm{Y}(\mathfrak{o}_2)$. Thus, taking $\delta=0$ in Corollary 4.4.6 we get a criterion of irreducibility of the restriction of the $\mathrm{Y}(\mathfrak{gl}_2)$-module (4.62) to the subalgebra $\mathrm{Y}(\mathfrak{o}_2)$.

(ii) The conditions (4.61) can be written in an equivalent form by replacing $\delta-1/2$ by $-\delta-1/2$. □

EXAMPLE 4.4.8. Consider the restriction of the representation $L(\alpha,\beta)$ of $\mathrm{Y}(\mathfrak{gl}_2)$ with $\alpha-\beta\in\mathbb{Z}_+$ to $\mathrm{Y}(\mathfrak{o}_2)$. By Corollary 4.4.6, it is reducible if and only if the string $S(\alpha,\beta)$ contains $-1/2$. In this case $\alpha-1/2$ and $-\beta-1/2$ are nonnegative integers. Denote the highest weight vector of $L(\alpha,\beta)$ by ξ and set $\xi_p=(E_{1,-1})^p\xi$. Suppose first that $\alpha+\beta\geqslant 0$. Then the following are $\mathrm{Y}(\mathfrak{o}_2)$-submodules of $L(\alpha,\beta)$:

$$L_1 = \text{ span of } \{\xi_0,\dots,\xi_{-\beta-1/2}\},$$
$$L_2 = \text{ span of } \{\xi_{\alpha+1/2},\dots,\xi_{\alpha-\beta}\}.$$

By Proposition 4.4.2, we have the isomorphisms

$$L_1\cong L(-1/2,\beta)\otimes V(-\alpha-1/2) \quad\text{and}\quad L_2\cong L(-1/2,\beta)\otimes V(\alpha+1/2).$$

The quotient $L(\alpha,\beta)/(L_1\oplus L_2)$ is isomorphic to $L(\alpha,-\beta+1)$. If $\alpha+\beta<0$, then $L(\alpha,\beta)$ contains the submodule

$$L = \text{ span of } \{\xi_{\alpha+1/2},\dots,\xi_{-\beta-1/2}\}$$

which is isomorphic to $L(-\beta,\alpha+1)$ and the quotient $L(\alpha,\beta)/L$ is isomorphic to the direct sum of $L(\alpha,1/2)\otimes V(\beta-1/2)$ and $L(\alpha,1/2)\otimes V(-\beta+1/2)$. □

PROPOSITION 4.4.9. *Given any pair $(P(u),\gamma)$, where $P(u)$ is a monic polynomial in u with $P(u)=P(-u+1)$ and $\gamma\in\mathbb{C}$ with $P(\gamma)\neq 0$, the irreducible $\mathrm{SY}(\mathfrak{o}_2)$-module of the form (4.60) corresponding to $(P(u),\gamma)$ is determined uniquely, up to a permutation of the tensor factors $L(\alpha_i,\beta_i)$ and replacements of factors $L(\alpha_j,\beta_j)$ by $L(-\beta_j,-\alpha_j)$, respectively, for any $j\in\{1,\dots,k\}$. Moreover, the value of δ is found by $\delta=\gamma-1/2$.*

PROOF. Consider the multiset of strings $\{S_1,\dots,S_k,S_1',\dots,S_k'\}$ where $S_i=S(\alpha_i,\beta_i)$ and $S_i'=S(-\beta_i,-\alpha_i)$. Using (4.56) we find that the polynomial $P(u)$

corresponding to the representation (4.60) is found by the formula (4.41), while $\gamma = \delta + 1/2$. By Proposition 4.3.6, the multiset of strings is uniquely determined by the polynomial $P(u)$, thus proving the claim. □

We conclude the discussion of representations of the twisted Yangian $\mathrm{Y}(\mathfrak{o}_2)$ by providing bases for a class of its irreducible representations. Consider the $\mathrm{Y}(\mathfrak{o}_2)$-module $V = L \otimes V(\delta)$, where $V(\delta)$ is a one-dimensional $\mathrm{Y}(\mathfrak{o}_2)$-module spanned by the basis vector ξ, and L is defined in (4.62). Suppose that all differences $\alpha_i - \beta_i$ are nonnegative integers. Using the vector space isomorphism

$$L \otimes V(\delta) \to L, \qquad \eta \otimes \xi \mapsto \eta, \qquad \eta \in L, \tag{4.63}$$

we can regard L as a $\mathrm{Y}(\mathfrak{o}_2)$-module. Using the defining relations (1.3) and the coproduct formula (4.23), we can write $s_{1,-1}(u)$ as an operator in L in the form

$$s_{1,-1}(u) = \frac{u - \delta - 1/2}{u}\, t_{1,-1}(u)\, t_{11}(-u) + \frac{u + \delta + 1/2}{u}\, t_{1,-1}(-u)\, t_{11}(u).$$

Recall the operators $T_{ij}(u)$ in L defined in (3.86). The operator in L defined by

$$\begin{aligned} S_{1,-1}(u) &= u^{2k}\, s_{1,-1}(u) \\ &= \frac{(-1)^k}{u}\Big((u - \delta - 1/2)\, T_{1,-1}(u)\, T_{11}(-u) \\ &\qquad\qquad + (u + \delta + 1/2)\, T_{1,-1}(-u)\, T_{11}(u)\Big) \end{aligned} \tag{4.64}$$

is an even polynomial in u of degree $\leqslant 2k - 2$. Let ζ_i be the highest vector of the $\mathfrak{gl}_2$-module $L(\alpha_i, \beta_i)$ for $i = 1, \dots, k$, and let $\zeta = \zeta_1 \otimes \dots \otimes \zeta_k$. Suppose that a k-tuple $\gamma = (\gamma_1, \dots, \gamma_k)$ of complex numbers satisfies the conditions

$$\alpha_i - \gamma_i \in \mathbb{Z}_+, \qquad \gamma_i - \beta_i \in \mathbb{Z}_+, \qquad i = 1, \dots, k. \tag{4.65}$$

Define the corresponding vector $\zeta_\gamma \in L$ by

$$\zeta_\gamma = \prod_{i=1}^{k} S_{1,-1}(\gamma_i - 1)\, S_{1,-1}(\gamma_i - 2) \dots S_{1,-1}(\beta_i)\, \zeta. \tag{4.66}$$

Note that the ordering of the factors in the product is irrelevant, as the operators $S_{1,-1}(u)$ and $S_{1,-1}(v)$ in L commute due to the defining relations in $\mathrm{Y}(\mathfrak{o}_2)$.

THEOREM 4.4.10. *Suppose that the $\mathrm{Y}(\mathfrak{o}_2)$-module $L \otimes V(\delta)$ is irreducible and the strings $S(\alpha_i, \beta_i)$, $i = 1, \dots, k$, satisfy the conditions*

$$S(\alpha_i, \beta_i) \cap S(\alpha_j, \beta_j) = \emptyset \qquad \textit{and} \qquad S(\alpha_i, \beta_i) \cap S(-\beta_j, -\alpha_j) = \emptyset$$

for all $i < j$. Then the vectors ζ_γ with γ satisfying (4.65) *form a basis of L.*

PROOF. Consider the basis $\{\eta_\gamma\}$ of L constructed in Theorem 3.3.8. Using (4.64) and the formulas for the action of the operators $T_{1,-1}(u)$ and $T_{11}(u)$ in this basis, provided by Theorem 3.3.8, we obtain

$$S_{1,-1}(\gamma_i)\, \eta_\gamma = 2\,(-\delta - \gamma_i - 1/2) \prod_{a=1,\, a \neq i}^{k} (-\gamma_i - \gamma_a)\, \eta_{\gamma + \delta_i}, \qquad i = 1, \dots, k.$$

The second condition on the strings implies that $\gamma_i + \gamma_a \neq 0$ for all $a \neq i$. Furthermore, since the $\mathrm{Y}(\mathfrak{o}_2)$-module $L \otimes V(\delta)$ is irreducible, the relation $-\delta - \gamma_i - 1/2 = 0$

is possible only if $\gamma_i = \alpha_i$; see Remark 4.4.7(ii). Therefore, each vector ζ_γ coincides with η_γ, up to a nonzero constant factor. The statement now follows from Theorem 3.3.8. □

We now aim to prove the classification theorem for finite-dimensional irreducible representations of the twisted Yangian $\mathrm{Y}(\mathfrak{o}_{2n})$, where $n \geqslant 2$ is a positive integer. Recall that the irreducible highest weight representations $V(\mu(u))$ of $\mathrm{Y}(\mathfrak{o}_{2n})$ are parameterized by n-tuples of formal series in u^{-1},

$$\mu(u) = (\mu_1(u), \dots, \mu_n(u)). \tag{4.67}$$

Due to the defining relations (4.3), (4.4) and the Poincaré–Birkhoff–Witt theorem for the algebra $\mathrm{Y}(\mathfrak{o}_{2n})$ (see Section 4.2) we may regard the twisted Yangian $\mathrm{Y}(\mathfrak{o}_2)$ as the subalgebra of $\mathrm{Y}(\mathfrak{o}_{2n})$ generated by the elements $s_{ij}^{(r)}$ with $r \geqslant 1$ and $i, j \in \{-1, 1\}$. Let ξ denote the highest vector of the $\mathrm{Y}(\mathfrak{o}_{2n})$-module $V(\mu(u))$.

LEMMA 4.4.11. *The* $\mathrm{Y}(\mathfrak{o}_2)$*-module* $V = \mathrm{Y}(\mathfrak{o}_2)\,\xi$ *is irreducible and isomorphic to the highest weight module* $V(\mu_1(u))$.

PROOF. Let J be the left ideal of $\mathrm{Y}(\mathfrak{o}_{2n})$ generated by the elements $s_{kl}^{(r)}$ with $r \geqslant 1$, $k < l$ and $(k, l) \neq (-1, 1)$. We claim that V is annihilated by J. In order to see this, note that by the Poincaré–Birkhoff–Witt theorem for the algebra $\mathrm{Y}(\mathfrak{o}_2)$, the module V is spanned by the elements

$$s_{1,-1}^{(r_1)} \dots s_{1,-1}^{(r_m)}\,\xi, \qquad m \geqslant 0.$$

However, $J\,\xi = 0$ and $J\,s_{1,-1}^{(r)} \subset J$ for any $r \geqslant 1$. The latter is a particular case of a more general relation immediate from the defining relations (4.3) and (4.4),

$$[s_{ij}(u), s_{kl}(v)] \equiv 0 \mod J, \tag{4.68}$$

where $i, j \in \{-1, 1\}$ and $k < l$ with $(k, l) \neq (-1, 1)$. Thus, $J\,V = \{0\}$.

Suppose now that V contains a nonzero submodule K. Then K contains a nonzero vector ζ such that $s_{-1,1}(u)\,\zeta = 0$. The vector ζ then satisfies $s_{kl}(u)\,\zeta = 0$ for all $-n \leqslant k < l \leqslant n$ since $J\,\zeta = 0$. By Corollary 4.2.7, ζ is proportional to ξ, and so $K = V$, proving that V is irreducible. Obviously, ξ is the highest vector of the $\mathrm{Y}(\mathfrak{o}_2)$-module V so that $V \cong V(\mu_1(u))$. □

Consider the permutation of the set $\{-n, \dots, -1, 1, \dots, n\}$ defined by the rule $i' = -i$ for $i = 1$ or -1 and $i' = i$ otherwise.

PROPOSITION 4.4.12. *The mapping*

$$s_{ij}(u) \mapsto s_{i'j'}(u), \qquad i, j \in \{-n, \dots, -1, 1, \dots, n\} \tag{4.69}$$

defines an automorphism of the algebra $\mathrm{Y}(\mathfrak{o}_{2n})$.

PROOF. This follows directly from the defining relations (4.3) and (4.4) of $\mathrm{Y}(\mathfrak{o}_{2n})$. Alternatively, this is also a particular case of the automorphism (2.18) rewritten in terms of the current presentation of $\mathrm{Y}(\mathfrak{o}_{2n})$ for an appropriate choice of the matrix B. □

Consider the finite-dimensional irreducible representation $V(\mu(u))$ of $\mathrm{Y}(\mathfrak{o}_2)$ with the highest weight $\mu(u)$. By Theorem 4.4.3, $V(\mu(u))$ is associated with a uniquely determined pair $(P(u), \gamma)$. Denote by $V(\mu(u))^\sharp$ the composition of the action of $\mathrm{Y}(\mathfrak{o}_2)$ on $V(\mu(u))$ with the automorphism (4.69).

LEMMA 4.4.13. *The* $\mathrm{Y}(\mathfrak{o}_2)$*-module* $V(\mu(u))^\sharp$ *is isomorphic to* $V(\mu^\sharp(u))$*, where the series* $\mu^\sharp(u)$ *is given by*

$$\mu^\sharp(u) = \mu(u) \cdot \frac{u-\gamma+1}{u+\gamma}.$$

In particular, the $\mathrm{Y}(\mathfrak{o}_2)$*-module* $V(\mu(u))^\sharp$ *is associated with the pair* $(P(u), -\gamma+1)$.

PROOF. By Propositions 4.4.1 and 4.4.2, we can find an even series $g(u)$ in u^{-1} such that the composition of the representation $V(\mu(u))$ of $\mathrm{Y}(\mathfrak{o}_2)$ with the automorphism (2.17) is isomorphic to a tensor product module of the form (4.60). Since the automorphisms (2.17) and (4.69) of $\mathrm{Y}(\mathfrak{o}_2)$ commute, we may assume without loss of generality, that the $\mathrm{Y}(\mathfrak{o}_2)$-module $V(\mu(u))$ is isomorphic to (4.60). Then Proposition 4.2.11 gives

$$\mu(u) = (1-\alpha_1 u^{-1}) \dots (1-\alpha_k u^{-1})(1+\beta_1 u^{-1}) \dots (1+\beta_k u^{-1}) \\ \times (1+(\delta+1/2)\, u^{-1})(1+1/2\, u^{-1})^{-1}.$$

Let ζ_i denote the highest vector of $L(\alpha_i, \beta_i)$ and let ξ be the basis vector of $V(\delta)$. Set

$$\zeta = (E_{1,-1})^{\alpha_1-\beta_1} \zeta_1 \otimes \dots \otimes (E_{1,-1})^{\alpha_k-\beta_k} \zeta_k \otimes \xi.$$

A simple calculation with the use of (4.23) gives the following relations for the action of $\mathrm{Y}(\mathfrak{o}_2)$ in the module (4.60): $s_{1,-1}(u)\,\zeta = 0$ and

$$s_{-1,-1}(u)\,\zeta = (1-\alpha_1 u^{-1}) \dots (1-\alpha_k u^{-1})(1+\beta_1 u^{-1}) \dots (1+\beta_k u^{-1}) \\ \times (1+(-\delta+1/2)\, u^{-1})(1+1/2\, u^{-1})^{-1}\, \zeta.$$

This means that the vector ζ, regarded as an element of the $\mathrm{Y}(\mathfrak{o}_2)$-module $V(\mu(u))^\sharp$, is the highest vector of this module with the highest weight given by

$$\mu^\sharp(u) = (1-\alpha_1 u^{-1}) \dots (1-\alpha_k u^{-1})(1+\beta_1 u^{-1}) \dots (1+\beta_k u^{-1}) \\ \times (1+(-\delta+1/2)\, u^{-1})(1+1/2\, u^{-1})^{-1}.$$

We have $\delta = \gamma - 1/2$ by Proposition 4.4.9, and so, comparing the above formulas for $\mu(u)$ and $\mu^\sharp(u)$, we derive the desired relation. The second part of the lemma now follows from (4.56). □

Let us call a series $\mu(u)$ *admissible* if the corresponding irreducible highest weight representation $V(\mu(u))$ of $\mathrm{Y}(\mathfrak{o}_2)$ is finite-dimensional; i.e., it satisfies the conditions of Theorem 4.4.3. In this case the corresponding series $\mu^\sharp(u)$ is found by Lemma 4.4.13. In the following theorem we use the notation (3.81) and (4.39).

THEOREM 4.4.14. *The irreducible highest weight representation* $V(\mu(u))$ *of the twisted Yangian* $\mathrm{Y}(\mathfrak{o}_{2n})$, $n \geqslant 2$, *is finite-dimensional if and only if the series* $\mu_1(u)$ *is admissible and at least one of the following four relations holds:*

$$\mu_1(-u) \Longrightarrow \mu_1(u) \longrightarrow \mu_2(u) \longrightarrow \dots \longrightarrow \mu_n(u), \tag{4.70}$$

$$\frac{u-1/2}{u+1/2} \cdot \mu_1(-u) \Longrightarrow \mu_1(u) \longrightarrow \mu_2(u) \longrightarrow \dots \longrightarrow \mu_n(u), \tag{4.71}$$

$$\mu_1^\sharp(-u) \Longrightarrow \mu_1^\sharp(u) \longrightarrow \mu_2(u) \longrightarrow \dots \longrightarrow \mu_n(u), \tag{4.72}$$

$$\frac{u-1/2}{u+1/2} \cdot \mu_1^\sharp(-u) \Longrightarrow \mu_1^\sharp(u) \longrightarrow \mu_2(u) \longrightarrow \dots \longrightarrow \mu_n(u). \tag{4.73}$$

PROOF. Suppose first that $\dim V(\mu(u)) < \infty$. Denote by $V(\mu(u))^\sharp$ the composition of the action of $\mathrm{Y}(\mathfrak{o}_{2n})$ on $V(\mu(u))$ with the automorphism (4.69). By Theorem 4.2.6, $V(\mu(u))^\sharp$ is a highest weight representation. In order to find its highest weight, consider the $\mathrm{Y}(\mathfrak{o}_2)$-module V introduced in Lemma 4.4.11. Since V is finite-dimensional, that lemma implies that the series $\mu_1(u)$ is admissible. By Lemma 4.4.13, the subspace $V \subset V(\mu(u))$ contains a nonzero vector ζ such that

$$s_{1,-1}(u)\,\zeta = 0 \qquad \text{and} \qquad s_{-1,-1}(u)\,\zeta = \mu_1^\sharp(u)\,\zeta,$$

where

$$\mu_1^\sharp(u) = \mu_1(u) \cdot \frac{u-\gamma+1}{u+\gamma} \tag{4.74}$$

and $(P(u), \gamma)$ is the pair associated with $\mu_1(u)$ by Theorem 4.4.3. Due to (4.68), the vector ζ is annihilated by the ideal J introduced in the proof of Lemma 4.4.13. Moreover, we have

$$s_{kk}(u)\,\zeta = \mu_k(u)\,\zeta, \qquad k = 2, \dots, n,$$

since (4.68) holds for the values $k = l \geqslant 2$ as well. This implies that the highest weight of the $\mathrm{Y}(\mathfrak{o}_{2n})$-module $V(\mu(u))^\sharp$ is the n-tuple $(\mu_1^\sharp(u), \mu_2(u), \dots, \mu_n(u))$. Applying Proposition 4.2.8, we find that

$$\mu_1(u) \longrightarrow \mu_2(u) \qquad \text{and} \qquad \mu_1^\sharp(u) \longrightarrow \mu_2(u).$$

Hence, by (3.81), there exist monic polynomials $Q(u)$ and $Q^\sharp(u)$ in u such that

$$\frac{\mu_1(u)}{\mu_2(u)} = \frac{Q(u+1)}{Q(u)} \qquad \text{and} \qquad \frac{\mu_1^\sharp(u)}{\mu_2(u)} = \frac{Q^\sharp(u+1)}{Q^\sharp(u)}.$$

Then by (4.74) we get

$$\frac{Q^\sharp(u+1)}{Q^\sharp(u)} = \frac{Q(u+1)}{Q(u)} \cdot \frac{u-\gamma+1}{u+\gamma}. \tag{4.75}$$

However, this is possible only if the parameter γ takes half-integer values. In order to derive this claim from (4.75), we may assume without loss of generality that the polynomials $Q(u)$ and $Q^\sharp(u)$ have no common roots. Furthermore, we may also assume that at least one of the polynomials, say, $Q(u)$, has a positive degree; otherwise the claim is obvious. Suppose now that $2\gamma \notin \mathbb{Z}$ and pick up a root u_0 of $Q(u)$ such that $u_0 + 1$ is not a root. Then (4.75) implies that $u_0 = \gamma - 1$. As $Q(u+1)$ is zero at $u = u_0 - 1$ while $u_0 + \gamma - 1 \neq 0$, this forces $Q(u_0 - 1) = 0$. This, in its turn, implies $Q(u_0 - 2) = 0$, $Q(u_0 - 3) = 0$, etc., which is impossible. The contradiction shows that $2\gamma \in \mathbb{Z}$.

Now we consider four cases. First, suppose that $\gamma \in 1/2 + \mathbb{Z}$ and $\gamma \leqslant 1/2$. By (4.56) applied to the series $\mu_1(u)$, we get

$$\frac{\mu_1(-u)}{\mu_1(u)} = \frac{P(u+1)}{P(u)} \cdot \frac{u+1/2}{u+\gamma} \cdot \frac{u-\gamma}{u-1/2}. \tag{4.76}$$

However, this can be written as

$$\frac{\mu_1(-u)}{\mu_1(u)} = \frac{P_1(u+1)}{P_1(u)}$$

for another monic polynomial $P_1(u)$ such that $P_1(u) = P_1(-u+1)$. Using Proposition 4.2.8, we conclude that in the case under consideration the relations (4.70) hold.

Next, let $\gamma \in 1/2 + \mathbb{Z}$ and $\gamma > 1/2$. Applying (4.56) to the series $\mu_1^\sharp(u)$ and using Lemma 4.4.13 we find that

$$\frac{\mu_1^\sharp(-u)}{\mu_1^\sharp(u)} = \frac{P(u+1)}{P(u)} \cdot \frac{u+1/2}{u-\gamma+1} \cdot \frac{u+\gamma-1}{u-1/2}. \tag{4.77}$$

Together with Proposition 4.2.8 applied to the $\mathrm{Y}(\mathfrak{o}_{2n})$-module $V(\mu(u))^\sharp$ this gives (4.72). Note that if $\gamma = 1/2$, then $\mu_1^\sharp(u) = \mu_1(u)$ so that both (4.70) and (4.72) hold.

Suppose now that $\gamma \in \mathbb{Z}$ and $\gamma \leqslant 0$. Then (4.76) implies that

$$\frac{u-1/2}{u+1/2} \cdot \frac{\mu_1(-u)}{\mu_1(u)} = \frac{P_1(u+1)}{P_1(u)}$$

for a monic polynomial $P_1(u)$ such that $P_1(u) = P_1(-u+1)$. Applying Proposition 4.2.8 we thus arrive at (4.71).

Similarly, in the remaining case with $\gamma \in \mathbb{Z}$ and $\gamma \geqslant 1$ we derive from (4.77) and Proposition 4.2.8 that (4.73) holds. This completes the proof of the "only if" part of the theorem.

Conversely, suppose now that an n-tuple of formal series (4.67) is given so that (4.70) holds. Then

$$\frac{\mu_1(-u)}{\mu_1(u)} = \frac{P_1(u+1)}{P_1(u)} \tag{4.78}$$

and

$$\frac{\mu_{i-1}(u)}{\mu_i(u)} = \frac{P_i(u+1)}{P_i(u)}, \qquad i = 2, \dots, n, \tag{4.79}$$

for certain monic polynomials $P_i(u)$ with $P_1(u) = P_1(-u+1)$. Then we can write

$$P_1(u) = Q(u)\, Q(-u+1)(-1)^{\deg Q}$$

for a monic polynomial $Q(u)$ in u.

Consider the irreducible highest weight representation $L(\lambda(u))$ of the Yangian $\mathrm{Y}(\mathfrak{gl}_{2n})$ with the highest weight $\lambda(u) = (\lambda_{-n}(u), \dots, \lambda_n(u))$ given by

$$\lambda_i(u) = u^{-k} Q(u)\, P_2(u) \dots P_i(u)\, P_{i+1}(u+1) \dots P_n(u+1)$$

and

$$\lambda_{-i}(u) = u^{-k} Q(u+1)\, P_2(u+1) \dots P_n(u+1)$$

for $i = 1, \dots, n$, where k denotes the sum of the degrees of the polynomials $Q(u)$, $P_2(u), \dots, P_n(u)$. By Theorem 3.4.1, the representation $L(\lambda(u))$ of $\mathrm{Y}(\mathfrak{gl}_{2n})$ is finite-dimensional. Due to Corollary 4.2.10, the cyclic $\mathrm{Y}(\mathfrak{o}_{2n})$-span of the highest vector of $L(\lambda(u))$ is a highest weight representation of $\mathrm{Y}(\mathfrak{o}_{2n})$ with the highest weight $\mu'(u) = (\mu'_1(u), \dots, \mu'_n(u))$ where $\mu'_i(u) = \lambda_i(u)\lambda_{-i}(-u)$. So, $\dim V(\mu'(u)) < \infty$. However, the components $\mu'_i(u)$ satisfy (4.78) and (4.79). Therefore, there exists an even formal series $g(u)$ such that $\mu_i(u) = g(u)\, \mu'_i(u)$ for all i. Hence, the composition of the representation of $\mathrm{Y}(\mathfrak{o}_{2n})$ on $V(\mu'(u))$ and the automorphism (2.17) is isomorphic to $V(\mu(u))$ proving that the latter is also finite-dimensional.

Similarly, if (4.72) holds, then the irreducible highest module over $\mathrm{Y}(\mathfrak{o}_{2n})$ with the highest weight $(\mu_1^\sharp(u), \mu_2(u), \dots, \mu_n(u))$ is finite-dimensional. However, as we have seen in the first part of the proof, the composition of this module with the automorphism (4.69) is isomorphic to the $\mathrm{Y}(\mathfrak{o}_{2n})$-module $V(\mu(u))$ since the automorphism (4.69) is involutive. This proves that $\dim V(\mu(u)) < \infty$.

If (4.71) holds, then (4.78) is now replaced by the relation

$$\frac{u-1/2}{u+1/2}\cdot\frac{\mu_1(-u)}{\mu_1(u)}=\frac{P_1(u+1)}{P_1(u)}$$

for a monic polynomial $P_1(u)$ such that $P_1(u) = P_1(-u+1)$, while (4.79) holds in the same form. Introduce the series $\lambda_i(u)$ exactly as above and consider the irreducible representation $L(\lambda(u))$ of $\mathrm{Y}(\mathfrak{gl}_{2n})$ which is finite-dimensional by Theorem 3.4.1. Furthermore, introduce the finite-dimensional irreducible representation $V(\mu_0)$ of the Lie algebra $\mathfrak{o}_{2n}$ with the highest weight $\mu_0 = (-1/2,\dots,-1/2)$ (in fact, $\dim V(\mu_0) = 2^{n-1}$, but this will not be used in this argument). Extend $V(\mu_0)$ to a module over the twisted Yangian $\mathrm{Y}(\mathfrak{o}_{2n})$ via the evaluation homomorphism (2.106) and consider the tensor product module $L(\lambda(u))\otimes V(\mu_0)$ over $\mathrm{Y}(\mathfrak{o}_{2n})$. Using formulas (4.20) and Proposition 4.2.9, we derive that the irreducible highest weight representation of $\mathrm{Y}(\mathfrak{o}_{2n})$ with the highest weight $\mu'(u) = (\mu'_1(u),\dots,\mu'_n(u))$ where

$$\mu'_i(u) = \lambda_i(u)\lambda_{-i}(-u)\,(1+1/2\,u^{-1})^{-1}$$

is finite-dimensional. The argument is now completed in the same way as for the case of (4.70) considered above.

Finally, if (4.73) holds, then by the previous case, the irreducible highest module over $\mathrm{Y}(\mathfrak{o}_{2n})$ with the highest weight $(\mu_1^\sharp(u),\mu_2(u),\dots,\mu_n(u))$ is finite-dimensional. Hence, so is the module $V(\mu(u))$. □

DEFINITION 4.4.15. The polynomials $P_i(u)$ with $i=1,\dots,n$ associated with a finite-dimensional $\mathrm{Y}(\mathfrak{o}_{2n})$-module $V(\mu(u))$ by one of the relations (4.70)–(4.73) are called the *Drinfeld polynomials* of $V(\mu(u))$. □

EXAMPLE 4.4.16. Suppose that the representation $V(\mu)$ of the Lie algebra $\mathfrak{o}_{2n}$, $n\geqslant 2$, is finite-dimensional and extend $V(\mu)$ to a $\mathrm{Y}(\mathfrak{o}_{2n})$-module via the evaluation homomorphism (2.106). Using (4.20) and Lemma 4.4.13 we find that

$$\mu_1(u)=\frac{u+\mu_1+1/2}{u+1/2}\qquad\text{and}\qquad \mu_1^\sharp(u)=\frac{u-\mu_1+1/2}{u+1/2}.$$

Then for the Drinfeld polynomials of the evaluation module $V(\mu)$ we find

$$P_i(u)=(u+\mu_i+1/2)(u+\mu_i+3/2)\dots(u+\mu_{i-1}-1/2)$$

for $i=3,\dots,n$,

$$P_2(u)=(u+\mu_2+1/2)(u+\mu_2+3/2)\dots(u-|\mu_1|-1/2),$$

while $P_1(u)$ is given by different formulas depending on the value of μ_1. We have

$$P_1(u)=(u-|\mu_1|+1/2)(u-|\mu_1|+3/2)\dots(u+|\mu_1|-3/2)$$

if $\mu_1\in 1/2+\mathbb{Z}$, and

$$P_1(u)=(u-|\mu_1|+1/2)(u-|\mu_1|+3/2)\dots(u-1/2)\\ \times(u-1/2)(u+1/2)\dots(u+|\mu_1|-3/2)$$

if $\mu_1\in\mathbb{Z}$. □

As we saw in the proof of Theorem 4.4.14, every finite-dimensional irreducible representation of the twisted Yangian $\mathrm{Y}(\mathfrak{o}_{2n})$ with $n\geqslant 2$ is associated with a parameter γ such that $2\gamma\in\mathbb{Z}$. The four cases (4.70)–(4.73) are distinguished by the four subsets of values of γ, with the value $\gamma=1/2$ corresponding to both cases (4.70) and (4.72) simultaneously. Accordingly, the set of values of γ can be partitioned into five subsets:

$$\{1/2\}, \qquad -1/2-\mathbb{Z}_+, \qquad 3/2+\mathbb{Z}_+, \qquad -\mathbb{Z}_+, \qquad 1+\mathbb{Z}_+. \tag{4.80}$$

Hence, taking into account Proposition 4.2.15, we obtain the following parametrization of the representations of the special twisted Yangian $\mathrm{SY}(\mathfrak{o}_{2n})$ with $n \geqslant 2$.

COROLLARY 4.4.17. *Finite-dimensional irreducible representations of the special twisted Yangian* $\mathrm{SY}(\mathfrak{o}_{2n})$ *are parameterized by the tuples* $\big(P_1(u),\dots,P_n(u)\big)$ *of monic polynomials in* u *such that* $P_1(u)=P_1(-u+1)$ *and a choice of the subset in* (4.80). □

4.5. Representations of $\mathrm{Y}(\mathfrak{o}_{2n+1})$

As in the previous two sections, the particular case $n=1$ plays a key role in the proof of the classification theorem for the representations of the twisted Yangians $\mathrm{Y}(\mathfrak{o}_{2n+1})$. We give necessary and sufficient conditions on the highest weight $\mu(u)$ for the irreducible highest weight representation $V(\mu(u))$ of $\mathrm{Y}(\mathfrak{o}_3)$ to be finite-dimensional. To prove that the conditions are sufficient we apply an argument similar to those used in the previous sections. However, a quite different argument is employed here to prove that the conditions are necessary. Our main instrument is an investigation of the restriction of $V(\mu(u))$ to the natural subalgebra $\mathrm{Y}(\mathfrak{o}_2)$ and the use of the results of the previous section, namely, Corollary 4.4.4.

By Definition 4.2.5, the irreducible highest weight representations $V(\mu(u))$ of $\mathrm{Y}(\mathfrak{o}_3)$ are parameterized by pairs of formal series $\mu(u)=(\mu_0(u),\mu_1(u))$, where

$$\mu_0(u)=1+\mu_0^{(2)}u^{-2}+\mu_0^{(4)}u^{-4}+\dots, \qquad \mu_0^{(2r)}\in\mathbb{C}, \tag{4.81}$$

and

$$\mu_1(u)=1+\mu_1^{(1)}u^{-1}+\mu_1^{(2)}u^{-2}+\dots, \qquad \mu_1^{(r)}\in\mathbb{C}. \tag{4.82}$$

PROPOSITION 4.5.1. *If* $\dim V(\mu(u))<\infty$, *then there exists a formal series*

$$f(u)=1+f_1u^{-1}+f_2u^{-2}+\dots, \qquad f_r\in\mathbb{C},$$

such that $f(u)\,\mu_0(u)$ *and* $f(u)\,\mu_1(u)$ *are polynomials in* u^{-1}.

PROOF. By twisting the action of $\mathrm{Y}(\mathfrak{o}_3)$ on $V(\mu(u))$ by the automorphism (2.17) with $g(u)=\mu_0(u)^{-1}$, we obtain a module over $\mathrm{Y}(\mathfrak{o}_3)$ which is isomorphic to the irreducible highest weight representation $V(1,g(u)\mu_1(u))$. So, we may assume without loss of generality that the highest weight of $V(\mu(u))$ has the form $\mu(u)=(1,\mu_1(u))$.

Let ξ denote the highest vector of $V(\mu(u))$. Since this representation is finite-dimensional, the vectors $s_{10}^{(i)}\xi\in V(\mu(u))$ with $i\geqslant 1$ are linearly dependent. Therefore, the corresponding Verma module $M(\mu(u))$ contains a nonzero vector η of the form

$$\eta=\sum_{i=1}^{m}c_i\,s_{10}^{(i)}\,1_{\mu(u)}, \qquad c_i\in\mathbb{C},$$

which belongs to the maximal proper submodule K of $M(\mu(u))$. Here m is a positive integer, and we may assume that $c_m\neq 0$. Then we have $s_{01}^{(r)}\eta=0$ for all $r \geqslant 1$ because otherwise the highest vector $1_{\mu(u)}$ would belong to K. By the defining relations (4.3), we get

$$(u-v)\, s_{01}(u)\, s_{10}(v)\, 1_{\mu(u)} = \big(\mu_1(u) - \mu_1(v)\big)\, 1_{\mu(u)}.$$

Dividing both sides by $u-v$ and taking the coefficients of $u^{-r}v^{-i}$ with the use of (3.35), we come to

$$s_{01}^{(r)}\, s_{10}^{(i)}\, 1_{\mu(u)} = -\mu_1^{(r+i-1)}\, 1_{\mu(u)}.$$

Hence, for all $r \geqslant 1$ we have the relations

$$\sum_{i=1}^{m} c_i\, \mu_1^{(r+i-1)} = 0.$$

They imply

$$\mu_1(u)\,(c_1 + c_2\, u + \dots + c_m\, u^{m-1}) = b_1 + b_2\, u + \dots + b_m\, u^{m-1}$$

for some coefficients $b_i \in \mathbb{C}$ with $b_m = c_m$. Thus, taking now

$$f(u) = c_m^{-1} \sum_{i=1}^{m} c_i\, u^{-m+i}$$

we conclude that both $f(u)1$ and $f(u)\,\mu_1(u)$ are polynomials in u^{-1}. □

By Proposition 4.5.1, if $V(\mu(u))$ is finite-dimensional, then for an appropriate series $f(u)$ we can write

$$f(u)\,\mu_0(u) = (1+\alpha_1 u^{-1})\dots(1+\alpha_k u^{-1}),$$
$$f(u)\,\mu_1(u) = (1+\beta_1 u^{-1})\dots(1+\beta_k u^{-1}),$$

for some constants $\alpha_i, \beta_i \in \mathbb{C}$ and some $k \geqslant 0$. Hence,

$$g(u)\,\mu_0(u) = (1-\alpha_1^2 u^{-2})\dots(1-\alpha_k^2 u^{-2}),$$
$$g(u)\,\mu_1(u) = (1-\alpha_1 u^{-1})\dots(1-\alpha_k u^{-1})\,(1+\beta_1 u^{-1})\dots(1+\beta_k u^{-1}),$$

where $g(u) = f(u)f(-u)\,\mu_0(u)$ is an even series in u^{-2}. Therefore, twisting the $\mathrm{Y}(\mathfrak{o}_3)$-module by an appropriate automorphism of the form (2.17), we can now concentrate on the representations $V(\mu(u))$ of $\mathrm{Y}(\mathfrak{o}_3)$ such that the highest weight $\mu(u) = (\mu_0(u), \mu_1(u))$ is given by

$$\begin{aligned} \mu_0(u) &= (1-\alpha_1^2 u^{-2})\dots(1-\alpha_k^2 u^{-2}), \\ \mu_1(u) &= (1-\alpha_1 u^{-1})\dots(1-\alpha_k u^{-1})\,(1+\beta_1 u^{-1})\dots(1+\beta_k u^{-1}). \end{aligned} \tag{4.83}$$

LEMMA 4.5.2. *Suppose that the irreducible representation $V(\mu(u))$ of the twisted Yangian $\mathrm{Y}(\mathfrak{o}_3)$ with the components of $\mu(u)$ given by (4.83) is finite-dimensional. Let $V(\widetilde{\mu}(u))$ be the irreducible highest weight representation of $\mathrm{Y}(\mathfrak{o}_3)$ such that the components of $\widetilde{\mu}(u)$ are obtained from $\mu_0(u)$ and $\mu_1(u)$ by replacing the parameters α_i and β_i by the rule*

$$\alpha_i \mapsto \alpha_i + l_i, \qquad \beta_i \mapsto \beta_i - m_i, \qquad i = 1, \dots, k,$$

where the l_i and m_i are some nonnegative integers. Then the representation $V(\widetilde{\mu}(u))$ is also finite-dimensional.

PROOF. For each $i \in \{1, \dots, k\}$ the representation

$$L(\alpha_i + l_i, \alpha_i + l_i, \alpha_i) \otimes V(\mu(u))$$

of $\mathrm{Y}(\mathfrak{o}_3)$ is finite-dimensional. By Proposition 4.2.9, the cyclic $\mathrm{Y}(\mathfrak{o}_3)$-span of the tensor product of the highest vectors of $L(\alpha_i+l_i, \alpha_i+l_i, \alpha_i)$ and $V(\mu(u))$ is a highest weight module with the highest weight $\mu'(u)$, where

$$\mu'_0(u) = \mu_0(u)\,(1 - (\alpha_i + l_i)^2\, u^{-2}),$$
$$\mu'_1(u) = \mu_1(u)\,(1 - (\alpha_i + l_i)\, u^{-1})\,(1 + \alpha_i\, u^{-1}).$$

By twisting this module by the automorphism (2.17), where the series $g(u)$ is given by $g(u) = (1-\alpha_i^2\, u^{-2})^{-1}$, we get a finite-dimensional representation with the highest weight obtained from $\mu(u)$ by replacing α_i with α_i+l_i. Similarly, considering tensor products of the form $L(\beta_i, \beta_i, \beta_i - m_i) \otimes V(\mu(u))$ we show that the representation with the highest weight obtained from $\mu(u)$ by replacing β_i with $\beta_i - m_i$ is also finite-dimensional. □

Let the series $\mu_0(u)$ and $\mu_1(u)$ be given by (4.83) with some complex parameters α_i and β_i. By Corollary 4.2.12, the $\mathrm{Y}(\mathfrak{o}_3)$-module $V(\mu(u))$ with $\mu(u) = (\mu_0(u), \mu_1(u))$ is isomorphic to a subquotient of the $\mathrm{Y}(\mathfrak{o}_3)$-module

$$L(\alpha_1, \alpha_1, \beta_1) \otimes \dots \otimes L(\alpha_k, \alpha_k, \beta_k). \tag{4.84}$$

Therefore, by Proposition 4.2.13(ii), the series

$$S_{ij}(u) = u^{2k}\, s_{ij}(u)$$

is a polynomial in u as an operator in $V(\mu(u))$. In particular, for the highest vector ξ of $V(\mu(u))$ we have $S_{ij}(u)\,\xi = 0$ for $i < j$ and

$$S_{00}(u)\,\xi = M_0(u)\,\xi, \qquad S_{11}(u)\,\xi = M_1(u)\,\xi,$$

where

$$M_0(u) = (u^2 - \alpha_1^2) \dots (u^2 - \alpha_k^2),$$
$$M_1(u) = (u - \alpha_1) \dots (u - \alpha_k)\,(u + \beta_1) \dots (u + \beta_k).$$

By Corollary 2.15.5, the mapping

$$S(u) \mapsto \widehat{S}(-u + 1/2) \tag{4.85}$$

defines an automorphism of the algebra $\mathrm{Y}(\mathfrak{o}_3)$. Here $\widehat{S}(u)$ is the Sklyanin comatrix for $S(u)$. The entries of $\widehat{S}(u)$ are given by Proposition 4.1.4. In particular, choosing an appropriate permutation (a_1, a_2, a_3) of the indices $(-1, 0, 1)$ we get

$$\begin{aligned} \widehat{s}_{00}(u) &= s_{11}(-u)\, s_{11}(u-1) - s_{1,-1}(-u)\, s_{-1,1}(u-1), \\ \widehat{s}_{-1,0}(u) &= s_{0,-1}(-u)\, s_{-1,1}(u-1) - s_{01}(-u)\, s_{11}(u-1), \\ \widehat{s}_{11}(u) &= s_{11}(-u)\, s_{00}(u-1) - s_{10}(-u)\, s_{-1,0}(u-1). \end{aligned} \tag{4.86}$$

Indeed, the respective choices are $(a_1, a_2, a_3) = (1, -1, 0)$, $(1, 0, -1)$ and $(0, -1, 1)$. By twisting the action of $\mathrm{Y}(\mathfrak{o}_3)$ on $V(\mu(u))$ with the automorphism (4.85) we obtain another $\mathrm{Y}(\mathfrak{o}_3)$-module structure on the same vector space. Using also an automorphism of the twisted Yangian of the form (2.17), we conclude that for any even formal series $g(u)$ in u^{-1} the operators on the vector space $V(\mu(u))$ given by

$$s^*_{ij}(u) = g(u)\, \widehat{s}_{ij}(-u + 1/2), \qquad i, j \in \{-1, 0, 1\}, \tag{4.87}$$

define a representation of $\mathrm{Y}(\mathfrak{o}_3)$.

LEMMA 4.5.3. *We have the following relations:*

$$\begin{aligned}
&\big(u^2-(v+3/2)^2\big)\,[s^*_{ij}(u),s_{kl}(v)]\\
&=(u-v-3/2)\Big(\delta_{kj}\sum_a s^*_{ia}(u)\,s_{al}(v)-\delta_{il}\sum_a s_{ka}(v)\,s^*_{aj}(u)\Big)\\
&\quad-(u+v+3/2)\Big(\delta_{i,-k}\sum_a s^*_{-a,j}(u)\,s_{al}(v)-\delta_{j,-l}\sum_a s_{ka}(v)\,s^*_{i,-a}(u)\Big)\\
&\qquad\qquad+\delta_{i,-k}\sum_a s^*_{-j,a}(u)\,s_{al}(v)-\delta_{j,-l}\sum_a s_{ka}(v)\,s^*_{a,-i}(u).
\end{aligned}$$

PROOF. The argument is similar to the one used for the proof of Proposition 2.12.2. Inverting both sides of the quaternary relation (2.102) for the twisted Yangian $Y(\mathfrak{o}_3)$ and multiplying from the left and right by $S_2(v)$ we obtain

$$R'(u+v+3)\,S_1^{-1}(u)\,R(v-u)\,S_2(v)=S_2(v)\,R(v-u)\,S_1^{-1}(u)\,R'(u+v+3).$$

Writing this in terms of the matrix elements, we get relations between the entries of the matrices $S^{-1}(u)$ and $S(v)$. Now use the definition of the Sklyanin comatrix (2.112) and the centrality of the Sklyanin determinant in $Y(\mathfrak{o}_3)$ to get the relations between the entries of the matrices $\widehat{S}(u)$ and $S(v)$. Finally apply the definition (4.87) of the operators $s^*_{ij}(u)$ to get the desired relations. □

By (4.86), we have the following formulas for the action of these operators on the highest vector ξ of $V(\mu(u))$:

$$\begin{aligned}
s^*_{-1,0}(u)\,\xi&=0,\\
s^*_{00}(u)\,\xi&=g(u)\,\mu_1(u-1/2)\,\mu_1(-u-1/2),\\
s^*_{11}(u)\,\xi&=g(u)\,\mu_1(u-1/2)\,\mu_0(-u-1/2).
\end{aligned}\tag{4.88}$$

By (4.17) and the symmetry relation (4.4), we obtain that $s^*_{ij}(u)\,\xi=0$ for all $i<j$. Hence, the action of the operators $s^*_{ij}(u)$ makes $V(\mu(u))$ into a highest weight module over $Y(\mathfrak{o}_3)$, where ξ is the highest vector. The highest weight $\mu^*(u)=(\mu^*_0(u),\mu^*_1(u))$ is found from the relations (4.88). We will now fix a particular choice of the series $g(u)$, namely,

$$g(u)=\mu_1(u-1/2)^{-1}\,\mu_1(-u-1/2)^{-1}\prod_{i=1}^{k}\Big(1-(\beta_i-1/2)\,u^{-2}\Big).$$

Then the components of the highest weight $\mu^*(u)$ will be given by

$$\begin{aligned}
\mu^*_0(u)&=(1-(\alpha^*_1)^2\,u^{-2})\dots(1-(\alpha^*_k)^2\,u^{-2}),\\
\mu^*_1(u)&=(1-\alpha^*_1\,u^{-1})\dots(1-\alpha^*_k\,u^{-1})(1+\beta^*_1\,u^{-1})\dots(1+\beta^*_k\,u^{-1}),
\end{aligned}\tag{4.89}$$

where

$$\alpha^*_i=-\beta_i+1/2,\qquad \beta^*_i=-\alpha_i+1/2,\qquad i=1,\dots,k.\tag{4.90}$$

As with the operators $s_{ij}(u)$, the series

$$S^*_{ij}(u)=u^{2k}\,s^*_{ij}(u)$$

is a polynomial in u as an operator in $V(\mu(u))$. For the highest vector ξ of $V(\mu(u))$ we have $S^*_{ij}(u)\,\xi=0$ for $i<j$ and

$$S^*_{00}(u)\,\xi=M^*_0(u)\,\xi,\qquad S^*_{11}(u)\,\xi=M^*_1(u)\,\xi,$$

where

$$M_0^*(u) = (u^2 - (\alpha_1^*)^2) \dots (u^2 - (\alpha_k^*)^2),$$
$$M_1^*(u) = (u - \alpha_1^*) \dots (u - \alpha_k^*)\,(u + \beta_1^*) \dots (u + \beta_k^*).$$

For the proof of the next lemma fix an index $i \in \{1, \dots, k\}$ and set $\alpha = \alpha_i$, $\beta^* = -\alpha_i + 1/2$. For any nonnegative integer p introduce the following vector in $V(\mu(u))$:

$$\eta_p = S_{10}(-\alpha + p - 1) \dots S_{10}(-\alpha)\,\xi.$$

In particular, $\eta_0 = \xi$.

LEMMA 4.5.4. *We have the relations*

$$S_{00}(u)\,\eta_p = \frac{u^2 - (\alpha - p)^2}{u^2 - \alpha^2}\,M_0(u)\,\eta_p, \tag{4.91}$$

$$S_{-1,0}(u)\,\eta_p = \frac{p\,(u + \alpha - p)}{u^2 - \alpha^2}\,M_0(u)\,M_1(-\alpha + p - 1)\,\eta_{p-1}, \tag{4.92}$$

and

$$S^*_{-1,1}(u)\,\eta_p = 0, \tag{4.93}$$

$$S^*_{11}(u)\,\eta_p = \frac{u + \beta^* + p}{u + \beta^*}\,M_1^*(u)\,\eta_p. \tag{4.94}$$

PROOF. We use induction on p to prove the four relations simultaneously. For $p = 0$ all of them are obvious. Suppose now that $p \geqslant 1$. By the defining relations (4.3), we have

$$\begin{aligned}[s_{00}(u), s_{10}(v)] &= \frac{u + v + 1}{u^2 - v^2}\,s_{10}(u)\,s_{00}(v) \\ &\quad - \frac{1}{u + v}\,s_{0,-1}(u)\,s_{00}(v) - \frac{2v + 1}{u^2 - v^2}\,s_{10}(v)\,s_{00}(u).\end{aligned} \tag{4.95}$$

By the induction hypothesis,

$$S_{00}(-\alpha + p - 1)\,\eta_{p-1} = 0.$$

Therefore,

$$\begin{aligned}S_{00}(u)\,\eta_p &= S_{00}(u)\,S_{10}(-\alpha + p - 1)\,\eta_{p-1} \\ &= \frac{u^2 - (\alpha - p)^2}{u^2 - (\alpha - p + 1)^2}\,S_{10}(-\alpha + p - 1)\,S_{00}(u)\,\eta_{p-1}.\end{aligned}$$

The induction hypothesis now implies (4.91).

By Lemma 4.5.3, we have

$$(u + v + 3/2)\,[s_{11}^*(u), s_{10}(v)] = \sum_a s_{1a}^*(u)\,s_{a0}(v),$$

and so,

$$\begin{aligned}s_{11}^*(u)\,s_{10}(v) &= \frac{u + v + 3/2}{u + v + 1/2}\,s_{10}(v)\,s_{11}^*(u) \\ &\quad + \frac{1}{u + v + 1/2}\Big(s_{10}^*(u)\,s_{00}(v) + s_{1,-1}^*(u)\,s_{-1,0}(v)\Big).\end{aligned}$$

Hence, using the induction hypothesis we can write

$$S^*_{11}(u)\,\eta_p = S^*_{11}(u)\,S_{10}(-\alpha+p-1)\,\eta_{p-1}$$
$$= \frac{u+\beta^*+p}{u+\beta^*+p-1}\,S_{10}(-\alpha+p-1)\,S^*_{11}(u)\,\eta_{p-1},$$

thus proving (4.94). The proof of (4.93) is the same; it suffices to apply Lemma 4.5.3 to the expansion of $[s^*_{-1,1}(u), s_{10}(v)]$. Finally, in order to prove (4.92), consider the following relation implied by (4.3):

$$(4.96)\quad [s_{-1,0}(u), s_{10}(v)] = \frac{1}{u-v}\Big(s_{10}(u)\,s_{-1,0}(v) - s_{10}(v)\,s_{-1,0}(u)\Big)$$
$$+\frac{u-v-1}{u^2-v^2}\,s_{11}(v)\,s_{00}(u) - \frac{1}{u+v}\,s_{-1,-1}(u)\,s_{00}(v) + \frac{1}{u^2-v^2}\,s_{11}(u)\,s_{00}(v).$$

Using this relation and the induction hypothesis, we obtain

$$S_{-1,0}(u)\,\eta_p = S_{-1,0}(u)\,S_{10}(-\alpha+p-1)\,\eta_{p-1}$$
$$= \frac{u+\alpha-p}{u+\alpha-p+1}\,S_{10}(-\alpha+p-1)\,S_{-1,0}(u)\,\eta_{p-1}$$
$$+\frac{u+\alpha-p}{u^2-(\alpha-p+1)^2}\,S_{11}(-\alpha+p-1)\,S_{00}(u)\,\eta_{p-1}.$$

Expanding further $S_{-1,0}(u)\,\eta_{p-1}$ and applying (4.91), we come to the relation

$$(4.97)\qquad S_{-1,0}(u)\,\eta_p = \frac{u+\alpha-p}{u^2-\alpha^2}\,M_0(u)\,\eta^{(1)}_p,$$

where

$$\eta^{(1)}_p = \sum_{q=1}^{p} S_{10}(-\alpha+p-1)\dots S_{11}(-\alpha+q-1)\dots S_{10}(-\alpha)\,\xi$$

and $S_{11}(-\alpha+q-1)$ takes the q-th position from the right.

On the other hand, by (4.86) and (4.87),

$$s^*_{11}(-u-1/2) = g(u+1/2)\,\Big(s_{11}(-u-1)\,s_{00}(u) - s_{10}(-u-1)\,s_{-1,0}(u)\Big).$$

Applying both sides of this relation to the vector η_{p-1} and using the induction hypothesis, we get

$$(4.98)\qquad M_1(-u-1)\,\eta_{p-1} = \frac{u-\alpha+p-1}{u-\alpha}\,S_{11}(-u-1)\,\eta_{p-1}$$
$$-\frac{1}{u-\alpha}\,S_{10}(-u-1)\,\eta^{(1)}_{p-1}.$$

Observe that

$$\eta^{(1)}_p = S_{11}(-\alpha+p-1)\,\eta_{p-1} + S_{10}(-\alpha+p-1)\,\eta^{(1)}_{p-1}.$$

Therefore, putting $u=\alpha-p$ into (4.98), we get

$$(4.99)\qquad \eta^{(1)}_p = p\,M_1(-\alpha+p-1)\,\eta_{p-1}.$$

Together with (4.97) this completes the proof of (4.92). □

For any nonnegative integers $p_1, \dots, p_k$ introduce vectors $\eta_{p_1,\dots,p_k} \in V(\mu(u))$ by

$$\eta_{p_1,\dots,p_k} = \prod_{i=1,\dots,k}^{\longrightarrow} \Big(S_{10}(-\alpha_i + p_i - 1)\dots S_{10}(-\alpha_i)\Big)\,\xi, \tag{4.100}$$

where the arrow indicates that the factors in the product are ordered according to increasing indices. In particular, the highest vector ξ coincides with $\eta_{0,\dots,0}$.

LEMMA 4.5.5. *For any nonnegative integers $p_1, \dots, p_k$ the expression*

$$\frac{S_{-1,0}(u)\,\eta_{p_1,\dots,p_k}}{(u+\alpha_1-p_1)\dots(u+\alpha_k-p_k)} \tag{4.101}$$

is a polynomial in u with values in $V(\mu(u))$. The value of this polynomial at $u = \alpha_i - p_i$ with $i \in \{1, \dots, k\}$ coincides with

$$-\,M_1(-\alpha_i + p_i - 1)\prod_{j=1}^{k}(\alpha_i - \alpha_j - p_i)\,\eta_{p_1,\dots,p_i-1,\dots,p_k}. \tag{4.102}$$

Moreover, we have the relation

$$S_{00}(u)\,\eta_{p_1,\dots,p_k} = \big(u^2 - (\alpha_1 - p_1)^2\big)\dots\big(u^2 - (\alpha_k - p_k)^2\big)\,\eta_{p_1,\dots,p_k}. \tag{4.103}$$

PROOF. The relation (4.103) is verified by induction with the use of (4.95); cf. the proof of (4.91) above. Furthermore, using (4.96) and (4.103), we obtain

$$\begin{aligned} S_{-1,0}(u)\,\eta_{p_1,\dots,p_k} &= (u+\alpha_1-p_1)\dots(u+\alpha_k-p_k) \\ &\times \sum_{j=1}^{k}(u-\alpha_1)\dots(u-\alpha_{j-1})(u-\alpha_{j+1}+p_{j+1})\dots(u-\alpha_k+p_k)\,\eta^{(j)}_{p_1,\dots,p_k}, \end{aligned} \tag{4.104}$$

where $\eta^{(j)}_{p_1,\dots,p_k}$ is the vector in $V(\mu(u))$ obtained from $\eta_{p_1,\dots,p_k}$ by replacing the product

$$S_{10}(-\alpha_j + p_j - 1)\dots S_{10}(-\alpha_j) \tag{4.105}$$

in (4.100) with the sum

$$\sum_{q=1}^{p_j} S_{10}(-\alpha_j + p_j - 1)\dots S_{11}(-\alpha_j + q - 1)\dots S_{10}(-\alpha_j);$$

the factor $S_{11}(-\alpha_j + q - 1)$ takes the q-th position from the right. This proves the first part of the lemma. Moreover, (4.99) and (4.104) imply that the value of the polynomial (4.101) at $u = \alpha_k - p_k$ coincides with (4.102) for $i = k$. Therefore, in order to complete the proof, it remains to show that any permutation of the products (4.105) in the expression (4.100) does not change its value. So, it suffices to verify the following auxiliary claim: if a vector $\eta \in V(\mu(u))$ satisfies

$$S_{00}(u)\,\eta = (u^2 - \gamma_1^2)\dots(u^2 - \gamma_k^2)\,\eta$$

with $k \geqslant 2$, then

$$S_{10}(\gamma_1)\,S_{10}(\gamma_2)\,\eta = S_{10}(\gamma_2)\,S_{10}(\gamma_1)\,\eta. \tag{4.106}$$

Indeed, by the defining relations (4.3),

$$(u+v)\,[s_{10}(u), s_{10}(v)] = -\,s_{1,-1}(u)\,s_{00}(v) + s_{1,-1}(v)\,s_{00}(u).$$

Therefore, if $\gamma_1 + \gamma_2 \neq 0$, then (4.106) holds because $S_{00}(\gamma_1)\,\eta = S_{00}(\gamma_2)\,\eta = 0$. If $\gamma_1 + \gamma_2 = 0$, then (4.106) follows since $S_{00}(u)\,\eta$ has double zero at $u = \gamma_1$. □

LEMMA 4.5.6. *For any nonnegative integers $p_1, \dots, p_k$ we have*

$$S^*_{-1,1}(u)\,\eta_{p_1,\dots,p_k} = 0 \tag{4.107}$$

and

$$S^*_{11}(u)\,\eta_{p_1,\dots,p_k} = (u - \alpha^*_1)\dots(u - \alpha^*_k) \times (u + \beta^*_1 + p_1)\dots(u + \beta^*_k + p_k)\,\eta_{p_1,\dots,p_k}. \tag{4.108}$$

PROOF. Both relations follow by induction with the use of Lemmas 4.5.3 and 4.5.5; cf. the proof of Lemma 4.5.4. □

We now aim to establish necessary conditions for the representation $V(\mu(u))$ to be finite-dimensional. The argument will be based upon Lemmas 4.5.2, 4.5.5 and 4.5.6. Suppose that $\dim V(\mu(u)) < \infty$. If the vector $\eta_{p_1,\dots,p_k}$ is nonzero, then by Lemma 4.5.6, the cyclic span $W = \mathrm{Y}(\mathfrak{o}_2)\,\eta_{p_1,\dots,p_k}$ is a highest weight module over the twisted Yangian $\mathrm{Y}(\mathfrak{o}_2)$, where the generators act via the operators $s^*_{ij}(u)$ with $i, j \in \{-1, 1\}$. The highest weight of W is the series

$$\big(1 - \alpha^*_1 u^{-1}\big)\dots\big(1 - \alpha^*_k u^{-1}\big)\big(1 + (\beta^*_1 + p_1)\,u^{-1}\big)\dots\big(1 + (\beta^*_k + p_k)\,u^{-1}\big).$$

Since $\dim W < \infty$, Corollary 4.4.4 implies that the multiset of the $2k+1$ complex numbers

$$\alpha^*_1,\ \dots,\ \alpha^*_k,\ -\beta^*_1 - p_1,\ \dots,\ -\beta^*_k - p_k,\ -1/2 \tag{4.109}$$

contains k disjoint pairs such that the sum of the numbers in each pair is a nonnegative integer. Using all possible values of the parameters p_i we get the desired conditions on the α^*_i and β^*_i. Lemma 4.5.2 gives us the freedom of shifting the numbers α_i and β_i by nonnegative integers to bring the original highest weight $\mu(u)$ to a simpler form, if necessary. Lemma 4.5.5 will be used to verify that the vector $\eta_{p_1,\dots,p_k}$ is nonzero. Namely, applying appropriate operators to the vectors of the form $\eta_{q_1,\dots,q_k}$ repeatedly and making sure that the corresponding coefficients in (4.102) are nonzero, we get the highest vector $\xi = \eta_{0,\dots,0}$ with a nonzero coefficient.

LEMMA 4.5.7. *Suppose that the representation $V(\mu(u))$ of $\mathrm{Y}(\mathfrak{o}_3)$ with the components of $\mu(u)$ given by (4.83) is finite-dimensional. Then there exists a renumbering of the α_i and a renumbering of the β_i such that one of the following two conditions holds:*

$$\alpha_i - \beta_i \in \mathbb{Z} \quad \textit{for} \quad i = 1, \dots, k, \quad \textit{or} \tag{4.110}$$

$$\alpha_i - \beta_i \in \mathbb{Z} \quad \textit{for} \quad i = 1, \dots, k-1 \quad \textit{and} \quad \alpha_k \in 1/2 + \mathbb{Z},\ \ \beta_k \in \mathbb{Z}. \tag{4.111}$$

PROOF. Shifting the α_i and β_i as in Lemma 4.5.2, if necessary, we may assume, without loss of generality, that the following conditions hold: for all $i, j \in \{1, \dots, k\}$

$$\alpha_i - \alpha_j \in \mathbb{Z} \implies \alpha_i = \alpha_j, \qquad \beta_i - \beta_j \in \mathbb{Z} \implies \beta_i = \beta_j,$$
$$\alpha_i + \beta_j \in \mathbb{Z} \implies \alpha_i + \beta_j \leqslant 0, \qquad \operatorname{Re} \alpha_i \geqslant 1/2.$$

Now define parameters $p_1, \dots, p_k$ as follows. If $\alpha_i + \alpha_j \in \mathbb{Z}$ for some indices i and j, then we set $p_i = p_j = \alpha_i + \alpha_j$. Due to the above conditions, for any i there is at most one value of the parameter α_j with this property, and so p_i and p_j are well-defined. For any i we set $p_i = 0$ if $\alpha_i + \alpha_j \notin \mathbb{Z}$ for all j. Clearly, all p_i are nonnegative integers.

Let us verify that the corresponding vector $\eta_{p_1,\dots,p_k}$ is nonzero. Indeed, if $p_i \geqslant 1$ for some i, then the coefficient of the vector $\eta_{p_1,\dots,p_i-1,\dots,p_k}$ in (4.102) equals

$$-\prod_{j=1}^{k}(\alpha_i+\alpha_j-p_i+1)\prod_{j=1}^{k}(\alpha_i-\beta_j-p_i+1)\prod_{j=1}^{k}(\alpha_i-\alpha_j-p_i).$$

By the definition of the p_i, this coefficient is nonzero. Moreover, it remains nonzero when p_i is replaced with $p_i-1, p_i-2, \dots, 1$. Thus, we can get the highest vector $\xi = \eta_{0,\dots,0}$ with a nonzero coefficient by acting on $\eta_{p_1,\dots,p_k}$ by an appropriate element of the algebra $\mathrm{Y}(\mathfrak{o}_3)$. This shows that $\eta_{p_1,\dots,p_k} \neq 0$.

By Corollary 4.4.4 and Lemma 4.5.6, the multiset of the $2k+1$ complex numbers (4.109) must contain k disjoint pairs such that the sum of the numbers in each pair is a nonnegative integer. However, recalling the formulas (4.90) for the α_i^* and β_i^*, we conclude that, by the choice of the parameters p_i, we have for all i and j,

$$-\beta_i^*-p_i-\beta_j^*-p_j \notin \mathbb{Z}_+, \qquad -\beta_i^*-p_i-1/2 \notin \mathbb{Z}_+.$$

Therefore, we must have

$$\begin{aligned}&\alpha_i^*-\beta_i^*-p_i \in \mathbb{Z}_+ &&\text{for}\quad i=1,\dots,k, \quad \text{or}\\ &\alpha_i^*-\beta_i^*-p_i \in \mathbb{Z}_+ &&\text{for}\quad i=1,\dots,k-1 \quad \text{and}\quad \alpha_k^* \in 1/2+\mathbb{Z}_+\end{aligned}$$

for some renumbering of the α_i^* and β_i^*. Since $\alpha_i^*-\beta_i^* = \alpha_i-\beta_i$, this implies (4.110) in the former case, while the latter implies

$$\alpha_i-\beta_i \in \mathbb{Z} \quad \text{for}\quad i=1,\dots,k-1 \quad\text{and}\quad \beta_k \in \mathbb{Z}. \tag{4.112}$$

In order to show that (4.110) or (4.111) holds in this case as well, note that the representation $V(\mu^*(u))$ with the highest weight given by (4.89) is also finite-dimensional. Therefore, repeating the previous argument for the parameters α_i^* and β_i^* instead of α_i and β_i, we obtain that

$$\alpha^*_{\sigma(i)}-\beta^*_{\tau(i)} \in \mathbb{Z} \qquad \text{for}\quad i=1,\dots,k-1 \quad\text{and}\quad \beta^*_{\tau(k)} \in \mathbb{Z},$$

for some permutations $\sigma, \tau \in \mathfrak{S}_k$. This is equivalent to the condition

$$\alpha_{\tau(i)}-\beta_{\sigma(i)} \in \mathbb{Z} \qquad \text{for}\quad i=1,\dots,k-1 \quad\text{and}\quad \alpha_{\tau(k)} \in 1/2+\mathbb{Z}.$$

If $\tau(k)=k$, then (4.111) follows. So suppose that $\tau(k)=l\neq k$. Then we have

$$\alpha_i-\beta_{\rho(i)} \in \mathbb{Z} \qquad \text{for}\quad i=1,\dots,l-1,\, l+1,\dots,k \quad\text{and}\quad \alpha_l \in 1/2+\mathbb{Z} \tag{4.113}$$

for some $\rho \in \mathfrak{S}_k$. Let d be the minimum positive integer such that $\rho^d(k)=k$. If $\rho^r(k)\neq l$ for all r, then using (4.113) for the values $i=k, \rho(k),\dots,\rho^{d-1}(k)$ we deduce from (4.112) that $\alpha_k-\beta_k \in \mathbb{Z}$. This implies (4.110). Suppose now that $\rho^r(k)=l$ for some positive integer r. Taking the minimum possible value of r and using (4.113) for $i=k, \rho(k),\dots,\rho^{r-1}(k)$ we find from (4.112) that $\alpha_k-\alpha_l \in \mathbb{Z}$. This yields (4.111). □

PROPOSITION 4.5.8. *The irreducible highest weight representation* $V(\mu(u))$ *of* $\mathrm{Y}(\mathfrak{o}_3)$ *with the components of* $\mu(u)$ *given by* (4.83) *is finite-dimensional if and only if there exists a renumbering of the* α_i *and a renumbering of the* β_i *such that one of the following two conditions holds:*

$$\alpha_i - \beta_i \in \mathbb{Z}_+ \quad \textit{for} \quad i = 1, \dots, k, \qquad \textit{or} \tag{4.114}$$

$$\alpha_i - \beta_i \in \mathbb{Z}_+ \quad \textit{for} \quad i = 1, \dots, k-1 \quad \textit{and} \quad \alpha_k \in 1/2 + \mathbb{Z}_+, \quad \beta_k \in -\mathbb{Z}_+. \tag{4.115}$$

PROOF. We start by proving that the conditions (4.114) are sufficient for the representation $V(\mu(u))$ to be finite-dimensional. By Corollary 4.2.12, the $\mathrm{Y}(\mathfrak{o}_3)$-module $V(\mu(u))$ is isomorphic to a subquotient of the finite-dimensional $\mathrm{Y}(\mathfrak{o}_3)$-module (4.84). Hence $\dim V(\mu(u)) < \infty$. Similarly, if (4.115) holds, then by Proposition 4.2.11, the $\mathrm{Y}(\mathfrak{o}_3)$-module $V(\mu(u))$ is isomorphic to a subquotient of the finite-dimensional $\mathrm{Y}(\mathfrak{o}_3)$-module

$$L(\alpha_1, \alpha_1, \beta_1) \otimes \dots \otimes L(\alpha_{k-1}, \alpha_{k-1}, \beta_{k-1}) \otimes L(\alpha_k, \alpha_k, 1/2) \otimes V(\beta_k - 1/2).$$

Thus, $\dim V(\mu(u)) < \infty$.

Conversely, suppose that $\dim V(\mu(u)) < \infty$. Our argument will follow the same pattern as the proof of Lemma 4.5.7. In particular, that proof shows that, in order to reach the desired conclusion, we may partition the multiset of complex numbers α_i and β_i into sub-multisets belonging to the same $\mathbb{Z}$-cosets in $\mathbb{C}$. First, consider all those numbers which belong to the union of two cosets $z + \mathbb{Z}$ and $-z + \mathbb{Z}$ for some fixed $z \in \mathbb{C}$ such that $z \not\equiv 0,\ 1/2 \mod \mathbb{Z}$. By Lemma 4.5.7, the number of α's and β's belonging to this union are the same. In order to simplify the notation, re-denote those numbers by α_i and β_i with $i = 1, \dots, k$. By the same lemma, renumbering the α_i and β_i, if necessary, we may further assume that for some $1 \leqslant r \leqslant k$,

$$\alpha_i, \beta_i \in z + \mathbb{Z}, \qquad \text{for} \quad i = 1, \dots, r, \quad \text{and}$$
$$\alpha_i, \beta_i \in -z + \mathbb{Z}, \qquad \text{for} \quad i = r+1, \dots, k,$$

where the α's and β's belonging to the same $\mathbb{Z}$-coset are ordered by their real parts; i.e. $\operatorname{Re} \alpha_1 \leqslant \dots \leqslant \operatorname{Re} \alpha_r$, etc. We will be proving that for any index $i = 1, \dots, k$ the number of indices $j \in \{1, \dots, k\}$ such that $\alpha_i - \beta_j \in \mathbb{Z}_+$ is not less than i. This will imply that (4.114) can be achieved by further renumbering the α_i and β_i. Suppose on the contrary that for some index $m \in \{1, \dots, r\}$ the number of indices $j \in \{1, \dots, r\}$ such that $\alpha_m - \beta_j \in \mathbb{Z}_+$ is $l < m$. Then, applying the shifts of Lemma 4.5.2, we may assume that the following conditions hold:

$$\alpha_1 = \dots = \alpha_m, \qquad \alpha_{m+1} = \dots = \alpha_r, \qquad \alpha_{r+1} = \dots = \alpha_k,$$
$$\beta_1 = \dots = \beta_l, \qquad \beta_{l+1} = \dots = \beta_r, \qquad \beta_{r+1} = \dots = \beta_k,$$

where

$$\operatorname{Re} \beta_1 < \operatorname{Re} \alpha_1 < \operatorname{Re} \beta_{l+1} < \operatorname{Re} \alpha_{m+1},$$

and also, if $r \leqslant k-1$, then

$$\alpha_1 + \alpha_{r+1} \geqslant 1 \qquad \text{and} \qquad \beta_1 + \alpha_{r+1} \leqslant 0.$$

Define the parameters $p_1, \dots, p_k$ by

$$p_1 = \dots = p_m = \alpha_1 + \alpha_{r+1} \qquad \text{and} \qquad p_{m+1} = \dots = p_k = 0.$$

Now, exactly as in the proof of Lemma 4.5.7, verifying that the coefficient in (4.102) is nonzero, we deduce that the vector $\eta_{p_1, \dots, p_k}$ is nonzero. However, we have for all i and j,

$$-\beta_i^* - p_i - \beta_j^* - p_j \notin \mathbb{Z}_+, \qquad -\beta_i^* - p_i - 1/2 \notin \mathbb{Z}_+.$$

Therefore, Corollary 4.4.4 implies that the $Y(\mathfrak{o}_2)$-module $Y(\mathfrak{o}_2)\,\eta_{p_1,\dots,p_k}$ with the action of the generators via the operators $s^*_{ij}(u)$ is infinite-dimensional. This contradiction completes the argument in the case under consideration.

It remains to consider those numbers α_i and β_i which belong to $\mathbb{Z}$ or $1/2+\mathbb{Z}$. In order to simplify the notation, we will assume that all these numbers with $i=1,\dots,k$ are contained in the union of the sets $\mathbb{Z}$ and $1/2+\mathbb{Z}$. By Lemma 4.5.7, we have the following two cases. Renumbering the α_i and β_i, if necessary, we may assume that either

$$\alpha_1,\dots,\alpha_r,\beta_1,\dots,\beta_r\in\mathbb{Z}\quad\text{and}\quad\alpha_{r+1},\dots,\alpha_k,\beta_{r+1},\dots,\beta_k\in 1/2+\mathbb{Z}\tag{4.116}$$

for some $r\in\{0,\dots,k\}$ or

$$\alpha_1,\dots,\alpha_{r-1},\beta_1,\dots,\beta_r\in\mathbb{Z}\quad\text{and}\quad\alpha_r,\dots,\alpha_k,\beta_{r+1},\dots,\beta_k\in 1/2+\mathbb{Z}\tag{4.117}$$

for some $r\in\{1,\dots,k\}$, where the sequences of α's and β's belonging to the same $\mathbb{Z}$-coset are increasing. Let (4.116) hold. We will be proving by contradiction that (4.114) holds. Suppose first that $r\geqslant 1$ and for some index $m\in\{1,\dots,r\}$ the number of indices $j\in\{1,\dots,r\}$ such that $\beta_j\leqslant\alpha_m$ is $l<m$. As in the previous case, applying the shifts of Lemma 4.5.2, we may assume that the following conditions hold:

$$\alpha_1=\dots=\alpha_m,\qquad\alpha_{m+1}=\dots=\alpha_r,$$
$$\beta_1=\dots=\beta_l,\qquad\beta_{l+1}=\dots=\beta_r,$$

where $\beta_1<\alpha_1<\beta_{l+1}<\alpha_{m+1}$. Now the argument is divided into two subcases. First, suppose that $\alpha_1\geqslant 0$. Shifting the numbers $\beta_1,\dots,\beta_l$ further to the left while preserving their equality, we may also assume that $\alpha_1+\beta_1\leqslant 0$. Define the parameters $p_1,\dots,p_k$ by

$$p_1=\dots=p_m=2\,\alpha_1\qquad\text{and}\qquad p_{m+1}=\dots=p_k=0.$$

Then using (4.102) we deduce that the vector $\eta_{p_1,\dots,p_k}$ is nonzero. However, we have

$$\alpha^*_1=\dots=\alpha^*_l=-\beta_1+1/2,\qquad\alpha^*_{l+1}=\dots=\alpha^*_r=-\beta_{l+1}+1/2,$$

while

$$-\beta^*_1-p_1=\dots=-\beta^*_m-p_m=-\alpha_1-1/2$$

and

$$-\beta^*_{m+1}-p_{m+1}=\dots=-\beta^*_r-p_r=\alpha_{m+1}-1/2.$$

By Corollary 4.4.4, the multiset of the $2k+1$ numbers (4.109) must contain k disjoint pairs such that the sum of the numbers in each pair is a nonnegative integer. However,

$$-\beta_{l+1}+1/2\leqslant -1/2\qquad\text{and}\qquad -\alpha_1-1/2\leqslant -1/2.$$

Therefore, since $l<m$, such k pairs are impossible to choose. This contradiction shows that the α_i and β_i with $i=1,\dots,r$ satisfy the condition $\alpha_i-\beta_i\in\mathbb{Z}_+$ for all $i\in\{1,\dots,r\}$.

Consider now the second subcase where $\alpha_1\leqslant -1$. Shifting the numbers $\alpha_{m+1},\dots,\alpha_r$ further to the right while keeping them equal to each other, we may assume that $\alpha_1+\alpha_{m+1}\geqslant 1$. Moreover, shifting the numbers $\beta_1,\dots,\beta_l$ further to the left while keeping them equal to each other, we may also assume that $\beta_1+\alpha_{m+1}\leqslant 0$. Define the parameters $p_1,\dots,p_k$ by

$$p_1=\dots=p_m=\alpha_1+\alpha_{m+1}\qquad\text{and}\qquad p_{m+1}=\dots=p_k=0.$$

We verify as above that $\eta_{p_1,\dots,p_k} \neq 0$ and note that

$$\alpha_1^* = \cdots = \alpha_l^* = -\beta_1 + 1/2, \qquad \alpha_{l+1}^* = \cdots = \alpha_r^* = -\beta_{l+1} + 1/2,$$

while

$$-\beta_1^* - p_1 = \cdots = -\beta_m^* - p_m = -\alpha_{m+1} - 1/2$$

and

$$-\beta_{m+1}^* - p_{m+1} = \cdots = -\beta_r^* - p_r = \alpha_{m+1} - 1/2.$$

Since $-\beta_{l+1} - \alpha_{m+1} < -\alpha_1 - \alpha_{m+1} \leqslant -1$ and $l < m$, this again contradicts Corollary 4.4.4. Hence, $\alpha_i - \beta_i \in \mathbb{Z}_+$ for all $i \in \{1, \dots, r\}$ in this subcase as well.

Similarly, if (4.116) holds for some $r \leqslant k - 1$, then, arguing by contradiction, we assume that for some index $m \in \{r+1, \dots, k\}$ the number of indices $j \in \{r+1, \dots, k\}$ such that $\beta_j \leqslant \alpha_m$ is $l < m$. The argument here is completed word by word as above, proving that (4.114) holds. The two subcases are distinguished here by the conditions $\alpha_{r+1} \geqslant 1/2$ and $\alpha_{r+1} \leqslant -1/2$.

Now let (4.117) hold. We will be proving that then (4.115) holds after a possible renumbering of the α_i and β_i. Again, we argue by contradiction. Exactly as in the above argument, we prove that there exists a renumbering of $\alpha_1, \dots, \alpha_{r-1}$ and a renumbering of $\beta_1, \dots, \beta_r$ such that $\alpha_i - \beta_i \in \mathbb{Z}_+$ for all $i = 1, \dots, r-1$. Furthermore, let us choose such a renumbering, where β_r takes the minimal possible value. We want to show that $\beta_r \leqslant 0$. Suppose on the contrary that $\beta_r \geqslant 1$. Using the shifts of Lemma 4.5.2, we may assume that for some $m \in \{0, \dots, r-1\}$ the following conditions hold:

$$\begin{aligned} \alpha_1 &= \cdots = \alpha_m, & \alpha_{m+1} &= \cdots = \alpha_{r-1}, \\ \beta_1 &= \cdots = \beta_m, & \beta_{m+1} &= \cdots = \beta_r, \end{aligned}$$

where $\beta_1 < \alpha_1 < \beta_{m+1} < \alpha_{m+1}$. Suppose first that $\alpha_1 \leqslant 0$. Then

$$\alpha_1^* = \cdots = \alpha_m^* = -\beta_1 + 1/2, \qquad \alpha_{m+1}^* = \cdots = \alpha_r^* = -\beta_{m+1} + 1/2,$$

while

$$-\beta_1^* = \cdots = -\beta_m^* = \alpha_1 - 1/2$$

and

$$-\beta_{m+1}^* = \cdots = -\beta_{r-1}^* = \alpha_{m+1} - 1/2.$$

However, by the assumption,

$$-\beta_{m+1} + 1/2 \leqslant -1/2 \qquad \text{and} \qquad \alpha_1 - 1/2 \leqslant -1/2.$$

Taking $p_1 = \cdots = p_k = 0$ in (4.109) and applying Corollary 4.4.4 we come to a contradiction, because it is impossible to choose k disjoint pairs from the $2k+1$ numbers (4.109) in such a way that the sum of the numbers in each pair would be a nonnegative integer.

Now let $\alpha_1 \geqslant 1$. We may assume that $\beta_1 \leqslant 0$. The proof in this case is completed by taking $p_1 = \cdots = p_m = \alpha_1$, $p_{m+1} = \cdots = p_k = 0$ and applying Corollary 4.4.4 once again.

Finally, since the representation $V(\mu(u))$ of $\mathrm{Y}(\mathfrak{o}_3)$ with the parameters α_i and β_i satisfying (4.117) is finite-dimensional, then so is the representation $V(\mu^*(u))$. We have

$$\alpha_{r+1}^*, \dots, \alpha_k^*, \beta_r^*, \dots, \beta_k^* \in \mathbb{Z} \quad \text{and} \quad \alpha_1^*, \dots, \alpha_r^*, \beta_1^*, \dots, \beta_{r-1}^* \in 1/2 + \mathbb{Z}.$$

Therefore, by the previous argument, there exists a renumbering of $\alpha_{r+1}^*, \dots, \alpha_k^*$ and a renumbering of $\beta_r^*, \dots, \beta_k^*$ such that $\alpha_i^* - \beta_i^* \in \mathbb{Z}_+$ for all $i = r+1, \dots, k$ and

$\beta_r^* \in -\mathbb{Z}_+$. This is equivalent to the conditions $\alpha_i - \beta_i \in \mathbb{Z}_+$ for $i = r+1, \dots, k$ and $\alpha_r \in 1/2 + \mathbb{Z}_+$. Together with the conditions $\alpha_i - \beta_i \in \mathbb{Z}_+$ for $i = 1, \dots, r-1$ and $\beta_r \in -\mathbb{Z}_+$ established above, this implies (4.115). □

The next theorem together with Theorem 4.2.6 provides a description of finite-dimensional irreducible representations of the twisted Yangian $\mathrm{Y}(\mathfrak{o}_{2n+1})$ with $n \geqslant 1$. We use notation (3.81).

THEOREM 4.5.9. *The irreducible highest weight representation $V(\mu(u))$ of the twisted Yangian $\mathrm{Y}(\mathfrak{o}_{2n+1})$ is finite-dimensional if and only if one of the following two relations holds:*

$$\mu_0(u) \longrightarrow \mu_1(u) \longrightarrow \dots \longrightarrow \mu_n(u), \tag{4.118}$$

$$\frac{u}{u+1/2} \cdot \mu_0(u) \longrightarrow \mu_1(u) \longrightarrow \dots \longrightarrow \mu_n(u). \tag{4.119}$$

PROOF. Suppose first that $\dim V(\mu(u)) < \infty$. Consider the homomorphism $\mathrm{Y}(\mathfrak{o}_3) \to \mathrm{Y}(\mathfrak{o}_{2n+1})$ which sends $s_{ij}(u)$ to the series with the same name in $\mathrm{Y}(\mathfrak{o}_{2n+1})$ for any $i, j \in \{-1, 0, 1\}$. Let $\mathrm{Y}(\mathfrak{o}_3)$ act on $V(\mu(u))$ via this homomorphism. The cyclic span $\mathrm{Y}(\mathfrak{o}_3)\xi$ of the highest vector ξ of $V(\mu(u))$ is a highest weight representation of $\mathrm{Y}(\mathfrak{o}_3)$ with the highest weight $(\mu_0(u), \mu_1(u))$. The irreducible quotient of this representation is finite-dimensional. So, as we have deduced from Proposition 4.5.1, the components of the highest weight $(\mu_0(u), \mu_1(u))$ can be assumed to be given by (4.83). Now, we apply Proposition 4.5.8. If (4.114) holds, then, as we saw in the proof of Theorem 3.3.3, we must have $\mu_0(u) \longrightarrow \mu_1(u)$. Similarly, if (4.115) holds, then $\mu_0(u) \longrightarrow (1 + 1/2\, u^{-1})\, \mu_1(u)$. Together with Proposition 4.2.8 this implies either (4.118) or (4.119).

Conversely, suppose that (4.118) holds. Then

$$\frac{\mu_{i-1}(u)}{\mu_i(u)} = \frac{P_i(u+1)}{P_i(u)}, \qquad i = 1, \dots, n, \tag{4.120}$$

for certain monic polynomials $P_i(u)$. Consider the irreducible highest weight representation $L(\lambda(u))$ of the Yangian $\mathrm{Y}(\mathfrak{gl}_{2n+1})$ with the highest weight $\lambda(u) = (\lambda_{-n}(u), \dots, \lambda_n(u))$ given by

$$\lambda_i(u) = u^{-k} P_1(u) \dots P_i(u)\, P_{i+1}(u+1) \dots P_n(u+1)$$

and

$$\lambda_{-i}(u) = u^{-k} P_1(u+1) \dots P_n(u+1)$$

for $i = 0, \dots, n$, where k denotes the sum of the degrees of the polynomials $P_1(u), \dots, P_n(u)$. By Theorem 3.4.1, the representation $L(\lambda(u))$ of $\mathrm{Y}(\mathfrak{gl}_{2n+1})$ is finite-dimensional. Due to Corollary 4.2.10, the cyclic $\mathrm{Y}(\mathfrak{o}_{2n+1})$-span of the highest vector of $L(\lambda(u))$ is a highest weight representation of $\mathrm{Y}(\mathfrak{o}_{2n+1})$ with the highest weight $\mu'(u) = (\mu'_1(u), \dots, \mu'_n(u))$ where $\mu'_i(u) = \lambda_i(u)\lambda_{-i}(-u)$. So, $\dim V(\mu'(u)) < \infty$. However, the components $\mu'_i(u)$ satisfy (4.120). Since both series $\mu_0(u)$ and $\mu'_0(u)$ are even, there exists an even formal series $g(u)$ such that $\mu_i(u) = g(u)\, \mu'_i(u)$ for all i. Hence, the composition of the representation of $\mathrm{Y}(\mathfrak{o}_{2n+1})$ on $V(\mu'(u))$ and the automorphism (2.17) is isomorphic to $V(\mu(u))$, proving that the latter is also finite-dimensional.

Similarly, if (4.119) holds, then there exist monic polynomials $P_1(u), \dots, P_n(u)$ in u such that (4.120) holds for $i = 2, \dots, n$ and

$$\frac{u}{u+1/2} \cdot \frac{\mu_0(u)}{\mu_1(u)} = \frac{P_1(u+1)}{P_1(u)}.$$

Introduce the series $\lambda_i(u)$ exactly as above and consider the finite-dimensional irreducible representation $L(\lambda(u))$ of $Y(\mathfrak{gl}_{2n+1})$. Furthermore, introduce the finite-dimensional irreducible representation $V(\mu_0)$ of the Lie algebra $\mathfrak{o}_{2n+1}$ with the highest weight $\mu_0 = (-1/2, \dots, -1/2)$ (in fact, $\dim V(\mu_0) = 2^n$, but this will not be used in this argument). Extend $V(\mu_0)$ to a module over the twisted Yangian $Y(\mathfrak{o}_{2n+1})$ via the evaluation homomorphism (2.106) and consider the tensor product module $L(\lambda(u)) \otimes V(\mu_0)$ over $Y(\mathfrak{o}_{2n+1})$. Using formulas (4.20) and Proposition 4.2.9, we derive that the irreducible highest weight representation of $Y(\mathfrak{o}_{2n+1})$ with the highest weight $\mu'(u) = (\mu'_0(u), \dots, \mu'_n(u))$, where

$$\mu'_i(u) = \lambda_i(u)\lambda_{-i}(-u)\,(1 + 1/2\,u^{-1})^{-1}, \qquad i = 1, \dots, n,$$

and $\mu'_0(u) = \lambda_0(u)\lambda_0(-u)$, is finite-dimensional. The argument is now completed in the same way as for the case of (4.118) considered above. □

DEFINITION 4.5.10. The polynomials $P_i(u)$ with $i = 1, \dots, n$ associated with a finite-dimensional $Y(\mathfrak{o}_{2n+1})$-module $V(\mu(u))$ by one of the relations (4.118) or (4.119) are called the *Drinfeld polynomials* of $V(\mu(u))$. □

EXAMPLE 4.5.11. Suppose that the representation $V(\mu)$ of the Lie algebra $\mathfrak{o}_{2n+1}$ is finite-dimensional and extend $V(\mu)$ to a $Y(\mathfrak{o}_{2n+1})$-module via the evaluation homomorphism (2.106). Then the Drinfeld polynomials of the evaluation module $V(\mu)$ defined in Section 4.2 can be easily found from (4.20) and are given by

$$P_i(u) = (u + \mu_i + 1/2)(u + \mu_i + 3/2) \dots (u + \mu_{i-1} - 1/2)$$

for $i = 2, \dots, n$, while $P_1(u)$ is given by different formulas depending on the value of μ_1. We have

$$P_1(u) = (u + \mu_1 + 1/2)(u + \mu_1 + 3/2) \dots (u - 1/2)$$

if $\mu_1 \in -\mathbb{Z}_+$, and

$$P_1(u) = (u + \mu_1 + 1/2)(u + \mu_1 + 3/2) \dots (u - 1)$$

if $\mu_1 \in -1/2 - \mathbb{Z}_+$. □

In accordance with Theorem 4.5.9, the finite-dimensional irreducible representations of the twisted Yangian $Y(\mathfrak{o}_{2n+1})$ are divided into two disjoint families depending on whether the highest weight $\mu(u)$ satisfies (4.118) or (4.119). Using Proposition 4.2.15, we thus obtain the following parametrization of the representations of the special twisted Yangian $SY(\mathfrak{o}_{2n+1})$.

COROLLARY 4.5.12. *Finite-dimensional irreducible representations of the special twisted Yangian* $SY(\mathfrak{o}_{2n+1})$ *are parameterized by the tuples* $\big(P_1(u), \dots, P_n(u)\big)$ *of monic polynomials in* u *and the choice of one of the two families corresponding to the conditions* (4.118) *or* (4.119). □

4.6. Examples

1. Consider the extended Drinfeld Yangian $X^D(\mathfrak{g}_N)$ which was introduced in Example 2.16.2 with the matrix G defined in (4.1). A representation L of the algebra $X^D(\mathfrak{g}_N)$ is called a *highest weight representation* if there exists a nonzero vector $\zeta \in L$ such that L is generated by ζ,

$$\begin{aligned} t_{ij}(u)\,\zeta &= 0 && \text{for} \quad -n \leqslant i < j \leqslant n, \qquad \text{and} \\ t_{ii}(u)\,\zeta &= \lambda_i(u)\,\zeta && \text{for} \quad -n \leqslant i \leqslant n, \end{aligned}$$

for some formal series

$$\lambda_i(u) = 1 + \lambda_i^{(1)}u^{-1} + \lambda_i^{(2)}u^{-2} + \dots, \qquad \lambda_i^{(r)} \in \mathbb{C},$$

where the value $i = 0$ occurs only in the case $\mathfrak{g}_N = \mathfrak{o}_{2n+1}$. The vector ζ is called the *highest vector* of L and the tuple $\lambda(u) = (\lambda_{-n}(u), \dots, \lambda_n(u))$ of the formal series is the *highest weight* of L.

Given any tuple $\lambda(u) = (\lambda_{-n}(u), \dots, \lambda_n(u))$ the *Verma module* $M(\lambda(u))$ is the quotient of $X^D(\mathfrak{g}_N)$ by the left ideal generated by all coefficients of the series $t_{ij}(u)$ with $-n \leqslant i < j \leqslant n$, and $t_{ii}(u) - \lambda_i(u)$ for $i = -n, \dots, n$.

The Verma module $M(\lambda(u))$ is nontrivial if and only if the components of the highest weight $\lambda(u)$ satisfy the conditions

$$\frac{\lambda_{-n+i-1}(u + \kappa - i)}{\lambda_{-n+i}(u + \kappa - i)} = \frac{\lambda_{n-i}(u)}{\lambda_{n-i+1}(u)} \tag{4.121}$$

for $i = 1, \dots, n-1$ if $\mathfrak{g}_N = \mathfrak{o}_{2n}$ or $\mathfrak{sp}_{2n}$, and for $i = 1, \dots, n$ if $\mathfrak{g}_N = \mathfrak{o}_{2n+1}$.

A nontrivial Verma module $M(\lambda(u))$ has a unique irreducible quotient denoted $L(\lambda(u))$. Every finite-dimensional irreducible $X^D(\mathfrak{g}_N)$-module is isomorphic to $L(\lambda(u))$ where the components of $\lambda(u)$ satisfy the conditions (4.121) and there exist monic polynomials $P_1(u), \dots, P_n(u)$ in u such that

$$\frac{\lambda_{i-1}(u)}{\lambda_i(u)} = \frac{P_i(u+1)}{P_i(u)}, \qquad \text{for} \quad i = 2, \dots, n$$

and also

$$\begin{aligned} \frac{\lambda_0(u)}{\lambda_1(u)} &= \frac{P_1(u+1/2)}{P_1(u)}, && \text{if} \quad \mathfrak{g}_N = \mathfrak{o}_{2n+1}, \\ \frac{\lambda_{-1}(u)}{\lambda_1(u)} &= \frac{P_1(u+2)}{P_1(u)}, && \text{if} \quad \mathfrak{g}_N = \mathfrak{sp}_{2n}, \\ \frac{\lambda_{-1}(u)}{\lambda_2(u)} &= \frac{P_1(u+1)}{P_1(u)}, && \text{if} \quad \mathfrak{g}_N = \mathfrak{o}_{2n}. \end{aligned}$$

Conversely, if (4.121) and the above conditions on the highest weight $\lambda(u)$ are satisfied, then $L(\lambda(u))$ is finite-dimensional.

The polynomials $P_1(u), \dots, P_n(u)$ are called the *Drinfeld polynomials* corresponding to the representation $L(\lambda(u))$. Any finite-dimensional irreducible representation of the Drinfeld Yangian $Y^D(\mathfrak{g}_N)$ is isomorphic to the restriction of a certain $X^D(\mathfrak{g}_N)$-module $L(\lambda(u))$ to the subalgebra $Y^D(\mathfrak{g}_N)$. In particular, such representations of $Y^D(\mathfrak{g}_N)$ are parameterized by the n-tuples of monic polynomials $\big(P_1(u), \dots, P_n(u)\big)$.

2. Recall the reflection equation algebra $\mathcal{B}(N, l)$ introduced in Example 2.16.4. A representation V of the algebra $\mathcal{B}(N, l)$ is called a *highest weight representation*

if there exists a nonzero vector $\xi \in V$ such that V is generated by ξ,

$$b_{ij}(u)\,\xi = 0 \qquad \text{for} \quad 1 \leqslant i < j \leqslant N, \qquad \text{and}$$
$$b_{ii}(u)\,\xi = \mu_i(u)\,\xi \qquad \text{for} \quad 1 \leqslant i \leqslant N,$$

for some formal series $\mu_i(u) \in \varepsilon_i + u^{-1}\mathbb{C}[[u^{-1}]]$. The vector ξ is called the *highest vector* of V and the set $\mu(u) = (\mu_1(u), \dots, \mu_N(u))$ is the *highest weight* of V.

Given an arbitrary N-tuple $\mu(u) = (\mu_1(u), \dots, \mu_N(u))$ of such form, we define the *Verma module* $M(\mu(u))$ as the quotient of $\mathcal{B}(N,l)$ by the left ideal generated by all coefficients of the series $b_{ij}(u)$ with $1 \leqslant i < j \leqslant N$, and $b_{ii}(u) - \mu_i(u)$ for $i = 1, \dots, N$.

The Verma module $M(\mu(u))$ is nontrivial if and only if

$$\mu_N(u)\,\mu_N(-u) = 1,$$

and for each $i = 1, \dots, N-1$ the following condition holds:

$$\widetilde{\mu}_i(u)\,\widetilde{\mu}_i(-u+N-i) = \widetilde{\mu}_{i+1}(u)\,\widetilde{\mu}_{i+1}(-u+N-i),$$

where

$$\widetilde{\mu}_i(u) = (2u - N + i)\,\mu_i(u) + \mu_{i+1}(u) + \dots + \mu_N(u).$$

A nontrivial Verma module $M(\mu(u))$ has a unique irreducible quotient denoted $V(\mu(u))$. Every finite-dimensional irreducible $\mathcal{B}(N,l)$-module is isomorphic to $V(\mu(u))$ for some N-tuple $\mu(u)$.

The $\mathcal{B}(N,0)$-module $V(\mu(u))$ is finite-dimensional if and only if there exist monic polynomials $P_1(u), \dots, P_{N-1}(u)$ in u such that $P_i(-u+N-i+1) = P_i(u)$ and

$$\frac{\widetilde{\mu}_i(u)}{\widetilde{\mu}_{i+1}(u)} = \frac{P_i(u+1)}{P_i(u)}, \qquad i = 1, \dots, N-1.$$

The module $V(\mu(u))$ over $\mathcal{B}(N,l)$ with $0 < l \leqslant k$ is finite-dimensional if and only if there exist monic polynomials $P_1(u), \dots, P_{N-1}(u)$ in u and $\gamma \in \mathbb{C}$ such that $P_i(-u+N-i+1) = P_i(u)$ with $P_k(\gamma) \neq 0$ and

$$\frac{\widetilde{\mu}_i(u)}{\widetilde{\mu}_{i+1}(u)} = \frac{P_i(u+1)}{P_i(u)}, \qquad i = 1, \dots, N-1, \quad i \neq k,$$

while

$$\frac{\widetilde{\mu}_k(u)}{\widetilde{\mu}_{k+1}(u)} = \frac{P_k(u+1)}{P_k(u)} \cdot \frac{\gamma - u}{\gamma + u - l}.$$

The polynomials $P_1(u), \dots, P_{N-1}(u)$ are called the *Drinfeld polynomials* corresponding to the finite-dimensional representation $V(\mu(u))$.

3. Consider the twisted quantized enveloping algebra $\mathrm{U}'_q(\mathfrak{sp}_{2n})$ introduced in Example 2.16.8 (the complex number q is nonzero and not a root of unity). Given an n-tuple of nonzero complex numbers $\lambda = (\lambda_1, \lambda_3, \dots, \lambda_{2n-1})$, the corresponding *Verma module* $M(\lambda)$ over $\mathrm{U}'_q(\mathfrak{sp}_{2n})$ is defined as the quotient of $\mathrm{U}'_q(\mathfrak{sp}_{2n})$ by the left ideal generated by the elements

$$s_{ij} \quad \text{with} \quad i = 1, 3, \dots, 2n-1, \quad j = 1, 2, \dots, i,$$

and

$$s_{i,i+1} - \lambda_i \quad \text{with} \quad i = 1, 3, \dots, 2n-1.$$

The Verma module $M(\lambda)$ has a unique irreducible quotient denoted $V(\lambda)$. Every finite-dimensional irreducible representation of $U'_q(\mathfrak{sp}_{2n})$ is isomorphic to $V(\lambda)$ where λ is an n-tuple of the form

$$\lambda = (\sigma_1 \, q^{m_1}, \dots, \sigma_n \, q^{m_n}),$$

with positive integers m_i satisfying $m_1 \leqslant m_2 \leqslant \dots \leqslant m_n$ and each σ_i is 1 or -1. In particular, the isomorphism classes of finite-dimensional irreducible representations of $U'_q(\mathfrak{sp}_{2n})$ are parameterized by such n-tuples.

Bibliographical notes

4.1–4.5. The exposition basically follows the author's papers [**294, 302**], although some modifications in the arguments were made. Apart from the simplest cases of $Y(\mathfrak{sp}_2)$ and $Y(\mathfrak{o}_2)$, the structure of an arbitrary finite-dimensional irreducible representation of $Y(\mathfrak{g}_N)$ largely remains unknown. In particular, no formulas are known for the dimensions or characters of these representations; see e.g. Arakawa [**6**], Brundan and Kleshchev [**53**], and Vasserot [**461**] for such formulas in the $\mathfrak{gl}_N$ case.

4.6. Example 1 is a reformulation of Drinfeld's classification theorem [**99**] for the finite-dimensional irreducible representations of the general Yangians; see also Chari and Pressley [**65**, Chapter 12]. An independent proof based on the R-matrix presentation of the extended Drinfeld Yangian $X^D(\mathfrak{g}_N)$ is given by Arnaudon, Molev and Ragoucy [**14**]. Example 2 is due to Molev and Ragoucy [**319**]. Representations of the twisted super-Yangians were described by Briot and Ragoucy [**48**]. Example 3 is due to the author [**312**]. A classification of the finite-dimensional irreducible representations of the twisted quantized enveloping algebra $U'_q(\mathfrak{o}_N)$ was obtained by Iorgov and Klimyk [**181**]. The result here is more complicated than in the symplectic case since the algebra $U'_q(\mathfrak{o}_N)$ does not have an analogue of the split presentation of the Lie algebra $\mathfrak{o}_N$.

CHAPTER 5

Gelfand–Tsetlin bases for representations of $\mathrm{Y}(\mathfrak{gl}_N)$

An explicit realization of a class of representations of the Yangian $\mathrm{Y}(\mathfrak{gl}_2)$ is provided by Theorem 3.3.8. In this chapter we generalize this construction and provide such a realization for representations of the Yangian $\mathrm{Y}(\mathfrak{gl}_N)$ for any N. As a particular case, we obtain the classical Gelfand–Tsetlin basis for representations of the Lie algebra $\mathfrak{gl}_N$. It will be convenient to work with a certain quotient of the algebra $\mathrm{Y}(\mathfrak{gl}_N)$. We start by describing some algebraic properties of this quotient.

5.1. Yangian of level p

As we saw in the proof of Theorem 3.4.1, given any finite-dimensional irreducible representation of the Yangian $\mathrm{Y}(\mathfrak{gl}_N)$, there exists an automorphism of $\mathrm{Y}(\mathfrak{gl}_N)$ of the form (1.20) such that its composition with the representation is isomorphic to a subquotient of a tensor product module

$$L(\lambda^{(1)}) \otimes \ldots \otimes L(\lambda^{(p)}), \tag{5.1}$$

where $L(\lambda^{(i)})$ is the irreducible representation of $\mathfrak{gl}_N$ with the highest weight $\lambda^{(i)}$. By Proposition 3.2.11, all generators $t_{ij}^{(r)}$ with $r \geqslant p+1$ act on (5.1) as zero operators. This motivates the following definition.

DEFINITION 5.1.1. For any positive integer p, the *Yangian of level* p is the quotient of the algebra $\mathrm{Y}(\mathfrak{gl}_N)$ by the ideal generated by all elements $t_{ij}^{(r)}$ with $r \geqslant p+1$ and $1 \leqslant i,j \leqslant N$. We denote this quotient by $\mathrm{Y}_p(\mathfrak{gl}_N)$. □

Thus, the composition of any finite-dimensional irreducible representation of $\mathrm{Y}(\mathfrak{gl}_N)$ with an appropriate automorphism (1.20) can be regarded as a representation of $\mathrm{Y}_p(\mathfrak{gl}_N)$ for some $p \geqslant 1$. Note that if $p=1$, then the algebra $\mathrm{Y}_1(\mathfrak{gl}_N)$ is isomorphic to the universal enveloping algebra $\mathrm{U}(\mathfrak{gl}_N)$; see Theorem 5.1.2 below.

We will keep the notation $t_{ij}^{(r)}$ for the image of the element $t_{ij}^{(r)} \in \mathrm{Y}(\mathfrak{gl}_N)$ in the quotient $\mathrm{Y}_p(\mathfrak{gl}_N)$. So, $\mathrm{Y}_p(\mathfrak{gl}_N)$ can be regarded as an algebra with generators $t_{ij}^{(r)}$ for $1 \leqslant r \leqslant p$ and $1 \leqslant i,j \leqslant N$, subject to the defining relations

$$(u-v)\,[T_{ij}(u), T_{kl}(v)] = T_{kj}(u)\,T_{il}(v) - T_{kj}(v)\,T_{il}(u), \tag{5.2}$$

where

$$T_{ij}(u) = \delta_{ij}\,u^p + t_{ij}^{(1)}\,u^{p-1} + \cdots + t_{ij}^{(p)}.$$

The following is an analogue of the Poincaré–Birkhoff–Witt theorem for the algebra $\mathrm{Y}_p(\mathfrak{gl}_N)$.

THEOREM 5.1.2. *Given an arbitrary linear ordering on the set of generators $t_{ij}^{(r)}$ with $1 \leqslant r \leqslant p$ and $1 \leqslant i,j \leqslant N$, any element of the algebra $\mathrm{Y}_p(\mathfrak{gl}_N)$ can be uniquely written as a linear combination of ordered monomials in these generators.*

PROOF. First, let us consider a particular linear ordering on the generators of $\mathrm{Y}(\mathfrak{gl}_N)$ defined by setting $t_{ij}^{(r)} \leqslant t_{kl}^{(s)}$ if $(r,i,j) \leqslant (s,k,l)$ in the lexicographical ordering. By Theorem 1.4.1, the ordered monomials

$$t_{i_1 j_1}^{(r_1)} \dots t_{i_q j_q}^{(r_q)}, \qquad t_{i_1 j_1}^{(r_1)} \leqslant \dots \leqslant t_{i_q j_q}^{(r_q)}, \tag{5.3}$$

form a basis in $\mathrm{Y}(\mathfrak{gl}_N)$. Let J_p denote the ideal introduced in Definition 5.1.1. We will show that monomials (5.3) with the additional condition $r_q \leqslant p$ form a basis of $\mathrm{Y}(\mathfrak{gl}_N)$ modulo the ideal J_p. Note that by the definition of the ordering, we have $r_1 \leqslant \dots \leqslant r_q$. Since any element of the ideal J_p is uniquely written as a linear combination of monomials (5.3), it suffices to prove that for each monomial occurring in this linear combination the inequality $r_q > p$ holds.

For each pair of generators $t_{ij}^{(r)} \geqslant t_{kl}^{(s)}$ with $r > p$, in the Yangian $\mathrm{Y}(\mathfrak{gl}_N)$ we have

$$t_{ij}^{(r)} t_{kl}^{(s)} = t_{kl}^{(s)} t_{ij}^{(r)} + \text{a linear combination of } t_{ab}^{(m)} t_{cd}^{(l)} \tag{5.4}$$

$$\text{with } \; t_{ab}^{(m)} < t_{cd}^{(l)}, \quad m + l < r + s, \quad \text{and} \quad l > p.$$

Indeed, by the defining relations (1.4),

$$[t_{ij}^{(r)}, t_{kl}^{(s)}] = \sum_{a=1}^{s} \Big(t_{kj}^{(a-1)} t_{il}^{(r+s-a)} - t_{kj}^{(r+s-a)} t_{il}^{(a-1)} \Big), \tag{5.5}$$

and (5.4) follows by induction on $r+s$. The ideal J_p is spanned by monomials of the form

$$t_{k_1 l_1}^{(s_1)} \dots t_{k_m l_m}^{(s_m)}, \qquad m \geqslant 1, \tag{5.6}$$

such that at least one of the indices $s_1, \dots, s_m$ is greater than p. Choose a maximal generator $t_{k_i l_i}^{(s_i)}$ in this monomial. Then we have $s_i > p$. Move this generator to the extreme right position, permuting it with the other generators using (5.4). Repeating this procedure and applying the induction on $s_1 + \dots + s_m$ we get an expression of the monomial (5.6) as a linear combination of the ordered monomials (5.3) with $r_q > p$. This proves the claim for the chosen ordering.

Finally, regarding (5.5) as a relation in the quotient $\mathrm{Y}_p(\mathfrak{gl}_N)$, we deduce that any two generators $t_{ij}^{(r)}$ and $t_{kl}^{(s)}$ commute modulo products $t_{ab}^{(m)} t_{cd}^{(l)}$ with the indices satisfying $m + l < r + s$. This reduces the proof for an arbitrary ordering on the generators $t_{ij}^{(r)}$ to the particular case considered above. □

Introduce two ascending filtrations on the algebra $\mathrm{Y}_p(\mathfrak{gl}_N)$ by setting

$$\deg t_{ij}^{(r)} = r \qquad \text{and} \qquad \deg' t_{ij}^{(r)} = r - 1, \qquad 1 \leqslant r \leqslant p;$$

cf. Sections 1.4 and 1.5. In order to describe the associated graded algebras $\operatorname{gr} \mathrm{Y}_p(\mathfrak{gl}_N)$ and $\operatorname{gr}' \mathrm{Y}_p(\mathfrak{gl}_N)$, consider the quotient of the polynomial current Lie algebra $\mathfrak{gl}_N[z]$ by the ideal $\sum_{k \geqslant p} \mathfrak{gl}_N z^k$. We denote this quotient by $\mathfrak{gl}_{N,p}$.

COROLLARY 5.1.3. (i) *The algebra* $\operatorname{gr} \mathrm{Y}_p(\mathfrak{gl}_N)$ *is isomorphic to the algebra of polynomials in* $p\,N^2$ *variables.*

(ii) *The algebra* $\operatorname{gr}' \mathrm{Y}_p(\mathfrak{gl}_N)$ *is isomorphic to the universal enveloping algebra* $\mathrm{U}(\mathfrak{gl}_{N,p})$.

PROOF. The graded algebra is commutative, so that the first statement is immediate from Theorem 5.1.2; cf. Corollary 1.4.2. Now let $\bar{t}_{ij}^{(r)}$ denote the image of $t_{ij}^{(r)}$ in the $(r-1)$-th component of $\mathrm{gr}'\,\mathrm{Y}_p(\mathfrak{gl}_N)$. By the defining relations (5.2) we get

$$[\bar{t}_{ij}^{(r)}, \bar{t}_{kl}^{(s)}] = \delta_{kj}\bar{t}_{il}^{(r+s-1)} - \delta_{il}\bar{t}_{kj}^{(r+s-1)},$$

where we assume that $\bar{t}_{ij}^{(r)} = 0$ for $r \geqslant p+1$. Hence the assignment $E_{ij}\, z^{r-1} \mapsto \bar{t}_{ij}^{(r)}$ for $1 \leqslant r \leqslant p$ defines a surjective homomorphism $\mathrm{U}(\mathfrak{gl}_{N,p}) \to \mathrm{gr}'\,\mathrm{Y}_p(\mathfrak{gl}_N)$. Its kernel is trivial by Theorem 5.1.2. □

The polynomial $T_{ij}(u)$ can be regarded as the image of the series $u^p\, t_{ij}(u)$ under the natural epimorphism $\mathrm{Y}(\mathfrak{gl}_N) \to \mathrm{Y}_p(\mathfrak{gl}_N)$. So we can extend the quantum minor formulas of Sections 1.6 and 1.7 to the algebra $\mathrm{Y}_p(\mathfrak{gl}_N)$. We define the quantum minors $T^{a_1\dots a_m}_{b_1\dots b_m}(u)$ by the following equivalent formulas, implied by (1.54) and (1.55),

$$T^{a_1\dots a_m}_{b_1\dots b_m}(u) = \sum_{\sigma\in\mathfrak{S}_m} \operatorname{sgn}\sigma \cdot T_{a_{\sigma(1)}b_1}(u)\dots T_{a_{\sigma(m)}b_m}(u-m+1) \tag{5.7}$$

$$= \sum_{\sigma\in\mathfrak{S}_m} \operatorname{sgn}\sigma \cdot T_{a_1 b_{\sigma(1)}}(u-m+1)\dots T_{a_m b_{\sigma(m)}}(u). \tag{5.8}$$

The *quantum determinant* is the polynomial $D(u) = T^{1\dots N}_{1\dots N}(u)$. Write

$$D(u) = u^{pN} + D_1\, u^{pN-1} + \dots + D_{pN}, \qquad D_k \in \mathrm{Y}_p(\mathfrak{gl}_N).$$

The following is an analogue of Theorem 1.7.5 for the algebra $\mathrm{Y}_p(\mathfrak{gl}_N)$.

THEOREM 5.1.4. *The coefficients $D_1, D_2, \dots, D_{pN}$ of the polynomial $D(u)$ belong to the center of the algebra $\mathrm{Y}_p(\mathfrak{gl}_N)$. Moreover, these elements are algebraically independent and generate the center of $\mathrm{Y}_p(\mathfrak{gl}_N)$.*

PROOF. The first part of the theorem follows from Corollary 1.7.2. For the proof of the second part, we use Corollary 5.1.3(ii). Observe that for any $a \in \mathbb{C}$, the highest degree components of the coefficients of the polynomial $T_{ij}(u-a)$ in the graded algebra $\mathrm{gr}'\,\mathrm{Y}_p(\mathfrak{gl}_N)$ are given by

$$\delta_{ij}\, u^p + \big(\bar{t}_{ij}^{(1)} - p\, a\, \delta_{ij}\big)\, u^{p-1} + \bar{t}_{ij}^{(2)}\, u^{p-2} + \dots + \bar{t}_{ij}^{(p)}.$$

By definition,

$$D(u) = \sum_{\sigma\in\mathfrak{S}_N} \operatorname{sgn}\sigma \cdot T_{\sigma(1),1}(u)\dots T_{\sigma(N),N}(u-N+1). \tag{5.9}$$

For any $k \in \{1, \dots, pN\}$ denote by $\overline{D}_k$ the highest degree component of the coefficient D_k of $D(u)$. We will identify $\mathrm{gr}'\,\mathrm{Y}_p(\mathfrak{gl}_N)$ with $\mathrm{U}(\mathfrak{gl}_{N,p})$ via the isomorphism $\bar{t}_{ij}^{(r)} \mapsto E_{ij}\, z^{r-1}$. Let us set

$$F_{ij}^{(r)} = E_{ij}\, z^r \qquad \text{for} \quad 1 \leqslant r \leqslant p-1, \qquad \text{and}$$
$$F_{ij}^{(0)} = E_{ij} - p\,(j-1)\,\delta_{ij}.$$

The (5.9) implies

$$\overline{D}_k = \sum_{\substack{1\leqslant i_1<\dots<i_s\leqslant N\\ j_1+\dots+j_s=k-s}} \sum_{\sigma\in\mathfrak{S}_s} \operatorname{sgn}\sigma \cdot F_{i_{\sigma(1)}i_1}^{(j_1)} \dots F_{i_{\sigma(s)}i_s}^{(j_s)}, \tag{5.10}$$

where the parameter $s \in \{1, \dots, N\}$ is uniquely determined by the decomposition $k = p\,(s-1)+r$ for $r \in \{1, \dots, p\}$, and the indices j_a run over nonnegative integers. Since the coefficients D_k are central in $\mathrm{Y}_p(\mathfrak{gl}_N)$, the elements $\overline{D}_k$ belong to the center of the universal enveloping algebra $\mathrm{U}(\mathfrak{gl}_{N,p})$. It will be sufficient to prove that the elements $\overline{D}_1, \overline{D}_2, \dots, \overline{D}_{pN}$ are algebraically independent and generate the center $\mathrm{Z}(\mathfrak{gl}_{N,p})$ of $\mathrm{U}(\mathfrak{gl}_{N,p})$. Let $\mathfrak{h}$, $\mathfrak{n}^+$ and $\mathfrak{n}^-$ denote the subalgebras of $\mathfrak{gl}_N$ consisting of all diagonal, upper triangular and lower triangular matrices, respectively. For any subalgebra $\mathfrak{a}$ of $\mathfrak{gl}_N$ denote by $\mathfrak{a}_p$ the corresponding subalgebra of $\mathfrak{gl}_{N,p}$ which is defined as the quotient of $\mathfrak{a}[z]$ by the ideal $\sum_{k\geqslant p} \mathfrak{a} z^k$. Let $\mathrm{U}(\mathfrak{gl}_{N,p})^{\mathfrak{h}}$ denote the centralizer of $\mathfrak{h}$ in $\mathrm{U}(\mathfrak{gl}_{N,p})$. Set

$$\mathrm{L} = \mathrm{U}(\mathfrak{gl}_{N,p})\, \mathfrak{n}_p^+ \cap \mathrm{U}(\mathfrak{gl}_{N,p})^{\mathfrak{h}}.$$

The following properties are implied by the Poincaré–Birkhoff–Witt theorem for $\mathrm{U}(\mathfrak{gl}_{N,p})$,

$$\mathrm{L} = \mathfrak{n}_p^-\, \mathrm{U}(\mathfrak{gl}_{N,p}) \cap \mathrm{U}(\mathfrak{gl}_{N,p})^{\mathfrak{h}};$$

$$\mathrm{L} \ \text{ is a two-sided ideal of } \ \mathrm{U}(\mathfrak{gl}_{N,p})^{\mathfrak{h}};$$

$$\mathrm{U}(\mathfrak{gl}_{N,p})^{\mathfrak{h}} = \mathrm{U}(\mathfrak{h}_p) \oplus \mathrm{L}.$$

Hence, the projection $\chi : \ \mathrm{U}(\mathfrak{gl}_{N,p})^{\mathfrak{h}} \to \mathrm{U}(\mathfrak{h}_p)$ with the kernel L is an algebra homomorphism. This is a generalization of the Harish-Chandra homomorphism for reductive Lie algebras.

Note that the elements $\overline{D}_1, \overline{D}_2, \dots, \overline{D}_{pN}$ belong to $\mathrm{U}(\mathfrak{gl}_{N,p})^{\mathfrak{h}}$. Their algebraic independence will follow from the algebraic independence of their images in $\mathrm{U}(\mathfrak{h}_p)$ under the homomorphism χ. Set

$$x_i^{(r)} = F_{ii}^{(r-1)}, \qquad 1 \leqslant r \leqslant p, \qquad 1 \leqslant i \leqslant N.$$

Then (5.10) implies

$$\chi(\overline{D}_k) = \sum_{\substack{1\leqslant i_1<\dots<i_s\leqslant N \\ j_1+\dots+j_s=k}} x_{i_1}^{(j_1)} \dots x_{i_s}^{(j_s)}, \tag{5.11}$$

where the indices j_a now run over the set of positive integers. Here, as before, we have used the bijection between the set $\{1, \dots, p\,N\}$ and the set of pairs (r, s), where $r \in \{1, \dots, p\}$ and $s \in \{1, \dots, N\}$, which is defined by the decomposition $k = p\,(s-1) + r$ for any $k \in \{1, \dots, p\,N\}$. Let us denote the polynomial (5.11) by $X_s^{(r)}$ and prove that the differentials $dX_s^{(r)}$ are linearly independent.

Note that $X_s^{(p)}$ is the elementary symmetric polynomial of degree s in the variables $x_1^{(p)}, \dots, x_N^{(p)}$. Therefore, the matrix $A = [a_{st}]_{s,t=1}^N$ defined by

$$dX_s^{(p)} = a_{s1}\, dx_1^{(p)} + \dots + a_{sN}\, dx_N^{(p)}$$

is nonsingular. Furthermore, for $1 \leqslant r < p$ we have

$$dX_s^{(r)} = a_{s1}\, dx_1^{(r)} + \dots + a_{sN}\, dx_N^{(r)}$$
$$+ \text{ a linear combination of } \ dx_q^{(m)} \ \text{ with } \ m > r.$$

Let us combine the coefficients of $dx_s^{(r)}$ in the expressions for the differentials $dX_s^{(r)}$ into a matrix and arrange its rows and columns in accordance with the lexicographical ordering on the pairs (r, s). Then the above relations imply that this matrix is block-triangular with p identical diagonal $N \times N$ blocks equal to the matrix A. This proves the linear independence of the differentials $dX_s^{(r)}$ and hence, the algebraic independence of the polynomials $X_s^{(r)}$.

Finally, it follows from a result of Rais and Tauvel [**399**, Théorème 4.5], that the center $\mathrm{Z}(\mathfrak{gl}_{N,p})$ possesses a system of algebraically independent homogeneous generators $P_s^{(r)}$ with $1 \leqslant r \leqslant p$ and $1 \leqslant s \leqslant N$ such that $P_s^{(r)}$ has degree s in the sense of the canonical filtration of the universal enveloping algebra $\mathrm{U}(\mathfrak{gl}_{N,p})$. Since this same property is shared by the element $\overline{D}_k$ with $k = p\,(s-1)+r$, we conclude that $\overline{D}_1, \dots, \overline{D}_{pN}$ are generators of $\mathrm{Z}(\mathfrak{gl}_{N,p})$. □

REMARK 5.1.5. In the course of the proof of Theorem 5.1.4 we constructed a family of algebraically independent generators of the center of the universal enveloping algebra $\mathrm{U}(\mathfrak{gl}_{N,p})$. A different family can be constructed by using an analogue of the series $z(u)$ for the algebra $\mathrm{Y}_p(\mathfrak{gl}_N)$; see Section 1.9. This construction is outlined in Example 5.5.1. □

5.2. Lowering and raising operators

As we pointed out in the beginning of the previous section, the composition of any finite-dimensional irreducible representation of the Yangian $\mathrm{Y}(\mathfrak{gl}_N)$ with an appropriate automorphism of the form (1.20) can be regarded as a representation of the Yangian of level p for some positive integer p. In other words, all components of the highest weight of such representation are polynomials in u^{-1} of degree $\leqslant p$. In this chapter we are working with the polynomials $T_{ij}(u)$ instead of the series $t_{ij}(u)$, and so it will be convenient to regard the highest weight as an N-tuple of monic polynomials in u of degree p, which are the eigenvalues of the highest vector for the action of the polynomials $T_{ii}(u)$ for $i = 1, \dots, N$. We will keep the notation $\lambda(u) = \big(\lambda_1(u), \dots, \lambda_N(u)\big)$ for the highest weight. Thus, rewriting (3.57) and (3.58) in terms of the polynomials $T_{ij}(u)$, for the highest vector ζ of the finite-dimensional representation $L = L(\lambda(u))$ we have

$$T_{ij}(u)\,\zeta = 0 \qquad \text{for} \quad 1 \leqslant i < j \leqslant N, \qquad \text{and} \tag{5.12}$$

$$T_{ii}(u)\,\zeta = \lambda_i(u)\,\zeta \qquad \text{for} \quad 1 \leqslant i \leqslant N, \tag{5.13}$$

where $\lambda_i(u)$ is a monic polynomial in u of degree p. Write

$$\lambda_i(u) = (u + \lambda_i^{(1)})\,(u + \lambda_i^{(2)}) \dots (u + \lambda_i^{(p)}), \qquad i = 1, \dots, N. \tag{5.14}$$

Using Theorem 3.4.1 and recalling the conditions (3.82), we can assume that the roots of the polynomials $\lambda_i(u)$ are numbered in such a way that for any $k \in \{1, \dots, p\}$ we have

$$\lambda_i^{(k)} - \lambda_{i+1}^{(k)} \in \mathbb{Z}_+, \qquad i = 1, \dots, N-1. \tag{5.15}$$

Proposition 3.2.9 then implies that the $\mathrm{Y}(\mathfrak{gl}_N)$-module $L(\lambda(u))$ is isomorphic to a subquotient of the tensor product module (5.1), where $\lambda^{(k)} = (\lambda_1^{(k)}, \dots, \lambda_N^{(k)})$. In particular, we have

$$\dim L(\lambda(u)) \leqslant \prod_{k=1}^{p} \dim L(\lambda^{(k)}). \tag{5.16}$$

Due to Theorem 5.1.2, the subalgebra of $\mathrm{Y}_p(\mathfrak{gl}_N)$ generated by the elements $t_{ij}^{(r)}$ with $1 \leqslant i, j \leqslant N-1$ and $r = 1, \dots, p$, is isomorphic to $\mathrm{Y}_p(\mathfrak{gl}_{N-1})$. We will identify $\mathrm{Y}_p(\mathfrak{gl}_{N-1})$ with this subalgebra.

Let $\mu_i(u)$ with $i = 1, \dots, N-1$ be monic polynomials in u of degree p. A vector $\eta \in L(\lambda(u))$ will be called a *singular vector of weight* $\mu(u) = \big(\mu_1(u), \dots, \mu_{N-1}(u)\big)$ (with respect to the subalgebra $\mathrm{Y}_p(\mathfrak{gl}_{N-1})$) if η satisfies

$$T_{ij}(u)\,\eta = 0 \qquad \text{for} \quad 1 \leqslant i < j \leqslant N-1, \qquad \text{and} \tag{5.17}$$

$$T_{ii}(u)\,\eta = \mu_i(u)\,\eta \qquad \text{for} \quad 1 \leqslant i \leqslant N-1. \tag{5.18}$$

If $\eta \neq 0$, then the cyclic span $\mathrm{Y}_p(\mathfrak{gl}_{N-1})\,\eta$ is a highest weight representation of $\mathrm{Y}_p(\mathfrak{gl}_{N-1})$ with the highest weight $\mu(u)$ and the highest vector η.

Using the quantum minors (5.7), for any indices $i < r$ introduce the polynomials

$$\tau_{ri}(u) = T^{\,i+1 \dots r}_{\,i \dots r-1}(u) \qquad \text{and} \qquad \tau_{ir}(u) = T^{\,1 \dots i}_{\,1 \dots i-1, r}(u). \tag{5.19}$$

The following properties of the elements (5.19) are implied by Corollary 1.7.2,

$$[\tau_{ri}(u), \tau_{rj}(v)] = 0 \qquad \text{and} \qquad [\tau_{ir}(u), \tau_{jr}(v)] = 0 \tag{5.20}$$

for all $i, j < r$.

Lemma 5.2.1. *Let a vector $\eta \in L(\lambda(u))$ be singular of weight*

$$\mu(u) = \big(\mu_1(u), \dots, \mu_{N-1}(u)\big)$$

and let $\mu_k(a) = 0$ for some $a \in \mathbb{C}$ and some $k \in \{1, \dots, N-1\}$. Then the vectors $\tau_{Nk}(a)\,\eta$ and $\tau_{kN}(a+k-1)\,\eta$ are also singular of weights

$$\Big(\mu_1(u), \dots, \frac{u-a-1}{u-a}\,\mu_k(u), \dots, \mu_{N-1}(u)\Big)$$

and

$$\Big(\mu_1(u), \dots, \frac{u-a+1}{u-a}\,\mu_k(u), \dots, \mu_{N-1}(u)\Big),$$

respectively.

Proof. Note that the quantum minor relations of Propositions 1.6.8 and 1.7.1 hold in the same form for the quantum minors (5.7). Therefore, using the first relation of Proposition 1.7.1, we obtain

$$\begin{aligned}(u-a)\,[T_{ij}(u), T^{\,k+1 \dots N}_{\,k \dots N-1}(a)] \\ = \sum_{l=k+1}^{N} T_{lj}(u)\, T^{\,k+1 \dots i \dots N}_{\,k \dots N-1}(a) - \sum_{l=k}^{N-1} T^{\,k+1 \dots N}_{\,k \dots j \dots N-1}(a)\, T_{il}(u),\end{aligned} \tag{5.21}$$

where the indices i and j in the quantum minors on the right-hand side replace the index l, respectively, in the first and second sum.

Now suppose that $1 \leqslant i \leqslant j \leqslant N-1$. If $1 \leqslant i \leqslant k$, then using the second relation of Proposition 1.6.8 and the skew-symmetry property of the quantum minors we can write

$$T^{\,k+1 \dots i \dots N}_{\,k \dots N-1}(a) = \sum_{m=k}^{N-1} (-1)^{l+m-1}\, T^{\,k+1 \dots \widehat{l} \dots N}_{\,k \dots \widehat{m} \dots N-1}(a-1)\, T_{im}(a).$$

However, by the assumption, $T_{im}(a)\,\eta = 0$, so that $T^{\,k+1 \dots i \dots N}_{\,k \dots N-1}(a)\,\eta = 0$. Furthermore, $T_{il}(u)\,\eta = 0$ unless $i = l = k$. Thus, if $1 \leqslant i \leqslant k$, then the right-hand side of (5.21) annihilates η unless $i = j = k$. If the latter is the case, then (5.21) gives

$$(u-a)\,T_{kk}(u)\,T^{\,k+1 \dots N}_{\,k \dots N-1}(a)\,\eta = (u-a-1)\,T^{\,k+1 \dots N}_{\,k \dots N-1}(a)\,T_{kk}(u)\,\eta.$$

If $k+1 \leqslant i \leqslant j$, then $T_{ij}(u)$ commutes with the quantum minor $T^{\,k+1\dots N}_{\,k\dots N-1}(a)$ by Corollary 1.7.2. This completes the proof for the vector $\tau_{Nk}(a)\,\eta$.

The argument for the vector $\tau_{kN}(a+k-1)\,\eta$ is similar. Applying the second relation of Proposition 1.7.1, we obtain

$$
\begin{aligned}
(5.22)\quad (u-a)\,[T_{ij}(u), T^{\,1\dots k}_{\,1\dots k-1,N}(a+k-1)] \\
= \sum_{l=1}^{k} T^{\,1\dots i\dots k}_{\,1\dots k-1,N}(a+k-1)\,T_{lj}(u) - \sum_{l=1}^{k-1} T_{il}(u)\,T^{\,1\dots k}_{\,1\dots j\dots k-1,N}(a+k-1) \\
- T_{iN}(u)\,T^{\,1\dots k}_{\,1\dots k-1,j}(a+k-1),
\end{aligned}
$$

where the indices i and j in the quantum minors on the right-hand side replace the index l, respectively, in the first and second sum.

Now suppose that $1 \leqslant i \leqslant j \leqslant N-1$. If $j \geqslant k$, then, applying the first relation of Proposition 1.6.8 to the quantum minors $T^{\,1\dots k}_{\,1\dots k-1,j}(a+k-1)$ and

$$
T^{\,1\dots k}_{\,1\dots j\dots k-1,N}(a+k-1) = (-1)^{k-l}\,T^{\,1\dots k}_{\,1\dots \widehat{l}\dots k-1,N,j}(a+k-1),
$$

we derive that the right-hand side of (5.22) annihilates η unless $i=j=k$. In the latter case, (5.22) gives

$$
\begin{aligned}
(u-a)\,T_{kk}(u)\,T^{\,1\dots k}_{\,1\dots k-1,N}(a+k-1)\,\eta \\
= (u-a+1)\,T^{\,1\dots k}_{\,1\dots k-1,N}(a+k-1)\,T_{kk}(u)\,\eta.
\end{aligned}
$$

Finally, if $i \leqslant j \leqslant k-1$, then the operator $T_{ij}(u)$ commutes with the quantum minor $T^{\,1\dots k}_{\,1\dots k-1,N}(a+k-1)$ by Corollary 1.7.2, completing the proof. □

We will call the operators $\tau_{Nk}(a)$ and $\tau_{kN}(a+k-1)$ introduced in Lemma 5.2.1, the *lowering* and *raising operators*, respectively.

We extend the notation (3.81) to pairs of monic polynomials in u of degree p. Namely, if $\lambda(u)$ and $\mu(u)$ are two such polynomials, then we write $\lambda(u) \longrightarrow \mu(u)$ if there exists a monic polynomial $P(u)$ in u such that

$$
\frac{\lambda(u)}{\mu(u)} = \frac{P(u+1)}{P(u)}.
$$

Equivalently, $\lambda(u) \longrightarrow \mu(u)$ if there exist decompositions

$$
\lambda(u) = (u+\alpha_1)\dots(u+\alpha_p), \qquad \mu(u) = (u+\beta_1)\dots(u+\beta_p),
$$

such that $\alpha_i - \beta_i \in \mathbb{Z}_+$ for all $i = 1,\dots,p$; see the proof of Theorem 3.3.3.

From now on, we will only consider a certain class of representations of $\mathrm{Y}_p(\mathfrak{gl}_N)$ by imposing a *generality condition* on the highest weights of the representations $L(\lambda(u))$. Namely, we will assume that

$$
(5.23)\qquad \lambda_i^{(k)} - \lambda_j^{(m)} \notin \mathbb{Z}, \qquad \text{for all } i,j \text{ and all } k \neq m.
$$

Suppose that $\mu(u) = (\mu_1(u),\dots,\mu_{N-1}(u))$ is a tuple of monic polynomials in u of degree p satisfying

$$
\lambda_i(u) \longrightarrow \mu_i(u), \qquad i = 1,\dots,N-1.
$$

By the generality condition, this implies that there exist uniquely determined decompositions

$$
(5.24)\qquad \mu_i(u) = (u+\mu_i^{(1)})\,(u+\mu_i^{(2)})\dots(u+\mu_i^{(p)}), \qquad i = 1,\dots,N-1,
$$

such that $\lambda_i^{(k)} - \mu_i^{(k)} \in \mathbb{Z}_+$ for all i and k. Let us define the corresponding vector $\zeta_\mu \in L(\lambda(u))$ by the formula

$$\zeta_\mu = \prod_{i=1}^{N-1} \prod_{k=1}^{p} \Big(\tau_{Ni}(-\mu_i^{(k)} - 1) \dots \tau_{Ni}(-\lambda_i^{(k)} + 1)\, \tau_{Ni}(-\lambda_i^{(k)}) \Big)\, \zeta. \tag{5.25}$$

Lemma 5.2.1 implies that the vector ζ_μ is singular of weight $\mu(u)$.

LEMMA 5.2.2. *Suppose that for some* $i \in \{1, \dots, N-1\}$ *and* $k \in \{1, \dots, p\}$ *the condition* $\lambda_i^{(k)} - \mu_i^{(k)} \geqslant 1$ *holds. Then we have*

$$\tau_{iN}(-\mu_i^{(k)} + i - 1)\, \zeta_\mu$$
$$= (-1)^{N-i} \lambda_1(-\mu_i^{(k)} + i - 1) \dots \lambda_N(-\mu_i^{(k)} + i - N)\, \zeta_{\mu+\delta_i^{(k)}},$$

where $\zeta_{\mu+\delta_i^{(k)}}$ *corresponds to the tuple of polynomials obtained from the tuple* $\mu(u)$ *by replacing* $\mu_i^{(k)}$ *by* $\mu_i^{(k)} + 1$.

PROOF. Applying (1.50) with $q = (1, \dots, i-1, N, i, \dots, N-1)$, we get

$$D(u)\, \zeta_{\mu+\delta_i^{(k)}} = (-1)^{N-i} \sum_{\sigma \in \mathfrak{S}_N} \operatorname{sgn} \sigma \cdot T_{\sigma(1),1}(u) \dots T_{\sigma(i-1),i-1}(u - i + 2)$$
$$\times T_{\sigma(i),N}(u - i + 1)\, T_{\sigma(i+1),i}(u - i) \dots T_{\sigma(N),N-1}(u - N + 1)\, \zeta_{\mu+\delta_i^{(k)}}.$$

Using the defining relations (5.2) we deduce that if $\sigma(l) < i$ for some $l \geqslant i+1$ or $\sigma(l) = i$ for some $l \geqslant i+2$, then the corresponding summand on the right-hand side of the formula is zero. Similarly, if $\sigma(i+1) = i$, then the value of the corresponding summand at $u = -\mu_i^{(k)} + i - 1$ is also zero, because $T_{ii}(-\mu_i^{(k)} + i - 1)\, \zeta_{\mu+\delta_i^{(k)}} = 0$. Hence, for this particular value of u the summation over $\sigma \in \mathfrak{S}_N$ in the above formula can be replaced with the summation over $\sigma = \sigma_1 \sigma_2$, were σ_1 and σ_2 run over the permutations of the sets $\{1, \dots, i\}$ and $\{i+1, \dots, N\}$, respectively. This implies

$$D(-\mu_i^{(k)} + i - 1)\, \zeta_{\mu+\delta_i^{(k)}} = (-1)^{N-i} \tau_{iN}(-\mu_i^{(k)} + i - 1)\, \tau_{Ni}(-\mu_i^{(k)} - 1)\, \zeta_{\mu+\delta_i^{(k)}}$$
$$= (-1)^{N-i} \tau_{iN}(-\mu_i^{(k)} + i - 1)\, \zeta_\mu,$$

where the second equality is implied by (5.20).

On the other hand, by Proposition 3.2.5, for the action of the quantum determinant $D(u)$ on the vector $\zeta_{\mu+\delta_i^{(k)}}$ we have

$$D(u)\, \zeta_{\mu+\delta_i^{(k)}} = \lambda_1(u) \dots \lambda_N(u - N + 1)\, \zeta_{\mu+\delta_i^{(k)}},$$

completing the proof. □

We will say that a tuple $\mu(u) = \big(\mu_1(u), \dots, \mu_{N-1}(u)\big)$ of monic polynomials in u of degree p satisfies the *betweenness conditions*, if the following relations hold:

$$\lambda_i(u) \longrightarrow \mu_i(u) \longrightarrow \lambda_{i+1}(u), \qquad i = 1, \dots, N-1. \tag{5.26}$$

We assume, as before, that (5.15) and the generality condition (5.23) hold.

COROLLARY 5.2.3. *If a tuple* $\mu(u)$ *satisfies the betweenness conditions* (5.26), *then the corresponding vector* $\zeta_\mu \in L(\lambda(u))$ *is nonzero.*

PROOF. The betweenness conditions mean that there are uniquely determined decompositions (5.24) such that

$$\lambda_i^{(k)} - \mu_i^{(k)} \in \mathbb{Z}_+ \qquad \text{and} \qquad \mu_i^{(k)} - \lambda_{i+1}^{(k)} \in \mathbb{Z}_+.$$

Therefore, if the assumption of Lemma 5.2.2 holds then applying the operator $\tau_{iN}(-\mu_i^{(k)} + i - 1)$ to the vector ζ_μ we obtain $\zeta_{\mu+\delta_i^{(k)}}$ with a nonzero coefficient. So, applying such operators repeatedly, we can get the highest vector ζ of $L(\lambda(u))$ with a nonzero coefficient. Thus, $\zeta_\mu \neq 0$. □

The *Gelfand–Tsetlin pattern* $\Lambda(u)$ (associated with the highest weight $\lambda(u)$) is an array of monic polynomials in u of degree p of the form

$$\begin{array}{ccccccc} \lambda_{N1}(u) & & \lambda_{N2}(u) & & \cdots & & \lambda_{NN}(u) \\ & \lambda_{N-1,1}(u) & & \cdots & & \lambda_{N-1,N-1}(u) & \\ & & \cdots & & \cdots & & \\ & & \lambda_{21}(u) & & \lambda_{22}(u) & & \\ & & & \lambda_{11}(u) & & & \end{array}$$

where $\lambda_{Ni}(u) = \lambda_i(u)$ for $i = 1, \dots, N$, so that the top row coincides with $\lambda(u)$, and the following conditions hold

$$\lambda_{r+1,i}(u) \longrightarrow \lambda_{ri}(u) \longrightarrow \lambda_{r+1,i+1}(u) \quad \text{for} \quad r = 1, \dots, N-1 \quad \text{and} \quad i = 1, \dots, r.$$

In other words, any two successive rows of the pattern obey the betweenness conditions (5.26). Due to (5.15) and the generality condition (5.23), there exist uniquely determined decompositions

$$\lambda_{ri}(u) = (u + \lambda_{ri}^{(1)}) \dots (u + \lambda_{ri}^{(p)}), \qquad 1 \leqslant i \leqslant r \leqslant N, \tag{5.27}$$

such that $\lambda_{Ni}^{(k)} = \lambda_i^{(k)}$,

$$\lambda_{r+1,i}^{(k)} - \lambda_{ri}^{(k)} \in \mathbb{Z}_+ \qquad \text{and} \qquad \lambda_{ri}^{(k)} - \lambda_{r+1,i+1}^{(k)} \in \mathbb{Z}_+ \tag{5.28}$$

for $k = 1, \dots, p$ and $1 \leqslant i \leqslant r \leqslant N-1$. For any pattern $\Lambda(u)$ introduce the corresponding vector $\zeta_\Lambda \in L(\lambda(u))$ by

$$\zeta_\Lambda = \prod_{r=2,\dots,N}^{\longrightarrow} \prod_{i=1}^{r-1} \prod_{k=1}^{p} \Big(\tau_{ri}(-\lambda_{r-1,i}^{(k)} - 1) \dots \tau_{ri}(-\lambda_{ri}^{(k)} + 1)\, \tau_{ri}(-\lambda_{ri}^{(k)}) \Big)\, \zeta.$$

A basis of Gelfand–Tsetlin type for the representation $L(\lambda(u))$ is provided by the following theorem.

THEOREM 5.2.4. *The vectors ζ_Λ parameterized by all patterns $\Lambda(u)$ associated with the highest weight $\lambda(u)$, form a basis of the representation $L(\lambda(u))$.*

PROOF. Observe that the factor corresponding to $r = N$ in the formula for ζ_Λ gives a singular vector with respect to the subalgebra $\mathrm{Y}_p(\mathfrak{gl}_{N-1})$ whose weight is the row $\big(\lambda_{N-1,1}(u), \dots, \lambda_{N-1,N-1}(u)\big)$ of the pattern $\Lambda(u)$. By Corollary 5.2.3, this vector is nonzero. Similarly, using the subalgebras of the chain

$$\mathrm{Y}_p(\mathfrak{gl}_1) \subset \mathrm{Y}_p(\mathfrak{gl}_2) \subset \dots \subset \mathrm{Y}_p(\mathfrak{gl}_N), \tag{5.29}$$

we deduce by induction that each vector ζ_Λ is nonzero. As in Section 1.13, consider the Gelfand–Tsetlin subalgebra of $\mathrm{Y}_p(\mathfrak{gl}_N)$ which is generated by the centers of the subalgebras $\mathrm{Y}_p(\mathfrak{gl}_r)$ of the chain (5.29). Equivalently, by Theorem 5.1.4, the Gelfand–Tsetlin subalgebra is generated by all coefficients of the polynomials $T^{1\dots r}_{1\dots r}(u)$ for $r = 1, \dots, N$. Moreover, by the same theorem, the lowering operators $\tau_{si}(v)$ commute with the minor $T^{1\dots r}_{1\dots r}(u)$ for all $s \leqslant r$. Since this minor coincides with the quantum determinant for the subalgebra $\mathrm{Y}_p(\mathfrak{gl}_r)$, arguing as in the proof of Proposition 3.2.5, we find that

$$T^{1\dots r}_{1\dots r}(u)\,\zeta_\Lambda = \lambda_{r1}(u)\dots\lambda_{rr}(u-r+1)\,\zeta_\Lambda. \tag{5.30}$$

So, ζ_Λ is an eigenvector for all operators $T^{1\dots r}_{1\dots r}(u)$ with distinct sets of eigenvalues. Indeed, the row $\big(\lambda_{N-1,1}(u), \dots, \lambda_{N-1,N-1}(u)\big)$ of the pattern $\Lambda(u)$ is uniquely determined by the eigenvalue of the operator $T^{1\dots N-1}_{1\dots N-1}(u)$ and the betweenness conditions for the rows of the pattern. Then by induction the entire pattern $\Lambda(u)$ is uniquely determined by the set of eigenvalues of the operators $T^{1\dots r}_{1\dots r}(u)$ with $r = 1, \dots, N-1$. This proves that the vectors ζ_Λ are linearly independent.

We complete the argument by counting the number of the vectors ζ_Λ and comparing it with $\dim L(\lambda(u))$. Observe that by conditions (5.28), for each fixed value of $k \in \{1, \dots, p\}$ the set of parameters $\big(\lambda^{(k)}_{ri}\big)$ with $1 \leqslant i \leqslant r \leqslant N$ forms a Gelfand–Tsetlin pattern associated with the highest weight $\lambda^{(k)}$ of the irreducible representation $L(\lambda^{(k)})$ of the Lie algebra $\mathfrak{gl}_N$ (see e.g. Section 5.4 below). This shows that the total number of patterns $\Lambda(u)$ coincides with the product of dimensions $\dim L(\lambda^{(k)})$ for $k = 1, \dots, N$. Comparing this with (5.16), we conclude that the number of patterns coincides with $\dim L(\lambda(u))$. □

The following is the branching rule for the restriction $\mathrm{Y}_p(\mathfrak{gl}_N) \downarrow \mathrm{Y}_p(\mathfrak{gl}_{N-1})$.

Corollary 5.2.5. *The restriction of the $\mathrm{Y}_p(\mathfrak{gl}_N)$-module $L(\lambda(u))$ to the subalgebra $\mathrm{Y}_p(\mathfrak{gl}_{N-1})$ is isomorphic to the direct sum of irreducible highest weight $\mathrm{Y}_p(\mathfrak{gl}_{N-1})$-modules $L'(\mu(u))$,*

$$L(\lambda(u))|_{\mathrm{Y}_p(\mathfrak{gl}_{N-1})} \cong \bigoplus_{\mu(u)} L'(\mu(u)),$$

where $\mu(u)$ runs over all tuples of monic polynomials $\mu(u) = \big(\mu_1(u), \dots, \mu_{N-1}(u)\big)$ of degree p satisfying the betweenness conditions (5.26).

Proof. Given a tuple $\mu(u)$ satisfying the betweenness conditions, consider the vector ζ_μ defined in (5.25). By Corollary 5.2.3, the vector ζ_μ is nonzero. As was shown in the proof of Theorem 5.2.4, the vectors ζ_μ are eigenvectors for the operator $T^{1\dots N-1}_{1\dots N-1}(u)$ with different eigenvalues. This implies that all pairwise intersections of the cyclic spans $\mathrm{Y}_p(\mathfrak{gl}_{N-1})\,\zeta_\mu$ are zero. Therefore, the restriction of $L(\lambda(u))$ to the subalgebra $\mathrm{Y}_p(\mathfrak{gl}_{N-1})$ contains the direct sum of these cyclic spans. On the other hand, by Theorem 5.2.4, we have the identity

$$\sum_{\mu(u)} \dim L'(\mu(u)) = \dim L(\lambda(u)),$$

summed over all tuples of monic polynomials $\mu(u) = \big(\mu_1(u), \dots, \mu_{N-1}(u)\big)$ of degree p satisfying the betweenness conditions (5.26). Since $L'(\mu(u))$ is isomorphic to a quotient of $\mathrm{Y}_p(\mathfrak{gl}_{N-1})\,\zeta_\mu$, this identity implies that the $\mathrm{Y}_p(\mathfrak{gl}_{N-1})$-module

$\mathrm{Y}_p(\mathfrak{gl}_{N-1})\,\zeta_\mu$ is irreducible. Hence this module is isomorphic to $L'(\mu(u))$ thus completing the proof. □

5.3. Action of the Drinfeld generators

In Section 3.1 we showed that the Yangian $\mathrm{Y}(\mathfrak{gl}_N)$ is generated by the coefficients of the series $h_r(u)$, $e_r(u)$ and $f_r(u)$; see also Section 1.11. Now we introduce a closely related family of generators by

$$a_r(u) = t^{\,1\dots r}_{\,1\dots r}(u), \qquad b_r(u) = t^{\,1\dots r}_{\,1\dots r-1,\,r+1}(u), \qquad c_r(u) = t^{\,1\dots r-1,\,r+1}_{\,1\dots r}(u). \tag{5.31}$$

PROPOSITION 5.3.1. *The coefficients of the series $a_r(u)$ for $r = 1,\dots,N$ and the series $b_r(u)$ and $c_r(u)$ for $r = 1,\dots,N-1$ generate the algebra $\mathrm{Y}(\mathfrak{gl}_N)$.*

PROOF. Recalling that $e_r(u) = e_{r,r+1}(u)$ and $f_r(u) = f_{r+1,r}(u)$, we find from Theorem 1.11.6 that

$$\begin{aligned} h_r(u) &= a_r(u+r-1)\,a_{r-1}(u+r-1)^{-1},\\ f_r(u) &= c_r(u+r-1)\,a_r(u+r-1)^{-1},\\ e_r(u) &= a_r(u+r-1)^{-1}\,b_r(u+r-1). \end{aligned}$$

The claim now follows from Theorem 3.1.5. □

LEMMA 5.3.2. *For any $1 \leqslant r \leqslant N-1$ we have the relations in $\mathrm{Y}(\mathfrak{gl}_N)$,*

$$a_r(u)\,c_r(v) = \frac{u-v-1}{u-v}\,c_r(v)\,a_r(u) + \frac{1}{u-v}\,c_r(u)\,a_r(v) \tag{5.32}$$

and

$$c_r(u-1)\,b_r(u) = d_r(u-1)\,a_r(u) - a_{r-1}(u-1)\,a_{r+1}(u), \tag{5.33}$$

where $a_0(u) = 1$ and $d_r(u) = t^{\,1\dots r-1,\,r+1}_{\,1\dots r-1,\,r+1}(u)$.

PROOF. Both relations follow from the quantum Sylvester theorem (Theorem 1.12.1) for the values $M = r-1$ and $N = 2$; relation (5.32) follows by the application of the homomorphism (1.93) to the defining relations in the Yangian $\mathrm{Y}(\mathfrak{gl}_2)$ while (5.33) is the identity for the quantum determinant of $T^\sharp(u)$. □

Returning to the Yangian of level p, introduce the polynomials with coefficients in $\mathrm{Y}_p(\mathfrak{gl}_N)$ by

$$A_r(u) = T^{\,1\dots r}_{\,1\dots r}(u), \qquad B_r(u) = T^{\,1\dots r}_{\,1\dots r-1,\,r+1}(u), \qquad C_r(u) = T^{\,1\dots r-1,\,r+1}_{\,1\dots r}(u). \tag{5.34}$$

COROLLARY 5.3.3. *The coefficients of the polynomials $A_r(u)$ for $r = 1,\dots,N$ and the polynomials $B_r(u)$ and $C_r(u)$ for $r = 1,\dots,N-1$ generate the algebra $\mathrm{Y}_p(\mathfrak{gl}_N)$.*

PROOF. Recalling that $T_{ij}(u) = u^p\,t_{ij}(u)$, we find that in the algebra $\mathrm{Y}_p(\mathfrak{gl}_N)$,

$$A_r(u) = \Big(u(u-1)\dots(u-r+1)\Big)^p\,a_r(u).$$

This implies that each coefficient of the series $a_r(u)$ is a linear combination of the coefficients of the polynomial $A_r(u)$. Similarly, each coefficient of the series $b_r(u)$ or $c_r(u)$ is a linear combination of the coefficients of the polynomial $B_r(u)$ or $C_r(u)$, respectively. The statement now follows from Proposition 5.3.1. □

We now describe the action of the operators $A_r(u)$, $B_r(u)$ and $C_r(u)$ on the basis $\{\zeta_\Lambda\}$ of the $\mathrm{Y}_p(\mathfrak{gl}_N)$-module $L(\lambda(u))$ provided by Theorem 5.2.4. Since $B_r(u)$ and $C_r(u)$ are polynomials in u of degree $< pr$, it suffices to find the values of these polynomials at pr different values of u. The polynomial can then be calculated by the Lagrange interpolation formula. For these values we take the numbers

$$l_{ri}^{(k)} = \lambda_{ri}^{(k)} - i + 1, \qquad k = 1, \dots, p \quad \text{and} \quad i = 1, \dots, r,$$

where the parameters $\lambda_{ri}^{(k)}$ are defined in (5.27). Set $\zeta_\Lambda = 0$ if the array $\Lambda(u)$ is not a pattern.

THEOREM 5.3.4. *We have*

$$A_r(u)\,\zeta_\Lambda = \lambda_{r1}(u)\dots\lambda_{rr}(u-r+1)\,\zeta_\Lambda, \tag{5.35}$$

for $r = 1, \dots, N$, *and*

$$B_r(-l_{ri}^{(k)})\,\zeta_\Lambda = -\lambda_{r+1,1}(-l_{ri}^{(k)})\dots\lambda_{r+1,r+1}(-l_{ri}^{(k)} - r)\,\zeta_{\Lambda+\delta_{ri}^{(k)}}, \tag{5.36}$$

$$C_r(-l_{ri}^{(k)})\,\zeta_\Lambda = \lambda_{r-1,1}(-l_{ri}^{(k)})\dots\lambda_{r-1,r-1}(-l_{ri}^{(k)} - r + 2)\,\zeta_{\Lambda-\delta_{ri}^{(k)}}, \tag{5.37}$$

for $r = 1, \dots, N-1$, *where* $\zeta_{\Lambda\pm\delta_{ri}^{(k)}}$ *corresponds to the pattern obtained from* $\Lambda(u)$ *by replacing* $\lambda_{ri}^{(k)}$ *by* $\lambda_{ri}^{(k)} \pm 1$, *and* $\lambda_{00}(u) \equiv 1$.

PROOF. The relation (5.35) was established in the proof of Theorem 5.2.4; see (5.30). For the proof of (5.36) observe that by Corollary 1.7.2, the operator $B_r(u)$ commutes with the lowering operators $\tau_{si}(v)$ for all $s \leqslant r$. Therefore, without loss of generality, we may only consider the case $r = N-1$.

Using the skew-symmetry property of the quantum minors and (1.55), for any $i \in \{1, \dots, N-1\}$ we can write

$$T^{\,1\dots\, N-1}_{\,1\dots\, N-2,\, N}(u) = \sum_q \operatorname{sgn} q \cdot T^{\,i+1\dots\, N-1}_{\,q_{i+1}\dots\, q_{N-1}}(u-i)\, T^{\,1\dots\, i}_{\,q_1\dots\, q_i}(u), \tag{5.38}$$

summed over the permutations $q = (q_1, \dots, q_{N-1})$ of the set $\{1, \dots, N-2, N\}$ such that $q_1 < \dots < q_i$ and $q_{i+1} < \dots < q_{N-1}$. Now we set $u = -l_{N-1,i}^{(k)}$ and apply the operator $B_{N-1}(u) = T^{\,1\dots\, N-1}_{\,1\dots\, N-2,\, N}(u)$ to the vector ζ_μ defined in (5.25), where $\mu(u)$ is the tuple $\big(\lambda_{N-1,1}(u), \dots, \lambda_{N-1,N-1}(u)\big)$. We claim that if $i \leqslant q_j \leqslant N-2$ for some $j \in \{1, \dots, i\}$, then

$$T^{\,1\dots\, i}_{\,q_1\dots\, q_i}(-l_{N-1,i}^{(k)})\,\zeta_\mu = 0.$$

Indeed, the vector ζ_μ is singular of weight $\mu(u)$, and so the claim follows by taking $b_m = q_j$ in the first formula of Proposition 1.6.8 and using the relation

$$T_{ii}(-\lambda_{N-1,i}^{(k)})\,\zeta_\mu = 0.$$

Therefore, the contribution of the term corresponding to a permutation q on the right-hand side of (5.38) is zero unless $q = (1, \dots, i-1, N, i, \dots, N-2)$. Hence

$$B_{N-1}(-l_{N-1,i}^{(k)})\,\zeta_\mu = (-1)^{N-i-1}\,\tau_{N-1,i}(-\lambda_{N-1,i}^{(k)} - 1)\,\tau_{iN}(-\lambda_{N-1,i}^{(k)} + i - 1)\,\zeta_\mu,$$

assuming $\tau_{N-1,i}(u) \equiv 1$ for $i = N-1$. Together with Lemma 5.2.2 this proves (5.36) in the case where $\lambda_{Ni}^{(k)} - \lambda_{N-1,i}^{(k)} \geqslant 1$. If $\lambda_{Ni}^{(k)} = \lambda_{N-1,i}^{(k)}$, then by Lemma 5.2.1 the vector $\tau_{iN}(-\lambda_{N-1,i}^{(k)} + i - 1)\,\zeta_\mu$ is singular, and its weight does not satisfy the betweenness conditions (5.26). Thus, this vector is zero by Corollary 5.2.5. This completes the proof of (5.36).

For the proof of (5.37) we apply Lemma 5.3.2. Suppose first that in the pattern $\Lambda(u)$ we have $\lambda_{ri}^{(k)} - \lambda_{r+1,i+1}^{(k)} \geqslant 1$. This implies that the vector $\zeta_{\Lambda-\delta_{ri}^{(k)}}$ is nonzero. Apply the operator $C_r(u-1)\, B_r(u)$ with $u = -l_{ri}^{(k)} + 1$ to this vector. By (5.33), we have

$$C_r(u-1)\, B_r(u) = D_r(u-1)\, A_r(u) - A_{r-1}(u-1)\, A_{r+1}(u),$$

where $A_0(u) = 1$ and $D_r(u) = T^{1\dots r-1,\, r+1}_{1\dots r-1,\, r+1}(u)$. Now (5.37) follows from (5.35) and (5.36) as

$$A_r(-l_{ri}^{(k)} + 1)\, \zeta_{\Lambda-\delta_{ri}^{(k)}} = 0.$$

It remains to show that if $\lambda_{ri}^{(k)} = \lambda_{r+1,i+1}^{(k)}$, then $C_r(-l_{ri}^{(k)})\, \zeta_\Lambda = 0$. Observe that the operator $A_s(u)$ commutes with $C_r(v)$ for any $s \neq r$ due to Corollary 1.7.2. Furthermore, by (5.32) we have

$$A_r(u)\, C_r(v) = \frac{u-v-1}{u-v}\, C_r(v)\, A_r(u) + \frac{1}{u-v}\, C_r(u)\, A_r(v).$$

Hence, using (5.35) we find that $C_r(-l_{ri}^{(k)})\, \zeta_\Lambda$ is a common eigenvector for all operators $A_s(u)$ with $s = 1, \dots, N$. So, if the vector $C_r(-l_{ri}^{(k)})\, \zeta_\Lambda$ is nonzero then it must be proportional to a certain basis vector of $L(\lambda(u))$. However, this is impossible because its eigenvalue with respect to the action of the operator $A_r(u)$ is given by

$$\frac{u + l_{ri}^{(k)} - 1}{u + l_{ri}^{(k)}}\, \lambda_{r1}(u) \dots \lambda_{rr}(u - r + 1),$$

thus completing the proof. □

REMARK 5.3.5. The generality condition (5.23) can be replaced by a less restrictive condition on the highest weight $\lambda(u)$ for the construction of the Gelfand–Tsetlin basis to remain valid; cf. Theorem 3.3.8. Namely, let the decompositions (5.14) be given so that conditions (5.15) hold. Then (5.23) can be replaced by the following condition:

$$\begin{aligned} &\text{if} \quad \lambda_i^{(k)} - \lambda_j^{(m)} \in \mathbb{Z} \qquad \text{for some } i, j \quad \text{and} \quad k \neq m, \\ &\text{then either} \quad \lambda_1^{(k)} - \lambda_N^{(m)} + N - 1 \leqslant 0 \quad \text{or} \quad \lambda_1^{(m)} - \lambda_N^{(k)} + N - 1 \leqslant 0. \end{aligned}$$

Theorems 5.2.4, 5.3.4 and Corollary 5.2.5 will then hold in the same form and the same arguments for their proofs apply. □

5.4. Gelfand–Tsetlin bases for representations of $\mathfrak{gl}_N$

As we pointed out in Section 5.1, the Yangian of level 1 is isomorphic to the universal enveloping algebra, $\mathrm{Y}_1(\mathfrak{gl}_N) \cong \mathrm{U}(\mathfrak{gl}_N)$. We will identify $\mathrm{Y}_1(\mathfrak{gl}_N)$ with $\mathrm{U}(\mathfrak{gl}_N)$ via the isomorphism given by $t_{ij}^{(1)} \mapsto E_{ij}$. The results of the previous sections thus provide a construction of the classical Gelfand–Tsetlin bases for representations of the Lie algebra $\mathfrak{gl}_N$. As before, we let $L(\lambda)$ denote a finite-dimensional irreducible representation of the Lie algebra $\mathfrak{gl}_N$ with the highest weight $\lambda = (\lambda_1, \dots, \lambda_N)$ so that $\lambda_i - \lambda_{i+1} \in \mathbb{Z}_+$ for all $i = 1, \dots, N-1$. Omitting the upper indices of the parameters $\lambda^{(1)}_{ri}$ we can define the *Gelfand–Tsetlin pattern* Λ (associated with the highest weight λ) is an array of row vectors of the form

$$\begin{array}{ccccccccc} \lambda_{N1} & & \lambda_{N2} & & \cdots & & & \lambda_{NN} & \\ & \lambda_{N-1,1} & & \cdots & & & \lambda_{N-1,N-1} & & \\ & & \cdots & & \cdots & & \cdots & & \\ & & & \lambda_{21} & & \lambda_{22} & & & \\ & & & & \lambda_{11} & & & & \end{array}$$

where $\lambda_{Ni} = \lambda_i$ for $i = 1, \dots, N$, so that the top row coincides with λ, and the following betweenness conditions hold

$$\lambda_{r+1,i} - \lambda_{ri} \in \mathbb{Z}_+ \qquad \text{and} \qquad \lambda_{ri} - \lambda_{r+1,i+1} \in \mathbb{Z}_+ \qquad \text{for} \quad 1 \leqslant i \leqslant r \leqslant N-1.$$

It can be shown directly that the number of patterns associated with a given highest weight λ coincides with the dimension of $L(\lambda)$ provided by the Weyl dimension formula. This formula is the only fact from the representation theory of $\mathfrak{gl}_N$ used in the proof of Theorem 5.2.4.

Now we have $T_{ij}(u) = \delta_{ij}\, u + E_{ij}$, and the quantum minors are defined by (5.7). Let ζ denote the highest vector of $L(\lambda)$. For any pattern Λ the corresponding vector $\zeta_\Lambda \in L(\lambda)$ is defined by

$$\zeta_\Lambda = \prod_{r=2,\dots,N}^{\longrightarrow} \prod_{i=1}^{r-1} \tau_{ri}(-\lambda_{r-1,i} - 1) \dots \tau_{ri}(-\lambda_{ri} + 1)\, \tau_{ri}(-\lambda_{ri})\, \zeta,$$

where the lowering operators $\tau_{ri}(u)$ are given in (5.19).

THEOREM 5.4.1. *The vectors ζ_Λ parameterized by all patterns Λ associated with the highest weight λ, form a basis of the representation $L(\lambda)$. Moreover, the action of generators of $\mathfrak{gl}_N$ in this basis is given by the formulas*

$$E_{rr}\, \zeta_\Lambda = \left(\sum_{i=1}^{r} \lambda_{ri} - \sum_{i=1}^{r-1} \lambda_{r-1,i} \right) \zeta_\Lambda, \tag{5.39}$$

$$E_{r,r+1}\, \zeta_\Lambda = -\sum_{i=1}^{r} \frac{(l_{r+1,1} - l_{ri}) \dots (l_{r+1,r+1} - l_{ri})}{(l_{r1} - l_{ri}) \dots \wedge_i \dots (l_{rr} - l_{ri})}\, \zeta_{\Lambda+\delta_{ri}}, \tag{5.40}$$

$$E_{r+1,r}\, \zeta_\Lambda = \sum_{i=1}^{r} \frac{(l_{r-1,1} - l_{ri}) \dots (l_{r-1,r-1} - l_{ri})}{(l_{r1} - l_{ri}) \dots \wedge_i \dots (l_{rr} - l_{ri})}\, \zeta_{\Lambda-\delta_{ri}}, \tag{5.41}$$

where $l_{ri} = \lambda_{ri} - i + 1$ and the symbol $\wedge_i$ indicates that the i-th factor in the denominator is skipped.

PROOF. The first statement is a reformulation of Theorem 5.2.4 for the case $p = 1$. For the proof of the second statement we use Theorem 5.3.4 which gives

$$A_r(u)\, \zeta_\Lambda = (u + l_{r1}) \dots (u + l_{rr})\, \zeta_\Lambda, \tag{5.42}$$

for $r = 1, \dots, N$, and

$$B_r(-l_{ri})\,\zeta_\Lambda = -\prod_{j=1}^{r+1} (l_{r+1,j} - l_{ri})\,\zeta_{\Lambda+\delta_{ri}}, \tag{5.43}$$

$$C_r(-l_{ri})\,\zeta_\Lambda = \prod_{j=1}^{r-1} (l_{r-1,j} - l_{ri})\,\zeta_{\Lambda-\delta_{ri}}, \tag{5.44}$$

for $r = 1, \dots, N-1$. By definition, $A_r(u)$ is a monic polynomial in u of degree r while the coefficient of u^{r-1} equals

$$E_{11} + \dots + E_{rr} - r(r-1)/2.$$

So, (5.39) follows from (5.42). Similarly, $B_r(u)$ and $C_r(u)$ are polynomials in u of degree $r-1$ and their coefficients of u^{r-1} are respectively equal to $E_{r,r+1}$ and $E_{r+1,r}$. By the Lagrange interpolation formula,

$$B_r(u) = \sum_{i=1}^{r} \frac{(u + l_{r1}) \dots \wedge_i \dots (u + l_{rr})}{(l_{r1} - l_{ri}) \dots \wedge_i \dots (l_{rr} - l_{ri})}\, B_r(-l_{ri}).$$

Thus, (5.43) implies (5.40). Similarly, (5.44) implies (5.41). □

Since the operator $C_r(-l_{ri})$ takes the basis vector ζ_Λ to another basis vector with a constant factor, it is possible to construct the Gelfand–Tsetlin basis of $L(\lambda)$ by applying these operators repeatedly to the highest vector ζ. For each pattern Λ define the vector $\kappa_\Lambda \in L(\lambda)$ by

$$\begin{aligned}\kappa_\Lambda = \prod_{r=1,\dots,N-1}^{\longrightarrow} \Big\{ & C_{N-1}(-l_{N-1,r} - 1) \dots C_{N-1}(-l_r + 1)\, C_{N-1}(-l_r) \\ & \times C_{N-2}(-l_{N-2,r} - 1) \dots C_{N-2}(-l_r + 1)\, C_{N-2}(-l_r) \\ & \qquad \times \dots \times C_r(-l_{rr} - 1) \dots C_r(-l_r + 1)\, C_r(-l_r) \Big\}\, \zeta,\end{aligned}$$

where we set $l_i = \lambda_i - i + 1$.

COROLLARY 5.4.2. *The vectors κ_Λ with Λ running over all patterns associated with λ form a basis of $L(\lambda)$.*

PROOF. We have $\kappa_\Lambda = K_\Lambda\, \zeta_\Lambda$ for a constant K_Λ whose value equals the product of the coefficients of the basis vectors ζ_Λ occurring in (5.44). Since all these coefficients are nonzero, then so is K_Λ, thus proving the claim. □

Consider the chain of natural Lie algebra embeddings

$$\mathfrak{gl}_1 \subset \mathfrak{gl}_2 \subset \dots \subset \mathfrak{gl}_N. \tag{5.45}$$

The *Gelfand–Tsetlin subalgebra* $\mathcal{Z}_N$ of $\mathrm{U}(\mathfrak{gl}_N)$ is the subalgebra generated by the centers $\mathrm{Z}(\mathfrak{gl}_m)$ of the universal enveloping algebras $\mathfrak{gl}_m$ of the chain (5.45) for all $m = 1, \dots, N$. Clearly, the subalgebra $\mathcal{Z}_N$ is commutative.

COROLLARY 5.4.3. *Each basis vector ζ_Λ is a common eigenvector for the action of the elements of the Gelfand–Tsetlin subalgebra $\mathcal{Z}_N$.*

PROOF. The construction of the basis $\{\zeta_\Lambda\}$ of $L(\lambda)$ is consistent with the consecutive restrictions of $L(\lambda)$ to the subalgebras of the chain (5.45). Therefore, the claim follows from the fact that the elements of $Z(\mathfrak{gl}_N)$ act in $L(\lambda)$ as multiplication by scalars. □

The polynomial $A_N(u)$ coincides with the Capelli determinant $\mathcal{C}(u)$ which will be introduced in Section 7.1. It is well known that the coefficients of $\mathcal{C}(u)$ generate the center of $U(\mathfrak{gl}_N)$; see also Theorem 7.1.1 below. Therefore, the Gelfand–Tsetlin subalgebra $\mathcal{Z}_N$ is generated by the coefficients of all polynomials $A_r(u)$ with $r = 1, \dots, N$. The corresponding eigenvalues of the basis vectors ζ_Λ with respect to these operators are provided by (5.42). Several other families of generators of the subalgebra $\mathcal{Z}_N$ will be constructed in Section 7.3.

5.5. Examples

1. In the notation of Section 5.1, for any $0 \leqslant m \leqslant p-1$ introduce the matrix $E^{(m)} = [E_{ij}^{(m)}]_{i,j=1}^N$, where $E_{ij}^{(m)} = E_{ij}\, z^m$. Represent any positive integer k in the form $k = p\,(s-1) + r$ with $s \geqslant 1$ and $1 \leqslant r \leqslant p$ and define the element θ_k by the formula

$$\theta_k = \sum_{r_1+\dots+r_s=k} r_s \cdot \operatorname{tr} E^{(r_1-1)} \dots E^{(r_s-1)},$$

where each r_i runs over the set $\{1, \dots, p\}$. Then all the elements θ_k belong to the center $Z(\mathfrak{gl}_{N,p})$ of the algebra $U(\mathfrak{gl}_{N,p})$. Moreover, the family $\{\theta_1, \dots, \theta_{pN}\}$ is algebraically independent and generates $Z(\mathfrak{gl}_{N,p})$.

In particular, in the case $p = 1$ the elements $\theta_k = \operatorname{tr} E^k$ with $E = E^{(0)}$ are the well known Gelfand invariants for the Lie algebra $\mathfrak{gl}_N$; see also Corollary 7.1.4 below.

2. The following generalization of Kostant's theorem for the Yangian of level p and the universal enveloping algebra $U(\mathfrak{gl}_{N,p})$ holds (see Section 5.1): each algebra $Y_p(\mathfrak{gl}_N)$ and $U(\mathfrak{gl}_{N,p})$ is a free module over its center.

3. Let Γ be the Gelfand–Tsetlin subalgebra of $Y_p(\mathfrak{gl}_N)$ which is generated by the coefficients of the polynomials $A_1(u), \dots, A_N(u)$.

Open problem: Is $Y_p(\mathfrak{gl}_N)$ a free left module over Γ? The answer is affirmative for $p = 1$ (i.e. for $U(\mathfrak{gl}_N)$) and for $N = 2$ and arbitrary p.

4. Consider the finite-dimensional irreducible representation $L(\lambda)$ of the quantized enveloping algebra $U_q(\mathfrak{gl}_N)$; see Example 3.5.5. This representation admits a q-analogue of the Gelfand–Tsetlin basis; cf. Theorem 5.4.1. In order to formulate the result, introduce another set of generators of $U_q(\mathfrak{gl}_N)$ by the formulas

$$t_r = t_{rr}, \qquad t_r^{-1} = \bar{t}_{rr}, \qquad e_r = -\frac{\bar{t}_{r,r+1}\, t_{rr}}{q - q^{-1}}, \qquad f_r = \frac{\bar{t}_{rr}\, t_{r+1,r}}{q - q^{-1}}.$$

For any integer m we set

$$[m] = \frac{q^m - q^{-m}}{q - q^{-1}}.$$

Define the Gelfand–Tsetlin pattern Λ (associated with λ) in the same way as in Section 5.4. There exists a basis $\{\zeta_\Lambda\}$ of $L(\lambda)$ parameterized by the patterns Λ such that the action of the generators of $\mathrm{U}_q(\mathfrak{gl}_N)$ is given by

$$t_r\,\zeta_\Lambda = q^{w_r}\,\zeta_\Lambda, \qquad w_r = \sum_{i=1}^{r} \lambda_{ri} - \sum_{i=1}^{r-1} \lambda_{r-1,i},$$

$$e_r\,\zeta_\Lambda = -\sum_{i=1}^{r} \frac{[\,l_{r+1,1} - l_{ri}]\dots[\,l_{r+1,r+1} - l_{ri}]}{[\,l_{r1} - l_{ri}]\dots\wedge_i\dots[\,l_{rr} - l_{ri}]}\,\zeta_{\Lambda+\delta_{ri}},$$

$$f_r\,\zeta_\Lambda = \sum_{i=1}^{r} \frac{[\,l_{r-1,1} - l_{ri}]\dots[\,l_{r-1,r-1} - l_{ri}]}{[\,l_{r1} - l_{ri}]\dots\wedge_i\dots[\,l_{rr} - l_{ri}]}\,\zeta_{\Lambda-\delta_{ri}}.$$

5. *Open problem*: Describe the composition factors for the restriction of an arbitrary irreducible finite-dimensional representation of $\mathrm{Y}(\mathfrak{gl}_N)$ to the subalgebra $\mathrm{Y}(\mathfrak{gl}_{N-1})$; cf. Corollary 5.2.5. A satisfactory solution of this problem would lead to a character formula for the irreducible finite-dimensional representations of $\mathrm{Y}(\mathfrak{gl}_N)$.

Bibliographical notes

5.1. The Yangian of level p was introduced by Cherednik [**74**]. The proof of the Poincaré–Birkhoff–Witt theorem follows the author's paper [**299**]. Another proof was given by Brundan and Kleshchev [**51**]. The first part of Theorem 5.1.4 is implied by the results of Kulish and Sklyanin [**251**] while the second part is due to Cherednik [**73**]; we have followed [**299**] where a different proof was given. The construction of the generalized Harish-Chandra homomorphism used in the proof of this theorem is due to Geoffriau [**138**]; cf. Dixmier [**89**, Chapter 7.4]. The Yangians of level p are a particular case of a more general class of algebras called the 'shifted Yangians' which were introduced by Brundan and Kleshchev [**52, 53**].

5.2–5.3. The exposition basically follows the author's paper [**295**]. Bases of Gelfand–Tsetlin type for a wider class of *tame* Yangian modules were constructed by Nazarov and Tarasov [**351**].

5.4. Up to a re-normalization, the basis of Theorem 5.4.1 is the original basis of Gelfand and Tsetlin (often spelt as Zetlin or Cetlin) [**134**]. Several different methods to construct this basis were developed in the literature (the original paper [**134**] does not contain a derivation of the formulas); see Baird and Biedenharn [**19**], Zhelobenko [**480, 481, 484**], Nagel and Moshinsky [**333**], Hou Pei-yu [**175**], Asherova, Smirnov and Tolstoy [**16**], Gould [**145**]. The Yangians were used for the construction of the basis by Nazarov and Tarasov [**350**] and the author [**295**]. In particular, the paper [**350**] contains an independent construction of the Gelfand–Tsetlin basis vectors κ_Λ with the use of the Drinfeld generators (Corollary 5.4.2). A discussion of various approaches to the construction of the Gelfand–Tsetlin basis and some historical remarks can be found in the review article [**310**].

5.5. Example 1 is due to the author [**299**]. A more general construction is due to Brown and Brundan [**50**], where a family of central elements for the shifted Yangians was produced; see also Brundan and Kleshchev [**52**]. The generalization of Kostant's theorem in Example 2 is due to Futorny and Ovsienko [**121**]. The particular cases of Example 3 are due to Ovsienko [**380**] ($p = 1$) and Futorny, Molev and Ovsienko [**120**] ($N = 2$). In a broader context of a new class of 'Galois algebras' these examples were considered by Futorny and Ovsienko [**122, 123**]. The q-analogue of the Gelfand–Tsetlin basis in Example 4 was constructed in different methods by several authors; see Jimbo [**199**], Ueno, Takebayashi and

Shibukawa [**449**], Nazarov and Tarasov [**350**], Tolstoy [**444**], Molev, Tolstoy and Zhang [**324**].

CHAPTER 6

Tensor products of evaluation modules for $\mathrm{Y}(\mathfrak{gl}_N)$

As we saw in Section 3.3, tensor products of evaluation modules of the Yangian $\mathrm{Y}(\mathfrak{sl}_2)$ exhaust all its finite-dimensional irreducible representations; see Corollary 3.3.4. The extension of this claim to the Yangians $\mathrm{Y}(\mathfrak{sl}_N)$ with $N \geqslant 3$ is false; see Remark 3.4.10. Nevertheless, those tensor products of evaluation modules which are irreducible provide explicit realizations of such modules. In this chapter we formulate explicit necessary and sufficient conditions for irreducibility of tensor products of evaluation modules for $\mathrm{Y}(\mathfrak{gl}_N)$ (and hence for $\mathrm{Y}(\mathfrak{sl}_N)$). The conditions for multiple tensor products will be reduced to the case of two factors via the *binary property*. In the case $N = 2$ the irreducibility criterion coincides with the one provided by Corollary 3.3.6. We start by considering tensor products of two evaluation modules.

6.1. Sufficient conditions

As before, by a $\mathfrak{gl}_N$-highest weight we will mean a tuple of complex numbers $\lambda = (\lambda_1, \ldots, \lambda_N)$ such that $\lambda_i - \lambda_{i+1} \in \mathbb{Z}_+$ for all $i = 1, \ldots, N-1$. Given another $\mathfrak{gl}_N$-highest weight $\mu = (\mu_1, \ldots, \mu_N)$ we will use the notation

$$l_i = \lambda_i - i + 1, \qquad m_i = \mu_i - i + 1, \qquad i = 1, \ldots, N.$$

For each pair of indices $1 \leqslant i < j \leqslant N$ introduce the corresponding subsets of $\mathbb{C}$ by

$$\langle\, l_j, l_i \,\rangle = \{l_j, l_j + 1, \ldots, l_i\} \setminus \{l_j, l_{j-1}, \ldots, l_i\},$$
$$\langle\, m_j, m_i \,\rangle = \{m_j, m_j + 1, \ldots, m_i\} \setminus \{m_j, m_{j-1}, \ldots, m_i\}.$$

In particular, if $\lambda_i = \lambda_{i+1} = \cdots = \lambda_j$, then $\langle\, l_j, l_i \,\rangle = \emptyset$. Recall the evaluation modules over $\mathrm{Y}(\mathfrak{gl}_N)$ of the form $L(\lambda)$ and their tensor products defined in Section 3.2.

THEOREM 6.1.1. *Let λ and μ be $\mathfrak{gl}_N$-highest weights. Suppose that for each pair of indices $1 \leqslant i < j \leqslant N$ we have*

$$m_j, m_i \notin \langle\, l_j, l_i \,\rangle \qquad \text{or} \qquad l_j, l_i \notin \langle\, m_j, m_i \,\rangle. \tag{6.1}$$

Then the $\mathrm{Y}(\mathfrak{gl}_N)$-module $L(\lambda) \otimes L(\mu)$ is irreducible.

PROOF. We will argue by induction on N. If $N = 1$, then the statement is trivial, as both $L(\lambda)$ and $L(\mu)$ are one-dimensional. For $N = 2$ the statement is a particular case of Corollary 3.3.6 although we do not need to use it. Suppose that $N \geqslant 2$ and let K be a nonzero $\mathrm{Y}(\mathfrak{gl}_N)$-submodule of $L(\lambda) \otimes L(\mu)$. Then K contains a nonzero vector ξ such that ξ is an eigenvector for the coefficients of all series $t_{ii}(u)$ for $i = 1, \ldots, N$ and

$$t_{ij}(u)\, \xi = 0, \qquad 1 \leqslant i < j \leqslant N.$$

Indeed, introduce the subspace K^0 of K as in (3.63) and then argue exactly as in the proof of Theorem 3.2.7. In an equivalent form, the conditions on ξ can be formulated as follows: ξ is annihilated by all coefficients of the series $b_1(u), \dots, b_{N-1}(u)$ and ξ is an eigenvector for each of the $a_1(u), \dots, a_N(u)$; see (5.31) for the definition of these series. The equivalence of the conditions is implied by Proposition 3.2.2 and the proof of Proposition 5.3.1.

Now let ζ and ζ' denote the highest vectors of the $\mathfrak{gl}_N$-modules $L(\lambda)$ and $L(\mu)$, respectively. The key part of the proof of the theorem is to show that ξ is proportional to $\zeta \otimes \zeta'$, that is,

$$\xi = \text{const} \cdot \zeta \otimes \zeta'. \tag{6.2}$$

Then considering dual modules we will also show that the vector $\zeta \otimes \zeta'$ is cyclic in the $\mathrm{Y}(\mathfrak{gl}_N)$-module $L(\lambda) \otimes L(\mu)$.

Observe that due to Proposition 3.2.10, the $\mathrm{Y}(\mathfrak{gl}_N)$-modules $L(\lambda) \otimes L(\mu)$ and $L(\mu) \otimes L(\lambda)$ are reducible or irreducible simultaneously. Hence, exchanging λ and μ if necessary, we may assume without loss of generality that for the pair (i, j) with $i = 1$ and $j = N$ the condition

$$m_1, m_N \notin \langle\, l_N, l_1 \,\rangle \tag{6.3}$$

holds. Consider the Gelfand–Tsetlin basis $\{\zeta_\Lambda\}$ of the $\mathfrak{gl}_N$-module $L(\lambda)$; see Section 5.4. The vector ξ is uniquely written in the form

$$\xi = \sum_{\Lambda} \zeta_\Lambda \otimes \eta_\Lambda, \tag{6.4}$$

summed over all patterns Λ associated with λ, and $\eta_\Lambda \in L(\mu)$.

We will be using the standard partial ordering on the weights of $L(\lambda)$ introduced in Section 3.2. Given two weights $v, w \in \mathfrak{h}^*$, we will write $v \prec w$ if $w - v$ is a $\mathbb{Z}_+$-linear combination of the simple roots $\varepsilon_i - \varepsilon_{i+1}$. It is convenient to reformulate this condition in an equivalent form: $v \prec w$ if and only if

$$w - v = p_1\, \varepsilon_1 + \dots + p_N\, \varepsilon_N, \tag{6.5}$$

with

$$p_1,\ p_1 + p_2,\ \dots,\ p_1 + \dots + p_{N-1} \in \mathbb{Z}_+, \qquad p_1 + \dots + p_N = 0.$$

The vector ξ is a $\mathfrak{gl}_N$-highest vector with respect to the natural action of $\mathfrak{gl}_N$ on $L(\lambda) \otimes L(\mu)$. This action coincides with the one obtained via the embedding (1.6). In particular, ξ is a weight vector. Since the basis $\{\zeta_\Lambda\}$ consists of weight vectors, each element $\eta_\Lambda \in L(\mu)$ in (6.4) is also a $\mathfrak{gl}_N$-weight vector. Moreover, all elements $\zeta_\Lambda \otimes \eta_\Lambda$ in (6.4) have the same $\mathfrak{gl}_N$-weight. We will denote the weight of the vector ζ_Λ by $w(\Lambda)$. Due to (5.39), we have

$$w(\Lambda) = w_1\, \varepsilon_1 + \dots + w_N\, \varepsilon_N, \qquad w_r = \sum_{i=1}^{r} \lambda_{ri} - \sum_{i=1}^{r-1} \lambda_{r-1,i}. \tag{6.6}$$

We will say that a pattern Λ *occurs* in the expansion (6.4) if $\eta_\Lambda \neq 0$. Consider the set of patterns occurring in (6.4) and suppose that Λ^0 is a minimal element of this set with respect to the partial ordering on the weights $w(\Lambda)$. In other words, if Λ occurs in (6.4) and $w(\Lambda) \prec w(\Lambda^0)$, then $w(\Lambda) = w(\Lambda^0)$. We will denote the entries of the pattern Λ^0 by λ^0_{ki}.

LEMMA 6.1.2. *The vector η_{Λ^0} coincides with ζ' up to a constant factor.*

PROOF. It follows from (1.54) that the vector ξ is annihilated by the quantum minors $t^{1\dots m}_{1\dots m-1,m+1}(u)$ for $m=1,\dots,N-1$. Therefore, by Proposition 1.6.9,

$$\sum_{c_1<\dots<c_m}\sum_{\Lambda} t^{1\dots m}_{c_1\dots c_m}(u)\,\zeta_\Lambda\otimes t^{c_1\dots c_m}_{1\dots m-1,m+1}(u)\,\eta_\Lambda=0. \tag{6.7}$$

Expand $t^{1\dots m}_{c_1\dots c_m}(u)\,\zeta_\Lambda$ as a linear combination of the basis vectors ζ_Λ and take the coefficient of ζ_{Λ^0} in (6.7). By (1.54) and (3.60), the weight of each coefficient of the series $t^{1\dots m}_{c_1\dots c_m}(u)\,\zeta_\Lambda$ equals

$$w'=w(\Lambda)+\varepsilon_1+\dots+\varepsilon_m-\varepsilon_{c_1}-\dots-\varepsilon_{c_m}.$$

Therefore $w'\succ w(\Lambda)$, and $w'=w(\Lambda)$ if and only if $c_i=i$ for each i. Since the weight of Λ^0 is minimal, the vector ζ_{Λ^0} can occur only in the expansion of

$$t^{1\dots m}_{1\dots m}(u)\,\zeta_{\Lambda^0}=a_m(u)\,\zeta_{\Lambda^0}. \tag{6.8}$$

Hence, (5.42) implies

$$t^{1\dots m}_{1\dots m-1,m+1}(u)\,\eta_{\Lambda^0}=b_m(u)\,\eta_{\Lambda^0}=0$$

for each $m=1,\dots,N-1$. Thus, η_{Λ^0} is a highest vector of $L(\mu)$, and so η_{Λ^0} coincides with ζ' up to a constant factor. □

LEMMA 6.1.3. *The pattern Λ^0 is determined uniquely. Hence, if a pattern Λ occurs in* (6.4), *then $w(\Lambda)\succ w(\Lambda^0)$.*

PROOF. We have

$$a_m(u)\,\xi=\alpha_m(u)\,\xi,\qquad m=1,\dots,N \tag{6.9}$$

for certain formal series $\alpha_m(u)$ in u^{-1}. On the other hand, Proposition 1.6.9 implies

$$a_m(u)\,\xi=\sum_{c_1<\dots<c_m}\sum_{\Lambda} t^{1\dots m}_{c_1\dots c_m}(u)\,\zeta_\Lambda\otimes t^{c_1\dots c_m}_{1\dots m}(u)\,\eta_\Lambda. \tag{6.10}$$

As we have seen in the proof of Lemma 6.1.2, the vector ζ_{Λ^0} can occur in (6.10) only in the expansion of (6.8). By (5.42) and Lemma 6.1.2, comparing the coefficients of ζ_{Λ^0}, we find that

$$\alpha_m(u)=\frac{(u+l^0_{m1})\dots(u+l^0_{mm})}{u(u-1)\dots(u-m+1)}\,\beta_m(u), \tag{6.11}$$

where $l^0_{mi}=\lambda^0_{mi}-i+1$ and the series $\beta_m(u)$ is defined by $a_m(u)\,\zeta'=\beta_m(u)\,\zeta'$. The equation (6.11) uniquely determines the parameters l^0_{mi}, since $l^0_{mi}-l^0_{m,i+1}$ is a positive integer for each i. Thus, the pattern Λ^0 is also determined uniquely. □

LEMMA 6.1.4. *For each entry λ_{ki} of any pattern Λ occurring in* (6.4) *we have*

$$\lambda_{ki}-\lambda^0_{ki}\in\mathbb{Z}_+\qquad \textit{for all}\quad 1\leqslant i\leqslant k\leqslant N-1.$$

PROOF. If Λ occurs in (6.4), then $w(\Lambda)\succ w(\Lambda^0)$ by Lemma 6.1.3. We use induction on $w(\Lambda)$. Fix $\Lambda\neq\Lambda^0$. Then there exists another pattern Λ' occurring in (6.4) such that $w(\Lambda^0)\prec w(\Lambda')\prec w(\Lambda)$ with $\Lambda'\neq\Lambda$, and for some m and some indices $c_1<\dots<c_m$ the expansion of $t^{1\dots m}_{c_1\dots c_m}(u)\,\zeta_{\Lambda'}$ contains ζ_Λ with a nonzero coefficient. Indeed, if this is not the case, then considering the coefficient of ζ_Λ in (6.7) we come to the conclusion that $b_m(u)\,\eta_\Lambda=0$ for all $m=1,\dots,N-1$, and so η_Λ is proportional to the highest vector of $L(\mu)$: see the proof of Lemma 6.1.2. This implies that Λ and Λ^0 must have the same weight. Due to Lemma 6.1.3, we have to conclude that $\Lambda=\Lambda^0$, a contradiction.

By Proposition 1.7.1, the operator $t^{1\dots m}_{c_1\dots c_m}(u)$ can be represented as the commutator

$$t^{1\dots m}_{c_1\dots c_m}(u) = [\dots[[t^{1\dots m}_{1\dots m}(u), E_{mc_m}], E_{m-1,c_{m-1}}], \dots, E_{pc_p}], \tag{6.12}$$

where p is the minimum index i such that $c_i \neq i$. Here, as before, we identify the elements E_{ij} and $t^{(1)}_{ij}$ using the embedding (1.6). The operator $t^{1\dots m}_{1\dots m}(u)$ acts on the basis vectors ζ_Λ by scalar multiplication; see (5.42). Furthermore, E_{ij} with $i<j$ can be written as a commutator of some elements $E_{k,k+1}$ with $k \in \{1,\dots,N-1\}$. By the Gelfand–Tsetlin formulas (5.40), the vector $E_{k,k+1}\,\zeta_{\Lambda'}$ is a linear combination of the basis vectors $\zeta_{\Lambda'+\delta_{ki}}$, where $i=1,\dots,k$. The proof is completed by the application of the induction hypothesis to the pattern Λ'. □

LEMMA 6.1.5. *In the pattern Λ^0 we have $\lambda^0_{N-1,i} = \lambda_i$ for all $i=1,\dots,N-1$.*

PROOF. We will be proving that for every $r=1,\dots,N-1$ we have the property: if $i \geqslant r$ and

$$\lambda^0_{N-1,i} = \lambda^0_{N-2,i-1} = \dots = \lambda^0_{N-r,i-r+1},$$

then either $\lambda^0_{N-1,i} = \lambda_i$ or $i \geqslant r+1$ and

$$\lambda^0_{N-1,i} = \lambda^0_{N-2,i-1} = \dots = \lambda^0_{N-r,i-r+1} = \lambda^0_{N-r-1,i-r}.$$

This clearly implies the statement. Arguing by contradiction, consider the minimum value of r for which the property fails. That is, there exists $i \geqslant r$ such that $\Lambda' := \Lambda^0 + \delta_{N-1,i} + \dots + \delta_{N-r,\,i-r+1}$ is a pattern. By (1.54) we have

$$t^{1\dots N-r}_{1\dots N-r-1,N}(u)\,\xi = 0. \tag{6.13}$$

Applying Proposition 1.6.9 once again, we come to

$$\sum_{c_1<\dots<c_{N-r}} \sum_{\Lambda} t^{1\dots N-r}_{c_1\dots c_{N-r}}(u)\,\zeta_\Lambda \otimes t^{c_1\dots c_{N-r}}_{1\dots N-r-1,N}(u)\,\eta_\Lambda = 0. \tag{6.14}$$

The coefficient of the vector $\zeta_{\Lambda'} \otimes \eta_{\Lambda^0}$ in the expansion of the left-hand side of (6.14) must be 0. Let us determine which patterns Λ yield a nontrivial contribution to this coefficient. Considering the weight of $t^{1\dots N-r}_{c_1\dots c_{N-r}}(u)\,\zeta_\Lambda$ we come to the relation

$$w(\Lambda) + \varepsilon_1 + \dots + \varepsilon_{N-r} - \varepsilon_{c_1} - \dots - \varepsilon_{c_{N-r}} = w(\Lambda^0) + \varepsilon_{N-r} - \varepsilon_N,$$

and hence

$$w(\Lambda) - w(\Lambda^0) = \varepsilon_{c_1} + \dots + \varepsilon_{c_{N-r}} - \varepsilon_1 - \dots - \varepsilon_{N-r-1} - \varepsilon_N.$$

Since $w(\Lambda) \succ w(\Lambda^0)$ by Lemma 6.1.3, we obtain from (6.5) that $c_i = i$ for all values $i=1,\dots,N-r-1$ while $c_{N-r} \in \{N-r,\dots,N\}$. Then $w(\Lambda) = w(\Lambda^0) + \varepsilon_{c_{N-r}} - \varepsilon_N$. By Lemma 6.1.4, Λ is obtained from Λ^0 by increasing exactly one entry by 1 in each of the rows $c_{N-r}, c_{N-r}+1, \dots, N-1$. On the other hand, by the minimality of r and the betweenness conditions for the patterns, the array Λ would not be a pattern unless $c_{N-r} = N-r$ or $c_{N-r} = N$. Thus, the coefficient in question can have a contribution from only two summands in (6.14), namely,

$$t^{1\dots N-r}_{1\dots N-r-1,N}(u)\,\zeta_{\Lambda^0} \otimes t^{1\dots N-r-1,N}_{1\dots N-r-1,N}(u)\,\eta_{\Lambda^0} \tag{6.15}$$

and

$$t^{1\dots N-r}_{1\dots N-r}(u)\,\zeta_{\Lambda'} \otimes t^{1\dots N-r}_{1\dots N-r-1,N}(u)\,\eta_{\Lambda'}. \tag{6.16}$$

We consider (6.15) first. Proposition 1.7.1 implies

$$t^{1\dots N-r}_{1\dots N-r-1,N}(u) = [t^{1\dots N-r}_{1\dots N-r}(u), E_{N-r,N}]. \tag{6.17}$$

As we have observed above, the minimality of r and the betweenness conditions imply that $E_{iN}\,\zeta_{\Lambda^0} = 0$ for $N-r < i \leqslant N-1$. Hence,

$$E_{N-r,N}\,\zeta_{\Lambda^0} = (-1)^{r-1} E_{N-1,N} E_{N-2,N-1} \dots E_{N-r,N-r+1}\,\zeta_{\Lambda^0}.$$

Therefore, by (5.40), the expansion of $E_{N-r,N}\,\zeta_{\Lambda^0}$ in terms of the basis vectors ζ_Λ contains $\zeta_{\Lambda'}$ with a nonzero coefficient. It will now be convenient to use polynomial quantum minor operators defined by

$$T^{a_1\dots a_m}_{b_1\dots b_m}(u) = u(u-1)\dots(u-m+1)\,t^{a_1\dots a_m}_{b_1\dots b_m}(u);$$

see (1.54). Using (5.42) we find from (6.17) that the coefficient of $\zeta_{\Lambda'}$ in the expansion of $T^{1\dots N-r}_{1\dots N-r-1,N}(u)\,\zeta_{\Lambda^0}$ equals

$$C\,(u+l^0_{N-r,1})\dots\wedge_j\dots(u+l^0_{N-r,N-r}), \qquad j := i-r+1,$$

where C is a nonzero constant. For the second factor in (6.15) we find from (1.54) and Lemma 6.1.2 that

$$T^{1\dots N-r-1,N}_{1\dots N-r-1,N}(u)\,\eta_{\Lambda^0} = (u+m_1)\dots(u+m_{N-r-1})(u+m_N+r)\,\eta_{\Lambda^0}.$$

Consider now the expression (6.16). By (5.42) we have

$$T^{1\dots N-r}_{1\dots N-r}(u)\,\zeta_{\Lambda'} = (u+l^0_{N-r,1})\dots(u+l^0_{N-r,j}+1)\dots(u+l^0_{N-r,N-r})\,\zeta_{\Lambda'}.$$

Since ξ is a $\mathfrak{gl}_N$-weight vector and $w(\Lambda') = w(\Lambda^0) + \varepsilon_{N-r} - \varepsilon_N$, the vector $\eta_{\Lambda'}$ is a linear combination of the vectors ζ'_M with $w(M) = \mu - \varepsilon_{N-r} + \varepsilon_N$, where $\{\zeta'_M\}$ is the Gelfand–Tsetlin basis of $L(\mu)$. This implies that the row $N-r$ of each of the patterns M is $(\mu_1, \dots, \mu_{N-r-1}, \mu_{N-r}-1)$. Therefore we have $E_{N-r,N}\,\eta_{\Lambda'} = C'\,\zeta'$ for a constant C', and so by (5.42) and (6.17)

$$T^{1\dots N-r}_{1\dots N-r-1,N}(u)\,\eta_{\Lambda'} = C'\,(u+m_1)\dots(u+m_{N-r-1})\,\zeta'.$$

Combining the results of the above calculations and taking the coefficient of the vector $\zeta_{\Lambda'}\otimes\eta_{\Lambda^0}$ in (6.13), we obtain

$$\begin{aligned} C\,(u+l^0_{N-r,1})\dots\wedge_j\dots(u+l^0_{N-r,N-r})(u+m_1)\dots(u+m_{N-r-1})(u+m_N+r) \\ +\,C'\,(u+l^0_{N-r,1})\dots(u+l^0_{N-r,j}+1)\dots(u+l^0_{N-r,N-r}) \\ \times\,(u+m_1)\dots(u+m_{N-r-1}) = 0. \end{aligned}$$

Reducing by the common factors gives

$$C\,(u+m_N+r) + C'\,(u+l^0_{N-r,j}+1) = 0.$$

Put $u = -l^0_{N-r,j}-1$ in this relation. Since C is nonzero we get $m_N = l^0_{N-r,j}-r+1$ and thus $m_N = l^0_{N-1,i}$. By the betweenness conditions for Λ^0 and Λ',

$$\lambda_i - \lambda^0_{N-1,i} > 0 \quad\text{and}\quad \lambda^0_{N-1,i} - \lambda_{i+1} \geqslant 0,$$

which implies that both differences $l_i - m_N$ and $m_N - l_{i+1}$ are positive integers. Thus $m_N \in \langle\, l_{i+1}, l_i\,\rangle \subseteq \langle\, l_N, l_1\,\rangle$, which contradicts (6.3). Therefore, our assumption that Λ' is a pattern must be wrong. □

Using Lemmas 6.1.4 and 6.1.5 we can now conclude that all vectors ζ_Λ which occur in (6.4) belong to the $U(\mathfrak{gl}_{N-1})$-span of the highest vector ζ of $L(\lambda)$. This span is isomorphic to the irreducible representation $L(\lambda_-)$ of $\mathfrak{gl}_{N-1}$ with the highest weight $\lambda_- = (\lambda_1, \dots, \lambda_{N-1})$. In particular, $E_{NN}\,\zeta_\Lambda = \lambda_N\,\zeta_\Lambda$ for each Λ. Furthermore, writing ξ as a linear combination of the vectors $\zeta_\Lambda \otimes \zeta'_M$, by Lemma 6.1.2 for the corresponding patterns we obtain

$$w(\Lambda) + w(M) = w(\Lambda^0) + \mu. \tag{6.18}$$

Therefore, $E_{NN}\,\zeta'_M = \mu_N\,\zeta'_M$ for all M, which implies that the row $N-1$ of each pattern M coincides with $(\mu_1, \dots, \mu_{N-1})$. In other words, each vector ζ'_M belongs to the $U(\mathfrak{gl}_{N-1})$-span of ζ' which is isomorphic to $L(\mu_-)$ where $\mu_- = (\mu_1, \dots, \mu_{N-1})$. Thus, ξ belongs to

$$L(\lambda_-) \otimes L(\mu_-). \tag{6.19}$$

Using (1.35), we derive that the $Y(\mathfrak{gl}_{N-1})$-module structure on (6.19) coincides with the one obtained by restriction from $Y(\mathfrak{gl}_N)$ to the subalgebra generated by the coefficients of the series $t_{ij}(u)$ with $1 \leqslant i, j \leqslant N-1$. The vector ξ is annihilated by the operators $b_1(u), \dots, b_{N-2}(u)$. By the assumption of the theorem, for each pair (i, j) such that $1 \leqslant i < j \leqslant N-1$ the condition (6.1) is satisfied. Therefore, the $Y(\mathfrak{gl}_{N-1})$-module (6.19) is irreducible by the induction hypothesis, and we may finally conclude from Corollary 3.2.8 that (6.2) holds.

In order to complete the proof of the theorem, we need to show that the $Y(\mathfrak{gl}_N)$-submodule of $L = L(\lambda) \otimes L(\mu)$ generated by the tensor product of the highest vectors $\zeta \otimes \zeta'$ coincides with L. By Proposition 3.2.12, the dual $Y(\mathfrak{gl}_N)$-module L^* is isomorphic to the tensor product $L(\widetilde{\lambda}) \otimes L(\widetilde{\mu})$. Now, if we assume that the vector $\zeta \otimes \zeta'$ generates a proper submodule K of L, then its annihilator

$$\operatorname{Ann} K = \{\omega \in L^* \mid \omega(v) = 0 \quad \text{for all} \quad v \in K\}$$

is a nonzero submodule of $L^* \cong L(\widetilde{\lambda}) \otimes L(\widetilde{\mu})$. However, the condition (6.3) holds for the highest weights $\widetilde{\lambda}$ and $\widetilde{\mu}$ as well, and so, by the previous argument, the submodule $\operatorname{Ann} K$ of $L(\widetilde{\lambda}) \otimes L(\widetilde{\mu})$ must contain the tensor product of the highest vectors of $L(\widetilde{\lambda})$ and $L(\widetilde{\mu})$. The highest vector ζ^* of $L(\widetilde{\lambda})$ can be identified with the element ζ^* of $L(\lambda)^*$ such that $\zeta^*(\zeta) = 1$ and $\zeta^*(\eta) = 0$ for all weight vectors $\eta \in L(\lambda)_\nu$, $\nu \neq \lambda$, and similarly for the element $\zeta'^* \in L(\mu)^*$. This leads to a contradiction (cf. the proof of Proposition 3.3.2), thus completing the proof of Theorem 6.1.1. □

The following corollary is immediate from Theorem 6.1.1.

COROLLARY 6.1.6. *The $Y(\mathfrak{gl}_N)$-module $L(\lambda) \otimes L(\lambda)$ is irreducible for any $\mathfrak{gl}_N$-highest weight λ.* □

6.2. Necessary conditions

As in the previous section, we assume that λ and μ are arbitrary complex $\mathfrak{gl}_N$-highest weights.

THEOREM 6.2.1. *Suppose that the $Y(\mathfrak{gl}_N)$-module $L(\lambda) \otimes L(\mu)$ is irreducible. Then for each pair of indices $1 \leqslant i < j \leqslant N$ we have*

$$m_j, m_i \notin \langle\, l_j, l_i \,\rangle \qquad \text{or} \qquad l_j, l_i \notin \langle\, m_j, m_i \,\rangle. \tag{6.20}$$

PROOF. We use induction on N. Given a $\mathfrak{gl}_N$-highest weight $\lambda = (\lambda_1, \dots, \lambda_N)$ set

$$\lambda_- = (\lambda_1, \dots, \lambda_{N-1}) \quad \text{and} \quad \lambda_+ = (\lambda_2, \dots, \lambda_N).$$

LEMMA 6.2.2. *Suppose that the* $\mathrm{Y}(\mathfrak{gl}_N)$*-module* $L(\lambda) \otimes L(\mu)$ *is irreducible. Then the* $\mathrm{Y}(\mathfrak{gl}_{N-1})$*-modules* $L(\lambda_-) \otimes L(\mu_-)$ *and* $L(\lambda_+) \otimes L(\mu_+)$ *are both irreducible.*

PROOF. We will identify $L(\lambda_-)$ and $L(\mu_-)$ with the $\mathrm{U}(\mathfrak{gl}_{N-1})$-spans of the highest vectors ζ in $L(\lambda)$ and ζ' in $L(\mu)$, respectively. Any generator E_{iN} of $\mathfrak{gl}_N$ with $i < N$ annihilates $L(\lambda_-)$ and $L(\mu_-)$. Hence, the subspace $L(\lambda_-) \otimes L(\mu_-)$ of $L(\lambda) \otimes L(\mu)$ is invariant with respect to the action of the subalgebra $\mathrm{Y}(\mathfrak{gl}_{N-1})$ of $\mathrm{Y}(\mathfrak{gl}_N)$. Moreover, this action of $\mathrm{Y}(\mathfrak{gl}_{N-1})$ coincides with the one obtained by regarding $L(\lambda_-)$ and $L(\mu_-)$ as evaluation modules over $\mathrm{Y}(\mathfrak{gl}_{N-1})$ and then using the comultiplication in $\mathrm{Y}(\mathfrak{gl}_{N-1})$; see (1.35).

Suppose that there is a nonzero submodule of $L(\lambda_-) \otimes L(\mu_-)$ which does not contain the vector $\zeta \otimes \zeta'$. Then this submodule contains a nonzero vector ξ annihilated by the coefficients of all series $t_{ij}(u)$ with $1 \leqslant i < j \leqslant N-1$. However, we also have $t_{iN}(u)\,\xi = 0$ for any $i < N$ which follows from (1.35). This implies that $L(\lambda) \otimes L(\mu)$ is not irreducible, a contradiction.

Suppose now that the $\mathrm{Y}(\mathfrak{gl}_{N-1})$-submodule of $L_- = L(\lambda_-) \otimes L(\mu_-)$ generated by $\zeta \otimes \zeta'$ is proper. Equip the dual space L_-^* with a $\mathrm{Y}(\mathfrak{gl}_{N-1})$-module structure by using the anti-automorphism t of $\mathrm{Y}(\mathfrak{gl}_{N-1})$ given by (1.26) so that

$$(y\omega)(v) = \omega(t(y)v), \qquad y \in \mathrm{Y}(\mathfrak{gl}_{N-1}), \quad \omega \in L_-^*, \quad v \in L_-.$$

We easily derive from (1.35) that the module L_-^* is isomorphic to $L(\mu_-) \otimes L(\lambda_-)$. Since $\zeta \otimes \zeta'$ generates a proper submodule of L_-, its annihilator in L_-^* is a nonzero submodule which does not contain the tensor product of the highest vectors of $L(\mu_-)$ and $L(\lambda_-)$; cf. the proof of Proposition 3.3.2. However, the $\mathrm{Y}(\mathfrak{gl}_N)$-module $L(\mu) \otimes L(\lambda)$ is irreducible by Proposition 3.2.10, thus leading again to a contradiction. This proves that the $\mathrm{Y}(\mathfrak{gl}_{N-1})$-module $L(\lambda_-) \otimes L(\mu_-)$ is irreducible.

Now consider the $\mathrm{Y}(\mathfrak{gl}_{N-1})$-module $L(\lambda_+) \otimes L(\mu_+)$. If the $\mathrm{Y}(\mathfrak{gl}_N)$-module $L = L(\lambda) \otimes L(\mu)$ is irreducible, then so is the dual module L^* defined in (3.73). By Proposition 3.2.12 the claim is now implied by the previous case. □

Lemma 6.2.2 and the induction hypothesis imply that conditions (6.20) are satisfied for all pairs $(i,j) \neq (1,N)$. We will argue by contradiction and suppose that these conditions are violated for the pair $(1,N)$. Then either $m_N \in \langle\, l_N, l_1 \,\rangle$ and $l_1 \in \langle\, m_N, m_1 \,\rangle$ or $l_N \in \langle\, m_N, m_1 \,\rangle$ and $m_1 \in \langle\, l_N, l_1 \,\rangle$. Using Proposition 3.2.10, if necessary, we may assume that $m_N \in \langle\, l_N, l_1 \,\rangle$ and $l_1 \in \langle\, m_N, m_1 \,\rangle$. Therefore, there exist indices $p, q \in \{1, \dots, N-1\}$ such that

$$m_N \in \langle\, l_{p+1}, l_p \,\rangle \qquad \text{and} \qquad l_1 \in \langle\, m_{q+1}, m_q \,\rangle. \tag{6.21}$$

If $p = N-1$, then by the conditions (6.20) for λ_+ and μ_+ we must have $q = 1$. Moreover, using (6.20) for λ_- and μ_- we conclude that there should exist indices r and s such that $1 \leqslant r \leqslant s \leqslant N-1$ and

$$m_2, \dots, m_r \in \{l_2, \dots, l_s\}, \qquad l_{s+1}, \dots, l_{N-1} \in \{m_{r+1}, \dots, m_{N-1}\}.$$

In particular, this implies that

$$l_i - m_i \in \mathbb{Z}_+ \qquad \text{for all} \quad i = 2, \dots, N-1. \tag{6.22}$$

Similarly, if $p \leqslant N-2$ in (6.21), then we must have $q = N-p$ and

$$l_{p-i+1} = m_{N-i} \qquad \text{for} \quad i = 1, \dots, p-1. \tag{6.23}$$

Our aim in the remaining part of the proof of Theorem 6.2.1 is to produce a nonzero singular vector ϑ of the $\mathrm{Y}(\mathfrak{gl}_N)$-module $L(\lambda) \otimes L(\mu)$ and thus prove that this module is not irreducible. Indeed, if $L(\lambda) \otimes L(\mu)$ were irreducible, then by Theorem 3.2.7 it would be isomorphic to an irreducible highest weight module $V(\lambda, \mu)$ whose highest weight is provided by Proposition 3.2.9. However, as we will see below, the image $\theta \in V(\lambda, \mu)$ of the vector ϑ under the isomorphism turns out to be zero, thus leading to a contradiction.

By Proposition 3.2.11, the tensor product $L(\lambda) \otimes L(\mu)$ can be regarded as a module over $\mathrm{Y}_2(\mathfrak{gl}_N)$, the Yangian of level 2; see Definition 5.1.1. Maintaining the notation of Section 5.1, we will use the quantum minors $T^{a_1 \dots a_m}_{b_1 \dots b_m}(u)$ defined in (5.7) and the polynomials $A_r(u)$ and $B_r(u)$ defined in (5.34). Recall also the raising and lowering operators $\tau_{ri}(u)$ and $\tau_{ir}(u)$ given by (5.19). We set $\tau_{ri}(u) = 1$ for $r \leqslant i$. For any nonnegative integer k introduce the product $\Omega_{ri}(u,k)$ of lowering operators by

$$\Omega_{ri}(u,k) = \tau_{ri}(u+k-1) \dots \tau_{ri}(u+1)\, \tau_{ri}(u) \qquad \text{for} \quad k \geqslant 1, \tag{6.24}$$

and set $\Omega_{ri}(u,0) = 1$.

With the parameter p defined in (6.21), the numbers

$$k_i = l_i - m_{N-p+i}, \qquad i = 1, \dots, p, \tag{6.25}$$

are positive integers by (6.22) and (6.23). Introduce the vector $\vartheta \in L(\lambda) \otimes L(\mu)$ by

$$\vartheta = \Omega_{N-p+1,1}(-\lambda_1, k_1)\, \Omega'_{N-p+2,2}(-\lambda_2, k_2) \dots \Omega'_{Np}(-\lambda_p, k_p)\, (\zeta \otimes \zeta'), \tag{6.26}$$

where ζ and ζ' are the highest vectors of the $\mathfrak{gl}_N$-modules $L(\lambda)$ and $L(\mu)$, respectively, and $\Omega'_{ra}(u, k_a)$ denotes the derivative of the polynomial $\Omega_{ra}(u, k_a)$ over u.

LEMMA 6.2.3. *The vector ϑ is nonzero.*

PROOF. Write the vector ϑ in the form

$$\vartheta = \sum_{\Lambda, M} c_{\Lambda, M}\, \zeta_\Lambda \otimes \zeta'_M, \qquad c_{\Lambda,M} \in \mathbb{C}, \tag{6.27}$$

where ζ_Λ and ζ'_M denote the Gelfand–Tsetlin basis vectors in $L(\lambda)$ and $L(\mu)$, respectively; see Section 5.4. It suffices to show that at least one coefficient $c_{\Lambda,M}$ is nonzero. Let us calculate the coefficients c_{Λ,M^0}, where M^0 is the pattern corresponding to the highest vector ζ' of $L(\mu)$. That is, the row k of the pattern M^0 coincides with $(\mu_1, \dots, \mu_k)$ for all k. For any $a = 2, \dots, p$ define the coefficients $c^{(a)}_{\Lambda,M} \in \mathbb{C}$ by the expansion

$$\Omega'_{N-p+a,a}(-\lambda_a, k_a) \dots \Omega'_{Np}(-\lambda_p, k_p)\, (\zeta \otimes \zeta') = \sum_{\Lambda, M} c^{(a)}_{\Lambda,M}\, \zeta_\Lambda \otimes \zeta'_M. \tag{6.28}$$

Furthermore, for any $a = 1, \dots, p$ define the pattern $\Lambda^{(a)}$ associated with λ as follows. The entry $\lambda^{(a)}_{ri}$ of $\Lambda^{(a)}$ coincides with λ_i unless $i = a, \dots, p$ and $r < N-p+i$. For these values of i and r we set $\lambda^{(a)}_{ri} = \lambda_i - k_i$. The betweenness conditions for

$\Lambda^{(a)}$ are guaranteed by (6.22) and (6.23). We will prove by a reverse induction on a that for each pattern M occurring in the expansion (6.28) we have

$$\mu - w(M) = \sum_{i=a}^{N-1} n_i\,(\varepsilon_i - \varepsilon_{i+1}), \qquad n_i \in \mathbb{Z}_+, \tag{6.29}$$

and $c^{(a)}_{\Lambda^{(a)},M^0} \neq 0$. Suppose that $a \leqslant p$ and denote the left-hand side of (6.28) by $\vartheta^{(a)}$. By Proposition 1.6.9, we have

$$\Delta(\tau_{ra}(v)) = \sum_{b_1<\dots<b_{N-p}} T^{\,a+1\dots r}_{\,b_1\dots b_{N-p}}(v) \otimes T^{\,b_1\dots b_{N-p}}_{\,a\dots r-1}(v), \qquad r = N-p+a,$$

where the quantum minor operators in $L(\lambda)$ and $L(\mu)$ are defined by the formulas (1.54) with the $t_{ij}(u)$ replaced by the polynomial operators $T_{ij}(u) = u\,t_{ij}(u)$. If w is a weight of the $\mathfrak{gl}_N$-module $L(\mu)$, then $w \prec \mu$. Therefore, by the induction hypothesis, if $b_1 < a$, then

$$T^{\,b_1\dots b_{N-p}}_{\,a\dots r-1}(v)\,\zeta'_M = 0$$

for each pattern M occurring in the expansion (6.28) for $\vartheta^{(a+1)}$. This proves (6.29). Hence, for any $k \in \mathbb{Z}_+$ the vectors $\zeta_\Lambda \otimes \zeta'_{M^0}$ which occur in the expansion of $\Omega_{ra}(v,k)\,\vartheta^{(a+1)}$ should have the form

$$\Omega_{ra}(v,k)\,\zeta_{\Lambda^{(a+1)}} \otimes T^{\,a\dots r-1}_{\,a\dots r-1}(v+k-1)\dots T^{\,a\dots r-1}_{\,a\dots r-1}(v)\,\zeta'. \tag{6.30}$$

The coefficient of ζ' in this expression equals

$$\prod_{j=1}^{k}(v+\mu_a+j-1)(v+\mu_{a+1}+j-2)\dots(v+\mu_{r-1}+j-N+p).$$

The conditions (6.22) and (6.23) imply that for $k = k_a$ there is a unique factor in this product which vanishes at $v = -\lambda_a$. Therefore, the derivative of (6.30) with $k = k_a$ at $v = -\lambda_a$ is, up to a nonzero constant factor,

$$\Omega_{N-p+a,a}(-\lambda_a, k_a)\,\zeta_{\Lambda^{(a+1)}} \otimes \zeta'.$$

By the definition of the basis vectors ζ_Λ (see Section 5.4), this coincides with the vector $\zeta_{\Lambda^{(a)}} \otimes \zeta'$, thus proving that the coefficient $c^{(a)}_{\Lambda^{(a)},M^0}$ in (6.28) is nonzero.

Finally, the application of the operator $\Omega_{N-p+1,1}(-\lambda_1,k_1)$ to the vector $\vartheta^{(2)}$ produces a linear combination (6.27). Here the coefficient $c_{\Lambda^{(1)},M^0}$ is the product of $c^{(2)}_{\Lambda^{(2)},M^0}$ and the factor

$$\prod_{j=1}^{k}(-\lambda_1+\mu_1+j-1)(-\lambda_1+\mu_2+j-2)\dots(-\lambda_1+\mu_{N-p}+j-N+p)$$

with $k = l_1 - m_{N-p+1}$, which comes from the expansion of the coefficient of ζ' in (6.30) for $a = 1$. It remains to observe that this factor is nonzero by (6.21). □

Now consider the irreducible highest weight module $V(\lambda,\mu)$ over $\mathrm{Y}_2(\mathfrak{gl}_N)$ generated by the highest vector ξ satisfying

$$T_{ii}(u)\,\xi = (u+\lambda_i)(u+\mu_i)\,\xi, \qquad i = 1,\dots,N. \tag{6.31}$$

Introduce the vector $\theta \in V(\lambda,\mu)$ by

$$\theta = \Omega_{N-p+1,1}(-\lambda_1,k_1)\,\Omega'_{N-p+2,2}(-\lambda_2,k_2)\dots\Omega'_{Np}(-\lambda_p,k_p)\,\xi; \tag{6.32}$$

cf. (6.26). Our aim is to prove that the vector θ is zero. We do this in Lemma 6.2.8 below, which together with Lemma 6.2.3 will complete the proof of Theorem 6.2.1. We will show that θ is annihilated by the operators $B_i(u)$ for all $i = 1, \dots, N-1$ and then apply Corollary 3.2.8, noting that θ is not proportional to ξ. This works directly in the case $p = 1$. However, if $p \geqslant 2$, then applying the operators $B_i(u)$ to θ, we come to a more general problem to prove that all vectors parameterized by a certain finite family of pattern-like arrays Λ are zero. We will rely on a few lemmas which generalize the techniques used in the construction of the Gelfand–Tsetlin bases; see Sections 5.2 and 5.3.

Let L be an arbitrary highest weight module over $\mathrm{Y}_2(\mathfrak{gl}_N)$ generated by the highest vector ξ satisfying (6.31). Sometimes we will also regard L as a module over the Yangian $\mathrm{Y}(\mathfrak{gl}_N)$ where all generators $t_{ij}^{(r)}$ with $r \geqslant 3$ act as the zero operators. Suppose that $\eta \in L$ is a singular vector for $\mathrm{Y}_2(\mathfrak{gl}_{N-1})$ of weight

$$\pi(u) = \big(\pi_1(u), \dots, \pi_{N-1}(u)\big);$$

see Section 5.2. That is, η is annihilated by $T_{ij}(u)$ for $1 \leqslant i < j \leqslant N-1$ and

$$T_{ii}(u)\,\eta = \pi_i(u)\,\eta, \qquad i = 1, \dots, N-1, \tag{6.33}$$

where the $\pi_i(u)$ are monic polynomials in u of degree two. Let a be a fixed element of the set $\{1, \dots, N-1\}$ and let k be a nonnegative integer.

LEMMA 6.2.4. *We have the following relations in L:*

$$T_{ii}(u)\,\Omega_{Na}(v,k)\,\eta = \pi_i(u)\,\Omega_{Na}(v,k)\,\eta, \tag{6.34}$$

for $1 \leqslant i \leqslant N-1$ and $i \neq a$, while

$$\begin{aligned} T_{aa}(u)\,\Omega_{Na}(v,k)\,\eta &= \frac{(u-v-k)\,\pi_a(u)}{u-v}\,\Omega_{Na}(v,k)\,\eta \\ &+ \frac{k}{u-v}\sum_{c=a+1}^{N} \pi_a(v)\,\pi_{a+1}(v-1)\dots\pi_{c-1}(v-c+a+1) \\ &\qquad \times T_{ca}(u)\,\Omega_{Na}(v+1,k-1)\,\tau_{Nc}(v-c+a)\,\eta. \end{aligned} \tag{6.35}$$

Moreover,

$$T_{i,i+1}(u)\,\Omega_{Na}(v,k)\,\eta = 0, \tag{6.36}$$

for $1 \leqslant i < N-1$ and $i \neq a$, while for $a < N-1$

$$\begin{aligned} &T_{a,a+1}(u)\,\Omega_{Na}(v,k)\,\eta \\ &= \frac{k}{u-v}\sum_{c=a+1}^{N} \pi_a(v)\,\pi_{a+1}(v-1)\dots\pi_{c-1}(v-c+a+1) \\ &\qquad \times T_{c,a+1}(u)\,\Omega_{Na}(v+1,k-1)\,\tau_{Nc}(v-c+a)\,\eta. \end{aligned} \tag{6.37}$$

PROOF. Note that the coefficients of the polynomial $\Omega_{Na}(v,k)$ are linear combinations of monomials in the generators $t_{rs}^{(j)}$ with $a \leqslant s \leqslant N-1$. Suppose that $i < a$. We have $T_{il}(u)\,\eta = 0$ for $a \leqslant l \leqslant N-1$. Therefore, applying (5.2) to the commutators $[T_{ii}(u), T_{rs}(v)]$ and $[T_{il}(u), T_{rs}(v)]$, we conclude by an easy induction that $T_{il}(u)\,\Omega_{Na}(v,k)\,\eta = 0$. This proves (6.34) and (6.36) for $i < a$. For $i > a$ both relations are immediate from Corollary 1.7.2. Furthermore, by Proposition 1.7.1,

$$T_{aa}(u)\,\Omega_{Na}(v,k) = \frac{u-v-k}{u-v-k+1}\,\tau_{Na}(v+k-1)\,T_{aa}(u)\,\Omega_{Na}(v,k-1)$$
$$+\frac{1}{u-v-k+1}\sum_{c=a+1}^{N} T_{ca}(u)\,T^{\,a\dots\widehat{c}\dots N}_{\,a\dots N-1}(v+k-1)\,\Omega_{Na}(v,k-1)\,(-1)^{c-a-1}.$$

The subalgebra Y_a of $\mathrm{Y}(\mathfrak{gl}_N)$ generated by the coefficients of the series $t_{rs}(u)$ with $a \leqslant r, s \leqslant N$ is naturally isomorphic to the Yangian $\mathrm{Y}(\mathfrak{gl}_{N-a+1})$. Applying the automorphism provided by Corollary 1.9.4 to this subalgebra, we derive from the defining relations (5.2) that

$$T^{\,a\dots\widehat{c}\dots N}_{\,a\dots N-1}(v+1)\,T^{\,a+1\dots N}_{\,a\dots N-1}(v) = T^{\,a+1\dots N}_{\,a\dots N-1}(v+1)\,T^{\,a\dots\widehat{c}\dots N}_{\,a\dots N-1}(v)$$

for every $c = a+1,\dots,N$. Hence,

$$T^{\,a\dots\widehat{c}\dots N}_{\,a\dots N-1}(v+k-1)\,\Omega_{Na}(v,k-1) = \Omega_{Na}(v+1,k-1)\,T^{\,a\dots\widehat{c}\dots N}_{\,a\dots N-1}(v).$$

Note that $T_{ca}(u)$ commutes with $\tau_{Na}(v+k-1)$ by Corollary 1.7.2. Therefore, the induction on k gives

$$T_{aa}(u)\,\Omega_{Na}(v,k) = \frac{u-v-k}{u-v}\,\Omega_{Na}(v,k)\,T_{aa}(u)$$
$$+\frac{k}{u-v}\sum_{c=a+1}^{N} T_{ca}(u)\,\Omega_{Na}(v+1,k-1)\,T^{\,a\dots\widehat{c}\dots N}_{\,a\dots N-1}(v)\,(-1)^{c-a-1}.$$

Similarly, for $a < N-1$ we have

$$T_{a,a+1}(u)\,\Omega_{Na}(v,k) = \frac{u-v-k}{u-v}\,\Omega_{Na}(v,k)\,T_{a,a+1}(u)$$
$$+\frac{k}{u-v}\sum_{c=a+1}^{N} T_{c,a+1}(u)\,\Omega_{Na}(v+1,k-1)\,T^{\,a\dots\widehat{c}\dots N}_{\,a\dots N-1}(v)\,(-1)^{c-a-1}.$$

Finally, by (5.8),

$$T^{\,a\dots\widehat{c}\dots N}_{\,a\dots N-1}(v) = \sum_{\sigma}\operatorname{sgn}\sigma\cdot T_{N,\sigma(N-1)}(v-N+a+1)\dots T_{c+1,\sigma(c)}(v-c+a)$$
$$\times\, T_{c-1,\sigma(c-1)}(v-c+a+1)\dots T_{a,\sigma(a)}(v),$$

summed over the permutations σ of the set $\{a,\dots,N-1\}$. Since $T_{ij}(u)\,\eta = 0$ for $1 \leqslant i < j \leqslant N-1$, we conclude from (6.33) that

$$T^{\,a\dots\widehat{c}\dots N}_{\,a\dots N-1}(v)\,\eta = \tau_{Nc}(v-c+a)\,\pi_a(v)\,\pi_{a+1}(v-1)\dots\pi_{c-1}(v-c+a+1)\,\eta,$$

completing the proof of (6.35) and (6.37). □

Note the following relations in L which are immediate from Lemma 6.2.4. If $\pi_a(-\beta) = 0$ for some $\beta \in \mathbb{C}$, then

$$T_{aa}(u)\,\Omega_{Na}(-\beta,k)\,\eta = \frac{(u+\beta-k)\,\pi_a(u)}{u+\beta}\,\Omega_{Na}(-\beta,k)\,\eta, \tag{6.38}$$

and for $a < N-1$

$$T_{a,a+1}(u)\,\Omega_{Na}(-\beta,k)\,\eta = 0. \tag{6.39}$$

In what follows we identify the elements $E_{ij} \in \mathfrak{gl}_N$ with their images $t_{ij}^{(1)}$ under the embedding of $\mathrm{U}(\mathfrak{gl}_N)$ into $\mathrm{Y}(\mathfrak{gl}_N)$; see (1.6).

LEMMA 6.2.5. *Let $1 \leqslant a < N-1$. Then we have the relations in* $\mathrm{Y}_2(\mathfrak{gl}_N)$

$$[E_{N-1,N}, \Omega_{Na}(v,k)] = -k\,\Omega_{Na}(v,k-1)\, T^{\,a+1\dots N}_{\,a\dots N-2,N}(v+k-1) \tag{6.40}$$

and

$$T^{\,a+1\dots N}_{\,a\dots N-2,N}(u)\, T^{\,a+1\dots N-1}_{\,a+1\dots N-1}(u) = \tau_{N-1,a}(u)\, T^{\,a+1\dots N}_{\,a+1\dots N}(u) + \tau_{Na}(u)\, T^{\,a+1\dots N-1}_{\,a+1\dots N-2,N}(u). \tag{6.41}$$

PROOF. Proposition 1.7.1 implies $[E_{N-1,N}, \tau_{Na}(v)] = -T^{\,a+1\dots N}_{\,a\dots N-2,N}(v)$. Hence

$$[E_{N-1,N}, \Omega_{Na}(v,k)] = -\sum_{i=1}^{k} \tau_{Na}(v)\dots T^{\,a+1\dots N}_{\,a\dots N-2,N}(v+i-1)\dots \tau_{Na}(v+k-1).$$

Applying the automorphism provided by Corollary 1.9.4 to the subalgebra Y_a introduced in the proof of Lemma 6.2.4, we derive from (5.2) that

$$T^{\,a+1\dots N}_{\,a\dots N-2,N}(v)\, \tau_{Na}(v+1) = \tau_{Na}(v)\, T^{\,a+1\dots N}_{\,a\dots N-2,N}(v+1). \tag{6.42}$$

This proves (6.40) by an easy induction.

Furthermore, note that the coefficients of all series involved in (6.41) are contained in the subalgebra Y_a. Hence, it suffices to prove the relation in the case $a = 1$. Proposition 1.7.1 gives

$$(u-v)\,[\,t_{11}(u), t^{\,N-1,N}_{\,1,N}(v)] = t_{N-1,1}(u)\, t^{\,1,N}_{\,1,N}(v) - t_{N,1}(u)\, t^{\,1,N-1}_{\,1,N}(v) - t^{\,N-1,N}_{\,1,N}(v)\, t_{11}(u).$$

So, taking $v = u$ we get

$$t_{N-1,1}(u)\, t^{\,1,N}_{\,1,N}(u) = t^{\,N-1,N}_{\,1,N}(u)\, t_{11}(u) + t_{N,1}(u)\, t^{\,1,N-1}_{\,1,N}(u).$$

The desired relation now follows by the application of Theorem 1.10.7; just replace all minors by their respective complements and use the fact that ω_N is an automorphism of $\mathrm{Y}(\mathfrak{gl}_N)$. □

We keep using the notation of Lemma 6.2.4. As before, ξ denotes the highest vector of L satisfying (6.31). We also regard L as a $\mathfrak{gl}_N$-module using (1.6).

LEMMA 6.2.6. *We have the following relations in L:*

$$E_{iN}\, \Omega_{Na}(v,k)\, \xi = 0 \qquad \text{if} \quad i < a. \tag{6.43}$$

If $a < i \leqslant N-1$, then

$$E_{iN}\, \Omega_{Na}(v,k)\, \xi = (-1)^{N-i}\, k \prod_{j=i+1}^{N} (v'+l_j)(v'+m_j)\, \tau_{ia}(v+k-1)\, \Omega_{Na}(v,k-1)\, \xi, \tag{6.44}$$

where $v' = v+a+k-1$. Moreover, if $1 \leqslant a \leqslant N-1$, then

$$E_{aN}\, \Omega_{Na}(v,k)\, \xi = (-1)^{N-a-1}\, k \times \Big(\Omega_{Na}(v+1,k-1)\, T^{\,a\dots N-1}_{\,a\dots N-1}(v) - \Omega_{Na}(v,k-1)\, T^{\,a+1\dots N}_{\,a+1\dots N}(v+k-1)\Big)\, \xi. \tag{6.45}$$

PROOF. The relation (6.43) follows from the fact that the coefficients of the polynomial $\Omega_{Na}(v,k)$ belong to the subalgebra Y_a; see the proof of Lemma 6.2.4. Furthermore, Proposition 1.7.1 implies

$$[E_{aN}, \tau_{Na}(v)] = (-1)^{N-a-1}\, T^{\,a\,\dots\,N-1}_{\,a\,\dots\,N-1}(v) - (-1)^{N-a-1}\, T^{\,a+1\,\dots\,N}_{\,a+1\,\dots\,N}(v).$$

Therefore,

$$E_{aN}\,\Omega_{Na}(v,k)\,\xi = (-1)^{N-a-1} \tag{6.46}$$
$$\times \sum_{i=1}^{k} \tau_{Na}(v)\dots\big(T^{\,a\,\dots\,N-1}_{\,a\,\dots\,N-1}(v+i-1) - T^{\,a+1\,\dots\,N}_{\,a+1\,\dots\,N}(v+i-1)\big)\dots\tau_{Na}(v+k-1)\,\xi.$$

We have the following analogue of (6.42) which is verified in the same way:

$$T^{\,a+1\,\dots\,N}_{\,a+1\,\dots\,N}(v)\,\tau_{Na}(v+1) = \tau_{Na}(v)\,T^{\,a+1\,\dots\,N}_{\,a+1\,\dots\,N}(v+1).$$

The right-hand side of (6.46) now takes the form

$$(-1)^{N-a-1}\sum_{i=1}^{k}\tau_{Na}(v)\dots\tau_{Na}(v+i-2)\,T^{\,a\,\dots\,N-1}_{\,a\,\dots\,N-1}(v+i-1)\,\Omega_{Na}(v+i,k-i)\,\xi$$
$$-\,(-1)^{N-a-1}\,k\,\Omega_{Na}(v,k-1)\,T^{\,a+1\,\dots\,N}_{\,a+1\,\dots\,N}(v+k-1)\,\xi.$$

Applying again Corollary 1.9.4 to Y_a we bring the sum here to the form

$$\sum_{i=1}^{k}\Big((k-i+1)\,\tau_{Na}(v+k-1)\dots\tau_{Na}\widehat{(v+i-1)}\dots\tau_{Na}(v)\,T^{\,a\,\dots\,N-1}_{\,a\,\dots\,N-1}(v+i-1)$$
$$-\,(k-i)\,\tau_{Na}(v+k-1)\dots\tau_{Na}\widehat{(v+i)}\dots\tau_{Na}(v)\,T^{\,a\,\dots\,N-1}_{\,a\,\dots\,N-1}(v+i)\Big)\,\xi,$$

which simplifies to $k\,\Omega_{Na}(v+1,k-1)\,T^{\,a\,\dots\,N-1}_{\,a\,\dots\,N-1}(v)\,\xi$, thus proving (6.45).

Finally, if $i > a$, then due to Proposition 1.7.1 we have

$$[E_{iN}, \tau_{Na}(v)] = (-1)^{N-i}\, T^{\,a+1\,\dots\,N}_{\,a\,\dots\,\widehat{i}\,\dots\,N}(v).$$

As in the proof of (6.40), this brings the left-hand side of (6.44) to the form

$$(-1)^{N-i}\,k\,\Omega_{Na}(v,k-1)\,T^{\,a+1\,\dots\,N}_{\,a\,\dots\,\widehat{i}\,\dots\,N}(v+k-1)\,\xi.$$

Using (5.7), we get

$$T^{\,a+1\,\dots\,N}_{\,a\,\dots\,\widehat{i}\,\dots\,N}(v+k-1)\,\xi = \prod_{j=i+1}^{N}(v'+l_j)(v'+m_j)\,\tau_{ia}(v+k-1)\,\xi,$$

completing the proof. □

Now return to the irreducible highest weight module $V(\lambda,\mu)$ over $\mathrm{Y}_2(\mathfrak{gl}_N)$. We will introduce a family of elements of $V(\lambda,\mu)$ which will include the vector θ defined in (6.32). Introduce the *admissible arrays* Λ associated with λ and μ as follows. Each Λ is a sequence of rows $\Lambda_r = (\lambda_{r1},\dots,\lambda_{rr})$ with $r = 1,\dots,N$ of the same shape as the patterns described in Section 5.4. The top row Λ_N coincides with λ, and for all $r \geqslant 2$ the following conditions hold:

$$\lambda_{ri} - \lambda_{r-1,i} \in \mathbb{Z}_+ \qquad \text{for} \quad i = 1,\dots,r-1.$$

Moreover, each entry λ_{ri} of Λ is equal to λ_i unless

$$i = 2,\dots,p \qquad \text{and} \qquad r < N-p+i, \tag{6.47}$$

and for all $i \in \{2, \ldots, p\}$ we have

$$\lambda_i - \lambda_{ii} \leqslant k_i; \tag{6.48}$$

see (6.25). This condition implies that $0 \leqslant \lambda_{ri} - \lambda_{r-1,i} \leqslant k_i$ for all i so that the set of admissible arrays is finite. By definition, only a part of an admissible array can vary with the remaining entries fixed, as illustrated:

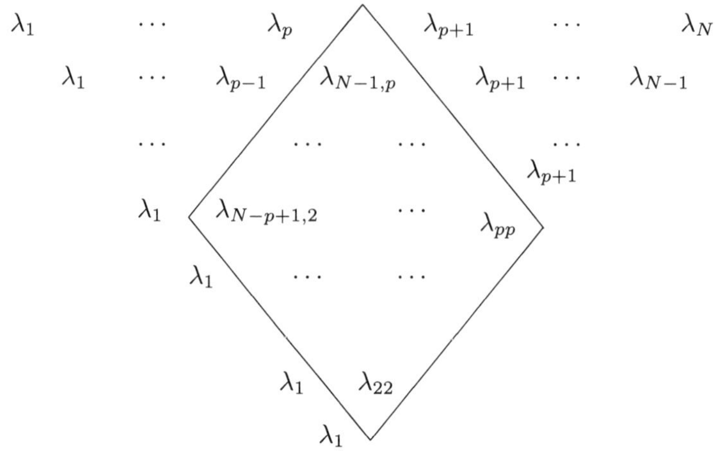

Given an admissible array Λ, introduce the corresponding vector $\theta_\Lambda \in V(\lambda, \mu)$ by

$$\theta_\Lambda = \prod_{r=3,\ldots,N}^{\longrightarrow} \Big(\prod_{i=2}^{p} \widetilde{\Omega}_{ri}(-\lambda_{ri}, \lambda_{ri} - \lambda_{r-1,i}) \Big)\, \xi, \tag{6.49}$$

where the polynomials $\widetilde{\Omega}_{ri}(v,k)$ are defined by

$$\widetilde{\Omega}_{ri}(v,k) = \begin{cases} \Omega'_{ri}(v,k) & \text{if } \ r = N-p+i \ \text{ and } \ k = k_i \\ \Omega_{ri}(v,k) & \text{otherwise;} \end{cases} \tag{6.50}$$

see (6.24). Note that the factors in the brackets in (6.49) commute for any index r due to Corollary 1.7.2. We have $\xi = \theta_{\Lambda^0}$ for the array Λ^0 with $\lambda^0_{ri} = \lambda_i$ for all r and i. Furthermore, $\theta = \Omega_{N-p+1,1}(-\lambda_1, k_1)\, \theta_\Lambda$ for the array Λ with $\lambda_{ri} = \lambda_i - k_i$ for all indices r and i satisfying (6.47). We define the *weight* $w(\Lambda)$ of an admissible array Λ by (6.6). We use the ordering on the weights described in Section 6.1. Given $\Lambda \neq \Lambda^0$, consider the minimum index $r = r(\Lambda)$ such that the difference $\lambda_{ra} - \lambda_{r-1,a}$ is a positive integer for some $2 \leqslant a \leqslant p$.

LEMMA 6.2.7. *Suppose that Λ is an admissible array with $r(\Lambda) \geqslant N - p + 2$. Then we have the following relations in $V(\lambda, \mu)$:*

$$T_{ii}(u)\, \theta_\Lambda = (u + \lambda_{r-1,i})(u + \mu_i)\, \theta_\Lambda \qquad \textit{for} \quad 1 \leqslant i \leqslant r-1, \tag{6.51}$$

$$T_{ij}(u)\, \theta_\Lambda = 0 \qquad \textit{for} \quad 1 \leqslant i < j \leqslant r-1, \tag{6.52}$$

$$B_i(u)\, \theta_\Lambda = \sum_{j=1}^{i} \beta_{ij}(u, \Lambda)\, \theta_{\Lambda + \delta_{ij}} \qquad \textit{for} \quad r-1 \leqslant i < N, \tag{6.53}$$

where $\beta_{ij}(u,\Lambda)$ are some polynomials in u, and we suppose that $\theta_\Lambda = 0$ if Λ is not an admissible array.

PROOF. Let $2 \leqslant a \leqslant p$ be the least index such that $k = \lambda_{ra} - \lambda_{r-1,a} > 0$. Then

$$\theta_\Lambda = \widetilde{\Omega}_{ra}(-\lambda_{ra}, k)\,\eta, \qquad \eta := \theta_{\Lambda'}, \tag{6.54}$$

where Λ' denotes the array obtained from Λ by increasing each of the entries $\lambda_{aa}, \dots, \lambda_{r-1,a}$ by k. We will use the reverse induction on the pairs (r, a) ordered lexicographically, with the induction base $\theta_\Lambda = \xi$. Note that by the definition of admissible arrays we must have $r \leqslant N - p + a$.

Identify the subalgebra of $\mathrm{Y}_2(\mathfrak{gl}_N)$ generated by the coefficients of the polynomials $T_{ij}(u)$ with $1 \leqslant i, j \leqslant r$ with $\mathrm{Y}_2(\mathfrak{gl}_r)$. By the induction hypothesis, η is a $\mathrm{Y}_2(\mathfrak{gl}_{r-1})$-singular vector such that (6.33) holds with N replaced by r, where

$$\pi_i(u) = \begin{cases} (u + \lambda_{ri})(u + \mu_i) & \text{if } i \leqslant a, \\ (u + \lambda_{r-1,i})(u + \mu_i) & \text{if } i > a. \end{cases}$$

Therefore, if $\widetilde{\Omega}_{ra}(-\lambda_{ra}, k) = \Omega_{ra}(-\lambda_{ra}, k)$, then (6.51) and (6.52) are immediate from Lemma 6.2.4.

Suppose now that $\widetilde{\Omega}_{ra}(-\lambda_{ra}, k) = \Omega'_{ra}(-\lambda_{ra}, k)$. Then $r = N - p + a$ and $k = k_a$. Therefore, $\lambda_{ra} = \lambda_a$ by (6.48). It is clear from Lemma 6.2.4 that (6.51) and (6.52) hold for $i \neq a$ so that we may assume $i = a$. In this case, due to (6.35) and (6.37), it suffices to show that

$$\Omega_{ra}(-\lambda_a, k_a)\,\eta = 0 \tag{6.55}$$

and that for every $c = a + 1, \dots, r$ the polynomial

$$\pi_a(v)\,\pi_{a+1}(v-1)\dots\pi_{c-1}(v-c+a+1)\,\Omega_{ra}(v+1, k_a - 1)\,\tau_{rc}(v - c + a)\,\eta \tag{6.56}$$

has zero of multiplicity at least two at $v = -\lambda_a$.

Suppose first that $p < N - 1$ and consider $\Omega_{ra}(-\lambda_a, k_a)\,\eta$. By (5.7) we have

$$\tau_{ra}(v) = \sum_\sigma \operatorname{sgn}\sigma \cdot T_{\sigma(a+1),a}(v) \dots T_{\sigma(r),r-1}(v - r + a + 1), \tag{6.57}$$

summed over the permutations σ of the set $\{a+1, \dots, r\}$. By the induction hypothesis we have $T_{\sigma(r),r-1}(u)\,\eta = 0$ if $\sigma(r) < r - 1$, while

$$T_{r-1,r-1}(u)\,\eta = (u + \lambda_{r-1,r-1})(u + \mu_{r-1})\,\eta.$$

The factor $u + \mu_{r-1}$ is zero if $u = -\lambda_a - r + a + 1$ by (6.23). Therefore, if $v = -\lambda_a$, then we may assume that $\sigma(r) = r$ in (6.57), which gives

$$\tau_{ra}(-\lambda_a)\,\eta = \tau_{r-1,a}(-\lambda_a)\,T_{r,r-1}(-\mu_{r-1})\,\eta.$$

Since $T_{r,r-1}(v) = \tau_{r,r-1}(v)$ is a lowering operator, we verify by an easy induction with the use of Corollary 1.7.2 and (6.38) that

$$\Omega_{ra}(-\lambda_a, k_a)\,\eta = \Omega_{r-1,a}(-\lambda_a, k_a)\,\Omega_{r,r-1}(-\mu_{r-1}, k_a)\,\eta.$$

We have $k_a = m_{r-1} - m_r$ by (6.23), and so the equality $\Omega_{ra}(-\lambda_a, k_a)\,\eta = 0$ in the case $p < N - 1$ will be implied by the fact that the vector

$$\widetilde{\eta} = T_{r,r-1}(-\mu_r) \dots T_{r,r-1}(-\mu_{r-1})\,\eta \tag{6.58}$$

is zero in $V(\lambda, \mu)$. Since the $\mathrm{Y}_2(\mathfrak{gl}_N)$-module $V(\lambda, \mu)$ is irreducible, it will be sufficient to show, due to Corollary 3.2.8, that the vector $\widetilde{\eta}$ is annihilated by all operators $B_i(u)$ with $i = 1, \dots, N-1$. Since $T_{r,r-1}(u)$ commutes with the lowering operators $\tau_{ri}(v)$, we may assume that the array Λ satisfies $\lambda_{ri} = \lambda_{r-1,i}$ for all $i \neq a$. In other words, $\eta = \theta_{\Lambda'}$ is a $\mathrm{Y}(\mathfrak{gl}_r)$-singular vector such that

$$T_{ii}(u)\,\eta = (u + \lambda_{ri})(u + \mu_i)\,\eta \qquad \text{for} \quad i = 1, \dots, r.$$

By (6.36), $\widetilde{\eta}$ is annihilated by the operators $T_{i,i+1}(u)$ with $i = 1, \dots, r-2$. Moreover, by (6.34) and (6.38), $\widetilde{\eta}$ is an eigenvector for the operators $T_{ii}(u)$ with $i = 1, \dots, r-1$. We have $T_{r-1,r}(u) = [T_{r-1,r-1}(u), E_{r-1,r}]$. By (6.45), $E_{r-1,r}\,\widetilde{\eta} = 0$ and therefore $T_{r-1,r}(u)\,\widetilde{\eta} = 0$. On the other hand, if $i \geqslant r$, then $B_i(u)$ commutes with the elements $T_{r,r-1}(v)$ by Corollary 1.7.2. Hence, by the induction hypothesis, $B_i(u)\,\widetilde{\eta}$ is a linear combination of the vectors

$$T_{r,r-1}(-\mu_r) \dots T_{r,r-1}(-\mu_{r-1})\,\theta_{\Lambda' + \delta_{ij}}$$

with $2 \leqslant j \leqslant p$. We conclude by induction on the weight of Λ' that $B_i(u)\,\widetilde{\eta} = 0$, thus proving (6.55).

Consider now the polynomial (6.56). Suppose first that $c = r$. Note that $r - 1 > a$ since $p < N - 1$. We have

$$\text{(6.59)} \qquad \pi_a(v) = (v + \lambda_a)(v + \mu_a), \qquad \pi_{r-1}(v) = (v + \lambda_{r-1,r-1})(v + \mu_{r-1}).$$

However, $\mu_{r-1} - r + a + 1 = \lambda_a$ by (6.23). This shows that the coefficient of η in (6.56) is divisible by $(v + \lambda_a)^2$. Now let $a + 1 \leqslant c < r$. By (5.7),

$$\tau_{rc}(-\lambda_a - c + a) = \sum_{\sigma} \operatorname{sgn}\sigma \cdot T_{\sigma(c+1),c}(-\lambda_a - c + a) \dots T_{\sigma(r),r-1}(-\lambda_a - r + a + 1),$$

summed over the permutations σ of the set $\{c + 1, \dots, r\}$. We can repeat the argument which we applied to the expression (6.57) to show that the polynomial $\Omega_{ra}(v + 1, k_a - 1)\,\tau_{rc}(v - c + a)\,\eta$ has zero at $v = -\lambda_a$. Together with (6.59) this completes the proof of (6.51) and (6.52) in the case $p < N - 1$.

In the case $p = N - 1$ we have $a = r - 1$ and

$$\Omega_{r,r-1}(-\lambda_{r-1}, k_{r-1})\,\eta = T_{r,r-1}(-\mu_r) \dots T_{r,r-1}(-\lambda_{r-1})\,\eta.$$

Note that the operators $T_{r,r-1}(u)$ and $T_{r,r-1}(v)$ commute. Therefore, due to (6.22) it suffices to show that the vector (6.58) is zero. The argument used in the case $p < N - 1$ applies here as well.

For $p = N - 1$ the polynomial (6.56) equals

$$\text{(6.60)} \qquad \pi_{r-1}(v)\,\Omega_{r,r-1}(-\lambda_{r-1} + 1, k_{r-1} + 1)\,\eta.$$

If $\lambda_{r-1} = \mu_{r-1}$, then $\pi_{r-1}(v) = (v + \lambda_{r-1})^2$. If $\lambda_{r-1} - \mu_{r-1} > 0$, then

$$\Omega_{r,r-1}(-\lambda_{r-1} + 1, k_{r-1} + 1)\,\eta = T_{r,r-1}(-\mu_r) \dots T_{r,r-1}(-\lambda_{r-1} + 1)\,\eta.$$

This vector is zero because the vector (6.58) is zero. In both cases (6.60) has zero of multiplicity at least two at $v = -\lambda_{r-1}$, proving (6.51) and (6.52).

To prove (6.53) we note that $B_i(u)$ commutes with $\Omega_{sa}(v, k)$ for $i \geqslant s$ by Corollary 1.7.2. Therefore, it suffices to consider the case $i = r - 1$. We derive from Proposition 1.7.1 that $B_{r-1}(u) = [A_{r-1}(u), E_{r-1,r}]$. Suppose first that $p < N - 1$. By (6.51) and (6.52) the operator $A_{r-1}(u)$ acts on θ_Λ as multiplication by a polynomial in u. So it suffices to prove that

$$\text{(6.61)} \qquad E_{r-1,r}\,\theta_\Lambda = \sum_{j=1}^{r-1} \beta_j(\Lambda)\,\theta_{\Lambda + \delta_{r-1,j}},$$

where $\beta_j(\Lambda)$ are some constants. Write θ_Λ in the form (6.54) and assume that $a < r-1$. We now use Lemma 6.2.5. By (6.40) we have

$$(6.62)\quad E_{r-1,r}\,\Omega_{ra}(v,k)\,\theta_{\Lambda'}$$
$$= \Omega_{ra}(v,k)\,E_{r-1,r}\,\theta_{\Lambda'} - k\,\Omega_{ra}(v,k-1)\,T^{\,a+1\dots r}_{\,a\dots r-2,r}(v+k-1)\,\theta_{\Lambda'}.$$

Furthermore, (6.41) gives

$$(6.63)\quad T^{\,a+1\dots r}_{\,a\dots r-2,r}(u)\,T^{\,a+1\dots r-1}_{\,a+1\dots r-1}(u)\,\theta_{\Lambda'}$$
$$= \tau_{r-1,a}(u)\,T^{\,a+1\dots r}_{\,a+1\dots r}(u)\,\theta_{\Lambda'} + \tau_{ra}(u)\,T^{\,a+1\dots r-1}_{\,a+1\dots r-2,r}(u)\,\theta_{\Lambda'}.$$

By Proposition 1.7.1 we have $T^{\,a+1\dots r-1}_{\,a+1\dots r-2,r}(u) = [T^{\,a+1\dots r-1}_{\,a+1\dots r-1}(u), E_{r-1,r}]$. Using the induction hypothesis we obtain

$$E_{r-1,r}\,\theta_{\Lambda'} = \sum_{j=a+1}^{r-1} \beta_j(\Lambda')\,\theta_{\Lambda'+\delta_{r-1,j}}.$$

Furthermore, using (6.51) and (6.52) we derive from (5.7) that

$$T^{\,a+1\dots r-1}_{\,a+1\dots r-1}(u)\,\theta_{\Lambda'} = \prod_{i=a+1}^{r-1} (u+l_{r-1,i}+a)(u+m_i+a)\,\theta_{\Lambda'},$$

while

$$T^{\,a+1\dots r}_{\,a+1\dots r}(u)\,\theta_{\Lambda'} = \prod_{i=a+1}^{r} (u+l_{ri}+a)(u+m_i+a)\,\theta_{\Lambda'}.$$

Here we have set $l_{ri} = \lambda_{ri} - i + 1$ and used the fact that $T^{\,a+1\dots r}_{\,a+1\dots r}(u)$ commutes with the lowering operators $\tau_{rb}(v)$. Thus, by (6.63)

$$(6.64)\quad T^{\,a+1\dots r}_{\,a\dots r-2,r}(u)\,\theta_{\Lambda'}$$
$$= (u+l_{rr}+a)(u+m_r+a)\prod_{i=a+1}^{r-1}\frac{u+l_{ri}+a}{u+l_{r-1,i}+a}\,\tau_{r-1,a}(u)\,\theta_{\Lambda'}$$
$$+ \sum_{j=a+1}^{r-1}\frac{\beta_j(\Lambda')}{u+l_{r-1,j}+a}\,\tau_{ra}(u)\,\theta_{\Lambda'+\delta_{r-1,j}}.$$

Now put $v = -\lambda_{ra}$ and $k = \lambda_{ra}-\lambda_{r-1,a}$ into (6.62). The denominator $u+l_{r-1,i}+a$ in (6.64) becomes $l_{r-1,i}-l_{r-1,a}$ at $u = v+k-1$. Due to the conditions (6.23) and (6.48) the difference $l_{r-1,i}-l_{r-1,a}$ can only be zero if $i = a+1$. Moreover, in this case $l_{r-1,a+1} = l_{r,a+1} = l_{a+1}$. Then $\Lambda'+\delta_{r-1,a+1}$ is not an admissible array so that the summand with $j = a+1$ does not occur in the sum in (6.64). The denominator $u+l_{r-1,i}+a$ with $i = a+1$ does not occur in the product either, since it cancels with $u+l_{r,a+1}+a$. Thus the substitution $u = v+k-1$ into (6.64) is well-defined. Using the fact that $\tau_{r-1,b}(v)$ commutes with $\tau_{ra}(u)$ if $b \geqslant a$ we complete the proof of (6.61) for the case $\widetilde{\Omega}_{ra}(v,k) = \Omega_{ra}(v,k)$ in (6.54).

Assume now that $\widetilde{\Omega}_{ra}(v,k) = \Omega'_{ra}(v,k)$. Then by (6.50) we must have $r = N-p+a$ and $k = k_a = l_a - m_r$. Moreover, we also have $\lambda_{ra} = \lambda_a$ by (6.48). Take the derivative with respect to v in (6.62) and put $v = -\lambda_a$. Note that the factor $u+m_r+a$ in (6.64) vanishes at $u = -\lambda_a+k_a-1$. Furthermore, as has been shown above, $\Omega_{ra}(-\lambda_a,k_a)\,\theta_{\Lambda'+\delta_{r-1,j}} = 0$; see (6.55). The application of the induction hypothesis finally proves (6.61) in the case $a < r-1$.

If $a = r - 1$ in (6.54), then $\eta = \theta_{\Lambda'}$ is a $\mathrm{Y}_2(\mathfrak{gl}_r)$-singular vector, so that we may use the relation (6.45) to prove (6.61). The same relation applies in the case $p = N - 1$, where we also use the fact that the polynomial (6.60) has zero of multiplicity at least two at $v = -\lambda_{r-1}$. □

Consider the vector $\theta \in V(\lambda, \mu)$ defined in (6.32).

LEMMA 6.2.8. *The vector θ is zero.*

PROOF. We will be proving by induction on the weight of Λ that

$$\Omega_{N-p+1,1}(-\lambda_1, k_1)\,\theta_\Lambda = 0 \tag{6.65}$$

for all admissible arrays Λ such that the parameter $r = r(\Lambda)$ satisfies $r \geqslant N - p + 2$. For the induction base we note that

$$\Omega_{N-p+1,1}(-\lambda_1, k_1)\,\xi = 0.$$

Indeed, using Corollary 1.7.2 we find that the vector on the left-hand side is annihilated by the operators $B_i(u)$ with $i = N - p + 1, \dots, N - 1$. On the other hand, by (6.39) it is also annihilated by $T_{i,i+1}(u)$ with $i = 1, \dots, N - p - 1$. Furthermore,

$$T_{N-p,N-p+1}(u) = [T_{N-p,N-p}(u), E_{N-p,N-p+1}],$$

and we find from (6.34) and (6.44) that the vector is also annihilated by the operator $E_{N-p,N-p+1}$. By Corollary 3.2.8 the vector must be zero.

Suppose now that $w(\Lambda) \prec \lambda$ and $\Lambda \neq \Lambda^0$. Denote the vector on the left-hand side of (6.65) by ϑ_Λ. We will show that $B_i(u)\,\vartheta_\Lambda = 0$ for all $i = 1, \dots, N - 1$. By Lemma 6.2.7, the $\mathrm{Y}_2(\mathfrak{gl}_{N-p+1})$-span of the vector θ_Λ is a highest weight module with the highest weight defined by (6.51) with $r = N - p + 2$. Exactly as above, we find that the vector $T_{N-p,N-p+1}(u)\,\theta_\Lambda$ is zero. Furthermore, the operators $B_i(u)$ with the values $i = N - p + 1, \dots, N - 1$ commute with $\Omega_{N-p+1,1}(-\lambda_1, k_1)$. Therefore, by (6.53), $B_i(u)\,\vartheta_\Lambda$ is a linear combination of the vectors $\Omega_{N-p+1,1}(-\lambda_1, k_1)\,\theta_{\Lambda+\delta_{ij}}$. If $i \geqslant N - p + 2$, then the arrays $\Lambda + \delta_{ij}$ satisfy the condition $r \geqslant N - p + 2$ on the parameter $r = r(\Lambda)$ used in Lemma 6.2.7, and we complete the proof in this case applying the induction hypothesis.

If $i = N - p + 1$, then

$$\theta_{\Lambda+\delta_{N-p+1,j}} = \tau_{N-p+1,j}(-\lambda_{N-p+1,j} - 1)\,\theta_{\Lambda'},$$

for some array Λ' for which the corresponding parameter $r' = r(\Lambda')$ satisfies $r' \geqslant N - p + 2$. However, $\tau_{N-p+1,j}(v)$ commutes with $\Omega_{N-p+1,1}(-\lambda_1, k_1)$, which again ensures that $B_i(u)\,\vartheta_\Lambda = 0$ by the induction hypothesis. □

Finally, if the $\mathrm{Y}(\mathfrak{gl}_N)$-module $L(\lambda) \otimes L(\mu)$ were irreducible, then it would be isomorphic to the highest weight module $V(\lambda, \mu)$. However, this is impossible due to Lemmas 6.2.3 and 6.2.8. So we have a contradiction with the assumption (6.21), thus completing the proof of Theorem 6.2.1. □

6.3. Irreducibility criterion

The composition of the evaluation homomorphism (1.5) and a shift automorphism (1.21) yields a generalized evaluation homomorphism $\mathrm{Y}(\mathfrak{gl}_N) \to \mathrm{U}(\mathfrak{gl}_N)$ given by

$$t_{ij}(u) \mapsto \delta_{ij} + E_{ij}\,(u - a)^{-1}, \tag{6.66}$$

where $a \in \mathbb{C}$ is a fixed constant. Every irreducible representation $L(\lambda)$ of $\mathfrak{gl}_N$ with the highest weight λ becomes a $\mathrm{Y}(\mathfrak{gl}_N)$-module defined via the homomorphism (6.66). We will denote this module by $L(\lambda)_a$. The evaluation module $L(\lambda)$ over $\mathrm{Y}(\mathfrak{gl}_N)$ introduced in Section 3.2 coincides with the module $L(\lambda)_0$. Obviously, $L(\lambda)_a$ is a highest weight representation of the Yangian with the components of the highest weight given by

$$\lambda_i(u) = 1 + \lambda_i(u-a)^{-1}, \qquad i = 1, \dots, N; \tag{6.67}$$

cf. (3.65). Similar to (3.66), we can regard tensor products of the form

$$L(\lambda^{(1)})_{a_1} \otimes L(\lambda^{(2)})_{a_2} \otimes \dots \otimes L(\lambda^{(k)})_{a_k}, \qquad a_i \in \mathbb{C}, \tag{6.68}$$

as Yangian modules. Let us denote the N-tuple $(1, 1, \dots, 1)$ by I. Suppose that all $\lambda^{(i)}$ with $i = 1, \dots, k$ are $\mathfrak{gl}_N$-highest weights so that all representations $L(\lambda^{(i)})$ of $\mathfrak{gl}_N$ are finite-dimensional.

PROPOSITION 6.3.1. *The* $\mathrm{Y}(\mathfrak{gl}_N)$*-module* (6.68) *is irreducible if and only if the* $\mathrm{Y}(\mathfrak{gl}_N)$*-module*

$$L(\lambda^{(1)} - a_1\, I) \otimes L(\lambda^{(2)} - a_2\, I) \otimes \dots \otimes L(\lambda^{(k)} - a_k\, I) \tag{6.69}$$

is irreducible.

PROOF. Suppose that the module (6.68) is irreducible. Let $\zeta^{(r)}$ denote the highest vector of the $\mathfrak{gl}_N$-module $L(\lambda^{(r)})$. As in the proof of Proposition 3.2.9, we find that $\zeta^{(1)} \otimes \dots \otimes \zeta^{(k)}$ is the highest vector of the $\mathrm{Y}(\mathfrak{gl}_N)$-module (6.68) with the highest weight $(\lambda_1(u), \dots, \lambda_N(u))$ where

$$\lambda_i(u) = \Big(1 + \frac{\lambda_i^{(1)}}{u - a_1}\Big) \dots \Big(1 + \frac{\lambda_i^{(k)}}{u - a_k}\Big). \tag{6.70}$$

Consider the automorphism (1.20) of $\mathrm{Y}(\mathfrak{gl}_N)$ where $f(u)$ is the formal series in u^{-1} given by

$$f(u) = (1 - a_1 u^{-1}) \dots (1 - a_k u^{-1}).$$

The composition of the module (6.68) with this automorphism is an irreducible $\mathrm{Y}(\mathfrak{gl}_N)$-module $\widetilde{L}$ with the highest weight $(\widetilde{\lambda}_1(u), \dots, \widetilde{\lambda}_N(u))$ where

$$\widetilde{\lambda}_i(u) = \big(1 + (\lambda_i^{(1)} - a_1)\, u^{-1}\big) \dots \big(1 + (\lambda_i^{(k)} - a_k)\, u^{-1}\big). \tag{6.71}$$

Therefore, $\widetilde{L}$ is isomorphic to a subquotient of (6.69) by Proposition 3.2.9. However, these two modules have the same dimension, and hence they are isomorphic. In particular, the module (6.69) is irreducible. The converse statement is verified by reversing the argument. □

PROPOSITION 6.3.2. *Let* $c \in \mathbb{C}$. *If the* $\mathrm{Y}(\mathfrak{gl}_N)$*-module* (3.66) *is irreducible, then so is the* $\mathrm{Y}(\mathfrak{gl}_N)$*-module obtained from* (3.66) *by the simultaneous shifts of the parameters*

$$\lambda_i^{(r)} \mapsto \lambda_i^{(r)} - c, \qquad r = 1, \dots, k, \quad i = 1, \dots, N.$$

PROOF. If the module (3.66) is irreducible, then so is the module L^c which is the composition of (3.66) with the shift automorphism of $\mathrm{Y}(\mathfrak{gl}_N)$ given by (1.21). The highest weight of L^c is $(\lambda_1^c(u), \dots, \lambda_N^c(u))$ with

$$\lambda_i^c(u) = \Big(1 + \frac{\lambda_i^{(1)}}{u - c}\Big) \dots \Big(1 + \frac{\lambda_i^{(k)}}{u - c}\Big).$$

The proof is completed by repeating the argument of the proof of Proposition 6.3.1 with the use of the automorphism (1.20) of $Y(\mathfrak{gl}_N)$ where $f(u) = (1 - cu^{-1})^k$. $\square$

Given complex numbers a, b, and the $\mathfrak{gl}_N$-highest weights λ and μ, we will now obtain an irreducibility criterion for the $Y(\mathfrak{gl}_N)$-module $L(\lambda)_a \otimes L(\mu)_b$. Due to Proposition 6.3.1, we may assume that the parameters a and b are both zero. Also, Theorem 6.1.1 implies that if the components of λ and μ belong to different $\mathbb{Z}$-cosets in $\mathbb{C}$, then the $Y(\mathfrak{gl}_N)$-module $L(\lambda) \otimes L(\mu)$ is irreducible. Hence, applying Proposition 6.3.2, we may assume that all components of λ and μ are integers.

We will call two disjoint finite subsets A and B of $\mathbb{Z}$ *crossing* if there exist elements $a_1, a_2 \in A$ and $b_1, b_2 \in B$ such that

$$a_1 < b_1 < a_2 < b_2 \qquad \text{or} \qquad b_1 < a_1 < b_2 < a_2.$$

Otherwise, A and B are called *noncrossing*. For any $\mathfrak{gl}_N$-highest weight λ with integer components introduce the subset $\mathcal{A}_\lambda \subset \mathbb{Z}$ by

$$\mathcal{A}_\lambda = \{\lambda_1, \lambda_2 - 1, \dots, \lambda_N - N + 1\}.$$

THEOREM 6.3.3. *Suppose that λ and μ are $\mathfrak{gl}_N$-highest weights with integer components. Then the $Y(\mathfrak{gl}_N)$-module $L(\lambda) \otimes L(\mu)$ is irreducible if and only if the sets $\mathcal{A}_\lambda \setminus \mathcal{A}_\mu$ and $\mathcal{A}_\mu \setminus \mathcal{A}_\lambda$ are noncrossing.*

PROOF. By Theorems 6.1.1 and 6.2.1, we only need to verify the following statement: the sets $\mathcal{A}_\lambda \setminus \mathcal{A}_\mu$ and $\mathcal{A}_\mu \setminus \mathcal{A}_\lambda$ are noncrossing if and only if for all pairs of indices $1 \leqslant i < j \leqslant N$ we have

$$m_j, m_i \notin \langle\, l_j, l_i \,\rangle \qquad \text{or} \qquad l_j, l_i \notin \langle\, m_j, m_i \,\rangle. \tag{6.72}$$

Let us write $\mathrm{Cond}(\mathcal{A}_\lambda, \mathcal{A}_\mu)$ for the condition that $\mathcal{A}_\lambda \setminus \mathcal{A}_\mu$ and $\mathcal{A}_\mu \setminus \mathcal{A}_\lambda$ are noncrossing. We use induction on N. In the case $N = 2$ the statement is obviously true. Let $N \geqslant 3$. Suppose first that (6.72) holds. Set

$$\mathcal{A}_\lambda^- = \{l_1, \dots, l_{N-1}\} \qquad \text{and} \qquad \mathcal{A}_\lambda^+ = \{l_2, \dots, l_N\}$$

and similarly define $\mathcal{A}_\mu^-$ and $\mathcal{A}_\mu^+$. By the induction hypothesis, both conditions $\mathrm{Cond}(\mathcal{A}_\lambda^-, \mathcal{A}_\mu^-)$ and $\mathrm{Cond}(\mathcal{A}_\lambda^+, \mathcal{A}_\mu^+)$ are satisfied. If $m_1 = l_1$, then $\mathrm{Cond}(\mathcal{A}_\lambda, \mathcal{A}_\mu)$ obviously holds. We may assume without loss of generality that $m_1 > l_1$. Let

$$\mathcal{A}_\lambda^- \setminus \mathcal{A}_\mu^- = \{l_{i_1}, \dots, l_{i_k}\}, \qquad \mathcal{A}_\mu^- \setminus \mathcal{A}_\lambda^- = \{m_{j_1}, \dots, m_{j_k}\},$$

where $1 \leqslant i_1 < \dots < i_k \leqslant N$ and $1 = j_1 < \dots < j_k \leqslant N$. We must have for some $a \in \{1, \dots, k\}$ that

$$m_{j_{a+1}} < l_{i_k} < \dots < l_{i_1} < m_{j_a},$$

where the leftmost inequality is ignored when $a = k$. If $2 \leqslant a \leqslant k-1$, then together with $\mathrm{Cond}(\mathcal{A}_\lambda^+, \mathcal{A}_\mu^+)$ this clearly ensures $\mathrm{Cond}(\mathcal{A}_\lambda, \mathcal{A}_\mu)$. Similarly, this is also true when $a = 1$ and $i_1 \geqslant 2$. So, if $a = 1$, then the only case where both $\mathrm{Cond}(\mathcal{A}_\lambda^-, \mathcal{A}_\mu^-)$ and $\mathrm{Cond}(\mathcal{A}_\lambda^+, \mathcal{A}_\mu^+)$ hold but $\mathrm{Cond}(\mathcal{A}_\lambda, \mathcal{A}_\mu)$ does not is the one with the following inequalities between the elements of $\mathcal{A}_\lambda \setminus \mathcal{A}_\mu$ and $\mathcal{A}_\mu \setminus \mathcal{A}_\lambda$:

$$l_N < m_N < m_{j_k} < \dots < m_{j_2} < l_{i_k} < \dots < l_{i_2} < l_1 < m_1.$$

However, in this case $m_N \in \langle\, l_N, l_1 \,\rangle$ and $l_1 \in \langle\, m_N, m_1 \,\rangle$, so that (6.72) is violated for $i = 1$ and $j = N$. An analogous argument shows that if $a = k$, then the only case where both $\mathrm{Cond}(\mathcal{A}_\lambda^-, \mathcal{A}_\mu^-)$ and $\mathrm{Cond}(\mathcal{A}_\lambda^+, \mathcal{A}_\mu^+)$ hold but $\mathrm{Cond}(\mathcal{A}_\lambda, \mathcal{A}_\mu)$ does not is

$$l_N < l_{i_k} < \dots < l_{i_2} < m_N < l_1 < m_{j_k} < \dots < m_{j_2} < m_1.$$

But then $m_N \in \langle l_N, l_1 \rangle$ and $l_1 \in \langle m_N, m_1 \rangle$, which contradicts (6.72) again.

Conversely, suppose that $\mathrm{Cond}(\mathcal{A}_\lambda, \mathcal{A}_\mu)$ holds. This condition clearly implies both $\mathrm{Cond}(\mathcal{A}^-_\lambda, \mathcal{A}^-_\mu)$ and $\mathrm{Cond}(\mathcal{A}^+_\lambda, \mathcal{A}^+_\mu)$, and so, by the induction hypothesis, (6.72) holds for all pairs $i < j$ with the possible exception of $(i, j) = (1, N)$. If the latter condition fails, then we have

$$\begin{cases} m_N \in \langle l_N, l_1 \rangle \\ \quad l_1 \in \langle m_N, m_1 \rangle \end{cases} \qquad \text{or} \qquad \begin{cases} l_N \in \langle m_N, m_1 \rangle \\ m_1 \in \langle l_N, l_1 \rangle. \end{cases}$$

However, in each of the two cases this contradicts $\mathrm{Cond}(\mathcal{A}_\lambda, \mathcal{A}_\mu)$. □

An irreducibility criterion for the $\mathrm{Y}(\mathfrak{gl}_N)$-modules of the form (6.68) will be obtained in Section 6.5. In order to prove the relevant theorem we need to develop an alternative approach to the Yangian evaluation modules based on the so-called fusion procedure for the symmetric group.

6.4. Fusion procedure for the symmetric group

Let us recall some well known facts about the representations of the symmetric group $\mathfrak{S}_k$; see e.g. James and Kerber [**194**] or Sagan [**411**]. A *partition* λ is a sequence $\lambda = (\lambda_1, \dots, \lambda_n)$ of integers such that $\lambda_1 \geqslant \dots \geqslant \lambda_n \geqslant 0$. We will identify a partition λ with its *diagram*, which is a left-justified array of rows of unit cells such that the top row contains λ_1 cells, the next row contains λ_2 cells, etc. The number of nonempty rows in the diagram is called the *length* of λ. The cells of a diagram will be identified by their row and column numbers so that the cell (i, j) is found at the intersection of row i and column j. The *content* of this cell is the number $j - i$. If $\lambda_1 + \dots + \lambda_n = k$, then λ is a *partition of* k, written $\lambda \vdash k$. A cell of λ is called *removable* if its removal leaves a diagram. Similarly, a cell is *addable* to λ if the union of λ and the cell is a diagram. We will write $\mu \nearrow \lambda$ if λ is obtained from μ by adding one cell. A *tableau* $\mathcal{U}$ *of shape* λ (or a λ-*tableau* $\mathcal{U}$) is obtained by filling in the cells of the diagram with the numbers $\{1, \dots, k\}$ so that each cell contains exactly one number. We write $\mathrm{sh}(\mathcal{U}) = \lambda$ if the shape of $\mathcal{U}$ is λ. A tableau $\mathcal{U}$ is called *standard* if its entries strictly increase along the rows and down the columns.

The irreducible representations of $\mathfrak{S}_k$ over $\mathbb{C}$ are parameterized by partitions of k. Given a partition λ of k denote the corresponding irreducible representation of $\mathfrak{S}_k$ by V_λ. The vector space V_λ is equipped with an $\mathfrak{S}_k$-invariant inner product $(\ ,\ )$. The orthonormal *Young basis* $\{v_\mathcal{U}\}$ of V_λ is parameterized by the set of standard λ-tableaux $\mathcal{U}$. The action of the generators $s_i = (i, i+1)$ of $\mathfrak{S}_k$ in the Young basis is described as follows. We denote by $c_j = c_j(\mathcal{U})$ the content of the cell occupied by the number j in a standard λ-tableau $\mathcal{U}$. Then for any $i \in \{1, \dots, k-1\}$ we have

$$s_i \cdot v_\mathcal{U} = d v_\mathcal{U} + \sqrt{1 - d^2}\, v_{s_i \mathcal{U}},$$

where $d = (c_{i+1} - c_i)^{-1}$, the tableau $s_i\mathcal{U}$ is obtained from $\mathcal{U}$ by swapping the entries i and $i+1$, and we assume $v_{s_i\mathcal{U}} = 0$ if the tableau $s_i\mathcal{U}$ is not standard.

For any $k \geqslant 2$ we regard $\mathfrak{S}_{k-1}$ as the natural subgroup of $\mathfrak{S}_k$ which consists of the permutations which fix k. The restriction of V_λ to the subgroup $\mathfrak{S}_{k-1}$ is described by the *branching rule*

$$V_\lambda\big|_{\mathfrak{S}_{k-1}} \cong \bigoplus_{\mu \nearrow \lambda} V'_\mu, \tag{6.73}$$

where V'_μ denotes the irreducible representation of $\mathfrak{S}_{k-1}$ associated with the diagram μ obtained from λ by removing one cell. The Young basis $\{v_{\mathcal{U}}\}$ of V_λ is consistent with the decomposition (6.73) in the following sense. For a fixed μ with $\mu \nearrow \lambda$ consider the subset of the standard λ-tableaux $\mathcal{U}$ such that the removal of the cell occupied by k leaves a standard tableau of shape μ. Then the linear span of the vectors $v_{\mathcal{U}}$ parameterized by the tableaux of this subset is a representation of $\mathfrak{S}_{k-1}$ isomorphic to V'_μ.

The group algebra $\mathbb{C}[\mathfrak{S}_k]$ is isomorphic to the direct sum of matrix algebras

$$\mathbb{C}[\mathfrak{S}_k] \cong \bigoplus_{\lambda \vdash k} \mathrm{Mat}_{f_\lambda}(\mathbb{C}),$$

where $f_\lambda = \dim V_\lambda$. The matrix units $e_{\mathcal{U}\mathcal{U}'} \in \mathrm{Mat}_{f_\lambda}(\mathbb{C})$ are parameterized by pairs of standard λ-tableaux $\mathcal{U}$ and $\mathcal{U}'$. We will identify $\mathbb{C}[\mathfrak{S}_k]$ with the direct sum of matrix algebras by the formulas

$$e_{\mathcal{U}\mathcal{U}'} = \frac{f_\lambda}{k!}\,\phi_{\mathcal{U}\mathcal{U}'}, \tag{6.74}$$

where $\phi_{\mathcal{U}\mathcal{U}'}$ is the matrix element corresponding to the basis vectors $v_{\mathcal{U}}$ and $v_{\mathcal{U}'}$ of the representation V_λ,

$$\phi_{\mathcal{U}\mathcal{U}'} = \sum_{s \in \mathfrak{S}_k} (s \cdot v_{\mathcal{U}}, v_{\mathcal{U}'}) \cdot s^{-1} \in \mathbb{C}[\mathfrak{S}_k]. \tag{6.75}$$

For the diagonal elements we will simply write $e_{\mathcal{U}} = e_{\mathcal{U}\mathcal{U}}$ and $\phi_{\mathcal{U}} = \phi_{\mathcal{U}\mathcal{U}}$.

Consider the *Jucys–Murphy* elements $x_1, \dots, x_k \in \mathbb{C}[\mathfrak{S}_k]$ defined by

$$x_1 = 0, \qquad x_i = (1\,i) + (2\,i) + \cdots + (i-1\,i), \quad i = 2, \dots, k.$$

These elements generate a commutative subalgebra of $\mathbb{C}[\mathfrak{S}_k]$. Moreover, x_k commutes with all elements of $\mathfrak{S}_{k-1}$. The vectors of the Young basis are eigenvectors for the action of x_i on V_λ. For any standard λ-tableau $\mathcal{U}$ we have

$$x_i \cdot v_{\mathcal{U}} = c_i(\mathcal{U})\, v_{\mathcal{U}}, \qquad i = 1, \dots, k. \tag{6.76}$$

The branching properties of the Young basis imply the following properties of the matrix units. If $\mathcal{V}$ is a given standard tableau with the entries $1, \dots, k-1$, then

$$e_{\mathcal{V}} = \sum_{\mathcal{V} \nearrow \mathcal{U}} e_{\mathcal{U}}, \tag{6.77}$$

where $\mathcal{V} \nearrow \mathcal{U}$ means that the standard tableau $\mathcal{U}$ is obtained from $\mathcal{V}$ by adding one cell with the entry k. The relations (6.76) imply

$$x_i\, e_{\mathcal{U}} = e_{\mathcal{U}}\, x_i = c_i(\mathcal{U})\, e_{\mathcal{U}}, \qquad i = 1, \dots, k \tag{6.78}$$

for any standard λ-tableau $\mathcal{U}$. In particular, we have the identity in $\mathbb{C}[\mathfrak{S}_k]$,

$$x_k = \sum_{\lambda \vdash k} \sum_{\mathrm{sh}(\mathcal{U}) = \lambda} c_k(\mathcal{U})\, e_{\mathcal{U}}, \tag{6.79}$$

so that x_k can be viewed as a diagonal matrix.

Given any diagonal matrix $D = \mathrm{diag}\,[d_1, \dots, d_m]$ and a function $g(v)$ of a complex variable v, the expression $g(D)$ will be understood, as usual, as the diagonal matrix $\mathrm{diag}\,[g(d_1), \dots, g(d_m)]$.

Now let $k \geqslant 2$ and let λ be a partition of k. Fix a standard λ-tableau $\mathcal{U}$ and denote by $\mathcal{V}$ the standard tableau obtained from $\mathcal{U}$ by removing the cell α occupied by k. Then the shape of $\mathcal{V}$ is a diagram which we denote by μ. Let u be a complex variable. Due to (6.79), the expression

$$e_{\mathcal{V}}\,\frac{u-c}{u-x_k}, \qquad c=c_k(T), \tag{6.80}$$

is a rational function in u with values in $\mathbb{C}[\mathfrak{S}_k]$.

PROPOSITION 6.4.1. *We have the relation in* $\mathbb{C}[\mathfrak{S}_k]$,

$$e_{\mathcal{U}}=e_{\mathcal{V}}\,\frac{(x_k-a_1)\dots(x_k-a_l)}{(c-a_1)\dots(c-a_l)}, \tag{6.81}$$

where $a_1,\dots,a_l$ *are the contents of all addable cells of* μ *except for* α, *while* c *is the content of the latter. Moreover, the rational function* (6.80) *is regular at* $u=c$, *and we have*

$$e_{\mathcal{U}}=e_{\mathcal{V}}\,\frac{u-c}{u-x_k}\Big|_{u=c}. \tag{6.82}$$

PROOF. Relation (6.81) follows by the application of (6.77) to $e_{\mathcal{V}}$ and then the use of (6.78). Similarly, by (6.77) and (6.78) we have

$$e_{\mathcal{V}}\,\frac{u-c}{u-x_k}=\sum_{\mathcal{V}\nearrow\mathcal{U}'}e_{\mathcal{U}'}\,\frac{u-c}{u-c_k(\mathcal{U}')}=e_{\mathcal{U}}+\sum_{\mathcal{V}\nearrow\mathcal{U}',\,\mathcal{U}'\neq\mathcal{U}}e_{\mathcal{U}'}\,\frac{u-c}{u-c_k(\mathcal{U}')}.$$

Since $c_k(\mathcal{U}')\neq c$ for all standard tableaux $\mathcal{U}'$ distinct from $\mathcal{U}$, the value of this rational function at $u=c$ is $e_{\mathcal{U}}$. □

Obviously, $e_{\mathcal{U}_0}=1$ if $\mathcal{U}_0$ is the (1)-tableau with the entry 1. Therefore, (6.81) yields an explicit expression for $e_{\mathcal{U}}$ in terms of the Jucys–Murphy elements $x_2,\dots,x_k$. In the following corollary we use the assumptions of Proposition 6.4.1.

COROLLARY 6.4.2. *We have*

$$\phi_{\mathcal{U}}=H_{\lambda,\mu}\,\phi_{\mathcal{V}}\,\frac{u-c}{u-x_k}\Big|_{u=c} \tag{6.83}$$

with

$$H_{\lambda,\mu}=\frac{(a_1-c)\dots(a_p-c)(c-a_{p+1})\dots(c-a_l)}{(b_1-c)\dots(b_q-c)(c-b_{q+1})\dots(c-b_r)}, \tag{6.84}$$

where the numbers $a_1,\dots,a_p,c,a_{p+1},\dots,a_l$ *are the contents of all addable cells of* μ *and* $b_1,\dots,b_q,c,b_{q+1},\dots,b_r$ *are the contents of all removable cells of* λ *with both sequences written in decreasing order.*

PROOF. We use (6.74) and note that by the hook length formula, the quotient $k!/f_\lambda$ equals the product of the hooks of λ. Hence the coefficient $H_{\lambda,\mu}$ is the ratio of the product of hooks of λ and the product of hooks of μ, which implies (6.84). □

REMARK 6.4.3. Consider the character χ_λ of V_λ,

$$\chi_\lambda=\sum_{s\in\mathfrak{S}_k}\chi_\lambda(s)\,s\in\mathbb{C}[\mathfrak{S}_k].$$

We have

$$\chi_\lambda=\sum_{\mathrm{sh}(\mathcal{U})=\lambda}\phi_{\mathcal{U}},$$

summed over all standard λ-tableaux $\mathcal{U}$. Formula (6.81) implies the following recurrence relation for the normalized characters $\widehat{\chi}_\lambda = f_\lambda\, \chi_\lambda/k!$:

$$\widehat{\chi}_\lambda = \sum_{\mu\nearrow\lambda} \widehat{\chi}_\mu \frac{(x_k - a_1)\dots(x_k - a_l)}{(c - a_1)\dots(c - a_l)},$$

summed over diagrams μ. Equivalently,

$$\chi_\lambda = \sum_{\mu\nearrow\lambda} \chi_\mu \frac{(a_1 - x_k)\dots(a_p - x_k)(x_k - a_{p+1})\dots(x_k - a_l)}{(b_1 - c)\dots(b_q - c)(c - b_{q+1})\dots(c - b_r)},$$

with the notation introduced in Corollary 6.4.2. □

For any distinct indices $i, j \in \{1, \dots, k\}$ introduce the rational function in two variables u, v with values in the group algebra $\mathbb{C}[\mathfrak{S}_k]$ by

$$\rho_{ij}(u, v) = 1 - \frac{(i\,j)}{u - v}. \tag{6.85}$$

Suppose that λ is a partition of k and let $\mathcal{U}$ be a standard λ-tableau. Set $c_i = c_i(\mathcal{U})$ for $i = 1, \dots, k$.

Proposition 6.4.4. *Let r be a fixed index, $r \geqslant k + 1$. We have the equalities of rational functions in u valued in $\mathbb{C}[\mathfrak{S}_r]$,*

$$\phi_{\mathcal{U}}\, \rho_{k,r}(-c_k, u)\dots\rho_{1r}(-c_1, u) = \rho_{1r}(-c_1, u)\dots\rho_{k,r}(-c_k, u)\, \phi_{\mathcal{U}} \tag{6.86}$$
$$= \phi_{\mathcal{U}}\Big(1 + \frac{(1\,r) + (2\,r) + \dots + (k\,r)}{u}\Big).$$

Proof. We argue by induction on k noting that the relations hold for $k = 1$. Suppose that $k \geqslant 2$. By (6.77) we can write $\phi_{\mathcal{U}}$ as the product

$$\phi_{\mathcal{U}} = \gamma \cdot \phi_{\mathcal{U}}\, \phi_{\mathcal{V}},$$

where $\mathcal{V}$ is the standard tableau obtained from $\mathcal{U}$ by removing the cell occupied by k and γ is a nonzero constant. Hence, using the induction hypothesis we can write the leftmost expression in (6.86) as

$$\begin{aligned}
&\gamma \cdot \phi_{\mathcal{U}}\, \phi_{\mathcal{V}}\, \rho_{k,r}(-c_k, u)\dots\rho_{1r}(-c_1, u)\\
&= \gamma \cdot \phi_{\mathcal{U}}\, \rho_{k,r}(-c_k, u)\, \phi_{\mathcal{V}}\, \rho_{k-1,r}(-c_{k-1}, u)\dots\rho_{1r}(-c_1, u)\\
&= \gamma \cdot \phi_{\mathcal{U}}\, \rho_{k,r}(-c_k, u)\, \phi_{\mathcal{V}}\Big(1 + \frac{(1\,r) + (2\,r) + \dots + (k-1\,r)}{u}\Big)
\end{aligned}$$

which equals

$$\phi_{\mathcal{U}}\Big(1 + \frac{(k\,r)}{u + c_k}\Big)\Big(1 + \frac{(1\,r) + (2\,r) + \dots + (k-1\,r)}{u}\Big). \tag{6.87}$$

Observe that

$$(k\,r)\big((1\,r) + (2\,r) + \dots + (k-1\,r)\big) = x_k\,(k\,r)$$

and recall that $\phi_{\mathcal{U}}\, x_k = c_k\, \phi_{\mathcal{U}}$ by (6.78). Hence, (6.87) simplifies to give the rightmost expression in (6.86).

The second equality in (6.86) is verified in the same way with the use of the decomposition $\phi_{\mathcal{U}} = \gamma \cdot \phi_{\mathcal{V}}\, \phi_{\mathcal{U}}$ implied by (6.77). □

Corollary 6.4.5. *In the notation of Proposition 6.4.4 we have the equality*

$$\phi_{\mathcal{U}}\, \rho_{1r}(u, -c_1)\dots\rho_{k,r}(u, -c_k) = \rho_{k,r}(u, -c_k)\dots\rho_{1r}(u, -c_1)\, \phi_{\mathcal{U}}.$$

PROOF. This follows from Proposition 6.4.4 with the use of the relation

$$\rho_{ij}(u,v)\,\rho_{ij}(v,u) = 1 - \frac{1}{(u-v)^2}, \tag{6.88}$$

which is immediate from (6.85). □

An expression for the matrix element $\phi_{\mathcal{U}}$ is provided by the *fusion procedure* described in the next theorem. Equip the set of all pairs (i,j) with $1 \leqslant i < j \leqslant k$ with the reverse lexicographical ordering so that (i,j) precedes (r,l) if $j < l$, or $j = l$ and $i < r$. Take k complex variables $u_1, \dots, u_k$ and consider the ordered product

$$\phi(u_1, \dots, u_k) = \prod_{1 \leqslant i < j \leqslant k}^{\longrightarrow} \rho_{ij}(u_i, u_j). \tag{6.89}$$

This product can be written in the form

$$\phi(u_1, \dots, u_k) = \sum_{s \in \mathfrak{S}_k} \phi_s(u_1, \dots, u_k) \cdot s,$$

where each $\phi_s(u_1, \dots, u_k)$ is a uniquely determined rational function in $u_1, \dots, u_k$ with values in $\mathbb{C}$. Suppose that λ is a partition of k and let $\mathcal{U}$ be a standard λ-tableau. As before, set $c_i = c_i(\mathcal{U})$ for $i = 1, \dots, k$. We would like to evaluate $\phi(u_1, \dots, u_k)$ at $u_i = c_i$ for all $i = 1, \dots, k$. However, the denominators of some rational functions $\phi_s(u_1, \dots, u_k)$ can vanish at these values of the u_i. The next theorem shows that the *consecutive* evaluations still make sense. At each step, the rational function which is to be evaluated at $u_i = c_i$ turns out to have a removable singularity at this point thus allowing us to define its value. We will express this by saying that the value is *well-defined*. Before formulating the general result let us consider an example.

EXAMPLE 6.4.6. Let $\lambda = (2^2)$ so that $k = 4$. Take the standard λ-tableau

$$\mathcal{U} = \begin{array}{|c|c|} \hline 1 & 2 \\ \hline 3 & 4 \\ \hline \end{array}$$

The contents are $c_1 = 0$, $c_2 = 1$, $c_3 = -1$, $c_4 = 0$. The rational function $\phi(u_1, u_2, u_3, u_4)$ is given by

$$\phi(u_1,u_2,u_3,u_4) = \rho_{12}(u_1,u_2)\,\rho_{13}(u_1,u_3)\,\rho_{23}(u_2,u_3) \\ \times \rho_{14}(u_1,u_4)\,\rho_{24}(u_2,u_4)\,\rho_{34}(u_3,u_4).$$

Multiplying the elements we find $\phi_{(14)}(u_1, u_2, u_3, u_4) = \alpha/\beta$, where

$$\alpha = -(u_1-u_2)(u_1-u_3)(u_2-u_3)(u_2-u_4)(u_3-u_4) - (u_1-u_2)(u_2-u_3)(u_2-u_4) \\ - (u_1-u_3)(u_2-u_3)(u_3-u_4) - u_1 + u_3 - u_2 + u_4$$

and

$$\beta = (u_1-u_2)(u_1-u_3)(u_2-u_3)(u_1-u_4)(u_2-u_4)(u_3-u_4).$$

The denominator β vanishes at $u_1 = 0$, $u_2 = 1$, $u_3 = -1$, $u_4 = 0$ so that the corresponding value of $\phi_{(14)}(u_1, u_2, u_3, u_4)$ is not defined. However,

$$\phi_{(14)}(0,1,-1,u) = \frac{2u^2+u}{2u^3-2u} = \frac{2u+1}{2u^2-2},$$

and this function is well-defined at $u = 0$ with the value $-1/2$. □

THEOREM 6.4.7. *The consecutive evaluations*

$$\phi(u_1, \dots, u_k)\big|_{u_1=c_1}\big|_{u_2=c_2} \cdots \big|_{u_k=c_k}$$

of the rational function $\phi(u_1, \dots, u_k)$ are well-defined. The corresponding value coincides with the matrix element $\phi_{\mathcal{U}}$.

PROOF. Clearly, it is sufficient to consider the last evaluation $u_k = c_k$. We argue by induction on k and suppose that $k \geqslant 2$. Using the induction hypothesis and setting $u = u_k$ we get

$$\phi(u_1, \dots, u_k)\big|_{u_1=c_1} \cdots \big|_{u_{k-1}=c_{k-1}} = \phi_{\mathcal{V}}\, \rho_{1k}(c_1, u) \dots \rho_{k-1,k}(c_{k-1}, u),$$

where the standard tableau $\mathcal{V}$ is obtained from $\mathcal{U}$ by removing the cell occupied by k. Let us verify that the expression on the right-hand side can be given by

$$\phi_{\mathcal{V}}\, \rho_{1k}(c_1, u) \dots \rho_{k-1,k}(c_{k-1}, u) = \prod_{i=1}^{k-1} \Big(1 - \frac{1}{(u - c_i)^2}\Big)\, \phi_{\mathcal{V}}\, (1 - x_k\, u^{-1})^{-1}. \tag{6.90}$$

Note that due to (6.79), the expression $(1 - x_k\, u^{-1})^{-1}$ is a well-defined rational function in u valued in $\mathbb{C}[\mathfrak{S}_k]$. Since x_k commutes with $\phi_{\mathcal{V}}$, relation (6.90) is equivalent to

$$\phi_{\mathcal{V}}\, \rho_{k-1,k}(-c_{k-1}, -u) \dots \rho_{1k}(-c_1, -u) = \phi_{\mathcal{V}}\, (1 - x_k\, u^{-1})$$

due to (6.88), where we also observed that $\rho_{ij}(v, u) = \rho_{ij}(-u, -v)$. However, the latter relation holds by Proposition 6.4.4. Now write the right-hand side of (6.90) as

$$\prod_{i=1}^{k-1} \Big(1 - \frac{1}{(u - c_i)^2}\Big) \frac{u}{u - c_k} \cdot \phi_{\mathcal{V}}\, \frac{u - c_k}{u - x_k}. \tag{6.91}$$

Note that the product

$$\prod_{i=1}^{k-1} \Big(1 - \frac{1}{(u - c_i)^2}\Big) \frac{u}{u - c_k}$$

depends only on the shape μ of $\mathcal{V}$ so we may choose a particular tableau $\mathcal{V}$ for its evaluation. We take $\mathcal{V}$ to be the *row tableau* $\mathcal{V}^r$ obtained by filling in the cells of μ with the numbers $1, \dots, k$ by consecutive rows from left to right in each row. A short calculation shows that this product is regular at $u = c_k$ and the value equals the number $H_{\lambda,\mu}$ given by (6.84) with $c = c_k$. Thus, due to (6.83), the value of (6.91) at $u = c_k$ is $\phi_{\mathcal{U}}$. □

EXAMPLE 6.4.8. Using the notation of Example 6.4.6, we have

$$\phi(0, 1, -1, u) = \Big(1 + (1\,2)\Big)\Big(1 - (1\,3)\Big)\Big(1 - \frac{(2\,3)}{2}\Big) \times \Big(1 + \frac{(1\,4)}{u}\Big)\Big(1 + \frac{(2\,4)}{u - 1}\Big)\Big(1 + \frac{(3\,4)}{u + 1}\Big).$$

By Theorem 6.4.7, this rational function is well-defined at $u = 0$. The corresponding value is

$$\phi_{\mathcal{U}} = \phi(0,1,-1,0) = \frac{1}{2}\Big(1+(1\,2)\Big)\Big(1-(1\,3)\Big)\Big(2-(2\,3)\Big) \\ \times\Big(2-(1\,4)-(2\,4)-(3\,4)\Big)\Big(2+(1\,4)+(2\,4)+(3\,4)\Big).$$

□

Given any standard λ-tableau $\mathcal{U}$, introduce the subgroup $\mathfrak{S}_{\mathcal{U}}$ (respectively, $\mathfrak{S}'_{\mathcal{U}}$) of $\mathfrak{S}_k$ which consist of the permutations preserving the sets of entries appearing in every row (resp. every column) of $\mathcal{U}$. Set

$$h_{\mathcal{U}} = \sum_{s\in\mathfrak{S}_{\mathcal{U}}} s, \qquad a_{\mathcal{U}} = \sum_{s\in\mathfrak{S}'_{\mathcal{U}}} \operatorname{sgn} s\cdot s. \tag{6.92}$$

These are elements of the group algebra $\mathbb{C}[\mathfrak{S}_k]$, and their product $h_{\mathcal{U}}a_{\mathcal{U}}$ is the *Young symmetrizer* associated with $\mathcal{U}$. The following well known relation for the matrix element $\phi_{\mathcal{U}^r}$ corresponding to the row tableau $\mathcal{U}^r$ of shape λ can be obtained from Weyl [**464**, Section IV.2]:

$$\phi_{\mathcal{U}^r} = \frac{h_{\mathcal{U}^r}a_{\mathcal{U}^r}h_{\mathcal{U}^r}}{\lambda_1!\dots\lambda_l!}. \tag{6.93}$$

Moreover, there exists an invertible element $h\in\mathbb{C}[\mathfrak{S}_k]$ such that

$$h_{\mathcal{U}^r}a_{\mathcal{U}^r}h_{\mathcal{U}^r} = h_{\mathcal{U}^r}a_{\mathcal{U}^r}h. \tag{6.94}$$

6.5. Multiple tensor products

The symmetric group $\mathfrak{S}_k$ acts naturally on the tensor product space

$$\mathbb{C}^N\otimes\mathbb{C}^N\otimes\dots\otimes\mathbb{C}^N, \qquad k \text{ factors}, \tag{6.95}$$

by permuting the factors. On the other hand, $\mathbb{C}^N$ carries the vector representation of the Lie algebra $\mathfrak{gl}_N$ so that the vector space (6.95) is a representation of $\mathfrak{gl}_N$. The mutually commuting actions of $\mathfrak{S}_k$ and $\mathfrak{gl}_N$ on the space (6.95) are described in the classical Schur–Weyl duality. In particular, this leads to a construction of the so-called polynomial representations of $\mathfrak{gl}_N$ as submodules of (6.95). More precisely, suppose that λ is a partition of k whose length does not exceed N. We will write $\lambda = (\lambda_1,\dots,\lambda_N)$, completing the N-tuple by zeros if necessary. Consider an arbitrary standard λ-tableau $\mathcal{U}$ and let $\Phi_{\mathcal{U}}\in\operatorname{End}(\mathbb{C}^N)^{\otimes k}$ denote the image of the matrix element $\phi_{\mathcal{U}}$ under the action of $\mathfrak{S}_k$ on (6.95). Then the subspace

$$L_{\mathcal{U}} = \Phi_{\mathcal{U}}(\mathbb{C}^N)^{\otimes k}$$

is a $\mathfrak{gl}_N$-submodule of (6.95). This submodule is irreducible and isomorphic to the highest weight representation $L(\lambda)$.

Note that if $\mathcal{U} = \mathcal{U}^r$ is the row tableau of shape λ, then by (6.93) and (6.94) the subspace $L_{\mathcal{U}^r}$ coincides with the image of the Young symmetrizer,

$$L_{\mathcal{U}^r} = H_{\mathcal{U}^r}A_{\mathcal{U}^r}(\mathbb{C}^N)^{\otimes k},$$

where $H_{\mathcal{U}^r}$ and $A_{\mathcal{U}^r}$ are the respective images in $\operatorname{End}(\mathbb{C}^N)^{\otimes k}$ of the elements $h_{\mathcal{U}^r}$ and $a_{\mathcal{U}^r}$ defined in (6.92).

Recall that the evaluation module $L(\lambda)$ over the Yangian $\mathrm{Y}(\mathfrak{gl}_N)$ is defined via the evaluation homomorphism (1.5). Let us write this homomorphism in a matrix form,

$$\pi_N : T(u) \mapsto 1 + E\, u^{-1},$$

interpreting the matrix E as the element

$$E = \sum_{i,j=1}^{N} e_{ij} \otimes E_{ij} \in \operatorname{End} \mathbb{C}^N \otimes \mathrm{U}(\mathfrak{gl}_N),$$

cf. (1.7). In the vector representation $\mathbb{C}^N$ of $\mathfrak{gl}_N$ we have $E_{ij} \mapsto e_{ij}$ and so the image of the matrix E under the action of $\mathfrak{gl}_N$ on (6.95) can be written as

$$\sum_{a=1}^{k} \sum_{i,j=1}^{N} e_{ij} \otimes 1^{\otimes(a-1)} \otimes e_{ij} \otimes 1^{\otimes(k-a)}.$$

This is an element of the algebra

$$\operatorname{End} \mathbb{C}^N \otimes \operatorname{End} (\mathbb{C}^N)^{\otimes k}, \tag{6.96}$$

and we identify the endomorphism algebra of the vector space (6.95) with the tensor product of the algebras $\operatorname{End} \mathbb{C}^N$. Labeling the tensor factors in (6.96) with the numbers $0, 1, \dots, k$ and using the permutation operator (1.11), we can write the image of the transposed matrix $T^t(u)$ in the representation $L_{\mathcal{U}}$ as

$$T^t(u) \mapsto 1 + \big(P_{01} + P_{02} + \dots + P_{0k}\big)\, u^{-1}. \tag{6.97}$$

In particular, if $k = 1$, then this takes the form

$$T^t(u) \mapsto R_{01}(-u), \tag{6.98}$$

where we have used the R-matrix (1.12). More generally, taking the composition of (6.98) with an appropriate shift automorphism (1.21) of $\mathrm{Y}(\mathfrak{gl}_N)$, for any complex number z we make the vector space $\mathbb{C}^N$ into a representation of $\mathrm{Y}(\mathfrak{gl}_N)$ by the assignment

$$T^t(u) \mapsto R_{01}(-u - z).$$

Furthermore, applying the comultiplication (1.42), we get a representation of $\mathrm{Y}(\mathfrak{gl}_N)$ in the vector space (6.95) defined by the assignment

$$T^t(u) \mapsto R_{01}(-u - z_1)\, R_{02}(-u - z_2) \dots R_{0k}(-u - z_k),$$

where $z_1, \dots, z_k$ are fixed complex numbers.

Now consider a standard λ-tableau $\mathcal{U}$ and for any index $r = 1, \dots, k$ denote by $c_r = c_r(\mathcal{U})$ the content of the cell of $\mathcal{U}$ occupied by r.

PROPOSITION 6.5.1. *The subspace $L_{\mathcal{U}}$ of the vector space* (6.95) *is stable under the action of $\mathrm{Y}(\mathfrak{gl}_N)$ defined by*

$$T^t(u) \mapsto R_{01}(-u - c_1)\, R_{02}(-u - c_2) \dots R_{0k}(-u - c_k). \tag{6.99}$$

Moreover, the representation of $\mathrm{Y}(\mathfrak{gl}_N)$ on $L_{\mathcal{U}}$ obtained by restriction is isomorphic to the evaluation module $L(\lambda)$.

PROOF. Observe that $R_{ij}(u-v)$ coincides with the image of the element $\rho_{ij}(u,v)$ defined in (6.85) under the action of the symmetric group $\mathfrak{S}_{k+1}$ on the tensor product of the vector spaces $\mathbb{C}^N$. Hence, applying Proposition 6.4.4 with r replaced by 0, we obtain

$$R_{01}(-u-c_1)\,R_{02}(-u-c_2)\dots R_{0k}(-u-c_k)\,\Phi_{\mathcal{U}} = \Phi_{\mathcal{U}}\Big(1+\frac{P_{01}+P_{02}+\dots+P_{0k}}{u}\Big).$$

This implies the first part of the proposition. The second part follows from (6.97) by taking into account that $P_{01}+P_{02}+\dots+P_{0k}$ commutes with $\Phi_{\mathcal{U}}$. □

Now consider the tensor product algebra

$$\operatorname{End}(\mathbb{C}^N)^{\otimes k}\otimes \mathrm{Y}(\mathfrak{gl}_N) \tag{6.100}$$

with k copies of the endomorphism algebra which we label with the numbers $1,\dots,k$; cf. (1.36). Given a standard λ-tableau $\mathcal{U}$ as above, introduce the corresponding formal series in u^{-1} with coefficients in the algebra (6.100) by

$$T_{\mathcal{U}}(u)=T_k^t(u-c_k)\dots T_1^t(u-c_1).$$

PROPOSITION 6.5.2. *All coefficients of the series $T_{\mathcal{U}}(u)$ preserve the subspace $L_{\mathcal{U}}$ of the vector space* (6.95).

PROOF. The image of the element (6.89) in the algebra $\operatorname{End}(\mathbb{C}^N)^{\otimes k}$ coincides with the element $R(u_1,\dots,u_k)$ given by (1.45) with $m=k$. Applying the automorphism (1.32) of $\mathrm{Y}(\mathfrak{gl}_N)$ to the relation of Proposition 1.6.1, we get

$$R(u_1,\dots,u_k)\,T_1^t(-u_1)\dots T_k^t(-u_k)=T_k^t(-u_k)\dots T_1^t(-u_1)\,R(u_1,\dots,u_k).$$

Hence, specifying the variables consecutively by $u_i=-u+c_i$ for $i=1,\dots,k$ and applying Theorem 6.4.7, we come to the relation

$$\Phi_{\mathcal{U}}\,T_1^t(u-c_1)\dots T_k^t(u-c_k)=T_k^t(u-c_k)\dots T_1^t(u-c_1)\,\Phi_{\mathcal{U}}$$

which implies the desired property of $T_{\mathcal{U}}(u)$. □

Thus, using the restriction to the subspace $L_{\mathcal{U}}$, we may regard the coefficients of the series $T_{\mathcal{U}}(u)$ as elements of the algebra

$$\operatorname{End}L_{\mathcal{U}}\otimes\mathrm{Y}(\mathfrak{gl}_N).$$

Now take a positive integer m and consider a partition μ of m whose length does not exceed N. Let $\mathcal{V}$ be a standard μ-tableau and let $d_1,\dots,d_m$ be the contents of the cells of $\mathcal{V}$ occupied by the numbers $1,\dots,m$, respectively. Consider the tensor product algebra

$$\operatorname{End}(\mathbb{C}^N)^{\otimes k}\otimes\operatorname{End}(\mathbb{C}^N)^{\otimes m}$$

and label the factors $\operatorname{End}\mathbb{C}^N$ with the numbers $1,\dots,k+m$. Introduce the rational function $R_{\mathcal{U}\mathcal{V}}(u)$ in a variable u with values in this algebra by

$$R_{\mathcal{U}\mathcal{V}}(u)=\prod_{i=1,\dots,k}^{\longleftarrow}\Big(\prod_{j=1,\dots,m}^{\longrightarrow}R_{i,k+j}(u+c_i-d_j)\Big),$$

where both products are taken in the orders indicated by the arrows: the values of the index j are increasing in the product while the values of i are decreasing. Similarly, we set

$$R^{(21)}_{\mathcal{V}\mathcal{U}}(u) = \overleftarrow{\prod_{j=1,\dots,m}} \Bigg(\overrightarrow{\prod_{i=1,\dots,k}} R_{k+j,i}(u + d_j - c_i) \Bigg).$$

PROPOSITION 6.5.3. *We have the identity*

$$R_{\mathcal{U}\mathcal{V}}(-u)\, R^{(21)}_{\mathcal{V}\mathcal{U}}(u) = \prod_{i=1}^{k} \prod_{j=1}^{m} \Big(1 - \frac{1}{(u - c_i + d_j)^2}\Big).$$

PROOF. This follows by the repeated application of the relation (1.13) and the fact that $R_{ij}(u)$ and $R_{rl}(v)$ commute if the indices i, j, r, l are distinct. □

Let us denote by $\mathfrak{S}^{\circ}_m$ the subgroup of the symmetric group $\mathfrak{S}_{k+m}$ which consists of the permutations which fix each of the indices $1, \dots, k$. This subgroup is isomorphic to the symmetric group $\mathfrak{S}_m$. We will introduce the respective objects associated with the standard tableau $\mathcal{V}$ by using the group $\mathfrak{S}^{\circ}_m$ instead of $\mathfrak{S}_m$ so that

$$L_{\mathcal{V}} = \Phi_{\mathcal{V}}(\mathbb{C}^N)^{\otimes m},$$

where the tensor factors $\mathbb{C}^N$ are labelled with the numbers $k+1, \dots, k+m$ and the matrix element $\Phi_{\mathcal{V}}$ belongs to the group algebra $\mathbb{C}[\mathfrak{S}^{\circ}_m]$.

PROPOSITION 6.5.4. *The subspace*

$$L_{\mathcal{U}} \otimes L_{\mathcal{V}} \subset (\mathbb{C}^N)^{\otimes k} \otimes (\mathbb{C}^N)^{\otimes m}$$

in stable under the operator $R_{\mathcal{U}\mathcal{V}}(u)$.

PROOF. Proposition 6.4.4 implies that for any $i \in \{1, \dots, k\}$ we have

$$R_{i,k+1}(u + c_i - d_1) \dots R_{i,k+m}(u + c_i - d_m)\, \Phi_{\mathcal{V}} \\ = \Phi_{\mathcal{V}}\, R_{i,k+m}(u + c_i - d_m) \dots R_{i,k+1}(u + c_i - d_1),$$

while for any $j \in \{1, \dots, m\}$ Corollary 6.4.5 gives

$$R_{k,k+j}(u + c_k - d_j) \dots R_{1,k+j}(u + c_1 - d_j)\, \Phi_{\mathcal{U}} \\ = \Phi_{\mathcal{U}}\, R_{1,k+j}(u + c_1 - d_j) \dots R_{k,k+j}(u + c_k - d_j).$$

The R-matrices $R_{ij}(u)$ and $R_{rl}(v)$ commute if the indices i, j, r, l are distinct. Therefore, we have

$$R_{\mathcal{U}\mathcal{V}}(u)\, \Phi_{\mathcal{U}}\, \Phi_{\mathcal{V}} = \Phi_{\mathcal{U}}\, \Phi_{\mathcal{V}} \overrightarrow{\prod_{i=1,\dots,k}} \Bigg(\overleftarrow{\prod_{j=1,\dots,m}} R_{i,k+j}(u + c_i - d_j) \Bigg),$$

proving the claim. □

Thus, by restriction to the subspace $L_{\mathcal{U}} \otimes L_{\mathcal{V}}$ we may regard $R_{\mathcal{U}\mathcal{V}}(u)$ as a rational function in u with values in the algebra

$$\operatorname{End} L_{\mathcal{U}} \otimes \operatorname{End} L_{\mathcal{V}}.$$

By Proposition 6.5.1, the vector space $L_{\mathcal{V}}$ carries an irreducible representation of the Yangian $\mathrm{Y}(\mathfrak{gl}_N)$ isomorphic to the evaluation module $L(\mu)$.

PROPOSITION 6.5.5. *The image of the series* $T_{\mathcal{U}}(u)$ *under the homomorphism* $\mathrm{Y}(\mathfrak{gl}_N) \to \mathrm{End}\, L_{\mathcal{V}}$ *coincides with* $R_{\mathcal{U}\mathcal{V}}(-u)$.

PROOF. This follows immediately from the definition of $T_{\mathcal{U}}(u)$ with the use of Proposition 6.5.1. □

Let $a \in \mathbb{C}$ be a given complex number. Expand the rational function $R_{\mathcal{U}\mathcal{V}}(u)$ into the Laurent series at $u = a$. Let $(u-a)^{-p} J_{\mathcal{U}\mathcal{V}}(a)$ be the leading term of the expansion. This defines a nonzero linear operator $J_{\mathcal{U}\mathcal{V}}(a)$ in the vector space $L_{\mathcal{U}} \otimes L_{\mathcal{V}}$. Note that if $a \notin \mathbb{Z}$, then $R_{\mathcal{U}\mathcal{V}}(u)$ is regular at $u = a$ so that in this case we have $p = 0$ and $J_{\mathcal{U}\mathcal{V}}(a) = R_{\mathcal{U}\mathcal{V}}(a)$.

Consider the composition of the Yangian action on the vector space $L_{\mathcal{U}}$ with the shift automorphism (1.21). This yields the Yangian module which we will denote by $L_{\mathcal{U},a}$. The $\mathrm{Y}(\mathfrak{gl}_N)$-module $L_{\mathcal{U},a}$ is isomorphic to the evaluation module $L(\lambda)_a$ obtained with the use of the evaluation homomorphism (6.66).

Now let another complex number $b \in \mathbb{C}$ be given. We will equip the vector space $L_{\mathcal{U}} \otimes L_{\mathcal{V}}$ with a structure of a $\mathrm{Y}(\mathfrak{gl}_N)$-module in two different ways. We denote by $L_{\mathcal{U},a} \otimes L_{\mathcal{V},b}$ the $\mathrm{Y}(\mathfrak{gl}_N)$-module obtained by taking the tensor product of the modules $L_{\mathcal{U},a}$ and $L_{\mathcal{V},b}$ with the use of the comultiplication Δ defined in (1.35). We will use the notation $L_{\mathcal{U},a} \otimes' L_{\mathcal{V},b}$ for the same vector space whose $\mathrm{Y}(\mathfrak{gl}_N)$-module structure is obtained by applying the opposite comultiplication Δ' defined in (1.42).

PROPOSITION 6.5.6. *The operator* $J_{\mathcal{U}\mathcal{V}}(b-a)$ *is an intertwining operator for the* $\mathrm{Y}(\mathfrak{gl}_N)$*-modules*

$$J_{\mathcal{U}\mathcal{V}}(b-a) : L_{\mathcal{U},a} \otimes L_{\mathcal{V},b} \to L_{\mathcal{U},a} \otimes' L_{\mathcal{V},b}.$$

PROOF. Recalling the notation of Section 1.5, for the image of the transposed matrix $T^t(u)$ under the comultiplication Δ we have

$$\Delta : T^t(u) \mapsto T^t_{[2]}(u)\, T^t_{[1]}(u), \tag{6.101}$$

which is implied by the definition (1.35). Hence, by (6.99) the image of $T^t(u)$ under the representation

$$\varrho : \mathrm{Y}(\mathfrak{gl}_N) \to \mathrm{End}\,(L_{\mathcal{U},a} \otimes L_{\mathcal{V},b})$$

is given by

$$T^t(u) \mapsto R_{0,\mathcal{V}}(-u+b)\, R_{0,\mathcal{U}}(-u+a),$$

where we have set

$$R_{0,\mathcal{V}}(u) = R_{0,k+1}(u-d_1) \dots R_{0,k+m}(u-d_m)$$

and

$$R_{0,\mathcal{U}}(u) = R_{01}(u-c_1) \dots R_{0k}(u-c_k).$$

Similarly, under the opposite comultiplication we have

$$\Delta' : T^t(u) \mapsto T^t_{[1]}(u)\, T^t_{[2]}(u),$$

and the image of $T^t(u)$ under the representation

$$\varrho' : \mathrm{Y}(\mathfrak{gl}_N) \to \mathrm{End}\,(L_{\mathcal{U},a} \otimes' L_{\mathcal{V},b})$$

is given by

$$T^t(u) \mapsto R_{0,\mathcal{U}}(-u+a)\, R_{0,\mathcal{V}}(-u+b).$$

Let us now verify the relation

$$(6.102)\quad R_{\mathcal{U}\mathcal{V}}(b-v)\, R_{0,\mathcal{V}}(-u+b)\, R_{0,\mathcal{U}}(-u+v) = R_{0,\mathcal{U}}(-u+v)\, R_{0,\mathcal{V}}(-u+b)\, R_{\mathcal{U}\mathcal{V}}(b-v),$$

where u and v are complex variables. Indeed, it follows easily by the repeated use of the relations

$$R_{i,k+j}(b-v+c_i-d_j)\, R_{0,k+j}(-u+b-d_j)\, R_{0\,i}(-u+v-c_i) = R_{0\,i}(-u+v-c_i)\, R_{0,k+j}(-u+b-d_j)\, R_{i,k+j}(b-v+c_i-d_j)$$

implied by (1.17) for any indices $i \in \{1,\dots,k\}$ and $j \in \{1,\dots,m\}$. Taking the leading terms of the Laurent expansions of both sides of (6.102) at $v=a$ we come to the relation

$$J_{\mathcal{U}\mathcal{V}}(b-a)\, R_{0,\mathcal{V}}(-u+b)\, R_{0,\mathcal{U}}(-u+a) = R_{0,\mathcal{U}}(-u+a)\, R_{0,\mathcal{V}}(-u+b)\, J_{\mathcal{U}\mathcal{V}}(b-a)$$

which yields

$$J_{\mathcal{U}\mathcal{V}}(b-a)\, \varrho(T^{t}(u)) = \varrho'(T^{t}(u))\, J_{\mathcal{U}\mathcal{V}}(b-a)$$

as required. □

Consider now a particular case where the standard tableaux $\mathcal{U}$ and $\mathcal{V}$ coincide. Denote by $P_{\mathcal{U}}$ the permutation operator in $L_{\mathcal{U}} \otimes L_{\mathcal{U}}$ defined by

$$P_{\mathcal{U}} : \eta \otimes \zeta \mapsto \zeta \otimes \eta, \qquad \eta, \zeta \in L_{\mathcal{U}}.$$

COROLLARY 6.5.7. *The intertwining operator*

$$J_{\mathcal{U}\mathcal{U}}(0) : L_{\mathcal{U},a} \otimes L_{\mathcal{U},a} \to L_{\mathcal{U},a} \otimes' L_{\mathcal{U},a}$$

equals $c \cdot P_{\mathcal{U}}$ *for some nonzero constant* c.

PROOF. Since the $Y(\mathfrak{gl}_N)$-module $L_{\mathcal{U},a}$ is isomorphic to the evaluation module $L(\lambda)_a$, the $Y(\mathfrak{gl}_N)$-module $L_{\mathcal{U},a} \otimes L_{\mathcal{U},a}$ is irreducible by Corollary 6.1.6 and Proposition 6.3.1. Hence the $Y(\mathfrak{gl}_N)$-module $L_{\mathcal{U},a} \otimes' L_{\mathcal{U},a}$ is also irreducible. Indeed, if ζ is the highest vector of $L(\lambda)_a$, then the vector $\zeta \otimes \zeta$ generates a highest weight submodule of $L(\lambda)_a \otimes' L(\lambda)_a$ whose highest weight coincides with that of $L(\lambda)_a \otimes L(\lambda)_a$; cf. the proof of Proposition 3.2.9. As both tensor product modules have the same dimension, we may conclude that they are isomorphic. Moreover, the permutation map

$$P_{\mathcal{U}} : L_{\mathcal{U},a} \otimes L_{\mathcal{U},a} \to L_{\mathcal{U},a} \otimes' L_{\mathcal{U},a}$$

is an isomorphism of the Yangian modules. Therefore the intertwining operator $J_{\mathcal{U}\mathcal{U}}$ must be equal to $c \cdot P_{\mathcal{U}}$ for a constant c. The constant is nonzero because $J_{\mathcal{U}\mathcal{U}}$ is nonzero. □

We are now in a position to prove the main theorem of this section, which can be called the *binary property* of the tensor products of the Yangian modules. Consider the tensor product of the $Y(\mathfrak{gl}_N)$ evaluation modules

$$(6.103)\qquad L = L(\lambda^{(1)})_{a_1} \otimes L(\lambda^{(2)})_{a_2} \otimes \dots \otimes L(\lambda^{(l)})_{a_l},$$

where each $\lambda^{(i)}$ is a partition of length not exceeding N and each a_i is a complex number. We will obtain an irreducibility criterion for such Yangian modules. Note that our assumptions on the parameters of the module L do not restrict generality since applying Propositions 6.3.1 and 6.3.2, if necessary, we can derive an irreducibility criterion for an arbitrary module of the form (6.68). Note also that the irreducibility criterion for the modules (6.103) with $l = 2$ is obtained from Theorem 6.3.3.

THEOREM 6.5.8. *The* $\mathrm{Y}(\mathfrak{gl}_N)$*-module* L *is irreducible if and only if the modules* $L(\lambda^{(i)})_{a_i} \otimes L(\lambda^{(j)})_{a_j}$ *are irreducible for all* $1 \leqslant i < j \leqslant l$.

PROOF. Suppose first that the module L is irreducible. We will argue by contradiction and suppose that the $\mathrm{Y}(\mathfrak{gl}_N)$-module $L(\lambda^{(i)})_{a_i} \otimes L(\lambda^{(j)})_{a_j}$ contains a nonzero proper submodule K for some $1 \leqslant i < j \leqslant l$. Using Propositions 3.2.10 and 6.3.1 if necessary, we may assume without loss of generality that $i = l-1$ and $j = l$. However, in this case the vector subspace

$$L(\lambda^{(1)})_{a_1} \otimes L(\lambda^{(2)})_{a_2} \otimes \ldots \otimes L(\lambda^{(l-2)})_{a_{l-2}} \otimes K$$

of L is a nonzero proper submodule. This contradiction completes the proof of the "only if" part of the theorem.

Now suppose that all $\mathrm{Y}(\mathfrak{gl}_N)$-modules $L(\lambda^{(i)})_{a_i} \otimes L(\lambda^{(j)})_{a_j}$ with $i < j$ are irreducible. Let

$$\pi : \mathrm{Y}(\mathfrak{gl}_N) \to \operatorname{End} L$$

be the homomorphism associated with the $\mathrm{Y}(\mathfrak{gl}_N)$-module L. Denote by P_L the permutation operator in $L \otimes L$ defined by

$$P_L : \eta \otimes \zeta \mapsto \zeta \otimes \eta, \qquad \eta, \zeta \in L.$$

So P_L is an element of the endomorphism algebra $\operatorname{End}(L \otimes L)$ which we identify with $\operatorname{End} L \otimes \operatorname{End} L$. Our argument will be based upon the following simple observation.

LEMMA 6.5.9. *If the vector space* $\operatorname{End} L \otimes \pi\big(\mathrm{Y}(\mathfrak{gl}_N)\big)$ *contains the permutation operator* P_L*, then the* $\mathrm{Y}(\mathfrak{gl}_N)$*-module* L *is irreducible.*

PROOF. For any element $A \in \operatorname{End} L$ we have

$$(A \otimes 1)\, P_L \in \operatorname{End} L \otimes \pi\big(\mathrm{Y}(\mathfrak{gl}_N)\big).$$

Let $\operatorname{tr} : \operatorname{End} L \to \mathbb{C}$ be the trace map. Now observe that

$$A = (\operatorname{tr} \otimes \operatorname{id})\big((A \otimes 1)\, P_L\big)$$

and so the image $\pi\big(\mathrm{Y}(\mathfrak{gl}_N)\big)$ contains the element A. Thus $\pi\big(\mathrm{Y}(\mathfrak{gl}_N)\big) = \operatorname{End} L$ and the $\mathrm{Y}(\mathfrak{gl}_N)$-module L is irreducible. □

For each $i = 1, \ldots, l$ choose a standard $\lambda^{(i)}$-tableau $\mathcal{U}_i$. Introduce the formal series in u^{-1} with coefficients in $\operatorname{End} L \otimes \mathrm{Y}(\mathfrak{gl}_N)$ by

$$T_L(u) = T^{(1)}_{\mathcal{U}_1}(u + a_1) \ldots T^{(l)}_{\mathcal{U}_l}(u + a_l),$$

where we identify the endomorphism algebra $\operatorname{End} L$ with the tensor product

$$\operatorname{End} L_{\mathcal{U}_1} \otimes \operatorname{End} L_{\mathcal{U}_2} \otimes \ldots \otimes \operatorname{End} L_{\mathcal{U}_l}$$

and label the factors with the numbers $1, \ldots, l$. We will now verify that the image

$$(\operatorname{id} \otimes \pi)\big(T_L(u)\big) \tag{6.104}$$

is a rational function in u with values in $\operatorname{End} L \otimes \pi\big(\mathrm{Y}(\mathfrak{gl}_N)\big)$ such that the leading coefficient of its Laurent expansion at $u = 0$ is proportional to P_L. In view of

Lemma 6.5.9, this will complete the proof. Due to (6.101) and Proposition 6.5.5, the image (6.104) is given by

$$(\mathrm{id}\otimes\pi)\big(T_L(u)\big) = \prod_{i=1,\dots,l}^{\longrightarrow}\left(\prod_{j=1,\dots,l}^{\longleftarrow} R^{(i,l+j)}_{\mathcal{U}_i\mathcal{U}_j}(-u-a_i+a_j)\right), \tag{6.105}$$

where we regard $\operatorname{End} L\otimes\operatorname{End} L$ as the tensor product

$$\operatorname{End} L_{\mathcal{U}_1}\otimes\dots\otimes\operatorname{End} L_{\mathcal{U}_l}\otimes\operatorname{End} L_{\mathcal{U}_1}\otimes\dots\otimes\operatorname{End} L_{\mathcal{U}_l}$$

with the factors labelled with the numbers $1,\dots,2l$. Now expand the product of the rational functions in (6.105) into the Laurent series at $u=0$. The leading coefficient of this expansion is found by

$$\prod_{i=1,\dots,l}^{\longrightarrow}\left(\prod_{j=1,\dots,l}^{\longleftarrow} J^{(i,l+j)}_{\mathcal{U}_i\mathcal{U}_j}(a_j-a_i)\right). \tag{6.106}$$

By Corollary 6.5.7, each operator $J^{(i,l+i)}_{\mathcal{U}_i\mathcal{U}_i}(0)$ with $i\in\{1,\dots,l\}$ is proportional to the permutation operator $P^{(i,l+i)}_{\mathcal{U}_i}$. Hence, up to a nonzero constant factor, (6.106) equals

$$P^{(1,l+1)}_{\mathcal{U}_1}\dots P^{(l,2l)}_{\mathcal{U}_l}\prod_{i=1,\dots,l}^{\longrightarrow} J^{(l+i,l)}_{\mathcal{U}_i\mathcal{U}_l}(a_l-a_i)\dots J^{(l+i,i+1)}_{\mathcal{U}_i\mathcal{U}_{i+1}}(a_{i+1}-a_i)$$
$$\times J^{(i,l+i-1)}_{\mathcal{U}_i\mathcal{U}_{i-1}}(a_{i-1}-a_i)\dots J^{(i,l+1)}_{\mathcal{U}_i\mathcal{U}_1}(a_1-a_i). \tag{6.107}$$

By Proposition 6.5.3, for all $1\leqslant i<j\leqslant l$ the product

$$R^{(l+i,j)}_{\mathcal{U}_i\mathcal{U}_j}(-u)\,R^{(j,l+i)}_{\mathcal{U}_j\mathcal{U}_i}(u)$$

is a rational function in u with values in $\mathbb{C}$. Therefore, the operator

$$J^{(l+i,j)}_{\mathcal{U}_i\mathcal{U}_j}(a_j-a_i)\,J^{(j,l+i)}_{\mathcal{U}_j\mathcal{U}_i}(a_i-a_j)$$

equals a constant times the identity operator. By the assumptions of the theorem, all the operators $J^{(l+i,j)}_{\mathcal{U}_i\mathcal{U}_j}(a_j-a_i)$ and $J^{(j,l+i)}_{\mathcal{U}_j\mathcal{U}_i}(a_i-a_j)$ are invertible so that the constant is nonzero. This implies that the product (6.107) equals

$$c\cdot P^{(1,l+1)}_{\mathcal{U}_1}\dots P^{(l,2l)}_{\mathcal{U}_l}=c\cdot P_L$$

where the constant c is nonzero. □

6.6. Examples

1. Suppose that ν and λ are two partitions such that the diagram of λ is contained in the diagram of ν. The *skew diagram* ν/λ is the set-theoretical difference of the diagrams of ν and λ. If the skew diagram ν/λ has k cells, then a *standard tableau* $\mathcal{U}$ *of shape* ν/λ is obtained by filling in the cells of the diagram bijectively with the numbers $\{1,\dots,k\}$ in such a way that the entries strictly increase along the rows and down the columns; cf. Section 6.4. The corresponding skew representation $V_{\nu/\lambda}$ of $\mathfrak{S}_k$ is spanned by the basis vectors $v_{\mathcal{U}}$ parameterized by all standard tableaux of shape ν/λ. The action of the generators $s_i=(i,i+1)$ of $\mathfrak{S}_k$ in the basis $\{v_{\mathcal{U}}\}$ is described by the same formulas as for the normal (nonskew) case; see Section 6.4. Let $\chi_{\nu/\lambda}$ denote the character of $V_{\nu/\lambda}$. The irreducible representation V_μ of $\mathfrak{S}_k$ occurs in the irreducible decomposition of $V_{\nu/\lambda}$ with the multiplicity

$c^{\nu}_{\lambda\mu}$ which is found by the classical *Littlewood–Richardson rule.* In terms of the characters,

$$\chi_{\nu/\lambda} = \sum_{\mu} c^{\nu}_{\lambda\mu}\, \chi_{\mu}.$$

The coefficients $c^{\nu}_{\lambda\mu}$ also describe the multiplication in the algebra of symmetric functions in the basis of Schur functions $s_{\lambda}(x)$,

$$s_{\lambda}(x)\, s_{\mu}(x) = \sum_{\nu} c^{\nu}_{\lambda\mu}\, s_{\nu}(x).$$

A version of the Littlewood–Richardson rule is provided by Example 8.10.6 below. Its particular case, where μ is a row or column diagram, is the *Pieri rule*; all coefficients in this case do not exceed 1.

2. Consider the evaluation modules of the form $L(\lambda)$ over the quantum affine algebra $\mathrm{U}_q(\widehat{\mathfrak{gl}}_N)$ as defined in Example 3.5.10. The irreducibility criterion for the tensor product $L(\lambda) \otimes L(\mu)$ of two evaluation modules has exactly the same form as for the Yangian modules; see Theorem 6.3.3.

3. The binary property of Theorem 6.5.8 holds for the evaluation modules over the quantum affine algebra $\mathrm{U}_q(\widehat{\mathfrak{gl}}_N)$ in the same form.

4. *Open problem*: Does the binary property of Theorem 6.5.8 extend to tensor products of arbitrary (finite-dimensional) irreducible Yangian modules?

Bibliographical notes

6.1–6.3. Theorems 6.1.1, 6.2.1 and 6.3.3 were proved in the author's paper [**306**]. The criterion of Theorem 6.3.3 was inspired by a similar irreducibility criterion for the induction products of the evaluation modules over the affine Hecke algebras of type A; see Leclerc, Nazarov and Thibon [**263**]. In the case where the induction products are reducible, a combinatorial description of the composition factors is given by Leclerc [**262**]. The application of the Drinfeld functor [**98**] (see also Arakawa [**6**]) to the induction products of [**263**] leads to an irreducibility criterion for the Yangian modules where the highest weights satisfy some additional conditions. Similarly, the results of [**262**] also provide a combinatorial description of the composition factors of the tensor product of two evaluation modules of the corresponding quantum affine algebra. Another proof of Theorem 6.3.3 was given by Brundan and Kleshchev [**53**]. In a particular case this theorem was proved in an earlier paper by Nazarov and Tarasov [**352**]. The Drinfeld Yangians in A type are exceptional in the sense that the evaluation homomorphisms exist only in this case, as pointed out by Drinfeld [**97**]; see also Chari and Pressley [**65**, Chapter 12] for details. For the simple Lie algebras of other types one may consider tensor products of the so-called ‘fundamental modules’. They coincide with the evaluation modules $L(\lambda)_a$ in the A type for fundamental weights λ. For arbitrary types, irreducibility conditions for tensor products of the fundamental modules over the quantum affine algebra $\mathrm{U}_q(\widehat{\mathfrak{g}})$ were conjectured by Akasaka and Kashiwara in [**4**] and proved there in the cases where $\widehat{\mathfrak{g}}$ is of type $A^{(1)}$ or $C^{(1)}$. In some other cases the conjecture was proved in different ways by Frenkel and Mukhin [**114**], and Varagnolo and Vasserot [**460**] before the general conjecture was settled by Kashiwara [**207**]. A generalization of this result for some other types of modules was given by Chari [**59**].

6.4. The 'fusion procedure' was used in the pioneering works of the St.-Petersburg school as a tool to produce new solutions of the Yang–Baxter equation out of the given ones; see for instance Kulish, Reshetikhin and Sklyanin [**249**], Kulish and Sklyanin [**251**], Kulish and Reshetikhin [**248**]. Theorem 6.4.7 goes back to Cherednik [**72, 73, 74**]; see also Jucys [**203**]. Cherednik's papers do not contain complete proofs, however. More details were given by Jimbo, Kuniba, Miwa and Okado [**200**, Lemmas 3.2 and A.1], while a complete proof of Theorem 6.4.7 in a slightly different form was given by Nazarov [**341**, Theorem 2.2]; see also Guizzi and Papi [**165**]. The arguments of the papers [**200**], [**341**] and [**165**] rely on the relationship of $\phi_{\mathcal{U}}$ with the Young symmetrizers in the case where $\mathcal{U}$ is the row or column tableau. In the present form Theorem 6.4.7 was proved in the author's paper [**313**]. The fusion procedure for the Hecke algebra was developed by Nazarov [**347**] where a detailed proof of a q-analogue of Theorem 6.4.7 can be found. A hook version of the fusion procedure for the symmetric group and the Hecke algebra was developed by Grime [**162, 163**]. The fusion procedure involving skew diagrams goes back to Cherednik's work [**72, 73**], complete details and proofs were given by Nazarov [**345, 346**].

6.5. The binary property of Theorem 6.5.8 was established by Nazarov and Tarasov [**353**, Theorem 4.9] in a greater generality. This property holds for the tensor products of the skew Yangian modules; see Section 8.5 below. The argument of [**353**] relies on an observation made by Kitanine, Maillet and Terras [**230, 282**]. We have generally followed [**353**] restricting the exposition to the case of normal (nonskew) diagrams and considering arbitrary standard tableaux. One exception is Corollary 6.5.7, whose proof in the skew case is different and requires additional arguments; see [**353**, Theorem 3.5].

6.6. The Littlewood–Richardson coefficients $c^{\nu}_{\lambda\mu}$ introduced in Example 1 can be calculated by various versions of the rule; see e.g. Fulton [**117**], Macdonald [**280**] or Sagan [**411**]. Example 2 is due to Molev, Tolstoy and Zhang [**324**]. Example 3 is due to Leclerc, Nazarov and Thibon [**263**]. A general binary cyclicity property was established by Chari [**59**] for tensor products of arbitrary irreducible finite-dimensional representations of $\mathrm{U}_q(\widehat{\mathfrak{g}})$ with $\mathfrak{g}$ of any type. It has been known as a "folklore theorem" that the finite-dimensional representation theories for the Drinfeld Yangian and the quantum affine algebras are essentially the same; see Varagnolo [**459**] for precise theorems.

CHAPTER 7

Casimir elements and Capelli identities

Here we discuss several constructions of families of Casimir elements for the classical Lie algebras implied by the results of the previous chapters. All of these constructions (some of them are well known) are related with the quantum determinant for the Yangian $\mathrm{Y}(\mathfrak{gl}_N)$ or the Sklyanin determinant for the twisted Yangian $\mathrm{Y}(\mathfrak{g}_N)$. We also show that the images of these determinants in the respective universal enveloping algebras provide the noncommutative characteristic polynomials associated with the generator matrices of the Lie algebras $\mathfrak{gl}_N$ or $\mathfrak{g}_N$. Moreover, we calculate the images of some Casimir elements in the representations of the corresponding Lie algebras by differential operators, thus providing analogues of the classical Capelli identity.

7.1. Newton's formulas

Consider the case of $\mathfrak{gl}_N$ first. We let $\mathfrak{h}$, $\mathfrak{n}^+$ and $\mathfrak{n}^-$ denote the subalgebras of $\mathfrak{gl}_N$ consisting of all diagonal, upper triangular and lower triangular matrices, respectively. The Cartan subalgebra $\mathfrak{h}$ is spanned by the elements E_{ii}. We will fix the basis $E_{11}, \dots, E_{NN}$ of $\mathfrak{h}$. As in Section 3.2, any weight $\lambda \in \mathfrak{h}^*$ will be identified with the N-tuple of complex numbers $\lambda = (\lambda_1, \dots, \lambda_N)$, where $\lambda_i = \lambda(E_{ii})$.

Let $\mathrm{U}(\mathfrak{gl}_N)^{\mathfrak{h}}$ denote the centralizer of $\mathfrak{h}$ in $\mathrm{U}(\mathfrak{gl}_N)$. The subspace

$$\mathrm{J}_N = \mathrm{U}(\mathfrak{gl}_N)\mathfrak{n}^+ \cap \mathrm{U}(\mathfrak{gl}_N)^{\mathfrak{h}}$$

is a two-sided ideal of $\mathrm{U}(\mathfrak{gl}_N)^{\mathfrak{h}}$ which can also be given by

$$\mathrm{J}_N = \mathfrak{n}^-\mathrm{U}(\mathfrak{gl}_N) \cap \mathrm{U}(\mathfrak{gl}_N)^{\mathfrak{h}}.$$

By the Poincaré–Birkhoff–Witt theorem, we have the vector space decomposition

$$\mathrm{U}(\mathfrak{gl}_N)^{\mathfrak{h}} = \mathrm{J}_N \oplus \mathrm{U}(\mathfrak{h}).$$

The projection

$$\chi : \mathrm{U}(\mathfrak{gl}_N)^{\mathfrak{h}} \to \mathrm{U}(\mathfrak{h})$$

with the kernel J_N is an algebra homomorphism called the *Harish-Chandra homomorphism*. We will identify the algebra $\mathrm{U}(\mathfrak{h}) = \mathrm{S}(\mathfrak{h})$ with the algebra of polynomial functions on the space $\mathfrak{h}^*$. Then the restriction of χ to the center $\mathrm{Z}(\mathfrak{gl}_N)$ of $\mathrm{U}(\mathfrak{gl}_N)$ is an isomorphism onto the algebra of symmetric polynomials in the variables $l_1, \dots, l_N$, where $l_i = \lambda_i - i + 1$. This restriction is called the *Harish-Chandra isomorphism*.

Equivalently, any element $z \in \mathrm{Z}(\mathfrak{gl}_N)$ acts as a multiplication by a scalar in any irreducible highest weight representation $L(\lambda)$ of $\mathfrak{gl}_N$ (see Section 3.2 for the definition of $L(\lambda)$). This scalar coincides with the value of the polynomial $\chi(z)$ at $(l_1, \dots, l_N)$.

Here and below we use expressions of the form $v+A$, where A is a square matrix and v is a variable or constant. Such expressions are understood by interpreting v as the scalar matrix of the same size as A which has v on the diagonal.

As before, we denote by E the $N \times N$ matrix whose ij-th entry is E_{ij}. Denote by $\mathcal{C}(u)$ the *Capelli determinant*

$$\mathcal{C}(u) = \sum_{p \in \mathfrak{S}_N} \operatorname{sgn} p \cdot (u+E)_{p(1),1} \cdots (u+E-N+1)_{p(N),N}. \tag{7.1}$$

This is a polynomial in u with coefficients in the universal enveloping algebra $\mathrm{U}(\mathfrak{gl}_N)$,

$$\mathcal{C}(u) = u^N + \mathcal{C}_1\, u^{N-1} + \cdots + \mathcal{C}_N, \qquad \mathcal{C}_i \in \mathrm{U}(\mathfrak{gl}_N). \tag{7.2}$$

THEOREM 7.1.1. *The coefficients $\mathcal{C}_1, \dots, \mathcal{C}_N$ belong to $\mathrm{Z}(\mathfrak{gl}_N)$. The image of $\mathcal{C}(u)$ under the Harish-Chandra isomorphism is given by*

$$\chi : \mathcal{C}(u) \mapsto (u+l_1) \dots (u+l_N), \tag{7.3}$$

so that $\chi(\mathcal{C}_k)$ is the elementary symmetric polynomial of degree k in $l_1, \dots, l_N$. Moreover, the algebra $\mathrm{Z}(\mathfrak{gl}_N)$ is generated by $\mathcal{C}_1, \dots, \mathcal{C}_N$.

PROOF. Due to Theorem 1.7.5, the coefficients of the quantum determinant $\operatorname{qdet} T(u)$ belong to the center of the Yangian $\mathrm{Y}(\mathfrak{gl}_N)$. Since the evaluation homomorphism π_N defined in (1.5) is surjective, the images of the coefficients of $\operatorname{qdet} T(u)$ under π_N belong to $\mathrm{Z}(\mathfrak{gl}_N)$. Using the formula (1.52) for $\operatorname{qdet} T(u)$ we obtain

$$\mathcal{C}(u) = u(u-1) \dots (u-N+1) \cdot \pi_N\big(\operatorname{qdet} T(u)\big), \tag{7.4}$$

proving the first part of the proposition. For the proof of the second part observe that, by the definition of the Harish-Chandra isomorphism, the only term in (7.1) which provides a nonzero contribution to $\chi(\mathcal{C}(u))$, corresponds to the identity permutation p. Finally, since the elementary symmetric polynomials generate the algebra of symmetric polynomials in the variables $l_1, \dots, l_N$, the elements $\mathcal{C}_1, \dots, \mathcal{C}_N$ are generators of the algebra $\mathrm{Z}(\mathfrak{gl}_N)$. □

The following alternative formulas for the Capelli determinant are immediate from (7.4) and Proposition 1.6.6.

COROLLARY 7.1.2. *For any permutation $q \in \mathfrak{S}_N$ we have*

$$\mathcal{C}(u) = \operatorname{sgn} q \sum_{p \in \mathfrak{S}_N} \operatorname{sgn} p \cdot (u+E)_{p(1),q(1)} \cdots (u+E-N+1)_{p(N),q(N)} \tag{7.5}$$

and

$$\mathcal{C}(u) = \operatorname{sgn} q \sum_{p \in \mathfrak{S}_N} \operatorname{sgn} p \cdot (u+E-N+1)_{q(1),p(1)} \cdots (u+E)_{q(N),p(N)}. \tag{7.6}$$

The *Gelfand invariants* are the elements of $\mathrm{U}(\mathfrak{gl}_N)$ defined by

$$\operatorname{tr} E^k = \sum_{i_1, i_2, \dots, i_k = 1}^{N} E_{i_1 i_2} E_{i_2 i_3} \dots E_{i_k i_1}, \qquad k = 0, 1, \dots.$$

The following can be regarded as a noncommutative analogue of the classical Newton formula which relates the elementary and power sums symmetric functions.

THEOREM 7.1.3. *We have the formula*

$$1+\sum_{k=0}^{\infty}\frac{(-1)^k\,\mathrm{tr}\,E^k}{(u-N+1)^{k+1}}=\frac{\mathcal{C}(u+1)}{\mathcal{C}(u)}.$$

PROOF. Using the definition (1.68) of the series $z(u)$ and applying the evaluation homomorphism (1.5), we find

$$\begin{aligned}\pi_N : z(-u+N)^{-1} &\mapsto \frac{1}{N}\,\mathrm{tr}\Big((1-E\,(u-N)^{-1})(1-E\,u^{-1})^{-1}\Big)\\ &=1-\frac{1}{u-N}\sum_{k=1}^{\infty}\mathrm{tr}\,E^k\,u^{-k}.\end{aligned}\tag{7.7}$$

The quantum Liouville formula (1.69) gives

$$z(u+1)^{-1}=\frac{\mathrm{qdet}\,T(u+1)}{\mathrm{qdet}\,T(u)}.$$

Applying the evaluation homomorphism (1.5) to both sides of this relation and using (7.4) and (7.7), we arrive at the desired formula. □

COROLLARY 7.1.4. *All Gelfand invariants* $\mathrm{tr}\,E^k$ *belong to* $\mathrm{Z}(\mathfrak{gl}_N)$. *Their images under the Harish-Chandra isomorphism are found by*

$$1+\sum_{k=0}^{\infty}\frac{(-1)^k\,\chi(\mathrm{tr}\,E^k)}{(u-N+1)^{k+1}}=\prod_{i=1}^{N}\frac{u+l_i+1}{u+l_i}.\tag{7.8}$$

PROOF. Both statements follow from Theorem 7.1.1 and Theorem 7.1.3. □

It can be derived from Corollary 7.1.4, or directly from the definition of the Gelfand invariants, that for each $k\geqslant 1$ the symmetric polynomial $\chi(\mathrm{tr}\,E^k)$ has degree k and its homogeneous component of degree k coincides with $l_1^k+\dots+l_N^k$. This implies that the first N Gelfand invariants $\mathrm{tr}\,E,\dots,\mathrm{tr}\,E^N$ generate the center of $\mathrm{U}(\mathfrak{gl}_N)$.

Now we turn to the orthogonal and symplectic Lie algebras. It will be convenient to use the split realization of the Lie algebra $\mathfrak{g}_N=\mathfrak{o}_N$ or $\mathfrak{sp}_N$ introduced in Section 4.2. Namely, we regard $\mathfrak{g}_N$ as a subalgebra of $\mathfrak{gl}_N$ spanned by the elements

$$F_{ij}=E_{ij}-\theta_{ij}\,E_{-j,-i},\qquad -n\leqslant i,j\leqslant n,\tag{7.9}$$

where we use the indices $-n,\dots,n$ to number the rows and columns of $N\times N$ matrices as in Section 4.1.

We let $\mathfrak{h}$, $\mathfrak{n}^+$ and $\mathfrak{n}^-$ denote the subalgebras of $\mathfrak{g}_N$ consisting of all diagonal, upper triangular and lower triangular matrices, respectively. The Cartan subalgebra $\mathfrak{h}$ is spanned by the elements F_{ii}. We will fix the basis $F_{11},\dots,F_{nn}$ of $\mathfrak{h}$.

As in Section 4.2, any weight $\lambda\in\mathfrak{h}^*$ will be identified with the n-tuple of complex numbers $\lambda=(\lambda_1,\dots,\lambda_n)$, where $\lambda_i=\lambda(F_{ii})$. For $i=1,\dots,n$ denote

$$\rho_{-i}=-\rho_i=\begin{cases} i-1 & \text{for}\quad \mathfrak{g}_N=\mathfrak{o}_{2n},\\ i-\frac{1}{2} & \text{for}\quad \mathfrak{g}_N=\mathfrak{o}_{2n+1},\\ i & \text{for}\quad \mathfrak{g}_N=\mathfrak{sp}_{2n},\end{cases}\tag{7.10}$$

and set $l_i=-l_{-i}=\lambda_i+\rho_i$. We also set $\rho_0=1/2$ in the case of $\mathfrak{g}_N=\mathfrak{o}_{2n+1}$.

Let $\mathrm{U}(\mathfrak{g}_N)^{\mathfrak{h}}$ denote the centralizer of $\mathfrak{h}$ in $\mathrm{U}(\mathfrak{g}_N)$. As in the $\mathfrak{gl}_N$ case, we have

$$\mathrm{U}(\mathfrak{g}_N)\mathfrak{n}^+ \cap \mathrm{U}(\mathfrak{g}_N)^{\mathfrak{h}} = \mathfrak{n}^-\mathrm{U}(\mathfrak{g}_N) \cap \mathrm{U}(\mathfrak{g}_N)^{\mathfrak{h}}.$$

Denote this subspace by J_N. Then J_N is a two-sided ideal of $\mathrm{U}(\mathfrak{g}_N)^{\mathfrak{h}}$, and we have the vector space decomposition

$$\mathrm{U}(\mathfrak{g}_N)^{\mathfrak{h}} = \mathrm{J}_N \oplus \mathrm{U}(\mathfrak{h}).$$

The projection

$$\chi : \mathrm{U}(\mathfrak{g}_N)^{\mathfrak{h}} \to \mathrm{U}(\mathfrak{h}) \tag{7.11}$$

with the kernel J_N is an algebra homomorphism called the *Harish-Chandra homomorphism*. We will identify the algebra $\mathrm{U}(\mathfrak{h}) = \mathrm{S}(\mathfrak{h})$ with the algebra of polynomial functions on the space $\mathfrak{h}^*$. In the cases $\mathfrak{g}_N = \mathfrak{o}_{2n+1}$ and $\mathfrak{g}_N = \mathfrak{sp}_{2n}$ the restriction of χ to the center $\mathrm{Z}(\mathfrak{g}_N)$ of $\mathrm{U}(\mathfrak{g}_N)$ is an isomorphism onto the algebra of symmetric polynomials in the variables $l_1^2, \dots, l_n^2$. In the case $\mathfrak{g}_N = \mathfrak{o}_{2n}$ the restriction of χ to $\mathrm{Z}(\mathfrak{g}_N)$ is an isomorphism onto the subalgebra of the algebra of polynomials in $l_1, \dots, l_n$ generated by the polynomial $l_1 \dots l_n$ and the symmetric polynomials in the variables $l_1^2, \dots, l_n^2$. In all the cases, this restriction is called the *Harish-Chandra isomorphism*.

Recall that for any n-tuple of complex numbers $\lambda = (\lambda_1, \dots, \lambda_n)$ the corresponding irreducible highest weight representation $V(\lambda)$ is generated by a nonzero vector ξ such that

$$\begin{aligned} &F_{ij}\,\xi = 0 \qquad \text{for} \quad -n \leqslant i < j \leqslant n, \qquad \text{and} \\ &F_{ii}\,\xi = \lambda_i\,\xi \qquad \text{for} \quad 1 \leqslant i \leqslant n; \end{aligned} \tag{7.12}$$

see (4.19). Any element $z \in \mathrm{Z}(\mathfrak{g}_N)$ acts in $V(\lambda)$ by multiplying each vector by a scalar. This scalar coincides with the value of the polynomial $\chi(z)$ at $(l_1, \dots, l_n)$.

As before, we denote by F the $N \times N$ matrix whose ij-th entry is F_{ij}. Introduce the *Capelli-type determinant*

$$\begin{aligned} \mathcal{C}(u) = (-1)^n \sum_{p \in \mathfrak{S}_N} \operatorname{sgn} pp' \cdot \big(u + \rho_{-n} + F\big)_{-a_{p(1)},\, a_{p'(1)}} \\ \times \dots \times \big(u + \rho_n + F\big)_{-a_{p(N)},\, a_{p'(N)}}, \end{aligned} \tag{7.13}$$

where $(a_1, \dots, a_N)$ is a fixed permutation of the indices $(-n, \dots, n)$ and p' is the image of p under the map (2.53). Then $\mathcal{C}(u)$ is a polynomial in u with coefficients in $\mathrm{U}(\mathfrak{g}_N)$.

EXAMPLE 7.1.5. If $\mathfrak{g}_N = \mathfrak{sp}_2$, then taking $(a_1, a_2) = (-1, 1)$ we get

$$\begin{aligned} \mathcal{C}(u) &= \big(u + F_{-1,-1} + 1\big)\big(u + F_{11} - 1\big) - F_{1,-1}\,F_{-1,1} \\ &= u^2 - (F_{11} - 1)^2 - F_{1,-1}\,F_{-1,1}. \end{aligned}$$

If $\mathfrak{g}_N = \mathfrak{o}_3$, then taking $(a_1, a_2, a_3) = (-1, 0, 1)$ we get

$$\begin{aligned} \mathcal{C}(u) &= \big(u + F_{-1,-1} + 1/2\big)\big(u + 1/2\big)\big(u + F_{11} - 1/2\big) \\ &\quad - F_{0,-1}\,F_{-1,0}\,\big(u + F_{11} - 1/2\big) - F_{10}\,\big(u + F_{-1,-1} + 1/2\big)\,F_{01} \\ &= \big(u + 1/2\big)\big(u^2 - (F_{11} - 1/2)^2 - 2\,F_{10}\,F_{01}\big). \end{aligned}$$

In both cases the coefficients of $\mathcal{C}(u)$ are Casimir elements for $\mathfrak{g}_N$. ☐

THEOREM 7.1.6. *The polynomial $\mathcal{C}(u)$ does not depend on the choice of the permutation $(a_1, \dots, a_N)$. All coefficients of $\mathcal{C}(u)$ belong to $\mathrm{Z}(\mathfrak{g}_N)$. Moreover, the image of $\mathcal{C}(u)$ under the Harish-Chandra isomorphism is given by*

$$\chi : \mathcal{C}(u) \mapsto \prod_{i=1}^{n} (u^2 - l_i^2), \qquad \text{if} \quad N = 2n,$$

and

$$\chi : \mathcal{C}(u) \mapsto \Big(u + \frac{1}{2}\Big) \prod_{i=1}^{n} (u^2 - l_i^2), \qquad \text{if} \quad N = 2n + 1.$$

PROOF. Due to Theorem 2.8.2 and Proposition 2.15.2, the coefficients of the Sklyanin determinant $\mathrm{sdet}\, S(u)$ belong to the center of the twisted Yangian $\mathrm{Y}(\mathfrak{g}_N)$. Since the evaluation homomorphism ϱ_N defined in (2.106) is surjective, the images of the coefficients of $\mathrm{sdet}\, S(u)$ under ϱ_N belong to $\mathrm{Z}(\mathfrak{g}_N)$. Using Proposition 4.1.3, we obtain

$$\mathcal{C}(u) = \prod_{i=-n}^{n} (u + \rho_i) \cdot \varrho_N\big(\mathbf{c}(u)\big), \tag{7.14}$$

where, as in Section 4.1, we set

$$\mathbf{c}(u) = \alpha_N\big(u + (N-1)/2\big)^{-1} \cdot \mathrm{sdet}\, S\big(u + (N-1)/2\big),$$

and the index $i = 0$ occurs in the product only if $N = 2n + 1$. This proves the first two statements. The third is verified by the same argument as in the proof of Proposition 4.2.4. Indeed, take $(a_1, \dots, a_N) = (-n, \dots, n)$. Then the only term in the expression for $\mathcal{C}(u)$ which provides a nonzero contribution to $\chi(\mathcal{C}(u))$ corresponds to the permutation $p = (N, N-1, \dots, 1)$. In this case $p' = (1, 2, \dots, N)$, thus yielding the desired formulas. □

The *Gelfand invariants* are the elements of $\mathrm{U}(\mathfrak{g}_N)$ defined by

$$\mathrm{tr}\, F^k = \sum_{i_1, i_2, \dots, i_k = -n}^{n} F_{i_1 i_2} F_{i_2 i_3} \dots F_{i_k i_1}, \qquad k = 0, 1, \dots.$$

The following is an analogue of Theorem 7.1.3 for the Lie algebra $\mathfrak{g}_N$.

THEOREM 7.1.7. *If $N = 2n$, then*

$$1 + \frac{2u+1}{2u+1 \mp 1} \sum_{k=0}^{\infty} \frac{(-1)^k \,\mathrm{tr}\, F^k}{(u + \rho_n)^{k+1}} = \frac{\mathcal{C}(u+1)}{\mathcal{C}(u)},$$

where the upper sign is taken in the orthogonal case and the lower sign in the symplectic case. If $N = 2n + 1$, then

$$1 + \frac{2u+1}{2u} \sum_{k=0}^{\infty} \frac{(-1)^k \,\mathrm{tr}\, F^k}{(u + \rho_n)^{k+1}} = \frac{\overline{\mathcal{C}}(u+1)}{\overline{\mathcal{C}}(u)},$$

where

$$\overline{\mathcal{C}}(u) = \frac{2u}{2u+1}\, \mathcal{C}(u).$$

PROOF. We use the quantum Liouville formula for the twisted Yangian; see Corollary 2.15.6. This formula gives

$$\mathbf{y}(u)^{-1} = \frac{\operatorname{sdet} S(u)}{\operatorname{sdet} S(u-1)}. \tag{7.15}$$

Apply the evaluation homomorphism ϱ_N defined in (2.106) to both sides of (7.15). To find the image of the series $\mathbf{y}(u)^{-1}$ we use the formula (2.113). Note that by (2.106), for the images under ϱ_N of the matrices occurring in (2.113) we have

$$S(-u) \mapsto 1 - \frac{F}{u \mp 1/2}, \qquad S'(-u) \mapsto 1 + \frac{F}{u \mp 1/2}$$

and

$$S^{-1}(u-N) \mapsto \sum_{k=0}^{\infty} \frac{(-1)^k F^k}{(u - N \pm 1/2)^k}.$$

Replacing u by $u + (N+1)/2$ in (7.15) and using (7.14), we get the desired formulas after a straightforward calculation. □

COROLLARY 7.1.8. *All Gelfand invariants* $\operatorname{tr} F^k$ *belong to* $\mathrm{Z}(\mathfrak{g}_N)$*. Their images under the Harish-Chandra isomorphism are found by*

$$1 + \frac{2u+1}{2u+1 \mp 1} \sum_{k=0}^{\infty} \frac{(-1)^k \chi(\operatorname{tr} F^k)}{(u+\rho_n)^{k+1}} = \prod_{i=-n}^{n} \frac{u + l_i + 1}{u + l_i}, \tag{7.16}$$

where the zero index is skipped in the product if $N = 2n$*, while for* $N = 2n+1$ *one should set* $l_0 = 0$.

PROOF. Both statements follow from Theorems 7.1.6 and 7.1.7. □

Most of the results of this chapter concerning the orthogonal and symplectic Lie algebras can be easily reformulated in terms of their presentations associated with arbitrary matrices G. In particular, in the presentation of $\mathfrak{g}_N$ used in Section 2.1, the Capelli-type determinant is defined by the formula

$$\begin{aligned} \mathcal{C}_G(u) &= \big(\det G\big)^{-1} \\ &\times \sum_{p \in \mathfrak{S}_N} \operatorname{sgn} pp' \cdot \big(G(u+\sigma_1) + F\big)_{p(1),p'(1)} \cdots \big(G(u+\sigma_N) + F\big)_{p(N),p'(N)}, \end{aligned} \tag{7.17}$$

where $(\sigma_1, \dots, \sigma_N) = (\rho_{-n}, \dots, \rho_n)$; cf. (7.13). Using Theorem 2.7.2 and the evaluation homomorphism (2.8), we derive that the polynomial $\mathcal{C}_G(u)$ is monic of degree N and all its coefficients belong to the center of $\mathrm{U}(\mathfrak{g}_N)$.

Suppose that G and G' are two nonsingular symmetric (respectively, skew-symmetric) $N \times N$-matrices. Consider the corresponding Lie algebras $\mathfrak{g}_N$ and $\mathfrak{g}'_N$ which are respectively spanned by the elements F_{ij} and F'_{ij}; see Section 2.1. Fix a matrix B such that $BGB^t = G'$. Using the embedding (2.9) and Corollary 2.3.2, we find that the mapping

$$F' \mapsto BFB^t \tag{7.18}$$

defines an isomorphism $\mathfrak{g}'_N \to \mathfrak{g}_N$, where F and F' denote the matrices with the entries F_{ij} and F'_{ij}, respectively. This isomorphism naturally extends to an isomorphism of the universal enveloping algebras $\mathrm{U}(\mathfrak{g}'_N) \to \mathrm{U}(\mathfrak{g}_N)$. Corollary 2.5.5 implies the following.

PROPOSITION 7.1.9. *The image of the polynomial $\mathcal{C}_{G'}(u)$ under the isomorphism* (7.18) *coincides with $\mathcal{C}_G(u)$.* □

Similarly, if the matrix G is chosen as in Section 4.1, then applying the isomorphism (2.104) and the corresponding isomorphism between the presentations of the respective Lie algebras, we conclude from Proposition 2.15.2 that the image of the polynomial $\mathcal{C}_G(u)$ coincides with $\mathcal{C}(u)$, where $\mathcal{C}(u)$ is defined in (7.13). Together with Theorem 7.1.6, this provides, in particular, the Harish-Chandra images of the coefficients of $\mathcal{C}_G(u)$.

7.2. Noncommutative Cayley–Hamilton theorem

The polynomials $\mathcal{C}(u)$ introduced in the previous section turn out to play the role of noncommutative characteristic polynomials for the matrices E and F. Note that since the coefficients of $\mathcal{C}(u)$ are central in the corresponding universal enveloping algebra, the substitution of the variable u by a matrix with entries in the universal enveloping algebra is well-defined.

Consider the case of $\mathfrak{gl}_N$ first. The following is a noncommutative analogue of the Cayley–Hamilton theorem for the matrix of generators of $\mathfrak{gl}_N$.

THEOREM 7.2.1. *We have the identities*

$$\mathcal{C}(-E+N-1)=0 \qquad \text{and} \qquad \mathcal{C}(-E^t)=0. \tag{7.19}$$

PROOF. Apply the evaluation homomorphism (1.5) to both sides of relation (1.65). By Proposition 1.9.2, for any $i,j\in\{1,\dots,N\}$ the expression

$$u(u-1)\dots(u-N+2)\cdot\pi_N\big(\widehat{t}_{ij}(u)\big)$$

is a polynomial in u with coefficients in $\mathrm{U}(\mathfrak{gl}_N)$. Hence, using (7.4), we obtain the relation

$$\mathcal{C}(u)=\widehat{\mathcal{C}}(u)\,(u+E-N+1),$$

where $\widehat{\mathcal{C}}(u)$ is a polynomial in u with coefficients in $\operatorname{End}\mathbb{C}^N\otimes\mathrm{U}(\mathfrak{gl}_N)$. So, the first relation in (7.19) follows by the substitution $u=-E+N-1$. In order to prove the second, note that the application of the evaluation homomorphism (1.5) to both sides of (1.67) yields

$$\mathcal{C}(u)=\widehat{\mathcal{C}}^{\,t}(u-1)\,(u+E^t).$$ □

Let L be a (not necessarily irreducible) highest weight representation of $\mathfrak{gl}_N$ with the highest weight $(\lambda_1,\dots,\lambda_N)$. That is, L is generated by a nonzero vector ζ satisfying the relations (3.64). Theorem 7.2.1 implies the *characteristic identities* for the image of the matrix E in L.

COROLLARY 7.2.2. *The image of the matrix E in L satisfies the identities*

$$\prod_{i=1}^{N}(E-l_i-N+1)=0 \qquad \text{and} \qquad \prod_{i=1}^{N}(E^t-l_i)=0,$$

where $l_i=\lambda_i-i+1$.

PROOF. The coefficients of the Capelli polynomial $\mathcal{C}(u)$ act in L as multiplications by scalars which are found by (7.3). So the image of $\mathcal{C}(u)$ in L is the product $(u+l_1)\dots(u+l_N)$. Now the identities are implied by (7.19). □

Now turn to the case of the Lie algebras $\mathfrak{g}_N$. As before, we consider the three cases simultaneously. The following is a noncommutative Cayley–Hamilton theorem for the matrix F.

THEOREM 7.2.3. *We have the identity*

$$\mathcal{C}(-F-\rho_n)=0. \tag{7.20}$$

PROOF. Replacing u by $u+(N-1)/2$ in (2.112), we get

$$\operatorname{sdet} S\big(u+(N-1)/2\big)=\widehat{S}\big(u+(N-1)/2\big)\cdot S\big(u-(N-1)/2\big).$$

Apply the evaluation homomorphism (2.106) to both sides of this relation. By Proposition 4.1.4, for any $a,b\in\{-n,\dots,n\}$ the expression

$$\prod_{i=-n}^{n-1}(u+\rho_i)\cdot\frac{\varrho_N\big(\widehat{s}_{ab}(u+(N-1)/2)\big)}{\alpha_N\big(u+(N-1)/2\big)}$$

is a polynomial in u with coefficients in $\mathrm{U}(\mathfrak{g}_N)$. Hence, using (7.14), we obtain the relation

$$\mathcal{C}(u)=\widehat{\mathcal{C}}(u)\,(u+F+\rho_n),$$

where $\widehat{\mathcal{C}}(u)$ is a polynomial in u with coefficients in $\operatorname{End}\mathbb{C}^N\otimes\mathrm{U}(\mathfrak{g}_N)$. The proof is completed by the substitution $u=-F-\rho_n$. □

Let V be a (not necessarily irreducible) highest weight representation of the Lie algebra $\mathfrak{g}_N$ with the highest weight $(\lambda_1,\dots,\lambda_n)$. That is, V is generated by a nonzero vector ξ satisfying the relations (7.12).

The following is the *characteristic identity* for the image of the matrix F in V.

COROLLARY 7.2.4. *The image of the matrix F in V satisfies the identity*

$$\prod_{i=-n}^{n}(F-l_i+\rho_n)=0,$$

where $l_i=-l_{-i}=\lambda_i+\rho_i$ for $i=1,\dots,n$. The zero index is skipped in the product if $N=2n$, while for $N=2n+1$ one should set $l_0=1/2$.

PROOF. The coefficients of the Capelli polynomial $\mathcal{C}(u)$ act in V as multiplications by scalars which are found by Theorem 7.1.6. The identity is now implied by Theorem 7.2.3. □

7.3. Graphical constructions of Casimir elements

Let $A=[a_{ij}]$ be an $N\times N$ matrix with entries from an arbitrary ring with 1 and let q be a formal variable. Fix an index $i\in\{1,\dots,N\}$. The noncommutative *elementary symmetric functions* $\Lambda_k^{(i)}$ associated with the matrix A and the index i are defined as the coefficients in the expansion of the following quasideterminant (see Definition 1.10.1):

$$1+\sum_{k=1}^{\infty}\Lambda_k^{(i)}q^k=\big|1+qA\big|_{ii}. \tag{7.21}$$

The *complete symmetric functions* $S_k^{(i)}$ are defined by

$$1+\sum_{k=1}^{\infty}S_k^{(i)}q^k=\big|1-qA\big|_{ii}^{-1}, \tag{7.22}$$

while the *power sums symmetric functions of the first kind* $\Psi_k^{(i)}$ and the *power sums symmetric functions of the second kind* $\Phi_k^{(i)}$ are defined by the respective formulas

$$\sum_{k=1}^{\infty} \Psi_k^{(i)} q^{k-1} = \left|1 - qA\right|_{ii} \frac{d}{dq} \left|1 - qA\right|_{ii}^{-1} \tag{7.23}$$

and

$$\sum_{k=1}^{\infty} \Phi_k^{(i)} q^{k-1} = -\frac{d}{dq} \log \left|1 - qA\right|_{ii}, \tag{7.24}$$

where the derivative over q and logarithm are regarded as formal operations on the formal power series in q with coefficients in the ring. In particular, the logarithm is well-defined on the series which begin with 1.

Let us consider the complete oriented graph $\mathcal{A}$ with N vertices $\{1, 2, \ldots, N\}$, the arrow from i to j being labelled by a_{ij}. Every directed path in the graph going from vertex i to vertex j defines a monomial of the form $a_{ir_1} a_{r_1 r_2} \cdots a_{r_{k-1} j}$ which is obtained by taking the product of the labels of the consecutive arrows of the path. The positive integer k is the *length* of the path. A *simple path* is a path such that $r_s \neq i, j$ for all $s \in \{1, \ldots, k-1\}$. We have the following description of the noncommutative symmetric functions defined above in terms of the graph $\mathcal{A}$.

PROPOSITION 7.3.1. (i) $(-1)^{k-1} \Lambda_k^{(i)}$ *is the sum of all monomials labeling simple paths in* $\mathcal{A}$ *of length* k *going from* i *to* i*;*

(ii) $S_k^{(i)}$ *is the sum of all monomials labeling paths in* $\mathcal{A}$ *of length* k *going from* i *to* i*;*

(iii) $\Psi_k^{(i)}$ *is the sum of all monomials labeling paths in* $\mathcal{A}$ *of length* k *going from* i *to* i*, the coefficient of each monomial being the length of the first return to vertex* i*;*

(iv) $\Phi_k^{(i)}$ *is the sum of all monomials labeling paths in* $\mathcal{A}$ *of length* k *going from* i *to* i*, the coefficient of each monomial being the ratio of* k *to the number of returns to* i*.*

PROOF. By Proposition 1.10.4, we have

$$\left|1 - qA\right|_{ii}^{-1} = \left[(1 - qA)^{-1}\right]_{ii} = 1 + \sum_{k=1}^{\infty} q^k \, (A^k)_{ii}.$$

Hence, $S_k^{(i)} = (A^k)_{ii}$, proving (ii). Furthermore, by Definition 1.10.1,

$$\left|1 + qA\right|_{ii} = 1 + q\, a_{ii} - q^2 \sum_{k,l \neq i} a_{ik} \left[(1 + qA^{ii})^{-1}\right]_{kl} a_{li},$$

which yields (i). Note that (7.23) can be written in the following equivalent form:

$$\sum_{k=1}^{\infty} \Psi_k^{(i)} q^{k-1} = \Big(-\frac{d}{dq} \left|1 - qA\right|_{ii}\Big) \left|1 - qA\right|_{ii}^{-1}.$$

Therefore, using (7.21) and (7.22) we get

$$\Psi_k^{(i)} = \sum_{r+p=k} (-1)^{r-1}\, r\, \Lambda_r^{(i)}\, S_p^{(i)}.$$

Now (iii) follows from (i) and (ii). Finally, using (7.21) and the formal expansion

$$-\log(1-x)=\sum_{s=1}^{\infty}\frac{x^s}{s},$$

we obtain from (7.24),

$$\Phi_k^{(i)}=\sum_{k_1+\dots+k_s=k}\frac{k}{s}\,\widetilde{\Lambda}_{k_1}^{(i)}\dots\widetilde{\Lambda}_{k_s}^{(i)},$$

summed over $s\geqslant 1$ and positive integers k_j, where $\widetilde{\Lambda}_k^{(i)}=(-1)^{k-1}\Lambda_k^{(i)}$. So, (iv) follows from (i). □

Note that the power sums symmetric functions $\Psi_k^{(i)}$ and $\Phi_k^{(i)}$ are different, in general, for $k\geqslant 3$.

We will now use the noncommutative symmetric functions to construct Casimir elements for the classical Lie algebras. Consider the case of $\mathfrak{gl}_N$ first.

For $1\leqslant m\leqslant N$ denote by $E^{(m)}$ the $m\times m$ matrix with the entries E_{ij}, where $i,j=1,\dots,m$. Introduce the elements $\Lambda_k^{(m)}$, $S_k^{(m)}$, $\Psi_k^{(m)}$ and $\Phi_k^{(m)}$ of the universal enveloping algebra $\mathrm{U}(\mathfrak{gl}_N)$ as the respective noncommutative symmetric functions associated with the matrix $E^{(m)}-m+1$ and the index m. Explicit expressions for these elements are provided by Proposition 7.3.1. For any $k\geqslant 1$ set

$$\Lambda_k=\sum_{k_1+\dots+k_N=k}\Lambda_{k_1}^{(1)}\dots\Lambda_{k_N}^{(N)},$$

$$S_k=\sum_{k_1+\dots+k_N=k}S_{k_1}^{(1)}\dots S_{k_N}^{(N)},$$

$$\Psi_k=\sum_{m=1}^{N}\Psi_k^{(m)},$$

$$\Phi_k=\sum_{m=1}^{N}\Phi_k^{(m)},$$

where the indices k_i run over nonnegative integers and $\Lambda_0^{(m)}=S_0^{(m)}=1$.

THEOREM 7.3.2. *All elements Λ_k, S_k, Ψ_k and Φ_k belong to the center $\mathrm{Z}(\mathfrak{gl}_N)$ of the universal enveloping algebra $\mathrm{U}(\mathfrak{gl}_N)$ for all $k\geqslant 1$. Moreover, $\Psi_k=\Phi_k$ for all k, and the images of Λ_k, S_k and Ψ_k under the Harish-Chandra isomorphism are, respectively, the elementary, complete and power sums symmetric polynomials of degree k in the variables $l_1,\dots,l_N$,*

$$\chi:\Lambda_k\mapsto\sum_{1\leqslant i_1<\dots<i_k\leqslant N}l_{i_1}\dots l_{i_k},$$

$$\chi:S_k\mapsto\sum_{1\leqslant i_1\leqslant\dots\leqslant i_k\leqslant N}l_{i_1}\dots l_{i_k},$$

$$\chi:\Psi_k\mapsto l_1^k+\dots+l_N^k.$$

PROOF. Consider the Capelli determinant $\mathcal{C}(u)$ introduced in (7.1) and set $\widetilde{\mathcal{C}}(q) = q^N \mathcal{C}(q^{-1})$. Then (7.2) gives

$$\widetilde{\mathcal{C}}(q) = 1 + q\,\mathcal{C}_1 + \dots + q^N \mathcal{C}_N. \tag{7.25}$$

Now apply the evaluation homomorphism (1.5) to both sides of the quasideterminant factorization of the quantum determinant $\operatorname{qdet} T(u)$ provided by Theorem 1.10.5. By (7.4),

$$\widetilde{\mathcal{C}}(q) = (1-q)\dots(1-(N-1)\,q)\cdot \pi_N\big(\operatorname{qdet} T(q^{-1})\big).$$

Furthermore, for any $1 \leqslant m \leqslant N$ we have

$$(1-(m-1)\,q)\cdot \pi_N\Big(\big|T^{(m)}(q^{-1}-m+1)\big|_{mm}\Big) = \big|1 + q\,(E^{(m)}-m+1)\big|_{mm}.$$

Thus, Theorem 1.10.5 implies the following decomposition of $\widetilde{\mathcal{C}}(q)$ in the algebra of formal power series in q with coefficients in $\mathrm{U}(\mathfrak{gl}_N)$:

$$\widetilde{\mathcal{C}}(q) = \big|1+q\,E^{(1)}\big|_{11}\cdots\big|1+q\,(E^{(N)}-N+1)\big|_{NN}, \tag{7.26}$$

and the factors pairwise commute. By the definition of the elements $\Lambda_k^{(m)}$,

$$1+\sum_{k=1}^{\infty}\Lambda_k^{(m)}q^k = \big|1+q\,(E^{(m)}-m+1)\big|_{mm}. \tag{7.27}$$

Therefore, (7.25) implies that $\Lambda_k = \mathcal{C}_k$ for $k=1,\dots,N$ and $\Lambda_k = 0$ for all values $k \geqslant N+1$. So, the statements of the theorem about the elements Λ_k now follow from Theorem 7.1.1.

Furthermore, by the definition of the elements $S_k^{(m)}$, we have

$$1+\sum_{k=1}^{\infty}S_k^{(m)}q^k = \big|1-q\,(E^{(m)}-m+1)\big|_{mm}^{-1},$$

which implies the relation

$$1+\sum_{k=1}^{\infty}S_k\,q^k = \widetilde{\mathcal{C}}(-q)^{-1}.$$

Hence, all elements S_k belong to $\mathrm{Z}(\mathfrak{gl}_N)$. Due to Theorem 7.1.1, their Harish-Chandra images are found by

$$1+\sum_{k=1}^{\infty}\chi(S_k)\,q^k = \prod_{i=1}^{N}\frac{1}{1-l_i\,q}.$$

This proves the statements about the elements S_k. Now observe that by the decomposition (7.26), the series (7.27) is the ratio of the polynomials $\widetilde{\mathcal{C}}(q)$ associated with the Lie algebras $\mathfrak{gl}_m$ and $\mathfrak{gl}_{m-1}$. Since the coefficients of $\widetilde{\mathcal{C}}(q)$ are central in the corresponding universal enveloping algebra, all elements $\Lambda_k^{(m)}$ pairwise commute. Using the definition of the elements $\Psi_k^{(m)}$ and $\Phi_k^{(m)}$, we come to the relations

$$\sum_{k=1}^{\infty}\Psi_k\,q^{k-1} = \widetilde{\mathcal{C}}(-q)\,\frac{d}{dq}\,\widetilde{\mathcal{C}}(-q)^{-1} \tag{7.28}$$

and

$$\sum_{k=1}^{\infty}\Phi_k\,q^{k-1} = -\frac{d}{dq}\,\log\widetilde{\mathcal{C}}(-q).$$

As the coefficients of the polynomial $\widetilde{C}(q)$ are central in $\mathrm{U}(\mathfrak{gl}_N)$, we deduce that $\Psi_k, \Phi_k \in \mathrm{Z}(\mathfrak{gl}_N)$. Since

$$\widetilde{C}(-q)\,\frac{d}{dq}\,\widetilde{C}(-q)^{-1} = -\frac{d}{dq}\,\log\widetilde{C}(-q),$$

we conclude that $\Psi_k = \Phi_k$. Finally, the Harish-Chandra images $\chi(\Psi_k)$ are found from (7.28) and Theorem 7.1.1. □

COROLLARY 7.3.3. *Each family* $\{\Lambda_k\}$, $\{S_k\}$ *and* $\{\Psi_k\}$ *with* $k = 1, 2, \dots, N$ *generates* $\mathrm{Z}(\mathfrak{gl}_N)$.

PROOF. It is well known that the first N elementary symmetric polynomials, complete symmetric polynomials, or power sums symmetric polynomials generate the algebra of symmetric polynomials in $l_1, \dots, l_N$. The claim then follows by the application of the Harish-Chandra isomorphism and Theorem 7.3.2. □

EXAMPLE 7.3.4. We have

$$\Psi_1 = \sum_{m=1}^{N} (E_{mm} - m + 1),$$

$$\Psi_2 = \sum_{m=1}^{N} (E_{mm} - m + 1)^2 + 2 \sum_{1\leqslant l<m\leqslant N} E_{ml}\, E_{lm}.$$ □

Recall the Gelfand–Tsetlin subalgebra $\mathcal{Z}_N$ of $\mathrm{U}(\mathfrak{gl}_N)$ introduced in Section 5.4.

COROLLARY 7.3.5. *The subalgebra* $\mathcal{Z}_N$ *is generated by each family of elements* $\{\Lambda_k^{(m)}\}$, $\{S_k^{(m)}\}$ *and* $\{\Psi_k^{(m)}\}$ *with* $1 \leqslant k \leqslant m \leqslant N$.

PROOF. We show first that these elements belong to the subalgebra $\mathcal{Z}_N$. As we noted in the proof of Theorem 7.3.2, the series (7.27) is the ratio of the polynomials $\widetilde{C}(q)$ associated with the Lie algebras $\mathfrak{gl}_m$ and $\mathfrak{gl}_{m-1}$. So, all elements $\Lambda_k^{(m)}$ belong to $\mathcal{Z}_N$. The same argument applied to the series $\widetilde{C}(-q)^{-1}$ instead of $\widetilde{C}(q)$ shows that all elements $S_k^{(m)}$ also belong to $\mathcal{Z}_N$. The element $\Psi_k^{(m)}$ is the difference of the central elements Ψ_k corresponding to the Lie algebras $\mathfrak{gl}_m$ and $\mathfrak{gl}_{m-1}$ and so belongs to $\mathcal{Z}_N$ as well.

Finally, by Theorem 7.3.2, each element of $\mathrm{Z}(\mathfrak{gl}_N)$ can be expressed as a polynomial in the elements of each of the families. This shows that each family $\{\Lambda_k^{(m)}\}$, $\{S_k^{(m)}\}$ and $\{\Psi_k^{(m)}\}$ generates $\mathcal{Z}_N$. □

Note that since $\Phi_k = \Psi_k$ for all k, using the argument of the proof of Corollary 7.3.5, we obtain $\Psi_k^{(m)} = \Phi_k^{(m)}$ for all k and m.

Consider the Gelfand–Tsetlin basis $\{\zeta_\Lambda\}$ of the representation $L(\lambda)$ of $\mathfrak{gl}_N$ constructed in Section 5.4. By Corollary 5.4.3, each basis vector ζ_Λ is a common eigenvector for all elements $\Lambda_k^{(m)}$, $S_k^{(m)}$ and $\Psi_k^{(m)}$. The corresponding eigenvalues can be easily found by using (5.42) and the relationship between the Capelli determinant and the central elements Λ_k, S_k and Ψ_k.

EXAMPLE 7.3.6. In the notation of Section 5.4, we have

$$\Psi_k^{(m)}\,\zeta_\Lambda = \Big(l_{m1}^k + \dots + l_{mm}^k - l_{m-1,1}^k - \dots - l_{m-1,m-1}^k\Big)\,\zeta_\Lambda.$$ □

Now turn to the case of the orthogonal and symplectic Lie algebras $\mathfrak{g}_N = \mathfrak{o}_N$ or $\mathfrak{g}_N = \mathfrak{sp}_N$. Let $1 \leqslant m \leqslant n$. Denote by $F^{(m)}$ the matrix with the entries F_{ij}, where $i, j = -m, -m+1, \dots, m$, and by $\widetilde{F}^{(m)}$ the submatrix of $F^{(m)}$ obtained by removing the row and column enumerated by $-m$.

Let $\Lambda_k^{(m)}$, $S_k^{(m)}$, $\Phi_k^{(m)}$ respectively denote the elementary, complete and power sums symmetric functions of the second kind associated with the matrix $F^{(m)} + \rho_m$ and the index m. Also, let $\widetilde{\Lambda}_k^{(m)}$, $\widetilde{S}_k^{(m)}$, $\widetilde{\Phi}_k^{(m)}$ denote the respective symmetric functions associated with the matrix $-\widetilde{F}^{(m)} - \rho_m$ and the index m. The explicit expressions of these elements in terms of the generators F_{ij} are provided by Proposition 7.3.1. For any $k \geqslant 1$ set

$$\Lambda_{2k} = \sum_{k_1+\dots+k_{2n}=2k} \widetilde{\Lambda}_{k_1}^{(1)} \Lambda_{k_2}^{(1)} \dots \widetilde{\Lambda}_{k_{2n-1}}^{(n)} \Lambda_{k_{2n}}^{(n)},$$

$$S_{2k} = \sum_{k_1+\dots+k_{2n}=2k} \widetilde{S}_{k_1}^{(1)} S_{k_2}^{(1)} \dots \widetilde{S}_{k_{2n-1}}^{(n)} S_{k_{2n}}^{(n)},$$

$$\Phi_{2k} = \sum_{m=1}^{n} \big(\widetilde{\Phi}_{2k}^{(m)} + \Phi_{2k}^{(m)}\big),$$

where the k_i run over nonnegative integers and $\Lambda_0^{(m)} = \widetilde{\Lambda}_0^{(m)} = S_0^{(m)} = \widetilde{S}_0^{(m)} = 1$.

THEOREM 7.3.7. *All elements Λ_{2k}, S_{2k} and Φ_{2k} belong to the center $\mathrm{Z}(\mathfrak{g}_N)$ of the universal enveloping algebra $\mathrm{U}(\mathfrak{g}_N)$ for all $k \geqslant 1$. Moreover, the images of $(-1)^k \Lambda_{2k}$, S_{2k} and $\Phi_{2k}/2$ under the Harish-Chandra isomorphism are, respectively, the elementary, complete and power sums symmetric polynomials of degree k in the variables $l_1^2, \dots, l_n^2$.*

PROOF. Consider the Capelli-type determinant $\mathcal{C}(u)$ introduced in (7.13) and set

$$\widetilde{\mathcal{C}}(q) = \begin{cases} q^N\, \mathcal{C}(q^{-1}) & \text{if} \quad N = 2n, \\ \dfrac{q^N}{1+q/2}\, \mathcal{C}(q^{-1}) & \text{if} \quad N = 2n+1. \end{cases}$$

Due to Theorem 7.1.6, $\widetilde{\mathcal{C}}(q)$ is an even polynomial in q of degree $2n$ with constant term 1,

$$\widetilde{\mathcal{C}}(q) = 1 + q^2\, \mathcal{C}_1 + \dots + q^{2n}\, \mathcal{C}_n. \tag{7.29}$$

Now apply the evaluation homomorphism (2.106) to both sides of the quasideterminant factorization of the Sklyanin determinant provided by Theorem 4.1.7. Arguing as in the proof of Theorem 7.3.2, we obtain the following decomposition of $\widetilde{\mathcal{C}}(q)$ in the algebra of formal power series in q with coefficients in $\mathrm{U}(\mathfrak{g}_N)$:

$$\begin{aligned} \widetilde{\mathcal{C}}(q) = \big|1 - q\,(\widetilde{F}^{(1)} + \rho_1)\big|_{11} \cdot \big|1 + q\,(F^{(1)} + \rho_1)\big|_{11} \\ \times \dots \big|1 - q\,(\widetilde{F}^{(n)} + \rho_n)\big|_{nn} \cdot \big|1 + q\,(F^{(n)} + \rho_n)\big|_{nn}, \end{aligned} \tag{7.30}$$

and the factors pairwise commute. By the definition of the elements $\Lambda_k^{(m)}$ and $\widetilde{\Lambda}_k^{(m)}$, we have

$$1 + \sum_{k=1}^{\infty} \Lambda_k^{(m)} q^k = \big|1 + q\,(F^{(m)} + \rho_m)\big|_{mm}$$

and

$$1+\sum_{k=1}^{\infty}\widetilde{\Lambda}_k^{(m)}q^k=\big|1-q\,(\widetilde{F}^{(m)}+\rho_m)\big|_{mm}.$$

Therefore, (7.29) implies that $\Lambda_{2k}=\mathcal{C}_k$ for $k=1,\dots,n$ and $\Lambda_{2k}=0$ for all values $k\geqslant n+1$. So, the statements of the theorem about the elements Λ_{2k} now follow from Theorem 7.1.6. Furthermore, using the definition of the elements $S_k^{(m)}$ and $\widetilde{S}_k^{(m)}$, we derive that

$$1+\sum_{k=1}^{\infty}S_{2k}\,q^{2k}=\widetilde{\mathcal{C}}(-q)^{-1}.$$

This implies the statements about the elements S_{2k}. Similarly, since the factors in the expansion (7.30) commute, using the definition of the elements $\Phi_k^{(m)}$ and $\widetilde{\Phi}_k^{(m)}$, we come to the relation

$$\sum_{k=1}^{\infty}\Phi_{2k}\,q^{2k-1}=-\frac{d}{dq}\,\log\widetilde{\mathcal{C}}(-q),$$

which implies the statements about the elements Φ_{2k}. □

Regarding a possible definition of the analogues of the elements Ψ_{2k} for the Lie algebra $\mathfrak{g}_N$, see Example 7.7.7.

Note that in the case $\mathfrak{g}_N=\mathfrak{o}_{2n}$ we have

$$\chi:\Lambda_{2n}\mapsto(-1)^n\,l_1^2\dots l_n^2.$$

Therefore, Λ_{2n} is the square of a certain element of $\mathrm{Z}(\mathfrak{o}_{2n})$ which we denote by $\sqrt{\Lambda_{2n}}$.

COROLLARY 7.3.8. *Each family $\{\Lambda_{2k}\}$, $\{S_{2k}\}$ and $\{\Phi_{2k}\}$ with $k=1,2,\dots,n$ generates $\mathrm{Z}(\mathfrak{g}_N)$ for $\mathfrak{g}_N=\mathfrak{o}_{2n+1}$ and $\mathfrak{g}_N=\mathfrak{sp}_{2n}$. In the case $\mathfrak{g}_N=\mathfrak{o}_{2n}$ the algebra $\mathrm{Z}(\mathfrak{o}_{2n})$ is generated by each family $\{\Lambda_{2k}\}$, $\{S_{2k}\}$ and $\{\Phi_{2k}\}$ with $k=1,2,\dots,n-1$ together with the element $\sqrt{\Lambda_{2n}}$.*

PROOF. The statement follows from Theorem 7.3.7 by the application of the Harish-Chandra isomorphism. □

EXAMPLE 7.3.9. We have

$$\Phi_2^{(m)}=(F_{mm}+\rho_m)^2+2\sum_{-m\leqslant i<m}F_{mi}F_{im},\qquad\text{and}$$

$$\widetilde{\Phi}_2^{(m)}=(F_{mm}+\rho_m)^2+2\sum_{-m<i<m}F_{mi}F_{im}.$$

Hence, if $\mathfrak{g}_N=\mathfrak{o}_N$, then the second order Casimir element can be written as

$$\Phi_2=2\sum_{m=1}^{n}\Big((F_{mm}+\rho_m)^2+2\sum_{-m<i<m}F_{mi}F_{im}\Big).$$

If $\mathfrak{g}_N=\mathfrak{sp}_{2n}$, then

$$\Phi_2=2\sum_{m=1}^{n}\Big((F_{mm}-m)^2+F_{m,-m}F_{-m,m}+2\sum_{-m<i<m}F_{mi}F_{im}\Big).$$

□

An alternative version of Theorem 7.3.7 for the presentation of the orthogonal Lie algebra $\mathfrak{o}_N$ corresponding to the matrix $G = 1$ is given in Example 7.7.8. It is based upon the quasideterminant decomposition of the Sklyanin determinant provided by Theorem 2.12.1.

7.4. Higher Capelli identities and quantum immanants

We will use a class of symmetric polynomials known as the double (or factorial) Schur polynomials, whose definition and basic properties are briefly reviewed below; see e.g. Macdonald [**279**], Okounkov [**365**], Okounkov and Olshanski [**368**].

Recall from Section 6.4 that a partition μ of length $\leqslant n$ is a sequence of integers $\mu = (\mu_1, \dots, \mu_n)$ such that $\mu_1 \geqslant \dots \geqslant \mu_n \geqslant 0$. As in that section, we will identify a partition with its diagram. Fix an arbitrary sequence $b = (b_1, b_2, \dots)$ of complex numbers and for each $k = 0, 1, 2, \dots$ introduce the corresponding k-th factorial power of a variable y by

$$(y\,|\,b)^0 = 1, \qquad (y\,|\,b)^k = (y - b_1)\dots(y - b_k), \quad k \geqslant 1.$$

For an n-tuple of variables $x = (x_1, \dots, x_n)$ consider the polynomial

$$s_\mu(x\,|\,b) = \frac{\det\big[(x_j\,|\,b)^{\mu_i+n-i}\big]}{\det\big[(x_j\,|\,b)^{n-i}\big]}$$

where the determinants are taken with respect to the indices $i, j = 1, \dots, n$. This polynomial is symmetric in $x_1, \dots, x_n$ and is called the *double Schur polynomial* (associated with the sequence b). Note that the denominator

$$\det\big[(x_j\,|\,b)^{n-i}\big] = \prod_{1\leqslant i<j\leqslant n} (x_i - x_j)$$

is the Vandermonde determinant. Thus it does not depend on the sequence b. If $b = (0, 0, \dots)$, then $s_\mu(x\,|\,b)$ is the ordinary Schur polynomial $s_\mu(x)$. Moreover, for an arbitrary sequence b the homogeneous component of $s_\mu(x\,|\,b)$ in x of the highest degree coincides with $s_\mu(x)$.

An equivalent combinatorial definition of the double Schur polynomials can be given in terms of semistandard μ-tableaux. A μ-tableau $\mathcal{T}$ (see Section 6.4) is called *semistandard* if its entries weakly increase along the rows and strictly increase down the columns. If $\alpha \in \mu$ is a cell in the diagram μ, then $\mathcal{T}(\alpha)$ will denote the entry of the tableau $\mathcal{T}$ at this cell. Recall that the content $c(\alpha)$ of a cell $\alpha = (i, j) \in \mu$ is the difference between its column and row numbers, $c(\alpha) = j - i$. The double Schur polynomial associated with μ is then given by the formula

$$s_\mu(x\,|\,b) = \sum_{\mathrm{sh}(\mathcal{T})=\mu} \prod_{\alpha\in\mu} \big(x_{\mathcal{T}(\alpha)} - b_{\mathcal{T}(\alpha)+c(\alpha)}\big),$$

summed over semistandard tableaux $\mathcal{T}$ of shape μ with entries in the set $\{1, \dots, n\}$.

The polynomials $s_\mu(x\,|\,b)$ possess the following vanishing property which can be derived from either of their two definitions. Assume that the sequence b is multiplicity-free; that is, $b_k \neq b_l$ for all $k \neq l$. For any partition $\lambda = (\lambda_1, \dots, \lambda_n)$ introduce the n-tuple b_λ of elements of b by

$$b_\lambda = (b_{\lambda_1+n}, \dots, b_{\lambda_n+1}).$$

We will write $|\lambda|$ for the degree $\lambda_1 + \dots + \lambda_n$ of the partition λ. Then

$$s_\mu(b_\lambda\,|\,b) = 0 \qquad \text{unless} \quad \mu \subset \lambda,$$

while $s_\mu(b_\mu\,|\,b) \neq 0$. In particular,

$$s_\mu(b_\lambda\,|\,b) = 0 \qquad \text{if} \quad |\lambda| < |\mu|.$$

It is straightforward to verify that these properties characterize the double Schur polynomials in the following sense. Let $f(x)$ be a symmetric polynomial of degree $|\mu|$ whose homogeneous component of the top degree is $s_\mu(x)$. Then we have:

$$\text{if } f(b_\lambda) = 0 \text{ for any } \lambda \text{ such that } |\lambda| < |\mu|, \text{ then } f(x) = s_\mu(x\,|\,b). \tag{7.31}$$

For an arbitrary sequence b, the *double elementary* and *double complete symmetric polynomials* are respectively given by

$$e_k(x\,|\,b) = s_{(1^k)}(x\,|\,b) = \sum_{i_1<\dots<i_k} (x_{i_1} - b_{i_1})(x_{i_2} - b_{i_2-1})\dots(x_{i_k} - b_{i_k-k+1})$$

and

$$h_k(x\,|\,b) = s_{(k)}(x\,|\,b) = \sum_{i_1\leqslant\dots\leqslant i_k} (x_{i_1} - b_{i_1})(x_{i_2} - b_{i_2+1})\dots(x_{i_k} - b_{i_k+k-1}).$$

The following identities of formal power series in q^{-1} take place:

$$1 + \sum_{k=1}^{n} \frac{(-1)^k e_k(x\,|\,b)}{(q - b_{n-k+1})\dots(q - b_n)} = \frac{(q - x_1)\dots(q - x_n)}{(q - b_1)\dots(q - b_n)} \tag{7.32}$$

and

$$1 + \sum_{k=1}^{\infty} \frac{h_k(x\,|\,b)}{(q - b_{n+1})\dots(q - b_{n+k})} = \frac{(q - b_1)\dots(q - b_n)}{(q - x_1)\dots(q - x_n)}. \tag{7.33}$$

We now need to generalize the matrix notation of Section 1.2. An arbitrary matrix $A = [a_{ij}]$ of size $p \times r$ whose entries are elements of an algebra $\mathcal{A}$ over $\mathbb{C}$ will be identified with the element of the tensor product

$$A = \sum_{i,j} e_{ij} \otimes a_{ij} \in \mathrm{Mat}_{p\times r} \otimes \mathcal{A},$$

where the e_{ij} with $i = 1, \dots, p$ and $j = 1, \dots, r$ are the standard matrix units which form a basis of the vector space $\mathrm{Mat}_{p\times r} = \mathrm{Mat}_{p\times r}(\mathbb{C})$. More generally, given k matrices $A^{(1)}, \dots, A^{(k)}$ of size $p\times r$ we define their tensor product $A^{(1)} \otimes \dots \otimes A^{(k)}$ as the element

$$\sum e_{i_1 j_1} \otimes \dots \otimes e_{i_k j_k} \otimes a^{(1)}_{i_1 j_1} \dots a^{(k)}_{i_k j_k} \in (\mathrm{Mat}_{p\times r})^{\otimes k} \otimes \mathcal{A}.$$

The matrix multiplication induces the natural map

$$\Big((\mathrm{Mat}_{p\times r})^{\otimes k} \otimes \mathcal{A}\Big) \times \Big((\mathrm{Mat}_{r\times s})^{\otimes k} \otimes \mathcal{A}\Big) \to (\mathrm{Mat}_{p\times s})^{\otimes k} \otimes \mathcal{A}.$$

In the case $p = r$ the vector space $\mathrm{Mat}_{p\times r}$ is an algebra isomorphic to $\mathrm{End}\,\mathbb{C}^p$ which we also denote by Mat_p.

Now we briefly review some properties of a representation of the universal enveloping algebra $\mathrm{U}(\mathfrak{gl}_N)$ by polynomial coefficient differential operators; see e.g. Howe [**177**] and Howe and Umeda [**178**] for more details. Suppose that m is a positive integer and consider the algebra of polynomials $\mathcal{P}$ in mN variables x_{ia} with

$i = 1, \dots, N$ and $a = 1, \dots, m$. Let $\partial_{ia} : \mathcal{P} \to \mathcal{P}$ denote the partial differentiation operator with respect to the variable x_{ia}. The mapping

$$E_{ij} \mapsto \sum_{a=1}^{m} x_{ia} \partial_{ja} \tag{7.34}$$

defines a representation of the Lie algebra $\mathfrak{gl}_N$ on $\mathcal{P}$. Introduce the $N \times m$ matrices $X = [x_{ia}]$ and $\mathcal{D} = [\partial_{ia}]$ with entries in the algebra of polynomial coefficient differential operators $\mathcal{PD}$ in the variables x_{ia} so that

$$X = \sum_{i,a} e_{ia} \otimes x_{ia}, \qquad \mathcal{D} = \sum_{i,a} e_{ia} \otimes \partial_{ia}$$

are also regarded as elements of $\mathrm{Mat}_{N \times m} \otimes \mathcal{PD}$. The mapping (7.34) can be written as

$$E \mapsto X \mathcal{D}^t, \tag{7.35}$$

where, as before,

$$E = \sum_{i,j=1}^{N} e_{ij} \otimes E_{ij} \in \mathrm{Mat}_N \otimes \mathrm{U}(\mathfrak{gl}_N),$$

while $\mathcal{D}^t$ is the transposed matrix

$$\mathcal{D}^t = \sum_{a,i} e_{ai} \otimes \partial_{ia} \in \mathrm{Mat}_{m \times N} \otimes \mathcal{PD},$$

summed over $a = 1, \dots, m$ and $i = 1, \dots, N$. The multiplication of the matrix X from the right by elements of the group GL_m defines an action of GL_m on $\mathcal{P}$ which naturally extends to the vector space $\mathcal{PD}$. If $m \geqslant N$, then the extension of the homomorphism (7.34) to the universal enveloping algebra yields an isomorphism

$$\mathrm{U}(\mathfrak{gl}_N) \to \mathcal{PD}^{GL_m} \tag{7.36}$$

of $\mathrm{U}(\mathfrak{gl}_N)$ and the algebra of GL_m-invariant polynomial coefficient differential operators. Furthermore, the restriction of the isomorphism (7.36) to the center of $\mathrm{Z}(\mathfrak{gl}_N)$ gives an isomorphism

$$\mathrm{Z}(\mathfrak{gl}_N) \to \mathcal{PD}^{GL_N \times GL_m} \tag{7.37}$$

of the center and the algebra of $GL_N \times GL_m$-invariants in $\mathcal{PD}$. The GL_N action here corresponds to the representation of $\mathfrak{gl}_N$ defined by (7.34) and is determined by the multiplication of the matrix X from the left by elements of the group GL_N.

As in Section 6.5, consider the natural action of the symmetric group $\mathfrak{S}_k$ on the tensor product of the k copies of $\mathbb{C}^N$; cf. (6.95). So the permutations will be represented as elements of the algebra $(\mathrm{Mat}_N)^{\otimes k}$. Furthermore, as in Section 6.4, for a partition μ of k and standard μ-tableaux $\mathcal{U}$ and $\mathcal{U}'$ with entries in $\{1, \dots, k\}$ consider the matrix element $\phi_{\mathcal{U}\mathcal{U}'}$ defined as in (6.75). Its image in $(\mathrm{Mat}_N)^{\otimes k}$ will be denoted by $\Phi_{\mathcal{U}\mathcal{U}'}$, and we set $\Phi_{\mathcal{U}} = \Phi_{\mathcal{U}\mathcal{U}}$; cf. Section 6.5. Recall also that $c_r(\mathcal{U}) = j - i$ if the cell $(i, j) \in \mu$ is occupied by the entry r of the tableau $\mathcal{U}$.

THEOREM 7.4.1. *For any standard μ-tableaux $\mathcal{U}$ and $\mathcal{U}'$, under the map* (7.35) *we have*

$$(E - c_1) \otimes \dots \otimes (E - c_k)\, \Phi_{\mathcal{U}\mathcal{U}'} \mapsto X^{\otimes k}\, (\mathcal{D}^t)^{\otimes k}\, \Phi_{\mathcal{U}\mathcal{U}'}, \tag{7.38}$$

where $c_r = c_r(\mathcal{U})$ for $r = 1, \dots, k$ and $E - c_r$ denotes the matrix obtained from E by subtracting c_r from all diagonal entries.

PROOF. We will argue by induction on k; cf. the proof of Proposition 6.4.4. The matrix elements $\phi_{\mathcal{U}\mathcal{U}'}$ are proportional to the matrix units $e_{\mathcal{U}\mathcal{U}'}$ (see Section 6.4), and so using (6.77) we can write

$$\Phi_{\mathcal{U}\mathcal{U}'} = \gamma \cdot \Phi_{\mathcal{V}}\, \Phi_{\mathcal{U}\mathcal{U}'},$$

where $\mathcal{V}$ is the standard tableau obtained from $\mathcal{U}$ by removing the cell occupied by k and γ is a nonzero constant. Using (7.35) and the induction hypothesis, we obtain

$$\begin{aligned}(E - c_1) \otimes \ldots \otimes (E - c_k)\, \Phi_{\mathcal{U}\mathcal{U}'} \\ &= \gamma \cdot (E - c_1) \otimes \ldots \otimes (E - c_{k-1})\, \Phi_{\mathcal{V}} \otimes (E - c_k)\, \Phi_{\mathcal{U}\mathcal{U}'} \\ &\mapsto \gamma \cdot X^{\otimes(k-1)}\, (\mathcal{D}^t)^{\otimes(k-1)}\, \Phi_{\mathcal{V}} \otimes (X\, \mathcal{D}^t - c_k)\, \Phi_{\mathcal{U}\mathcal{U}'} \\ &= X^{\otimes(k-1)} \cdot (\mathcal{D}^t)^{\otimes(k-1)} \otimes (X\, \mathcal{D}^t - c_k)\, \Phi_{\mathcal{U}\mathcal{U}'}.\end{aligned}$$

Expand the last expression in terms of the matrix elements to get

$$\sum e_{i_1 j_1} \otimes \ldots \otimes e_{i_k j_k} \otimes x_{i_1 a_1} \ldots x_{i_{k-1} a_{k-1}} \\ \times \partial_{j_1 a_1} \ldots \partial_{j_{k-1} a_{k-1}} \Big(x_{i_k a_k} \partial_{j_k a_k} - \frac{\delta_{i_k j_k}\, c_k}{N} \Big)\, \Phi_{\mathcal{U}\mathcal{U}'}.$$

Now we transform this expression using the relations $\partial_{ia}\, x_{jb} = x_{jb}\, \partial_{ia} + \delta_{ij}\, \delta_{ab}$ to obtain

$$\sum e_{i_1 j_1} \otimes \ldots \otimes e_{i_k j_k} \otimes x_{i_1 a_1} \ldots x_{i_k a_k}\, \partial_{j_1 a_1} \ldots \partial_{j_k a_k}\, \Phi_{\mathcal{U}\mathcal{U}'} \\ + \sum e_{i_1 j_1} \otimes \ldots \otimes e_{i_{k-1} j_{k-1}} \otimes 1 \otimes x_{i_1 a_1} \ldots x_{i_{k-1} a_{k-1}}\, \partial_{j_1 a_1} \ldots \partial_{j_{k-1} a_{k-1}} \\ \times \big(P_{1k} + \cdots + P_{k-1,k} - c_k \big)\, \Phi_{\mathcal{U}\mathcal{U}'},$$

where P_{ik} is the image of the transposition $(i\, k) \in \mathfrak{S}_k$ in the algebra $(\mathrm{Mat}_N)^{\otimes k}$. Note that $P_{1k} + \cdots + P_{k-1,k}$ is the image of the Jucys–Murphy element x_k so that

$$\big(P_{1k} + \cdots + P_{k-1,k} \big)\, \Phi_{\mathcal{U}\mathcal{U}'} = c_k\, \Phi_{\mathcal{U}\mathcal{U}'},$$

which is implied by (6.79). The second summand in the above expression vanishes, thus proving the theorem. □

Consider the trace map

$$\mathrm{tr} : (\mathrm{Mat}_N)^{\otimes k} \to \mathbb{C}$$

which is a linear map defined on the basis elements by

$$\mathrm{tr}\, (e_{i_1 j_1} \otimes \ldots \otimes e_{i_k j_k}) = \delta_{i_1 j_1} \ldots \delta_{i_k j_k}.$$

Recall that f_μ denotes the dimension of the irreducible representation V_μ of $\mathfrak{S}_k$ associated with the partition μ. The ratio $k!/f_\mu$ equals the product of hooks of μ, which we denote by H_μ. The *quantum immanant* $\mathbb{S}_\mu$ associated with μ is the element of $\mathrm{U}(\mathfrak{gl}_N)$ defined by

$$\mathbb{S}_\mu = \frac{1}{H_\mu}\, \mathrm{tr}\, (E - c_1) \otimes \ldots \otimes (E - c_k) \cdot \Phi_{\mathcal{U}}, \tag{7.39}$$

where $\mathcal{U}$ is a standard μ-tableau, $c_r = c_r(\mathcal{U})$ for $r = 1, \ldots, k$, and the trace is taken over all matrix factors. We will see below that $\mathbb{S}_\mu$ depends only on the partition

μ and does not depend on the standard tableau $\mathcal{U}$, thus justifying the notation. Recall that

$$\chi_\mu = \sum_{s\in\mathfrak{S}_k} \chi_\mu(s)\cdot s \in \mathbb{C}[\mathfrak{S}_k]$$

denotes the character of the irreducible representation V_μ of $\mathfrak{S}_k$. The *higher Capelli identities* are provided by the following corollary.

COROLLARY 7.4.2. *The image of the quantum immanant in the representation* (7.35) *is given by*

$$\mathbb{S}_\mu \mapsto \frac{1}{k!}\,\mathrm{tr}\, X^{\otimes k}\,(\mathcal{D}^t)^{\otimes k}\,\chi_\mu. \tag{7.40}$$

PROOF. Take the trace of both sides of (7.38) with $\mathcal{U}' = \mathcal{U}$. On the left-hand side we get $H_\mu\,\mathbb{S}_\mu$, while for the right-hand side we have

$$\begin{aligned}\mathrm{tr}\, X^{\otimes k}\,(\mathcal{D}^t)^{\otimes k}\,\Phi_{\mathcal{U}} &= \frac{1}{k!}\sum_{s\in\mathfrak{S}_k} \mathrm{tr}\Big(s\, X^{\otimes k}\,(\mathcal{D}^t)^{\otimes k}\,\Phi_{\mathcal{U}}\, s^{-1}\Big)\\ &= \frac{1}{f_\mu}\,\mathrm{tr}\, X^{\otimes k}\,(\mathcal{D}^t)^{\otimes k}\,\chi_\mu.\end{aligned}$$

Here we have used the invariance of $X^{\otimes k}$ and $(\mathcal{D}^t)^{\otimes k}$ under the conjugations by the elements $s\in\mathfrak{S}_k$ and the following well known identity in the group algebra of $\mathfrak{S}_k$:

$$\frac{1}{k!}\sum_{s\in\mathfrak{S}_k} s\,\Phi_{\mathcal{U}}\,s^{-1} = \frac{1}{f_\mu}\,\chi_\mu.$$

In order to verify the latter, we use the notation of Section 6.4 and note that the representation V_μ admits a basis of the form

$$\sigma_1\cdot v_{\mathcal{U}},\ \dots,\ \sigma_f\cdot v_{\mathcal{U}},\qquad \sigma_i\in\mathfrak{S}_k,\qquad f = f_\mu.$$

Since the character is constant on the elements of a conjugacy class of $\mathfrak{S}_k$, for any $\sigma\in\mathfrak{S}_k$ we have

$$\begin{aligned}\chi_\mu(\sigma) &= \sum_{i=1}^{f}\big(\sigma\,\sigma_i\cdot v_{\mathcal{U}},\,\sigma_i\cdot v_{\mathcal{U}}\big) = \sum_{i=1}^{f}\big(\sigma_i^{-1}\sigma\,\sigma_i\cdot v_{\mathcal{U}},\,v_{\mathcal{U}}\big)\\ &= \frac{1}{k!}\sum_{s\in\mathfrak{S}_k}\sum_{i=1}^{f}\big(\sigma_i^{-1}s^{-1}\sigma\,s\,\sigma_i\cdot v_{\mathcal{U}},\,v_{\mathcal{U}}\big) = \frac{f_\mu}{k!}\sum_{s\in\mathfrak{S}_k}\big(s^{-1}\sigma\,s\cdot v_{\mathcal{U}},\,v_{\mathcal{U}}\big),\end{aligned}$$

as desired. □

In particular, taking $m \geqslant N$ and using the isomorphism (7.36), we obtain from Corollary 7.4.2 that the element $\mathbb{S}_\mu$ is independent of the μ-tableau $\mathcal{U}$. Take $\mathcal{U} = \mathcal{U}^r$, the row tableau of shape μ. Then by (6.93) we have

$$\Phi_{\mathcal{U}^r} = \frac{1}{\mu!}\,H_{\mathcal{U}^r}A_{\mathcal{U}^r}H_{\mathcal{U}^r},\qquad \mu! = \mu_1!\dots\mu_l!,$$

where l is the length of μ, and $H_{\mathcal{U}^r}$ and $A_{\mathcal{U}^r}$ are the respective images in $\mathrm{End}\,(\mathbb{C}^N)^{\otimes k}$ of the elements $h_{\mathcal{U}^r}$ and $a_{\mathcal{U}^r}$ defined in (6.92).

COROLLARY 7.4.3. *We have*

$$\mathbb{S}_\mu = \frac{1}{H_\mu} \operatorname{tr} (E - c_1) \otimes \ldots \otimes (E - c_k)\, H_{\mathcal{U}^r} A_{\mathcal{U}^r}, \tag{7.41}$$

where $c_i = c_i(\mathcal{U}^r)$ *for* $i = 1, \ldots, k$.

PROOF. Propositions 1.6.1 and 1.6.3 imply

$$(E - c_1) \otimes \ldots \otimes (E - c_k)\, H_{\mathcal{U}^r} = H_{\mathcal{U}^r} (E - c_1') \otimes \ldots \otimes (E - c_k'), \tag{7.42}$$

where $c_1', \ldots, c_k'$ denote the contents of the cells of μ listed in the reverse row order, so that the cells are read from right to left in each row, starting from the top row. Indeed, since $H_{\mathcal{U}^r}$ equals the product of the symmetrizers of the rows of $\mathcal{U}^r$, it suffices to verify (7.42) in the case where μ is a row tableau. In that case, (7.42) follows by applying the evaluation homomorphism (1.5) to both sides of the relation of Proposition 1.6.1. Hence, using the cyclic property of trace and the relation $H_{\mathcal{U}^r}^2 = \mu!\, H_{\mathcal{U}^r}$, we can write

$$\begin{aligned} \mathbb{S}_\mu &= \frac{1}{H_\mu\, \mu!} \operatorname{tr} (E - c_1) \otimes \ldots \otimes (E - c_k)\, H_{\mathcal{U}^r} A_{\mathcal{U}^r} H_{\mathcal{U}^r} \\ &= \frac{1}{H_\mu\, \mu!} \operatorname{tr} H_{\mathcal{U}^r} (E - c_1') \otimes \ldots \otimes (E - c_k')\, A_{\mathcal{U}^r} H_{\mathcal{U}^r} \\ &= \frac{1}{H_\mu} \operatorname{tr} H_{\mathcal{U}^r} (E - c_1') \otimes \ldots \otimes (E - c_k')\, A_{\mathcal{U}^r}, \end{aligned}$$

which yields (7.41). □

EXAMPLE 7.4.4. For $\mu = (2, 1)$ we have

$$\mathbb{S}_{(2,1)} = \frac{1}{3} \operatorname{tr} E \otimes (E - 1) \otimes (E + 1) \cdot (1 + P_{12})(1 - P_{13}),$$

or, more explicitly,

$$\begin{aligned} \mathbb{S}_{(2,1)} = \frac{1}{3} \sum_{i_1, i_2, i_3} \Big(& E_{i_1 i_1} (E_{i_2 i_2} - 1)(E_{i_3 i_3} + 1) + E_{i_1 i_2} (E_{i_2 i_1} - \delta_{i_2 i_1})(E_{i_3 i_3} + 1) \\ & - E_{i_1 i_3} (E_{i_2 i_2} - 1)(E_{i_3 i_1} + \delta_{i_3 i_1}) - E_{i_1 i_2} (E_{i_2 i_3} - \delta_{i_2 i_3})(E_{i_3 i_1} + \delta_{i_3 i_1}) \Big), \end{aligned}$$

summed over the indices $i_1, i_2, i_3 \in \{1, \ldots, N\}$. □

In order to get some alternative formulas for $\mathbb{S}_\mu$, for an arbitrary standard μ-tableau $\mathcal{U}$ write

$$\Phi_{\mathcal{U}} = \sum_{s \in \mathfrak{S}_k} \Phi_{\mathcal{U}}(s) \cdot s^{-1} \in \mathbb{C}[\mathfrak{S}_k].$$

For any multiset $\{a_1, \ldots, a_k\}$ with entries from $\{1, \ldots, N\}$ we denote by α_i the multiplicity of the element $i \in \{1, \ldots, N\}$ in the multiset.

COROLLARY 7.4.5. *The quantum immanant* $\mathbb{S}_\mu$ *can be given by the formulas*

$$\begin{aligned} \mathbb{S}_\mu &= \sum_{a_1 \leqslant \cdots \leqslant a_k} \frac{1}{\alpha_1! \ldots \alpha_N!} \sum_{\operatorname{sh}(\mathcal{U}) = \mu} \sum_{s \in \mathfrak{S}_k} \Phi_{\mathcal{U}}(s)\, (E - c_1)_{a_1, a_{s(1)}} \cdots (E - c_k)_{a_k, a_{s(k)}} \\ &= \sum_{a_1 \geqslant \cdots \geqslant a_k} \frac{1}{\alpha_1! \ldots \alpha_N!} \sum_{\operatorname{sh}(\mathcal{U}) = \mu} \sum_{s \in \mathfrak{S}_k} \Phi_{\mathcal{U}}(s)\, (E - c_1)_{a_1, a_{s(1)}} \cdots (E - c_k)_{a_k, a_{s(k)}}, \end{aligned}$$

where the second summation in each formula is taken over all standard μ*-tableaux* $\mathcal{U}$ *and* $c_r = c_r(\mathcal{U})$ *for* $r = 1, \ldots, k$.

PROOF. Using the definition (7.39) and the relation $H_\mu f_\mu = k!$ we can write

$$\mathbb{S}_\mu = \frac{1}{k!} \sum_{a_1,\dots,a_k} \sum_{\mathrm{sh}(\mathcal{U})=\mu} \sum_{s\in\mathfrak{S}_k} \Phi_{\mathcal{U}}(s)\,(E-c_1)_{a_1,a_{s(1)}} \cdots (E-c_k)_{a_k,a_{s(k)}}. \tag{7.43}$$

Now take $m \geqslant N$. By Theorem 7.4.1, under the isomorphism (7.36) we have

$$\sum_{\mathrm{sh}(\mathcal{U})=\mu} (E-c_1)\otimes\dots\otimes(E-c_k)\,\Phi_{\mathcal{U}} \mapsto X^{\otimes k}\,(\mathcal{D}^t)^{\otimes k}\,\chi_\mu.$$

Therefore the element on the left-hand side is stable under the conjugations by the permutations $\sigma \in \mathfrak{S}_k$. This implies that for any $\sigma \in \mathfrak{S}_k$ the summands in (7.43) corresponding to the k-tuples $(a_1,\dots,a_k)$ and $(a_{\sigma(1)},\dots,a_{\sigma(k)})$ are equal. Hence we may restrict the summation to either the increasing or decreasing sets of k-tuples. Taking into account the number $k!/(\alpha_1!\dots\alpha_N!)$ of equal summands we get the required formulas. $\square$

Consider now the double Schur polynomial $s_\mu(x\,|\,b)$ in the N-tuple of variables $x = (l_1+N-1,\dots,l_N+N-1)$, which is associated with the sequence of parameters $b = (0,1,2,\dots)$.

THEOREM 7.4.6. *All quantum immanants $\mathbb{S}_\mu$ belong to the center $\mathrm{Z}(\mathfrak{gl}_N)$ of $\mathrm{U}(\mathfrak{gl}_N)$. These elements parameterized by all partitions μ of length not exceeding N form a basis of $\mathrm{Z}(\mathfrak{gl}_N)$. Moreover, the image of $\mathbb{S}_\mu$ under the Harish-Chandra isomorphism is given by*

$$\chi : \mathbb{S}_\mu \mapsto s_\mu(x\,|\,b).$$

PROOF. We will be proving that the element $\mathbb{S}_\mu$ is invariant with respect to the adjoint action of the general linear group GL_N. Recall that GL_N acts on $\mathfrak{gl}_N$ by the rule

$$\mathrm{Ad}\,\mathbf{g} : X \mapsto \mathbf{g}X\mathbf{g}^{-1}, \qquad X \in \mathfrak{gl}_N, \quad \mathbf{g}\in GL_N.$$

This extends to an action of GL_N on the universal enveloping algebra $\mathrm{U}(\mathfrak{gl}_N)$ so that elements of GL_N act on $\mathrm{U}(\mathfrak{gl}_N)$ as automorphisms. The center $\mathrm{Z}(\mathfrak{gl}_N)$ coincides with the subalgebra of GL_N-invariants in $\mathrm{U}(\mathfrak{gl}_N)$. Calculating the image of the basis element E_{ij} under $\mathrm{Ad}\,\mathbf{g}$ we find that the image of the matrix E is given by

$$\mathrm{Ad}\,\mathbf{g} : E \mapsto \mathbf{g}^t\,E\,(\mathbf{g}^t)^{-1}.$$

So it suffices to verify that the element $\mathbb{S}_\mu$ remains unchanged if E is replaced by $\mathbf{h}E\mathbf{h}^{-1}$ for any $\mathbf{h}\in GL_N$. The matrix $E-c_r$ is then replaced by $\mathbf{h}(E-c_r)\mathbf{h}^{-1}$, and we have

$$\mathbf{h}(E-c_1)\mathbf{h}^{-1}\otimes\dots\otimes\mathbf{h}(E-c_k)\mathbf{h}^{-1} = \mathbf{h}^{\otimes k}(E-c_1)\otimes\dots\otimes(E-c_k)\,(\mathbf{h}^{-1})^{\otimes k}.$$

By the cyclic property of trace,

$$\begin{aligned} &\mathrm{tr}\,\mathbf{h}^{\otimes k}(E-c_1)\otimes\dots\otimes(E-c_k)\,(\mathbf{h}^{-1})^{\otimes k}\,\Phi_{\mathcal{U}} \\ &\qquad = \mathrm{tr}\,(E-c_1)\otimes\dots\otimes(E-c_k)\,(\mathbf{h}^{-1})^{\otimes k}\,\Phi_{\mathcal{U}}\,\mathbf{h}^{\otimes k}. \end{aligned} \tag{7.44}$$

However,

$$(\mathbf{h}^{-1})^{\otimes k}\,\Phi_{\mathcal{U}}\,\mathbf{h}^{\otimes k} = \Phi_{\mathcal{U}}$$

since the actions of GL_N and $\mathfrak{S}_k$ commute, and so (7.44) equals $H_\mu\,\mathbb{S}_\mu$, proving that $\mathbb{S}_\mu$ is central.

Next, we calculate the Harish-Chandra image of the quantum immanant $\mathbb{S}_\mu$. Applying the second formula of Corollary 7.4.5 we find

$$\chi(\mathbb{S}_\mu) = \sum_{a_1 \geqslant \dots \geqslant a_k} \frac{1}{\alpha_1! \dots \alpha_N!} \sum_{\mathrm{sh}(\mathcal{U})=\mu} \sum_{s \in \mathfrak{S}_{(a)}} \Phi_{\mathcal{U}}(s) \left(\lambda_{a_1} - c_1(\mathcal{U})\right) \dots \left(\lambda_{a_k} - c_k(\mathcal{U})\right),$$

where $\mathfrak{S}_{(a)}$ denotes the subgroup of $\mathfrak{S}_k$ which consists of the permutations stabilizing the k-tuple $(a_1, \dots, a_k)$. Obviously, this subgroup can be identified with the direct product $\mathfrak{S}_{\alpha_1} \times \dots \times \mathfrak{S}_{\alpha_N}$. For each standard μ-tableau $\mathcal{U}$ denote by $\mathcal{T} = a(\mathcal{U})$ the tableau obtained from $\mathcal{U}$ by replacing the entry r with the number a_r for $r = 1, \dots, k$. So, the entries of $\mathcal{T}$ weakly decrease along the rows and down the columns. Changing the order of summation in the formula for $\chi(\mathbb{S}_\mu)$ we can write it as

$$\chi(\mathbb{S}_\mu) = \sum_{a_1 \geqslant \dots \geqslant a_k} \frac{1}{\alpha_1! \dots \alpha_N!} \sum_{\mathrm{sh}(\mathcal{T})=\mu} \prod_{\alpha \in \mu} \left(\lambda_{\mathcal{T}(\alpha)} - c(\alpha)\right) \Psi_{\mathcal{T}}, \tag{7.45}$$

summed over the tableaux $\mathcal{T}$ with the entries $a_1, \dots, a_k$, where we have set

$$\Psi_{\mathcal{T}} = \sum_{a(\mathcal{U})=\mathcal{T}} \sum_{s \in \mathfrak{S}_{(a)}} \Phi_{\mathcal{U}}(s). \tag{7.46}$$

Given a μ-tableau $\mathcal{T}$ with entries in $\{1, \dots, N\}$ such that the entries of $\mathcal{T}$ weakly decrease along the rows and down the columns, introduce the skew diagrams $\omega_1, \dots, \omega_N$ as follows. The diagram ω_r is the union of the cells of $\mathcal{T}$ which are occupied by $N - r + 1$ (in particular, ω_r can be empty). Then (7.46) can be rewritten as

$$\Psi_{\mathcal{T}} = \prod_{r=1}^{N} \sum_{s_r \in \mathfrak{S}_{\alpha_r}} \chi_{\omega_r}(s_r),$$

where χ_{ω_r} denotes the skew character of $\mathfrak{S}_{\alpha_r}$ associated with ω_r; see Example 6.6.1. However, the expression

$$\frac{1}{\alpha_r!} \sum_{s \in \mathfrak{S}_{\alpha_r}} \chi_{\omega_r}(s)$$

coincides with the standard inner product of χ_{ω_r} with the trivial character of $\mathfrak{S}_{\alpha_r}$. This is therefore the multiplicity of the trivial representation in the skew representation of $\mathfrak{S}_{\alpha_r}$ associated with ω_r. By the Pieri rule (see e.g. Examples 6.6.1 and 8.10.6) the multiplicity is nonzero only if ω_r does not contain two cells in the same column, in which case the multiplicity is 1. This implies that the summation in (7.45) can be restricted to those tableaux $\mathcal{T}$ whose columns strictly decrease, and we have

$$\chi(\mathbb{S}_\mu) = \sum_{\mathrm{sh}(\mathcal{T})=\mu} \prod_{\alpha \in \mu} \left(\lambda_{\mathcal{T}(\alpha)} - c(\alpha)\right), \tag{7.47}$$

summed over all *reverse* μ-tableaux $\mathcal{T}$ with entries in $\{1, \dots, N\}$ such that the entries of $\mathcal{T}$ weakly decrease along the rows and strictly decrease down the columns. The set of such μ-tableaux is in bijection with the set of semistandard μ-tableaux. The bijection is obtained by replacing the entry r of $\mathcal{T}$ with $N - r + 1$ for each $r = 1, \dots, N$. Clearly, when (7.47) is rewritten in terms of the semistandard μ-tableaux it becomes $s_\mu(x \,|\, b)$ with $x = (l_1 + N - 1, \dots, l_N + N - 1)$ and $b = (0, 1, 2, \dots)$.

Finally, as the double Schur polynomials $s_\mu(x \,|\, b)$ with μ running over all partitions with the length not exceeding N form a basis of the algebra of symmetric polynomials in x, the corresponding quantum immanants form a basis of $\mathrm{Z}(\mathfrak{gl}_N)$ due to the Harish-Chandra isomorphism. □

REMARK 7.4.7. The polynomials in $\lambda_1, \dots, \lambda_N$ defined in (7.47) are known as the *shifted Schur polynomials*; see also Section 8.2 below.

An alternative calculation of the image of $\mathbb{S}_\mu$ under the Harish-Chandra isomorphism is based upon the characteristic properties of the double Schur polynomials (7.31). One can verify that these properties are satisfied by the image $\chi(\mathbb{S}_\mu)$ with the use of Corollary 7.4.2 and regarding the irreducible representations of $\mathfrak{gl}_N$ as submodules of $\mathcal{P}$; cf. Section 7.6. □

We now consider two particular cases where μ is either a column or row diagram. We set

$$C_k = \mathbb{S}_{(1^k)} \qquad \text{and} \qquad D_k = \mathbb{S}_{(k)}.$$

Corollary 7.4.5 gives

$$C_k = \sum_{a_1<\dots<a_k} \sum_{p\in\mathfrak{S}_k} \operatorname{sgn} p \cdot E_{a_1,a_{p(1)}} \dots (E+k-1)_{a_k,a_{p(k)}}. \tag{7.48}$$

In particular, $C_k = 0$ for $k > N$. Equivalently, C_k can be defined with the use of the quantum minors $t^{a_1\dots a_k}_{b_1\dots b_k}(u)$; see (1.55). Namely, C_k is the value at $u = k-1$ of the polynomial

$$u(u-1)\dots(u-k+1)\cdot \sum_{1\leqslant a_1<\dots<a_k\leqslant N} \pi_N\Big(t^{a_1\dots a_k}_{a_1\dots a_k}(u)\Big),$$

where π_N is the evaluation homomorphism defined in (1.5).

Furthermore, by Corollary 7.4.5,

$$D_k = \sum_{a_1\leqslant\dots\leqslant a_k} \frac{1}{\alpha_1!\dots\alpha_N!} \sum_{p\in\mathfrak{S}_k} E_{a_1,a_{p(1)}} \dots (E-k+1)_{a_k,a_{p(k)}}. \tag{7.49}$$

By Theorem 7.4.6 the elements C_k and D_k belong to $\mathrm{Z}(\mathfrak{gl}_N)$, and due to (7.47) their images under the Harish-Chandra isomorphism can be written as

$$\chi : C_k \mapsto \sum_{1\leqslant a_1<\dots<a_k\leqslant N} \big(\lambda_{a_1}+k-1\big)\big(\lambda_{a_2}+k-2\big)\dots\,\lambda_{a_k}$$

and

$$\chi : D_k \mapsto \sum_{1\leqslant a_1\leqslant\dots\leqslant a_k\leqslant N} \big(\lambda_{a_1}-k+1\big)\big(\lambda_{a_2}-k+2\big)\dots\,\lambda_{a_k}.$$

We can now establish a relationship between the Capelli determinant $\mathcal{C}(u)$ defined by relation (7.1) and the elements C_k and D_k.

COROLLARY 7.4.8. *We have the relations*

$$\frac{\mathcal{C}(u)}{u(u-1)\dots(u-N+1)} = 1 + \sum_{k=1}^{N} \frac{C_k}{u(u-1)\dots(u-k+1)}$$

and

$$\left(\frac{\mathcal{C}(u)}{u(u-1)\dots(u-N+1)}\right)^{-1} = 1 + \sum_{k=1}^{\infty} \frac{(-1)^k D_k}{(u+1)\dots(u+k)}.$$

PROOF. By Theorem 7.4.6 the polynomials $\chi(C_k)$ and $\chi(D_k)$ coincide with $e_k(x\,|\,b)$ and $h_k(x\,|\,b)$, respectively, for $x = (l_1 + N - 1, \dots, l_N + N - 1)$ and $b = (0, 1, 2, \dots)$. Now the desired relations are implied by Theorem 7.1.1 and the identities (7.32) and (7.33), respectively, by taking $q = -u + N - 1$. □

We let $\operatorname{per} A$ stand for the permanent of a square matrix $A = [a_{ij}]$, so that

$$\operatorname{per} A = \sum_{\sigma \in \mathfrak{S}_N} a_{\sigma(1)1} \dots a_{\sigma(N)N}. \tag{7.50}$$

By Corollary 7.4.2, the images of the elements C_k and D_k in the representation (7.34) are given by the formulas

$$C_k \mapsto \sum_{i_1 < \dots < i_k} \sum_{a_1 < \dots < a_k} \det [\, x_{i_p a_q} \,] \det [\, \partial_{i_p a_q} \,] \tag{7.51}$$

and

$$D_k \mapsto \sum_{i_1 \leqslant \dots \leqslant i_k} \sum_{a_1 \leqslant \dots \leqslant a_k} \frac{\operatorname{per} [\, x_{i_p a_q} \,] \operatorname{per} [\, \partial_{i_p a_q} \,]}{f_1 ! \dots f_N ! \; \alpha_1 ! \dots \alpha_m !}, \tag{7.52}$$

where the summation indices $i_1, \dots, i_k$ and $a_1, \dots, a_k$ run over the sets $\{1, \dots, N\}$ and $\{1, \dots, m\}$ respectively, while $f_1, \dots, f_N$ are the multiplicities of the numbers $1, \dots, N$ in the sequence $(i_1, \dots, i_k)$ and $\alpha_1, \dots, \alpha_m$ are the multiplicities of the numbers $1, \dots, m$ in the sequence $(a_1, \dots, a_k)$.

REMARK 7.4.9. The relation (7.51) is the classical Capelli identity. Its dual form can be obtained by considering a representation of the Lie algebra $\mathfrak{gl}_m$ on the same space $\mathcal{P}$. The basis elements E'_{ab} of $\mathfrak{gl}_m$ act by the rule

$$E'_{ab} \mapsto \sum_{i=1}^{N} x_{ia} \partial_{ib}.$$

The images of the corresponding central elements C'_k and D'_k are given by the same differential operators which occur in (7.51) and (7.52), respectively. The differential operator in (7.51) is understood to be zero if $k > \min\{m, N\}$. □

7.5. Noncommutative Pfaffians and Hafnians

Now we turn to the orthogonal case. As in Section 2.1, let $G = [g_{ij}]$ be a symmetric nonsingular $N \times N$ matrix. We consider the orthogonal group O_N as the group of complex matrices preserving the bilinear symmetric form on $\mathbb{C}^N$ defined by G, so that

$$O_N = \{\, \mathbf{h} \in \operatorname{Mat}_N(\mathbb{C}) \mid \mathbf{h}^t G\, \mathbf{h} = G \}.$$

The corresponding orthogonal Lie algebra $\mathfrak{o}_N$ is then given by

$$\mathfrak{o}_N = \{ X \in \operatorname{Mat}_N(\mathbb{C}) \mid X^t G + G X = 0 \}.$$

This presentation of $\mathfrak{o}_N$ is consistent with the one used in Section 2.1. In particular, $\mathfrak{o}_N$ is spanned by the elements F_{ij} defined in (2.1). The adjoint action of the group O_N on $\mathfrak{o}_N$ is given by the rule

$$\operatorname{Ad} \mathbf{h} : X \mapsto \mathbf{h} X \mathbf{h}^{-1}, \qquad X \in \mathfrak{o}_N, \quad \mathbf{h} \in O_N.$$

This extends to an action of O_N on the universal enveloping algebra $\mathrm{U}(\mathfrak{o}_N)$ so that elements of O_N act on $\mathrm{U}(\mathfrak{o}_N)$ as automorphisms. If $N = 2n + 1$, then the center $\mathrm{Z}(\mathfrak{o}_N)$ coincides with the subalgebra of O_N-invariants in $\mathrm{U}(\mathfrak{o}_N)$. If $N = 2n$, then the subalgebra of O_N-invariants in $\mathrm{U}(\mathfrak{o}_N)$ is a proper subalgebra of the center $\mathrm{Z}(\mathfrak{o}_N)$. The image of this subalgebra under the Harish-Chandra isomorphism is the algebra of symmetric polynomials in the variables $l_1^2, \dots, l_n^2$; see Section 7.1.

In the case of $\mathfrak{o}_{2n}$, the *Pfaffian* $\operatorname{Pf} F$ of the matrix $F = [F_{ij}]$ is the element of $\mathrm{U}(\mathfrak{o}_{2n})$ defined by

$$\operatorname{Pf} F = \frac{1}{2^n n!} \sum_{\sigma \in \mathfrak{S}_{2n}} \operatorname{sgn} \sigma \cdot F_{\sigma(1),\sigma(2)} \cdots F_{\sigma(2n-1),\sigma(2n)}. \tag{7.53}$$

This is a noncommutative generalization of the ordinary Pfaffian of a numerical skew-symmetric matrix.

Now let N be arbitrary and let k be a positive integer such that $2k \leqslant N$. For any subset $I = \{i_1, \dots, i_{2k}\}$ of $\{1, \dots, N\}$ denote by F_I the $2k \times 2k$ submatrix of F whose rows and columns are enumerated by the elements of I. Then the Pfaffian $\operatorname{Pf} F_I$ of the matrix F_I will be defined as in (7.53).

Given two subsets $I = \{i_1, \dots, i_m\}$ and $J = \{j_1, \dots, j_m\}$ of $\{1, \dots, N\}$ denote by G^{IJ} the submatrix of G^{-1} which corresponds to the rows $i_1, \dots, i_m$ and columns $j_1, \dots, j_m$. Introduce the element C_k of $\mathrm{U}(\mathfrak{o}_N)$ by the formula

$$C_k = \sum_{I,J} \det G^{IJ} \cdot \operatorname{Pf} F_I \cdot \operatorname{Pf} F_J, \tag{7.54}$$

summed over all $2k$-subsets $I, J \subset \{1, \dots, N\}$.

THEOREM 7.5.1. *All elements C_k with $1 \leqslant k \leqslant [N/2]$ are O_N-invariants. In particular, $C_k \in \mathrm{Z}(\mathfrak{o}_N)$.*

PROOF. Consider the dual vector space $V = (\mathbb{C}^N)^*$ with the basis $\{\varepsilon_1, \dots, \varepsilon_N\}$ such that $\varepsilon_i(e_j) = \delta_{ij}$. Then V is a representation of O_N so that an element $\mathbf{h} \in O_N$ acts on $v \in V$ by

$$v \mapsto (\mathbf{h}^t)^{-1}\, v.$$

This action uniquely extends to an action by automorphisms on the exterior algebra $\Lambda(V)$. The tensor product algebra $\Lambda(V) \otimes \mathrm{U}(\mathfrak{o}_N)$ is then also a representation of O_N. Observe that the element

$$\Phi = \sum_{i<j} (\varepsilon_i \wedge \varepsilon_j) \otimes F_{ij} = \frac{1}{2} \cdot \sum_{i,j} (\varepsilon_i \wedge \varepsilon_j) \otimes F_{ij} \in \Lambda(V) \otimes \mathrm{U}(\mathfrak{o}_N) \tag{7.55}$$

is invariant with respect to the action of O_N. By the definition of the Pfaffian, we have an equality in the algebra $\Lambda(V) \otimes \mathrm{U}(\mathfrak{o}_N)$,

$$\frac{\Phi^k}{k!} = \sum_{i_1 < \cdots < i_{2k}} \big(\varepsilon_{i_1} \wedge \cdots \wedge \varepsilon_{i_{2k}}\big) \otimes \operatorname{Pf} F_I,$$

where $I = \{i_1, \dots, i_{2k}\}$.

Denote by g^{ij} the entries of the inverse matrix $G^{-1} = [g^{ij}]$. The symmetric bilinear form on V defined by

$$\langle \varepsilon_i, \varepsilon_j \rangle = g^{ij} \tag{7.56}$$

is invariant with respect to the O_N-action on V. Furthermore, extend this form to an invariant bilinear form on $\Lambda(V)$ as follows. The homogeneous components of $\Lambda(V)$ of different degrees are pairwise orthogonal while

$$\langle \varepsilon_{i_1} \wedge \cdots \wedge \varepsilon_{i_m}, \varepsilon_{j_1} \wedge \cdots \wedge \varepsilon_{j_m} \rangle = \det G^{IJ},$$

for $I = \{i_1, \dots, i_m\}$ and $J = \{j_1, \dots, j_m\}$. Extend it further to the $\mathbb{C}$-bilinear form on $\Lambda(V) \otimes \mathrm{U}(\mathfrak{o}_N)$ valued in $\mathrm{U}(\mathfrak{o}_N)$,

$$\langle \xi \otimes X, \eta \otimes Y \rangle = \langle \xi, \eta \rangle \cdot XY.$$

The latter form is O_N-equivariant; that is,

$$\langle \mathbf{h}x, \mathbf{h}y \rangle = \mathbf{h}\langle x, y \rangle$$

for any $\mathbf{h} \in O_N$ and $x, y \in \Lambda(V) \otimes \mathrm{U}(\mathfrak{o}_N)$. This implies O_N-invariance of the element

$$\Big\langle \frac{\Phi^k}{k!}, \frac{\Phi^k}{k!} \Big\rangle = \sum_{I,J} \det G^{IJ} \cdot \operatorname{Pf} F_I \cdot \operatorname{Pf} F_J,$$

which coincides with C_k. □

In the next section we will find the Harish-Chandra images of the elements C_k and their symplectic counterparts D_k defined below with the use of certain representations of the orthogonal and symplectic Lie algebras in a space of polynomials.

Note that in the particular case $G = 1$ the formulas for the elements C_k have the simpler form

$$C_k = \sum_I \big(\operatorname{Pf} F_I\big)^2, \tag{7.57}$$

summed over all $2k$-subsets $I \subset \{1, \dots, N\}$.

Now consider the symplectic case. For an even $N = 2n$ we let $G = [g_{ij}]$ be a skew-symmetric nonsingular $N \times N$ matrix. The symplectic group Sp_N consists of the complex matrices preserving the bilinear skew-symmetric form on $\mathbb{C}^N$ defined by G, so that

$$Sp_N = \{\mathbf{h} \in \mathrm{Mat}_N(\mathbb{C}) \mid \mathbf{h}^t G \,\mathbf{h} = G\}.$$

The corresponding symplectic Lie algebra $\mathfrak{sp}_N$ is then given by

$$\mathfrak{sp}_N = \{X \in \mathrm{Mat}_N(\mathbb{C}) \mid X^t G + GX = 0\}.$$

This presentation of $\mathfrak{sp}_N$ is consistent with the one used in Section 2.1. In particular, $\mathfrak{sp}_N$ is spanned by the elements F_{ij} defined in (2.1). The adjoint action of the group Sp_N on $\mathfrak{sp}_N$ is given by the rule

$$\mathrm{Ad}\,\mathbf{h} : X \mapsto \mathbf{h}X\mathbf{h}^{-1}, \qquad X \in \mathfrak{sp}_N, \quad \mathbf{h} \in Sp_N.$$

This extends to an action of Sp_N on the universal enveloping algebra $\mathrm{U}(\mathfrak{sp}_N)$ so that elements of Sp_N act on $\mathrm{U}(\mathfrak{sp}_N)$ as automorphisms. The center $\mathrm{Z}(\mathfrak{sp}_N)$ coincides with the subalgebra of Sp_N-invariants in $\mathrm{U}(\mathfrak{sp}_N)$.

For any $k \geqslant 1$ let $I = \{i_1, \dots, i_{2k}\}$ be a multiset whose elements belong to $\{1, \dots, N\}$. Denote by F_I the $2k \times 2k$ matrix whose (a, b) entry is $F_{i_a i_b}$. The *Hafnian* $\operatorname{Hf} F$ of the symmetric matrix F_I is the element of $\mathrm{U}(\mathfrak{sp}_N)$ defined by

$$\operatorname{Hf} F_I = \frac{1}{2^k k!} \sum_{\sigma \in \mathfrak{S}_{2k}} F_{i_{\sigma(1)}, i_{\sigma(2)}} \dots F_{i_{\sigma(2k-1)}, i_{\sigma(2k)}}. \tag{7.58}$$

Denote by g^{ij} the entries of the inverse matrix $G^{-1} = [g^{ij}]$. Given two multisets $I = \{i_1, \dots, i_m\}$ and $J = \{j_1, \dots, j_m\}$ with elements in $\{1, \dots, N\}$, denote by G^{IJ} the $m \times m$ matrix whose (a, b) entry is $g^{i_a j_b}$. Furthermore, let α_l and β_l denote the multiplicities of the index l in the multisets I and J, respectively. Introduce the element D_k of $\mathrm{U}(\mathfrak{sp}_N)$ by the formula

$$D_k = \sum_{I,J} \frac{\operatorname{per} G^{IJ}}{\alpha_1! \dots \alpha_N! \, \beta_1! \dots \beta_N!} \cdot \operatorname{Hf} F_I \cdot \operatorname{Hf} F_J, \tag{7.59}$$

summed over all $2k$-multisets I, J with elements in $\{1, \dots, N\}$, where $\operatorname{per} A$ denotes the permanent of a matrix A; see (7.50).

THEOREM 7.5.2. *All elements D_k with $k \geqslant 1$ belong to $\mathrm{Z}(\mathfrak{sp}_N)$.*

PROOF. Consider the dual vector space $V = (\mathbb{C}^N)^*$ with the basis $\{\varepsilon_1, \dots, \varepsilon_N\}$ such that $\varepsilon_i(e_j) = \delta_{ij}$. Then V is a representation of Sp_N so that an element $\mathbf{h} \in Sp_N$ acts on $v \in V$ by

$$v \mapsto (\mathbf{h}^t)^{-1} v.$$

This action is uniquely extended to an action by automorphisms on the symmetric algebra $S(V)$. The tensor product algebra $S(V) \otimes \mathrm{U}(\mathfrak{sp}_N)$ is then also a representation of Sp_N. Observe that the element

$$\Psi = \frac{1}{2} \cdot \sum_{i,j} \varepsilon_i \, \varepsilon_j \otimes F_{ij} \in S(V) \otimes \mathrm{U}(\mathfrak{sp}_N)$$

is invariant with respect to the action of Sp_N. By the definition of the Hafnian, we have an equality in the algebra $S(V) \otimes \mathrm{U}(\mathfrak{sp}_N)$,

$$\frac{\Psi^k}{k!} = \sum_{I} \frac{1}{\alpha_1! \dots \alpha_N!} \, \varepsilon_{i_1} \dots \varepsilon_{i_{2k}} \otimes \operatorname{Hf} F_I,$$

summed over multisets $I = \{i_1, \dots, i_{2k}\}$.

Furthermore, the form on V defined by the formula (7.56) is invariant with respect to the action of Sp_N on V. Extend this form to an invariant bilinear form on $S(V)$ as follows. The homogeneous components of $S(V)$ of different degrees are pairwise orthogonal while

$$\langle \varepsilon_{i_1} \dots \varepsilon_{i_k}, \varepsilon_{j_1} \dots \varepsilon_{j_k} \rangle = \operatorname{per} G^{IJ},$$

for $I = \{i_1, \dots, i_m\}$ and $J = \{j_1, \dots, j_m\}$. Extend it further to the $\mathbb{C}$-bilinear form on $S(V) \otimes \mathrm{U}(\mathfrak{sp}_N)$ valued in $\mathrm{U}(\mathfrak{sp}_N)$,

$$\langle \xi \otimes X, \eta \otimes Y \rangle = \langle \xi, \eta \rangle \cdot XY.$$

The latter form is Sp_N-equivariant; that is,

$$\langle \mathbf{h}x, \mathbf{h}y \rangle = \mathbf{h} \langle x, y \rangle$$

for any $\mathbf{h} \in Sp_N$ and $x, y \in S(V) \otimes \mathrm{U}(\mathfrak{sp}_N)$. This implies Sp_N-invariance of the element

$$\left\langle \frac{\Psi^k}{k!}, \frac{\Psi^k}{k!} \right\rangle = \sum_{I,J} \frac{\operatorname{per} G^{IJ}}{\alpha_1! \dots \alpha_N! \, \beta_1! \dots \beta_N!} \cdot \operatorname{Hf} F_I \cdot \operatorname{Hf} F_J$$

which coincides with D_k. □

We conclude this section by establishing an invariance property of the elements C_k and D_k with respect to the matrix G. We consider the orthogonal and symplectic cases simultaneously. Suppose that G and G' are two nonsingular symmetric (respectively, skew-symmetric) $N \times N$-matrices and consider the isomorphism

$\mathfrak{g}'_N \to \mathfrak{g}_N$ given by (7.18). Denote by C'_k and D'_k the respective elements defined in (7.54) and (7.59) for the Lie algebra $\mathfrak{g}'_N$.

PROPOSITION 7.5.3. *The images of the elements C'_k and D'_k under the isomorphism (7.18) coincide with C_k and D_k, respectively.*

PROOF. We will use primes to indicate the respective objects associated with the Lie algebra $\mathfrak{g}'_N$. In the orthogonal case consider the element Φ' which is defined by (7.55),

$$\Phi' = \frac{1}{2} \cdot \sum_{i,j} (\varepsilon'_i \wedge \varepsilon'_j) \otimes F'_{ij} \in \Lambda(V') \otimes \mathrm{U}(\mathfrak{o}'_N).$$

Applying the isomorphism (7.18) to the elements F'_{ij} we find that the image of Φ' will be given by

$$\Phi' \mapsto \frac{1}{2} \cdot \sum_{i,j} (\varepsilon'_i \wedge \varepsilon'_j) \otimes \sum_{r,s} b_{ir}\, b_{js}\, F_{rs} = \frac{1}{2} \cdot \sum_{r,s} (\widetilde{\varepsilon}_r \wedge \widetilde{\varepsilon}_s) \otimes F_{rs},$$

where $\widetilde{\varepsilon}_r = \sum_i \varepsilon'_i\, b_{ir}$ and the b_{ij} are the matrix elements of $B = [b_{ij}]$. Now observe that duc to the relation $B^t (G')^{-1} B = G^{-1}$, we have

$$\langle \widetilde{\varepsilon}_r, \widetilde{\varepsilon}_s \rangle = \sum_{i,j} g'^{\,ij}\, b_{ir}\, b_{js} = g^{rs}.$$

This implies that the image of the element

$$C'_k = \Big\langle \frac{(\Phi')^k}{k!}, \frac{(\Phi')^k}{k!} \Big\rangle'$$

under the isomorphism (7.18) is C_k. In the symplectic case the argument is exactly the same. □

7.6. Capelli identities for $\mathfrak{o}_N$ and $\mathfrak{sp}_{2n}$

Consider the orthogonal Lie algebra $\mathfrak{o}_N$ first. We will work with the presentation of $\mathfrak{o}_N$ used in Section 7.1, so $\mathfrak{o}_N$ is spanned by the elements F_{ij} defined in (7.9). For any subset I of $\{-n, \dots, n\}$ containing $2k$ elements $i_1 < \dots < i_{2k}$, the $2k \times 2k$ matrix $[F_{i_p, -i_q}]$ is skew-symmetric. We denote its Pfaffian by

$$\Phi_I = \mathrm{Pf}\,[F_{i_p,-i_q}], \qquad p, q = 1, \dots, 2k; \tag{7.60}$$

see (7.53). The matrix G is given by (4.1) so that $G^{-1} = G$. Writing the elements C_k defined in (7.54) for the current presentation of $\mathfrak{o}_N$ we come to the expression

$$C_k = (-1)^k \cdot \sum_I \Phi_I\, \Phi_{I^*}, \tag{7.61}$$

summed over $2k$-subsets I, where $I^* = \{-i_{2k}, \dots, -i_1\}$. Indeed, the determinant $\det G^{IJ}$ occurring in (7.54) is nonzero only if $J = I^*$, in which case it equals $(-1)^k$.

Introduce the algebra of polynomials $\mathcal{P}$ in mN commutative variables x_{ai} with $i = -n, \dots, n$ and $a = 1, \dots, m$. As before, ∂_{ai} will denote the partial differentiation operator with respect to the variable x_{ai}. The mapping

$$F_{ij} \mapsto \sum_{a=1}^{m} \big(x_{ai}\, \partial_{aj} - x_{a,-j}\, \partial_{a,-i} \big) \tag{7.62}$$

defines a representation of $\mathfrak{o}_N$ on $\mathcal{P}$. Furthermore, let $A = \{a_1, \dots, a_k\}$ be a subset of $\{1, \dots, m\}$. Introduce the differential operator

$$\Omega_{AI} = \sum_J \operatorname{sgn} JJ' \cdot \det\left[x_{a_p j_q}\right] \det\left[\partial_{a_p, -j'_q}\right], \tag{7.63}$$

where the sum is taken over all partitions of I into two subsets $J = \{j_1, \dots, j_k\}$ and $J' = \{j'_1, \dots, j'_k\}$. Here the determinants are taken with respect to the indices $p, q = 1, \dots, k$ and $\operatorname{sgn} JJ'$ is the sign of the permutation $(j_1, j'_1, \dots, j_k, j'_k)$ of the sequence $(i_1, \dots, i_{2k})$.

THEOREM 7.6.1. *The image of the element C_k in the representation* (7.62) *is given by the formula*

$$C_k \mapsto (-1)^k \sum_{I,A,B} \Omega_{AI}\, \Omega_{BI^*},$$

summed over k-element subsets A and B of $\{1, \dots, m\}$ and $2k$-element subsets I of $\{-n, \dots, n\}$. Moreover, the image of C_k under the Harish-Chandra isomorphism is given by

$$\chi : C_k \mapsto (-1)^k \sum_{1 \leqslant i_1 < \dots < i_k \leqslant n} (l_{i_1}^2 - \rho_{i_1}^2) \dots (l_{i_k}^2 - \rho_{i_k - k+1}^2).$$

PROOF. Due to (7.61), the first part of the theorem will follow if we prove that the image of the Pfaffian in the representation (7.62) is given by

$$\Phi_I \mapsto \sum_A \Omega_{AI}. \tag{7.64}$$

By definition, Φ_I equals

$$\sum_\sigma \frac{\operatorname{sgn} \sigma}{2^k\, k!} \cdot \left(E_{i_{\sigma(1)}, -i_{\sigma(2)}} - E_{i_{\sigma(2)}, -i_{\sigma(1)}}\right) \dots \left(E_{i_{\sigma(2k-1)}, -i_{\sigma(2k)}} - E_{i_{\sigma(2k)}, -i_{\sigma(2k-1)}}\right)$$

$$= \sum_\sigma \frac{\operatorname{sgn} \sigma}{k!} \cdot E_{i_{\sigma(1)}, -i_{\sigma(2)}} \dots E_{i_{\sigma(2k-1)}, -i_{\sigma(2k)}}$$

in $\mathrm{U}(\mathfrak{o}_N) \subset \mathrm{U}(\mathfrak{gl}_N)$. Therefore the image of Φ_I in the representation (7.62) equals

$$\sum_\sigma \sum_{a_1, \dots, a_k = 1}^{m} \frac{\operatorname{sgn} \sigma}{k!} \cdot x_{a_1 i_{\sigma(1)}} \partial_{a_1, -i_{\sigma(2)}} \dots x_{a_k i_{\sigma(2k-1)}} \partial_{a_k, -i_{\sigma(2k)}}. \tag{7.65}$$

Move each operator $\partial_{a_p, -i_{\sigma(2p)}}$ in every monomial of (7.65) to the right, commuting it consecutively with the multiplication operators $x_{a_q i_{\sigma(2q-1)}}$ for $q = p+1, \dots, k$. Let us start with the operator $\partial_{a_1, -i_{\sigma(2)}}$. Take the commutator

$$\left[\partial_{a_1, -i_{\sigma(2)}}, x_{a_2 i_{\sigma(3)}}\right] = \delta_{a_1 a_2}\, \delta_{-i_{\sigma(2)}, i_{\sigma(3)}}.$$

The sum over σ in (7.65) involves two commutators of this form with opposite signs. Let $\bar\sigma \in \mathfrak{S}_{2k}$ be such that $\bar\sigma(r) = \sigma(r)$ for $r \neq 2, 3$ while $\bar\sigma(2) = \sigma(3)$ and $\bar\sigma(3) = \sigma(2)$. Then $\operatorname{sgn} \bar\sigma = -\operatorname{sgn} \sigma$ whilst $\left[\partial_{a_1, -i_{\bar\sigma(2)}}, x_{a_2 i_{\bar\sigma(3)}}\right] = \left[\partial_{a_1, -i_{\sigma(2)}}, x_{a_2 i_{\sigma(3)}}\right]$. So by repeating this argument we get

$$\Phi_I \mapsto \sum_\sigma \sum_{a_1, \dots, a_k = 1}^{m} \frac{\operatorname{sgn} \sigma}{k!} \cdot x_{a_1 i_{\sigma(1)}} \dots x_{a_k i_{\sigma(2k-1)}} \partial_{a_1, -i_{\sigma(2)}} \dots \partial_{a_k, -i_{\sigma(2k)}}.$$

This can obviously be rewritten as

$$\Phi_I \mapsto \sum_J \sum_{a_1,\dots,a_k=1}^{m} \frac{\operatorname{sgn} JJ'}{k!} \cdot \det\big[\,x_{a_p j_q}\big] \det\big[\,\partial_{a_p,-j'_q}\big],$$

where J and J' are the same sets as in (7.63). We now obtain the desired relation (7.64) since both determinants $\det\big[\,x_{a_p j_q}\big]$ and $\det\big[\,\partial_{a_p,-j'_q}\big]$ are skew-symmetric with respect to permutations of the sequence $(a_1,\dots,a_k)$.

In order to prove the second part of the theorem, consider the finite-dimensional irreducible representations $V(\lambda)$ of the Lie algebra $\mathfrak{o}_N$, where the components λ_i of the highest weight $\lambda=(\lambda_1,\dots,\lambda_n)$ are nonpositive integers; see (7.12). In this case, $\widetilde{\lambda}=(-\lambda_n,\dots,-\lambda_1)$ is a partition. It will be sufficient to show that the eigenvalue of C_k in any such representation $V(\lambda)$ coincides with the corresponding value of the double elementary symmetric polynomial $(-1)^k\, e_k(l_1^2,\dots,l_n^2\,|\,b)$, where $l_i=\lambda_i+\rho_i$ and the sequence b is defined by

$$b=\big(\varepsilon^2,(\varepsilon+1)^2,(\varepsilon+2)^2,\dots\big), \tag{7.66}$$

with $\varepsilon=0$ or $1/2$ for $N=2n$ or $2n+1$, respectively; see Section 7.5.

Suppose that $m\geqslant n$. Then every $\mathfrak{o}_N$-module $V(\lambda)$ is contained in $\mathcal{P}$ as a submodule. Indeed, an $\mathfrak{o}_N$-singular vector in $\mathcal{P}$ of weight λ can be constructed as follows. For each $p=1,\dots,n$ set

$$\Delta_p=\det[\,x_{ai}\,],\qquad a=1,\dots,p,\qquad i=-n,\dots,-n+p-1.$$

An easy direct calculation shows that the polynomial

$$\xi=\Delta_1^{\lambda_{n-1}-\lambda_n}\cdots\Delta_{n-1}^{\lambda_1-\lambda_2}\,\Delta_n^{-\lambda_1}\in\mathcal{P} \tag{7.67}$$

satisfies conditions (7.12) with respect to the action of $\mathfrak{o}_N$ defined in (7.62). Note that the polynomial ξ has degree $|\widetilde{\lambda}|$. So by the first part of the theorem, the operator C_k vanishes in every module $V(\lambda)$ with $|\widetilde{\lambda}|<k$. Now apply the characterization property (7.31) taking into account the observation that the n-tuple $b_{\widetilde{\lambda}}$ coincides with $(l_n^2,\dots,l_1^2)$. Therefore, the proof will be completed if we show that the leading component of $(-1)^k\,\chi(C_k)$, regarded as a polynomial in $\lambda_1,\dots,\lambda_k$, is the elementary symmetric polynomial $e_k(\lambda_1^2,\dots,\lambda_n^2)$. The leading terms of this polynomial arise from the summands of Φ_I and Φ_{I^*} which depend only on the generators F_{ii}. Hence we may restrict the sum in (7.61) to the sets I of the form $\{-p_k,\dots,-p_1,p_1,\dots,p_k\}$ where $1\leqslant p_1<\dots<p_k\leqslant n$, so that $I^*=I$. For any of these sets I there are $2^k\,k!$ summands in (7.60) depending only on the F_{ii}. All these summands are equal to each other. So the leading terms of our polynomial add up to

$$\sum_{1\leqslant p_1<\dots<p_k\leqslant n}\chi\big(F_{p_1p_1}^2\cdots F_{p_kp_k}^2\big)=\sum_{1\leqslant p_1<\dots<p_k\leqslant n}\lambda_{p_1}^2\cdots\lambda_{p_k}^2=e_k(\lambda_1^2,\dots,\lambda_n^2),$$

completing the proof. □

We now prove an analogue of Theorem 7.6.1 for the symplectic Lie algebra. As in the orthogonal case, we will work with the presentation of $\mathfrak{sp}_{2n}$ used in Section 7.1, so $\mathfrak{sp}_{2n}$ is spanned by the elements F_{ij} defined in (7.9). Let us set $\widetilde{F}_{ij}=\operatorname{sgn} i\cdot F_{ij}$. Then we have $\widetilde{F}_{i,-j}=\widetilde{F}_{j,-i}$. Let I be any sequence of length $2k$ of elements from the set $\{-n,\dots,n\}$. Suppose these elements are $i_1\leqslant\dots\leqslant i_{2k}$.

We will denote the multiplicity of an element i in I by α_i. Denote the Hafnian of the symmetric matrix $[\widetilde{F}_{i_p,-i_q}]$ by

$$\Psi_I = \mathrm{Hf}\,[\widetilde{F}_{i_p,-i_q}], \qquad p,q=1,\dots,2k; \tag{7.68}$$

see (7.58). The matrix G is given by (4.1) so that $G^{-1} = -G$. The elements D_k defined in (7.59) now take the form

$$D_k = \sum_I \frac{\mathrm{sgn}\,(i_1 \dots i_{2k})}{\alpha_{-n}! \dots \alpha_n!} \cdot \Psi_I \, \Psi_{I^*}, \tag{7.69}$$

summed over $2k$-sequences I, where $I^* = \{-i_{2k},\dots,-i_1\}$. Indeed, the permanent $\mathrm{per}\, G^{IJ}$ occurring in (7.59) is nonzero only if $J = I^*$ in which case it equals $\mathrm{sgn}\,(i_1 \dots i_{2k})$.

As in (7.62), the mapping

$$F_{ij} \mapsto \sum_{a=1}^{m} \big(x_{ai}\,\partial_{aj} - \mathrm{sgn}\, ij \cdot x_{a,-j}\,\partial_{a,-i}\big) \tag{7.70}$$

defines a representation of $\mathfrak{sp}_{2n}$ on the space of polynomials $\mathcal{P}$ in the variables x_{ai} with $i = -n,\dots,n$ and $a = 1,\dots,m$. Let $A = \{a_1,\dots,a_k\}$ be a sequence of elements from the set $\{1,\dots,m\}$ and let $d_1,\dots,d_m$ be the multiplicities of $1,\dots,m$ in A. Introduce the differential operator

$$\Theta_{AI} = \frac{1}{d_1! \dots d_m!} \sum_J \mathrm{sgn}\,(j_1 \dots j_k) \cdot \mathrm{per}\,[x_{a_p j_q}]\, \mathrm{per}\,[\partial_{a_p,-j'_q}], \tag{7.71}$$

where the sum is taken over all partitions of I into subsequences $J = (j_1,\dots,j_k)$ and $J' = (j'_1,\dots,j'_k)$ each of length k. So $j_1 \leqslant \dots \leqslant j_k$ and $j'_1 \leqslant \dots \leqslant j'_k$. Here the permanents are taken with respect to the indices $p,q = 1,\dots,k$.

THEOREM 7.6.2. *The image of the element D_k in the representation* (7.70) *is given by the formula*

$$D_k \mapsto \sum_{I,A,B} \frac{\mathrm{sgn}\,(i_1 \dots i_{2k})}{\alpha_{-n}! \dots \alpha_n!} \cdot \Theta_{AI}\, \Theta_{BI^*},$$

summed over k-element sequences A and B from the set $\{1,\dots,m\}$ and $2k$-element sequences I from the set $\{-n,\dots,n\}$. Moreover, the image of D_k under the Harish-Chandra isomorphism is given by

$$\chi : D_k \mapsto (-1)^k \sum_{1 \leqslant i_1 \leqslant \dots \leqslant i_k \leqslant n} (l_{i_1}^2 - i_1^2) \dots (l_{i_k}^2 - (i_k + k - 1)^2).$$

PROOF. Due to (7.69), the first part of the theorem will follow if we prove that the image of the Hafnian in the representation (7.70) is given by

$$\Psi_I \mapsto \sum_A \Theta_{AI}. \tag{7.72}$$

We can transform the formula for Ψ_I as

$$\begin{aligned}
&\frac{1}{2^k\, k!} \cdot \sum_\sigma \big(\mathrm{sgn}\, i_{\sigma(1)} \cdot E_{i_{\sigma(1)},-i_{\sigma(2)}} + \mathrm{sgn}\, i_{\sigma(2)} \cdot E_{i_{\sigma(2)},-i_{\sigma(1)}}\big) \\
&\qquad \times \dots \times \big(\mathrm{sgn}\, i_{\sigma(2k-1)} \cdot E_{i_{\sigma(2k-1)},-i_{\sigma(2k)}} + \mathrm{sgn}\, i_{\sigma(2k)} \cdot E_{i_{\sigma(2k)},-i_{\sigma(2k-1)}}\big) \\
&= \frac{1}{k!} \cdot \sum_\sigma \mathrm{sgn}\,\big(i_{\sigma(1)}\, i_{\sigma(3)} \dots i_{\sigma(2k-1)}\big) \cdot E_{i_{\sigma(1)},-i_{\sigma(2)}} \dots E_{i_{\sigma(2k-1)},-i_{\sigma(2k)}}
\end{aligned}$$

in $\mathrm{U}(\mathfrak{sp}_{2n}) \subset \mathrm{U}(\mathfrak{gl}_{2n})$. Therefore the image of Ψ_I in the representation (7.70) equals

$$\sum_{\sigma} \sum_{a_1,\dots,a_k=1}^{m} \frac{\operatorname{sgn}(i_{\sigma(1)} \dots i_{\sigma(2k-1)})}{k!} \cdot x_{a_1 i_{\sigma(1)}} \partial_{a_1,-i_{\sigma(2)}} \dots x_{a_k i_{\sigma(2k-1)}} \partial_{a_k,-i_{\sigma(2k)}}.$$

Move to the right each operator $\partial_{a_p,-i_{\sigma(2p)}}$ in every monomial above, commuting it consecutively with the multiplication operators $x_{a_q i_{\sigma(2q-1)}}$ for $q = p+1, \dots, k$. Let us start with the operator $\partial_{a_1,-i_{\sigma(2)}}$. Take the commutator

$$\left[\partial_{a_1,-i_{\sigma(2)}}, x_{a_2 i_{\sigma(3)}}\right] = \delta_{a_1 a_2}\, \delta_{-i_{\sigma(2)}, i_{\sigma(3)}}.$$

The above sum over σ contains two such commutators with opposite signs. Indeed, if $\bar\sigma \in \mathfrak{S}_{2k}$ is such that $\bar\sigma(r) = \sigma(r)$ for $r \neq 2, 3$ while $\bar\sigma(2) = \sigma(3)$ and $\bar\sigma(3) = \sigma(2)$, then

$$\left[\partial_{a_1,-i_{\bar\sigma(2)}}, x_{a_2 i_{\bar\sigma(3)}}\right] = \left[\partial_{a_1,-i_{\sigma(2)}}, x_{a_2 i_{\sigma(3)}}\right]$$

whilst

$$\operatorname{sgn}(i_{\bar\sigma(1)} i_{\bar\sigma(3)} \dots i_{\bar\sigma(2k-1)}) = -\operatorname{sgn}(i_{\sigma(1)} i_{\sigma(3)} \dots i_{\sigma(2k-1)})$$

for $i_{\sigma(2)} = -i_{\sigma(3)}$. So by repeating this argument we bring the image of Ψ_I to the form

$$\sum_{\sigma} \sum_{a_1,\dots,a_k=1}^{m} \frac{\operatorname{sgn}(i_{\sigma(1)} \dots i_{\sigma(2k-1)})}{k!} \cdot x_{a_1 i_{\sigma(1)}} \dots x_{a_k i_{\sigma(2k-1)}} \partial_{a_1,-i_{\sigma(2)}} \dots \partial_{a_k,-i_{\sigma(2k)}}.$$

The latter sum can be rewritten as

$$\gamma(\Psi_I) = \sum_{J} \sum_{a_1,\dots,a_k=1}^{m} \frac{\operatorname{sgn}(j_1 \dots j_k)}{k!} \cdot \operatorname{per}\left[x_{a_p j_q}\right] \operatorname{per}\left[\partial_{a_p,-j'_q}\right]$$

where J and J' are the same as in (7.71). We now obtain the desired relation (7.72) since both permanents $\operatorname{per}[x_{a_p j_q}]$ and $\operatorname{per}[\partial_{a_p,-j'_q}]$ are symmetric with respect to permutations of the sequence $(a_1, \dots, a_k)$.

For the proof of the second part of the theorem, consider the finite-dimensional irreducible representations $V(\lambda)$ of the Lie algebra $\mathfrak{sp}_{2n}$, so that the components λ_i of the highest weight $\lambda = (\lambda_1, \dots, \lambda_n)$ are nonpositive integers. Then the n-tuple $\widetilde{\lambda} = (-\lambda_n, \dots, -\lambda_1)$ is a partition. It will be sufficient to show that the eigenvalue of $(-1)^k D_k$ in any such representation $V(\lambda)$ coincides with the corresponding value of the double complete symmetric polynomial $h_k(l_1^2, \dots, l_n^2\,|\,b)$, where $l_i = \lambda_i + \rho_i$ and the sequence b is defined by (7.66) with $\varepsilon = 1$.

Suppose that $m \geqslant n$. Then every $\mathfrak{sp}_{2n}$-module $V(\lambda)$ is contained in $\mathcal{P}$ as a submodule. The corresponding $\mathfrak{sp}_{2n}$-singular vector ξ in $\mathcal{P}$ of weight λ is given by the formula (7.67). Since the polynomial ξ has degree $|\widetilde{\lambda}|$, the first part of the theorem implies that the operator D_k vanishes in every module $V(\lambda)$ with $|\widetilde{\lambda}| < k$. Now apply the characterization property (7.31) taking into account the observation that the n-tuple $b_{\widetilde{\lambda}}$ coincides with $(l_n^2, \dots, l_1^2)$. So, the proof will be completed if we show that the leading component of $(-1)^k \chi(D_k)$, regarded as a polynomial in $\lambda_1, \dots, \lambda_k$, is the complete symmetric polynomial $h_k(\lambda_1^2, \dots, \lambda_n^2)$.

The leading terms of this polynomial arise from the summands of Ψ_I and Ψ_{I^*} which depend only on the generators F_{ii}. So we may restrict the sum in (7.69) to the sequences I of the form $\{-p_k, \dots, p_1, p_1, \dots, p_k\}$ where $1 \leqslant p_1 \leqslant \dots \leqslant p_k \leqslant n$. Then we have $I^* = I$. For any of these sequences I there are $2^k k! \cdot \alpha_1! \dots \alpha_n!$

summands in (7.68) depending only on F_{ii}. All these summands are equal to each other. So the leading terms of our polynomial add up to

$$\sum_{1\leqslant p_1\leqslant\cdots\leqslant p_k\leqslant n} \chi(F^2_{p_1p_1}\dots F^2_{p_kp_k}) = \sum_{1\leqslant p_1\leqslant\cdots\leqslant p_k\leqslant n} \lambda^2_{p_1}\dots\lambda^2_{p_k} = h_k(\lambda_1^2,\dots,\lambda_n^2),$$

completing the argument. □

Finally we prove an analogue of Corollary 7.4.8 for the orthogonal and symplectic Lie algebras. We now return to the presentation of $\mathfrak{g}_N = \mathfrak{o}_N$ or $\mathfrak{sp}_N$ introduced in Section 2.1. For an arbitrary nonsingular symmetric or skew-symmetric matrix G consider the Capelli-type determinant $\mathcal{C}_G(u)$ defined by (7.17) and the elements C_k and D_k given by (7.54) and (7.59), respectively.

COROLLARY 7.6.3. *We have the following relations. If* $\mathfrak{g}_N = \mathfrak{o}_N$, *then*

$$\frac{\mathcal{C}_G(u)}{(u+\rho_{-n})\dots(u+\rho_n)} = 1 + \sum_{k=1}^{n} \frac{C_k}{(u^2-\rho^2_{n-k+1})\dots(u^2-\rho_n^2)}.$$

If $\mathfrak{g}_N = \mathfrak{sp}_N$, *then*

$$\left(\frac{\mathcal{C}_G(u)}{(u+\rho_{-n})\dots(u+\rho_n)}\right)^{-1} = 1 + \sum_{k=1}^{\infty} \frac{(-1)^k D_k}{(u^2-(n+1)^2)\dots(u^2-(n+k)^2)}.$$

PROOF. By Propositions 7.1.9 and 7.5.3, it is sufficient to prove the relations for a particular choice of the matrix G. In other words, is suffices to verify the corresponding relations for the Harish-Chandra images of the Capelli type determinant and the elements C_k and D_k. These are provided by Theorems 7.1.6, 7.6.1 and 7.6.2, respectively, so the relations follow from (7.32) and (7.33). □

The following is a noncommutative analogue of the relation $(\operatorname{Pf} A)^2 = \det A$ which holds for a numerical skew-symmetric $2n \times 2n$ matrix A; see also Example 7.7.9.

COROLLARY 7.6.4. *In the case* $\mathfrak{g}_N = \mathfrak{o}_{2n}$ *we have the identity*

$$\big(\operatorname{Pf} F\big)^2 = \sum_{p\in\mathfrak{S}_{2n}} \operatorname{sgn} pp' \cdot \big(\sigma_1 G + F\big)_{p(1),p'(1)} \cdots \big(\sigma_{2n} G + F\big)_{p(2n),p'(2n)},$$

where $\sigma_i = n - i$ *for* $i = 1,\dots,n$ *and* $\sigma_i = n - i + 1$ *for* $i = n+1,\dots,2n$.

PROOF. Observe that the expression on the right-hand side of the relation coincides with $\det G \cdot \mathcal{C}_G(0)$; see (7.17). Therefore, the relation is implied by the first expansion formula in Corollary 7.6.3 for $N = 2n$ together with the observation that $\rho_1 = 0$ and $\big(\operatorname{Pf} F\big)^2 = \det G \cdot C_n$. □

7.7. Examples

1. An analogue of the Capelli determinant (7.1) for the quantized enveloping algebra $\mathrm{U}_q(\mathfrak{gl}_N)$ (see Example 1.15.1) can be introduced by the formula

$$\mathcal{C}(u) = \sum_{s\in\mathfrak{S}_N} (-q)^{-l(s)} \cdot \big(uT - \overline{T}\big)_{s(1)1} \big(uT - q^2\overline{T}\big)_{s(2)2} \cdots \big(uT - q^{2N-2}\overline{T}\big)_{s(N)N},$$

where $l(s)$ is the length of a reduced decomposition of s. All coefficients of $\mathcal{C}(u)$ belong to the center of $\mathrm{U}_q(\mathfrak{gl}_N)$. This can be deduced with the use of Examples 1.15.5 and 1.15.6.

The following q-analogues of the noncommutative Cayley–Hamilton theorem (Theorem 7.2.1) take place:

$$\mathcal{C}\big(q^{2N-2}T^{-1}\overline{T}\big) = 0 \qquad \text{and} \qquad \mathcal{C}\big((T^t)^{-1}\overline{T}^t\big) = 0.$$

2. Consider the twisted quantized enveloping algebra $\mathrm{U}'_q(\mathfrak{o}_N)$ introduced in Example 2.16.8. A q-analogue of the Capelli-type determinant (7.17) can be introduced by the formula

$$\mathcal{C}(u) = \sum_{p\in\mathfrak{S}_N} (-q)^{-l(p)+l(p')}\big(u\,\overline{S} + q\,S\big)_{p'(1),p(1)}\cdots\big(u\,\overline{S} + q^{2n-1}S\big)_{p'(n),p(n)}$$
$$\times\big(u\,S + q^{2n-1}\,\overline{S}\big)_{p(n+1),p'(n+1)}\cdots\big(u\,S + q^{2N-3}\,\overline{S}\big)_{p(N),p'(N)}.$$

The polynomial $\mathcal{C}(u)$ is monic of degree N and all its coefficients belong to the center of $\mathrm{U}'_q(\mathfrak{o}_N)$.

The corresponding version of the noncommutative Cayley–Hamilton theorem (Theorem 7.2.3) has the form

$$\mathcal{C}(-q^{2N-3}\,\overline{S}\,S^{-1}) = 0.$$

3. In the limit $q \to 1$, central elements of the algebra $\mathrm{U}'_q(\mathfrak{o}_N)$ give rise to invariants of the Poisson bracket on the algebra of polynomials $\mathcal{P}_N$ in the variables a_{ij}; see Example 2.16.11. We combine the variables a_{ij} into the lower triangular matrix $A = [a_{ij}]$ where we set $a_{ii} = 1$ for all i and $a_{ij} = 0$ for $i < j$. It can be derived from Example 2 that the coefficients of the polynomial

$$\det(A + \lambda A^t) = f_0 + f_1\,\lambda + \cdots + f_N\,\lambda^N$$

are invariants of the Poisson algebra $\mathcal{P}_N$. Moreover, $f_0 = f_N = 1$ and $f_{N-i} = f_i$ for all i. If $N = 2n+1$ is odd, then the coefficients $f_1,\dots,f_n$ are algebraically independent generators of the algebra of invariants of $\mathcal{P}_N$. If $N = 2n$ is even, then

$$\det(A - A^t) = \mathrm{Pf}\,(A - A^t)^2,$$

and a family of algebraically independent generators of the algebra of invariants of $\mathcal{P}_N$ is obtained by replacing any one of the elements $f_1,\dots,f_n$ with $\mathrm{Pf}\,(A - A^t)$.

4. The elements

$$\mathrm{tr}\,(A^{-1}A^t)^k, \qquad k = 1,2,\dots,$$

are invariants of $\mathcal{P}_N$. This can be derived from Example 3 by using the Liouville formula

$$\sum_{k=1}^{\infty}(-1)^{k-1}\lambda^{k-1}\mathrm{tr}\,H^k = \frac{d}{d\lambda}\ln\det(1+\lambda H),$$

where $H = A^{-1}A^t$, and observing that $\det(A+\lambda A^t) = \det(1+\lambda H)$ since $\det A = 1$.

5. Given a lower triangular $N\times N$ matrix B and a $2k$-element subset I of $\{1,\dots,N\}$, denote by $\mathrm{Pf}_I(B)$ the Pfaffian of the $2k\times 2k$ submatrix $(B^t - B)_I$ of $B^t - B$ whose rows and columns are determined by the elements of I. For each positive integer k such that $2k \leqslant N$ the element

$$c_k = (-1)^k \sum_{I,\,|I|=2k} \mathrm{Pf}_I(A)\,\mathrm{Pf}_I(A^{-1})$$

is an invariant of the Poisson bracket on $\mathcal{P}_N$. Moreover, in the case $N = 2n$, both $\mathrm{Pf}_{I_0}(A)$ and $\mathrm{Pf}_{I_0}(A^{-1})$ with $I_0 = \{1,\dots,2n\}$ are also invariants, and we have the relation $\mathrm{Pf}_{I_0}(A^{-1}) = (-1)^n\,\mathrm{Pf}_{I_0}(A)$.

The algebra of invariants of $\mathcal{P}_N$ is generated by $c_1, \dots, c_n$ for $N = 2n+1$, and by $c_1, \dots, c_{n-1}, \operatorname{Pf}_{I_0}(A)$ if $N = 2n$. In both cases, the families of generators are algebraically independent.

6. Consider the Poisson algebra $\mathcal{P}_{2n}$ introduced in Example 2.16.12. Similar to Example 3, combine the variables a_{ij} into a matrix $A = [a_{ij}]$ which has a block-triangular form with n diagonal 2×2-blocks.

The elements

$$a_{i+1,i+1}\, a_{ii} - a_{i+1,i}\, a_{i,i+1}, \qquad i = 1, 3, \dots, 2n-1,$$

and the coefficients of the polynomial

$$\det(A + \lambda A^t) = f_0 + f_1 \lambda + \cdots + f_{2n} \lambda^{2n}$$

are invariants of the Poisson bracket on $\mathcal{P}_{2n}$.

Open problem: Find an algebraically independent family of generators of the algebra of invariants of $\mathcal{P}_{2n}$.

7. By analogy with the $\mathfrak{gl}_N$ case, introduce the power sums symmetric functions of the first kind $\Psi_k^{(m)}$ and $\widetilde{\Psi}_k^{(m)}$ associated, respectively, with the matrices $F^{(m)} + \rho_m$ and $-\widetilde{F}^{(m)} - \rho_m$ and the index m; see Section 7.3. For all $k \geqslant 1$ introduce the elements

$$\Psi_{2k} = \sum_{m=1}^{n} \big(\widetilde{\Psi}_{2k}^{(m)} + \Psi_{2k}^{(m)} \big).$$

Open problem: Do the elements Ψ_{2k} belong to the center of $\mathrm{U}(\mathfrak{g}_N)$? If so, is it true that $\Psi_{2k} = \Phi_{2k}$ for all k?

8. Consider the presentation of the orthogonal Lie algebra $\mathfrak{o}_N$ corresponding to the matrix $G = 1$. Thus, the elements of $\mathfrak{o}_N$ are skew-symmetric matrices with the usual matrix commutator. The elements $F_{ij} = E_{ij} - E_{ji}$ span $\mathfrak{o}_N$. For any $2 \leqslant m \leqslant N$ denote by $F^{(m)}$ the $m \times m$ matrix with the entries F_{ij}, where $i, j = 1, \dots, m$. Let $\Lambda_k^{(m)}$, $S_k^{(m)}$, $\Psi_k^{(m)}$ and $\Phi_k^{(m)}$ respectively denote the elementary, complete, and power sums noncommutative symmetric functions of the first and second kind associated with the matrix $F^{(m)} + N/2 - m + 1$ and the index m; see Section 7.3. For any $k \geqslant 1$ set

$$\Lambda_k = \sum_{k_2 + \cdots + k_N = k} \Lambda_{k_2}^{(2)} \dots \Lambda_{k_N}^{(N)},$$

$$S_k = \sum_{k_2 + \cdots + k_N = k} S_{k_2}^{(2)} \dots S_{k_N}^{(N)},$$

$$\Psi_k = \sum_{m=2}^{N} \Psi_k^{(m)},$$

$$\Phi_k = \sum_{m=2}^{N} \Phi_k^{(m)},$$

where the indices k_i run over nonnegative integers and $\Lambda_0^{(m)} = S_0^{(m)} = 1$. All elements Λ_k, S_k, Ψ_k and Φ_k belong to the center $\mathrm{Z}(\mathfrak{o}_N)$ of the universal enveloping algebra $\mathrm{U}(\mathfrak{o}_N)$. These elements are zero for odd values of k. Moreover, $\Psi_{2k} = \Phi_{2k}$ for all $k \geqslant 1$, and the images of $(-1)^k \Lambda_{2k}$, S_{2k} and $\Psi_{2k}/2$ under the Harish-Chandra isomorphism are, respectively, the elementary, complete and power sums symmetric polynomials of degree k in the variables $l_1^2, \dots, l_n^2$; see Proposition 7.1.9.

9. Using the presentation of the orthogonal Lie algebra $\mathfrak{o}_N$ as in Example 8, set

$$\mathcal{D}(u) = \sum_{p \in \mathfrak{S}_N} \operatorname{sgn} p \cdot \big(u + F + N/2\big)_{p(1),1} \big(u + F + N/2 - 1\big)_{p(2),2} \\ \times \dots \times \big(u + F - N/2 + 1\big)_{p(N),N}.$$

Then all coefficients of the polynomial $\mathcal{D}(u)$ belong to the center of $\mathrm{U}(\mathfrak{o}_N)$. Moreover, we have the identity

$$\mathcal{D}(u) = \sum_{k=0}^{n} C_k \, \big(u + N/2 - k\big)\big(u + N/2 - k - 1\big) \dots \big(u - N/2 + k + 1\big),$$

where the elements C_k are defined in (7.57) with $C_0 = 1$. In particular, the following analogue of Corollary 7.6.4 holds for $N = 2n$:

$$\big(\operatorname{Pf} F\big)^2 = \sum_{p \in \mathfrak{S}_{2n}} \operatorname{sgn} p \cdot \big(F + n\big)_{p(1),1} \big(F + n - 1\big)_{p(2),2} \cdots \big(F - n + 1\big)_{p(2n),2n}.$$

10. In the presentation of the orthogonal Lie algebra $\mathfrak{o}_N$ used in Section 7.1, introduce the polynomial

$$\mathcal{W}(u) = \sum_{p \in \mathfrak{S}_N} \operatorname{sgn} p \cdot (u + \rho_{-n} + F)_{a_{p(1)}, a_1} \cdots (u + \rho_n + F)_{a_{p(N)}, a_N},$$

where $(a_1, \dots, a_N) = (-n, \dots, n)$ and ρ_0 is replaced by 0 if N is odd. All coefficients of $\mathcal{W}(u)$ belong to the center of $\mathrm{U}(\mathfrak{o}_N)$.

The Harish-Chandra images of the coefficients of $\mathcal{W}(u)$ are given by

$$\chi : \mathcal{W}(u) \mapsto \prod_{i=1}^{n} (u^2 - l_i^2), \qquad \text{if} \quad N = 2n,$$

and

$$\chi : \mathcal{W}(u) \mapsto u \prod_{i=1}^{n} (u^2 - l_i^2), \qquad \text{if} \quad N = 2n + 1;$$

cf. Theorem 7.1.6.

11. Consider the restriction of the representation (7.34) to the orthogonal Lie algebra $\mathfrak{o}_N$ corresponding to the matrix $G = 1$; see Example 8. The elements F_{ij} act by the rule

$$F_{ij} \mapsto \sum_{a=1}^{m} \big(x_{ia} \partial_{ja} - x_{ja} \partial_{ia}\big).$$

Now suppose that $N = 2m$. The image of the Pfaffian $\operatorname{Pf} F$ is then given by

$$\operatorname{Pf} F \mapsto \det \begin{pmatrix} x_{11} & \partial_{11} & \cdots & x_{1m} & \partial_{1m} \\ x_{21} & \partial_{21} & \cdots & x_{2m} & \partial_{2m} \\ \vdots & \vdots & \ddots & \vdots & \vdots \\ x_{N1} & \partial_{N1} & \cdots & x_{Nm} & \partial_{Nm} \end{pmatrix}.$$

12. *Open problem*: Construct analogues of the quantum immanants for the Lie algebras $\mathfrak{o}_N$ and $\mathfrak{sp}_{2n}$.

Bibliographical notes

7.1. More details on the properties of the Harish-Chandra isomorphism can be found, e.g., in the books by Dixmier [**89**] or Humphreys [**179**]. The construction of the Gelfand invariants follows the general method of Gelfand [**128**]. The first direct proof of the centrality of the Capelli determinant (7.1) was given by Howe and Umeda [**178**]; see also Howe [**176**]. The Capelli-type determinant (7.13) was introduced by the author in [**296**], where Theorem 7.1.6 was proved. The images of the Gelfand invariants under the Harish-Chandra isomorphism (Corollaries 7.1.4 and 7.1.8) were first found by Perelomov and Popov [**387, 388, 389**]. The observation that the formula (7.8) is related to the theory of Yangians was made by Cherednik [**74**]. A derivation of the Perelomov–Popov formulas from the quantum Liouville formulas (Theorems 1.9.5 and 2.11.2) is contained in the author's paper [**298**]. Different proofs of Newton's formulas without using the Yangians were given by Umeda [**455**] and Itoh [**184, 185**] for the general linear and orthogonal Lie algebras. Some q-analogues of the Gelfand invariants were given by Gould, Zhang and Bracken [**154, 479**].

7.2. Theorem 7.2.1 was formulated in this form by Nazarov and Tarasov [**350**]. On the other hand, it is also implied by the characteristic identities of Green [**160**] with the use of the Harish-Chandra image of the Capelli determinant. Theorem 7.2.3 was proved by the author [**296**]. The identity (7.20) was independently obtained by Nazarov (1991, unpublished), where he used (7.14) as a definition of the polynomial $\mathcal{C}(u)$. His proof relies on an alternative calculation of the Harish-Chandra image of $\mathcal{C}(u)$; see Molev and Nazarov [**315**, Theorem 6.2]. Some analogues of (7.20) were given by Itoh [**185, 186**] with the use of the determinant $\mathcal{D}(u)$ introduced in Example 9 instead of $\mathcal{C}(u)$. A noncommutative Cayley–Hamilton theorem for a generic matrix was proved by Gelfand *et al.* [**131**] by using the quasideterminants; see Section 1.10. Various versions of this theorem for quantum algebras were given by Gurevich, Pyatov and Saponov [**166**], Isaev, Ogievetsky and Pyatov [**183**], Ogievetsky and Pyatov [**363**], and Chervov and Talalaev [**77**]. Corollaries 7.2.2 and 7.2.4 are the characteristic identities discovered by Green [**160**] and Bracken and Green [**46**]. More general identities were obtained by Gould [**148**] for the semi-simple Lie algebras and by Gould and Jarvis [**151**] for the Kac–Moody algebras. A new class of 'family algebras' associated with the semi-simple Lie algebras was introduced by Kirillov [**222, 223**]. The family algebras provide new tools for the study of the Lie algebras and their representations, in particular the characteristic identities and Casimir elements; see also Rozhkovskaya [**407, 408**]. Super and color analogues of the characteristic identities were given by Jarvis and Green [**161, 195**] and Jarvis and Murray [**196**]. Gould and Stoilova [**153**] gave characteristic identities for $\mathfrak{gl}_\infty$.

7.3. The noncommutative symmetric functions associated with a matrix were introduced by Gelfand *et al.* [**131**], where Proposition 7.3.1 was proved. Theorem 7.3.2 was proved in [**131**, Section 7.5]. Theorem 7.3.7 is due to the author [**297**].

7.4. The higher Capelli identities and the quantum immanants were discovered by Okounkov [**365**], where Corollaries 7.4.2, 7.4.5 and Theorem 7.4.6 were first proved; see also Okounkov and Olshanski [**368**]. Originally, these results were motivated by the R-matrix techniques and the fusion procedure; cf. Sections 6.4 and 6.5. Theorem 7.4.1 was proved in a subsequent paper by Okounkov [**366**] and also by Nazarov [**341**]. The proof in [**341**] employs the relationship with the Yangian representation theory. We basically followed [**366**] and the author's work [**301**], where a simplified version of the argument of [**366**] was found. The elements (7.48) are the classical Capelli elements, and relation (7.51) is due to Capelli [**55, 56**]; see also Weyl [**464**], Howe [**176**], Howe and Umeda [**178**] and Umeda [**456**]. The elements (7.49) and relation (7.52) were discovered by Nazarov [**338**]. The double (or factorial) Schur polynomials $s_\mu(x|b)$ were first introduced by Biedenharn and Louck [**33, 34**] for a particular sequence b. The general definition is due to Macdonald [**279**] and Goulden and Greene [**156**]. It was observed by Lascoux [**259**] that the double Schur polynomials are recovered as a particular case of the double Schubert polynomials. More details on this relationship can be found in Chen, Li and Louck [**70**]. The vanishing and characterization properties of the polynomials $s_\mu(x|b)$ were discovered by Okounkov [**365**]; see also Knop [**234, 235**], Sahi [**412**], Okounkov and Olshanski [**368**]. The generating series expansions (7.32) and (7.33) can be found in Okounkov and Olshanski [**368**]; cf. Abderrezzak [**1**]. Analogues of the Capelli identities for some Lie superalgebras were proved by Nazarov [**340**]. The higher Capelli identities for the Lie superalgebra $\mathfrak{gl}_{M|N}$ were obtained in the author's paper [**301**].

7.5. The exposition follows Molev and Nazarov [**315**]. The term 'Hafnian' was devised by Caianiello [**54**] for a symmetric matrix with commuting entries; see also Kuperberg [**256**, Section 2]. Analogues of the Casimir elements C_k for the twisted quantized enveloping algebra $\mathrm{U}'_q(\mathfrak{o}_N)$ were constructed by Gavrilik and Iorgov [**126**]. Some central permanent-type elements in the universal enveloping algebra $\mathrm{U}(\mathfrak{sp}_{2n})$ were given by Itoh [**191**]. The "generalized symmetrization map" constructed by Olshanski [**377**] can be combined with the results of Okounkov and Olshanski [**369**, Section 5] to produce explicit expressions for bases of the centers of $\mathrm{U}(\mathfrak{o}_N)$ and $\mathrm{U}(\mathfrak{sp}_{2n})$; see also a related work of Nazarov [**343**].

7.6. Theorems 7.6.1 and 7.6.2 are due to Molev and Nazarov [**315**]. Some other analogues, generalizations and connections of the classical Capelli identity can be found in the literature; see e.g. Howe [**176, 177**], Howe and Umeda [**178**], Itoh [**187, 189**], Knop [**235**], Kostant and Sahi [**243, 244**], Lee, Nishiyama and Wachi [**265**], Nazarov [**341, 343**], Noumi, Umeda and Wakayama [**359, 360, 361**], Mukhin, Tarasov and Varchenko [**329**].

7.7. The q-analogue of the Cayley–Hamilton theorem in Example 1 is due to Nazarov and Tarasov [**350**]. In a different form, some q-analogues of the characteristic identities were given by Gould, Zhang and Bracken [**154, 479**]. Example 2 is due to Molev, Ragoucy and Sorba [**321**]. The invariants of the Poisson algebra $\mathcal{P}_N$ given in Example 3 were first constructed by Nelson and Regge [**356**]. The algebraic independence of the generators was established by Bondal [**39**]. The invariants of Example 4 can be obtained by using the Casimir elements of the algebra $\mathrm{U}'_q(\mathfrak{o}_N)$ constructed by Noumi, Umeda and Wakayama [**361**]. The invariants of Example 5 were produced in Molev and Ragoucy [**320**] by using the Casimir elements constructed by Gavrilik and Iorgov [**126**]. Example 6 is contained in [**320**]. Example 8 is due to the author [**300**]. The centrality of the determinant $\mathcal{D}(u)$ in Example 9 was first proved by Howe and Umeda [**178**]. Its relationship with the Pfaffians was established by Itoh and Umeda [**192**]. Example 10 is due to Wachi [**463**]. Another proof was given by Itoh [**190**] with the use of a 'symmetrized determinant' associated with the matrix F. Example 11 provides an alternative form of (7.64) and was found independently by Ochiai [**362**].

CHAPTER 8

Centralizer construction

In this chapter we explore further the relationship between the classical Lie algebras and the associated (twisted) Yangians. We show that the Yangian and the twisted Yangians can be realized as subalgebras of some projective limit algebras of a sequence of centralizers in the classical enveloping algebras. This construction gives rise to a class of 'skew representations' over the twisted Yangians. We describe the skew representations of $\mathrm{Y}(\mathfrak{gl}_N)$ by calculating their highest weights, Drinfeld polynomials and Gelfand–Tsetlin characters. Some analogues of these results for the twisted Yangians are discussed in Chapter 9.

8.1. Olshanski algebra associated with $\mathfrak{gl}_\infty$

For any positive integer N consider the general linear Lie algebra $\mathfrak{gl}_N$. The Lie algebra $\mathfrak{gl}_{N-1}$ can be identified with the subalgebra of $\mathfrak{gl}_N$ spanned by the basis elements E_{ij} with $1 \leqslant i, j \leqslant N-1$. Fix a nonnegative integer M such that $M \leqslant N$ and denote by $\mathfrak{gl}_{N,M}$ the subalgebra in $\mathfrak{gl}_N$ spanned by the basis elements E_{ij} with $M+1 \leqslant i, j \leqslant N$. The subalgebra $\mathfrak{gl}_{N,M}$ is isomorphic to $\mathfrak{gl}_{N-M}$. Let $\mathrm{A}_M(N)$ denote the centralizer of $\mathfrak{gl}_{N,M}$ in the universal enveloping algebra $\mathrm{U}(\mathfrak{gl}_N)$. Let $\mathrm{A}(N)^0$ denote the centralizer of the element E_{NN} in $\mathrm{U}(\mathfrak{gl}_N)$ and let $\mathrm{I}(N)$ be the left ideal of $\mathrm{U}(\mathfrak{gl}_N)$ generated by the elements E_{iN}, $i = 1, \dots, N$. Then the Poincaré–Birkhoff–Witt theorem for the Lie algebra $\mathfrak{gl}_N$ implies that $\mathrm{I}(N)^0 := \mathrm{I}(N) \cap \mathrm{A}(N)^0$ is a two-sided ideal of $\mathrm{A}(N)^0$, and one has the vector space decomposition

$$\mathrm{A}(N)^0 = \mathrm{I}(N)^0 \oplus \mathrm{U}(\mathfrak{gl}_{N-1}).$$

Therefore, the projection of $\mathrm{A}(N)^0$ onto $\mathrm{U}(\mathfrak{gl}_{N-1})$ with the kernel $\mathrm{I}(N)^0$ is an algebra homomorphism. If $M < N$, then its restriction to the subalgebra $\mathrm{A}_M(N)$ defines a homomorphism

$$o_N : \mathrm{A}_M(N) \to \mathrm{A}_M(N-1). \tag{8.1}$$

Note that the algebra $\mathrm{A}_M(N)$ inherits the natural filtration of $\mathrm{U}(\mathfrak{gl}_N)$ and the homomorphism o_N is filtration-preserving.

Definition 8.1.1. The algebra A_M is defined as the projective limit of the sequence of the algebras $\mathrm{A}_M(N)$, $N \geqslant M$, with respect to the homomorphisms

$$\mathrm{A}_M(M) \xleftarrow{o_{M+1}} \mathrm{A}_M(M+1) \xleftarrow{o_{M+2}} \dots \xleftarrow{o_N} \mathrm{A}_M(N) \xleftarrow{o_{N+1}} \dots,$$

where the limit is taken in the category of filtered associative algebras. □

In other words, an element of the algebra A_M is a sequence of the form $a = (a_M, a_{M+1}, \dots, a_N, \dots)$ with $a_N \in \mathrm{A}_M(N)$, $o_N(a_N) = a_{N-1}$ for $N > M$, and

$$\deg a := \sup_{N \geqslant M} \deg a_N < \infty, \tag{8.2}$$

where $\deg a_N$ denotes the degree of a_N in the universal enveloping algebra $\mathrm{U}(\mathfrak{gl}_N)$. If $b = (b_M, b_{M+1}, \dots, b_N, \dots)$ is another element of A_M, then the product ab is the sequence

$$ab = (a_M b_M, a_{M+1} b_{M+1}, \dots, a_N b_N, \dots).$$

We define the Lie algebra $\mathfrak{gl}_\infty$ as the inductive limit of the Lie algebras $\mathfrak{gl}_N$ with respect to the natural embeddings $\mathfrak{gl}_N \hookrightarrow \mathfrak{gl}_{N+1}$,

$$\mathfrak{gl}_\infty = \bigcup_{N \geqslant 1} \mathfrak{gl}_N.$$

Equivalently, $\mathfrak{gl}_\infty$ is the Lie algebra of all complex matrices $B = [b_{ij}]$, where the indices i and j run over the set of positive integers such that the number of nonzero entries b_{ij} is finite. So, $\mathfrak{gl}_\infty$ has a basis $\{E_{ij}\}$, where $i, j \geqslant 1$ and the E_{ij} satisfy the standard commutation relations (0.1).

By definition, the algebra $\mathrm{A}_0(N)$ coincides with the center $\mathrm{Z}(\mathfrak{gl}_N)$ of the universal enveloping algebra $\mathrm{U}(\mathfrak{gl}_N)$. The elements of the algebra A_0 can therefore be regarded as *virtual Casimir elements* (or *virtual Laplace operators*) for the Lie algebra $\mathfrak{gl}_\infty$. We will give a more detailed description of this algebra in the next section.

Observe that the homomorphisms o_N are compatible with the natural embeddings $\mathrm{A}_M(N) \hookrightarrow \mathrm{A}_{M+1}(N)$; that is, the following diagram commutes:

$$\begin{array}{ccc} \mathrm{A}_M(N) & \longrightarrow & \mathrm{A}_{M+1}(N) \\ {\scriptstyle o_N}\downarrow & & \downarrow{\scriptstyle o_N} \\ \mathrm{A}_M(N-1) & \longrightarrow & \mathrm{A}_{M+1}(N-1). \end{array}$$

Therefore we can define an embedding $\mathrm{A}_M \hookrightarrow \mathrm{A}_{M+1}$ by

$$(a_M, a_{M+1}, a_{M+2}, \dots) \mapsto (a_{M+1}, a_{M+2}, \dots).$$

DEFINITION 8.1.2. The *Olshanski algebra* $\mathrm{A} = \mathrm{A}(\mathfrak{gl}_\infty)$ *associated with* $\mathfrak{gl}_\infty$ is the inductive limit of the filtered algebras A_M,

$$\mathrm{A} = \bigcup_{M \geqslant 0} \mathrm{A}_M.$$

□

Note that the universal enveloping algebra $\mathrm{U}(\mathfrak{gl}_\infty)$ is canonically embedded into the Olshanski algebra A. The image of an element $x \in \mathrm{U}(\mathfrak{gl}_\infty)$ in A is the sequence $(x, x, \dots)$. This is a well-defined map because x belongs to $\mathrm{U}(\mathfrak{gl}_M)$ for some M and hence to $\mathrm{A}_M(N)$ for all $N > M$.

PROPOSITION 8.1.3. *The center of the Olshanski algebra* A *coincides with* A_0.

PROOF. It is clear that A_0 is contained in the center. Conversely, suppose that $a = (a_M, a_{M+1}, \dots) \in \mathrm{A}$ is a central element. Then it commutes with $\mathfrak{gl}_\infty \subset \mathrm{A}$, and so a_N lies in the center of $\mathrm{U}(\mathfrak{gl}_N)$ for all $N \geqslant M$, which implies that $a \in \mathrm{A}_0$. □

REMARK 8.1.4. It is well known that the center of the universal enveloping algebra $\mathrm{U}(\mathfrak{gl}_\infty)$ is trivial, so that $\mathfrak{gl}_\infty$ has no "genuine" Casimir elements. In contrast to $\mathrm{U}(\mathfrak{gl}_\infty)$, the Olshanski algebra A has a large center A_0 consisting of the virtual Casimir elements. As we will see in the next section, these elements act as scalars in the natural class of highest weight $\mathfrak{gl}_\infty$-modules. From this perspective, the algebra A appears to be a more natural analogue of the enveloping algebras $\mathrm{U}(\mathfrak{gl}_N)$ than $\mathrm{U}(\mathfrak{gl}_\infty)$. □

8.2. Virtual Casimir elements and highest weight modules for $\mathfrak{gl}_\infty$

Due to the Harish-Chandra isomorphism (see Section 7.1), the center $\mathrm{A}_0(N)$ of $\mathrm{U}(\mathfrak{gl}_N)$ is naturally isomorphic to the algebra $\Lambda^*(N)$ of polynomials in N variables $\lambda_1, \dots, \lambda_N$ which are symmetric in the shifted variables $\lambda_1, \lambda_2 - 1, \dots, \lambda_N - N + 1$. Moreover, the Harish-Chandra isomorphism preserves the filtrations on $\mathrm{A}_0(N)$ and $\Lambda^*(N)$, where the former is inherited from the natural filtration on $\mathrm{U}(\mathfrak{gl}_N)$, while the latter is determined by the usual degrees of polynomials. Thus, in the case $M = 0$ the homomorphism o_N given in (8.1) can be interpreted as the specialization homomorphism $o_N : \Lambda^*(N) \to \Lambda^*(N-1)$ such that

$$o_N : f(\lambda_1, \dots, \lambda_N) \mapsto f(\lambda_1, \dots, \lambda_{N-1}, 0). \tag{8.3}$$

The corresponding projective limit in the category of filtered algebras is called the *algebra of shifted symmetric functions* and is denoted by Λ^*. This construction of the algebra Λ^* is quite similar to that of the algebra Λ of symmetric functions, and both algebras are indeed closely related to each other. Namely, it is easily seen that Λ is naturally isomorphic to the graded algebra $\operatorname{gr} \Lambda^*$ associated with the filtered algebra Λ^*.

Summarizing the previous discussion, we obtain the following analogue of the Harish-Chandra isomorphism for the Lie algebra $\mathfrak{gl}_\infty$.

PROPOSITION 8.2.1. *The algebra of virtual Casimir elements A_0 is isomorphic to the algebra of shifted symmetric functions Λ^*. The isomorphism $\chi : \mathrm{A}_0 \to \Lambda^*$ is determined by the Harish-Chandra isomorphisms $\chi_N : \mathrm{A}_0(N) \to \Lambda^*(N)$ with $N \geqslant 0$.* □

Some families of generators of the algebra A_0 can be constructed by using the results of Chapter 7 and taking an appropriate limit as $N \to \infty$. Here we will denote by E the infinite matrix $[E_{ij}]$ whose rows and columns are numbered by positive integers while $E^{(N)}$ will denote its submatrix determined by the first N rows and columns. The series

$$\operatorname{qdet}(1 + E^{(N)} u^{-1}) = \sum_{p \in \mathfrak{S}_N} \operatorname{sgn} p \cdot (1 + E u^{-1})_{p(1),1} \cdots (1 + E(u - N + 1)^{-1})_{p(N),N}$$

in the image of the quantum determinant $\operatorname{qdet} T(u)$ under the evaluation homomorphism (1.5). Write

$$\operatorname{qdet}(1 + E^{(N)} u^{-1}) = 1 + \mathcal{E}_1^{(N)} u^{-1} + \mathcal{E}_2^{(N)} u^{-2} + \dots, \qquad \mathcal{E}_k^{(N)} \in \mathrm{U}(\mathfrak{gl}_N).$$

Due to relation (7.4) and Theorem 7.1.1, we see that all coefficients $\mathcal{E}_k^{(N)}$ belong to $\mathrm{Z}(\mathfrak{gl}_N)$. Moreover, by the definition of the homomorphism o_N we have

$$o_N : \operatorname{qdet}(1 + E^{(N)} u^{-1}) \mapsto \operatorname{qdet}(1 + E^{(N-1)} u^{-1}).$$

Hence, for any $k \geqslant 1$ we may define a virtual Casimir element $\mathcal{E}_k \in \mathrm{A}_0$ as the sequence

$$\mathcal{E}_k = (\mathcal{E}_k^{(0)}, \mathcal{E}_k^{(1)}, \mathcal{E}_k^{(2)}, \dots).$$

Define the *virtual quantum determinant* as the formal power series

$$\operatorname{qdet}(1 + E u^{-1}) = 1 + \mathcal{E}_1 u^{-1} + \mathcal{E}_2 u^{-2} + \dots.$$

In order to get an alternative expression for the virtual Casimir elements $\mathcal{E}_k$, denote by $\mathfrak{S}_\infty$ the group of finite permutations of the set of positive integers so that for any $p \in \mathfrak{S}_\infty$ we have $p(l) = l$ for sufficiently large l. Using the expression

$$\operatorname{qdet}(1 + Eu^{-1}) = \sum_{p \in \mathfrak{S}_\infty} \operatorname{sgn} p \cdot (1 + Eu^{-1})_{p(1),1} (1 + E(u-1)^{-1})_{p(2),2} \cdots,$$

we can regard each coefficient $\mathcal{E}_k$ as a formal series of elements of $\mathrm{U}(\mathfrak{gl}_\infty)$. For instance,

$$\mathcal{E}_1 = \sum_{i=1}^{\infty} E_{ii},$$

$$\mathcal{E}_2 = \sum_{i=1}^{\infty} (i-1) E_{ii} + \sum_{1 \leqslant i < j} (E_{ii}E_{jj} - E_{ji}E_{ij}).$$

PROPOSITION 8.2.2. *The elements $\mathcal{E}_1, \mathcal{E}_2, \dots$ are algebraically independent and generate the algebra A_0. Their images under the isomorphism χ are found by*

$$\chi : \operatorname{qdet}(1 + Eu^{-1}) \mapsto \prod_{l=1}^{\infty} \frac{u + \lambda_l - l}{u - l}.$$

PROOF. The second statement is immediate from the definition of the virtual quantum determinant. This implies that $\chi(\mathcal{E}_k)$ is the shifted symmetric function found by

$$1 + \sum_{k=1}^{\infty} \chi(\mathcal{E}_k) u^{-k} = \prod_{l=1}^{\infty} \frac{u + \lambda_l - l}{u - l}.$$

It is clear that $\chi(\mathcal{E}_k)$ has degree k as an element of Λ^* and the homogeneous component of $\chi(\mathcal{E}_k)$ of degree k coincides with the k-th elementary symmetric function in the variables λ_i. Since the elementary symmetric functions are algebraically independent generators of the algebra Λ, the first statement follows. □

Virtual analogues of the Gelfand invariants (see Section 7.1) can be constructed as follows. Set

$$\mathcal{G}_k^{(N)} = \operatorname{tr}\Big(E^{(N)}(E^{(N)} - N)^{k-1}\Big), \qquad k \geqslant 1.$$

PROPOSITION 8.2.3. *For any $k \geqslant 1$ the sequence*

$$\mathcal{G}_k = (\mathcal{G}_k^{(0)}, \mathcal{G}_k^{(1)}, \mathcal{G}_k^{(2)}, \dots)$$

is an element of the algebra A_0. Moreover, the elements $\mathcal{G}_1, \mathcal{G}_2, \dots$ are algebraically independent and generate A_0. Their images under the isomorphism χ are found by

$$\sum_{k=1}^{\infty} \chi(\mathcal{G}_k) u^{-k} = u - u \cdot \prod_{l=1}^{\infty} \frac{u - \lambda_l + l - 1}{u + l - 1} \cdot \frac{u + l}{u - \lambda_l + l}.$$

PROOF. Corollary 7.1.4 implies that the elements $\mathcal{G}_k^{(N)}$ belong to $\mathrm{Z}(\mathfrak{gl}_N)$. We need to verify only that for any $N \geqslant 1$,

$$o_N : \mathcal{G}_k^{(N)} \mapsto \mathcal{G}_k^{(N-1)}.$$

We do this by regarding o_N as the specialization homomorphism (8.3) and considering the Harish-Chandra images of the elements $\mathcal{G}_k^{(N)}$ and $\mathcal{G}_k^{(N-1)}$. By the definition of the elements $\mathcal{G}_k^{(N)}$, we have

$$\sum_{k=1}^{\infty} \mathcal{G}_k^{(N)} u^{-k} = \sum_{k=1}^{\infty} \frac{\operatorname{tr}\big(E^{(N)}\big)^k}{(u+N)^k}.$$

Using Corollary 7.1.4, we derive the following relation for the Harish-Chandra images $\chi_N\big(\mathcal{G}_k^{(N)}\big) \in \Lambda^*(N)$:

$$\sum_{k=1}^{\infty} \chi_N\big(\mathcal{G}_k^{(N)}\big)\, u^{-k} = u - (u+N)\cdot \frac{(u-\lambda_1)\dots(u-\lambda_N+N-1)}{(u-\lambda_1+1)\dots(u-\lambda_N+N)}. \tag{8.4}$$

Therefore, we have

$$o_N : \chi_N\big(\mathcal{G}_k^{(N)}\big) \mapsto \chi_{N-1}\big(\mathcal{G}_k^{(N-1)}\big),$$

completing the proof of the first statement. Furthermore, the element $\mathcal{G}_k \in \mathrm{A}_0$ clearly has degree $\leqslant k$. An easy calculation shows that the homogeneous component of degree k of the shifted symmetric function $\chi(\mathcal{G}_k)$ coincides with $\sum_{i=1}^{\infty} \lambda_i^k$. Since the power sums symmetric functions are algebraically independent generators of the algebra Λ, the second statement follows. The formula for the images $\chi(\mathcal{G}_k)$ is implied by (8.4). □

EXAMPLE 8.2.4. The first two elements $\mathcal{G}_1$ and $\mathcal{G}_2$ can be written as

$$\mathcal{G}_1 = \mathcal{E}_1 = \sum_{i=1}^{\infty} E_{ii},$$

$$\mathcal{G}_2 = \mathcal{E}_1^2 - \mathcal{E}_1 - 2\,\mathcal{E}_2 = \sum_{i=1}^{\infty} E_{ii}\,(E_{ii} - 2i + 1) + 2 \sum_{1\leqslant i<j} E_{ji}E_{ij}.$$

□

Note that apart from the elements $\chi(\mathcal{E}_k)$ and $\chi(\mathcal{G}_k)$, the algebra of shifted symmetric functions Λ^* admits some other families of algebraically independent generators. In particular, analogues of the power sums symmetric functions can be given by

$$\sum_{i=1}^{\infty} \big((\lambda_i - i)^k - (-i)^k\big), \qquad k = 1, 2, \dots,$$

while analogues of the complete symmetric functions can be defined as the coefficients in the expansion of the series

$$\prod_{l=1}^{\infty} \frac{u+l}{u-\lambda_l+l}$$

in the powers of u^{-1}. We also remark that the elements of Λ^* can be viewed as functions on the set of all sequences $\lambda = (\lambda_1, \lambda_2, \dots)$ which contain only a finite number of nonzero terms.

Explicit linear bases of the algebras A_0 and Λ^* can be constructed with the use of the quantum immanants and the shifted Schur polynomials; see Section 7.4. In order to indicate the dependence on N we will now denote the quantum immanants for $\mathfrak{gl}_N$ defined in (7.39) by $\mathbb{S}_{\mu|N}$. We set $\mathbb{S}_{\mu|N} = 0$ if the length of the partition μ exceeds N.

PROPOSITION 8.2.5. *Given any partition μ, we have*

$$o_N : \mathbb{S}_{\mu|N} \mapsto \mathbb{S}_{\mu|N-1}. \tag{8.5}$$

PROOF. By (7.47), the image of $\mathbb{S}_{\mu|N}$ under the Harish-Chandra isomorphism is found by

$$\chi_N : \mathbb{S}_{\mu|N} \mapsto s^*_{\mu|N},$$

where $s^*_{\mu|N}$ is the shifted Schur polynomial

$$s^*_{\mu|N} = \sum_{\mathrm{sh}(\mathcal{T})=\mu} \prod_{\alpha\in\mu} \big(\lambda_{\mathcal{T}(\alpha)} - c(\alpha)\big),$$

summed over all reverse μ-tableaux $\mathcal{T}$ with entries in $\{1,\dots,N\}$ such that the entries of $\mathcal{T}$ weakly decrease along the rows and strictly decrease down the columns. We may assume that the length of μ does not exceed N. If a reverse μ-tableau contains entries equal to N, then N must occur in the $(1,1)$ cell. Hence, the evaluation $\lambda_N = 0$ amounts to restricting the entries of μ-tableaux to the set $\{1,\dots,N-1\}$. This shows that

$$o_N : s^*_{\mu|N} \mapsto s^*_{\mu|N-1},$$

which implies (8.5). □

Given any partition μ, Proposition 8.2.5 allows us to define the corresponding *virtual quantum immanant* $\mathbb{S}_\mu \in \mathrm{A}_0$ as the sequence

$$\mathbb{S}_\mu = (\,\mathbb{S}_{\mu|N} \mid N \geqslant 0\,).$$

Similarly, define the *shifted Schur function* $s^*_{\mu|N} \in \Lambda^*$ as the sequence

$$s^*_\mu = (\,s^*_{\mu|N} \mid N \geqslant 0\,).$$

COROLLARY 8.2.6. *As μ runs over the set of all partitions, the virtual quantum immanants $\mathbb{S}_\mu$ form a basis of the algebra A_0, while the shifted Schur functions s^*_μ form a basis of the algebra Λ^*. Moreover,*

$$\chi : \mathbb{S}_\mu \mapsto s^*_\mu.$$

PROOF. This is implied by Theorem 7.4.6 with the use of the fact that the shifted Schur polynomial $s^*_{\mu|N}$ coincides with the respective double Schur polynomial $s_\mu(x\,|\,b)$ with $x = (l_1 + N - 1, \dots, l_N + N - 1)$ and $b = (0, 1, 2, \dots)$. □

REMARK 8.2.7. The multiplication in the algebra of virtual Casimir elements A_0 and in the algebra of shifted symmetric functions Λ^* can be described by calculating the structure constants $f^\nu_{\lambda\mu}$ in the respective bases $\{\mathbb{S}_\mu\}$ and $\{s^*_\mu\}$,

$$\mathbb{S}_\lambda\,\mathbb{S}_\mu = \sum_\nu f^\nu_{\lambda\mu}\,\mathbb{S}_\nu, \qquad s^*_\lambda\,s^*_\mu = \sum_\nu f^\nu_{\lambda\mu}\,s^*_\nu.$$

The constants $f^\nu_{\lambda\mu}$ turn out to be nonnegative integers, and a combinatorial rule for their calculation is provided by Example 8.10.7 below.

A generalized family of rings Λ^a of symmetric functions is introduced in Example 8.10.2 below. The family is parameterized by an infinite set of variables, and both algebras Λ and Λ^* can be obtained as specializations of the extension of Λ^a over the field of complex numbers. □

Now we introduce a category Ω of modules over the Lie algebra $\mathfrak{gl}_\infty$ analogous to the Bernstein–Gelfand–Gelfand category $\mathcal{O}$. We let $\mathfrak{h}$ denote the Cartan subalgebra of $\mathfrak{gl}_\infty$ which consists of diagonal matrices and let $\mathfrak{n}^+$ (respectively, $\mathfrak{n}^-$) denote the subalgebra of upper (respectively, lower) triangular matrices. We have the triangular decomposition

$$\mathfrak{gl}_\infty = \mathfrak{n}^- \oplus \mathfrak{h} \oplus \mathfrak{n}^+.$$

For a linear functional $\lambda \in \mathfrak{h}^*$ set $\lambda_i = \lambda(E_{ii})$. We will identify λ with the sequence $(\lambda_1, \lambda_2, \dots)$.

A $\mathfrak{gl}_\infty$-module V is said to be *highest weight* if it has a nonzero cyclic vector v satisfying $\mathfrak{n}^+ v = \{0\}$ and there exists $\lambda \in \mathfrak{h}^*$ such that $hv = \lambda(h)v$ for any $h \in \mathfrak{h}$. The functional λ is the *highest weight* of V, and v is the *highest weight vector* of V; it is unique up to scalar multiples. The universal $\mathfrak{gl}_\infty$-module $M(\lambda)$ with the highest weight $\lambda \in \mathfrak{h}^*$ (the *Verma module*) may be defined as the quotient of the universal enveloping algebra $\mathrm{U}(\mathfrak{gl}_\infty)$ by the left ideal generated by $\mathfrak{n}^+$ and the elements $h - \lambda(h)$, $h \in \mathfrak{h}$. Denote by $L(\lambda)$ the unique irreducible quotient of $M(\lambda)$. The $\mathfrak{gl}_\infty$-module $L(\lambda)$ can be regarded as an inductive limit of irreducible highest weight $\mathfrak{gl}_N$-modules. Namely, for a positive integer N denote by $L(\lambda)^{(N)}$ the $\mathfrak{gl}_N$-cyclic span of the highest weight vector $v \in L(\lambda)$. Obviously, $L(\lambda)^{(N)}$ is a highest weight module with the highest weight $(\lambda_1, \dots, \lambda_N)$. We have the following.

PROPOSITION 8.2.8. *For any $N \geqslant 1$ the $\mathfrak{gl}_N$-module $L(\lambda)^{(N)}$ is irreducible.*

PROOF. Suppose that a nonzero vector $w \in L(\lambda)^{(N)}$ satisfies

$$E_{ij}\, w = 0 \qquad \text{for} \quad 1 \leqslant i < j \leqslant N.$$

The vector w can be written as a linear combination of vectors of the form

$$E_{j_1 i_1} \dots E_{j_k i_k}\, v, \qquad 1 \leqslant i_a < j_a \leqslant N.$$

The commutation relations of $\mathfrak{gl}_\infty$ imply that w is annihilated by the entire algebra $\mathfrak{n}^+$. Since $L(\lambda)$ is irreducible, w must be proportional to v. This implies that $L(\lambda)^{(N)}$ is irreducible. □

Obviously, $L(\lambda)^{(N)} \subset L(\lambda)^{(N+1)}$ so that we can regard $L(\lambda)$ as the inductive limit

$$L(\lambda) = \bigcup_{N=1}^{\infty} L(\lambda)^{(N)}.$$

For a $\mathfrak{gl}_\infty$-module V and a functional $\mu = (\mu_1, \mu_2, \dots) \in \mathfrak{h}^*$ set

$$\begin{aligned} V_\mu(N) = \{v \in V \mid E_{ii} v = \mu_i v \quad &\text{for} \quad i > N \quad \text{and} \\ E_{ij} v = 0 \quad &\text{for} \quad 1 \leqslant i < j; \ \ j > N\}. \end{aligned}$$

Obviously, $V_\mu(N) \subset V_\mu(N+1)$. Set

$$V_\mu(\infty) = \bigcup_{N=1}^{\infty} V_\mu(N).$$

Note that $V_\mu(\infty)$ is a submodule of V since $V_\mu(N)$ is $\mathfrak{gl}_N$-invariant for every N. We will write $\mu \sim \mu'$ if $\mu_i = \mu'_i$ for all sufficiently large i. Then $\mu \sim \mu'$ implies $V_\mu(\infty) = V_{\mu'}(\infty)$.

DEFINITION 8.2.9. Define Ω as the category of $\mathfrak{gl}_\infty$-modules V satisfying the condition $V_{\mathbf{o}}(\infty) = V$, where $\mathbf{o} = (0, 0, \dots) \in \mathfrak{h}^*$. □

PROPOSITION 8.2.10. *Let V be a $\mathfrak{gl}_\infty$-module with the highest weight λ, where $\lambda \sim \mathbf{o}$. Then V belongs to Ω. In particular, Ω contains the modules $M(\lambda)$ and $L(\lambda)$ with $\lambda \sim \mathbf{o}$.*

PROOF. Suppose that $\lambda_i = 0$ for $i \geqslant N + 1$. Then the highest weight vector v lies in the subspace $V_{\mathbf{o}}(N)$. Since v is a cyclic vector of V, the submodule $V_{\mathbf{o}}(\infty)$ coincides with V. □

The following theorem shows that the Olshanski algebra A may be regarded as a 'true' analogue of the universal enveloping algebra $\mathrm{U}(\mathfrak{gl}_\infty)$.

THEOREM 8.2.11. *Any $\mathfrak{gl}_\infty$-module $V \in \Omega$ has a natural* A*-module structure.*

PROOF. Let $v \in V$ and $a = (a_M, a_{M+1}, \dots) \in \mathrm{A}$. By the definition of the Olshanski algebra A (Definition 8.1.2), $a_N - a_{N-1} \in \mathrm{I}(N)$. Hence, if N is large enough, $a_N v$ does not depend on N and this value can be taken as the definition of the vector av. Let us verify that this definition yields an action of A. Indeed, if $a = (a_N)$ and $b = (b_N)$ are elements of A, then abv is equal to $a_k b_l v$, where k and l are large enough. On the other hand, $(ab)v$ is $(a_N b_N)v = a_N b_N v$ for a sufficiently large N, and so $abv = (ab)v$. This action of A clearly extends the action of its subalgebra $\mathrm{U}(\mathfrak{gl}_\infty)$. □

Recall the isomorphism $\chi : \mathrm{A}_0 \to \Lambda^*$ provided by Proposition 8.2.1.

PROPOSITION 8.2.12. *Let V be a highest weight $\mathfrak{gl}_\infty$-module with the highest weight $\lambda \sim \mathbf{o}$. Then every element $a \in \mathrm{A}_0$ acts on V as multiplication by the scalar $\chi(a)_\lambda$ obtained by evaluating $\chi(a) \in \Lambda^*$ at λ.*

PROOF. Let v be the highest weight vector of the module V and $a = (a_N)$ be an arbitrary element of A_0. By the definition of χ, $a_N v = \chi_N(a)v$ for a sufficiently large N. Hence $av = \chi(a)_\lambda v$. Since v is a cyclic vector and a belongs to the center of the algebra A, the same is true for all vectors of the module V. □

REMARK 8.2.13. Given $c \in \mathbb{C}$ we can define $\Omega(c)$ as the category of modules V over $\mathfrak{gl}_\infty$ satisfying the condition $V_{\mathbf{c}}(\infty) = V$, where $\mathbf{c} = (c, c, \dots) \in \mathfrak{h}^*$; cf. Definition 8.2.9. The category $\Omega(c)$ contains a distinguished one-dimensional module $L(\mathbf{c})$ and the mapping $V \mapsto V \otimes L(\mathbf{c})$ establishes a category isomorphism $\Omega \to \Omega(c)$. So, all the categories $\Omega(c)$ are canonically isomorphic to each other.

The results of Sections 8.1 and 8.2 can be carried over to the Lie algebra $\mathfrak{gl}_{2\infty+1}$ of all complex matrices $A = [A_{ij}]$ where $i, j \in \mathbb{Z}$ and only a finite number of the A_{ij} is nonzero. In particular, here $\Omega(c)$ is replaced with the category $\Omega(c_-, c_+)$ of modules which 'stabilize' in both negative and positive directions. In contrast to the category $\Omega(c)$ for $\mathfrak{gl}_\infty$, the category $\Omega(c_-, c_+)$ for $\mathfrak{gl}_{2\infty+1}$ involves a substantial continuous parameter, namely, the difference $c_- - c_+$. □

8.3. Polynomial invariants for $\mathfrak{gl}_N$

Here, for the use in the next section, we give a description of the structure of the graded algebras $\operatorname{gr} \mathrm{A}_M$ and $\operatorname{gr} \mathrm{A}$ with respect to the filtration defined by (8.2). Keeping the notation of Section 8.1, denote by $\mathrm{P}_M(N)$ the subalgebra of the symmetric algebra $\mathrm{S}(\mathfrak{gl}_N)$ which consists of the elements invariant under the adjoint action of the subalgebra $\mathfrak{gl}_{N,M}$. We let $\mathrm{I}'(N)$ denote the ideal of $\mathrm{S}(\mathfrak{gl}_N)$ generated by the elements E_{iN}, $i = 1, \dots, N$.

Replacing respectively $\mathrm{A}_M(N)$ by $\mathrm{P}_M(N)$ and $\mathrm{I}(N)$ by $\mathrm{I}'(N)$ in the construction of Section 8.1, we get a commutative graded algebra P_M which is the projective limit of the commutative graded algebras $\mathrm{P}_M(N)$, $N \geqslant M$. The commutative analogue P of the Olshanski algebra A is then defined as the inductive limit of the algebras P_M. Note that the algebra P contains the symmetric algebra of $\mathfrak{gl}_\infty$.

The role of the algebra Λ^* is played by the algebra Λ of symmetric functions. In more detail, $\mathrm{P}_0(N)$ coincides with the algebra of invariants of $N \times N$ matrices and may be identified with the algebra of symmetric polynomials in N variables $\lambda_1, \dots, \lambda_N$. Then P_0 may be identified with the algebra of symmetric functions in infinitely many variables $\lambda_1, \lambda_2, \dots$.

PROPOSITION 8.3.1. *We have the isomorphisms*

$$\operatorname{gr} \mathrm{A}_M \cong \mathrm{P}_M \qquad \textit{and} \qquad \operatorname{gr} \mathrm{A} \cong \mathrm{P}.$$

PROOF. Let $\big(\mathrm{U}^k(\mathfrak{gl}_N)\big)$, $k = 0, 1, 2, \dots$ stand for the canonical filtration of the universal enveloping algebra $\mathrm{U}(\mathfrak{gl}_N)$, and let $\mathrm{S}(\mathfrak{gl}_N) = \oplus \mathrm{S}^k(\mathfrak{gl}_N)$ stand for the canonical gradation of the symmetric algebra $\mathrm{S}(\mathfrak{gl}_N)$. Set

$$\mathrm{A}^k_M(N) = \mathrm{A}_M(N) \cap \mathrm{U}^k(\mathfrak{gl}_N), \qquad \mathrm{P}^k_M(N) = \mathrm{P}_M(N) \cap \mathrm{S}^k(\mathfrak{gl}_N).$$

The canonical isomorphism

$$\mathrm{U}^k(\mathfrak{gl}_N)/\mathrm{U}^{k-1}(\mathfrak{gl}_N) \cong \mathrm{S}^k(\mathfrak{gl}_N) \tag{8.6}$$

commutes with the adjoint action of $\mathfrak{gl}_N$ and hence with that of $\mathfrak{gl}_{N,M}$. Since all the spaces in (8.6) are semisimple $\mathfrak{gl}_{N,M}$-modules, we obtain the isomorphisms

$$\mathrm{A}^k_M(N)/\mathrm{A}^{k-1}_M(N) \cong \mathrm{P}^k_M(N),$$

which allows us to identify $\operatorname{gr} \mathrm{A}_M(N)$ with $\mathrm{P}_M(N)$.

Furthermore, for any $N > M$ the diagram

$$\begin{array}{ccc} \mathrm{A}^k_M(N) & \longrightarrow & \mathrm{P}^k_M(N) \\ \downarrow & & \downarrow \\ \mathrm{A}^k_M(N-1) & \longrightarrow & \mathrm{P}^k_M(N-1) \end{array}$$

is commutative, which follows immediately from the definition of the ideals $\mathrm{I}(N)$ and $\mathrm{I}'(N)$. This yields isomorphisms $\operatorname{gr} \mathrm{A}_M \cong \mathrm{P}_M$ for all $M \geqslant 0$ and, hence, an isomorphism $\operatorname{gr} \mathrm{A} \cong \mathrm{P}$. □

We will identify $\mathrm{S}(\mathfrak{gl}_N)$ with the algebra of polynomial functions on $\mathfrak{gl}_N$. We will regard the elements of this algebra as polynomials in N^2 variables x_{ij} with $i, j \in \{1, \dots, N\}$ which will be combined into the matrix $X = [x_{ij}]$. Observe that the coefficients of the following polynomials in an indeterminate q belong to the algebra $\mathrm{P}_M(N)$:

$$\det(1 + qX)_{\mathcal{B}\mathcal{B}} \qquad \text{and} \qquad \det(1 + qX)_{\mathcal{B}_i\mathcal{B}_j}, \qquad 1 \leqslant i, j \leqslant M, \tag{8.7}$$

where $\mathcal{B}_i = \{i, M+1, M+2, \dots, N\}$ and $\mathcal{B} = \{M+1, \dots, N\}$. Here, as before, we use the notation $Y_{\mathcal{P}\mathcal{Q}}$ for the submatrix of a matrix Y whose rows and columns are enumerated by the elements of sets $\mathcal{P}$ and $\mathcal{Q}$, respectively.

PROPOSITION 8.3.2. *The algebra $\mathrm{P}_M(N)$ is generated by the coefficients of the polynomials* (8.7).

PROOF. Recalling the definition of the quasideterminants from Section 1.10 and using formula (1.75) we can write

$$\det(1 + qX)_{\mathcal{B}_i\mathcal{B}_j} = \det(1 + qX)_{\mathcal{B}\mathcal{B}} \, \big|(1 + qX)_{\mathcal{B}_i\mathcal{B}_j}\big|_{ij}. \tag{8.8}$$

Furthermore, Proposition 7.3.1(i) gives

$$\big|(1 + qX)_{\mathcal{B}_i\mathcal{B}_j}\big|_{ij} = \delta_{ij} + \sum_{r=1}^{\infty} (-1)^{r-1} \Lambda^{(r)}_{ij} q^r, \tag{8.9}$$

where $\Lambda^{(r)}_{ij} = \sum x_{ik_1}x_{k_1k_2}\dots x_{k_{r-1}j}$, summed over the indices $k_l \in \{M+1,\dots,N\}$. Write X as a block matrix

$$X = \begin{bmatrix} A & B \\ C & D \end{bmatrix}$$

in accordance with the decomposition $N = M + (N - M)$. Denote by $GL_{N,M}$ the subgroup of GL_N which consists of the matrices stabilizing each of the basis vectors e_i with $1 \leqslant i \leqslant M$ in $\mathbb{C}^N$ and the space spanned by the remaining basis vectors. This subgroup is isomorphic to GL_{N-M}. It suffices to prove that any polynomial $\phi(X)$ satisfying the invariance condition

(8.10)
$$\phi(X) \equiv \phi(A, B, C, D) = \phi(A, B\mathbf{g}^{-1}, \mathbf{g}C, \mathbf{g}D\mathbf{g}^{-1}), \qquad \text{for all} \quad \mathbf{g} \in GL_{N,M},$$

is a polynomial function of the invariants $\Lambda^{(r)}_{ij}$ and the coefficients of the polynomial $\det(1 + qD)$. However, the coefficients of this polynomial generate the full set of invariants of the matrix D (this is the well known case $M = 0$ of the proposition) so that we can replace them by the invariants $\operatorname{tr} D^r$ with $r \geqslant 1$. Furthermore, note that $\Lambda^{(1)}_{ij} = a_{ij}$, while for any $r \geqslant 2$ we have $\Lambda^{(r)}_{ij} = (BD^{r-2}C)_{ij}$.

To simplify the notation, we will assume below that $M = 1$; the generalization to the case $M > 1$ will then be obvious. We can ignore the element a_{11}, so we have to show that any polynomial invariant $\phi(B, C, D)$, where C is an element of the GL_{N-1}-module V of column vectors of the length $N - 1$, B is an element of the dual module V^* of row vectors, and $D \in V \otimes V^*$, can be expressed in terms of $\operatorname{tr} D^r$ and $BD^{r-2}C$. Any such invariant can be decomposed into a sum of expressions of the form

$$\psi(\underbrace{B, \dots, B}_{p}, \underbrace{C, \dots, C}_{q}, \underbrace{D, \dots, D}_{m}),$$

where ψ is a multilinear invariant. In its turn, ψ is determined by a multilinear invariant of the form

(8.11)
$$\chi(b_1, \dots, b_p, c_1, \dots, c_q, u_1, \dots, u_m, v_1, \dots, v_m),$$

where $b_1, \dots, b_p, u_1, \dots, u_m \in V^*$ and $c_1, \dots, c_q, v_1, \dots, v_m \in V$. The multilinear invariants are described by the First Fundamental Theorem of invariants for the general linear group; see H. Weyl [**464**, Section II.6]. First of all, nonzero invariants exist only when the number of vector arguments is equal to the number of covector arguments. So in (8.11) we must have $p = q$. Furthermore, any invariant can be uniquely written as a polynomial in 'elementary invariants' that come from the canonical pairing $V^* \otimes V \to \mathbb{C}$. In our notation, this means that χ may be represented as a linear combination of monomials in bilinear invariants $b_i c_j$, $b_i v_j$, $u_i c_j$, $u_i v_j$ such that in each of the monomials each letter occurs exactly once. Hence, each monomial can be written as a product of closed chains of the form

(8.12)
$$(u_{k_1}v_{k_2})(u_{k_2}v_{k_3})\dots(u_{k_r}v_{k_1}),$$

and open chains of the form

(8.13)
$$(b_k v_{l_1})(u_{l_1}v_{l_2})\dots(u_{l_{r-1}}c_n).$$

However, under the passage $\chi \mapsto \psi \mapsto \phi$, each chain (8.12) gives rise to the invariant $\operatorname{tr} D^r$, while each chain (8.13) corresponds to an invariant of the form $BD^{r-2}C$. □

Let integers $K \geqslant 1$ and $M \geqslant 0$ be fixed.

PROPOSITION 8.3.3. *For a sufficiently large N the coefficients of $q, q^2, \dots, q^K$ of all polynomials* (8.7) *are algebraically independent.*

PROOF. Using (8.8) and arguing as in the proof of Proposition 8.3.2, we see that the claim will follow if we prove that for a sufficiently large N the elements $\operatorname{tr} D^r$ and $\Lambda_{ij}^{(r)}$ with $1 \leqslant r \leqslant K$ and $1 \leqslant i, j \leqslant M$ are algebraically independent. We will modify slightly the argument used in the proof of Theorem 1.4.1. For any triple of indices (i, j, r) satisfying the conditions choose a subset

$$\mathcal{O}_{ij}^{(r)} \subset \{M+1, M+2, \dots\}$$

of cardinality $r-1$ in such a way that all these subsets are disjoint. Let N be so large that all of them are contained in $\{M+1, M+2, \dots, N-K\}$. Let

$$y_i, \quad 1 \leqslant i \leqslant K \qquad \text{and} \qquad y_{ij}^{(r)}, \quad 1 \leqslant i, j \leqslant M, \quad 1 \leqslant r \leqslant K$$

be complex parameters. Define a linear operator $x_{ij}^{(r)}$ in $\mathbb{C}^N$ depending on $y_{ij}^{(r)}$ as follows. Let $a_1 < \dots < a_{r-1}$ be all the elements of the set $\mathcal{O}_{ij}^{(r)}$. Then put

$$x_{ij}^{(r)}: \quad e_j \mapsto y_{ij}^{(r)} e_{a_{r-1}}, \quad e_{a_{r-1}} \mapsto e_{a_{r-2}}, \quad \dots, \quad e_{a_1} \mapsto e_i,$$
$$e_k \mapsto 0 \quad \text{for} \quad k \notin \{j\} \cup \mathcal{O}_{ij}^{(r)}.$$

In the case $r = 1$, the first line above reads as $x_{ij}^{(1)} : e_j \mapsto y_{ij}^{(1)} e_i$. Now define a linear operator X in $\mathbb{C}^N$ by

$$X e_k = \begin{cases} \sum_{i,j,r} x_{ij}^{(r)} e_k & \text{if} \quad k \leqslant N-K, \\ y_{k-N+K}\, e_k & \text{if} \quad k > N-K. \end{cases} \tag{8.14}$$

Regarding X as a matrix, we get

$$\operatorname{tr} D^r = y_1^r + \dots + y_K^r,$$
$$\Lambda_{ij}^{(r)} = y_{ij}^{(r)}.$$

Hence the polynomials $\operatorname{tr} D^r$ and $\Lambda_{ij}^{(r)}$ are algebraically independent even if they are restricted to the affine subspace of matrices of the form (8.14). □

Obviously, setting $x_{iN} = 0$ for $i = 1, \dots, N$ in the determinants (8.7) we get the corresponding elements of the algebra $\mathrm{P}_M(N-1)$. Therefore, we can define the *virtual determinants*

$$\det(1+qX)_{\mathcal{B}\mathcal{B}} \qquad \text{and} \qquad \det(1+qX)_{\mathcal{B}_i\mathcal{B}_j}, \qquad 1 \leqslant i, j \leqslant M, \tag{8.15}$$

of the submatrices of the infinite matrix $1 + qX$, where $X = [x_{ij}]_{i,j=1}^{\infty}$, as formal series in q with coefficients in P_M. Here $\mathcal{B}$ and $\mathcal{B}_i$ respectively denote the infinite sets $\{M+1, M+2, \dots\}$ and $\{i, M+1, M+2, \dots\}$.

PROPOSITION 8.3.4. *For any fixed $M \geqslant 0$ the coefficients of $q, q^2, \dots$ of all series* (8.15) *are algebraically independent generators of the algebra P_M.*

PROOF. Let $p = (p_N |\ N \geqslant M) \in \mathrm{P}_M$. By Proposition 8.3.2, for any $N \geqslant M$ the element $p_N \in \mathrm{P}_M(N)$ can be represented as a polynomial ϕ_N in the coefficients of the polynomials (8.7). By Proposition 8.3.3, ϕ_N does not depend on N for sufficiently large N. This proves that the coefficients of $q, q^2, \dots$ of all series (8.15) generate the algebra P_M. Their algebraic independence is implied by Proposition 8.3.3. □

Now we will construct an alternative family of generators of the algebra P_M.

PROPOSITION 8.3.5. *The algebra* $\mathrm{P}_M(N)$ *is generated by the polynomials*

$$p_N^{(r)}(X) = \operatorname{tr} X^r \tag{8.16}$$

and

$$p_{ij|N}^{(r)}(X) = (X^r)_{ij}, \tag{8.17}$$

where $1 \leqslant i, j \leqslant M$ *and* $r \geqslant 1$.

PROOF. The elements (8.17) occur only for $M \geqslant 1$, while for $M = 0$ the claim is well known. So we will assume $M \geqslant 1$. In the notation of the proof of Proposition 8.3.2, we need to show that any polynomial $\phi(X)$ satisfying the invariance condition (8.10) is a polynomial function of the invariants $\operatorname{tr} X^r$ and $(X^r)_{ij}$ with $1 \leqslant i, j \leqslant M$. Note that these invariants can be replaced by the invariants of the form $\operatorname{tr} D^r$ with $r \geqslant 1$, $(BD^{r-2}C)_{ij}$ with $r \geqslant 2$, and a_{ij}. Indeed, we have $x_{ij} = a_{ij}$, and for $r \geqslant 2$ the invariant $(X^r)_{ij} - (BD^{r-2}C)_{ij}$ can be expressed in terms of the invariants a_{kl} and $(BD^{s-2}C)_{kl}$ with $s < r$. Hence, instead of $(X^r)_{ij}$ we can take $(BD^{r-2}C)_{ij}$ for $r \geqslant 2$. Moreover, $\operatorname{tr} X^r - \operatorname{tr} D^r$ can be expressed in terms of the invariants of the form a_{kl}, $\operatorname{tr} D^s$ with $s < r$, and $(BD^{s-2}C)_{kl}$ with $s \leqslant r$. Hence, instead of $\operatorname{tr} X^r$ we can take $\operatorname{tr} D^r$. However, for these invariants the desired property was established in the proof of Proposition 8.3.2. □

Fix integers $K \geqslant 1$ and $M \geqslant 0$ and assume that the indices i, j, r satisfy the conditions $1 \leqslant i, j \leqslant M$ and $1 \leqslant r \leqslant K$.

PROPOSITION 8.3.6. *For a sufficiently large* N *the elements* $p_{ij|N}^{(r)}$, $p_N^{(r)}$ *of the algebra* $\mathrm{P}_M(N)$ *with the indices satisfying the above conditions are algebraically independent.*

PROOF. Define a linear operator X in $\mathbb{C}^N$ in exactly the same way as in the proof of Proposition 8.3.3; see (8.14). Regarding X as a matrix, we get

$$p_N^{(r)}(X) = y_1^r + \cdots + y_K^r + \phi^{(r)},$$
$$p_{ij|N}^{(r)}(X) = y_{ij}^{(r)} + \psi_{ij}^{(r)},$$

for some polynomials $\phi^{(r)}$ and $\psi_{ij}^{(r)}$ in the $y_{kl}^{(s)}$ such that $\phi^{(r)}$ does not depend on $y_1, \ldots, y_K$, while $\psi_{ij}^{(r)}$ depends only on $y_{kl}^{(s)}$ with $s < r$. Hence the polynomials $p_N^{(r)}$ and $p_{ij|N}^{(r)}$ are algebraically independent even if they are restricted to the affine subspace of matrices of the form (8.14). □

It is straightforward from the definition of the polynomials $p_N^{(r)}$ and $p_{ij|N}^{(r)}$ that for any $r \geqslant 1$ and $N \geqslant 1$ we have

$$p_N^{(r)} - p_{N-1}^{(r)} \in \mathrm{I}'(N), \tag{8.18}$$

$$p_{ij|N}^{(r)} - p_{ij|N-1}^{(r)} \in \mathrm{I}'(N), \quad 1 \leqslant i, j \leqslant N-1. \tag{8.19}$$

We claim that the sequence

$$p^{(r)} = (p_N^{(r)}), \qquad N \geqslant 1,$$

is an element of P_0. Indeed, $p_N^{(r)}$ is a homogeneous element of $\mathrm{P}_0(N)$ of degree r so that the claim is implied by (8.18). Similarly, (8.19) implies that for any $1 \leqslant i, j \leqslant M$ and $r \geqslant 1$ the sequence

$$p_{ij}^{(r)} = (p_{ij|N}^{(r)}), \qquad N \geqslant M,$$

is an element of P_M.

PROPOSITION 8.3.7. *For any fixed $M \geqslant 0$ the elements $p^{(r)}$ and $p_{ij}^{(r)}$ with $r \geqslant 1$ and $1 \leqslant i, j \leqslant M$ are algebraically independent generators of the algebra P_M.*

PROOF. Let $p = (p_N|\ N \geqslant M) \in \mathrm{P}_M$. By Proposition 8.3.5, for any $N \geqslant M$ the element $p_N \in \mathrm{P}_M(N)$ can be represented as a polynomial ϕ_N in the variables $p_N^{(r)}$, $p_{ij|N}^{(r)}$, where $M \leqslant \deg p$ and $1 \leqslant i, j \leqslant M$. By Proposition 8.3.6, ϕ_N does not depend on N for sufficiently large N. This proves that the elements $p^{(r)}$ and $p_{ij}^{(r)}$ generate the algebra P_M. Their algebraic independence is implied by Proposition 8.3.6. □

COROLLARY 8.3.8. *The algebra P is isomorphic to the algebra of polynomials in countably many variables $p^{(r)}$, $p_{ij}^{(r)}$, where $r \geqslant 1$ and $i, j \geqslant 1$.*

PROOF. This is immediate from Proposition 8.3.7. □

8.4. Algebraic structure of $\mathrm{A}(\mathfrak{gl}_\infty)$

We are now in a position to establish a relationship between the Olshanski algebra $\mathrm{A} = \mathrm{A}(\mathfrak{gl}_\infty)$ and the Yangian for the general linear Lie algebras.

As in Section 1.12, for any $1 \leqslant M \leqslant N$ introduce the series

$$t_{ij}^{\flat}(u) = t\,{}^{i,M+1\dots N}_{j,M+1\dots N}(u), \qquad 1 \leqslant i, j \leqslant M,$$

with coefficients in $\mathrm{Y}(\mathfrak{gl}_N)$. Theorem 1.12.2 provides the homomorphism $\mathrm{Y}(\mathfrak{gl}_M) \to \mathrm{Y}(\mathfrak{gl}_N)$ given by $t_{ij}(u) \mapsto t_{ij}^{\flat}(u)$. The composition of this homomorphism with the evaluation map (1.5) is a homomorphism $\psi_N : \mathrm{Y}(\mathfrak{gl}_M) \to \mathrm{U}(\mathfrak{gl}_N)$ which can be given by

$$\psi_N : t_{ij}(u) \mapsto (1 + Eu^{-1})^{i,M+1\dots N}_{j,M+1\dots N}, \tag{8.20}$$

where E is the infinite matrix $[E_{ij}]$, as in Section 8.2. Using the explicit formulas (1.54) for the quantum minors we can also write (8.20) in the form

$$\psi_N : t_{ij}(u) \mapsto \mathrm{qdet}\,(1 + Eu^{-1})_{\mathcal{B}_i\mathcal{B}_j}, \tag{8.21}$$

regarding the quantum minor in (8.20) as the quantum determinant of the submatrix of $1 + Eu^{-1}$ corresponding to the rows $\mathcal{B}_i$ and columns $\mathcal{B}_j$, where $\mathcal{B}_i$ denotes the set $\{i, M+1, \dots, N\}$. By Corollary 1.7.2, the image of ψ_N commutes with the Lie algebra $\mathfrak{gl}_{N,M}$. Hence, ψ_N defines a homomorphism

$$\psi_N : \mathrm{Y}(\mathfrak{gl}_M) \to \mathrm{A}_M(N).$$

By the explicit formula for $\mathrm{qdet}\,(1 + Eu^{-1})_{\mathcal{B}_i\mathcal{B}_j}$ provided by (1.54), the diagram

$$\begin{array}{ccccccccc}
\mathrm{Y}(\mathfrak{gl}_M) & = & \mathrm{Y}(\mathfrak{gl}_M) & = & \cdots & = & \mathrm{Y}(\mathfrak{gl}_M) & = & \cdots \\
\Big\downarrow{\scriptstyle \psi_M} & & \Big\downarrow{\scriptstyle \psi_{M+1}} & & & & \Big\downarrow{\scriptstyle \psi_N} & & \\
\mathrm{A}_M(M) & \xleftarrow[\ o_{M+1}\ ]{} & \mathrm{A}_M(M+1) & \longleftarrow & \cdots & \xleftarrow[\ o_N\ ]{} & \mathrm{A}_M(N) & \xleftarrow[\ o_{N+1}\ ]{} & \cdots
\end{array}$$

is commutative. Furthermore, the image of $t_{ij}^{(r)}$ under ψ_N clearly has degree $\leqslant r$ for any N. Hence the sequence of homomorphisms $(\psi_N \mid N \geqslant M)$ defines an algebra homomorphism $\psi : \mathrm{Y}(\mathfrak{gl}_M) \to \mathrm{A}_M$ which can be written in terms of the virtual quantum determinants by

$$\psi : t_{ij}(u) \mapsto \mathrm{qdet}\,(1 + Eu^{-1})_{\mathcal{B}_i\mathcal{B}_j},$$

where $\mathcal{B}_i$ now denotes the infinite set $\{i, M+1, M+2, \dots\}$; see Section 8.2. Also, denote by $\widetilde{\mathrm{A}}_0$ the commutative subalgebra of A_M generated by the coefficients of the virtual quantum determinant $\mathrm{qdet}\,(1+Eu^{-1})_{\mathcal{B}\mathcal{B}}$ with $\mathcal{B} = \{M+1, M+2, \dots\}$. By Proposition 8.2.1, the algebra $\widetilde{\mathrm{A}}_0$ is isomorphic to the algebra of shifted symmetric functions in the variables $\lambda_{M+1}, \lambda_{M+2}, \dots$.

THEOREM 8.4.1. *The homomorphism ψ is an algebra embedding of $\mathrm{Y}(\mathfrak{gl}_M)$ into the algebra A_M. Moreover, we have the tensor product decomposition*

$$\mathrm{A}_M = \widetilde{\mathrm{A}}_0 \otimes \mathrm{Y}(\mathfrak{gl}_M), \tag{8.22}$$

where the Yangian $\mathrm{Y}(\mathfrak{gl}_M)$ is identified with its image under the embedding ψ.

PROOF. The coefficient of u^{-r} of the series $\mathrm{qdet}\,(1 + Eu^{-1})_{\mathcal{B}_i\mathcal{B}_j}$ is an element of A_M of degree $\leqslant r$. Recall that A_M is a filtered algebra with $\mathrm{gr}\,\mathrm{A}_M \cong \mathrm{P}_M$; see Proposition 8.3.1. Considering the highest order term of the coefficient we find that its image in the r-th component of P_M coincides with the coefficient of u^{-r} of the virtual determinant $\det(1 + Xu^{-1})_{\mathcal{B}_i\mathcal{B}_j}$. Since the algebra $\mathrm{gr}\,\mathrm{A}_M = \mathrm{P}_M$ is commutative, we obtain from Proposition 8.3.4 that the elements $\psi(t_{ij}^{(r)})$ of the algebra A_M satisfy a Poincaré–Birkhoff–Witt-type condition: for any linear ordering on the set

$$\{\psi(t_{ij}^{(r)}) \mid 1 \leqslant i, j \leqslant M, \quad r \geqslant 1\},$$

each element of the subalgebra of A_M generated by this set has a unique representation as a linear combination of ordered monomials in the $\psi(t_{ij}^{(r)})$. Applying Theorem 1.4.1, we conclude that the homomorphism ψ is injective.

Similarly, the image of the coefficient of u^{-r} of the series $\mathrm{qdet}\,(1 + Eu^{-1})_{\mathcal{B}\mathcal{B}}$ in the r-th component of P_M coincides with the coefficient of u^{-r} of the virtual determinant $\det(1 + Xu^{-1})_{\mathcal{B}\mathcal{B}}$. Therefore, the decomposition (8.22) is implied by Proposition 8.3.4. □

We will now construct a different embedding of the Yangian $\mathrm{Y}(\mathfrak{gl}_M)$ into the algebra A_M and prove the corresponding analogue of Theorem 8.4.1.

Consider the automorphisms of the Yangian $\mathrm{Y}(\mathfrak{gl}_N)$ given by

$$T(u) \mapsto T(u+N) \qquad \text{and} \qquad T(u) \mapsto T^{-1}(-u);$$

see (1.21) and (1.31). Their composition gives another automorphism

$$T(u) \mapsto T^{-1}(-u-N).$$

Its composition with the evaluation homomorphism (1.5) yields a homomorphism $\varphi_N : \mathrm{Y}(\mathfrak{gl}_N) \to \mathrm{U}(\mathfrak{gl}_N)$ given by

$$\varphi_N : T(u) \mapsto \Big(1 - \frac{E}{u+N}\Big)^{-1}. \tag{8.23}$$

By Corollary 1.4.3, for any $M \leqslant N$ the Yangian $\mathrm{Y}(\mathfrak{gl}_M)$ can be regarded as a natural subalgebra of $\mathrm{Y}(\mathfrak{gl}_N)$.

PROPOSITION 8.4.2. *The image of the restriction of the homomorphism φ_N to the subalgebra $\mathrm{Y}(\mathfrak{gl}_M)$ is contained in the centralizer $\mathrm{A}_M(N)$.*

PROOF. Taking the coefficient of v^0 in (1.3) we deduce that

$$[t_{kl}^{(1)}, t_{ij}(u)] = \delta_{il}\, t_{kj}(u) - \delta_{kj}\, t_{il}(u). \tag{8.24}$$

In particular,

$$[t_{kl}^{(1)}, t_{ij}(u)] = 0 \qquad \text{for} \quad i, j \leqslant M < k, l.$$

On the other hand, (8.23) implies

$$\varphi_N : t_{kl}^{(1)} \mapsto E_{kl}. \tag{8.25}$$

Hence, for any $i, j \leqslant M$ the coefficients of the series $\varphi_N(t_{ij}(u))$ commute with all elements of the subalgebra $\mathfrak{gl}_{N,M}$. □

Proposition 8.4.2 implies that for any $N \geqslant M$ we have a homomorphism of the algebra $\mathrm{Y}(\mathfrak{gl}_M)$ to $\mathrm{A}_M(N)$ obtained by the restriction of φ_N.

THEOREM 8.4.3. *For any fixed $M \geqslant 1$ the sequence $(\varphi_N \mid N \geqslant M)$ defines an algebra embedding $\varphi : \mathrm{Y}(\mathfrak{gl}_M) \hookrightarrow \mathrm{A}_M$. Moreover, we have the tensor product decomposition*

$$\mathrm{A}_M = \mathrm{A}_0 \otimes \mathrm{Y}(\mathfrak{gl}_M), \tag{8.26}$$

where the Yangian $\mathrm{Y}(\mathfrak{gl}_M)$ is identified with its image under the embedding φ.

PROOF. In order to prove that the sequence of homomorphisms $(\varphi_N \mid N \geqslant M)$ defines an algebra homomorphism $\varphi : \mathrm{Y}(\mathfrak{gl}_M) \to \mathrm{A}_M$ we need to verify that the following diagram is commutative:

$$\begin{array}{ccccccccc}
\mathrm{Y}(\mathfrak{gl}_M) & = & \mathrm{Y}(\mathfrak{gl}_M) & = & \cdots & = & \mathrm{Y}(\mathfrak{gl}_M) & = & \cdots \\
{\scriptstyle\varphi_M}\downarrow & & {\scriptstyle\varphi_{M+1}}\downarrow & & & & {\scriptstyle\varphi_N}\downarrow & & \\
\mathrm{A}_M(M) & \xleftarrow[o_{M+1}]{} & \mathrm{A}_M(M+1) & \longleftarrow & \cdots & \xleftarrow[o_N]{} & \mathrm{A}_M(N) & \xleftarrow[o_{N+1}]{} & \cdots .
\end{array}$$

Denote the image of $t_{ij}(u)$ under the homomorphism (8.23) by $\tau_{ij|N}(u)$ and prove by induction on r that for the coefficients of this series the following holds:

$$\tau_{iN|N}^{(r)} \in \mathrm{I}(N), \qquad 1 \leqslant i \leqslant N, \quad r \geqslant 1, \tag{8.27}$$

$$\tau_{ij|N}^{(r)} - \tau_{ij|N-1}^{(r)} \in \mathrm{I}(N), \qquad 1 \leqslant i, j \leqslant N-1, \quad r \geqslant 1. \tag{8.28}$$

Introduce the matrix $\mathcal{T}(u) = [\tau_{ij|N}(u)]_{i,j=1}^N$. We have

$$\mathcal{T}(u)(u + N - E) = u + N.$$

Hence

$$u\,\mathcal{T}(u) = u + N + \mathcal{T}(u)(E - N). \tag{8.29}$$

Write $\mathcal{T}(u) = \mathcal{T}^{(0)} + \mathcal{T}^{(1)}u^{-1} + \dots$. Then by (8.29), $\mathcal{T}^{(0)} = 1$ while

$$\mathcal{T}^{(1)} = E,$$

$$\mathcal{T}^{(r)} = \mathcal{T}^{(r-1)}(E - N), \quad r \geqslant 2. \tag{8.30}$$

Therefore,

$$\tau_{iN|N}^{(1)} = E_{iN} \in \mathrm{I}(N), \quad 1 \leqslant i \leqslant N,$$

and

$$\tau_{ij|N}^{(1)} - \tau_{ij|N-1}^{(1)} = 0, \quad 1 \leqslant i,j \leqslant N-1.$$

So, we have verified (8.27) and (8.28) for $r = 1$. For $r > 1$ we obtain from (8.30) that

$$\begin{aligned}\tau_{iN|N}^{(r)} &= \sum_{a=1}^{N} \tau_{ia|N}^{(r-1)} (E_{aN} - \delta_{aN} N) \\ &= \sum_{a=1}^{N-1} \tau_{ia|N}^{(r-1)} E_{aN} + \tau_{iN|N}^{(r-1)} E_{NN} - N \tau_{iN|N}^{(r-1)}.\end{aligned}$$

By the induction hypotheses, this expression lies in $\mathrm{I}(N)$, which proves (8.27). Using again (8.30), we obtain for $1 \leqslant i,j \leqslant N-1$ that

$$\tau_{ij|N}^{(r)} - \tau_{ij|N-1}^{(r)} = \sum_{a=1}^{N} \tau_{ia|N}^{(r-1)} \big(E_{aj} - \delta_{aj} N\big) - \sum_{a=1}^{N-1} \tau_{ia|N-1}^{(r-1)} \big(E_{aj} - \delta_{aj}(N-1)\big),$$

which can be rewritten as

$$\sum_{a=1}^{N-1} \big(\tau_{ia|N}^{(r-1)} - \tau_{ia|N-1}^{(r-1)}\big) E_{aj} - (N-1)\big(\tau_{ij|N}^{(r-1)} - \tau_{ij|N-1}^{(r-1)}\big) + \tau_{iN|N}^{(r-1)} E_{Nj} - \tau_{ij|N}^{(r-1)}.$$

Since φ_N is an algebra homomorphism, we obtain from (8.24) and (8.25) that

$$\begin{aligned}\big(\tau_{ia|N}^{(r-1)} - \tau_{ia|N-1}^{(r-1)}\big) E_{aj} = E_{aj} \big(\tau_{ia|N}^{(r-1)} - \tau_{ia|N-1}^{(r-1)}\big) \\ + \tau_{ij|N}^{(r-1)} - \tau_{ij|N-1}^{(r-1)} - \delta_{ij}\big(\tau_{aa|N}^{(r-1)} - \tau_{aa|N-1}^{(r-1)}\big),\end{aligned}$$

while

$$\begin{aligned}\tau_{iN|N}^{(r-1)} E_{Nj} - \tau_{ij|N}^{(r-1)} &= E_{Nj} \tau_{iN|N}^{(r-1)} + [\tau_{iN|N}^{(r-1)}, E_{Nj}] - \tau_{ij|N}^{(r-1)} \\ &= E_{Nj} \tau_{iN|N}^{(r-1)} - \delta_{ij} \tau_{NN|N}^{(r-1)}.\end{aligned}$$

Hence, using (8.27) and the induction hypothesis, we complete the proof of (8.28). In particular, we have proved that for $i,j \leqslant M$ the sequence $(\tau_{ij|N}^{(r)} \mid N \geqslant M)$ is an element of the algebra A_M, and so the homomorphism φ is well-defined.

As the next step of the argument, we verify that φ is an embedding; that is, its kernel is trivial. As in the proof of Theorem 8.4.1 we make use of the filtration on A_M and pass to the corresponding graded algebra P_M; see Proposition 8.3.1. Consider the highest degree component of the sequence $\tau_{ij}^{(r)} = (\tau_{ij|N}^{(r)} \mid N \geqslant M)$. By the definition of $\tau_{ij|N}(u)$, $\tau_{ij|N}^{(r)}$ has the form

$$\tau_{ij|N}^{(r)} = \big(E^r\big)_{ij} + \sum_{k \geqslant 1} c_k \big(E^{r-k}\big)_{ij}, \qquad c_k \in \mathbb{C}.$$

This proves that the image of $\tau_{ij|N}^{(r)}$ in the r-th component of the graded algebra $\mathrm{S}(\mathfrak{gl}_N) = \operatorname{gr} \mathrm{U}(\mathfrak{gl}_N)$ is $p_{ij|N}^{(r)}$. Hence, the highest order term of $\tau_{ij}^{(r)}$ coincides with $p_{ij}^{(r)}$. As in the proof of Theorem 8.4.1, we obtain from Proposition 8.3.7 that the homomorphism φ taking $t_{ij}^{(r)}$ to $\tau_{ij}^{(r)}$ is injective.

Finally, the previous argument shows that the graded algebra $\operatorname{gr} \mathrm{Y}(\mathfrak{gl}_M)$ can be identified with the subalgebra of $\operatorname{gr} \mathrm{A}_M = \mathrm{P}_M$ generated by the elements $p_{ij}^{(r)}$. So, the decomposition (8.26) follows from Proposition 8.3.7. □

Using the embeddings $\mathrm{Y}(\mathfrak{gl}_M) \hookrightarrow \mathrm{Y}(\mathfrak{gl}_{M+1})$ provided by Corollary 1.4.3, define the *Yangian for* $\mathfrak{gl}_\infty$ as the inductive limit algebra

$$\mathrm{Y}(\mathfrak{gl}_\infty) = \bigcup_{M \geqslant 1} \mathrm{Y}(\mathfrak{gl}_M).$$

The following corollary which is immediate from Theorem 8.4.3 yields a description of the Olshanski algebra A.

COROLLARY 8.4.4. *We have the isomorphism* $\mathrm{A} \cong \mathrm{A}_0 \otimes \mathrm{Y}(\mathfrak{gl}_\infty)$. □

REMARK 8.4.5. Since the Yangian $\mathrm{Y}(\mathfrak{gl}_M)$ has a nontrivial center (see Theorem 1.7.5), the center of the algebra A_M strictly contains A_0. However, the center of the algebra $\mathrm{Y}(\mathfrak{gl}_\infty)$ is trivial, since by Proposition 8.1.3 the center of the algebra A coincides with A_0.

Note also that in contrast to the algebra $\mathrm{Y}(\mathfrak{gl}_M)$, the 'infinite' Yangian $\mathrm{Y}(\mathfrak{gl}_\infty)$ apparently does not possess a Hopf algebra structure. □

8.5. Skew representations of $\mathrm{Y}(\mathfrak{gl}_N)$

In this section we employ a dual form of the homomorphism $t_{ij}(u) \mapsto t^{\flat}_{ij}(u)$ used in Section 8.4. More precisely, we adopt the notation of Section 1.11 and for $M \geqslant 0$ consider the homomorphism $\mathrm{Y}(\mathfrak{gl}_N) \to \mathrm{Y}(\mathfrak{gl}_{M+N})$ given by

$$t_{ij}(u) \mapsto \big(t^{\,1\dots\, M}_{\,1\dots\, M}(u)\big)^{-1} \cdot t^{\,1\dots\, M, M+i}_{\,1\dots\, M, M+j}(u);$$

see (1.21) and Corollary 1.11.4. Its composition with the evaluation homomorphism (1.5) is a homomorphism $\mathrm{Y}(\mathfrak{gl}_M) \to \mathrm{U}(\mathfrak{gl}_{M+N})$ which can be given by

$$t_{ij}(u) \mapsto \mathrm{qdet}\,\big(1 + E\,u^{-1}\big)^{-1}_{\mathcal{A}\mathcal{A}} \cdot \mathrm{qdet}\,\big(1 + E\,u^{-1}\big)_{\mathcal{A}_i\mathcal{A}_j}, \tag{8.31}$$

where $\mathcal{A} = \{1, \dots, M\}$ and $\mathcal{A}_i = \{1, \dots, M, M+i\}$. By Lemma 1.11.2, the image of $t^{(1)}_{ij}$ under the homomorphism (8.31) is $E_{M+i,M+j}$. Therefore, Corollary 1.7.2 implies that the image of this homomorphism is contained in the centralizer $\mathrm{U}(\mathfrak{gl}_{M+N})^{\mathfrak{gl}_M}$, where $\mathfrak{gl}_M$ is regarded as a natural subalgebra of $\mathfrak{gl}_{M+N}$.

Consider now a $\mathfrak{gl}_{M+N}$-highest weight $\lambda = (\lambda_1, \dots, \lambda_{M+N})$ so that the λ_i are complex numbers satisfying the condition $\lambda_i - \lambda_{i+1} \in \mathbb{Z}_+$ for all values of the index $i = 1, \dots, M+N-1$. Furthermore, let $\mu = (\mu_1, \dots, \mu_M)$ be a $\mathfrak{gl}_M$-highest weight. The vector space $\mathrm{Hom}_{\mathfrak{gl}_M}\big(L(\mu), L(\lambda)\big)$ is isomorphic to the subspace $L(\lambda)^+_\mu$ of $L(\lambda)$ which consists of $\mathfrak{gl}_M$-highest vectors of weight μ,

$$\begin{aligned} L(\lambda)^+_\mu = \{\eta \in L(\lambda) \mid E_{ij}\,\eta &= 0 \qquad 1 \leqslant i < j \leqslant M \qquad \text{and} \\ E_{ii}\,\eta &= \mu_i\,\eta \qquad i = 1, \dots, M\}. \end{aligned}$$

The subspace $L(\lambda)^+_\mu$ is nonzero if and only if

$$\lambda_i - \mu_i \in \mathbb{Z}_+ \qquad \text{and} \qquad \mu_i - \lambda_{i+N} \in \mathbb{Z}_+ \qquad \text{for} \quad i = 1, \dots, M. \tag{8.32}$$

This follows easily from the properties of the Gelfand–Tsetlin basis of $L(\lambda)$; see Section 5.4. We will suppose that the conditions (8.32) hold. Observe that $L(\lambda)^+_\mu$ is a natural module over the centralizer $\mathrm{U}(\mathfrak{gl}_{M+N})^{\mathfrak{gl}_M}$. We now make $L(\lambda)^+_\mu$ into a module over the Yangian $\mathrm{Y}(\mathfrak{gl}_N)$ via the homomorphism (8.31).

PROPOSITION 8.5.1. *The* $\mathrm{Y}(\mathfrak{gl}_N)$*-module* $L(\lambda)^+_\mu$ *is irreducible.*

PROOF. It is well known that $L(\lambda)^+_\mu$ is an irreducible representation of the centralizer $\mathrm{U}(\mathfrak{gl}_{M+N})^{\mathfrak{gl}_M}$; see Dixmier [**89**, Section 9.1]. Since the elements of the center of $\mathrm{U}(\mathfrak{gl}_M)$ act on $L(\lambda)^+_\mu$ as scalar operators, the claim will follow if we show that the centralizer is generated by the image of the homomorphism (8.31) and the center of $\mathrm{U}(\mathfrak{gl}_M)$. However, Theorem 7.1.1 implies that the center of $\mathrm{U}(\mathfrak{gl}_M)$ is generated by the coefficients of the series $\operatorname{qdet}(1+E\,u^{-1})_{\mathcal{A}\mathcal{A}}$. Therefore, the obvious modification of Proposition 8.3.2 implies that the coefficients of the series

$$\operatorname{qdet}(1+E\,u^{-1})_{\mathcal{A}\mathcal{A}} \qquad \text{and} \qquad \operatorname{qdet}(1+E\,u^{-1})_{\mathcal{A}_i\mathcal{A}_j}, \qquad 1\leqslant i,j\leqslant N,$$

generate the centralizer $\mathrm{U}(\mathfrak{gl}_{M+N})^{\mathfrak{gl}_M}$; cf. the arguments used in Section 8.4 for the proof of Theorem 8.4.1. This implies the statement. □

The subspace $L(\lambda)^+_\mu$ admits a basis $\{\zeta_\Lambda\}$ labelled by the trapezium Gelfand–Tsetlin patterns Λ of the form

$$\begin{array}{cccccccccc}
\lambda_1 & & \lambda_2 & & \lambda_3 & & \cdots & & \lambda_{M+N-1} & & \lambda_{M+N} \\
& \lambda_{M+N-1,1} & & \lambda_{M+N-1,2} & & & \cdots & & \lambda_{M+N-1,M+N-1} & \\
& & \ddots & & \ddots & & \cdots & & \cdot^{\,\cdot^{\,\cdot}} & \\
& & & \lambda_{M+1,1} & & \lambda_{M+1,2} & \cdots & \lambda_{M+1,M+1} & & \\
& & & & \mu_1 & \mu_2 & \cdots & \mu_M & &
\end{array}$$

These arrays are formed by complex numbers λ_{ki} satisfying the betweenness conditions

$$\lambda_{ki}-\lambda_{k+1,i+1}\in\mathbb{Z}_+ \qquad \text{and} \qquad \lambda_{k+1,i}-\lambda_{ki}\in\mathbb{Z}_+ \tag{8.33}$$

for $k=M,M+1,\dots,M+N-1$ and $1\leqslant i\leqslant k$, where we have set

$$\lambda_{Mi}=\mu_i, \quad i=1,\dots,M \qquad \text{and} \qquad \lambda_{M+N,j}=\lambda_j, \quad j=1,\dots,M+N.$$

Note that for fixed λ and μ all entries of Λ belong to the same $\mathbb{Z}$-coset of $\mathbb{C}$. Ordering the elements of the coset by their real parts we can rewrite (8.33) in an equivalent form as

$$\lambda_{k+1,i+1}\leqslant\lambda_{ki}\leqslant\lambda_{k+1,i}.$$

Using this convention, introduce the trapezium array Λ^0 with the entries given by

$$\lambda^0_{M+k,i}=\min\{\lambda_i,\mu_{i-k}\}, \qquad k=1,\dots,N-1, \quad i=1,\dots,M+k, \tag{8.34}$$

where we assume that μ_j is sufficiently large if $j\leqslant 0$. It is easy to check that Λ^0 is a pattern.

PROPOSITION 8.5.2. *The basis vector ζ_{Λ^0} is the highest vector of the* $\mathrm{Y}(\mathfrak{gl}_N)$-*module $L(\lambda)^+_\mu$.*

PROOF. The vector space $L(\lambda)^+_\mu$ is a natural module over the subalgebra $\mathrm{U}(\mathfrak{gl}_N)$ of $\mathrm{Y}(\mathfrak{gl}_N)$. The action of $\mathfrak{gl}_N$ on the basis vectors is given by Theorem 5.4.1, where $\mathfrak{gl}_N$ is identified with the subalgebra of $\mathfrak{gl}_{M+N}$ spanned by the elements $E_{M+i,M+j}$ with $1\leqslant i,j\leqslant N$. In particular, the $\mathfrak{gl}_N$-weights of the basis vectors ζ_Λ are found by the formulas (5.39). These formulas imply that ζ_{Λ^0} has a maximal weight with respect to the standard partial ordering on the set of weights of the $\mathfrak{gl}_N$-module $L(\lambda)^+_\mu$; see Section 3.2. Hence, by (3.60), the vector ζ_{Λ^0} is annihilated by $t_{ij}(u)$ with $1\leqslant i<j\leqslant N$. The proof is now completed by the application of Proposition 8.5.1 and Corollary 3.2.8. □

REMARK 8.5.3. An explicit expression for the highest vector can be produced with the use of the Mickelsson algebra $\mathrm{S}(\mathfrak{gl}_{M+N}, \mathfrak{gl}_M)$; see Example 9.7.2 below. □

Given three elements i, j, k of a $\mathbb{Z}$-coset in $\mathbb{C}$, we will denote by $\mathrm{mid}\{i, j, k\}$ that element which is between the other two. For instance, $\mathrm{mid}\{1, 1, 2\} = 1$.

THEOREM 8.5.4. *The* $\mathrm{Y}(\mathfrak{gl}_N)$*-module* $L(\lambda)^+_\mu$ *is isomorphic to the highest weight representation* $L(\nu(u))$*, where the components of the highest weight* $\nu(u)$ *are found by*

$$\nu_k(u) = \frac{(u+\nu_k^{(1)})(u+\nu_k^{(2)}-1)\dots(u+\nu_k^{(M+1)}-M)}{(u+\mu_1)(u+\mu_2-1)\dots(u+\mu_M-M+1)(u-M)}$$

where $k = 1, \dots, N$ *and*

$$\nu_k^{(i)} = \mathrm{mid}\{\mu_{i-1}, \mu_i, \lambda_{k+i-1}\}, \tag{8.35}$$

assuming μ_0 *is sufficiently large and* μ_{m+1} *is sufficiently small.*

PROOF. By Propositions 8.5.1 and 8.5.2, we only need to calculate the highest weight of the $\mathrm{Y}(\mathfrak{gl}_N)$-module $L(\lambda)^+_\mu$. As ζ_{Λ^0} is the highest vector, by (1.54) we have for any $1 \leqslant k \leqslant N$

$$t^{1\dots k}_{1\dots k}(u)\,\zeta_{\Lambda^0} = \nu_1(u)\,\nu_2(u-1)\dots\nu_k(u-k+1)\,\zeta_{\Lambda^0}.$$

On the other hand, by Lemma 1.11.3, the action of this quantum minor can also be found by applying the evaluation homomorphism (1.5) to the following series with coefficients in $\mathrm{Y}(\mathfrak{gl}_{M+N})$:

$$\big(t^{1\dots M}_{1\dots M}(u)\big)^{-1}\cdot t^{1\dots M,\,M+1\dots M+k}_{1\dots M,\,M+1\dots M+k}(u), \tag{8.36}$$

and then applying the image to the vector ζ_{Λ^0}. Using (5.42) we come to the relation

$$\nu_1(u)\,\nu_2(u-1)\dots\nu_k(u-k+1)$$
$$= \frac{(1+\lambda^0_{M+k,1}u^{-1})\dots(1+\lambda^0_{M+k,M+k}(u-M-k+1)^{-1})}{(1+\mu_1u^{-1})\dots(1+\mu_M(u-M+1)^{-1})}.$$

Replacing here k by $k-1$ we derive a formula for $\nu_k(u-k+1)$ which implies

$$\nu_k(u) = \frac{(1+\lambda^0_{M+k,1}(u+k-1)^{-1})\dots(1+\lambda^0_{M+k,M+k}(u-M)^{-1})}{(1+\lambda^0_{M+k-1,1}(u+k-1)^{-1})\dots(1+\lambda^0_{M+k-1,M+k-1}(u-M+1)^{-1})}.$$

Note that

$$\lambda^0_{M+k,i} = \lambda^0_{M+k-1,i} = \lambda_i \qquad \text{for} \quad i = 1, \dots, k-1$$

and

$$\frac{1+\lambda^0_{M+k,k+j-1}(u-j+1)^{-1}}{1+\lambda^0_{M+k-1,k+j-1}(u-j+1)^{-1}} = \frac{1+\nu_k^{(j)}(u-j+1)^{-1}}{1+\mu_j(u-j+1)^{-1}}$$

for $j = 1, \dots, M$. Since $\lambda^0_{M+k,M+k} = \nu_k^{(M+1)}$ this gives the desired formula. □

In the case where the components of λ and μ are nonnegative integers, we may regard them as partitions; see Section 6.4 for the definitions. The first condition in (8.32) implies that the diagram of μ is contained in the diagram of λ. The *skew diagram* λ/μ is the set-theoretical difference of the diagrams of λ and μ. As in Section 6.4, by the content of a cell $\alpha \in \lambda/\mu$ we mean the number $c(\alpha) = j - i$ if α occurs in row i and column j.

COROLLARY 8.5.5. *The Drinfeld polynomials $P_1(u), \dots, P_{N-1}(u)$ of the representation $L(\lambda)^+_\mu$ of $\mathrm{Y}(\mathfrak{gl}_N)$ are given by*

$$P_k(u) = \prod_{i=1}^{M+1} (u + \nu^{(i)}_{k+1} - i + 1)(u + \nu^{(i)}_{k+1} - i + 2) \dots (u + \nu^{(i)}_k - i),$$

for $k = 1, \dots, N-1$. If λ and μ are partitions, then the formula can also be written as

$$P_k(u) = \prod_\alpha (u + c(\alpha)),$$

where α runs over the top cells of the columns of height k in the diagram of λ/μ.

PROOF. The formulas follow from Theorem 8.5.4 and the relations between the highest weight and Drinfeld polynomials of a representation of $\mathrm{Y}(\mathfrak{gl}_N)$; see (3.98). □

EXAMPLE 8.5.6. If $\lambda = (10, 8, 5, 4, 2)$ and $\mu = (6, 3)$, then the corresponding skew diagram is

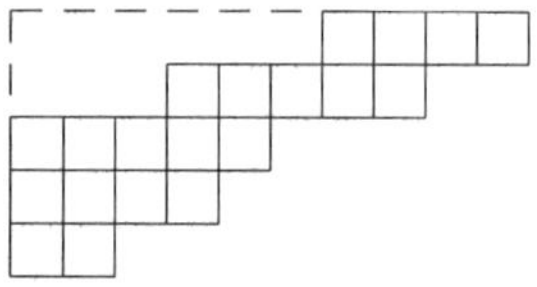

The Drinfeld polynomials of the $\mathrm{Y}(\mathfrak{gl}_3)$-module $L(\lambda)^+_\mu$ are

$$P_1(u) = (u+4)(u+8)(u+9), \qquad P_2(u) = u(u+3)(u+6)(u+7). \qquad \square$$

Note that in the case $M = 0$ the skew representation $L(\lambda)^+_\mu$ of $\mathrm{Y}(\mathfrak{gl}_N)$ coincides with the evaluation module $L(\lambda)$. Obviously, the formulas for the Drinfeld polynomials in this case agree with those given in Example 3.4.4.

For a given N consider the set $\mathcal{P}_N$ of tuples $\lambda(u) = \big(\lambda_1(u), \dots, \lambda_N(u)\big)$ where each $\lambda_i(u)$ is a formal series in u^{-1} of the form (3.56). Then $\mathcal{P}_N$ is an abelian group with respect to the component-wise multiplication of the tuples. Consider the group ring $\mathbb{Z}[\mathcal{P}_N]$ of the abelian group $\mathcal{P}_N$ whose elements are finite linear combinations of the form $\sum m_{\lambda(u)}[\lambda(u)]$, where $m_{\lambda(u)} \in \mathbb{Z}$.

Now recall the Gelfand–Tsetlin subalgebra H_N of $\mathrm{Y}(\mathfrak{gl}_N)$ generated by the coefficients $h^{(r)}_i$ of the series $h_1(u), \dots, h_N(u)$ defined in Theorem 1.11.6; see also Section 1.13.

DEFINITION 8.5.7. Suppose that V is a finite-dimensional representation of the Yangian $\mathrm{Y}(\mathfrak{gl}_N)$. For any $\lambda(u) \in \mathcal{P}_N$, the corresponding *Gelfand–Tsetlin subspace* $V_{\lambda(u)}$ consists of the vectors $v \in V$ with the property that for each $i = 1, \dots, N$ and each $r \geqslant 1$ there exists $p \geqslant 1$ such that $\big(h^{(r)}_i - \lambda^{(r)}_i\big)^p\, v = 0$. Then the *Gelfand–Tsetlin character* of V is the element of $\mathbb{Z}[\mathcal{P}_N]$ defined by

$$\mathrm{ch}\, V = \sum_{\lambda(u) \in \mathcal{P}_N} \big(\dim V_{\lambda(u)}\big)\, [\lambda(u)].$$

□

We conclude this section with a calculation of the Gelfand–Tsetlin character of the skew representation $L(\lambda)^+_\mu$ of the Yangian. It follows from Theorem 8.5.4 that the composition of the $\mathrm{Y}(\mathfrak{gl}_N)$-module $L(\lambda)^+_\mu$ with a shift automorphism (1.21) is isomorphic to the skew representation $L(\lambda')^+_{\mu'}$ with the shifted components of λ and μ,

$$\lambda'_i = \lambda_i - c, \qquad \mu'_i = \mu_i - c.$$

So we may assume without loss of generality that λ and μ are partitions. The formulas in the general case can then be obtained by applying an appropriate shift in u.

Introduce the following special elements of the group ring $\mathbb{Z}[\mathcal{P}_N]$ by

$$x_{i,a} = \big[(1,\dots,1+(u+a+i-1)^{-1},\dots,1)\big], \qquad 1 \leqslant i \leqslant N, \quad a \in \mathbb{C},$$

where the only series not equal to 1 is the i-th component of the tuple.

A *semistandard λ/μ-tableau $\mathcal{T}$* is obtained by writing the numbers $1,\dots,N$ into the cells of the skew diagram λ/μ in such a way that the elements in each row weakly increase while the elements in each column strictly increase. By $\mathcal{T}(\alpha)$ we denote the entry of $\mathcal{T}$ in the cell $\alpha \in \lambda/\mu$.

COROLLARY 8.5.8. *The Gelfand–Tsetlin character of the* Y($\mathfrak{gl}_N$)*-module $L(\lambda)^+_\mu$ is given by*

$$\operatorname{ch} L(\lambda)^+_\mu = \sum_{\mathcal{T}} \prod_{\alpha\in\lambda/\mu} x_{\mathcal{T}(\alpha),c(\alpha)},$$

summed over all semistandard λ/μ-tableaux $\mathcal{T}$. In particular, in the case $M = 0$ this gives the Gelfand–Tsetlin character of the evaluation module for Y($\mathfrak{gl}_N$).

PROOF. We start with the particular case $M = 0$ so that $L(\lambda)^+_\mu$ coincides with the evaluation module $L(\lambda)$. The coefficients of the quantum determinant of Y($\mathfrak{gl}_N$) act on $L(\lambda)$ as scalar operators found from Proposition 3.2.5, so that

$$\operatorname{qdet} T(u)|_{L(\lambda)} = \big(1+\lambda_1 u^{-1}\big)\dots\big(1+\lambda_N(u-N+1)^{-1}\big).$$

Observe that, regarding λ as a diagram, we can write the product here as

$$\prod_{\alpha\in\lambda} \frac{u+c(\alpha)+1}{u+c(\alpha)}.$$

Note also that by Corollary 1.11.8, we have the decomposition

$$\operatorname{qdet} T(u) = h_1(u)\,h_2(u-1)\dots h_N(u-N+1).$$

Now we employ the well known bijection between the patterns associated with λ and the semistandard λ-tableaux; see Section 5.4. Namely, the pattern Λ can be viewed as the sequence of diagrams

$$\Lambda_1 \subset \Lambda_2 \subset \dots \subset \Lambda_N = \lambda,$$

where the $\Lambda_k = (\lambda_{k1},\dots,\lambda_{kk})$ are the rows of Λ. The corresponding semistandard λ-tableau is obtained by placing the entry k into each cell of Λ_k/Λ_{k-1}. Hence, using the properties of the Gelfand–Tsetlin basis of $L(\lambda)$ (see Section 5.4), for any basis vector $\zeta_\Lambda \in L(\lambda)$ and any $1 \leqslant k \leqslant N$ we have

$$h_1(u)\,h_2(u-1)\dots h_k(u-k+1)\,\zeta_\Lambda = \prod_{\alpha\in\Lambda_k} \frac{u+c(\alpha)+1}{u+c(\alpha)}\,\zeta_\Lambda.$$

This implies

$$h_k(u-k+1)\,\zeta_\Lambda = \prod_{\alpha\in\Lambda_k/\Lambda_{k-1}} \frac{u+c(\alpha)+1}{u+c(\alpha)}\,\zeta_\Lambda. \tag{8.37}$$

Thus, the element of $\mathbb{Z}[\mathcal{P}_n]$ corresponding to the eigenvalue of $h_k(u)$ on ζ_Λ coincides with the product

$$\prod_{\alpha\in\Lambda_k/\Lambda_{k-1}} x_{k,c(\alpha)}.$$

This completes the proof in the case $M=0$.

In the skew case ($M \geqslant 1$) we note that the action of the quantum minor $t^{1\dots k}_{1\dots k}(u)$ on the basis vector $\zeta_\Lambda \in L(\lambda)^+_\mu$ is found by the application of the evaluation homomorphism (1.5) to the series (8.36). Hence the formula (8.37) remains valid in the skew case as well, where Λ_k now denotes the row $(\lambda_{M+k,1},\dots,\lambda_{M+k,M+k})$ of the trapezium pattern Λ, and we use the natural bijection between the trapezium patterns and the semistandard λ/μ-tableaux. □

8.6. Olshanski algebra associated with $\mathfrak{g}_\infty$

We will regard the Lie algebra $\mathfrak{g}_N = \mathfrak{o}_N$ or $\mathfrak{sp}_N$ as the subalgebra of $\mathfrak{gl}_N$ spanned by the elements F_{ij} defined in (7.9). The Lie algebra $\mathfrak{g}_{N-2}$ can be identified with the subalgebra of $\mathfrak{g}_N$ spanned by the elements F_{ij} with the indices satisfying $-n+1 \leqslant i,j \leqslant n-1$. Fix a nonnegative integer M such that $N-M$ is even. So, if $N=2n$ or $N=2n+1$, then $M=2m$ or $M=2m+1$, respectively, for some $m \leqslant n$. Denote by $\mathfrak{g}_{N,M}$ the subalgebra in $\mathfrak{g}_N$ spanned by the elements F_{ij} with $m+1 \leqslant |i|,|j| \leqslant n$. The subalgebra $\mathfrak{g}_{N,M}$ is isomorphic to $\mathfrak{g}_{N-M}$. Let $\mathrm{A}_M(N)$ denote the centralizer of $\mathfrak{g}_{N,M}$ in the universal enveloping algebra $\mathrm{U}(\mathfrak{g}_N)$. Let $\mathrm{A}(N)^0$ denote the centralizer of the element F_{nn} in $\mathrm{U}(\mathfrak{g}_N)$ and let $\mathrm{I}(N)$ be the left ideal of $\mathrm{U}(\mathfrak{g}_N)$ generated by the elements F_{in}, $i=-n,\dots,n$. Then the Poincaré–Birkhoff–Witt theorem for the Lie algebra $\mathfrak{g}_N$ implies that $\mathrm{I}(N)^0 = \mathrm{I}(N)\cap \mathrm{A}(N)^0$ is a two-sided ideal of $\mathrm{A}(N)^0$ which coincides with the intersection $\mathrm{J}(N)\cap\mathrm{A}(N)^0$, where $\mathrm{J}(N)$ is the right ideal of $\mathrm{U}(\mathfrak{g}_N)$ generated by the elements F_{ni}, $i=-n,\dots,n$. We have the vector space decomposition

$$\mathrm{A}(N)^0 = \mathrm{I}(N)^0 \oplus \mathrm{U}(\mathfrak{g}_{N-2}).$$

Therefore, the projection of $\mathrm{A}(N)^0$ onto $\mathrm{U}(\mathfrak{g}_{N-2})$ with the kernel $\mathrm{I}(N)^0$ is an algebra homomorphism. If $M<N$, then its restriction to the subalgebra $\mathrm{A}_M(N)$ defines a homomorphism

$$o_N : \mathrm{A}_M(N) \to \mathrm{A}_M(N-2). \tag{8.38}$$

Note that the algebra $\mathrm{A}_M(N)$ inherits the natural filtration of $\mathrm{U}(\mathfrak{g}_N)$ and the homomorphism o_N is filtration-preserving.

DEFINITION 8.6.1. The algebra A_M is defined as the projective limit of the sequence of the algebras $\mathrm{A}_M(N)$ with $N \geqslant M$, and $N-M$ even, with respect to the homomorphisms

$$\mathrm{A}_M(M) \xleftarrow{o_{M+2}} \mathrm{A}_M(M+2) \xleftarrow{o_{M+4}} \dots \xleftarrow{o_N} \mathrm{A}_M(N) \xleftarrow{o_{N+2}} \dots,$$

where the limit is taken in the category of filtered associative algebras. □

As in Section 8.1, an element of the algebra A_M is a sequence of the form $a=(a_M,a_{M+2},\dots,a_N,\dots)$ with $a_N\in\mathrm{A}_M(N)$, $o_N(a_N)=a_{N-2}$ for $N>M$, and

$$\deg a := \sup_N \deg a_N < \infty, \tag{8.39}$$

where $\deg a_N$ denotes the degree of a_N in the universal enveloping algebra $\mathrm{U}(\mathfrak{g}_N)$.

We define the Lie algebra $\mathfrak{g}_\infty$ as the inductive limit of the Lie algebras $\mathfrak{g}_N$ with respect to the natural embeddings $\mathfrak{g}_N \hookrightarrow \mathfrak{g}_{N+2}$, so that

$$\mathfrak{g}_\infty = \mathfrak{o}_{2\infty}, \quad \mathfrak{sp}_{2\infty}, \quad \text{or} \quad \mathfrak{o}_{2\infty+1},$$

respectively, for $\mathfrak{g}_N = \mathfrak{o}_{2n}$, $\mathfrak{sp}_{2n}$ or $\mathfrak{o}_{2n+1}$. In other words, the Lie algebra $\mathfrak{g}_\infty$ is spanned by the elements F_{ij} given by (7.9), where the indices i, j run over the set of all integers with the zero value excluded in the cases $\mathfrak{g}_\infty = \mathfrak{o}_{2\infty}$ and $\mathfrak{sp}_{2\infty}$.

DEFINITION 8.6.2. The *Olshanski algebra* $\mathrm{A} = \mathrm{A}(\mathfrak{g}_\infty)$ *associated with* $\mathfrak{g}_\infty$ is the inductive limit of the filtered algebras A_M with respect to the embeddings $\mathrm{A}_M \hookrightarrow \mathrm{A}_{M+2}$ given by

$$(a_M, a_{M+2}, a_{M+4}, \dots) \mapsto (a_{M+2}, a_{M+4}, \dots)$$

so that

$$\mathrm{A} = \bigcup_{m \geqslant 0} \mathrm{A}_{2m}$$

if $N = 2n$, and

$$\mathrm{A} = \bigcup_{m \geqslant 0} \mathrm{A}_{2m+1}$$

if $N = 2n + 1$. □

The universal enveloping algebra $\mathrm{U}(\mathfrak{g}_\infty)$ is canonically embedded into the Olshanski algebra A so that the image of an element $x \in \mathrm{U}(\mathfrak{g}_\infty)$ in A is the sequence $(x, x, \dots)$.

In the case $\mathfrak{g}_N = \mathfrak{o}_{2n+1}$ we also consider the center $\mathrm{A}_0(N)$ of the universal enveloping algebra $\mathrm{U}(\mathfrak{g}_N)$. Then using the homomorphisms (8.38) with $M = 0$ we define the algebra A_0, extending Definition 8.6.1 to this case. Observe that $\mathrm{A}_0(N) \subset \mathrm{A}_M(N)$ for any odd positive integer $M \leqslant N$ so that A_0 is a subalgebra of A_M. Indeed, the image of a sequence $(a_1, a_3, a_5, \dots) \in \mathrm{A}_0$ under this embedding is the sequence $(a_M, a_{M+2}, \dots) \in \mathrm{A}_M$.

Returning to the general case, we have the following analogue of Proposition 8.1.3 which is proved by the same argument.

PROPOSITION 8.6.3. *The center of the Olshanski algebra* $\mathrm{A} = \mathrm{A}(\mathfrak{g}_\infty)$ *coincides with* A_0. □

We will call the elements of the algebra A_0 the *virtual Casimir elements* for the Lie algebra $\mathfrak{g}_\infty$.

8.7. Virtual Casimir elements and highest weight modules for $\mathfrak{g}_\infty$

Recall that the center $\mathrm{A}_0(N)$ of $\mathrm{U}(\mathfrak{g}_N)$ is isomorphic to the subalgebra $\Lambda^*(n)$ of the algebra of polynomials in n variables $\lambda_1, \dots, \lambda_n$ under the Harish-Chandra isomorphism; see Section 7.1. For $\mathfrak{g}_N = \mathfrak{o}_{2n+1}$ and $\mathfrak{g}_N = \mathfrak{sp}_{2n}$, the elements of $\Lambda^*(n)$ are the polynomials which are symmetric in the variables $l_1^2, \dots, l_n^2$, where $l_i = \lambda_i + \rho_i$; see (7.10). For $\mathfrak{g}_N = \mathfrak{o}_{2n}$ the subalgebra $\Lambda^*(n)$ is generated by the polynomial $l_1 \dots l_n$ and the symmetric polynomials in the variables $l_1^2, \dots, l_n^2$. Note that the Harish-Chandra isomorphism preserves the filtrations on $\mathrm{A}_0(N)$ and $\Lambda^*(n)$, where the former is inherited from the natural filtration on $\mathrm{U}(\mathfrak{g}_N)$ while the latter is determined by the usual degrees of polynomials. So, the homomorphism o_N

given in (8.38) with $M = 0$ can be interpreted as the specialization homomorphism $\mathrm{o}_N : \Lambda^*(n) \to \Lambda^*(n-1)$ such that

$$\mathrm{o}_N : f(\lambda_1, \ldots, \lambda_n) \mapsto f(\lambda_1, \ldots, \lambda_{n-1}, 0). \tag{8.40}$$

The corresponding projective limit in the category of filtered algebras is denoted by Λ^*. We have the following analogue of Proposition 8.2.1.

PROPOSITION 8.7.1. *The algebra of virtual Casimir elements* A_0 *is isomorphic to the algebra* Λ^*. *The isomorphism* $\chi : \mathrm{A}_0 \to \Lambda^*$ *is determined by the Harish-Chandra isomorphisms* $\chi_N : \mathrm{A}_0(N) \to \Lambda^*(n)$ *with* $n \geqslant 0$. □

Consider the double elementary symmetric polynomials $e_k(l_1^2, \ldots, l_n^2 \,|\, a)$ with the sequence $a = (\rho_1^2, \rho_2^2, \ldots)$, where the numbers ρ_i are defined by (7.10) for any i; see Section 7.4. The relation (7.32) implies that

$$\mathrm{o}_N : e_k(l_1^2, \ldots, l_n^2 \,|\, a) \mapsto e_k(l_1^2, \ldots, l_{n-1}^2 \,|\, a). \tag{8.41}$$

Hence, for any $k \geqslant 1$ the sequence $(e_k(l_1^2, \ldots, l_n^2 \,|\, a) \mid n \geqslant k)$ determines an element $e_k \in \Lambda^*$ of degree $2k$.

PROPOSITION 8.7.2. *The elements* $e_1, e_2, \ldots$ *are algebraically independent and generate the algebra* Λ^*.

PROOF. Suppose that $f \in \Lambda^*$ is an element of degree $\leqslant p$ so that

$$f = (f^{(0)}, f^{(1)}, \ldots), \qquad \mathrm{o}_N : f^{(n)} \mapsto f^{(n-1)},$$

and $f^{(n)}$ is an element of $\Lambda^*(n)$ of degree $\leqslant p$ for each $n \geqslant 0$. Note that the top degree component of $e_k(l_1^2, \ldots, l_n^2 \,|\, a)$ is the elementary symmetric polynomial $e_k(l_1^2, \ldots, l_n^2)$. Therefore, for any $n > p$ the element $f^{(n)}$ can be uniquely represented as a polynomial in the $e_k(l_1^2, \ldots, l_n^2 \,|\, a)$ with $k = 1, \ldots, n$. (The polynomial $l_1 \ldots l_n$ in the case $\mathfrak{g}_N = \mathfrak{o}_{2n}$ will not occur, as its degree is $n > p$). Using (8.41), we conclude that the element f can be uniquely represented as a polynomial in the e_k with $k \geqslant 1$. □

We now construct a family of generators of the algebra A_0. Denote by F the infinite matrix $[F_{ij}]$ whose rows and columns are numbered by all integers with the zero value excluded for $\mathfrak{g}_\infty = \mathfrak{o}_{2\infty}$ and $\mathfrak{sp}_{2\infty}$. By $F^{(n)}$ we will denote the submatrix of F whose rows and columns are numbered by the indices $\{-n, \ldots, n\}$. Consider the series

$$\mathcal{C}^{(n)}(u) = (-1)^n \sum_{p \in \mathfrak{S}_N} \operatorname{sgn} pp' \cdot \Big(1 + \frac{F}{u + \rho_{-n}}\Big)_{-a_{p(1)}, a_{p'(1)}} \times \cdots \times \Big(1 + \frac{F}{u + \rho_n}\Big)_{-a_{p(N)}, a_{p'(N)}}, \tag{8.42}$$

where $(a_1, \ldots, a_N)$ is a fixed permutation of the indices $(-n, \ldots, n)$ and p' is the image of p under the map (2.53). Comparing this expression with (7.13), we find that $\mathcal{C}^{(n)}(u)$ is related to the Capelli-type determinant $\mathcal{C}(u)$ by the formula

$$\mathcal{C}(u) = \mathcal{C}^{(n)}(u) \cdot \prod_{i=-n}^{n} (u + \rho_i).$$

Due to Theorem 7.1.6, the coefficients of the series $\mathcal{C}^{(n)}(u)$ are central in $\mathrm{U}(\mathfrak{g}_N)$, and its image under the Harish-Chandra isomorphism is given by

$$\chi_N : \mathcal{C}^{(n)}(u) \mapsto \prod_{i=1}^{n} \frac{u^2 - l_i^2}{u^2 - \rho_i^2}. \tag{8.43}$$

Write

$$\mathcal{C}^{(n)}(u) = 1 + \mathcal{F}_1^{(n)} u^{-2} + \mathcal{F}_2^{(n)} u^{-4} + \dots, \qquad \mathcal{F}_k^{(n)} \in \mathrm{A}_0(N).$$

By (8.40), we have

$$o_N : \mathcal{C}^{(n)}(u) \mapsto \mathcal{C}^{(n-1)}(u).$$

Hence, for any $k \geqslant 1$ we may define a virtual Casimir element $\mathcal{F}_k \in \mathrm{A}_0$ as the sequence

$$\mathcal{F}_k = (\mathcal{F}_k^{(0)}, \mathcal{F}_k^{(1)}, \mathcal{F}_k^{(2)}, \dots).$$

PROPOSITION 8.7.3. *The elements $\mathcal{F}_1, \mathcal{F}_2, \dots$ are algebraically independent and generate the algebra A_0. Their images under the isomorphism χ are found by*

$$1 + \sum_{k=1}^{\infty} \chi(\mathcal{F}_k) u^{-k} = \prod_{i=1}^{\infty} \frac{u^2 - l_i^2}{u^2 - \rho_i^2}.$$

PROOF. The Harish-Chandra image $\chi_N(\mathcal{F}_k^{(n)})$ coincides with the double elementary symmetric polynomial $(-1)^k e_k(l_1^2, \dots, l_n^2 \,|\, a)$ associated with the sequence $a = (\rho_1^2, \rho_2^2, \dots)$; see (7.32). Hence, $\chi(\mathcal{F}_k) = (-1)^k e_k$. So, the claims follow from Proposition 8.7.2. □

Note that the algebra Λ^* admits some other families of algebraically independent generators. In particular, analogues of the power sums symmetric functions can be given by

$$\sum_{i=1}^{\infty} (l_i^{2k} - \rho_i^{2k}), \qquad k = 1, 2, \dots,$$

while analogues of the complete symmetric functions can be defined as the coefficients in the expansion of the series

$$\prod_{i=1}^{\infty} \frac{u^2 - \rho_i^2}{u^2 - l_i^2}$$

in the powers of u^{-2}. We also remark that the elements of Λ^* are well-defined functions on the set of all sequences $\lambda = (\lambda_1, \lambda_2, \dots)$ which contain only a finite number of nonzero terms.

Now we introduce a category Ω of modules over the Lie algebra $\mathfrak{g}_\infty$; cf. Section 8.2. We let $\mathfrak{h}$ denote the Cartan subalgebra of $\mathfrak{g}_\infty$ which consists of diagonal matrices (finite linear combinations of the elements F_{ii}) and let $\mathfrak{n}^+$ (respectively, $\mathfrak{n}^-$) denote the subalgebra of upper (respectively, lower) triangular matrices. We have the triangular decomposition

$$\mathfrak{g}_\infty = \mathfrak{n}^- \oplus \mathfrak{h} \oplus \mathfrak{n}^+.$$

For a linear functional $\lambda \in \mathfrak{h}^*$ set $\lambda_i = \lambda(F_{ii})$ for $i \geqslant 1$. We will identify λ with the sequence $(\lambda_1, \lambda_2, \dots)$.

A $\mathfrak{g}_\infty$-module V is said to be *highest weight* if it has a nonzero cyclic vector v satisfying $\mathfrak{n}^+ v = \{0\}$ and there exists $\lambda \in \mathfrak{h}^*$ such that $hv = \lambda(h)v$ for any $h \in \mathfrak{h}$. The functional λ is the *highest weight* of V, and v is the *highest weight* *vector* of V; it is unique up to scalar multiples. The universal $\mathfrak{g}_\infty$-module $M(\lambda)$ with the highest weight $\lambda \in \mathfrak{h}^*$ (the *Verma module*) may be defined as the quotient of the universal enveloping algebra $\mathrm{U}(\mathfrak{g}_\infty)$ by the left ideal generated by $\mathfrak{n}^+$ and the elements $h - \lambda(h)$, $h \in \mathfrak{h}$. Denote by $L(\lambda)$ the unique irreducible quotient of $M(\lambda)$. The $\mathfrak{g}_\infty$-module $L(\lambda)$ can be regarded as an inductive limit of irreducible highest weight $\mathfrak{g}_N$-modules. Namely, for a positive integer $N = 2n$ or $N = 2n+1$ denote by $L(\lambda)^{(n)}$ the $\mathfrak{g}_N$-cyclic span of the highest weight vector $v \in L(\lambda)$. Obviously, $L(\lambda)^{(n)}$ is a highest weight module with the highest weight $(\lambda_1, \dots, \lambda_n)$. The following is an analogue of Proposition 8.2.8 which is proved by a similar argument.

PROPOSITION 8.7.4. *The $\mathfrak{g}_N$-module $L(\lambda)^{(n)}$ is irreducible.* □

So, $L(\lambda)$ may be regarded as the inductive limit

$$L(\lambda) = \bigcup_{n=1}^{\infty} L(\lambda)^{(n)}.$$

For a $\mathfrak{g}_\infty$-module V and a functional $\mu = (\mu_1, \mu_2, \dots) \in \mathfrak{h}^*$ set

$$\begin{aligned} V_\mu(n) = \{v \in V \mid F_{ii}v = \mu_i v \quad &\text{for} \quad i > n \quad \text{and} \\ F_{ij}v = 0 \quad &\text{for} \quad i < j; \;\; j > n\}. \end{aligned}$$

Since $V_\mu(n) \subset V_\mu(n+1)$, we may set

$$V_\mu(\infty) = \bigcup_{n=1}^{\infty} V_\mu(n).$$

Note that $V_\mu(\infty)$ is a submodule of V since $V_\mu(n)$ is $\mathfrak{g}_N$-invariant for every N. We will write $\mu \sim \mu'$ if $\mu_i = \mu'_i$ for all sufficiently large i. Then $\mu \sim \mu'$ implies $V_\mu(\infty) = V_{\mu'}(\infty)$.

DEFINITION 8.7.5. Define Ω as the category of $\mathfrak{g}_\infty$-modules V satisfying the condition $V_{\mathbf{o}}(\infty) = V$, where $\mathbf{o} = (0, 0, \dots) \in \mathfrak{h}^*$. □

The following statements are respective counterparts of Proposition 8.2.10, Theorem 8.2.11, and Proposition 8.2.12. We omit the proofs as they are essentially the same as the corresponding arguments in Section 8.2.

PROPOSITION 8.7.6. *Let V be a $\mathfrak{g}_\infty$-module with the highest weight λ, where $\lambda \sim \mathbf{o}$. Then V belongs to Ω. In particular, Ω contains the modules $M(\lambda)$ and $L(\lambda)$ with $\lambda \sim \mathbf{o}$.* □

THEOREM 8.7.7. *Any $\mathfrak{g}_\infty$-module $V \in \Omega$ has a natural A-module structure.* □

PROPOSITION 8.7.8. *Let V be a highest weight $\mathfrak{g}_\infty$-module with the highest weight $\lambda \sim \mathbf{o}$. Then every element $a \in \mathrm{A}_0$ acts on V as multiplication by the scalar $\chi(a)_\lambda$ obtained by evaluating $\chi(a) \in \Lambda^*$ at λ.* □

8.8. Polynomial invariants for $\mathfrak{g}_N$

We will describe the structure of the graded algebras $\operatorname{gr} \mathrm{A}_M$ and $\operatorname{gr} \mathrm{A}$ with respect to the filtration defined by (8.39). Keeping the notation of Section 8.6, denote by $\mathrm{P}_M(N)$ the subalgebra of the symmetric algebra $\mathrm{S}(\mathfrak{g}_N)$ which consists of the elements invariant under the adjoint action of the subalgebra $\mathfrak{g}_{N,M}$. We let $\mathrm{I}'(N)$ denote the ideal of $\mathrm{S}(\mathfrak{g}_N)$ generated by the elements F_{in}, $i = -n, \dots, n$.

As in Section 8.1, replace respectively $\mathrm{A}_M(N)$ by $\mathrm{P}_M(N)$ and $\mathrm{I}(N)$ by $\mathrm{I}'(N)$ and repeat the arguments of Section 8.6 to construct a commutative graded algebra P_M which is the projective limit of the commutative graded algebras $\mathrm{P}_M(N)$, $N \geqslant M$, $N - M$ is even. The algebra P is then defined as the inductive limit of the algebras P_M. Note that the algebra P contains the symmetric algebra of $\mathfrak{g}_\infty$. Moreover, in the case $N = 2n+1$ we also consider the algebra $\mathrm{P}_0(N)$ of $\mathfrak{g}_N$-invariants in $\mathrm{S}(\mathfrak{g}_N)$ and the corresponding projective limit algebra P_0.

The proof of the next proposition is not essentially different from that of Proposition 8.3.1.

PROPOSITION 8.8.1. *We have the isomorphisms*

$$\mathrm{gr}\,\mathrm{A}_M \cong \mathrm{P}_M \qquad \textit{and} \qquad \mathrm{gr}\,\mathrm{A} \cong \mathrm{P}. \qquad \square$$

Elements of $\mathrm{S}(\mathfrak{g}_N)$ can be naturally identified with polynomials in matrix elements of an $N \times N$ matrix $X = [x_{ij}]_{i,j=-n}^n$ such that $X' = -X$, where the matrix transposition is defined in (4.2). Observe that the coefficients of the following polynomials in an indeterminate q belong to the algebra $\mathrm{P}_M(N)$,

$$\det(1+qX)_{\mathcal{B}\mathcal{B}} \qquad \text{and} \qquad \det(1+qX)_{\mathcal{B}_i\mathcal{B}_j}, \qquad -m \leqslant i,j \leqslant m, \tag{8.44}$$

where

$$\mathcal{B} = \{-n,\dots,-m-1,m+1,\dots,n\}, \qquad \mathcal{B}_i = \{-n,\dots,-m-1,i,m+1,\dots,n\}.$$

Here, as before, we use the notation $Y_{\mathcal{P}\mathcal{Q}}$ for the submatrix of a matrix Y whose rows and columns are enumerated by the elements of sets $\mathcal{P}$ and $\mathcal{Q}$, respectively.

PROPOSITION 8.8.2. *Every element of the algebra $\mathrm{P}_M(N)$ of degree $< n - m$ can be represented as a polynomial in the coefficients of the polynomials* (8.44).

PROOF. We will follow the argument of the proof of Proposition 8.3.2. Observe that the relations (8.8) and (8.9) will hold in the same form for the determinants (8.44), where $\Lambda_{ij}^{(r)} = \sum x_{ik_1}x_{k_1k_2}\cdots x_{k_{r-1}j}$, summed over the indices $k_l \in \mathcal{B}$.

Let us introduce submatrices A, B, C, D of the matrix X according to the decomposition $n = m + (n-m)$. Namely, A and D are the square matrices whose rows and columns are numbered by the indices $i,j \in \{-m,\dots,m\}$ and $i,j \notin \{-m,\dots,m\}$, respectively, while B corresponds to the subsets of indices $i \in \{-m,\dots,m\}$ and $j \notin \{-m,\dots,m\}$, and $C = -B'$.

We let G_N denote the classical Lie group corresponding to the Lie algebra $\mathfrak{g}_N$, namely $G_N = SO_N$ and SP_N for $\mathfrak{g}_N = \mathfrak{o}_N$ and $\mathfrak{sp}_N$, respectively. Furthermore, let $G_{N,M}$ denote the subgroup of G_N which consists of the matrices stabilizing each of the basis vectors e_i with $-m \leqslant i \leqslant m$ in $\mathbb{C}^N$ and the subspace spanned by the remaining basis vectors. This subgroup is isomorphic to G_{N-M}. We need to prove that any polynomial $\phi(X)$ of degree $< n-m$ satisfying the invariance condition

$$\phi(X) \equiv \phi(A,B,C,D) = \phi(A, B\mathbf{g}^{-1}, \mathbf{g}C, \mathbf{g}D\mathbf{g}^{-1}), \qquad \text{for all} \quad \mathbf{g} \in G_{N,M}, \tag{8.45}$$

is a polynomial function of the invariants $\Lambda_{ij}^{(r)}$ and the coefficients of the polynomial $\det(1+qD)$. Arguing as in the proof of Proposition 8.3.2, we can replace the coefficients with $\mathrm{tr}\,D^r$ with $r \geqslant 1$ and note that $\Lambda_{ij}^{(1)} = a_{ij}$, while $\Lambda_{ij}^{(r)} = (BD^{r-2}C)_{ij}$ for $r \geqslant 2$. Using the relation $B = -C'$, we reduce the task to describing the polynomial invariants $\phi(C,D)$, where C lies in the direct sum of several copies of the $G_{N,M}$-module V of column vectors of the length $N-M$, and D is an element of the adjoint module $\mathfrak{g}_{N,M}$. Any such invariant can be decomposed into a sum of expressions of the form

$$\psi(C, \dots, C, D, \dots, D),$$

where ψ is a multilinear invariant. That is, ψ is a morphism of $G_{N,M}$-modules

$$\psi: \underbrace{V \otimes \dots \otimes V}_{q} \otimes \underbrace{\mathfrak{g}_{N,M} \otimes \dots \otimes \mathfrak{g}_{N,M}}_{r} \to \mathbb{C}, \tag{8.46}$$

for some q and r, where we consider $\mathbb{C}$ as the trivial module. Note that the adjoint module $\mathfrak{g}_{N,M}$ is isomorphic to $\Lambda^2(V)$ in the orthogonal case and to $S^2(V)$ in the symplectic case. Let us identify the exterior or symmetric square of the module V with a submodule in the tensor product $V \otimes V$. Then any morphism (8.46) can be obtained as the restriction of a morphism of $G_{N,M}$-modules of the form

$$\eta: \underbrace{V \otimes \dots \otimes V}_{q} \otimes \underbrace{V \otimes \dots \otimes V}_{2r} \to \mathbb{C}.$$

Consider the invariant nondegenerate bilinear form on the space V which is preserved by the action of $G_{N,M}$:

$$\langle e_i, e_j \rangle = \delta_{i,-j} \qquad \text{in the orthogonal case,} \tag{8.47}$$

$$\langle e_i, e_j \rangle = \operatorname{sgn} i \cdot \delta_{i,-j} \quad \text{in the symplectic case,} \tag{8.48}$$

where $\{e_i\}$ is the canonical basis of $V \simeq \mathbb{C}^{N-M}$, $i = -n, \dots, -m-1, m+1, \dots, n$. Using the classical invariant theory for the orthogonal and symplectic groups (see H. Weyl [**464**, Theorems 2.9.A and 6.1.A]), we obtain that the invariant

$$\eta(c_1, \dots, c_q, u_1, v_1, \dots, u_r, v_r), \qquad c_i, u_i, v_i \in V$$

can be represented as a sum of monomials in bilinear invariants $\langle c_i, c_j \rangle$, $\langle c_i, u_j \rangle$, $\langle c_i, v_j \rangle$, $\langle u_i, u_j \rangle$, $\langle u_i, v_j \rangle$, and $\langle v_i, v_j \rangle$ such that in each of the monomials each letter appears exactly once. Here we have used the assumption $\deg \phi < n - m$, which implies that $q + r < n - m$ and hence $q + 2r < 2(n - m)$. This has allowed us to eliminate the invariants of the form $\det [w_1, \dots, w_{2(n-m)}]$, $w_i \in V$, which could occur in the orthogonal case. (In the symplectic case this assumption is not essential.) Note that if for some i we permute the letters u_i and v_i in such a monomial, then its restriction will still determine, up to a sign, the same invariant of the form (8.46). Therefore we may only consider the monomials which are products of submonomials of the form

$$\langle v_{s_1}, u_{s_2} \rangle \langle v_{s_2}, u_{s_3} \rangle \dots \langle v_{s_k}, u_{s_1} \rangle \tag{8.49}$$

and

$$\langle c_\alpha, u_{s_1} \rangle \langle v_{s_1}, u_{s_2} \rangle \dots \langle v_{s_k}, c_\beta \rangle \qquad \text{(in particular, } \langle c_\alpha, c_\beta \rangle\text{)}. \tag{8.50}$$

Let us check that the monomials (8.49) and (8.50) determine the invariants of the form $\operatorname{tr} D^k$ and $(C' D^k C)_{ij}$, respectively.

Consider first the orthogonal case and calculate $\eta(D)$, where the invariant η is determined by a monomial of the form (8.49). We can obviously assume that $(s_1, \dots, s_k) = (1, \dots, k)$. That is, $\eta: V^{\otimes 2k} \to \mathbb{C}$ with

$$\eta: u_1 \otimes v_1 \otimes \dots \otimes u_k \otimes v_k \mapsto \langle v_1, u_2 \rangle \langle v_2, u_3 \rangle \dots \langle v_k, u_1 \rangle. \tag{8.51}$$

We will regard the matrix D as an element of $V \otimes V$ using the vector space isomorphism $\mathfrak{g}_{N,M} \to \Lambda^2(V)$ defined by the formula $F_{ij} \mapsto e_i \otimes e_{-j} - e_{-j} \otimes e_i$. We have $D = \sum D_{ij} E_{ij} = \frac{1}{2} \sum D_{ij} F_{ij}$, because $D_{ij} = -D_{-j,-i}$. Therefore,

$$\underbrace{D \otimes \cdots \otimes D}_{k} = \frac{1}{2^k} \sum D_{i_1 j_1} \dots D_{i_k j_k} F_{i_1 j_1} \otimes \cdots \otimes F_{i_k j_k} \mapsto$$

$$\frac{1}{2^k} \sum D_{i_1 j_1} \dots D_{i_k j_k} (e_{i_1} \otimes e_{-j_1} - e_{-j_1} \otimes e_{i_1}) \otimes \ldots \otimes (e_{i_k} \otimes e_{-j_k} - e_{-j_k} \otimes e_{i_k})$$

$$= \sum D_{i_1 j_1} \dots D_{i_k j_k} (e_{i_1} \otimes e_{-j_1}) \otimes \ldots \otimes (e_{i_k} \otimes e_{-j_k}).$$

So, by (8.51)

$$\eta(D) = \sum D_{i_1 j_1} \dots D_{i_k j_k} \langle e_{-j_1}, e_{i_2} \rangle \langle e_{-j_2}, e_{i_3} \rangle \dots \langle e_{-j_k}, e_{i_1} \rangle$$

$$= \sum D_{i_1 j_1} \dots D_{i_k j_k} \delta_{j_1 i_2} \delta_{j_2 i_3} \dots \delta_{j_k i_1} = \sum D_{i_1 i_2} D_{i_2 i_3} \dots D_{i_k i_1} = \operatorname{tr} D^k.$$

In the symplectic case the vector space isomorphism $\mathfrak{g}_{N,M} \to S^2(V)$ can be defined by the formula $F_{ij} \mapsto \operatorname{sgn} j \, (e_i \otimes e_{-j} + e_{-j} \otimes e_i)$. An analogous calculation shows that here for an invariant η of the form (8.49) one has $\eta(D) = (-1)^k \operatorname{tr} D^k$.

For the monomials of the form (8.50) the calculation is quite similar and will be omitted. □

Observe that since the matrix X satisfies $X' = -X$, we have the relations

$$\det(1 + qX)_{\mathcal{B}\mathcal{B}} = \det(1 - qX)_{\mathcal{B}\mathcal{B}}$$

and

$$\Lambda_{ij}^{(r)} = (-1)^r \, \theta_{ij} \, \Lambda_{-j,-i}^{(r)}.$$

Hence, we can write

$$\det(1 + qX)_{\mathcal{B}\mathcal{B}} = 1 + \sum_{r=1}^{\infty} \Lambda^{(2r)} \, q^{2r}$$

for some polynomials $\Lambda^{(2r)}$ in the matrix elements of X. For the elements $\Lambda_{ij}^{(r)}$ we will impose the following restrictions on i, j, r:

$$i + j > 0 \quad \text{for } r \text{ odd}, \qquad i + j \geqslant 0 \quad \text{for } r \text{ even}$$

in the orthogonal case, and

$$i + j > 0 \quad \text{for } r \text{ even}, \qquad i + j \geqslant 0 \quad \text{for } r \text{ odd}$$

in the symplectic case.

Fix a positive integer K and suppose that the index r satisfies $1 \leqslant r \leqslant K$ while the indices i, j range over $\{-m, \dots, m\}$.

PROPOSITION 8.8.3. *There exists a sufficiently large value of N such that the polynomials $\Lambda^{(r)}$ for even r and $\Lambda_{ij}^{(r)}$ with the above restrictions on i, j, r are algebraically independent.*

PROOF. Arguing as in the proof of Proposition 8.8.2, we see that the polynomials $\Lambda^{(r)}$ can be replaced with the elements $\operatorname{tr} D^r$ for even r.

We will consider only the orthogonal case. The proof in the symplectic case could then be obtained by an obvious adjustment.

For any triple (i, j, r) satisfying our assumptions we choose a subset

$$\mathcal{O}_{ij}^{(r)} \subset \{m+1, m+2, \dots\}$$

of cardinality $r-1$ in such a way that all these subsets are disjoint. Let n be so large that all of them are contained in the set $\{m+1, m+2, \dots, n-K\}$. Introduce complex parameters

$$y_1, \dots, y_K \qquad \text{and} \qquad y_{ij}^{(r)}, \qquad -m \leqslant i, j \leqslant m, \qquad 1 \leqslant r \leqslant K.$$

Let us now define a linear operator $x_{ij}^{(r)}$ in $\mathbb{C}^N$ depending on $y_{ij}^{(r)}$ as follows. Let $(e_{-n}, \dots, e_n)$ be the canonical basis of $\mathbb{C}^N$ and let $a_1 < \cdots < a_{r-1}$ be all the elements of $\mathcal{O}_{ij}^{(r)}$. Then for $i + j \neq 0$

$$\begin{aligned} x_{ij}^{(r)}: \quad & e_j \mapsto y_{ij}^{(r)} e_{a_{r-1}}, \quad e_{a_{r-1}} \mapsto e_{a_{r-2}}, \quad \dots, \quad e_{a_1} \mapsto e_i, \\ & e_{-i} \mapsto -e_{-a_1}, \quad e_{-a_1} \mapsto -e_{-a_2}, \quad \dots, \quad e_{-a_{r-1}} \mapsto -y_{ij}^{(r)} e_{-j}, \\ & e_k \mapsto 0 \quad \text{for} \quad k \notin \{-i, j\} \cup \mathcal{O}_{ij}^{(r)} \cup -\mathcal{O}_{ij}^{(r)}; \end{aligned}$$

while for $i + j = 0$

$$\begin{aligned} x_{i,-i}^{(r)}: \quad & e_{-i} \mapsto y_{i,-i}^{(r)} e_{a_{r-1}} - \tfrac{1}{2} e_{-a_1}, \quad e_{a_{r-1}} \mapsto e_{a_{r-2}}, \quad \dots, \quad e_{a_1} \mapsto \tfrac{1}{2} e_i, \\ & e_{-a_1} \mapsto -e_{-a_2}, \quad \dots, \quad e_{-a_{r-1}} \mapsto -y_{i,-i}^{(r)} e_i, \\ & e_k \mapsto 0 \quad \text{for} \quad k \notin \{-i\} \cup \mathcal{O}_{i,-i}^{(r)} \cup -\mathcal{O}_{i,-i}^{(r)}, \end{aligned}$$

where $-\mathcal{O}_{ij}^{(r)}$ denotes the set of the elements of $\mathcal{O}_{ij}^{(r)}$ taken with the negative sign. Note that $(x_{ij}^{(r)})' = -x_{ij}^{(r)}$. Now define a linear operator X on $\mathbb{C}^N$ depending on all the variables by setting

$$X\, e_k = \begin{cases} \sum_{i,j,r} x_{ij}^{(r)} e_k & \text{if} \quad |k| \leqslant n - K, \\ \operatorname{sgn} k \cdot y_{|k|-(n-K)} e_k & \text{if} \quad n - K < |k| \leqslant n. \end{cases} \tag{8.52}$$

Regarding X as a matrix, we get

$$\begin{aligned} \operatorname{tr} D^r &= 2\,(y_1^r + \cdots + y_K^r) \quad \text{if } r \text{ is even}, \\ \Lambda_{ij}^{(r)} &= y_{ij}^{(r)} + (-1)^r y_{-j,-i}^{(r)} \quad \text{if} \quad i + j \neq 0, \\ \Lambda_{i,-i}^{(r)} &= y_{i,-i}^{(r)} \quad \text{if } r \text{ is even}. \end{aligned}$$

Thus our polynomials with the indices i, j, r satisfying the assumptions of the proposition are algebraically independent even if they are restricted to the affine subspace of matrices of the form (8.52). □

Setting $x_{in} = 0$ for $i = -n, \dots, n$ in the determinants (8.44) we get the corresponding elements of the algebra $\mathrm{P}_M(N-2)$. Therefore, we can define the *virtual determinants*

$$\det(1 + qX)_{\mathcal{B}\mathcal{B}} \qquad \text{and} \qquad \det(1 + qX)_{\mathcal{B}_i \mathcal{B}_j}, \qquad -m \leqslant i, j \leqslant m, \tag{8.53}$$

of the submatrices of the infinite matrix $1 + qX$, where $X = [x_{ij}]_{i,j=-\infty}^{\infty}$, as formal series in q with coefficients in P_M. Here $\mathcal{B}$ and $\mathcal{B}_i$ respectively denote the infinite sets $\{\dots, -m-2, -m-1, m+1, m+2, \dots\}$ and $\{\dots, -m-2, -m-1, i, m+1, m+2, \dots\}$. Extending the notation $\Lambda^{(r)}$ and $\Lambda_{ij}^{(r)}$ with $-m \leqslant i, j \leqslant m$ to the infinite case, we come to the following analogue of Proposition 8.3.4.

PROPOSITION 8.8.4. *The elements $\Lambda^{(r)}$ with r even are algebraically independent generators of the algebra P_0. Moreover, for any $m \geqslant 1$, the following elements are algebraically independent generators of the algebra P_M:*

$$\Lambda^{(r)} \quad \textit{with} \quad r \quad \textit{even},$$

$$\Lambda^{(r)}_{ij} \quad \textit{with} \quad i+j>0 \quad \textit{for} \quad r \quad \textit{odd}, \qquad \textit{and} \quad i+j\geqslant 0 \quad \textit{for} \quad r \quad \textit{even}$$

in the orthogonal case; and

$$\Lambda^{(r)} \quad \textit{with} \quad r \quad \textit{even},$$

$$\Lambda^{(r)}_{ij} \quad \textit{with} \quad i+j>0 \quad \textit{for} \quad r \quad \textit{even}, \qquad \textit{and} \quad i+j\geqslant 0 \quad \textit{for} \quad r \quad \textit{odd}$$

in the symplectic case.

PROOF. This follows from Propositions 8.8.2 and 8.8.3. □

We will also need another family of generators of the algebra P_M.

PROPOSITION 8.8.5. *Every element of the algebra $\mathrm{P}_M(N)$ of degree $< n-m$ can be represented as a polynomial in the functions*

$$p^{(r)}_N(X) = \operatorname{tr} X^r \tag{8.54}$$

and

$$p^{(r)}_{ij|N}(X) = (X^r)_{ij}, \tag{8.55}$$

where $-m \leqslant i, j \leqslant m$ and $r \geqslant 1$.

PROOF. The argument of the proof of Proposition 8.3.5 shows that the claim follows from Proposition 8.8.2. □

Since the matrix X satisfies $X' = -X$, we have

$$p^{(r)}_N(X) = (-1)^r p^{(r)}_N(X) \qquad \text{and} \qquad p^{(r)}_{ij|N}(X) = (-1)^r\, \theta_{ij}\, p^{(r)}_{-j,-i|N}(X).$$

Hence $p^{(r)}_N(X)$ vanishes for odd r, whereas for the elements $p^{(r)}_{ij|N}(X)$ we may impose the following restrictions on i, j, r:

$$i+j>0 \quad \text{for} \quad r \quad \text{odd}, \qquad i+j\geqslant 0 \quad \text{for} \quad r \quad \text{even}$$

in the orthogonal case, and

$$i+j>0 \quad \text{for} \quad r \quad \text{even}, \qquad i+j\geqslant 0 \quad \text{for} \quad r \quad \text{odd}$$

in the symplectic case.

Fix a positive integer K and suppose that the index r satisfies $1 \leqslant r \leqslant K$ while the indices i, j range over $\{-m, \dots, m\}$.

PROPOSITION 8.8.6. *There exists a sufficiently large value of N such that the polynomials $p^{(r)}_N$ for even r, and $p^{(r)}_{ij|N}$ with the above restrictions on i, j, r are algebraically independent.*

PROOF. Define a linear operator X in $\mathbb{C}^N$ in exactly the same way as in the proof of Proposition 8.8.3 (see (8.52)), where we consider only the orthogonal case. Regarding X as a matrix, we get

$$\begin{aligned} p_N^{(r)}(X) &= 2\,(y_1^r + \cdots + y_K^r) + \phi^{(r)} \quad \text{if } r \text{ is even}, \\ p_{ij|N}^{(r)}(X) &= y_{ij}^{(r)} + (-1)^r y_{-j,-i}^{(r)} + \psi_{ij}^{(r)} \quad \text{if} \quad i+j \neq 0, \\ p_{i,-i|N}^{(r)}(X) &= y_{i,-i}^{(r)} + \psi_{ij}^{(r)} \quad \text{if } r \text{ is even} \end{aligned}$$

for some polynomials $\phi^{(r)}$ and $\psi_{ij}^{(r)}$ in the $y_{kl}^{(s)}$ such that $\phi^{(r)}$ does not depend on $y_1, \dots, y_K$, while $\psi_{ij}^{(r)}$ depends only on $y_{kl}^{(s)}$ with $s < r$. Hence the polynomials $p_N^{(r)}$ and $p_{ij|N}^{(r)}$ are algebraically independent even if they are restricted to the affine subspace of matrices of the form (8.52). □

The definition of the polynomials $p_N^{(r)}$ and $p_{ij|N}^{(r)}$ implies that for any $r \geqslant 1$ and $N \geqslant 2$ we have

$$\begin{aligned} p_N^{(r)} - p_{N-2}^{(r)} &\in \mathrm{I}'(N), \\ p_{ij|N}^{(r)} - p_{ij|N-2}^{(r)} &\in \mathrm{I}'(N), \quad -n+1 \leqslant i,j \leqslant n-1. \end{aligned}$$

Hence the sequence

$$p^{(r)} = \begin{cases} p_{2n}^{(r)}, & n \geqslant 0, \quad \text{for} \quad \mathfrak{g}_N = \mathfrak{o}_{2n} \quad \text{or} \quad \mathfrak{sp}_{2n} \\ p_{2n+1}^{(r)}, & n \geqslant 0, \quad \text{for} \quad \mathfrak{g}_N = \mathfrak{o}_{2n+1} \end{cases}$$

is a well-defined element of P_0. Similarly, for a fixed value of m, if $-m \leqslant i, j \leqslant m$, then for any $r \geqslant 1$ the sequence

$$p_{ij}^{(r)} = \big(p_{ij|N}^{(r)} \mid N \geqslant M, \quad N - M \quad \text{is even}\big)$$

is a well-defined element of P_M.

PROPOSITION 8.8.7. *The elements $p^{(r)}$ with r even are algebraically independent generators of the algebra P_0. Moreover, for any $m \geqslant 1$, the following elements are algebraically independent generators of the algebra P_M:*

$$\begin{aligned} &p^{(r)} \quad \textit{with} \quad r \quad \textit{even}, \\ &p_{ij}^{(r)} \quad \textit{with} \quad i+j>0 \quad \textit{for} \quad r \quad \textit{odd} \quad \textit{and} \quad i+j \geqslant 0 \quad \textit{for} \quad r \quad \textit{even} \end{aligned}$$

in the orthogonal case; and

$$\begin{aligned} &p^{(r)} \quad \textit{with} \quad r \quad \textit{even}, \\ &p_{ij}^{(r)} \quad \textit{with} \quad i+j>0 \quad \textit{for} \quad r \quad \textit{even} \quad \textit{and} \quad i+j \geqslant 0 \quad \textit{for} \quad r \quad \textit{odd} \end{aligned}$$

in the symplectic case.

PROOF. This follows from Propositions 8.8.5 and 8.8.6. □

Note that dropping the restrictions $-m \leqslant i, j \leqslant m$ for the indices i, j in Propositions 8.8.4 and 8.8.7, we obtain the respective families of algebraically independent generators of the algebra P.

8.9. Algebraic structure of $\mathrm{A}(\mathfrak{g}_\infty)$

Here we establish a relationship between the Olshanski algebra $\mathrm{A} = \mathrm{A}(\mathfrak{g}_\infty)$ and the twisted Yangians.

Recall the evaluation homomorphism $\varrho_N : \mathrm{Y}(\mathfrak{g}_N) \to \mathrm{U}(\mathfrak{g}_N)$ given by (2.106) and denote by $f_{ij}(u)$ the image of $s_{ij}(u)$ under ϱ_N. Furthermore, we let $f^{\,a_1\dots a_k}_{\,b_1\dots b_k}(u)$ denote the image of the Sklyanin minor under ϱ_N so that

$$\varrho_N : s^{\,a_1\dots a_k}_{\,b_1\dots b_k}(u) \mapsto f^{\,a_1\dots a_k}_{\,b_1\dots b_k}(u).$$

Take the composition of the homomorphism $\mathrm{Y}(\mathfrak{g}_M) \to \mathrm{Y}(\mathfrak{g}_N)$ provided by Proposition 4.1.10 and the evaluation homomorphism ϱ_N. This yields the homomorphism $\psi_N : \mathrm{Y}(\mathfrak{g}_M) \to \mathrm{U}(\mathfrak{g}_N)$ given by

$$s_{ij}(u) \mapsto \alpha_{M-N}(u) \cdot f^{-n\dots -m-1,\, i,\, m+1\dots n}_{-n\dots -m-1,\, j,\, m+1\dots n}(u+n-m), \tag{8.56}$$

where $i, j \in \{-m, \dots, m\}$. By Corollary 4.1.9, the image of ψ_N is contained in the centralizer $\mathrm{A}_M(N)$, so that $\psi_N : \mathrm{Y}(\mathfrak{g}_M) \to \mathrm{A}_M(N)$.

PROPOSITION 8.9.1. *For any fixed value of M, the sequence of homomorphisms ψ_N with $N \geqslant M$ such that $N - M$ is even defines a homomorphism*

$$\psi : \mathrm{Y}(\mathfrak{g}_M) \to \mathrm{A}_M.$$

PROOF. We have to verify that the homomorphisms ψ_N are compatible with the sequence of homomorphisms (8.38); that is, the following diagram is commutative:

$$\begin{array}{ccccccccc}
\mathrm{Y}(\mathfrak{g}_M) & = & \mathrm{Y}(\mathfrak{g}_M) & = & \cdots & = & \mathrm{Y}(\mathfrak{g}_M) & = & \cdots \\
\downarrow{\scriptstyle \psi_M} & & \downarrow{\scriptstyle \psi_{M+2}} & & & & \downarrow{\scriptstyle \psi_N} & & \\
\mathrm{A}_M(M) & \xleftarrow[o_{M+2}]{} & \mathrm{A}_M(M+2) & \longleftarrow & \cdots & \xleftarrow[o_N]{} & \mathrm{A}_M(N) & \longleftarrow & \cdots .
\end{array}$$

Let us calculate the image of the series $\psi_N(s_{ij}(u))$ under the homomorphism o_N. Applying Proposition 4.1.1, we obtain

$$f^{-n\dots -m-1,\, i,\, m+1\dots n}_{-n\dots -m-1,\, j,\, m+1\dots n}(u+n-m) = \sum_c \check{f}^{-n\dots -m-1,\, i,\, m+1\dots n}_{-n\dots -m-1,\, j,\, m+1\dots n-1,\, c}(u+n-m)\, f_{cn}(u-n+m),$$

where by $\check{f}^{\,a_1\dots a_k}_{\,b_1\dots b_{k-1},c}(u)$ we denote the image of the auxiliary minor $\check{s}^{\,a_1\dots a_k}_{\,b_1\dots b_{k-1},c}(u)$ under the evaluation homomorphism (2.106). Now observe that $f_{cn}(u-n+m)$ belongs to the left ideal $\mathrm{I}(N)$ unless $c = n$. In this case $f_{nn}(u-n+m) \equiv 1 \mod \mathrm{I}(N)$. Furthermore, by Proposition 4.1.2,

$$\check{f}^{-n\dots -m-1,\, i,\, m+1\dots n}_{-n\dots -m-1,\, j,\, m+1\dots n}(v) = \frac{2v+1}{2v\pm 1} \sum_{i=1}^{2n-2m} (-1)^{i-1} f'_{a_i,-n}(-v)\, f^{\,a_1\dots \widehat{a}_i \dots a_{2n-2m}}_{-n+1\dots -m-1,\, j,\, m+1\dots n-1}(v-1),$$

where we have set $(a_1, \dots, a_{2n-2m}) = (-n \dots -m-1,\, i,\, m+1 \dots n-1)$ and $v = u+n-m$. Now $f'_{a_i,-n}(-v) = \theta_{-n,a_i} f_{n,-a_i}(-v)$ belongs to the right ideal $\mathrm{J}(N)$ unless $a_i = -n$, that is, $i = 1$. In this case $f_{nn}(-v) \equiv 1 \mod \mathrm{J}(N)$. Note that

$$\alpha_{M-N}(u)\, \frac{2v+1}{2v\pm 1} = \alpha_{M-N+2}(u).$$

Hence, o_N takes $\psi_N(s_{ij}(u))$ to the series

$$\alpha_{M-N+2}(u)\cdot f^{-n+1\dots -m-1,\,i,\,m+1\dots n-1}_{-n+1\dots -m-1,\,j,\,m+1\dots n-1}(u+n-m-1)$$

which coincides with $\psi_{N-2}(s_{ij}(u))$. Thus, the sequence of coefficients at each power of u^{-1} in $\psi_N(s_{ij}(u))$ with $N=M,M+2,\dots$ defines an element of A_M. □

Corollary 4.1.9 implies that all coefficients of the series

$$\alpha_{M-N}(u)\cdot f^{-n\dots -m-1,\,m+1\dots n}_{-n\dots -m-1,\,m+1\dots n}(u+n-m) \tag{8.57}$$

belong to the center of the universal enveloping algebra $\mathrm{U}(\mathfrak{g}_{N,M})$. The argument of the proof of Proposition 8.9.1 shows that the image of the series (8.57) under the homomorphism o_N coincides with

$$\alpha_{M-N+2}(u)\cdot f^{-n+1\dots -m-1,\,m+1\dots n-1}_{-n+1\dots -m-1,\,m+1\dots n-1}(u+n-m-1).$$

Therefore, for each r the sequence of coefficients of u^{-r} in the series (8.57) with $n=m+1,m+2,\dots$ determines an element of the projective limit algebra A_M. Denote by $\widetilde{\mathrm{A}}_0$ the subalgebra of A_M generated by all these elements with $r\geqslant 1$. By Proposition 8.7.1, the algebra $\widetilde{\mathrm{A}}_0$ is isomorphic to the algebra of shifted symmetric functions in the variables $\lambda_{m+1},\lambda_{m+2},\dots$.

THEOREM 8.9.2. *The homomorphism ψ is an algebra embedding of $\mathrm{Y}(\mathfrak{g}_M)$ into the algebra A_M. Moreover, we have the tensor product decomposition*

$$\mathrm{A}_M=\widetilde{\mathrm{A}}_0\otimes \mathrm{Y}(\mathfrak{g}_M), \tag{8.58}$$

where the Yangian $\mathrm{Y}(\mathfrak{g}_M)$ is identified with its image under the embedding ψ.

PROOF. For any $r\geqslant 1$, the sequence of the coefficients of u^{-r} of the series (8.57) with $n=m,m+1,\dots$ determines an element of the algebra A_M of degree $\leqslant r$. Recall that A_M is a filtered algebra with $\mathrm{gr}\,\mathrm{A}_M\cong \mathrm{P}_M$; see Proposition 8.8.1. Let us calculate the highest degree term of the coefficient of u^{-r} of (8.57). An explicit expression for the series (8.57) is provided by Proposition 4.1.3, where the series $s_{ij}(u)$ should be replaced by their images $f_{ij}(u)$ under the evaluation homomorphism (2.106). Observe that the image of $s'_{ij}(-u)$ coincides with $f_{ij}(u\mp 1)$. Since we are interested only in the highest degree component of the coefficient at each power of u^{-1}, we may replace each expression of type $f_{ij}(u+c)$ by $\delta_{ij}+F_{ij}u^{-1}$. Hence, denoting the elements of the set $\mathcal{B}=\{-n,\dots,-m-1,m+1,\dots,n\}$ by $a_1,\dots,a_{N-M}$ we can write, modulo lower degree terms at each power of u,

$$\begin{aligned} f^{-a_1\dots -a_{N-M}}_{\;a_1\dots a_{N-M}}(u+n-m)&\equiv \alpha_{N-M}(u+n-m)\\ &\times\sum_{p\in\mathfrak{S}_M}\mathrm{sgn}\,pp'\cdot(1+Fu^{-1})_{-a_{p(1)},a_{p'(1)}}\cdots(1+Fu^{-1})_{-a_{p(N-M)},a_{p'(N-M)}},\end{aligned}$$

where F denotes the $(N-M)\times(N-M)$ matrix whose rows and columns are numbered by the elements of the set $\mathcal{B}$ and whose (i,j) entry is F_{ij}. Since these matrix elements commute modulo lower degree terms, taking into account the identity

$$\alpha_{M-N}(u)\,\alpha_{N-M}(u+n-m)=1,$$

we can conclude from Lemma 2.7.5 that the image of the series (8.57) in $\mathrm{P}_M(N)$ is the determinant $\det(1+Xu^{-1})_{\mathcal{B}\mathcal{B}}$. The same argument shows that the image of the series (8.56) in $\mathrm{P}_M(N)$ is $\det(1+Xu^{-1})_{\mathcal{B}_i\mathcal{B}_j}$, where

$$\mathcal{B}_i=\{-n,\dots,-m-1,i,m+1,\dots,n\}.$$

The statements of the theorem now follow from Proposition 8.8.4; cf. the proof of Theorem 8.4.1. □

Now we construct a different embedding of the twisted Yangian $\mathrm{Y}(\mathfrak{g}_M)$ into the algebra A_M and prove the corresponding analogue of Theorem 8.9.2.

Consider the automorphism of the twisted Yangian $\mathrm{Y}(\mathfrak{g}_N)$ provided by Corollary 2.15.5 and take its composition with the evaluation homomorphism (2.106). This yields a homomorphism $\varphi_N : \mathrm{Y}(\mathfrak{g}_N) \to \mathrm{U}(\mathfrak{g}_N)$. Using the definition (2.112) of the Sklyanin comatrix, we can write for the image of $S(u)$,

$$\varphi_N : S(u) \mapsto \varsigma_N(u)\Big(1 - \frac{F}{u+\kappa_N}\Big)^{-1}, \tag{8.59}$$

where $\kappa_N = (N \mp 1)/2$ and

$$\varsigma_N(u) = \alpha_N(u)\, \varrho_N\Big(\operatorname{sdet} S(-u + N/2 - 1)\Big).$$

By Theorem 7.1.6, the coefficients of the series $\varsigma_N(u)$ belong to the center of the universal enveloping algebra $\mathrm{U}(\mathfrak{g}_N)$. Moreover, its image under the Harish-Chandra isomorphism is found by

$$\chi : \varsigma_N(u) \mapsto \prod_{i=1}^{n} \frac{(u+1/2)^2 - l_i^2}{(u+1/2)^2 - \rho_i^2}. \tag{8.60}$$

Corollary 2.4.4 implies that for any $M \leqslant N$ such that $N - M$ is even, the twisted Yangian $\mathrm{Y}(\mathfrak{g}_M)$ can be regarded as a natural subalgebra of $\mathrm{Y}(\mathfrak{g}_N)$.

PROPOSITION 8.9.3. *The image of the restriction of the homomorphism φ_N to the subalgebra $\mathrm{Y}(\mathfrak{g}_M)$ is contained in the centralizer $\mathrm{A}_M(N)$.*

PROOF. We deduce from the defining relations (4.3) that

$$[s_{kl}^{(1)}, s_{ij}(u)] = \delta_{il} s_{kj}(u) - \delta_{kj} s_{il}(u) - \theta_{i,-l}\delta_{k,-i} s_{-l,j}(u) + \theta_{k,-j}\delta_{-j,l} s_{i,-k}(u). \tag{8.61}$$

In particular,

$$[s_{kl}^{(1)}, s_{ij}(u)] = 0 \quad \text{for} \quad |i|, |j| \leqslant m < |k|, |l|.$$

On the other hand, using (8.60) we derive that

$$\varphi_N : s_{kl}^{(1)} \mapsto F_{kl}. \tag{8.62}$$

Hence, for any $-m \leqslant i, j \leqslant m$ the coefficients of the series $\varphi_N(s_{ij}(u))$ commute with all elements of the subalgebra $\mathfrak{g}_{N,M}$. □

The proof of Proposition 8.9.3 implies the following well known relations.

COROLLARY 8.9.4. *In the universal enveloping algebra $\mathrm{U}(\mathfrak{g}_N)$ we have*

$$[F_{kl}, (F^r)_{ij}] = \delta_{il}(F^r)_{kj} - \delta_{kj}(F^r)_{il} - \theta_{i,-l}\delta_{k,-i}(F^r)_{-l,j} + \theta_{k,-j}\delta_{-j,l}(F^r)_{i,-k}.$$

PROOF. This follows from (8.61) by using (8.59) and the expansion

$$(1 - Fv)^{-1} = 1 + \sum_{r=1}^{\infty} F^r v^r,$$

where $v = (u + \kappa_N)^{-1}$. □

By Proposition 8.9.3, for any $N \geqslant M$ such that $N - M$ is even, we have a homomorphism of the twisted Yangian $\mathrm{Y}(\mathfrak{g}_M)$ to the centralizer $\mathrm{A}_M(N)$ obtained by the restriction of φ_N.

THEOREM 8.9.5. *For any fixed $M \geqslant 1$ the sequence of homomorphisms φ_N with $N \geqslant M$ such that $N - M$ is even defines an algebra embedding $\varphi : \mathrm{Y}(\mathfrak{g}_M) \hookrightarrow \mathrm{A}_M$. Moreover, we have the tensor product decomposition*

$$\mathrm{A}_M = \mathrm{A}_0 \otimes \mathrm{Y}(\mathfrak{g}_M), \tag{8.63}$$

where the Yangian $\mathrm{Y}(\mathfrak{g}_M)$ is identified with its image under the embedding φ.

PROOF. In order to prove that the sequence of homomorphisms φ_N defines an algebra homomorphism $\varphi : \mathrm{Y}(\mathfrak{g}_M) \to \mathrm{A}_M$ we need to verify that the following diagram is commutative:

$$\begin{array}{ccccccccc}
\mathrm{Y}(\mathfrak{g}_M) & = & \mathrm{Y}(\mathfrak{g}_M) & = & \cdots & = & \mathrm{Y}(\mathfrak{g}_M) & = & \cdots \\
\varphi_M \downarrow & & \varphi_{M+2} \downarrow & & & & \varphi_N \downarrow & & \\
\mathrm{A}_M(M) & \xleftarrow[o_{M+2}]{} & \mathrm{A}_M(M+2) & \longleftarrow & \cdots & \xleftarrow[o_N]{} & \mathrm{A}_M(N) & \longleftarrow & \cdots .
\end{array}$$

Let us set

$$\mathfrak{S}(u) = \Big(1 - \frac{F}{u + \kappa_N}\Big)^{-1}$$

and denote the matrix elements of $\mathfrak{S}(u)$ by $\sigma_{ij|N}(u)$. We need to prove that

$$\varsigma_N(u)\, \sigma_{ij|N}(u) - \varsigma_{N-2}(u)\, \sigma_{ij|N-2}(u) \in \mathrm{I}(N)[[u^{-1}]], \qquad -n+1 \leqslant i,j \leqslant n-1.$$

However, by (8.60),

$$\varsigma_N(u) - \varsigma_{N-2}(u) \in \mathrm{I}(N)[[u^{-1}]];$$

see Section 8.7. Therefore, it suffices to demonstrate that

$$\sigma_{ij|N}(u) - \sigma_{ij|N-2}(u) \in \mathrm{I}(N)[[u^{-1}]], \qquad -n+1 \leqslant i,j \leqslant n-1.$$

We will prove by induction on r that for the coefficients of these series the following holds:

$$\sigma^{(r)}_{in|N} \in \mathrm{I}(N), \qquad -n \leqslant i \leqslant n, \quad r \geqslant 1, \tag{8.64}$$

$$\sigma^{(r)}_{ij|N} - \sigma^{(r)}_{ij|N-2} \in \mathrm{I}(N), \qquad -n+1 \leqslant i,j \leqslant n-1, \quad r \geqslant 1. \tag{8.65}$$

By definition of $\mathfrak{S}(u)$,

$$\mathfrak{S}(u)(u + \kappa_N - F) = u + \kappa_N.$$

Hence

$$u\, \mathfrak{S}(u) = u + \kappa_N + \mathfrak{S}(u)(F - \kappa_N).$$

Write $\mathfrak{S}(u) = \mathfrak{S}^{(0)} + \mathfrak{S}^{(1)} u^{-1} + \dots$. Then $\mathfrak{S}^{(0)} = 1$ and

$$\mathfrak{S}^{(1)} = F, \tag{8.66}$$

$$\mathfrak{S}^{(r)} = \mathfrak{S}^{(r-1)}(F - \kappa_N), \quad r \geqslant 2. \tag{8.67}$$

So,

$$\sigma^{(1)}_{in|N} = F_{in} \in \mathrm{I}(N), \quad -n \leqslant i \leqslant n,$$

and

$$\sigma^{(1)}_{ij|N} - \sigma^{(1)}_{ij|N-2} = 0, \quad -n+1 \leqslant i,j \leqslant n-1.$$

This gives (8.64) and (8.65) for $r = 1$. For $r > 1$ we obtain from (8.67) that

$$\begin{aligned}\sigma^{(r)}_{in|N} &= \sum_{a=-n}^{n} \sigma^{(r-1)}_{ia|N}(F_{an} - \delta_{an}\kappa_N)\\ &= \sum_{a=-n}^{n-1} \sigma^{(r-1)}_{ia|N} F_{an} + \sigma^{(r-1)}_{in|N} F_{nn} - \kappa_N \sigma^{(r-1)}_{in|N}.\end{aligned}$$

By the induction hypotheses, this expression lies in I(N), which proves (8.64). Furthermore, if $-n+1 \leqslant i, j \leqslant n-1$, then

$$\sigma^{(r)}_{ij|N} - \sigma^{(r)}_{ij|N-2} = \sum_{a=-n}^{n} \sigma^{(r-1)}_{ia|N}(F_{aj} - \delta_{aj}\kappa_N) - \sum_{a=-n+1}^{n-1} \sigma^{(r-1)}_{ia|N-2}(F_{aj} - \delta_{aj}\kappa_{N-2}),$$

which can be rewritten as

$$\sum_{a=-n+1}^{n-1} \big(\sigma^{(r-1)}_{ia|N} - \sigma^{(r-1)}_{ia|N-2}\big) F_{aj} - \kappa_N \sigma^{(r-1)}_{ij|N} + \kappa_{N-2} \sigma^{(r-1)}_{ij|N-2} + \sigma^{(r-1)}_{in|N} F_{nj} + \sigma^{(r-1)}_{i,-n|N} F_{-n,j}.$$

Since φ_N is an algebra homomorphism and the coefficients of $\varsigma_N(u)$ belong to Z($\mathfrak{g}_N$), we obtain from (8.61) and (8.62) that

$$\begin{aligned}\big(\sigma^{(r-1)}_{ia|N} - \sigma^{(r-1)}_{ia|N-2}\big) F_{aj} &= F_{aj}\big(\sigma^{(r-1)}_{ia|N} - \sigma^{(r-1)}_{ia|N-2}\big)\\ &+ \sigma^{(r-1)}_{ij|N} - \sigma^{(r-1)}_{ij|N-2} - \delta_{ij}\big(\sigma^{(r-1)}_{aa|N} - \sigma^{(r-1)}_{aa|N-2}\big)\\ &- \theta_{a,-a}\delta_{i,-a}\big(\sigma^{(r-1)}_{-a,j|N} - \sigma^{(r-1)}_{-a,j|N-2}\big) + \theta_{i,-j}\delta_{-j,a}\big(\sigma^{(r-1)}_{a,-i|N} - \sigma^{(r-1)}_{a,-i|N-2}\big).\end{aligned}$$

Furthermore, since $\kappa_N - \kappa_{N-2} = 1$ we have

$$\begin{aligned}&- \kappa_N \sigma^{(r-1)}_{ij|N} + \kappa_{N-2} \sigma^{(r-1)}_{ij|N-2} + \sigma^{(r-1)}_{in|N} F_{nj}\\ &= -\kappa_{N-2}\big(\sigma^{(r-1)}_{ij|N} - \sigma^{(r-1)}_{ij|N-2}\big) + \sigma^{(r-1)}_{in|N} F_{nj} - \sigma^{(r-1)}_{ij|N}\\ &= -\kappa_{N-2}\big(\sigma^{(r-1)}_{ij|N} - \sigma^{(r-1)}_{ij|N-2}\big) + F_{nj}\sigma^{(r-1)}_{in|N} - \delta_{ij}\sigma^{(r-1)}_{nn|N},\end{aligned}$$

while

$$\sigma^{(r-1)}_{i,-n|N} F_{-n,j} = -\theta_{-j,n}\sigma^{(r-1)}_{i,-n|N} F_{-j,n}.$$

Hence, using (8.64) and the induction hypothesis, we complete the proof of (8.65).

Thus, we have proved that the homomorphism φ is well-defined. Next we show that the kernel of φ is trivial. We make use of the filtration on A_M and pass to the corresponding graded algebra P_M; see Proposition 8.8.1. Consider the highest degree component of the sequence $\varphi(s^{(r)}_{ij}) = \big(\varphi_N(s^{(r)}_{ij}) \mid N \geqslant M,\ N - M \text{ is even}\big)$. By (8.59), the highest degree component of $\varphi_N(s^{(r)}_{ij})$ coincides with that of the element

$$(F^r)_{ij} + (F^{r-1})_{ij}\, \varsigma^{(1)}_N + \dots + \delta_{ij}\, \varsigma^{(r)}_N,$$

where $\varsigma^{(r)}_N$ denotes the coefficient of u^{-r} of the series $\varsigma_N(u)$. Hence, using notation of Section 8.8, we can identify the image of $\varphi_N(s^{(r)}_{ij})$ in the r-th component of the graded algebra $\mathrm{S}(\mathfrak{g}_N) = \operatorname{gr}\mathrm{U}(\mathfrak{g}_N)$ with the element

$$p^{(r)}_{ij|N} + p^{(r-1)}_{ij|N}\, \bar{\varsigma}^{(1)}_N + \dots + \delta_{ij}\, \bar{\varsigma}^{(r)}_N, \tag{8.68}$$

where $\bar{\varsigma}_N^{(r)}$ is the image of $\varsigma_N^{(r)}$ in the r-th component of $\mathrm{S}(\mathfrak{g}_N)$. The elements (8.68) with $N = M, M+2, \dots$ thus determine an element

$$p_{ij}^{(r)} + p_{ij}^{(r-1)}\, \bar{\varsigma}^{(1)} + \dots + \delta_{ij}\, \bar{\varsigma}^{(r)} \in \mathrm{P}_M \tag{8.69}$$

which is the highest degree component of the image $\varphi(s_{ij}^{(r)}) \in \mathrm{A}_M$.

On the other hand, by Proposition 8.8.7, the elements $p_{ij}^{(r)} \in \mathrm{P}_M$ with appropriate restrictions on the indices i, j, r are algebraically independent over the subalgebra P_0. Applying the Poincaré–Birkhoff–Witt theorem for the algebra $\mathrm{Y}(\mathfrak{g}_N)$ (Corollary 4.2.3), we can finally conclude that the kernel of φ is trivial.

The previous argument also shows that the graded algebra $\operatorname{gr} \mathrm{Y}(\mathfrak{g}_M)$ can be identified with the subalgebra of $\operatorname{gr} \mathrm{A}_M = \mathrm{P}_M$ generated by the elements (8.69) with appropriate restrictions on the indices i, j, r. Therefore, the decomposition (8.63) follows from Proposition 8.8.7. □

Using the embeddings $\mathrm{Y}(\mathfrak{g}_M) \hookrightarrow \mathrm{Y}(\mathfrak{g}_{M+2})$, define the *twisted Yangian for* $\mathfrak{g}_\infty$ as the inductive limit algebra

$$\mathrm{Y}(\mathfrak{o}_{2\infty}) = \bigcup_{m \geqslant 1} \mathrm{Y}(\mathfrak{o}_{2m}), \qquad \mathrm{Y}(\mathfrak{o}_{2\infty+1}) = \bigcup_{m \geqslant 1} \mathrm{Y}(\mathfrak{o}_{2m+1}),$$

$$\mathrm{Y}(\mathfrak{sp}_{2\infty}) = \bigcup_{m \geqslant 1} \mathrm{Y}(\mathfrak{sp}_{2m}).$$

The following corollary which is immediate from Theorem 8.9.5 yields a description of the Olshanski algebra A.

COROLLARY 8.9.6. *We have the isomorphism* $\mathrm{A} \cong \mathrm{A}_0 \otimes \mathrm{Y}(\mathfrak{g}_\infty)$. □

REMARK 8.9.7. Since the Yangian $\mathrm{Y}(\mathfrak{g}_M)$ has a nontrivial center (see Theorem 2.8.2), the center of the algebra A_M strictly contains A_0. However, the center of the algebra $\mathrm{Y}(\mathfrak{g}_\infty)$ is trivial, since by Proposition 8.6.3 the center of the algebra A coincides with A_0. □

8.10. Examples

1. Consider the quantized enveloping algebra $\mathrm{U}_q(\mathfrak{gl}_N)$ introduced in Example 1.15.1. Here we consider q as a formal variable so that $\mathrm{U}_q(\mathfrak{gl}_N)$ is regarded as an algebra over $\mathbb{C}(q)$. Denote by $\widetilde{\mathrm{U}}_q(\mathfrak{gl}_N)$ the subalgebra of $\mathrm{U}_q(\mathfrak{gl}_N)$ generated by the elements

$$\tau_{ij} = t_{ij}\bar{t}_{jj}, \quad i > j \qquad \text{and} \qquad \bar{\tau}_{ij} = \bar{t}_{ij}\bar{t}_{jj}, \quad i \leqslant j.$$

Fix a nonnegative integer M such that $M \leqslant N$ and denote by $\widetilde{\mathrm{U}}_q(\mathfrak{gl}_{N,M})$ the subalgebra of $\widetilde{\mathrm{U}}_q(\mathfrak{gl}_N)$ generated by the elements τ_{ij} and $\bar{\tau}_{ij}$ with $M+1 \leqslant i, j \leqslant N$. Let $\mathrm{A}_M(N)$ denote the centralizer of $\widetilde{\mathrm{U}}_q(\mathfrak{gl}_{N,M})$ in $\widetilde{\mathrm{U}}_q(\mathfrak{gl}_N)$. The algebra $\mathrm{A}_M(N)$ inherits the filtration of $\widetilde{\mathrm{U}}_q(\mathfrak{gl}_N)$ defined by

$$\deg \tau_{ij} = 0, \quad i > j, \qquad \text{and} \qquad \deg \bar{\tau}_{ij} = 1, \quad i \leqslant j.$$

For any $M < N$ there exists a filtration-preserving homomorphism

$$o_N : \mathrm{A}_M(N) \to \mathrm{A}_M(N-1).$$

The algebra A_M is defined as the projective limit in the category of filtered associative algebras of the sequence of the algebras $\mathrm{A}_M(N)$, $N \geqslant M$, with respect to the homomorphisms o_N, $N > M$.

Now consider the q-Yangian $\mathrm{Y}_q(\mathfrak{gl}_M)$ defined in Example 1.15.3 and denote by $\widetilde{\mathrm{Y}}_q(\mathfrak{gl}_M)$ the subalgebra of $\mathrm{Y}_q(\mathfrak{gl}_M)$ generated by the coefficients of the series

$$\tau_{ij}(u) = t_{ij}(u)\, \bar{t}^{(0)}_{jj}, \qquad 1 \leqslant i, j \leqslant M.$$

Then the algebra A_M contains a subalgebra isomorphic to $\widetilde{\mathrm{Y}}_q(\mathfrak{gl}_M)$.

Open problem: Find analogues of the decompositions (8.22) and (8.26) for the algebra A_M.

2. Let $a = (a_i)$, $i \in \mathbb{Z}$ be a sequence of variables. Consider the ring of polynomials $\mathbb{Z}[a]$ in the variables a_i with integer coefficients. Introduce another infinite set of variables $x = (x_1, x_2, \dots)$ and for each nonnegative integer n denote by Λ^a_n the ring of symmetric polynomials in $x_1, \dots, x_n$ with coefficients in $\mathbb{Z}[a]$. The ring Λ^a_n is filtered by the usual degrees of polynomials in $x_1, \dots, x_n$. The evaluation map

$$\varphi_n : \Lambda^a_n \to \Lambda^a_{n-1}, \qquad P(x_1, \dots, x_n) \mapsto P(x_1, \dots, x_{n-1}, a_n)$$

is a homomorphism of filtered rings so that we can define the inverse limit ring Λ^a by

$$\Lambda^a = \lim_{\longleftarrow} \Lambda^a_n, \qquad n \to \infty,$$

where the limit is taken with respect to the homomorphisms φ_n in the category of filtered rings.

3. Let λ be a partition of length $\leqslant n$. As in Section 7.4, by a *reverse λ-tableau* $\mathcal{T}$ we will mean the tableau obtained by filling in the cells of λ with the numbers $1, 2, \dots, n$ in such a way that the entries weakly decrease along the rows and strictly decrease down the columns. If $\alpha = (i, j)$ is a cell of λ, we let $\mathcal{T}(\alpha) = \mathcal{T}(i, j)$ denote the entry of $\mathcal{T}$ occupying the cell α. For the sequence a as in Example 2 and $x = (x_1, \dots, x_n)$ set

$$s_\lambda(x \,\|\, a) = \sum_{\mathcal{T}} \prod_{\alpha \in \lambda} (x_{\mathcal{T}(\alpha)} - a_{\mathcal{T}(\alpha) - c(\alpha)}),$$

summed over the reverse λ-tableaux $\mathcal{T}$. These polynomials coincide with the double Schur polynomials defined in Section 7.4,

$$s_\lambda(x \,\|\, a) = s_\lambda(x|b),$$

for the sequences a and b related by $a_{n-i+1} = b_i$ with $i = 1, 2, \dots$. For this reason we will use the same name for the polynomials $s_\lambda(x \,\|\, a)$.

The double Schur polynomials are compatible with respect to the homomorphisms φ_n,

$$\varphi_n : s_\lambda(x_1, \dots, x_n \,\|\, a) \mapsto s_\lambda(x_1, \dots, x_{n-1} \,\|\, a),$$

so that we can define the *double Schur function* $s_\lambda(x \,\|\, a)$ with $x = (x_1, x_2, \dots)$ as the sequence of the double Schur polynomials $(s_\lambda(x_1, \dots, x_n \,\|\, a) \mid n \geqslant 0)$.

4. As λ runs over the set of all partitions, the double Schur functions $s_\lambda(x \,\|\, a)$ form a basis of the ring Λ^a over $\mathbb{Z}[a]$. Introduce the *Littlewood–Richardson polynomials* $c^\nu_{\lambda\mu}(a)$ as the structure coefficients of the ring Λ^a in the basis of double Schur

functions,

$$s_\lambda(x\,\|\,a)\,s_\mu(x\,\|\,a)=\sum_\nu c_{\lambda\mu}^{\nu}(a)\,s_\nu(x\,\|\,a).$$

The Littlewood–Richardson polynomials can be calculated by the following combinatorial rule. Let R denote a sequence of diagrams

$$\mu=\rho^{(0)}\nearrow\rho^{(1)}\nearrow\dots\nearrow\rho^{(l-1)}\nearrow\rho^{(l)}=\nu,\tag{8.70}$$

where $\rho\nearrow\sigma$ means that σ is obtained from ρ by adding one cell. Let r_i denote the row number of the cell added to the diagram $\rho^{(i-1)}$. The sequence $r_1r_2\dots r_l$ is called the *Yamanouchi symbol* of R. Introduce the ordering on the set of cells of a diagram λ by reading them by columns from left to right and from bottom to top in each column. We call this the *column order*. We shall write $\alpha\prec\beta$ if α (strictly) precedes β with respect to the column order. Given a sequence R, construct the set $\mathcal{T}(\lambda,R)$ of *barred* reverse λ-tableaux $\mathcal{T}$ with entries from $\{1,2,\dots\}$ such that $\mathcal{T}$ contains cells $\alpha_1,\dots,\alpha_l$ with

$$\alpha_1\prec\dots\prec\alpha_l\qquad\text{and}\qquad\mathcal{T}(\alpha_i)=r_i,\quad 1\leqslant i\leqslant l.$$

We will distinguish the entries in $\alpha_1,\dots,\alpha_l$ by barring each of them. Thus, an element of $\mathcal{T}(\lambda,R)$ is a pair consisting of a reverse λ-tableau and a chosen sequence of barred entries compatible with R. We will keep the notation $\mathcal{T}$ for such a pair. For each cell α with $\alpha_i\prec\alpha\prec\alpha_{i+1}$, $0\leqslant i\leqslant l$, set $\rho(\alpha)=\rho^{(i)}$. A reverse λ-tableau $\mathcal{T}$ will be called *ν-bounded* if

$$\mathcal{T}(1,j)\leqslant\nu'_j\qquad\text{for all}\quad j=1,\dots,\lambda_1,$$

where ν'_j is the number of cells in column j of ν. Then the polynomial $c_{\lambda\mu}^{\nu}(a)$ is zero unless $\lambda,\mu\subset\nu$. If $\lambda,\mu\subset\nu$, then

$$c_{\lambda\mu}^{\nu}(a)=\sum_R\sum_{\mathcal{T}}\prod_{\substack{\alpha\in\lambda\\ \mathcal{T}(\alpha)\ \text{unbarred}}}\Big(a_{\mathcal{T}(\alpha)-\rho(\alpha)_{\mathcal{T}(\alpha)}}-a_{\mathcal{T}(\alpha)-c(\alpha)}\Big),$$

where the sums are taken over all sequences R of the form (8.70) and all ν-bounded reverse λ-tableaux $\mathcal{T}\in\mathcal{T}(\lambda,R)$. Moreover, for each factor occurring in the formula we have $\rho(\alpha)_{\mathcal{T}(\alpha)}>c(\alpha)$.

5. For the product of the double Schur functions $s_{(2)}(x\,\|\,a)$ and $s_{(2,1)}(x\,\|\,a)$ we have

$$\begin{aligned}s_{(2)}(x\,\|\,a)\,&s_{(2,1)}(x\,\|\,a)\\&=s_{(4,1)}(x\,\|\,a)+s_{(3,2)}(x\,\|\,a)+s_{(3,1,1)}(x\,\|\,a)+s_{(2,2,1)}(x\,\|\,a)\\&+\big(a_{-1}-a_2+a_{-2}-a_0\big)\,s_{(3,1)}(x\,\|\,a)+\big(a_{-1}-a_2\big)\,s_{(2,2)}(x\,\|\,a)\\&+\big(a_{-1}-a_0\big)\,s_{(2,1,1)}(x\,\|\,a)+\big(a_{-1}-a_2\big)\big(a_{-1}-a_0\big)\,s_{(2,1)}(x\,\|\,a).\end{aligned}$$

6. When a is specialized to the sequence of zeros, $s_\lambda(x\,\|\,a)$ becomes the Schur function $s_\lambda(x)$. Hence, if $|\nu|=|\lambda|+|\mu|$, the polynomial $c_{\lambda\mu}^{\nu}(a)$ coincides with the *Littlewood–Richardson coefficient* $c_{\lambda\mu}^{\nu}$. By Example 4, the coefficient $c_{\lambda\mu}^{\nu}$ equals the number of ν-bounded reverse λ-tableaux $\mathcal{T}$ whose column word coincides with the Yamanouchi symbol of a certain sequence R of the form (8.70).

Equivalently, $c_{\lambda\mu}^{\nu}$ counts the cardinality of the intersection of two finite sets: the set of column words of ν-bounded reverse λ-tableaux and the set of Yamanouchi symbols of the sequences of the form (8.70).

7. The shifted Schur functions s^*_μ defined in Section 8.2 can be regarded as specializations of the double Schur functions $s_\mu(x\,\|\,a)$ for the sequence a given by $a_i = -i$ for $i \in \mathbb{Z}$, and $x_i = \lambda_i - i$ for $i = 1, 2, \dots$. Hence, the structure constants $f^\nu_{\lambda\mu}$ introduced in Remark 8.2.7 can be calculated by the formula

$$f^\nu_{\lambda\mu} = \sum_R \sum_{\mathcal{T}} \prod_{\substack{\alpha\in\lambda \\ \mathcal{T}(\alpha)\ \text{unbarred}}} \Big(\rho(\alpha)_{\mathcal{T}(\alpha)} - c(\alpha)\Big),$$

where the sums are taken over all sequences R of the form (8.70) and all ν-bounded reverse λ-tableaux $\mathcal{T} \in \mathcal{T}(\lambda, R)$; see Example 4. In particular, the numbers $f^\nu_{\lambda\mu}$ are nonnegative integers.

8. We use the notation of Examples 3.5.5 and 3.5.6. Suppose that M is a nonnegative integer and let $\lambda = (\lambda_1, \dots, \lambda_{M+N})$ and $\mu = (\mu_1, \dots, \mu_M)$ be tuples of integers such that

$$\lambda_1 \geqslant \dots \geqslant \lambda_{M+N} \qquad \text{and} \qquad \mu_1 \geqslant \dots \geqslant \mu_M.$$

Consider the finite-dimensional irreducible representation $L(\lambda)$ of the quantized enveloping algebra $\mathrm{U}_q(\mathfrak{gl}_{M+N})$ with the highest weight λ and set

$$L(\lambda)^+_\mu = \{\eta \in L(\lambda) \mid \bar t_{ij}\,\eta = 0 \qquad 1 \leqslant i < j \leqslant M \qquad \text{and} \\ t_{ii}\,\eta = q^{\mu_i}\,\eta \qquad i = 1, \dots, M\}.$$

The vector space $L(\lambda)^+_\mu$ is nonzero if and only if

$$\lambda_i \geqslant \mu_i \geqslant \lambda_{i+N} \qquad \text{for} \quad i = 1, \dots, M.$$

We will assume that these inequalities hold. In that case, $L(\lambda)^+_\mu$ admits a basis $\{\zeta_\Lambda\}$ parameterized by the trapezium Gelfand–Tsetlin patterns Λ of the same form as in Section 8.5. We make $L(\lambda)^+_\mu$ into a module over the quantum affine algebra $\mathrm{U}_q(\widehat{\mathfrak{gl}}_N)$ by setting

$$t_{ij}(u) \mapsto \Big(\mathrm{qdet}\,(T - \overline{T}u^{-1})_{\mathcal{A}\mathcal{A}}\Big)^{-1} \cdot \mathrm{qdet}\,(T - \overline{T}u^{-1})_{\mathcal{A}_i\mathcal{A}_j},$$

$$\bar t_{ij}(u) \mapsto \Big(\mathrm{qdet}\,(\overline{T} - Tu)_{\mathcal{A}\mathcal{A}}\Big)^{-1} \cdot \mathrm{qdet}\,(\overline{T} - Tu)_{\mathcal{A}_i\mathcal{A}_j},$$

where $\mathcal{A} = \{1, \dots, M\}$ and $\mathcal{A}_i = \{1, \dots, M, M+i\}$, while T and $\overline{T}$ denote the generator matrices for the algebra $\mathrm{U}_q(\widehat{\mathfrak{gl}}_{M+N})$. We call $L(\lambda)^+_\mu$ the *skew representation* of $\mathrm{U}_q(\widehat{\mathfrak{gl}}_N)$.

9. The $\mathrm{U}_q(\widehat{\mathfrak{gl}}_N)$-module $L(\lambda)^+_\mu$ is isomorphic to the highest weight representation $L(\nu(u), \bar\nu(u))$, where the components of the highest weight are found by

$$\nu_k(u) = \frac{(q^{\nu_k^{(1)}} - q^{-\nu_k^{(1)}}u^{-1}) \dots (q^{\nu_k^{(M+1)}} - q^{-\nu_k^{(M+1)}+2M}u^{-1})}{(q^{\mu_1} - q^{-\mu_1}u^{-1}) \dots (q^{\mu_M} - q^{-\mu_M+2M-2}u^{-1})}$$

and

$$\bar\nu_k(u) = \frac{(q^{-\nu_k^{(1)}} - q^{\nu_k^{(1)}}u) \dots (q^{-\nu_k^{(M+1)}} - q^{\nu_k^{(M+1)}-2M}u)}{(q^{-\mu_1} - q^{\mu_1}u) \dots (q^{-\mu_M} - q^{\mu_M-2M+2}u)},$$

where $k = 1, \dots, N$ and the $\nu_k^{(i)}$ are defined in (8.35).

10. Assume that λ and μ are partitions and use the notation of Section 8.5. The Drinfeld polynomials $P_1(u), \dots, P_{N-1}(u)$ of the $\mathrm{U}_q(\widehat{\mathfrak{gl}}_N)$-module $L(\lambda)^+_\mu$ are given by

$$P_k(u) = \prod_\alpha (1 - q^{2\,c(\alpha)}u), \qquad k = 1, \dots, N-1,$$

where α runs over the top cells of the columns of height k in the diagram of λ/μ.

11. *Open problem*: Find irreducibility conditions for the tensor product of two skew representations $L(\lambda)^+_\mu \otimes L(\lambda')^+_{\mu'}$ of $\mathrm{Y}(\mathfrak{gl}_N)$; cf. Theorem 6.3.3.

Bibliographical notes

8.1–8.4. The connection between centralizers in the classical enveloping algebras and the (twisted) Yangians (the 'centralizer construction') was originally discovered by Olshanski [**372, 373, 374, 376**]. In particular, proofs of Proposition 8.2.3 and Theorem 8.4.3 are contained in [**374**]. The modified version of the construction involving determinants is due to the author; Theorem 8.4.1 was proved in [**307**]. The virtual Gelfand invariants were also found by Gould and Stoilova [**153**].

8.5. The skew representations of the Yangians and quantum affine algebras were first defined by Cherednik [**73**], [**74**] in a slightly different way. The Drinfeld polynomials of the skew representations (Corollary 8.5.5) were first calculated by Nazarov and Tarasov [**351**] using a different approach. Another calculation is contained in the author's paper [**307**]. We have followed Hopkins and Molev [**174**], where the Drinfeld polynomials for the skew representations of the quantum affine algebra were found; see Example 10. A more general class of the *tame* Yangian modules was introduced by Nazarov and Tarasov [**351**]. These are irreducible finite-dimensional modules of $\mathrm{Y}(\mathfrak{gl}_N)$ with a semisimple action of the Gelfand–Tsetlin subalgebra H_N. By the main theorem of [**351**], every tame module is isomorphic to a tensor product of skew representations, up to twisting by an automorphism of the form (1.20). The skew representations of the Yangians and quantum affine algebras were studied in the literature from various viewpoints providing, in particular, different interpretations of the character formula of Corollary 8.5.8; see e.g. Bazhanov and Reshetikhin [**23**], Khoroshkin and Nazarov [**210, 211**], Kirillov, Kuniba and Nakanishi [**225, 226**], Nakai and Nakanishi [**334**], and Nazarov [**345**].

The characters of representations of the Drinfeld Yangians were introduced by Knight [**233**]. He also proved, in particular, that $\mathrm{ch}\,(V \otimes W) = \mathrm{ch}\,V \cdot \mathrm{ch}\,W$ for finite-dimensional representation V and W. In the A type the characters of Knight essentially coincide with the Gelfand–Tsetlin characters (Definition 8.5.7); this terminology was used by Brundan and Kleshchev [**53**, Section 6.2]. Using the tensor product decompositions of the irreducible finite-dimensional $\mathrm{Y}(\mathfrak{gl}_2)$-modules provided by Corollary 3.3.4 and the formula for the Gelfand–Tsetlin character of the module $L(\alpha, \beta)$ implied by Corollary 8.5.8, one can get a formula for the Gelfand–Tsetlin character of any irreducible finite-dimensional $\mathrm{Y}(\mathfrak{gl}_2)$-module. This formula (in its $\mathfrak{sl}_2$-version) was given by Chari and Pressley [**66**]. The *q-characters* of representations of the quantum affine algebras were introduced by Frenkel and Reshetikhin [**116**], which is a q-version of the Knight characters or the Gelfand–Tsetlin characters. For the evaluation modules, a calculation similar to the one given in the proof of Corollary 8.5.8 can be found e.g. in Frenkel and Mukhin [**115**,

Section 4.5] and Brundan and Kleshchev [**53**, Section 7.4]. For further developments and generalizations of the q-characters see e.g. Chari and Moura [**61**], Nakajima [**336**], and Hernandez [**173**].

8.6–8.9. A proof of Theorem 8.9.5 was sketched in Olshanski's paper [**376**]. A detailed argument with some modifications was given in Molev and Olshanski [**317**]. Theorem 8.9.2 was proved in the author's paper [**311**]. The centralizer construction was the starting point for Khoroshkin and Nazarov [**210, 211, 212**] to develop a generalization of the Howe duality in the context of the (twisted) Yangians. Their work also provides a correspondence between intertwining operators on certain (twisted) Yangian modules and the Mickelsson–Zhelobenko algebras (the definition of these algebras is given in Section 9.1 below); see also Khoroshkin and Ogievetsky [**213**].

8.10. Example 1 is a q-version of the centralizer construction; see Hopkins and Molev [**174**]. An analogue of the construction for the symmetric groups was found by Olshanski [**372**] with detailed arguments given in Molev and Olshanski [**318**]. A super-version of the centralizer construction is given by Nazarov and Sergeev [**349**] which produces the Yangian of the queer Lie superalgebra; see also Stukopin [**430**], where a Drinfeld-type presentation of this Yangian is produced. Examples 2–7 are contained in the author's paper [**314**]. An independent proof of the formula for $c_{\lambda\mu}^{\nu}(a)$ was given by Kreiman [**245**]. In a certain specialization of the sequence a, the Littlewood–Richardson polynomials become the structure coefficients for the equivariant cohomology ring on the Grassmannian in the basis of the equivariant Schubert classes. A multiplication rule for the double Schur polynomials and the quantum immanants was given in an earlier paper by Molev and Sagan [**323**]. However, the formula of [**323**] lacks the positivity property established in a more general context by Graham [**159**] (the $c_{\lambda\mu}^{\nu}(a)$ are polynomials in the differences $a_i - a_j$, $i < j$, with positive integer coefficients). The first manifestly positive formula was given by Knutson and Tao [**236**] by using combinatorics of 'puzzles'. The structure coefficients provided by the rules of [**323**] and [**236**] depend on the number of variables and do not exhibit the stability property (as $n \to \infty$) implied by the formula of Example 4. As observed by Fulton [**118**], this property can also be derived from the puzzle rule of [**236**]. A weight-preserving bijection between the puzzles and tableaux was constructed by Kreiman [**245**]. The coefficients $f_{\lambda\mu}^{\nu}$ calculated in Example 7 also determine the multiplication rule for the 'virtual Capelli operators', which are defined as the sequences of the images of the quantum immanants in the representation (7.35) provided by Corollary 7.4.2; see Okounkov and Olshanski [**368**]. Examples 8–10 are taken from [**174**].

CHAPTER 9

Weight bases for representations of $\mathfrak{g}_N$

In this chapter we use the representations of the twisted Yangians to construct analogues of the Gelfand–Tsetlin bases for all finite-dimensional irreducible representations of the classical Lie algebras $\mathfrak{g}_N = \mathfrak{o}_{2n}$, $\mathfrak{o}_{2n+1}$ and $\mathfrak{sp}_N$; cf. Section 5.4. We will use the restrictions of the representations to the subalgebras of the chains

$$\mathfrak{o}_2 \subset \mathfrak{o}_4 \subset \cdots \subset \mathfrak{o}_{2n}, \qquad \mathfrak{o}_3 \subset \mathfrak{o}_5 \subset \cdots \subset \mathfrak{o}_{2n+1}, \qquad \text{and}$$
$$\mathfrak{sp}_2 \subset \mathfrak{sp}_4 \subset \cdots \subset \mathfrak{sp}_{2n},$$

respectively. It is well known that the reductions $\mathfrak{g}_N \downarrow \mathfrak{g}_{N-2}$ are not multiplicity-free in general. In order to "separate" these multiplicities, we employ the bases of the corresponding $\mathrm{Y}(\mathfrak{g}_2)$-modules constructed in Chapter 4.

9.1. The Mickelsson algebra theory

Here we reproduce some basic results of the theory of Mickelsson algebras which will be used later in this chapter. The proofs can be found in the book by Zhelobenko [**487**] and his expository paper [**486**].

Let $\mathfrak{g}$ be a Lie algebra over $\mathbb{C}$ and $\mathfrak{k}$ be its subalgebra reductive in $\mathfrak{g}$. This means that the adjoint $\mathfrak{k}$-module $\mathfrak{g}$ is completely reducible. In particular, $\mathfrak{k}$ is a reductive Lie algebra. Fix a Cartan subalgebra $\mathfrak{h}$ of $\mathfrak{k}$ and a triangular decomposition

$$\mathfrak{k} = \mathfrak{k}^- \oplus \mathfrak{h} \oplus \mathfrak{k}^+.$$

The subalgebras $\mathfrak{k}^-$ and $\mathfrak{k}^+$ are respectively spanned by the negative and positive root vectors $e_{-\alpha}$ and e_α with α running over the set of positive roots Δ^+ of $\mathfrak{k}$ with respect to $\mathfrak{h}$. The root vectors will be assumed to be normalized in such a way that

$$[e_\alpha, e_{-\alpha}] = h_\alpha, \qquad \alpha(h_\alpha) = 2 \tag{9.1}$$

for all $\alpha \in \Delta^+$.

Let $\mathrm{J} = \mathrm{U}(\mathfrak{g})\,\mathfrak{k}^+$ be the left ideal of $\mathrm{U}(\mathfrak{g})$ generated by $\mathfrak{k}^+$. Its normalizer $\mathrm{Norm}\,\mathrm{J}$ is a subalgebra of $\mathrm{U}(\mathfrak{g})$ defined by

$$\mathrm{Norm}\,\mathrm{J} = \{u \in \mathrm{U}(\mathfrak{g}) \mid \mathrm{J}\,u \subseteq \mathrm{J}\}.$$

Then J is a two-sided ideal of Norm J and the *Mickelsson algebra* $\mathrm{S}(\mathfrak{g}, \mathfrak{k})$ is defined as the quotient

$$\mathrm{S}(\mathfrak{g}, \mathfrak{k}) = \mathrm{Norm}\,\mathrm{J}/\mathrm{J}.$$

Let $\mathrm{R}(\mathfrak{h})$ denote the field of fractions of the commutative algebra $\mathrm{U}(\mathfrak{h})$. In what follows it will be convenient to consider the extension $\mathrm{U}'(\mathfrak{g})$ of the universal enveloping algebra $\mathrm{U}(\mathfrak{g})$ defined by

$$\mathrm{U}'(\mathfrak{g}) = \mathrm{U}(\mathfrak{g}) \otimes_{\mathrm{U}(\mathfrak{h})} \mathrm{R}(\mathfrak{h}). \tag{9.2}$$

We will identify $\mathrm{U}(\mathfrak{g})$ with the subalgebra $\mathrm{U}(\mathfrak{g}) \otimes 1$ of $\mathrm{U}'(\mathfrak{g})$. Let $\mathrm{J}' = \mathrm{U}'(\mathfrak{g})\,\mathfrak{k}^+$ be the left ideal of $\mathrm{U}'(\mathfrak{g})$ generated by $\mathfrak{k}^+$. Exactly as with the ideal J above, J' is a two-sided ideal of the normalizer Norm J′ and the *Mickelsson–Zhelobenko algebra* $\mathrm{Z}(\mathfrak{g},\mathfrak{k})$ is defined as the quotient

$$\mathrm{Z}(\mathfrak{g},\mathfrak{k}) = \mathrm{Norm}\,\mathrm{J}'/\mathrm{J}'.$$

Clearly, $\mathrm{Z}(\mathfrak{g},\mathfrak{k})$ is an extension of the Mickelsson algebra $\mathrm{S}(\mathfrak{g},\mathfrak{k})$,

$$\mathrm{Z}(\mathfrak{g},\mathfrak{k}) = \mathrm{S}(\mathfrak{g},\mathfrak{k}) \otimes_{\mathrm{U}(\mathfrak{h})} \mathrm{R}(\mathfrak{h}).$$

The Mickelsson–Zhelobenko algebra is an algebra over $\mathbb{C}$ as well as a left and right $\mathrm{R}(\mathfrak{h})$-module. An equivalent definition of $\mathrm{Z}(\mathfrak{g},\mathfrak{k})$ can be given by using the quotient space

$$\mathrm{M}(\mathfrak{g},\mathfrak{k}) = \mathrm{U}'(\mathfrak{g})/\mathrm{J}'.$$

The algebra $\mathrm{Z}(\mathfrak{g},\mathfrak{k})$ coincides with the subspace of $\mathfrak{k}$-highest vectors in $\mathrm{M}(\mathfrak{g},\mathfrak{k})$,

$$\mathrm{Z}(\mathfrak{g},\mathfrak{k}) = \mathrm{M}(\mathfrak{g},\mathfrak{k})^{+},$$

where

$$\mathrm{M}(\mathfrak{g},\mathfrak{k})^{+} = \{v \in \mathrm{M}(\mathfrak{g},\mathfrak{k}) \mid \mathfrak{k}^{+} v = 0\}.$$

The algebraic structure of $\mathrm{Z}(\mathfrak{g},\mathfrak{k})$ can be described with the use of the 'extremal projector' for the Lie algebra $\mathfrak{k}$. In order to define it, suppose that the positive roots are $\Delta^{+} = \{\alpha_1,\dots,\alpha_m\}$. Consider the vector space $\mathrm{F}_{\mu}(\mathfrak{k})$ of formal series of monomials

$$e_{-\alpha_1}^{k_1} \dots e_{-\alpha_m}^{k_m} e_{\alpha_m}^{r_m} \dots e_{\alpha_1}^{r_1}$$

of weight μ with coefficients in $\mathrm{R}(\mathfrak{h})$, where

$$(r_1 - k_1)\,\alpha_1 + \dots + (r_m - k_m)\,\alpha_m = \mu.$$

Introduce the vector space $\mathrm{F}(\mathfrak{k})$ as the direct sum

$$\mathrm{F}(\mathfrak{k}) = \bigoplus_{\mu} \mathrm{F}_{\mu}(\mathfrak{k}).$$

That is, the elements of $\mathrm{F}(\mathfrak{k})$ are finite sums $\sum x_{\mu}$ with $x_{\mu} \in \mathrm{F}_{\mu}(\mathfrak{k})$. It can be shown that $\mathrm{F}(\mathfrak{k})$ is an algebra with respect to the natural multiplication of formal series. The algebra $\mathrm{F}(\mathfrak{k})$ is equipped with a Hermitian anti-involution (conjugate-linear involutive anti-automorphism) defined by

$$e_{\alpha}^{*} = e_{-\alpha}, \qquad \alpha \in \Delta^{+}.$$

Furthermore, call a linear ordering of the positive roots *normal* if any composite root lies between its components. For instance, there are precisely two normal orderings for the root system of type B_2,

$$\Delta^{+} = \{\alpha, \alpha+\beta, \alpha+2\beta, \beta\} \qquad \text{and} \qquad \Delta^{+} = \{\beta, \alpha+2\beta, \alpha+\beta, \alpha\},$$

where α and β are the simple roots. In general, the number of normal orderings coincides with the number of reduced decompositions of the longest element of the corresponding Weyl group.

For any $\alpha \in \Delta^{+}$ introduce the element of $\mathrm{F}(\mathfrak{k})$ by

$$p_{\alpha} = 1 + \sum_{k=1}^{\infty} e_{-\alpha}^{k}\, e_{\alpha}^{k}\, \frac{(-1)^k}{k!\,(h_{\alpha} + \rho(h_{\alpha}) + 1) \dots (h_{\alpha} + \rho(h_{\alpha}) + k)}, \tag{9.3}$$

where h_{α} is defined in (9.1) and ρ is the half sum of the positive roots. Finally, define the *extremal projector* $p = p_{\mathfrak{k}}$ by

$$p = p_{\alpha_1} \cdots p_{\alpha_m}$$

with the product taken in a normal ordering of the positive roots α_i.

THEOREM 9.1.1. *The element* $p \in \mathrm{F}(\mathfrak{k})$ *does not depend on the normal ordering on* Δ^+ *and satisfies the conditions*

$$e_\alpha\, p = p\, e_{-\alpha} = 0 \qquad \text{for all} \qquad \alpha \in \Delta^+. \tag{9.4}$$

Moreover, $p^* = p$ *and* $p^2 = p$. □

In fact, the relations (9.4) uniquely determine the element p, up to a factor from $\mathrm{R}(\mathfrak{h})$. The extremal projector naturally acts on the vector space $\mathrm{M}(\mathfrak{g}, \mathfrak{k})$.

THEOREM 9.1.2. *The extremal projector* p *projects* $\mathrm{M}(\mathfrak{g}, \mathfrak{k})$ *onto* $\mathrm{Z}(\mathfrak{g}, \mathfrak{k})$ *with the kernel* $\mathfrak{k}^-\mathrm{M}(\mathfrak{g}, \mathfrak{k})$. *In particular,* $\mathrm{Z}(\mathfrak{g}, \mathfrak{k}) = p\,\mathrm{M}(\mathfrak{g}, \mathfrak{k})$ *and*

$$\mathrm{M}(\mathfrak{g}, \mathfrak{k}) = \mathrm{Z}(\mathfrak{g}, \mathfrak{k}) \oplus \mathfrak{k}^-\mathrm{M}(\mathfrak{g}, \mathfrak{k}).$$ □

Similarly, for any $\mathfrak{g}$-module V set

$$V^+ = \{v \in V \mid \mathfrak{k}^+ v = 0\}.$$

For any $\mu \in \mathfrak{h}^*$ we let V_μ denote the corresponding $\mathfrak{k}$-weight subspace of V,

$$V_\mu = \{\eta \in V \mid h\,\eta = \mu(h)\,\eta, \qquad h \in \mathfrak{h}\}$$

and set $V^+_\mu = V_\mu \cap V^+$. Suppose now that $\dim V < \infty$ and μ has the property that $\mu(h_\alpha) + \rho(h_\alpha) + k \neq 0$ for all positive integers k and all positive roots α. Then the extremal projector p can be regarded as an operator on V_μ. Moreover, p projects V_μ onto V^+_μ, annihilating the subspace $V_\mu \cap \mathfrak{k}^- V$. That is, if $e_{-\alpha}\eta \in V_\mu$ for some $\alpha \in \Delta^+$ and $\eta \in V$, then $p e_{-\alpha}\eta = 0$.

In order to get a more explicit description of the algebra $\mathrm{Z}(\mathfrak{g}, \mathfrak{k})$, consider a $\mathfrak{k}$-module decomposition

$$\mathfrak{g} = \mathfrak{k} \oplus \mathfrak{p}.$$

Choose a weight basis $e_1, \dots, e_n$ (with respect to the adjoint action of $\mathfrak{h}$) of the complementary module $\mathfrak{p}$.

THEOREM 9.1.3. *The elements*

$$a_i = p\, e_i, \qquad i = 1, \dots, n$$

are generators of the Mickelsson–Zhelobenko algebra $\mathrm{Z}(\mathfrak{g}, \mathfrak{k})$. *Moreover, the monomials*

$$a_1^{k_1} \dots a_n^{k_n}, \qquad k_i \in \mathbb{Z}_+,$$

form a basis of $\mathrm{Z}(\mathfrak{g}, \mathfrak{k})$. □

It can be proved that the generators a_i of $\mathrm{Z}(\mathfrak{g}, \mathfrak{k})$ satisfy quadratic defining relations. For the pairs $(\mathfrak{g}, \mathfrak{k})$ relevant to the constructions of bases of the Gelfand–Tsetlin type, the relations will be explicitly written down below in this chapter.

Regarding $\mathrm{Z}(\mathfrak{g}, \mathfrak{k})$ as a right $\mathrm{R}(\mathfrak{h})$-module, it is possible to introduce the normalized elements

$$z_i = a_i\, \pi_i, \qquad \pi_i \in \mathrm{U}(\mathfrak{h})$$

by multiplying a_i by its *right denominator* π_i in such a way that the z_i can be viewed as elements of the Mickelsson algebra $\mathrm{S}(\mathfrak{g}, \mathfrak{k})$.

THEOREM 9.1.4. *Let* $V = \mathrm{U}(\mathfrak{g})\, v$ *be a cyclic* $\mathrm{U}(\mathfrak{g})$*-module generated by an element* $v \in V^+$. *Then the subspace* V^+ *is linearly spanned by the elements*

$$z_1^{k_1} \dots z_n^{k_n}\, v, \qquad k_i \in \mathbb{Z}_+.$$ □

9.2. Mickelsson–Zhelobenko algebra $\mathrm{Z}(\mathfrak{g}_N, \mathfrak{g}_{N-2})$

Whenever possible we consider the three cases $\mathfrak{g}_N = \mathfrak{o}_{2n+1}$, $\mathfrak{sp}_{2n}$ and $\mathfrak{o}_{2n}$ simultaneously, unless otherwise stated. They will often be referred to as the B, C and D cases, respectively. The rows and columns of $2n \times 2n$ matrices will be numbered by the indices $-n, \dots, -1, 1, \dots, n$, while the rows and columns of $(2n+1) \times (2n+1)$ matrices will be numbered by the indices $-n, \dots, -1, 0, 1, \dots, n$. Accordingly, the index 0 will usually be skipped in the former case. The Lie algebra $\mathfrak{g}_N$ is spanned by the elements F_{ij}, $-n \leqslant i, j \leqslant n$, defined in (7.9). We will regard $\mathfrak{g}_{N-2}$ as the subalgebra of $\mathfrak{g}_N$ spanned by the elements F_{ij} with $-n+1 \leqslant i, j \leqslant n-1$. This subalgebra is nonzero for $n \geqslant 2$. In what follows, the trivial case $n = 1$ (and $n = 2$ in the D case) will usually be excluded.

We let $\mathfrak{h}_N$ denote the Cartan subalgebra of $\mathfrak{g}_N$ which consists of the diagonal matrices. The elements $F_{11}, \dots, F_{nn}$ form a basis of $\mathfrak{h}_N$. As in Section 4.2, the vectors of the dual basis of $\mathfrak{h}_N^*$ will be denoted by $\varepsilon_1, \dots, \varepsilon_n$. Any element of the dual space $\lambda = \lambda_1 \varepsilon_1 + \cdots + \lambda_n \varepsilon_n \in \mathfrak{h}_N^*$ will be identified with the n-tuple of complex numbers $\lambda = (\lambda_1, \dots, \lambda_n)$. We choose the set of positive roots Δ^+ for $\mathfrak{g}_N$ as in Section 4.2 so that the half-sum of the positive roots is $\rho = (\rho_1, \dots, \rho_n)$, where the ρ_i are defined in (7.10).

We apply the definitions and constructions of Section 9.1 to the pair of Lie algebras $\mathfrak{g} = \mathfrak{g}_N$ and $\mathfrak{k} = \mathfrak{g}_{N-2}$. In particular, J and J$'$ will denote the left ideals of the algebras $\mathrm{U}(\mathfrak{g}_N)$ and $\mathrm{U}'(\mathfrak{g}_N)$, respectively, generated by the elements F_{ij} with $-n < i < j < n$. The following simple observation will be frequently used. The Poincaré–Birkhoff–Witt theorem for the algebra $\mathrm{U}(\mathfrak{g}_N)$ implies

$$\mathrm{U}(\mathfrak{k}^-)\mathfrak{p} \cap \mathrm{J} = \{0\}, \tag{9.5}$$

where $\mathfrak{k}^-$ is the subalgebra of $\mathfrak{g}_N$ spanned by the elements F_{ji} with $-n < i < j < n$, while $\mathfrak{p}$ is the vector subspace of $\mathfrak{g}_N$ spanned by the elements F_{in} and $F_{i,-n}$ with $-n \leqslant i \leqslant n$.

We will write $\mathfrak{h} = \mathfrak{h}_{N-2}$ and let p denote the extremal projector for the Lie algebra $\mathfrak{g}_{N-2}$. The projector has the properties

$$F_{ij}\, p = 0 \qquad \text{and} \qquad p\, F_{ji} = 0 \qquad \text{for} \quad -n < i < j < n \tag{9.6}$$

implied by Theorem 9.1.1. By Theorem 9.1.3, the elements

$$F_{nn}, \qquad pF_{ia}, \qquad a = -n, n, \quad i = -n+1, \dots, n-1 \tag{9.7}$$

are generators of the Mickelsson–Zhelobenko algebra $\mathrm{Z}(\mathfrak{g}_N, \mathfrak{g}_{N-2})$ in the orthogonal case. In the symplectic case, the algebra $\mathrm{Z}(\mathfrak{g}_N, \mathfrak{g}_{N-2})$ is generated by the elements (9.7) together with $F_{n,-n}$ and $F_{-n,n}$. In order to write down explicit formulas for the generators, introduce the elements $f_i \in \mathrm{U}(\mathfrak{h})$ by

$$f_i = F_{ii} + \rho_i, \qquad f_{-i} = -f_i$$

for $i = 1, \dots, n$. Moreover, in the case of $\mathfrak{o}_{2n+1}$ also set $f_0 = -1/2$. The generators pF_{ia} can be given by a uniform expression in all three cases. Introduce the elements $\check{z}_{ia} \in \mathrm{U}'(\mathfrak{g}_N)$ by

$$\check{z}_{ia} = F_{ia} + \sum_{i > i_1 > \cdots > i_s > -n} F_{ii_1} F_{i_1 i_2} \dots F_{i_{s-1} i_s} F_{i_s a} \frac{1}{(f_i - f_{i_1}) \dots (f_i - f_{i_s})}, \tag{9.8}$$

summed over $s \geqslant 1$.

PROPOSITION 9.2.1. *For any $a \in \{-n, n\}$ and $i \in \{-n+1, \dots, n-1\}$ we have the relation in $\mathrm{U}'(\mathfrak{g}_N)$ modulo the left ideal J',*

$$pF_{ia} = \check{z}_{ia}. \tag{9.9}$$

PROOF. Let us verify that $\check{z}_{ia}$ belongs to the normalizer $\mathrm{Norm}\,\mathrm{J}'$ of the left ideal J'. It suffices to check that $F_{j,j+1}\check{z}_{ia}$ with $j = 1, \dots, n-2$ and also $F_{01}\check{z}_{ia}$, $F_{-1,1}\check{z}_{ia}$ and $F_{-1,2}\check{z}_{ia}$ in the B, C and D cases, respectively, are contained in J'. Indeed, suppose first that $-j < i \leqslant 0$. Then for the indices

$$i > i_1 > \dots > i_k > -j > -j-1 > i_l > \dots > i_s > -n$$

we have the following relations modulo J':

$$\begin{aligned} F_{j,j+1} \cdot F_{ii_1} \dots F_{i_k,-j} F_{-j,-j-1} F_{-j-1,i_l} \dots F_{i_s a} \\ = -F_{ii_1} \dots F_{i_k,-j} F_{-j-1,i_l} \dots F_{i_s a}(f_j - f_{j+1}), \end{aligned}$$

$$F_{j,j+1} \cdot F_{ii_1} \dots F_{i_k,-j} F_{-j,i_l} \dots F_{i_s a} = -F_{ii_1} \dots F_{i_k,-j} F_{-j-1,i_l} \dots F_{i_s a},$$

and

$$F_{j,j+1} \cdot F_{ii_1} \dots F_{i_k,-j-1} F_{-j-1,i_l} \dots F_{i_s a} = F_{ii_1} \dots F_{i_k,-j} F_{-j-1,i_l} \dots F_{i_s a}.$$

Since

$$-\frac{f_j - f_{j+1}}{(f_i - f_{-j})(f_i - f_{-j-1})} - \frac{1}{f_i - f_{-j}} + \frac{1}{f_i - f_{-j-1}} = 0,$$

the monomial

$$F_{ii_1} \dots F_{i_k,-j} F_{-j-1,i_l} \dots F_{i_s a}$$

occurs in the expansion of $F_{j,j+1}\check{z}_{ia}$ with the zero coefficient. Hence, $F_{j,j+1}\check{z}_{ia} \in \mathrm{J}'$. A similar calculation proves the claim in the remaining cases.

Thus, $p\check{z}_{ia} = \check{z}_{ia} \mod \mathrm{J}'$. On the other hand, due to the second equality in (9.6), we have $p\check{z}_{ia} = pF_{ia} \mod \mathrm{J}'$. □

We also need a relation dual to (9.8) provided by the following lemma.

LEMMA 9.2.2. *Let $a \in \{-n, n\}$ and $i \in \{-n+1, \dots, n-1\}$. Then we have the relation in $\mathrm{U}'(\mathfrak{g}_N)$,*

$$F_{ia} = \check{z}_{ia} + \sum_{i > i_1 > \dots > i_s > -n} F_{ii_1} F_{i_1 i_2} \dots F_{i_{s-1} i_s} \check{z}_{i_s,a} \frac{1}{(f_{i_s} - f_i)(f_{i_s} - f_{i_1}) \dots (f_{i_s} - f_{i_{s-1}})},$$

summed over $s \geqslant 1$.

PROOF. Expanding $\check{z}_{i_s,a}$, we write the right-hand side as a linear combination of monomials of the form

$$F_{ii_1} F_{i_1 i_2} \dots F_{i_{s-1} i_s} F_{i_s i_{s+1}} \dots F_{i_r,a} \tag{9.10}$$

with coefficients in $\mathrm{R}(\mathfrak{h})$, where $i > i_1 > \dots > i_s > i_{s+1} > \dots > i_r > -n$. Observe that F_{ia} occurs on the right-hand side with the coefficient 1, while the coefficient of any other monomial of the form (9.10) with $r \geqslant 1$ is

$$\sum_{k=0}^{r} \frac{1}{(f_{i_k} - f_{i_0}) \dots \wedge_k \dots (f_{i_k} - f_{i_r})} = 0,$$

where we set $i_0 = i$. Here we have used the well known identity

$$\sum_{i=1}^{n} \frac{1}{(x_i - x_1)\dots\wedge_i\dots(x_i - x_n)} = 0 \tag{9.11}$$

which holds for $n \geqslant 2$ and arbitrary variables $x_1, \dots, x_n$. The left-hand side of (9.11) can be written as the ratio of two polynomials. The denominator is the Vandermonde polynomial, while the numerator is a skew-symmetric polynomial in $x_1, \dots, x_n$ of degree less than $n(n-1)/2$ and hence is zero. □

In what follows it will be convenient to work with normalized generators of the Mickelsson-Zhelobenko algebra. Set

$$z_{ia} = \check{z}_{ia}\,(f_i - f_{i-1})\dots(f_i - f_{-n+1}) \tag{9.12}$$

in the B, C cases, and

$$z_{ia} = \check{z}_{ia}\,(f_i - f_{i-1})\dots\widehat{(f_i - f_{-i})}\dots(f_i - f_{-n+1}) \tag{9.13}$$

in the D case, where $\check{z}_{ia}$ is defined in (9.8) and the hat indicates the factor to be omitted; this factor occurs only for positive values of i. As usual, the zero index is skipped in the C and D cases.

COROLLARY 9.2.3. *For all $a \in \{-n, n\}$ and $i \in \{-n+1, \dots, n-1\}$ the elements z_{ia} are contained in the universal enveloping algebra $\mathrm{U}(\mathfrak{g}_N)$. In particular, they may be regarded as elements of the Mickelsson algebra $\mathrm{S}(\mathfrak{g}_N, \mathfrak{g}_{N-2})$.*

PROOF. In the B and C cases the claim is immediate from relation (9.8). In the D case, this relation only implies the claim for $i \leqslant 1$. In order to prove it for $i \geqslant 2$, we use the explicit formula for the extremal projector p given in Section 9.1. As can easily be seen, the sequence of roots

$$\Delta_i = \{\varepsilon_{i-1} - \varepsilon_i,\ \dots,\ \varepsilon_1 - \varepsilon_i,\ -\varepsilon_1 - \varepsilon_i,\ \dots,\ -\varepsilon_{i-1} - \varepsilon_i\}$$

can be completed to a normally ordered sequence containing all positive roots Δ^+ of $\mathfrak{o}_{2n-2}$. Hence, the extremal projector p can be written as the ordered product

$$p = \prod_{\alpha \in \Delta_i} p_\alpha \cdot p',$$

where p' is the product of the p_α with $\alpha \in \Delta^+ \setminus \Delta_i$. Now, we have $p' F_{ia} = F_{ia}$ modulo J' so that

$$pF_{ia} = p_{i-1,i}\, p_{i-2,i} \cdots p_{1,i}\, p_{-1,i} \cdots p_{-i+1,i}\, F_{ia}, \tag{9.14}$$

where $p_{l,i} = p_\alpha$ for $\alpha = \varepsilon_l - \varepsilon_i$ and $p_{-l,i} = p_\alpha$ for $\alpha = -\varepsilon_l - \varepsilon_i$. Due to (9.3), for $1 \leqslant l < i$ we have

$$p_{l,i} = 1 + \sum_{k=1}^{\infty} F_{il}^k\, F_{li}^k\, \frac{1}{k!\,(f_i - f_l - 1)\dots(f_i - f_l - k)}$$

and

$$p_{-l,i} = 1 + \sum_{k=1}^{\infty} F_{i,-l}^k\, F_{-l,i}^k\, \frac{1}{k!\,(f_i + f_l - 1)\dots(f_i + f_l - k)}.$$

Hence, (9.14) yields a formula for $pF_{ia} \in \mathrm{U}'(\mathfrak{o}_{2n})/\mathrm{J}'$ as a linear combination of elements of $\mathrm{U}(\mathfrak{k}^-)\mathfrak{p}$ with coefficients in $\mathrm{R}(\mathfrak{h})$. Moreover, as the above formulas for

$p_{l,i}$ and $p_{-l,i}$ show, the coefficients of this linear combination are rational functions in $f_1, \dots, f_{n-1}$ which are regular at $f_i = 0$.

Now choose an arbitrary basis of $\mathrm{U}(\mathfrak{k}^-)\mathfrak{p}$ and use (9.13) to write z_{ia} as a linear combination of the basis elements with coefficients in $\mathrm{R}(\mathfrak{h})$. Then each coefficient is either a polynomial in $f_1, \dots, f_{n-1}$ or the ratio of a polynomial in $f_1, \dots, f_{n-1}$ and f_i. However, in the latter case, the ratio must be a rational function regular at $f_i = 0$ due to Proposition 9.2.1. Thus, this ratio is a polynomial in $f_1, \dots, f_{n-1}$. □

EXAMPLE 9.2.4. If $a \in \{-3, 3\}$, then in the Mickelsson–Zhelobenko algebra $\mathrm{Z}(\mathfrak{o}_6, \mathfrak{o}_4)$ we have

$$z_{-1,a} = F_{-1,a}(f_{-1} - f_{-2}) + F_{-1,-2}F_{-2,a}, \qquad z_{-2,a} = F_{-2,a},$$
$$z_{1,a} = F_{1,a}(f_1 - f_{-2}) + F_{1,-2}F_{-2,a},$$

while

$$\begin{aligned} z_{2,a} &= F_{2,a}(f_2 - f_1)(f_2 - f_{-1}) + F_{2,1}F_{1,a}(f_2 - f_{-1}) + F_{2,-1}F_{-1,a}(f_2 - f_1) \\ &+ F_{2,1}F_{1,-2}F_{-2,a}\frac{f_2 - f_{-1}}{f_2 - f_{-2}} + F_{2,-1}F_{-1,-2}F_{-2,a}\frac{f_2 - f_1}{f_2 - f_{-2}}, \end{aligned}$$

which can also be written as

$$\begin{aligned} z_{2,a} &= F_{2,a}(f_2 - f_1)(f_2 - f_{-1}) + F_{2,1}F_{1,a}(f_2 - f_{-1}) + F_{2,-1}F_{-1,a}(f_2 - f_1) \\ &+ F_{2,1}F_{1,-2}F_{-2,a}. \end{aligned}$$

□

As we mentioned in Section 9.1, the elements z_{ia} satisfy some quadratic relations which can be shown to be the defining relations of the algebra $\mathrm{Z}(\mathfrak{g}_N, \mathfrak{g}_{N-2})$. We will use only a part of these relations given by the next proposition.

PROPOSITION 9.2.5. *For any $a, b \in \{-n, n\}$ and $i, j \in \{-n+1, \dots, n-1\}$ such that $i + j \neq 0$ we have in $\mathrm{Z}(\mathfrak{g}_N, \mathfrak{g}_{N-2})$,*

$$z_{ia}z_{jb} + z_{ja}z_{ib}(f_i - f_j - 1) = z_{ib}z_{ja}(f_i - f_j). \tag{9.15}$$

In particular, z_{ia} and z_{ja} commute for $i + j \neq 0$. Also, z_{ia} and z_{ib} commute for $i \neq 0$ and all values $a, b \in \{-n, n\}$.

PROOF. Suppose first that $i > j$ and simplify the product $pF_{ja}pF_{ib}$ modulo J' by using the expression for pF_{ib} provided by Proposition 9.2.1. Observe that by (9.6), if $i > i_1 > i_2 > -n$, then $pF_{ja}F_{ii_1} = 0$ unless $i_1 = j$. In that case $pF_{ja}F_{ij} = -pF_{ia}$ and $pF_{ia}F_{ji_2} = 0$. Therefore, modulo J',

$$pF_{ja}pF_{ib} = pF_{ja}\Big(F_{ib} + F_{ij}F_{jb}\frac{1}{f_i - f_j}\Big) = pF_{ja}F_{ib} - pF_{ia}F_{jb}\frac{1}{f_i - f_j},$$

while $pF_{ia}pF_{jb} = pF_{ia}F_{jb} = pF_{jb}F_{ia}$. Hence,

$$pF_{ja}pF_{ib} = pF_{ib}pF_{ja} - pF_{ia}pF_{jb}\frac{1}{f_i - f_j}.$$

Now, for $i > j$ the relation (9.15), regarded modulo J', follows from (9.12) and (9.13). This also implies that z_{ia} and z_{ja} commute. Furthermore, if $a \neq b$, then swapping a and b we get the relation

$$z_{ib}z_{ja} + z_{jb}z_{ia}(f_i - f_j - 1) = z_{ia}z_{jb}(f_i - f_j). \tag{9.16}$$

Regarding (9.15) and (9.16) with $i > j$ as a system of equations on the unknowns $z_{ia}z_{jb}$ and $z_{ib}z_{ja}$, we can express $z_{ia}z_{jb}$ as a linear combination of $z_{jb}z_{ia}$ and $z_{ja}z_{ib}$. The resulting expression yields (9.15) with $i < j$.

The last claim of the proposition is verified by the same argument. □

Apart from the z_{ia}, we will also use the elements z_{ai} defined by

$$z_{ai} = (-1)^{n-i}\, z_{-i,-a} \qquad \text{and} \qquad z_{ai} = (-1)^{n-i}\, \operatorname{sgn} a \cdot z_{-i,-a}$$

in the orthogonal and symplectic case, respectively. The next proposition provides alternative expressions for the z_{ai}. As usual, the zero index is skipped in the C and D cases.

PROPOSITION 9.2.6. *For any $a \in \{-n, n\}$ and $i \in \{-n+1, \dots, n-1\}$ the element $z_{ai} \in \mathrm{U}(\mathfrak{g}_N)$ can be given by the formula*

$$z_{ai} = \sum_{n > i_1 > \cdots > i_s > i} (f_i - f_{j_1}) \dots (f_i - f_{j_r})\, F_{ai_1} F_{i_1 i_2} \dots F_{i_s i} \tag{9.17}$$

in the B, C cases. The same formula applies in the D case for $i \geqslant 1$, while

$$z_{ai} = \sum_{n > i_1 > \cdots > i_s > i} \frac{(f_i - f_{j_1}) \dots (f_i - f_{j_r})}{2 f_i}\, F_{ai_1} F_{i_1 i_2} \dots F_{i_s i} \tag{9.18}$$

for $i \leqslant -1$. Here $s = 0, 1, \dots$ and $\{j_1, \dots, j_r\}$ is the complement to the subset $\{i_1, \dots, i_s\}$ in the set $\{i+1, \dots, n-1\}$.

PROOF. Denote by z'_{ai} the element of $\mathrm{U}'(\mathfrak{g}_N)$ which occurs on the right-hand side of (9.17) or (9.18). A direct calculation similar to the one used in the proof of Proposition 9.2.1 shows that z'_{ai} belongs to the normalizer $\operatorname{Norm} \mathrm{J}'$ of the left ideal J'. Hence z'_{ai} is stable under the application of the extremal projector p for the subalgebra $\mathfrak{g}_{N-2}$ modulo J'. On the other hand, $p z'_{ai}$ can be calculated as follows. Suppose first that $i \geqslant 0$. By the second equality in (9.6), in the quotient $\mathrm{U}'(\mathfrak{g}_N)/\mathrm{J}'$ we have

$$p F_{ai_1} F_{i_1 i_2} \dots F_{i_s i} = p F_{ai}.$$

Therefore,

$$p z'_{ai} = (f_i - f_{i+1} + 1) \dots (f_i - f_{n-1} + 1)\, p F_{ai}.$$

Note that $F_{ai} = -F_{-i,-a}$ and $F_{ai} = -\operatorname{sgn} a \cdot F_{-i,-a}$ in the orthogonal and symplectic cases, respectively. Since $f_j = -f_{-j}$ for $j \neq 0$ and $f_0 + 1 = -f_0$, we conclude from (9.12) and (9.13) that $z'_{ai} = z_{ai} \mod \mathrm{J}'$ in the case under consideration.

Now suppose $i \leqslant -1$ and consider the D case first. Exactly as in the above argument, we find that

$$p z'_{ai} = (f_i - f_{i+1}) \dots \widehat{(f_i - f_{-i})} \dots (f_i - f_{n-1}) \sum \frac{1}{(f_i - f_{i_1}) \dots (f_i - f_{i_s})}\, p F_{ai},$$

summed over $s \geqslant 0$ and the indices $n > i_1 > \cdots > i_s > i$ such that $i_p + i_{p+1} \neq 0$ for $p = 1, \dots, s$ with $i_{s+1} = i$. Using the relation

$$\begin{aligned} 1 + \frac{1}{f_i - f_k} + \frac{1}{f_i - f_{-k}} &+ \frac{1}{(f_i - f_{-i})(f_i - f_k)} \\ &+ \frac{1}{(f_i - f_{-i})(f_i - f_{-k})} = \frac{(f_i - f_k + 1)(f_i - f_{-k} + 1)}{(f_i - f_k)(f_i - f_{-k})}, \end{aligned} \tag{9.19}$$

which holds for any (nonzero) index k such that $i < k < -i$, we get

$$\begin{aligned} p z'_{ai} &= (f_i - f_{i+1} + 1) \dots \widehat{(f_i - f_{-i} + 1)} \dots (f_i - f_{n-1} + 1)\, p F_{ai} \\ &= p F_{ai}\, (f_i - f_{i+1}) \dots \widehat{(f_i - f_{-i})} \dots (f_i - f_{n-1}). \end{aligned}$$

Again, as in the case $i \geqslant 1$, we conclude that $z'_{ai} = z_{ai} \mod \mathrm{J}'$.

Similarly, in the C case, we obtain

$$pz'_{ai} = (f_i - f_{i+1}) \dots (f_i - f_{n-1}) \sum \frac{c(i_1, \dots, i_s)}{(f_i - f_{i_1}) \dots (f_i - f_{i_s})} pF_{ai},$$

summed over $s \geqslant 0$ and the indices $n > i_1 > \cdots > i_s > i$, where $c(i_1, \dots, i_s) = 2$ if there is a pair of indices i_p and i_{p+1} such that $i_p + i_{p+1} = 0$ for $p \in \{1, \dots, s\}$ with $i_{s+1} = i$, and $c(i_1, \dots, i_s) = 1$ otherwise. Using the relation (9.19), we come to

$$\begin{aligned} pz'_{ai} &= (f_i - f_{i+1} + 1) \dots (f_i - f_{-i} + 2) \dots (f_i - f_{n-1} + 1)\, pF_{ai} \\ &= pF_{ai}\, (f_i - f_{i+1}) \dots (f_i - f_{-i}) \dots (f_i - f_{n-1}) \end{aligned}$$

which gives $z'_{ai} = z_{ai} \mod \mathrm{J}'$.

Finally, in the B case,

$$pz'_{ai} = (f_i - f_{i+1}) \dots (f_i - f_{n-1}) \sum \frac{1}{(f_i - f_{i_1}) \dots (f_i - f_{i_s})} pF_{ai},$$

summed over $s \geqslant 0$ and the indices $n > i_1 > \cdots > i_s > i$ such that $i_p + i_{p+1} \neq 0$ for $p = 1, \dots, s$ with $i_{s+1} = i$. We divide the sum here into two parts, where the first is taken over the sets of indices $\{i_1, \dots, i_s\}$ containing the zero index, while the second is taken over the remaining sets. Both sums are easily calculated with the use of (9.19) for the second. Then the relation

$$1 + \frac{f_i - f_{-i} + 1}{(f_i - f_0)(f_i - f_{-i})} = \frac{f_i - f_{-i} + 2}{f_i - f_{-i}}$$

allows us to simplify the whole sum and leads to

$$\begin{aligned} pz'_{ai} &= (f_i - f_{i+1} + 1) \dots (f_i - f_0) \dots (f_i - f_{-i} + 2) \dots (f_i - f_{n-1} + 1)\, pF_{ai} \\ &= pF_{ai}\, (f_i - f_{i+1}) \dots (f_i - f_0 - 1) \dots (f_i - f_{-i}) \dots (f_i - f_{n-1}). \end{aligned}$$

Since $f_0 + 1 = -f_0$, we get $z'_{ai} = z_{ai} \mod \mathrm{J}'$.

Thus, we have showed that in all cases $z'_{ai} - z_{ai} \in \mathrm{J}'$. Since the subspace $\mathfrak{p}$ of $\mathfrak{g}_N$ is invariant under the adjoint action of $\mathfrak{k}^-$, we find that

$$z'_{ai} \in \mathrm{U}(\mathfrak{k}^-)\, \mathfrak{p} \otimes_{\mathrm{U}(\mathfrak{h})} \mathrm{R}(\mathfrak{h}).$$

By (9.5), we may now conclude that $z'_{ai} = z_{ai}$. □

9.3. Twisted Yangian and Mickelsson–Zhelobenko algebra

For all $a, b \in \{-n, n\}$ introduce the polynomials $Z_{ab}(u)$ in a variable u with coefficients in $\mathrm{U}'(\mathfrak{g}_N)$ by

$$Z_{ab}(u) = -\Big(\delta_{ab}(u + \rho_n + 1/2) + F_{ab}\Big) \prod_{i=-n+1}^{n-1} (u + g_i) + \sum_{i=-n+1}^{n-1} z_{ai} z_{ib} \prod_{j=-n+1,\, j\neq i}^{n-1} \frac{u + g_j}{g_i - g_j} \tag{9.20}$$

in the B case, and

$$Z_{ab}(u) = \Big(\delta_{ab}(u+\rho_n+1/2)+F_{ab}\Big)\prod_{i=-n+1}^{n-1}(u+g_i) - \sum_{i=-n+1}^{n-1} z_{ai}z_{ib}\prod_{j=-n+1,\, j\neq i}^{n-1}\frac{u+g_j}{g_i-g_j} \tag{9.21}$$

in the C case, while in the D case we introduce the rational function in u by

$$Z_{ab}(u) = -\Bigg(\Big(\delta_{ab}(u+\rho_n+1/2)+F_{ab}\Big)\prod_{i=-n+1}^{n-1}(u+g_i) - \sum_{i=-n+1}^{n-1} z_{ai}z_{ib}\,(u+g_{-i})\prod_{j=-n+1,\, j\neq\pm i}^{n-1}\frac{u+g_j}{g_i-g_j}\Bigg)\frac{1}{2u+1}, \tag{9.22}$$

where $g_i = f_i + 1/2$ for all i, and ρ_n is defined in (7.10). In the B and C cases, the coefficients of the polynomial $Z_{ab}(u)$ belong to the normalizer Norm J$'$ and so they can be regarded as elements of the Mickelsson–Zhelobenko algebra $\mathrm{Z}(\mathfrak{g}_N, \mathfrak{g}_{N-2})$. Similarly, in the D case we may regard $Z_{ab}(u)$ as a Laurent series in u^{-1} with coefficients in $\mathrm{Z}(\mathfrak{g}_N, \mathfrak{g}_{N-2})$.

Consider the split presentation of the twisted Yangian $\mathrm{Y}(\mathfrak{g}_2)$ as defined in Section 4.1. It will be convenient in this chapter to label the generator series of $\mathrm{Y}(\mathfrak{g}_2)$ by $s_{ab}(u)$ with a and b taking values in $\{-n, n\}$ instead of $\{-1, 1\}$.

THEOREM 9.3.1. (i) *The assignment*

$$s_{ab}(u) \mapsto -u^{-2n}\, Z_{ab}(u), \qquad a, b \in \{-n, n\} \tag{9.23}$$

defines an algebra homomorphism $\mathrm{Y}(\mathfrak{o}_2) \to \mathrm{Z}(\mathfrak{o}_{2n+1}, \mathfrak{o}_{2n-1})$.

(ii) *The assignment*

$$s_{ab}(u) \mapsto (u+1/2)\, u^{-2n}\, Z_{ab}(u), \qquad a, b \in \{-n, n\} \tag{9.24}$$

defines an algebra homomorphism $\mathrm{Y}(\mathfrak{sp}_2) \to \mathrm{Z}(\mathfrak{sp}_{2n}, \mathfrak{sp}_{2n-2})$.

(iii) *The assignment*

$$s_{ab}(u) \mapsto -2\, u^{-2n+2}\, Z_{ab}(u), \qquad a, b \in \{-n, n\} \tag{9.25}$$

defines an algebra homomorphism $\mathrm{Y}(\mathfrak{o}_2) \to \mathrm{Z}(\mathfrak{o}_{2n}, \mathfrak{o}_{2n-2})$.

PROOF. Consider the homomorphism $\mathrm{Y}(\mathfrak{g}_2) \to \mathrm{Y}(\mathfrak{g}_N)$ provided by Proposition 4.1.11 with $M = N-2$, so that

$$s_{ab}(u) \mapsto \alpha_{-N+2}(u) \cdot s^{-n+1\dots n-1,\, a}_{-n+1\dots n-1,\, b}(u+N/2-1). \tag{9.26}$$

Let us set

$$g(u) = \prod_{i=1}^{n-1}\big(1-(\rho_i-1/2)^2u^{-2}\big) \qquad \text{for} \quad \mathfrak{g}_N = \mathfrak{o}_N$$

and

$$g(u) = \prod_{i=1}^{n}\big(1-(\rho_i+1/2)^2u^{-2}\big) \qquad \text{for} \quad \mathfrak{g}_N = \mathfrak{sp}_{2n},$$

with the ρ_i defined in (7.10). It will be convenient to twist the homomorphism (9.26) by the automorphism of $\mathrm{Y}(\mathfrak{g}_2)$ given by

$$s_{ab}(u) \mapsto g(u)\, s_{ab}(u), \qquad a, b \in \{-n, n\},$$

so that for the images of the generators we have

$$s_{ab}(u) \mapsto \alpha_{-N+2}(u) \cdot g(u) \cdot s^{-n+1\dots n-1,\, a}_{-n+1\dots n-1,\, b}\,(u + N/2 - 1).$$

Now take the composition of this homomorphism with the evaluation homomorphism ϱ_N defined in (2.106). By Corollary 4.1.9, the image of this composition is contained in the centralizer $\mathrm{U}(\mathfrak{g}_N)^{\mathfrak{g}_{N-2}}$ of the subalgebra $\mathfrak{g}_{N-2}$ in the universal enveloping algebra $\mathrm{U}(\mathfrak{g}_N)$. This gives a homomorphism

$$\mathrm{Y}(\mathfrak{g}_2) \to \mathrm{U}(\mathfrak{g}_N)^{\mathfrak{g}_{N-2}}. \tag{9.27}$$

However, the centralizer is a subalgebra of the normalizer Norm J; see Section 9.2. Hence, we have a natural homomorphism

$$\mathrm{U}(\mathfrak{g}_N)^{\mathfrak{g}_{N-2}} \to \mathrm{S}(\mathfrak{g}_N, \mathfrak{g}_{N-2}) \tag{9.28}$$

obtained by the restriction of the canonical epimorphism Norm J $\to \mathrm{S}(\mathfrak{g}_N, \mathfrak{g}_{N-2})$. The composition of (9.27) and (9.28) yields a homomorphism from the twisted Yangian $\mathrm{Y}(\mathfrak{g}_2)$ to the Mickelsson algebra $\mathrm{S}(\mathfrak{g}_N, \mathfrak{g}_{N-2})$. We will demonstrate that under the homomorphism

$$\mathrm{Y}(\mathfrak{g}_2) \to \mathrm{S}(\mathfrak{g}_N, \mathfrak{g}_{N-2}) \tag{9.29}$$

constructed in this way, the images of the generator series $s_{ab}(u)$ will be expressed in terms of the elements z_{ai} and z_{ib} by the desired formulas.

Note that due to Corollary 4.1.12, the image of $s_{ab}(u)$ under the homomorphism (9.29) can also be found with the use of the map (4.9). For fixed $a, b \in \{-n, n\}$, denote by $\widetilde{F}$ the matrix with the rows numbered by $-n+1, \dots, n-1, a$ and columns numbered by $-n+1, \dots, n-1, b$ such that its (i,j) entry is F_{ij}. Then by Lemma 4.1.6, under the homomorphism (9.29) we have

$$s_{ab}(u) \mapsto g(u) \left| 1 + \frac{\widetilde{F}}{u - N/2 + 3/2} \right|_{ab} \varrho_N\big(\mathrm{sdet}\, S^{(n-1)}(-u + N/2 - 2)\big)$$

in the orthogonal case, and

$$s_{ab}(u) \mapsto g(u)\, \frac{2u+1}{2u - 2n + 3} \left| 1 + \frac{\widetilde{F}}{u - n + 1/2} \right|_{ab} \varrho_N\big(\mathrm{sdet}\, S^{(n-1)}(-u + n - 2)\big)$$

in the symplectic case, regarding the images of the coefficients of the series $s_{ab}(u)$ in $\mathrm{U}'(\mathfrak{g}_N)$ modulo the left ideal J'. We will use the notation

$$F^{(r)}_{kl} = \sum_{i_1, \dots, i_{r-1}} F_{ki_1} F_{i_1 i_2} \dots F_{i_{r-1} l}, \tag{9.30}$$

where $r \geqslant 1$ and the indices $i_1, \dots, i_{r-1}$ run over the set $\{-n+1, \dots, n-1\}$. By the definition of the quasideterminant (Definition 1.10.1), for a variable v we have

$$\big|1 - \widetilde{F}\, v\big|_{ab} = \delta_{ab} - F_{ab}\, v - \sum_{r=2}^{\infty} F^{(r)}_{ab}\, v^r. \tag{9.31}$$

Note that the following relations in $\mathrm{U}(\mathfrak{g}_N)$ hold:

$$[F^{(r-1)}_{ai}, F_{kl}] = \delta_{ki} F^{(r-1)}_{al} - \theta_{kl}\, \delta_{i,-l} F^{(r-1)}_{a,-k}, \tag{9.32}$$

where $i, k, l \in \{-n+1, \dots, n-1\}$. Indeed, they are easily deduced from Corollary 8.9.4 by taking into account that

$$F_{ai}^{(r-1)} = \sum_{j=-n+1}^{n-1} F_{aj} (\check{F}^{r-2})_{ji},$$

where $\check{F}$ denotes the generator matrix F associated with $\mathfrak{g}_{N-2}$. Similarly, since for $r \geqslant 2$ we have

$$F_{ab}^{(r)} = \sum_{i=-n+1}^{n-1} F_{ai}^{(r-1)} F_{ib}, \tag{9.33}$$

relation (9.32) implies that $F_{ab}^{(r)}$ belongs to the centralizer $\mathrm{U}(\mathfrak{g}_N)^{\mathfrak{g}_{N-2}}$. Therefore, $F_{ab}^{(r)} \in \mathrm{Norm}\, \mathrm{J}'$, so that we have $F_{ab}^{(r)} = p F_{ab}^{(r)}$ modulo J', where p denotes the extremal projector for $\mathfrak{g}_{N-2}$.

Now replace F_{ib} in (9.33) by the expression provided by Lemma 9.2.2. Using (9.6) and (9.32), for $i > i_1 > \dots > i_s > -n$ we obtain modulo J'

$$pF_{ai}^{(r-1)} F_{ii_1} F_{i_1 i_2} \dots F_{i_{s-1} i_s} pF_{i_s b} = pF_{ai_s}^{(r-1)} pF_{i_s b}, \tag{9.34}$$

unless $i_p + i_{p+1} = 0$ for some $p \in \{0, 1, \dots, s-1\}$, where $i_0 := i$. If the latter is the case, then the left-hand side of (9.34) is zero in the orthogonal case, while

$$pF_{ai}^{(r-1)} F_{ii_1} F_{i_1 i_2} \dots F_{i_{s-1} i_s} pF_{i_s b} = 2\, pF_{ai_s}^{(r-1)} pF_{i_s b}$$

in the symplectic case. Hence,

$$F_{ab}^{(r)} = \sum_{j=-n+1}^{n-1} pF_{aj}^{(r-1)} pF_{jb} \sum \frac{c(i_1, \dots, i_s)}{(f_j - f_{i_1}) \dots (f_j - f_{i_s})}, \tag{9.35}$$

where the second sum is taken over $s \geqslant 0$ and the indices $n > i_1 > \dots > i_s > j$, with $c(i_1, \dots, i_s) = 1$ unless there is a pair of indices i_p and i_{p+1} such that $i_p + i_{p+1} = 0$ for $p \in \{1, \dots, s\}$ with $i_{s+1} = j$. In the latter case $c(i_1, \dots, i_s)$ equals 2 or 0 in the symplectic or orthogonal case, respectively. Now observe that exactly the same sum was calculated in the proof of Proposition 9.2.6. Furthermore, arguing as above, for any index $j \in \{-n+1, \dots, n-1\}$ with the exception of the value $j = 0$ in the B case, we obtain

$$pF_{aj}^{(r-1)} = pF_{aj} (f_j - \rho_n - 1)^{r-2}$$

modulo J'. In the B case we have $pF_{a0}^{(r-1)} = pF_{a0} (n-1)^{r-2}$ and hence, in all the cases,

$$pF_{aj}^{(r-1)} pF_{jb} = pF_{aj}\, pF_{jb} (f_j - \rho_n)^{r-2}.$$

Thus, writing (9.35) in terms of the elements z_{ai} and z_{ib} (see Section 9.2) we come to the relation (modulo J')

$$F_{ab}^{(r)} = \sum_{j=-n+1}^{n-1} z_{aj} z_{jb} \frac{(f_j - \rho_n)^{r-2}}{(f_j - f_{-n+1}) \dots \wedge_j \dots (f_j - f_{n-1})} \tag{9.36}$$

in the B, C cases and

$$F_{ab}^{(r)} = \sum_{j=-n+1}^{n-1} z_{aj} z_{jb} \frac{(f_j - \rho_n)^{r-2}}{(f_j - f_{-n+1}) \dots \wedge_{-j,j} \dots (f_j - f_{n-1})} \tag{9.37}$$

in the D case.

Now, we obviously have

$$\varrho_N\big(\operatorname{sdet} S^{(n-1)}(-u+N/2-2)\big)=\varrho_{N-2}\big(\operatorname{sdet} S^{(n-1)}(-u+N/2-2)\big),$$

and the coefficients of this series belong to the center of the universal enveloping algebra $\mathrm{U}(\mathfrak{g}_{N-2})$; see Theorem 7.1.6. Therefore, the images of these coefficients in the Mickelsson algebra $\mathrm{S}(\mathfrak{g}_N,\mathfrak{g}_{N-2})$ coincide with their images under the Harish-Chandra homomorphism (7.11). They can be found from Theorem 7.1.6 and are given by

$$\varrho_{N-2}\big(\operatorname{sdet} S^{(n-1)}(-u+N/2-2)\big)=\frac{1}{\alpha_{N-2}(u)}\prod_{i=1}^{n-1}\frac{(u+1/2)^2-f_i^2}{(u+1/2)^2-\rho_i^2}.$$

So, it remains to put $v=-(u+\rho_n+1/2)^{-1}$ in (9.31), then use the identity

$$\sum_{r=2}^{\infty}(f_j-\rho_n)^{r-2}\,v^{r-1}=-\frac{1}{u+f_j+1/2}$$

and replace the coefficients f_i by $g_i-1/2$ everywhere to arrive at the desired relations. □

COROLLARY 9.3.2. *For any $a\in\{-n,n\}$ we have*

$$Z_{a,-a}(-u)=Z_{a,-a}(u).$$

Moreover, in the orthogonal case

$$Z_{-a,-a}(u)=\frac{2u-1}{2u}\,Z_{a,a}(-u)+\frac{1}{2u}\,Z_{a,a}(u),$$

while in the symplectic case

$$Z_{-a,-a}(u)=\frac{1-2u}{2u}\,Z_{a,a}(-u)-\frac{1}{2u}\,Z_{a,a}(u).$$

PROOF. All relations are immediate from Theorem 9.3.1 and the symmetry relation (4.4). □

COROLLARY 9.3.3. *For any $a\in\{-n,n\}$ in $\mathrm{Z}(\mathfrak{g}_N,\mathfrak{g}_{N-2})$ we have*

$$Z_{a,-a}(u)\,Z_{a,-a}(v)=Z_{a,-a}(v)\,Z_{a,-a}(u),$$

where u and v are variables.

PROOF. This follows from Theorem 9.3.1 and the defining relations (4.3) of $\mathrm{Y}(\mathfrak{g}_2)$ as they give $[s_{a,-a}(u),s_{a,-a}(v)]=0$. □

COROLLARY 9.3.4. *For any $i\in\{-n+1,\dots,n-1\}$ and $a\in\{-n,n\}$, in $\mathrm{Z}(\mathfrak{g}_N,\mathfrak{g}_{N-2})$ we have*

$$[z_{ai},Z_{a,-a}(u)]=0.$$

PROOF. Consider the series $g(u)$ defined in the proof of Theorem 9.3.1 and take the composition of the automorphism of $\mathrm{Y}(\mathfrak{g}_N)$ given by $S(u)\mapsto g(u)\,S(u)$ with the homomorphism $\varphi_N:\mathrm{Y}(\mathfrak{g}_N)\to\mathrm{U}(\mathfrak{g}_N)$ given by (8.59). This composition is the homomorphism $\psi:\mathrm{Y}(\mathfrak{g}_N)\to\mathrm{U}(\mathfrak{g}_N)$ such that

$$\psi: S(u)\mapsto \varsigma_N(u)\cdot g(u)\cdot\Big(1-\frac{F}{u+\kappa_N}\Big)^{-1},$$

with $\varsigma_N(u)$ and κ_N defined in (8.59). Then (8.62) gives

$$\psi: s_{kl}^{(1)}\mapsto F_{kl},\qquad k,l\in\{-n,\dots,n\}.$$

Hence, by (8.61) for any $i, j \in \{-n+1, \dots, n-1\}$ and $a \in \{-n, n\}$ we have

$$\big[F_{ai}, \psi\big(s_{a,-a}(u)\big)\big] = 0, \qquad \big[F_{ij}, \psi\big(s_{a,-a}(u)\big)\big] = 0.$$

Using the expressions (9.17) and (9.18) for z_{ai} we derive that

$$\big[z_{ai}, \psi\big(s_{a,-a}(u)\big)\big] = 0.$$

Finally, the coefficients of the series $\psi\big(s_{a,-a}(u)\big)$ belong to the centralizer $\mathrm{U}(\mathfrak{g}_N)^{\mathfrak{g}_{N-2}}$ and the image of $-\psi\big(s_{a,-a}(u)\big)$ or $\psi\big(s_{a,-a}(u)\big)$ in the orthogonal or symplectic case, respectively, in the Mickelsson–Zhelobenko algebra $\mathrm{Z}(\mathfrak{g}_N, \mathfrak{g}_{N-2})$ coincides with the image under the homomorphism provided by Theorem 9.3.1. □

We will need to regard $\mathrm{U}'(\mathfrak{g}_N)$ as a subalgebra of the extended algebra

$$\mathrm{U}(\mathfrak{g}_N) \otimes_{\mathrm{U}(\mathfrak{h}_N)} \mathrm{R}(\mathfrak{h}_N), \tag{9.38}$$

where $\mathfrak{h}_N$ is the Cartan subalgebra of $\mathfrak{g}_N$ spanned by the elements $F_{11}, \dots, F_{nn}$ and $\mathrm{R}(\mathfrak{h}_N)$ denotes the field of fractions of the commutative algebra $\mathrm{U}(\mathfrak{h}_N)$. Accordingly, we consider the extended Mickelsson–Zhelobenko algebra defined by

$$\mathrm{S}(\mathfrak{g}_N, \mathfrak{g}_{N-2}) \otimes_{\mathrm{U}(\mathfrak{h}_N)} \mathrm{R}(\mathfrak{h}_N). \tag{9.39}$$

It has natural structures of the left and right $\mathrm{R}(\mathfrak{h}_N)$-modules.

Suppose that $Q(u)$ is a linear combination

$$Q(u) = c_1\, q_1(u) + \dots + c_k\, q_k(u), \qquad c_i \in \mathrm{U}(\mathfrak{g}_N),$$

where the $q_i(u)$ are rational functions in u with coefficients in $\mathrm{R}(\mathfrak{h}_N)$. We will need to consider the values of $Q(h)$ where $h \in \mathrm{U}(\mathfrak{h}_N)$. In order to make such evaluations unambiguous, we will agree to write the coefficients c_i to the left of the rational functions $q_i(u)$ so that the element $Q(h)$ of the algebra (9.38) is defined by

$$Q(h) = c_1\, q_1(h) + \dots + c_k\, q_k(h).$$

The same convention applies to linear combinations $Q(u)$ where the coefficients c_i belong to the Mickelsson algebra $\mathrm{S}(\mathfrak{g}_N, \mathfrak{g}_{N-2})$.

Observe that under this convention, the following relations are immediate from the definition of $Z_{ab}(u)$: for any $i \in \{-n+1, \dots, n-1\}$ we have

$$Z_{ab}(-g_i) = z_{ai} z_{ib}, \qquad g_i = f_i + 1/2. \tag{9.40}$$

Now for any $a \in \{-n, n\}$ we set

$$z_{a,-a} = Z_{a,-a}(-g_n), \qquad g_n = f_n + 1/2. \tag{9.41}$$

Then $z_{a,-a}$ belongs to $\mathrm{U}'(\mathfrak{g}_N)$, and it can also be regarded as an element of the Mickelsson–Zhelobenko algebra $\mathrm{Z}(\mathfrak{g}_N, \mathfrak{g}_{N-2})$ in the B and C cases, while in the D case $z_{a,-a}$ belongs to the algebra (9.38) and can be regarded as an element of the extended Mickelsson–Zhelobenko algebra (9.39). An explicit expression for $z_{a,-a}$ in terms of the generators F_{ij} of $\mathfrak{g}_N$ is provided by the next proposition.

PROPOSITION 9.3.5. *For any $a \in \{-n, n\}$ we have the relations*

$$z_{a,-a} = \sum_{n > i_1 > \dots > i_s > -n} F_{ai_1} F_{i_1 i_2} \dots F_{i_s,-a}\, (f_n - f_{j_1}) \dots (f_n - f_{j_r}) \tag{9.42}$$

in the B, C cases and

$$z_{a,-a} = \sum_{n > i_1 > \dots > i_s > -n} F_{ai_1} F_{i_1 i_2} \dots F_{i_s,-a}\, \frac{(f_n - f_{j_1}) \dots (f_n - f_{j_r})}{2 f_n} \tag{9.43}$$

in the D case, where $s = 0, 1, \dots$ *and* $\{j_1, \dots, j_r\}$ *is the complement to the subset* $\{i_1, \dots, i_s\}$ *in the set* $\{-n+1, \dots, n-1\}$.

PROOF. We take (9.42) or (9.43), respectively, as a definition of $z_{a,-a}$ and show that (9.41) holds. Let us verify the following relations in the algebra (9.38):

$$(9.44) \qquad \sum_{i=-n+1}^{n} z_{ai}\, z_{i,-a} \frac{1}{(f_i - f_{-n+1}) \dots \wedge_i \dots (f_i - f_n)} = F_{a,-a}$$

in the B, C cases and

$$(9.45) \qquad \sum_{i=-n+1}^{n-1} z_{ai}\, z_{i,-a} \frac{1}{(f_i - f_{-n+1}) \dots \wedge_{-i,i} \dots (f_i - f_n)} + z_{a,-a} \frac{2f_n}{(f_n - f_{-n+1}) \dots (f_n - f_{n-1})} = 0$$

in the D case, where $z_{nn} = 1$ and the wedge indicates the indices for which the corresponding factors should be skipped. Write the expressions which occur on the left-hand side of the relations in terms of the generators F_{ij}. We use Proposition 9.2.6 in order to expand z_{ai}, and (9.12) and (9.13) to expand $z_{i,-a}$. For a fixed value of i, the expression will contain a linear combination of monomials of the form

$$(9.46) \qquad F_{a,i_1} F_{i_1 i_2} \dots F_{i_s i}\, F_{i k_1} F_{k_1 k_2} \dots F_{k_p,-a}$$

with $n = i_0 > i_1 > \dots > i_s > i > k_1 > \dots > k_p > -n$. Recall that $F_{a,-a} = 0$ in the B and D cases, while in the C case $F_{a,-a}$ occurs with the coefficient 1 in the summand corresponding to the value $i = n$. The coefficient of any other monomial of the form (9.46), written to the right of the monomial, is equal to

$$\frac{1}{(f_i - f_{i_0}) \dots (f_i - f_{i_s})(f_i - f_{k_1}) \dots (f_i - f_{k_p})}.$$

Taking the sum over i we find that the coefficient of the monomial

$$F_{a,l_1} F_{l_1 l_2} \dots F_{l_r,-a}, \qquad n = l_0 > l_1 > \dots > l_r > -n, \qquad r \geqslant 1,$$

in the entire expression equals

$$\sum_{j=0}^{r} \frac{1}{(f_{l_j} - f_{l_0}) \dots \wedge_j \dots (f_{l_j} - f_{l_r})} = 0,$$

where we have used the identity (9.11).

It remains to observe that (9.44) and (9.45), respectively, are equivalent to the relation $z_{a,-a} = Z_{a,-a}(-g_n)$. □

COROLLARY 9.3.6. *For any* $a \in \{-n, n\}$ *the element* $z_{a,-a}$ *is contained in the universal enveloping algebra* $\mathrm{U}(\mathfrak{g}_N)$ *so that it may be regarded as an element of the Mickelsson algebra* $\mathrm{S}(\mathfrak{g}_N, \mathfrak{g}_{N-2})$.

PROOF. No proof is required in the B and C cases, so we consider only the D case. Since $z_{a,-a}$ belongs to the algebra (9.39), the application of the extremal projector p for $\mathfrak{o}_{2n-2}$ to $z_{a,-a}$ does not change it modulo J$'$. On the other hand, using the second equality in (9.6), we get

$$p\, z_{a,-a} = \sum_{j=-n+1}^{n-1} pF_{a,j}F_{j,-a}\, c_j = \sum_{j=1}^{n-1} pF_{a,j}F_{j,-a}\, d_j, \qquad c_j \in \mathrm{R}(\mathfrak{h}_N), \tag{9.47}$$

where we have used the relation $F_{a,-j}F_{-j,-a} = F_{a,j}F_{j,-a}$ and set $d_j = c_j + c_{-j}$. For $j = 1, \dots, n-1$ the coefficients c_j are found by

$$c_j = \frac{(f_n - f_{n-1} + 1)\dots(f_n - f_{j+1} + 1)}{(f_n - f_{n-1})\dots(f_n - f_j)}\, d$$

and

$$c_{-j} = \frac{(f_n - f_{n-1} + 1)\dots(f_n - f_{j+1} + 1)}{(f_n - f_{n-1})\dots(f_n - f_{j+1})(f_n - f_{-j})}\, d + \sum_{k=1}^{j-1}(c_k + c_{-k})\, \frac{1}{f_n - f_{-j}},$$

with

$$d = \frac{(f_n - f_{n-1})\dots(f_n - f_{-n+1})}{2f_n}.$$

So, we have the following recurrence relation for the coefficients d_j:

$$\begin{aligned} d_j = {} & (f_n^2 - f_1^2)\dots(f_n^2 - f_{j-1}^2)(f_n + f_{j+1})\dots(f_n + f_{n-1}) \\ & \times (f_n - f_{j+1} + 1)\dots(f_n - f_{n-1} + 1) + \frac{1}{f_n + f_j}\sum_{k=1}^{j-1} d_k \end{aligned} \tag{9.48}$$

with $j = 1, \dots, n-1$ and $d_0 := 0$. An easy induction shows that each coefficient d_j in the expansion (9.47) is a polynomial in $f_1, \dots, f_n$.

The argument is now completed in a way similar to the proof of Corollary 9.2.3. We have the following analogue of (9.5):

$$\mathrm{U}(\mathfrak{k}^-)\,\mathfrak{p}^{(2)} \cap \mathrm{J} = \{0\}, \tag{9.49}$$

where $\mathfrak{p}^{(2)}$ is the subspace of $\mathrm{U}(\mathfrak{o}_{2n})$ spanned by the elements $F_{ia}F_{ja}$ such that $i, j \in \{-n+1, \dots, n-1\}$ and $a \in \{-n, n\}$. Choose an arbitrary basis of $\mathrm{U}(\mathfrak{k}^-)\,\mathfrak{p}^{(2)}$ and use (9.43) to write $z_{a,-a}$ as a linear combination of the basis elements with coefficients in $\mathrm{R}(\mathfrak{h}_N)$. Then each coefficient is the ratio of a polynomial in $f_1, \dots, f_n$ and f_n. However, f_n does not occur in the formula for the extremal projector p, and so by (9.47) and (9.49) the ratio must be a rational function regular at $f_n = 0$. Thus, this ratio is a polynomial in $f_1, \dots, f_n$. □

COROLLARY 9.3.7. *For any $a \in \{-n, n\}$ we have*

$$Z_{a,-a}(u) = \sum_{i=1}^{n} z_{ai} z_{i,-a} \prod_{j=1,\, j\neq i}^{n} \frac{u^2 - g_j^2}{g_i^2 - g_j^2} \tag{9.50}$$

in the B, C cases and

$$Z_{a,-a}(u) = \sum_{i=1}^{n-1} z_{ai} z_{i,-a} \prod_{j=1,\, j\neq i}^{n-1} \frac{u^2 - g_j^2}{g_i^2 - g_j^2} \tag{9.51}$$

in the D case, where $g_i = f_i + 1/2$ for all i.

PROOF. By Corollary 9.3.2, in the B and C cases $Z_{a,-a}(u)$ is a polynomial in u^2 of degree $n-1$. On the other hand, the values of $Z_{a,-a}(u)$ at $u=-g_i$ for $i=1,\dots,n$ are found from (9.40) and (9.41). Hence, the desired expression for $Z_{a,-a}(u)$ follows from the Lagrange interpolation formula.

For the same argument to work in the D case, we need to verify only that $Z_{a,-a}(u)$ is a polynomial in u. Indeed, if that is the case, then $Z_{a,-a}(u)$ is a polynomial in u^2 of degree $n-2$, so that the application of the Lagrange interpolation formula with the values $u=-g_i$ for $i=1,\dots,n-1$ gives the desired expression.

We will use the definition (9.22) of $Z_{a,-a}(u)$ and show that the polynomial $(2u+1)\,Z_{a,-a}(u)$ is zero at $u=-1/2$. That is, we need to verify the identity in $\mathrm{U}'(\mathfrak{o}_{2n})$,

$$\sum_{i=-n+1}^{n-1} z_{ai}\, z_{i,-a}\frac{1}{(f_i-f_{-n+1})\dots\wedge_i\dots(f_i-f_{n-1})}=0. \tag{9.52}$$

As in the proof of Proposition 9.3.5, we expand z_{ai} using Proposition 9.2.6 and expand $z_{i,-a}$ using (9.13). The left-hand side of (9.52) will be written as a linear combination of monomials

$$F_{a,l_1}F_{l_1l_2}\dots F_{l_r,-a},\qquad n>l_1>\dots>l_r>-n,\qquad r\geqslant 1.$$

The coefficient of such a monomial (written to the right of it) equals

$$\sum_{i=1}^{r}\frac{1}{2f_{l_i}\,(f_{l_i}-f_{l_1})\dots\wedge_i\dots(f_{l_i}-f_{l_r})}=\frac{(-1)^{r-1}}{2\,f_{l_1}\dots f_{l_r}}.$$

Here we have applied the identity (9.11) to the variables specialized to $0,f_{l_1},\dots,f_{l_r}$. Now we need to verify that

$$\sum_{n>l_1>\dots>l_r>-n} F_{a,l_1}F_{l_1l_2}\dots F_{l_r,-a}\,\frac{(-1)^{r-1}}{f_{l_1}\dots f_{l_r}}=0, \tag{9.53}$$

summed over $r\geqslant 1$. Note that the left-hand side is an element of the normalizer $\mathrm{Norm}\,\mathrm{J}'$ and so it is stable, modulo the left ideal J', under the application of the extremal projector p for the subalgebra $\mathfrak{o}_{2n-2}$. On the other hand, the application of p to this element with the use of (9.6) gives the expression

$$\sum_{j=-n+1}^{n-1} pF_{aj}F_{j,-a}\,c_j \mod \mathrm{J}', \tag{9.54}$$

where

$$c_j=\sum_{n>l_1>\dots>l_{r-1}>j}\frac{(-1)^{r-1}}{f_{l_1}\dots f_{l_{r-1}}f_j}$$

summed over $r\geqslant 1$ and $l_a+l_{a+1}\neq 0$ for $a=1,\dots,r-1$ with $l_r:=j$. Now observe that $F_{a,-j}F_{-j,-a}=F_{aj}F_{j,-a}$ so that (9.54) equals

$$\sum_{j=1}^{n-1} pF_{aj}F_{j,-a}\,(c_j+c_{-j}) \mod \mathrm{J}'.$$

However, $c_j+c_{-j}=0$ which follows easily by recalling that $f_{-i}=-f_i$ as all the terms in the expression for c_j+c_{-j} pairwise cancel.

Thus, we have verified that the left-hand side of (9.53) belongs to the left ideal J'. Therefore the identity (9.53) in $\mathrm{U}'(\mathfrak{o}_{2n})$ holds due to (9.49). □

LEMMA 9.3.8. *The element $z_{n,-n} \in \mathrm{U}(\mathfrak{g}_N)$ can be given by*

$$z_{n,-n} = \sum_{n>i_1>\cdots>i_s>-n} (f_{-n} - f_{j_1} + 1) \dots (f_{-n} - f_{j_r} + 1)\, F_{ni_1} F_{i_1 i_2} \dots F_{i_s,-n}$$

in the B, C cases and

$$z_{n,-n} = \sum_{n>i_1>\cdots>i_s>-n} \frac{(f_{-n} - f_{j_1} + 1) \dots (f_{-n} - f_{j_r} + 1)}{2f_{-n} + 2} F_{ni_1} F_{i_1 i_2} \dots F_{i_s,-n}$$

in the D case, where $s = 0, 1, \dots$ and $\{j_1, \dots, j_r\}$ is the complement to the subset $\{i_1, \dots, i_s\}$ in the set $\{-n+1, \dots, n-1\}$.

PROOF. By Corollary 9.3.2, the definition (9.41) for $a = n$ can be equivalently written as $z_{n,-n} = Z_{n,-n}(g_n)$. Then the proof of Proposition 9.3.5 can be repeated word by word to produce explicit expressions for $z_{n,-n}$ in the form (9.42) or (9.43), respectively, where f_n is replaced with $f_{-n} - 1$. This yields the desired expressions in all cases. □

PROPOSITION 9.3.9. *For $a \in \{-n, n\}$ we have the following relations in $\mathrm{U}'(\mathfrak{g}_N)$:*

$$F_{n-1,a} = \sum_{i=-n+1}^{n-1} z_{n-1,i}\, z_{ia} \frac{1}{(f_i - f_{-n+1}) \dots \wedge_i \dots (f_i - f_{n-1})}$$

in the B, C cases and

$$F_{n-1,a} = \sum_{i=-n+1}^{n-1} z_{n-1,i}\, z_{ia} \frac{1}{(f_i - f_{-n+1}) \dots \wedge_{-i,i} \dots (f_i - f_{n-1})}$$

in the D case, where $z_{n-1,n-1} = 1$ and the wedge indicates the indices for which the corresponding factors should be skipped.

PROOF. Let us write the expressions which occur on the right-hand side of the relations in terms of the generators F_{ij}. We use Lemma 9.3.8 and Proposition 9.2.6 in order to expand $z_{n-1,i}$, then use (9.12) and (9.13) to expand z_{ia}. For any fixed value of i, the expression will contain a linear combination of monomials of the form

$$F_{n-1,i_1} F_{i_1 i_2} \dots F_{i_s i}\, F_{ik_1} F_{k_1 k_2} \dots F_{k_p,a} \tag{9.55}$$

with $n - 1 = i_0 > i_1 > \dots > i_s > i > k_1 > \dots > k_p > -n$. Observe that the monomial $F_{n-1,a}$ occurs with the coefficient 1 in the summand corresponding to the value $i = n - 1$. The coefficient of any other monomial of the form (9.55), written to the right of the monomial, is equal to

$$\frac{1}{(f_i - f_{i_0}) \dots (f_i - f_{i_s})(f_i - f_{k_1}) \dots (f_i - f_{k_p})}.$$

Proceeding as in the proof of Proposition 9.3.5, we find that the coefficient of the monomial in the entire expression is zero. □

9.4. Yangian action on the multiplicity space

For any n-tuple of complex numbers $\lambda = (\lambda_1, \dots, \lambda_n)$ we denote by $V(\lambda)$ the irreducible representation of the Lie algebra $\mathfrak{g}_N$ with the highest weight λ. That is, $V(\lambda)$ is generated by a nonzero vector ξ such that

$$
\begin{aligned}
&F_{ij}\,\xi = 0 \qquad \text{for} \quad -n \leqslant i < j \leqslant n, \qquad \text{and}\\
&F_{ii}\,\xi = \lambda_i\,\xi \qquad \text{for} \quad 1 \leqslant i \leqslant n.
\end{aligned}
\tag{9.56}
$$

The representation $V(\lambda)$ is finite-dimensional if and only if

$$\lambda_i - \lambda_{i+1} \in \mathbb{Z}_+ \qquad \text{for} \quad i = 1, \dots, n-1$$

and

$$
\begin{aligned}
-\lambda_1 - \lambda_2 \in \mathbb{Z}_+ \qquad &\text{if} \quad \mathfrak{g}_N = \mathfrak{o}_{2n},\\
-\lambda_1 \in \mathbb{Z}_+ \qquad &\text{if} \quad \mathfrak{g}_N = \mathfrak{sp}_{2n},\\
-2\,\lambda_1 \in \mathbb{Z}_+ \qquad &\text{if} \quad \mathfrak{g}_N = \mathfrak{o}_{2n+1}.
\end{aligned}
$$

In particular, all components λ_i are nonpositive integers in the C case, while in the B and D cases, all components λ_i are either all integers or half-integers (elements of the set $1/2 + \mathbb{Z}$). We will work only with finite-dimensional representations $V(\lambda)$ in this chapter so that we will usually assume that the above conditions on λ are satisfied. Such n-tuples λ will be referred to as the *$\mathfrak{g}_N$-highest weights*.

Any weight of the module $V(\lambda)$ has the form $\lambda - \omega$, where ω is a linear combination of the positive roots with nonnegative integer coefficients; see Section 4.2.

Denote by $V(\lambda)^+$ the subspace of $\mathfrak{g}_{N-2}$-highest vectors in $V(\lambda)$:

$$V(\lambda)^+ = \{\eta \in V(\lambda) \mid F_{ij}\,\eta = 0, \qquad -n < i < j < n\}.$$

Given a tuple $\mu = (\mu_1, \dots, \mu_{n-1})$ we denote by $V(\lambda)_\mu$ the corresponding $\mathfrak{g}_{N-2}$-weight subspace in $V(\lambda)$:

$$V(\lambda)_\mu = \{\eta \in V(\lambda) \mid F_{ii}\,\eta = \mu_i\,\eta, \qquad i = 1, \dots, n-1\}.$$

The subspace is nonzero only if all components μ_i of μ are simultaneously integers or half-integers together with the λ_i. By the Weyl complete reducibility theorem, the restriction of $V(\lambda)$ to the subalgebra $\mathfrak{g}_{N-2}$ is given by

$$V(\lambda)|_{\mathfrak{g}_{N-2}} \cong \bigoplus_{\mu} \operatorname{mult}(\mu)\, V(\mu), \tag{9.57}$$

where $V(\mu)$ denotes the irreducible finite-dimensional representation of $\mathfrak{g}_{N-2}$ with the highest weight μ. The multiplicity $\operatorname{mult}(\mu)$ is known from the classical branching rules for the reduction $\mathfrak{g}_N \downarrow \mathfrak{g}_{N-2}$. In the orthogonal case, the rules can be obtained by considering the two-step reduction: $\mathfrak{o}_N \downarrow \mathfrak{o}_{N-1}$ followed by $\mathfrak{o}_{N-1} \downarrow \mathfrak{o}_{N-2}$. The exact values of $\operatorname{mult}(\mu)$ will be used in the proofs of Theorems 9.4.11, 9.4.13 and 9.4.15 below. Furthermore, $\operatorname{mult}(\mu)$ coincides with the dimension of the subspace

$$V(\lambda)^+_\mu = V(\lambda)_\mu \cap V(\lambda)^+,$$

which we will call the *multiplicity space*. Every nonzero element of $V(\lambda)^+_\mu$ generates a $\mathfrak{g}_{N-2}$-submodule of $V(\lambda)$ isomorphic to $V(\mu)$. So, we have a vector space isomorphism

$$V(\lambda) \cong \bigoplus_{\mu} V(\lambda)^+_\mu \otimes V(\mu). \tag{9.58}$$

Therefore, using induction on n, we can reduce the problem of constructing a basis of $V(\lambda)$ to the same problem for the multiplicity space $V(\lambda)^+_\mu$.

The centralizer $\mathrm{U}(\mathfrak{g}_N)^{\mathfrak{g}_{N-2}}$ naturally acts on $V(\lambda)^+_\mu$, and so the homomorphism (9.27) gives rise to a representation of $\mathrm{Y}(\mathfrak{g}_2)$ on $V(\lambda)^+_\mu$. Our next goal is to describe this representation relying on the classification results of Sections 4.3 and 4.4.

The arguments of the proof of Theorem 9.3.1 can be used to calculate the action of the generators of $\mathrm{Y}(\mathfrak{g}_2)$ on the multiplicity space. In fact, only the derivation of relations (9.36) and (9.37) requires a modification, with the elements $F^{(r)}_{ab}$ regarded as operators on $V(\lambda)^+_\mu$ rather than elements of the Mickelsson-Zhelobenko algebra. Proposition 9.2.1 and Corollary 9.2.3 imply that the elements z_{ia} and z_{ai} naturally act in the space $V(\lambda)^+$ by *raising* or *lowering* the weights. Namely, we have for $i = 1, \dots, n-1$:

$$z_{ia} : V(\lambda)^+_\mu \to V(\lambda)^+_{\mu+\delta_i}, \qquad z_{ai} : V(\lambda)^+_\mu \to V(\lambda)^+_{\mu-\delta_i}, \tag{9.59}$$

where $\mu \pm \delta_i$ is obtained from μ by replacing μ_i with $\mu_i \pm 1$. Hence, for any $a, b \in \{-n, n\}$ and $i \in \{-n+1, \dots, n-1\}$ the subspace $V(\lambda)^+_\mu$ is preserved by the operator $z_{ai} z_{ib}$. Moreover, we also have

$$z_{n,-n} : V(\lambda)^+_\mu \to V(\lambda)^+_\mu.$$

In the B case the operators z_{0a} also preserve each subspace $V(\lambda)^+_\mu$.

Note that if $\eta \in V(\lambda)_\mu$ is a weight vector, then η is an eigenvector of any element of $\mathrm{U}(\mathfrak{h})$ regarded as an operator on $V(\lambda)$. If $P/Q \in \mathrm{R}(\mathfrak{h})$ is a rational function with $P, Q \in \mathrm{U}(\mathfrak{h})$ and the eigenvalue of Q on the vector η is nonzero, then the action of P/Q on η is well-defined. In particular, if μ is a $\mathfrak{g}_{N-2}$-highest weight, then the extremal projector p for $\mathfrak{g}_{N-2}$ is a well-defined operator on the weight space $V(\lambda)_\mu$ and p projects $V(\lambda)_\mu$ onto $V(\lambda)^+_\mu$ with the kernel $V(\lambda)_\mu \cap \mathfrak{k}^- V(\lambda)$; see Section 9.1. Furthermore, p acts as the identity operator on $V(\lambda)^+_\mu$. Also, for any $i \in \{1, \dots, n-1\}$ the element $f_i \in \mathrm{U}(\mathfrak{h})$ acts on $V(\lambda)^+_\mu$ as multiplication by the scalar $\mu_i + \rho_i$. Thus, the resulting formulas for the action of the generators of $\mathrm{Y}(\mathfrak{g}_2)$ on $V(\lambda)^+_\mu$ will coincide with those provided by Theorem 9.3.1, except for the B case with $\mu_1 = 0$; here some coefficients may contain the factor $g_0 - g_1$ in the denominator whose value is zero on $V(\lambda)^+_\mu$. In this case we have the following.

LEMMA 9.4.1. *Let $\mathfrak{g}_N = \mathfrak{o}_{2n+1}$ and let the first component μ_1 of the highest weight μ be zero. Then for any $a \in \{-n, n\}$ both z_{0a} and z_{1a} are the zero operators on $V(\lambda)^+_\mu$.*

PROOF. Observe that the element $F_{10} \in \mathfrak{o}_{2n+1}$ acts as the zero operator on $V(\lambda)^+_\mu$ (assuming $n \geqslant 2$, as the case $n = 1$ is trivial). Indeed, the cyclic $\mathfrak{o}_{2n-1}$-span of any nonzero vector $\eta \in V(\lambda)^+_\mu$ is isomorphic to the irreducible representation $V(\mu)$ of $\mathfrak{o}_{2n-1}$ with the highest weight μ. On the other hand, the vector $F_{10}\,\eta$ is annihilated by all elements F_{ij} with $-n+1 \leqslant i < j \leqslant n-1$, which implies $F_{10}\,\eta = 0$.

By (9.12), the operator z_{0a} on $V(\lambda)^+_\mu$ can be written as

$$z_{0a} = \sum_{0 > i_1 > \dots > i_s > -n} F_{0 i_1} F_{i_1 i_2} \dots F_{i_s a} (f_0 - f_{j_1}) \dots (f_0 - f_{j_r}),$$

where $s = 0, 1, \dots$ and $\{j_1, \dots, j_r\}$ is the complement to the subset $\{i_1, \dots, i_s\}$ in the set $\{-n+1, \dots, -1\}$. Note that since $F_{0,-1} = -F_{10}$, if $i_1 = -1$, then we can replace the product $F_{0i_1} F_{i_1 i_2}$ by $F_{0 i_2}$. On the other hand, $f_0 - f_{-1} = -1$, which shows that all terms in the formula for z_{0a} pairwise cancel. Furthermore, by (9.12) the operator z_{1a} coincides with $F_{10}\, z_{0a}$, and hence is also zero. □

Lemma 9.4.1 motivates the following definition of the operators in $V(\lambda)^+_\mu$:

$$Z_{ab}(u) : V(\lambda)^+_\mu \to V(\lambda)^+_\mu, \qquad a, b \in \{-n, n\},$$

which depend on a complex parameter u. Namely, $Z_{ab}(u)$ is given by (9.20), (9.21) and (9.22) for the B, C, and D cases, respectively, excluding the B case with $\mu_1 = 0$, where for any $a, b \in \{-n, n\}$ we define this operator by

$$Z_{ab}(u) = -\Big(\delta_{ab}(u + \rho_n + 1/2) + F_{ab}\Big) \prod_{i=-n+1}^{n-1} (u + g_i) + \sum_{i=-n+1,\ i\neq 0,1}^{n-1} z_{ai} z_{ib} \prod_{j=-n+1,\, j\neq i}^{n-1} \frac{u + g_j}{g_i - g_j}. \tag{9.60}$$

The next corollary justifies these definitions.

COROLLARY 9.4.2. *With the above definitions of the operators $Z_{ab}(u)$, the action of the twisted Yangian* $\mathrm{Y}(\mathfrak{g}_2)$ *on the multiplicity space* $V(\lambda)^+_\mu$ *is given by* (9.23), (9.24) *and* (9.25), *respectively, for the B, C and D cases.*

PROOF. We need to consider only the B case with $\mu_1 = 0$. As we showed in the proof of Lemma 9.4.1, the operator $F_{10} = -F_{0,-1}$ annihilates $V(\lambda)^+_\mu$. Using the notation of the proof of Theorem 9.3.1, observe that each element $F^{(r)}_{ab}$ preserves the subspace $V(\lambda)^+_\mu$. Now (9.33) gives the following relation of operators on $V(\lambda)^+_\mu$:

$$F^{(r)}_{ab} = \sum_{i=-n+1,\ i\neq 0,\pm 1}^{n-1} F^{(r-1)}_{ai} F_{ib} + \big(F^{(r-1)}_{a,-1} + F^{(r-1)}_{a,0} F_{0,-1} + F^{(r-1)}_{a,1} F_{10} F_{0,-1}\big)\, F_{-1,b}. \tag{9.61}$$

Furthermore, since $F^{(r)}_{ab} = pF^{(r)}_{ab}$ as operators on $V(\lambda)^+_\mu$, applying (9.32) we can write

$$F^{(r)}_{ab} = \sum_{i=-n+1,\ i\neq 0,\pm 1}^{n-1} pF^{(r-1)}_{ai} F_{ib} + 3p\, F^{(r-1)}_{a,-1} F_{-1,b}. \tag{9.62}$$

For all $i \neq 0, 1$ the action of the element $\check{z}_{ia}$ on the space $V(\lambda)^+_\mu$ is well-defined; see (9.8). Furthermore, for $i \geqslant 2$, using the relation $F_{10}\, V(\lambda)^+_\mu = 0$, we obtain the following modified formula for $\check{z}_{ia}$, regarded as an operator on $V(\lambda)^+_\mu$:

$$\check{z}_{ia} = F_{ia} + \sum_{i>i_1>\cdots>i_s>-n} F_{ii_1} F_{i_1 i_2} \dots F_{i_{s-1} i_s} F_{i_s a} \frac{1}{(f_i - f_{i_1}) \dots (f_i - f_{i_s})}, \tag{9.63}$$

summed over $s \geqslant 1$, where none of the indices $i_1, \dots, i_s$ is equal to 0 or 1, and the factor of the form $F_{j,-1}$, if it occurs, should be replaced with the expression

$$F_{j,-1} + F_{j,0} F_{0,-1} + F_{j,1} F_{1,0} F_{0,-1}. \tag{9.64}$$

Indeed, note the following identities for the operators on $V(\lambda)^+_\mu$: for any $j \geqslant 2$ and $k \leqslant -2$ we have

$$F_{j1} F_{1k} = F_{j1} F_{10} F_{0,-1} F_{-1,k}, \qquad F_{j1} F_{10} F_{0k} = F_{j1} F_{10} F_{0,-1} F_{-1,k}$$

and $F_{j0}F_{0k} = F_{j0}F_{0,-1}F_{-1,k}$. Then (9.63) is implied by the relations

$$\frac{1}{f_i - f_1} + \frac{1}{(f_i - f_1)(f_i - f_0)} + \frac{1}{(f_i - f_1)(f_i - f_0)(f_i - f_{-1})} = \frac{1}{f_i - f_{-1}} \tag{9.65}$$

and

$$\frac{1}{f_i - f_0} + \frac{1}{(f_i - f_0)(f_i - f_{-1})} = \frac{1}{f_i - f_{-1}}, \tag{9.66}$$

which hold since $f_1 = f_0 = -1/2$ and $f_{-1} = 1/2$ on $V(\lambda)^+_\mu$.

Now, arguing as in the proof of Lemma 9.2.2, we get the following relation for operators in $V(\lambda)^+_\mu$: for $i \geqslant 2$,

$$F_{ia} = \check{z}_{ia} + \sum_{i>i_1>\dots>i_s>-n} F_{ii_1}F_{i_1i_2}\dots F_{i_{s-1}i_s}\,\check{z}_{i_s,a}\frac{1}{(f_{i_s} - f_i)(f_{i_s} - f_{i_1})\dots(f_{i_s} - f_{i_{s-1}})},$$

summed over $s \geqslant 1$, where none of the indices $i_1, \dots, i_s$ is equal to 0 or 1, and the factor of the form $F_{j,-1}$, if it occurs, should be replaced with the expression (9.64). Hence, using this relation and Lemma 9.2.2 with $i \leqslant -1$, we can bring (9.62) to the form

$$F^{(r)}_{ab} = \sum_{j=-n+1,\ j\neq 0,1}^{n-1} pF^{(r-1)}_{aj}\,\check{z}_{jb}\sum \frac{c(i_1, \dots, i_s, j)}{(f_j - f_{i_1})\dots(f_j - f_{i_s})}, \tag{9.67}$$

where the second sum is taken over $s \geqslant 0$ and the indices $n > i_1 > \dots > i_s > j$, none of which is 0 or 1, and $i_p + i_{p+1} \neq 0$ for $p \in \{1, \dots, s\}$ with $i_{s+1} = j$. The coefficient $c(i_1, \dots, i_s, j)$ equals 1 unless $i_p = -1$ for some $p \in \{1, \dots, s+1\}$; in the latter case $c(i_1, \dots, i_s, j)$ equals 3. This sum is not difficult to calculate; cf. the proof of Proposition 9.2.6. Indeed, the case $j \leqslant -1$ was done in that proof, while for $j \geqslant 2$ we need to verify only that the sum coincides with

$$\sum \frac{1}{(f_j - f_{i_1})\dots(f_j - f_{i_s})},$$

summed over $s \geqslant 0$ and the indices $n > i_1 > \dots > i_s > j$ with $i_p + i_{p+1} \neq 0$ for $p \in \{1, \dots, s\}$. However, this follows easily by recalling that $f_1 = f_0 = -1/2$ and $f_{-1} = 1/2$ on $V(\lambda)^+_\mu$ and using the identity

$$\frac{1}{f_j - f_1} + \frac{1}{f_j - f_0} + \frac{1}{f_j - f_{-1}} + \frac{1}{(f_j - f_1)(f_j - f_0)} + \frac{1}{(f_j - f_0)(f_j - f_{-1})} + \frac{1}{(f_j - f_1)(f_j - f_0)(f_j - f_{-1})} = \frac{3}{f_j - f_{-1}}.$$

Now the argument is completed exactly as in the proof of Theorem 9.3.1 by showing that (9.36) holds in the same form with the indices $j = 0, 1$ skipped in the sum and with both sides regarded as operators on $V(\lambda)^+_\mu$. □

Introduce the parameters m_i associated with μ by

$$m_i = \mu_i + \rho_i + 1/2, \qquad i = 1, \dots, n-1. \tag{9.68}$$

The element g_i acts on $V(\lambda)^+_\mu$ as multiplication by m_i for $i = 1, \dots, n-1$, while g_{-i} acts as multiplication by $-m_i + 1$. Moreover, in the B case we have $g_0 = 0$. Furthermore, the space $V(\lambda)^+_\mu$ is the direct sum of the eigenspaces for the operator

$g_n = f_n + 1/2$. A rational function P/Q in g_n is a well-defined operator on $V(\lambda)^+_\mu$ provided that Q does not vanish as g_n is evaluated at the eigenvalues.

COROLLARY 9.4.3. *For any $a \in \{-n, n\}$ the operator $Z_{a,-a}(u)$ on $V(\lambda)^+_\mu$ can be given by the formula (9.51) in the D case. Moreover, if the operator $g_i^2 - g_n^2$ on $V(\lambda)^+_\mu$ does not have zero eigenvalues for all $i = 1, \dots, n-1$, then the operator $Z_{a,-a}(u)$ can be given by the formula (9.50) in the B and C cases.*

PROOF. We generally follow the arguments of the proof of Corollary 9.3.7 with slight modifications in some special cases. Namely, in the D case with $\mu_1 = 0$ both f_1 and f_{-1} are zero, as operators on $V(\lambda)^+_\mu$. Therefore, the value of the operator $(2u+1)\, Z_{a,-a}(u)$ on $V(\lambda)^+_\mu$ at $u = -1/2$ is zero. In the B case with $\mu_1 = 0$ we use (9.60) for the definition of the operator $Z_{a,-a}(u)$. Then its value at $u = -g_1$ is zero, which agrees with (9.50) due to Lemma 9.4.1. □

Since f_i acts on $V(\lambda)^+_\mu$ as multiplication by the scalar $m_i - 1/2$ for each value $i = 1, \dots, n-1$, the formulas of Proposition 9.3.9 can be used to calculate the action of $F_{n-1,a}$ on the subspace $V(\lambda)^+_\mu$ of $V(\lambda)$, except for the B case with $\mu_1 = 0$ and $n \geqslant 2$. If the latter holds, the denominator $f_1 - f_0$ vanishes. In this case Lemma 9.4.1 motivates the following modification of the formulas.

COROLLARY 9.4.4. *For any $a \in \{-n, n\}$ the action of the operator $F_{n-1,a}$ on $V(\lambda)^+_\mu$ can be found by the formulas of Proposition 9.3.9 except for the B case with $\mu_1 = 0$ and $n \geqslant 2$, where the action is found by*

$$F_{n-1,a} = \sum_{i=-n+1,\, i\neq 0,1}^{n-1} z_{n-1,i}\, z_{ia} \frac{1}{(f_i - f_{-n+1}) \dots \wedge_i \dots (f_i - f_{n-1})}.$$

PROOF. We only need to modify the proof of Proposition 9.3.9 in the exceptional case. Represent the right-hand side of the above relation as a linear combination of monomials of the form (9.55). If neither 0 nor 1 occurs amongst the indices of the monomial, then the coefficient of this monomial is zero, as shown in the proof of Proposition 9.3.9. Now suppose that 0 occurs amongst the indices while 1 does not. As we observed in the proof of Lemma 9.4.1, the operator $F_{10} = -F_{0,-1}$ acts as zero on $V(\lambda)^+_\mu$. Hence, any product of the form $F_{0,-1}F_{-1,j}$ occurring in (9.55) can be replaced with $F_{0,j}$. Therefore, using the relation (9.66) and $f_{-1} - f_0 = 1$ for the operators on $V(\lambda)^+_\mu$ and calculating the coefficient of the monomial with the use of (9.11), we conclude that it is zero. Similarly, if 1 is one of the indices of the monomial (9.55), then for any $j \leqslant -2$ we can replace the products $F_{10}F_{0,j}$ and $F_{10}F_{0,-1}F_{-1,j}$ occurring in the monomial by $F_{1,j}$. The calculation is completed in the same way as above with the use of the relation (9.65). □

Given a $\mathfrak{g}_N$-highest weight λ, suppose that μ is a $\mathfrak{g}_{N-2}$-highest weight such that all components μ_i of μ are simultaneously integers of half-integers together with the λ_i. Introduce the vector $\xi_\mu \in V(\lambda)$ by

$$\xi_\mu = \prod_{i=1}^{n-1} \left(z_{ni}^{\max\{\lambda_i,\mu_i\}-\mu_i}\, z_{i,-n}^{\max\{\lambda_i,\mu_i\}-\lambda_i} \right) \xi, \tag{9.69}$$

where ξ is the highest vector of $V(\lambda)$. Since $\xi \in V(\lambda)^+_{\lambda'}$ with $\lambda' = (\lambda_1, \dots, \lambda_{n-1})$, using the properties (9.59) of the elements z_{ia} and z_{bi}, we easily find that the vector ξ_μ belongs to $V(\lambda)^+_\mu$.

In the next three lemmas we consider the cases B, C and D separately. In the B and C cases we set

$$\alpha_i = \min\{\lambda_{i-1}, \mu_{i-1}\} + \rho_i + 1/2,$$
$$\beta_i = \max\{\lambda_i, \mu_i\} + \rho_i + 1/2, \qquad i = 1, \dots, n,$$

assuming that $\lambda_0 = \mu_0 = 0$ and $\max\{\lambda_n, \mu_n\}$ is understood as being equal to λ_n.

LEMMA 9.4.5. *Let $\mathfrak{g}_N = \mathfrak{sp}_{2n}$. Then*

$$Z_{nn}(u)\,\xi_\mu = (u - \alpha_2)\dots(u - \alpha_n)(u + \beta_1)\dots(u + \beta_n)\,\xi_\mu. \tag{9.70}$$

Moreover, for any $i \in \{1, \dots, n-1\}$ we have

$$z_{in}\,\xi_\mu = -(m_i + \alpha_2)\dots(m_i + \alpha_n)(m_i - \beta_1)\dots(m_i - \beta_n)\,\xi_{\mu+\delta_i}$$

and

$$z_{-n,i}\,\xi_\mu = -(m_i - \alpha_2 - 1)\dots(m_i - \alpha_n - 1) \times (m_i + \beta_1 - 1)\dots(m_i + \beta_n - 1)\,\xi_{\mu-\delta_i}.$$

PROOF. Suppose first that $\lambda_i \leqslant \mu_i$ for some $i \in \{1, \dots, n-1\}$. Then $\beta_i = m_i$. By Proposition 9.2.5, $z_{in}z_{i,-n} - z_{i,-n}z_{in}$ and

$$z_{in}z_{j,-n}(f_i - f_j) = z_{i,-n}z_{jn} + z_{j,-n}z_{in}(f_i - f_j - 1)$$

for $j \in \{-n+1, \dots, n-1\}$. Since $z_{jn}\,\xi = 0$ for all j, we deduce that $z_{in}\,\xi_\mu = 0$ by an easy induction on the parameter $\ell = |\lambda_1 - \mu_1| + \dots + |\lambda_{n-1} - \mu_{n-1}|$ while keeping λ fixed. A similar argument shows that if $\lambda_i \geqslant \mu_i$ for some $i \in \{1, \dots, n-1\}$, then $z_{-n,i}\,\xi_\mu = 0$, which again agrees with the desired relation as $\alpha_{i+1} = m_i - 1$. We will now prove the remaining statements simultaneously by induction on the parameter ℓ. The value $\ell = 0$ will be taken as the induction base. In this case $\xi_\mu = \xi$, and the statements are immediate from (9.21) with $a = b = n$. Now suppose that $\lambda_i > \mu_i$ for some $i \in \{1, \dots, n-1\}$. By Proposition 9.2.5 we can write $\xi_\mu = z_{ni}\,\xi_{\mu+\delta_i}$. Observe that $g_{-i}\,\xi_{\mu+\delta_i} = -m_i\,\xi_{\mu+\delta_i}$. Hence, using (9.21) with $a = b = -n$ we get

$$z_{in}\,\xi_\mu = z_{in}\,z_{ni}\,\xi_{\mu+\delta_i} = -z_{-n,-i}\,z_{-i,-n}\,\xi_{\mu+\delta_i} = -Z_{-n,-n}(m_i)\,\xi_{\mu+\delta_i}.$$

By Corollary 9.3.2,

$$Z_{-n,-n}(m_i) = \frac{1 - 2m_i}{2m_i}\,Z_{nn}(-m_i) - \frac{1}{2m_i}\,Z_{nn}(m_i).$$

Note that the value of the parameter α_{i+1} associated with $\mu + \delta_i$ coincides with m_i, and so, by the induction hypothesis, $Z_{nn}(m_i)\,\xi_{\mu+\delta_i} = 0$ and

$$\frac{1 - 2m_i}{2m_i}\,Z_{nn}(-m_i)\,\xi_{\mu+\delta_i} = -(-m_i - \alpha_2)\dots(-m_i - \alpha_n) \times (-m_i + \beta_1)\dots(-m_i + \beta_n)\,\xi_{\mu+\delta_i},$$

where α_{i+1} stands for the value $m_i - 1$ associated with μ. This gives the desired expression for $z_{in}\,\xi_\mu$.

Next, suppose that $\lambda_i < \mu_i$ for some $i \in \{1, \dots, n-1\}$. By Proposition 9.2.5 we can write $\xi_\mu = z_{i,-n}\,\xi_{\mu-\delta_i}$. Observe that $g_i\,\xi_{\mu-\delta_i} = (m_i - 1)\,\xi_{\mu-\delta_i}$. Hence, using (9.21) with $a = b = -n$ we get

$$z_{-n,i}\,\xi_\mu = z_{-n,i}\,z_{i,-n}\,\xi_{\mu-\delta_i} = Z_{-n,-n}(-m_i + 1)\,\xi_{\mu-\delta_i}.$$

This time, the value of the parameter β_i associated with $\mu-\delta_i$ coincides with m_i-1 so that the calculation is completed exactly as above by using Corollary 9.3.2, providing the desired expression for $z_{-n,i}\,\xi_\mu$.

Finally, due to (9.21) with $a=b=n$, we have $Z_{nn}(u)\,\xi_\mu = A(u)\,\xi_\mu$ for a certain monic polynomial $A(u)$ in u of degree $2n-1$. However, such a polynomial is uniquely determined by its values at $2n-2$ distinct points and the sum of its roots. Indeed, if

$$A(u) = (u-\gamma_1)\dots(u-\gamma_{2n-1})$$

and the complex numbers $z_1,\dots,z_{2n-2}$ are distinct, then

$$A(u) = \sum_{i=1}^{2n-2} A(z_i) \prod_{j=1,\, j\neq i}^{2n-2} \frac{u-z_j}{z_i-z_j} + (u-z_0)\prod_{i=1}^{2n-2}(u-z_i), \tag{9.71}$$

where $z_0=\gamma_1+\dots+\gamma_{2n-1}-z_1-\dots-z_{2n-2}$; the latter is obtained by comparing the coefficients of u^{2n-2} on both sides.

Now, the numbers $-m_i$ and m_i-1 with $i=1,\dots,n-1$ are obviously distinct. We have $g_i\,\xi_\mu = m_i\,\xi_\mu$ so that using the expression for $z_{in}\,\xi_\mu$ proved above, we get

$$Z_{nn}(-m_i)\,\xi_\mu = z_{ni}\,z_{in}\,\xi_\mu = (-m_i-\alpha_2)\dots(-m_i-\alpha_n)(-m_i+\beta_1)\dots(-m_i+\beta_n)\,\xi_\mu,$$

which agrees with (9.70) for $u=-m_i$. Similarly, $g_{-i}\,\xi_\mu=(-m_i+1)\,\xi_\mu$ and

$$Z_{nn}(m_i-1)\,\xi_\mu = z_{n,-i}\,z_{-i,n}\,\xi_\mu = -z_{i,-n}\,z_{-n,i}\,\xi_\mu.$$

Using the expression for $z_{-n,i}\,\xi_\mu$ proved above, we get (9.70) for $u=m_i-1$. Note that the sum of all numbers $-m_i$ and m_i-1 is $-n+1$, so that by (9.21) with $a=b=n$, we need to verify only the relation

$$(F_{nn}-n+1/2)\,\xi_\mu = (\beta_1+\dots+\beta_n-\alpha_2-\dots-\alpha_n-n+1)\,\xi_\mu. \tag{9.72}$$

However, by the definition of ξ_μ,

$$F_{nn}\,\xi_\mu = \Big(\lambda_n + \sum_{i=1}^{n-1}\big(2\max\{\lambda_i,\mu_i\}-\lambda_i-\mu_i\big)\Big)\,\xi_\mu.$$

Since $\max\{\lambda_i,\mu_i\}=\beta_i+i-1/2$ and $\lambda_i+\mu_i=\beta_i+\alpha_{i+1}+2i$, the relation (9.72) follows. □

In the D case, in addition to the parameters m_i defined in (9.68), we set

$$\alpha_1=\min\{-|\lambda_1|,-|\mu_1|\}-1/2, \qquad \alpha_0=\alpha_1+|\lambda_1+\mu_1|,$$

and

$$\alpha_i=\min\{\lambda_i,\mu_i\}-i+1/2, \qquad i=2,\dots,n-1,$$
$$\beta_i=\max\{\lambda_{i+1},\mu_{i+1}\}-i+1/2, \qquad i=1,\dots,n-1,$$

where $\max\{\lambda_n,\mu_n\}$ is understood as being equal to λ_n. Also, let

$$\alpha_1'=\min\{\lambda_1,\mu_1\}-1/2, \qquad \beta_0=\max\{\lambda_1,\mu_1\}+1/2.$$

Note that the unordered pair $\{\alpha_1',-\beta_0\}$ coincides with $\{\alpha_0,\alpha_1\}$.

LEMMA 9.4.6. *Let* $\mathfrak{g}_N=\mathfrak{o}_{2n}$. *Then*

$$(2u+1)\,Z_{nn}(u)\,\xi_\mu = -(u-\alpha_0)\dots(u-\alpha_{n-1})(u+\beta_1)\dots(u+\beta_{n-1})\,\xi_\mu. \tag{9.73}$$

Moreover, for any $i \in \{1, \dots, n-1\}$ we have

$$z_{in}\, \xi_\mu = -(m_i + \alpha_1') \dots \widehat{(m_i + \alpha_i)} \dots (m_i + \alpha_{n-1}) \times (m_i - \beta_0) \dots (m_i - \beta_{n-1})\, \xi_{\mu+\delta_i}$$

and

$$z_{-n,i}\, \xi_\mu = -(m_i - \alpha_1' - 1) \dots (m_i - \alpha_{n-1} - 1) \times (m_i + \beta_0 - 1) \dots \widehat{(m_i + \beta_{i-1} - 1)} \dots (m_i + \beta_{n-1} - 1)\, \xi_{\mu-\delta_i},$$

where the hats indicate the factors to be omitted.

PROOF. We use the same argument as in the proof of Lemma 9.4.5. A slight modification is required for the calculation of $Z_{nn}(u)\, \xi_\mu$, where we need to distinguish two cases. If $\mu_1 \neq 0$, then $-(2u+1)\, Z_{nn}(u)\, \xi_\mu = A(u)\, \xi_\mu$ for a certain monic polynomial in u of degree $2n-1$. This polynomial is calculated with the use of (9.71) by taking $-m_i$ and $m_i - 1$ with $i = 1, \dots, n-1$ as the values of u. If $\mu_1 = 0$, then $g_1\, \xi_\mu = g_{-1}\, \xi_\mu = 1/2\, \xi_\mu$ so that $-2\, Z_{nn}(u)\, \xi_\mu = A(u)\, \xi_\mu$ for a certain monic polynomial $A(u)$ in u of degree $2n-2$. The argument is completed by using (9.71) and taking the distinct numbers $-m_i$ with $i = 2, \dots, n-1$ and $m_i - 1$ with $i = 1, \dots, n-1$ as the values of u. □

In the B case we introduce another element ξ_μ' of $V(\lambda)_\mu^+$ by $\xi_\mu' = z_{n0}\, \xi_\mu$. Since $z_{n0} = (-1)^n z_{0,-n}$, due to Lemma 9.4.1 we have $\xi_\mu' = 0$ if $\mu_1 = 0$. Furthermore, by Proposition 9.2.5, z_{n0} commutes with z_{ni} for any nonzero $i \in \{-n+1, \dots, n-1\}$. Hence, we also have $\xi_\mu' = 0$ if $\lambda_1 = 0$ because $z_{n0}\, \xi = 0$ by Lemma 9.4.1.

LEMMA 9.4.7. *Let $\mathfrak{g}_N = \mathfrak{o}_{2n+1}$. Then*

$$Z_{nn}(u)\, \xi_\mu = -(u - \alpha_1) \dots (u - \alpha_n)(u + \beta_1) \dots (u + \beta_n)\, \xi_\mu. \tag{9.74}$$

Moreover, $z_{0n}\, \xi_\mu = 0$ and for any $i \in \{1, \dots, n-1\}$ we have

$$z_{in}\, \xi_\mu = -(m_i + \alpha_1) \dots (m_i + \alpha_n)(m_i - \beta_1) \dots (m_i - \beta_n)\, \xi_{\mu+\delta_i}$$

and

$$z_{-n,i}\, \xi_\mu = -(m_i - \alpha_1 - 1) \dots (m_i - \alpha_n - 1) \times (m_i + \beta_1 - 1) \dots (m_i + \beta_n - 1)\, \xi_{\mu-\delta_i}.$$

Also,

$$Z_{nn}(u)\, \xi_\mu' = -(u+1)(u - \alpha_2) \dots (u - \alpha_n)(u + \beta_1) \dots (u + \beta_n)\, \xi_\mu' \tag{9.75}$$

and

$$z_{0n}\, \xi_\mu' = (-1)^n\, \alpha_2 \dots \alpha_n \beta_1 \dots \beta_n\, \xi_\mu.$$

PROOF. The relation $z_{0n}\, \xi_\mu = 0$ follows from Proposition 9.2.5, which also implies that $z_{in}\, \xi_\mu = 0$ for $\lambda_i \leqslant \mu_i$ and $z_{-n,i}\, \xi_\mu = 0$ for $\lambda_i \geqslant \mu_i$; cf. the proof of Lemma 9.4.5. The remaining statements are now verified simultaneously by induction almost exactly as in the C case. A slight modification is required for the calculation of the action of the operator $Z_{nn}(u)$ in the case where $\mu_1 = 0$, as the relation (9.60) should be used to define this operator instead of (9.20). Namely, in this case $g_1\, \xi_\mu = g_0\, \xi_\mu = 0$ so that $Z_{nn}(u)\, \xi_\mu = -u^2\, A(u)\, \xi_\mu$ for a certain monic polynomial in u of degree $2n-2$. Since $\alpha_1 = \beta_1 = 0$ we need to show that

$$A(u) = (u - \alpha_2) \dots (u - \alpha_n)(u + \beta_2) \dots (u + \beta_n).$$

This is verified by applying (9.71) as in the proof of Lemma 9.4.5, taking $-m_i$ with $i = 2, \dots, n-1$ and $m_i - 1$ with $i = 1, \dots, n-1$ as the values of u which are obviously distinct.

Furthermore, in order to calculate $z_{0n}\,\xi'_\mu$ we may assume that $\mu_1 \neq 0$. Note that $g_0 = 0$ and so

$$z_{0n}\,\xi'_\mu = z_{0n}\,z_{n0}\,\xi_\mu = z_{-n,0}\,z_{0,-n}\,\xi_\mu = Z_{-n,-n}(0)\,\xi_\mu.$$

By Corollary 9.3.2,

$$Z_{-n,-n}(0) = Z_{nn}(0) + Z'_{nn}(0),$$

where $Z'_{nn}(u)$ denotes the derivative of the polynomial $Z_{nn}(u)$ over u. Since $\alpha_1 = 0$ the desired relation now follows from (9.74). Finally, (9.75) is verified in the same way as (9.74). □

As before, we suppose that λ is a $\mathfrak{g}_N$-highest weight and μ is a $\mathfrak{g}_{N-2}$-highest weight such that all components μ_i of μ are simultaneously integers of half-integers together with the λ_i. Recall that $V(\lambda)^+_\mu$ is a module over the twisted Yangian $\mathrm{Y}(\mathfrak{g}_2)$ with the action provided by Corollary 9.4.2.

PROPOSITION 9.4.8. *Let $\mathfrak{g}_N = \mathfrak{sp}_{2n}$. Suppose that the components of λ and μ satisfy the inequalities*

$$(9.76)\qquad \lambda_i \geqslant \mu_{i+1}, \quad i = 1, \dots, n-2 \qquad \text{and} \qquad \mu_i \geqslant \lambda_{i+1}, \quad i = 1, \dots, n-1.$$

Then $\xi_\mu \neq 0$ and the cyclic span $\mathrm{Y}(\mathfrak{sp}_2)\,\xi_\mu$ is a highest weight module over $\mathrm{Y}(\mathfrak{sp}_2)$ with the highest weight

$$\mu(u) = (1 - \alpha_1 u^{-1}) \dots (1 - \alpha_n u^{-1})(1 + \beta_1 u^{-1}) \dots (1 + \beta_n u^{-1}).$$

PROOF. If $\lambda_i > \mu_i$ for some $i \in \{1, \dots, n-1\}$, then we find from Lemma 9.4.5 that $z_{in}\,\xi_\mu$ coincides with the vector $\xi_{\mu+\delta_i}$ multiplied by a nonzero constant. Similarly, if $\lambda_i < \mu_i$ for some $i \in \{1, \dots, n-1\}$, then $z_{-n,i}\,\xi_\mu$ equals the vector $\xi_{\mu-\delta_i}$ multiplied by a nonzero constant. As the highest vector ξ of $V(\mu)$ is nonzero, the first statement follows by induction on the parameter $|\lambda_1 - \mu_1| + \dots + |\lambda_{n-1} - \mu_{n-1}|$ while keeping λ fixed.

By Definition 4.2.1, in order to prove the second statement, we need to verify that $s_{nn}(u)\,\xi_\mu = \mu(u)\,\xi_\mu$ and $s_{-n,n}(u)\,\xi_\mu = 0$. However, this follows from Lemma 9.4.5 with the use of the formula (9.21) with $a = -n$, $b = n$, and the relation $[F_{-n,n}, z_{ni}] = 2\,z_{-n,i}$ which holds in $\mathrm{U}(\mathfrak{sp}_{2n})$ for any $i \in \{-n+1, \dots, n-1\}$. □

PROPOSITION 9.4.9. *Let $\mathfrak{g}_N = \mathfrak{o}_{2n}$. Suppose that the components of λ and μ satisfy the inequalities $-|\lambda_1| \geqslant \mu_2$ and $-|\mu_1| \geqslant \lambda_2$ together with*

$$(9.77)\qquad \lambda_i \geqslant \mu_{i+1}, \quad i = 2, \dots, n-2 \qquad \text{and} \qquad \mu_i \geqslant \lambda_{i+1}, \quad i = 2, \dots, n-1.$$

Then $\xi_\mu \neq 0$ and the cyclic span $\mathrm{Y}(\mathfrak{o}_2)\,\xi_\mu$ is a highest weight module over $\mathrm{Y}(\mathfrak{o}_2)$ with the highest weight

$$\mu(u) = (1 - \alpha_0 u^{-1}) \dots (1 - \alpha_{n-1} u^{-1}) \times (1 + \beta_1 u^{-1}) \dots (1 + \beta_{n-1} u^{-1})(1 + 1/2\, u^{-1})^{-1}.$$

PROOF. This follows from Lemma 9.4.6 with the use of the formula (9.22) with $a = -n$, $b = n$; cf. the proof of Proposition 9.4.8. □

PROPOSITION 9.4.10. *Let $\mathfrak{g}_N = \mathfrak{o}_{2n+1}$. Suppose that the components of λ and μ satisfy the inequalities*

$$\lambda_i \geqslant \mu_{i+1}, \quad i = 1, \dots, n-2 \qquad \text{and} \qquad \mu_i \geqslant \lambda_{i+1}, \quad i = 1, \dots, n-1. \tag{9.78}$$

Then $\xi_\mu \neq 0$ and the cyclic span $\mathrm{Y}(\mathfrak{o}_2)\,\xi_\mu$ is a highest weight module over $\mathrm{Y}(\mathfrak{o}_2)$ with the highest weight

$$\mu(u) = (1 - \alpha_2 u^{-1}) \dots (1 - \alpha_n u^{-1})(1 + \beta_1 u^{-1}) \dots (1 + \beta_n u^{-1}).$$

Moreover, if $\beta_1 \neq 0$, then $\xi'_\mu \neq 0$ and the cyclic span $\mathrm{Y}(\mathfrak{o}_2)\,\xi'_\mu$ is a highest weight module over $\mathrm{Y}(\mathfrak{o}_2)$ with the highest weight $(1 + u^{-1})\,\mu(u)$.

PROOF. The first part follows from Lemma 9.4.7 with the use of the formula (9.20) with $a = -n$, $b = n$. For the second part we use the same lemma together with the relations

$$z_{ni}\, z_{in}\, \xi'_\mu = z_{n0}\, z_{ni}\, z_{in}\, \frac{f_i - f_0 - 1}{f_i - f_0}\, \xi_\mu.$$

They hold for any nonzero $i \in \{-n+1, \dots, n-1\}$ due to Proposition 9.2.5, since $z_{0n}\xi_\mu = 0$. □

The structure of the $\mathrm{Y}(\mathfrak{g}_2)$-module $V(\lambda)^+_\mu$ (see Corollary 9.4.2) will be described in the following three theorems. Their proofs rely on the results of Chapter 4. We use the notation of Sections 4.3, 4.4 and 4.5, respectively. By the classical branching rules, the multiplicity $\mathrm{mult}(\mu)$ in the restriction (9.57) is zero unless the respective inequalities (9.76), (9.77) or (9.78) hold together with $-|\lambda_1| \geqslant \mu_2$ and $-|\mu_1| \geqslant \lambda_2$ in the D case. We will now assume that λ and μ satisfy these inequalities.

THEOREM 9.4.11. *Let $\mathfrak{g}_N = \mathfrak{sp}_{2n}$. The $\mathrm{Y}(\mathfrak{sp}_2)$-module $V(\lambda)^+_\mu$ is irreducible, and we have an isomorphism*

$$V(\lambda)^+_\mu \cong L(\alpha_1, \beta_1) \otimes \dots \otimes L(\alpha_n, \beta_n).$$

PROOF. Applying Corollary 4.3.5, we find that the $\mathrm{Y}(\mathfrak{sp}_2)$-module

$$V = L(\alpha_1, \beta_1) \otimes \dots \otimes L(\alpha_n, \beta_n) \tag{9.79}$$

is irreducible. Its highest weight is easily calculated with the use of Corollary 4.2.12, and it coincides with the highest weight $\mu(u)$ of the $\mathrm{Y}(\mathfrak{sp}_2)$-module $\mathrm{Y}(\mathfrak{sp}_2)\,\xi_\mu$; see Proposition 9.4.8. Hence V is isomorphic to the irreducible quotient of $\mathrm{Y}(\mathfrak{sp}_2)\,\xi_\mu$. However, $\mathrm{Y}(\mathfrak{sp}_2)\,\xi_\mu$ is a submodule of $V(\lambda)^+_\mu$, while the branching rule for the reduction $\mathfrak{sp}_{2n} \downarrow \mathfrak{sp}_{2n-2}$ gives

$$\dim V(\lambda)^+_\mu = \prod_{i=1}^{n} (\alpha_i - \beta_i + 1).$$

Since this coincides with $\dim V$, the statement follows. □

COROLLARY 9.4.12. *The Drinfeld polynomial of the $\mathrm{Y}(\mathfrak{sp}_2)$-module $V(\lambda)^+_\mu$ is given by $P(u) = Q(u)\,Q(-u+1)(-1)^{\deg Q}$ where*

$$Q(u) = \prod_{i=1}^{n} (u + \beta_i)(u + \beta_i + 1) \dots (u + \alpha_i - 1).$$

PROOF. This is immediate from (4.38). □

THEOREM 9.4.13. *Let* $\mathfrak{g}_N = \mathfrak{o}_{2n}$. *The* $\mathrm{Y}(\mathfrak{o}_2)$*-module* $V(\lambda)^+_\mu$ *is irreducible, and we have an isomorphism*

$$V(\lambda)^+_\mu \cong L(\alpha_1, \beta_1) \otimes \ldots \otimes L(\alpha_{n-1}, \beta_{n-1}) \otimes V(-\alpha_0 - 1/2).$$

PROOF. Applying Corollary 4.4.6 and using the relation $\alpha_1 \leqslant \alpha_0 < -\alpha_1$, we find that the $\mathrm{Y}(\mathfrak{o}_2)$-module

$$V = L(\alpha_1, \beta_1) \otimes \ldots \otimes L(\alpha_{n-1}, \beta_{n-1}) \otimes V(-\alpha_0 - 1/2) \tag{9.80}$$

is irreducible. Its highest weight is easy to calculate by using Proposition 4.2.11, and it coincides with the highest weight $\mu(u)$ of the $\mathrm{Y}(\mathfrak{o}_2)$-module $\mathrm{Y}(\mathfrak{o}_2)\,\xi_\mu$; see Proposition 9.4.9. Hence V is isomorphic to the irreducible quotient of $\mathrm{Y}(\mathfrak{o}_2)\,\xi_\mu$. However, $\mathrm{Y}(\mathfrak{o}_2)\,\xi_\mu$ is a submodule of $V(\lambda)^+_\mu$, while the branching rule for the reduction $\mathfrak{o}_{2n} \downarrow \mathfrak{o}_{2n-2}$ gives

$$\dim V(\lambda)^+_\mu = \prod_{i=1}^{n-1} (\alpha_i - \beta_i + 1).$$

Since this coincides with $\dim V$, the statement follows. □

COROLLARY 9.4.14. *The Drinfeld polynomial of the* $\mathrm{Y}(\mathfrak{o}_2)$*-module* $V(\lambda)^+_\mu$ *is given by* $P(u) = Q(u)\,Q(-u+1)(-1)^{\deg Q}$ *where*

$$Q(u) = \prod_{i=1}^{n-1} (u + \beta_i)(u + \beta_i + 1) \ldots (u + \alpha_i - 1)$$

and the module is associated with the pair $(P(u), -\alpha_0)$.

PROOF. This is immediate from (4.56). □

THEOREM 9.4.15. *Let* $\mathfrak{g}_N = \mathfrak{o}_{2n+1}$. *If* $\beta_1 \neq 0$, *then the* $\mathrm{Y}(\mathfrak{o}_2)$*-module* $V(\lambda)^+_\mu$ *is isomorphic to the direct sum of two irreducible modules,* $V(\lambda)^+_\mu \cong U \oplus U'$, *where*

$$U = L(0, \beta_1) \otimes L(\alpha_2, \beta_2) \otimes \ldots \otimes L(\alpha_n, \beta_n),$$
$$U' = L(-1, \beta_1) \otimes L(\alpha_2, \beta_2) \otimes \ldots \otimes L(\alpha_n, \beta_n)$$

if the λ_i *and* μ_i *are integers or*

$$U = L(-1/2, \beta_1) \otimes L(\alpha_2, \beta_2) \otimes \ldots \otimes L(\alpha_n, \beta_n) \otimes V(-1/2),$$
$$U' = L(-1/2, \beta_1) \otimes L(\alpha_2, \beta_2) \otimes \ldots \otimes L(\alpha_n, \beta_n) \otimes V(1/2)$$

if the λ_i *and* μ_i *are half-integers. Moreover, if* $\beta_1 = 0$, *then* $\mathrm{Y}(\mathfrak{o}_2)$*-module* $V(\lambda)^+_\mu$ *is irreducible, and we have an isomorphism*

$$V(\lambda)^+_\mu \cong L(\alpha_2, \beta_2) \otimes \ldots \otimes L(\alpha_n, \beta_n).$$

PROOF. Suppose first that $\beta_1 \neq 0$. Then applying Corollary 4.4.6 (see also Remark 4.4.7(i)), we find that the $\mathrm{Y}(\mathfrak{o}_2)$-modules U and U' are irreducible. Their highest weights are easy to calculate by using Proposition 4.2.11, and they respectively coincide with the highest weights $\mu(u)$ and $(1+u^{-1})\,\mu(u)$ of the $\mathrm{Y}(\mathfrak{o}_2)$-modules $\mathrm{Y}(\mathfrak{o}_2)\,\xi_\mu$ and $\mathrm{Y}(\mathfrak{o}_2)\,\xi'_\mu$; see Proposition 9.4.10. Hence U and U' are respectively isomorphic to the irreducible quotients of $\mathrm{Y}(\mathfrak{o}_2)\,\xi_\mu$ and $\mathrm{Y}(\mathfrak{o}_2)\,\xi'_\mu$. On the other hand, the branching rule for the reduction $\mathfrak{o}_{2n+1} \downarrow \mathfrak{o}_{2n-1}$ gives

$$\dim V(\lambda)^+_\mu = (-2\beta_1 + 1) \prod_{i=2}^{n} (\alpha_i - \beta_i + 1).$$

This coincides with $\dim U + \dim U'$ so that we need to verify only that the intersection of the submodules $\mathrm{Y}(\mathfrak{o}_2)\,\xi_\mu$ and $\mathrm{Y}(\mathfrak{o}_2)\,\xi'_\mu$ of $V(\lambda)^+_\mu$ is zero. In order to do this note that each of these submodules is a direct sum of eigenspaces for the operator F_{nn}. Moreover, ξ_μ and ξ'_μ are eigenvectors for F_{nn} and the difference between the respective eigenvalues equals -1. This implies that the difference between an arbitrary eigenvalue of F_{nn} in $\mathrm{Y}(\mathfrak{o}_2)\,\xi_\mu$ and an arbitrary eigenvalue of F_{nn} in $\mathrm{Y}(\mathfrak{o}_2)\,\xi'_\mu$ is an odd integer. Hence, these submodules cannot have a nonzero intersection, completing the proof of the first part of the theorem.

The second part is verified in the same way where we need to recall that if $\beta_1 = 0$, then the vector ξ'_μ is zero. (This also follows from the proof of the first part, as the submodule $\mathrm{Y}(\mathfrak{o}_2)\,\xi'_\mu$ must be zero by the dimension argument.) □

REMARK 9.4.16. (i) The pairs $(P(u),\gamma)$ associated with the $\mathrm{Y}(\mathfrak{o}_2)$-modules U and U' can be calculated from (4.56) as in Corollary 9.4.14.

(ii) It is well known that $V(\lambda)^+_\mu$ is an irreducible representation of the centralizer $\mathrm{U}(\mathfrak{g}_N)^{\mathfrak{g}_{N-2}}$; see e.g. Dixmier [**89**, Section 9.1]. Hence Theorem 9.4.15 implies that in the B case the centralizer $\mathrm{U}(\mathfrak{g}_N)^{\mathfrak{g}_{N-2}}$ is not generated by the image of the homomorphism (9.27) and the center of $\mathrm{U}(\mathfrak{g}_N)$; cf. Theorems 8.9.2 and 8.9.5. Otherwise the $\mathrm{Y}(\mathfrak{g}_2)$-module $V(\lambda)^+_\mu$ would be irreducible. It can easily be seen that the element $F_{21}F_{10} - F_{2,-1}F_{-1,0} - F_{20}F_{11}$ belongs to the centralizer $\mathrm{U}(\mathfrak{o}_5)^{\mathfrak{o}_3}$ but is neither central in $\mathrm{U}(\mathfrak{o}_5)$ nor belongs to the image of (9.27). □

9.5. Basis of the multiplicity space

By Theorems 9.4.11, 9.4.13 and 9.4.15, the multiplicity space $V(\lambda)^+_\mu$ can be equipped with an action of the Yangian $\mathrm{Y}(\mathfrak{gl}_2)$. We use this fact together with Theorem 3.3.8 to construct a basis of $V(\lambda)^+_\mu$. As before, we suppose that λ is a $\mathfrak{g}_N$-highest weight and μ is a $\mathfrak{g}_{N-2}$-highest weight such that all components μ_i of μ are simultaneously integers of half-integers together with the λ_i. We consider each of the B, C and D cases separately.

C **type case.** If $\mathfrak{g}_N = \mathfrak{sp}_{2n}$, then $\dim V(\lambda)^+_\mu$ equals the number of n-tuples of integers $\nu = (\nu_1,\dots,\nu_n)$ satisfying the betweenness conditions

$$0 \geqslant \nu_1 \geqslant \lambda_1 \geqslant \nu_2 \geqslant \lambda_2 \geqslant \cdots \geqslant \nu_{n-1} \geqslant \lambda_{n-1} \geqslant \nu_n \geqslant \lambda_n,$$
$$0 \geqslant \nu_1 \geqslant \mu_1 \geqslant \nu_2 \geqslant \mu_2 \geqslant \cdots \geqslant \nu_{n-1} \geqslant \mu_{n-1} \geqslant \nu_n.$$

Let us set

$$\gamma_i = \nu_i + \rho_i + 1/2, \qquad l_i = \lambda_i + \rho_i + 1/2 \tag{9.81}$$

for $i = 1,\dots,n$ where the ρ_i are defined in (7.10). These relations provide a one-to-one correspondence between the n-tuples ν satisfying the betweenness conditions and the n-tuples $\gamma = (\gamma_1,\dots,\gamma_n)$ satisfying the conditions (3.88) with $k = n$. For each such n-tuple ν introduce the vector $\xi_\nu \in V(\lambda)^+_\mu$ by

$$\xi_\nu = \prod_{i=1}^{n-1} z_{ni}^{\nu_i-\mu_i}\, z_{i,-n}^{\nu_i-\lambda_i} \cdot \prod_{k=l_n}^{\gamma_n-1} Z_{n,-n}(k)\,\xi. \tag{9.82}$$

The order of the factors $Z_{n,-n}(k)$ is irrelevant due to Corollary 9.3.3.

THEOREM 9.5.1. *The vectors ξ_ν parameterized by the n-tuples ν satisfying the betweenness conditions form a basis of the multiplicity space $V(\lambda)^+_\mu$.*

PROOF. Let us apply Theorem 9.4.11 to identify the vector space $V(\lambda)^+_\mu$ with the tensor product space V defined in (9.79). The space V is a module over the Yangian $\mathrm{Y}(\mathfrak{gl}_2)$ whose generators will now be denoted by $t_{ab}^{(r)}$ with $a, b \in \{-n, n\}$ and $r \geqslant 1$. The operators $T_{ab}(u) = u^n\, t_{ab}(u)$ in V are polynomials in u; cf. (3.86). The action of the twisted Yangian $\mathrm{Y}(\mathfrak{sp}_2)$ on $V(\lambda)^+_\mu$ is given by (9.24); see Corollary 9.4.2. On the other hand, the generators of $\mathrm{Y}(\mathfrak{sp}_2)$ and $\mathrm{Y}(\mathfrak{gl}_2)$ are related by (4.22). Hence, for all $a, b \in \{-n, n\}$ we have the following equality of operators on $V(\lambda)^+_\mu \cong V$:

$$Z_{ab}(u) = \frac{(-1)^n}{u + 1/2}\Big(\theta_{-n,b}\, T_{a,-n}(u)\, T_{-b,n}(-u) + \theta_{n,b}\, T_{a,n}(u)\, T_{-b,-n}(-u)\Big). \tag{9.83}$$

Furthermore, using the Yangian defining relations (1.3), we also obtain

$$Z_{a,-a}(u) = \frac{(-1)^n}{u}\Big(T_{a,-a}(u)\, T_{a,a}(-u) - T_{a,-a}(-u)\, T_{a,a}(u)\Big) \tag{9.84}$$

for any $a \in \{-n, n\}$. Now observe that the strings $S(\alpha_i, \beta_i)$ with $i = 1, \dots, n$ satisfy the assumptions of Theorem 3.3.8 with $k = n$. Therefore, this theorem provides a basis $\{\eta_\gamma\}$ of $V(\lambda)^+_\mu$. We will complete the proof by verifying that each basis vector η_γ coincides with ξ_ν, up to a nonzero constant factor, where γ and ν are related by (9.81); cf. the proof of Theorem 4.3.7. Indeed, the highest vector η_β of V is proportional to ξ_μ; see the proof of Theorem 9.4.11. Furthermore, using (9.84) with $a = n$ and the formulas for the action of the operators $T_{ab}(u)$ in this basis, provided by Theorem 3.3.8, we obtain

$$Z_{n,-n}(\gamma_i)\, \eta_\gamma = 2 \prod_{j=1,\, j\neq i}^{n} (-\gamma_i - \gamma_j)\, \eta_{\gamma+\delta_i}, \qquad i = 1, \dots, n. \tag{9.85}$$

Since $\gamma_i + \gamma_j \neq 0$ for all $i \neq j$ we can conclude that the vectors

$$\prod_{i=1}^{n} Z_{n,-n}(\gamma_i - 1) \dots Z_{n,-n}(\beta_i + 1)\, Z_{n,-n}(\beta_i)\, \xi_\mu \tag{9.86}$$

with ν satisfying the betweenness conditions form a basis of $V(\lambda)^+_\mu$. It remains to verify that the vector (9.86) coincides with ξ_ν. We will do this by induction on $\sum_i (\gamma_i - \beta_i)$ keeping λ and μ fixed. If $\gamma_i = \beta_i$ for all $i = 1, \dots, n$, then $\nu_i = \max\{\lambda_i, \mu_i\}$ and $\xi_\nu = \xi_\mu$. It is now sufficient to show that for an arbitrary ν and any $j \in \{1, \dots, n\}$ we have $Z_{n,-n}(\gamma_j)\, \xi_\nu = \xi_{\nu+\delta_j}$. If $j = n$, then this is immediate from Corollary 9.3.4. If $1 \leqslant j \leqslant n-1$, then using Corollary 9.3.4 again, we can write

$$Z_{n,-n}(\gamma_j)\, \xi_\nu = \prod_{i=1}^{j-1} z_{ni}^{\nu_i - \mu_i}\, z_{i,-n}^{\nu_i - \lambda_i}$$
$$\times\, z_{nj}^{\nu_j - \mu_j}\, Z_{n,-n}(\gamma_j)\, z_{j,-n}^{\nu_j - \lambda_j} \prod_{i=j+1}^{n-1} z_{ni}^{\nu_i - \mu_i}\, z_{i,-n}^{\nu_i - \lambda_i} \cdot \prod_{k=l_n}^{\gamma_n - 1} Z_{n,-n}(k)\, \xi.$$

Observe that $Z_{n,-n}(\gamma_j)$ here can be replaced with $Z_{n,-n}(g_j)$ because

$$g_j\, z_{j,-n}^{\nu_j - \lambda_j}\, \xi = \gamma_j\, z_{j,-n}^{\nu_j - \lambda_j}\, \xi.$$

On the other hand, Corollary 9.3.7 implies that $Z_{n,-n}(g_j) = z_{nj}\, z_{j,-n}$ and hence $Z_{n,-n}(\gamma_j)\, \xi_\nu = \xi_{\nu+\delta_j}$, as desired. □

D type case. If $\mathfrak{g}_N = \mathfrak{o}_{2n}$, then $\dim V(\lambda)^+_\mu$ equals the number of $(n-1)$-tuples $\nu = (\nu_1, \dots, \nu_{n-1})$ satisfying the betweenness conditions

$$-|\lambda_1| \geqslant \nu_1 \geqslant \lambda_2 \geqslant \nu_2 \geqslant \lambda_3 \geqslant \dots \geqslant \lambda_{n-1} \geqslant \nu_{n-1} \geqslant \lambda_n,$$
$$-|\mu_1| \geqslant \nu_1 \geqslant \mu_2 \geqslant \nu_2 \geqslant \mu_3 \geqslant \dots \geqslant \mu_{n-1} \geqslant \nu_{n-1}$$

with all the ν_i being simultaneously integers or half-integers together with the λ_i. Set $\nu_0 = \max\{\lambda_1, \mu_1\}$ and for each such $(n-1)$-tuple ν introduce the vector $\xi_\nu \in V(\lambda)^+_\mu$ by

$$\xi_\nu = \prod_{i=1}^{n-1} z_{ni}^{\nu_{i-1}-\mu_i} z_{i,-n}^{\nu_{i-1}-\lambda_i} \cdot \prod_{k=l_n}^{\gamma_{n-1}-1} Z_{n,-n}(k)\, \xi \tag{9.87}$$

where we have used the notation

$$\gamma_i = \nu_i - i + 1/2, \qquad l_i = \lambda_i - i + 3/2 \tag{9.88}$$

for $i = 1, \dots, n$. The order of the factors $Z_{n,-n}(k)$ in (9.87) is irrelevant due to Corollary 9.3.3.

THEOREM 9.5.2. *The vectors ξ_ν parameterized by the $(n-1)$-tuples ν satisfying the betweenness conditions form a basis of the multiplicity space $V(\lambda)^+_\mu$.*

PROOF. Let us apply Theorem 9.4.13 to identify the vector space $V(\lambda)^+_\mu$ with the tensor product space V defined in (9.80). We will regard V as a module over the Yangian $\mathrm{Y}(\mathfrak{gl}_2)$ by using the vector space isomorphism

$$V \cong L(\alpha_1, \beta_1) \otimes \dots \otimes L(\alpha_{n-1}, \beta_{n-1})$$

as in (4.63). Then the operators $T_{ab}(u) = u^{n-1}\, t_{ab}(u)$ in V are polynomials in u; cf. (3.86). The action of the twisted Yangian $\mathrm{Y}(\mathfrak{o}_2)$ on $V(\lambda)^+_\mu$ is given by (9.25); see Corollary 9.4.2. On the other hand, the actions of the generators of $\mathrm{Y}(\mathfrak{o}_2)$ and $\mathrm{Y}(\mathfrak{gl}_2)$ on V are related by the coproduct formula (4.23). Hence, recalling the definition of the one-dimensional $\mathrm{Y}(\mathfrak{o}_2)$-module $V(\delta)$ (see Section 4.2), for all $a, b \in \{-n, n\}$ we have the following equality of operators on $V(\lambda)^+_\mu \cong V$:

$$Z_{ab}(u) = \frac{(-1)^n}{2u+1} \Big(T_{a,-n}(u)\, T_{-b,n}(-u)\, (u + \alpha_0 + 1) + T_{a,n}(u)\, T_{-b,-n}(-u)\, (u - \alpha_0) \Big). \tag{9.89}$$

Furthermore, using the Yangian defining relations (1.3), we also obtain

$$Z_{a,-a}(u) = \frac{(-1)^n}{2u} \Big(T_{a,-n}(u)\, T_{a,n}(-u)\, (u + \alpha_0) + T_{a,-n}(-u)\, T_{a,n}(u)\, (u - \alpha_0) \Big) \tag{9.90}$$

for any $a \in \{-n, n\}$. The strings $S(\alpha_i, \beta_i)$ with $i = 1, \dots, n-1$ satisfy the assumptions of Theorem 3.3.8 with $k = n-1$, and so we have a basis $\{\eta_\gamma\}$ of $V(\lambda)^+_\mu$. Note that the vector η_β is proportional to ξ_μ; see the proof of Theorem 9.4.13. Furthermore, using (9.90) with $a = n$ and Theorem 3.3.8, we obtain

$$Z_{n,-n}(\gamma_i)\, \eta_\gamma = (\gamma_i - \alpha_0) \prod_{j=1,\, j\neq i}^{n-1} (-\gamma_i - \gamma_j)\, \eta_{\gamma+\delta_i}, \qquad i = 1, \dots, n-1. \tag{9.91}$$

Note that $\gamma_i + \gamma_j \neq 0$ for all $i \neq j$, while $\gamma_i - \alpha_0$ can only be zero if $i = 1$ and $|\lambda_1 + \mu_1| = 0$. However, in this case $\nu_1 = \min\{-|\lambda_1|, -|\mu_1|\}$ so that the vector $\eta_{\gamma+\delta_1}$ is zero. Hence we can conclude that the vectors

$$\prod_{i=1}^{n-1} Z_{n,-n}(\gamma_i - 1) \dots Z_{n,-n}(\beta_i + 1)\, Z_{n,-n}(\beta_i)\, \xi_\mu \tag{9.92}$$

with ν satisfying the betweenness conditions form a basis of $V(\lambda)^+_\mu$. Now the argument is completed in exactly the same way as in the proof of Theorem 9.5.1 by verifying that the vector (9.92) coincides with ξ_ν for the values of the parameters γ and ν related by (9.88). □

B type case. If $\mathfrak{g}_N = \mathfrak{o}_{2n+1}$, then $\dim V(\lambda)^+_\mu$ equals the number of $(n+1)$-tuples $\nu = (\sigma, \nu_1, \dots, \nu_n)$ satisfying the betweenness conditions

$$0 \geqslant \nu_1 \geqslant \lambda_1 \geqslant \nu_2 \geqslant \lambda_2 \geqslant \dots \geqslant \nu_{n-1} \geqslant \lambda_{n-1} \geqslant \nu_n \geqslant \lambda_n,$$
$$0 \geqslant \nu_1 \geqslant \mu_1 \geqslant \nu_2 \geqslant \mu_2 \geqslant \dots \geqslant \nu_{n-1} \geqslant \mu_{n-1} \geqslant \nu_n$$

with all the ν_i being simultaneously integers or half-integers together with the λ_i, while σ takes any of the two values 0 or 1 if $\nu_1 \neq 0$, and $\sigma = 0$ if $\nu_1 = 0$. For each such $(n+1)$-tuple ν introduce the vector $\xi_\nu \in V(\lambda)^+_\mu$ by

$$\xi_\nu = z_{n0}^{\sigma} \prod_{i=1}^{n-1} z_{ni}^{\nu_i - \mu_i} z_{i,-n}^{\nu_i - \lambda_i} \cdot \prod_{k=l_n}^{\gamma_n - 1} Z_{n,-n}(k)\, \xi, \tag{9.93}$$

where we have used the notation (9.81). The order of the factors $Z_{n,-n}(k)$ in (9.93) is irrelevant due to Corollary 9.3.3.

THEOREM 9.5.3. *The vectors ξ_ν parameterized by the $(n+1)$-tuples ν satisfying the betweenness conditions form a basis of the multiplicity space $V(\lambda)^+_\mu$.*

PROOF. The argument is similar to the proof of Theorem 9.5.2. Suppose first that $\beta_1 \neq 0$ and apply Theorem 9.4.15 to identify $V(\lambda)^+_\mu$ with the direct sum $U \oplus U'$. Each of the $\mathrm{Y}(\mathfrak{o}_2)$-modules U and U' has the form

$$L(\widetilde{\alpha}_1, \beta_1) \otimes L(\alpha_2, \beta_2) \otimes \dots \otimes L(\alpha_n, \beta_n) \otimes V(\delta), \tag{9.94}$$

for an appropriate choice of the parameters $\widetilde{\alpha}_1$ and δ (in particular, $\delta = 0$ in the case where the λ_i are integers). We regard (9.94) as a module over the Yangian $\mathrm{Y}(\mathfrak{gl}_2)$ in the same way as in the proof of Theorem 9.5.2. In particular, using (9.23) and Corollary 9.4.2, for all $a, b \in \{-n, n\}$ we derive the following equalities of operators on the space (9.94):

$$Z_{ab}(u) = \frac{(-1)^{n+1}}{u + 1/2} \Big(T_{a,-n}(u)\, T_{-b,n}(-u)\, (u - \delta + 1/2) + T_{a,n}(u)\, T_{-b,-n}(-u)\, (u + \delta + 1/2) \Big), \tag{9.95}$$

and for any $a \in \{-n, n\}$ we have

$$Z_{a,-a}(u) = \frac{(-1)^{n+1}}{u} \Big(T_{a,-n}(u)\, T_{a,n}(-u)\, (u - \delta - 1/2) + T_{a,-n}(-u)\, T_{a,n}(u)\, (u + \delta + 1/2) \Big). \tag{9.96}$$

Applying Theorem 3.3.8 we derive that the vectors

$$z_{n0}^{\sigma} \prod_{i=1}^{n} Z_{n,-n}(\gamma_i - 1) \dots Z_{n,-n}(\beta_i + 1)\, Z_{n,-n}(\beta_i)\, \xi_{\mu} \tag{9.97}$$

form a basis of $V(\lambda)^+_\mu$, where we have used the fact that ξ_μ and $\xi'_\mu = z_{n0}\, \xi_\mu$ are the highest vectors of the $\mathrm{Y}(\mathfrak{o}_2)$-modules U and U', respectively. Finally, the argument is completed by verifying in exactly the same way as in the proof of Theorem 9.5.1 that the vector (9.97) coincides with ξ_ν for the values of the parameters γ and ν related by (9.81).

In the case $\beta_1 = 0$ the same argument is used where only the value $\sigma = 0$ should be considered, as $\xi'_\mu = 0$. □

9.6. Basis of $V(\lambda)$

Using decomposition (9.58) and the basis of the multiplicity space $V(\lambda)^+_\mu$ constructed in Section 9.5, we can now construct a basis of the entire representation space $V(\lambda)$ by induction on n. We also provide explicit formulas for the action of the generators of $\mathfrak{g}_N$ in this basis. Again, we consider the B, C and D cases separately. We suppose that λ is an arbitrary fixed $\mathfrak{g}_N$-highest weight.

C **type case.** Define the *C type pattern* Λ associated with λ as an array of the form

$$\begin{array}{ccccccccccc}
 & \lambda_{n1} & & \lambda_{n2} & & \lambda_{n3} & & \cdots & & \lambda_{nn} & \\
\lambda'_{n1} & & \lambda'_{n2} & & \lambda'_{n3} & & \cdots & & \lambda'_{nn} & & \\
 & \lambda_{n-1,1} & & \lambda_{n-1,2} & & \cdots & & \lambda_{n-1,n-1} & & & \\
\lambda'_{n-1,1} & & \lambda'_{n-1,2} & & \cdots & & \lambda'_{n-1,n-1} & & & & \\
 & \cdots & & \cdots & & \cdots & & & & & \\
 & \lambda_{11} & & & & & & & & & \\
\lambda'_{11} & & & & & & & & & &
\end{array}$$

such that $\lambda_{ni} = \lambda_i$ for $i = 1, \dots, n$, the remaining entries are all nonpositive integers and the following inequalities hold:

$$\lambda'_{k1} \geqslant \lambda_{k1} \geqslant \lambda'_{k2} \geqslant \lambda_{k2} \geqslant \cdots \geqslant \lambda'_{k,k-1} \geqslant \lambda_{k,k-1} \geqslant \lambda'_{kk} \geqslant \lambda_{kk}$$

for $k = 1, \dots, n$, and

$$\lambda'_{k1} \geqslant \lambda_{k-1,1} \geqslant \lambda'_{k2} \geqslant \lambda_{k-1,2} \geqslant \cdots \geqslant \lambda'_{k,k-1} \geqslant \lambda_{k-1,k-1} \geqslant \lambda'_{kk}$$

for $k = 2, \dots, n$. We will be using the notation

$$l_{ki} = \lambda_{ki} + \rho_i + 1/2, \qquad l'_{ki} = \lambda'_{ki} + \rho_i + 1/2, \tag{9.98}$$

where the ρ_i are defined in (7.10).

COROLLARY 9.6.1. *The vectors*

$$\xi_\Lambda = \prod_{k=1,\dots,n}^{\longrightarrow} \left(\prod_{i=1}^{k-1} z_{ki}^{\lambda'_{ki} - \lambda_{k-1,i}}\, z_{i,-k}^{\lambda'_{ki} - \lambda_{ki}} \cdot \prod_{j=l_{kk}}^{l'_{kk}-1} Z_{k,-k}(j) \right) \xi$$

parameterized by all patterns Λ associated with λ form a basis of the representation $V(\lambda)$ of $\mathfrak{sp}_{2n}$.

PROOF. Using induction on n and applying Theorem 9.5.1 and the decomposition (9.58), we reduce the proof to the case $n = 1$. However, the claim obviously holds in this case as the vectors given by

$$\xi_\Lambda = \prod_{j=l_{11}}^{l'_{11}-1} Z_{1,-1}(j)\,\xi = F_{1,-1}^{\lambda'_{11}-\lambda_{11}}\,\xi$$

form a basis of $V(\lambda)$. □

Note that the elements $F_{k,-k}$, $F_{-k,k}$ with $k = 1, \dots, n$ and $F_{k-1,-k}$ with $k = 2, \dots, n$ generate $\mathfrak{sp}_{2n}$ as a Lie algebra. Therefore, the action of these elements on the basis vectors of $V(\lambda)$ determines the action of any element of $\mathfrak{sp}_{2n}$. In order to simplify the formulas for the matrix elements of the generators, we introduce the normalized basis vectors ζ_Λ of $V(\lambda)$ by

$$\zeta_\Lambda = N_\Lambda\, \xi_\Lambda,$$

where

$$N_\Lambda = \prod_{k=2}^{n} \prod_{1 \leqslant i < j \leqslant k} (-l'_{ki} - l'_{kj})!.$$

The basis vectors ζ_Λ are obtained by the same induction procedure as the one used in the proof of Corollary 9.6.1, where the basis vectors ξ_ν of $V(\lambda)^+_\mu$ are replaced with the normalized vectors

$$\zeta_\nu = \prod_{1 \leqslant i < j \leqslant n} (-\gamma_i - \gamma_j)!\, \xi_\nu. \tag{9.99}$$

THEOREM 9.6.2. *The action of the generators of $\mathfrak{sp}_{2n}$ in the basis $\{\zeta_\Lambda\}$ of $V(\lambda)$ is given by the formulas*

$$F_{kk}\,\zeta_\Lambda = \Big(2\sum_{i=1}^{k} \lambda'_{ki} - \sum_{i=1}^{k} \lambda_{ki} - \sum_{i=1}^{k-1} \lambda_{k-1,i}\Big)\zeta_\Lambda,$$

$$F_{k,-k}\,\zeta_\Lambda = \sum_{i=1}^{k} A_{ki}\,\zeta_{\Lambda+\delta'_{ki}},$$

$$F_{-k,k}\,\zeta_\Lambda = \sum_{i=1}^{k} B_{ki}\,\zeta_{\Lambda-\delta'_{ki}},$$

$$F_{k-1,-k}\,\zeta_\Lambda = \sum_{i=1}^{k-1} C_{ki}\,\zeta_{\Lambda-\delta_{k-1,i}} + \sum_{i=1}^{k}\sum_{j,m=1}^{k-1} D_{kijm}\,\zeta_{\Lambda+\delta'_{ki}+\delta_{k-1,j}+\delta'_{k-1,m}},$$

where

$$A_{ki} = \prod_{a=1,\, a\neq i}^{k} \frac{1}{l'_{ka} - l'_{ki}},$$

$$B_{ki} = 2\,A_{ki}\,(2\,l'_{ki} - 1) \prod_{a=1}^{k} (l_{ka} - l'_{ki}) \prod_{a=1}^{k-1} (l_{k-1,a} - l'_{ki}),$$

$$C_{ki} = \frac{1}{2\,l_{k-1,i} - 1} \prod_{a=1,\, a\neq i}^{k-1} \frac{1}{(l_{k-1,i} - l_{k-1,a})(l_{k-1,i} + l_{k-1,a} - 1)},$$

and

$$D_{kijm} = A_{ki} A_{k-1,m} C_{kj} \prod_{a=1,\, a\neq i}^{k} \left(l_{k-1,j}^2 - l_{ka}'^2\right) \prod_{a=1,\, a\neq m}^{k-1} \left(l_{k-1,j}^2 - l_{k-1,a}'^2\right).$$

The arrays $\Lambda \pm \delta_{ki}$ and $\Lambda \pm \delta'_{ki}$ are obtained from Λ by replacing λ_{ki} and λ'_{ki} by $\lambda_{ki} \pm 1$ and $\lambda'_{ki} \pm 1$ respectively. The vector ζ_Λ is zero if the array Λ is not a pattern.

PROOF. Note that the elements F_{kk}, $F_{k,-k}$, $F_{-k,k}$ belong to the centralizer of the subalgebra $\mathfrak{sp}_{2k-2}$ in $\mathrm{U}(\mathfrak{sp}_{2k})$. Therefore, these operators preserve the subspace of $\mathfrak{sp}_{2k-2}$-highest vectors in $V(\lambda)$. So, it suffices to compute the action of these operators with $k = n$ in the basis $\{\zeta_\nu\}$ of the space $V(\lambda)^+_\mu$; see Theorem 9.5.1 and (9.99). The definition of the vectors ξ_ν implies

$$F_{nn}\, \zeta_\nu = \left(2 \sum_{i=1}^{n} \nu_i - \sum_{i=1}^{n} \lambda_i - \sum_{i=1}^{n-1} \mu_i\right) \zeta_\nu.$$

Furthermore, as we saw in the proof of Theorem 9.5.1,

$$Z_{n,-n}(\gamma_i)\, \xi_\nu = \xi_{\nu+\delta_i}, \qquad i = 1, \dots, n. \tag{9.100}$$

However, $Z_{n,-n}(u)$ is a polynomial in u^2 of degree $n-1$; see Corollary 9.3.7. By (9.21), the leading coefficient of this polynomial is $F_{n,-n}$. Applying the Lagrange interpolation formula with the interpolation points γ_i, $i = 1, \dots, n$, we obtain

$$Z_{n,-n}(u)\, \xi_\nu = \sum_{i=1}^{n} \prod_{a=1,\, a\neq i}^{n} \frac{u^2 - \gamma_a^2}{\gamma_i^2 - \gamma_a^2}\, \xi_{\nu+\delta_i}.$$

Taking here the coefficient at u^{2n-2} we get

$$F_{n,-n}\, \xi_\nu = \sum_{i=1}^{n} \prod_{a=1,\, a\neq i}^{n} \frac{1}{\gamma_i^2 - \gamma_a^2}\, \xi_{\nu+\delta_i},$$

thus proving that

$$F_{n,-n}\, \zeta_\nu = \sum_{i=1}^{n} A_i\, \zeta_{\nu+\delta_i}, \qquad A_i = \prod_{a=1,\, a\neq i}^{n} \frac{1}{\gamma_a - \gamma_i}.$$

Similarly, by Corollary 9.3.7 and (9.21), $Z_{-n,n}(u)$ is a polynomial in u^2 of degree $n-1$ with the leading coefficient $F_{-n,n}$. Hence, applying (9.84) with $a = -n$ we find that $F_{-n,n}$, as an operator in $V(\lambda)^+_\mu$, coincides with $2\, t^{(1)}_{-n,n}$. However, $t^{(1)}_{-n,n}$ is the leading coefficient of the polynomial $T_{-n,n}(u)$, which has degree $n-1$. Comparing (9.85) and (9.100), we derive from Theorem 3.3.8 that

$$T_{n,-n}(-\gamma_i)\, \zeta_\nu = \frac{1}{2}\, \zeta_{\nu+\delta_i}$$

and hence

$$T_{-n,n}(-\gamma_i)\, \zeta_\nu = -2 \prod_{a=1}^{n} (\alpha_a - \gamma_i + 1)(\beta_a - \gamma_i)\, \zeta_{\nu-\delta_i}$$

for all $i = 1, \dots, n$. By the definition of the parameters α_i and β_i (see Section 9.4), we have

$$\prod_{a=1}^{n} (\alpha_a - \gamma_i + 1)(\beta_a - \gamma_i) = (1/2 - \gamma_i) \prod_{a=1}^{n} (l_a - \gamma_i) \prod_{a=1}^{n-1} (m_a - \gamma_i);$$

see (9.68) and (9.81). Applying the Lagrange interpolation formula to the polynomial $T_{-n,n}(u)$ with the interpolation points $-\gamma_i$ for $i = 1, \dots, n$ and taking the coefficient at u^{n-1}, we thus prove that

$$F_{-n,n}\,\zeta_\nu = \sum_{i=1}^{n} B_i\,\zeta_{\nu-\delta_i}, \qquad B_i = 2\,A_i\,(2\gamma_i - 1)\prod_{a=1}^{n}(l_a - \gamma_i)\prod_{a=1}^{n-1}(m_a - \gamma_i).$$

In order to calculate the action of the elements $F_{k-1,-k}$ it is obviously sufficient to consider the case $k = n$. The operator $F_{n-1,-n}$ on $V(\lambda)$ commutes with the action of all elements of the subalgebra $\mathfrak{sp}_{2n-4}$. Therefore, in the calculation of the expansion of $F_{n-1,-n}\,\zeta_\Lambda$ we may consider only the top part of Λ of the form

$$\begin{array}{ccccccccccc}
 & & \lambda_1 & & \lambda_2 & & \lambda_3 & & \cdots & & \lambda_n \\
 & \nu_1 & & \nu_2 & & \nu_3 & & \cdots & & \nu_n & \\
 & & \mu_1 & & \mu_2 & & \cdots & & \mu_{n-1} & & \\
 & \nu'_1 & & \nu'_2 & & \cdots & & \nu'_{n-1} & & & \\
 & & \mu'_1 & & \cdots & & \mu'_{n-2} & & & &
\end{array}$$

We set $\nu' = (\nu'_1, \dots, \nu'_{n-1})$ and $\gamma'_i = \nu'_i - i + 1/2$. We will calculate the expansion of $F_{n-1,-n}\,\xi_{\nu\mu\nu'}$ in terms of the basis vectors of $V(\lambda)$, where

$$\xi_{\nu\mu\nu'} = X_{\mu\nu'}\,\xi_{\nu\mu}$$

with

$$X_{\mu\nu'} = \prod_{i=1}^{n-2} z_{n-1,i}^{\nu'_i-\mu'_i}\, z_{i,-n+1}^{\nu'_i-\mu_i} \cdot \prod_{r=m_{n-1}}^{\gamma'_{n-1}-1} Z_{n-1,-n+1}(r)$$

and

$$\xi_{\nu\mu} = \prod_{i=1}^{n-1} z_{ni}^{\nu_i-\mu_i}\, z_{i,-n}^{\nu_i-\lambda_i} \cdot \prod_{k=l_n}^{\gamma_n-1} Z_{n,-n}(k)\,\xi.$$

Note that $\xi_{\nu\mu}$ coincides with the basis vector $\xi_\nu \in V(\lambda)^+_\mu$ defined in (9.82). As $F_{n-1,-n}$ is permutable with the elements $z_{n-1,i}$ and $Z_{n-1,-n+1}(u)$, we can write

$$F_{n-1,-n}\,\xi_{\nu\mu\nu'} = X_{\mu\nu'}\,F_{n-1,-n}\,\xi_{\nu\mu}.$$

By Corollary 9.4.4, we have

$$F_{n-1,-n}\,\xi_{\nu\mu\nu'} = \sum_{i=1}^{n-1} C_i\Big(X_{\mu\nu'}\,z_{i,-n+1}\,z_{ni}\,\xi_{\nu\mu} + X_{\mu\nu'}\,z_{n-1,i}\,z_{i,-n}\,\xi_{\nu\mu}\Big), \tag{9.101}$$

where

$$C_i = \frac{1}{2m_i - 1}\prod_{a=1,\,a\neq i}^{n-1}\frac{1}{(m_i - m_a)(m_i + m_a - 1)}.$$

Furthermore, let us verify that for $i = 1, \dots, n-1$ we have

$$X_{\mu\nu'}\,z_{i,-n+1}\,z_{ni}\,\xi_{\nu\mu} = \xi_{\nu,\mu-\delta_i,\nu'}.$$

Indeed, $z_{ni}\,\xi_{\nu\mu} = \xi_{\nu,\mu-\delta_i}$, while for $i < n-1$ we have $X_{\mu\nu'}\,z_{i,-n+1} = X_{\mu-\delta_i,\nu'}$, where we have used Proposition 9.2.5 and Corollary 9.3.4. If $i = n-1$, then by (9.41),

$$z_{n-1,-n+1} = Z_{n-1,-n+1}(g_{n-1}).$$

However,

$$g_{n-1}\,\xi_{\nu,\mu-\delta_{n-1}} = (m_{n-1} - 1)\,\xi_{\nu,\mu-\delta_{n-1}},$$

and so

$$X_{\mu\nu'}\, z_{n-1,-n+1} = X_{\mu\nu'} Z_{n-1,-n+1}(m_{n-1} - 1) = X_{\mu-\delta_{n-1},\nu'},$$

as desired.

Now for $j = 1, \dots, n-1$ consider the expression

$$X_{\mu\nu'}\, z_{n-1,j}\, z_{j,-n}\, \xi_{\nu\mu}.$$

Assume first that $\nu_j - \mu_j \geqslant 1$. Using (9.40) we obtain

$$z_{j,-n}\, \xi_{\nu\mu} = z_{j,-n}\, z_{nj}\, \xi_{\nu,\mu+\delta_j} = z_{n,-j}\, z_{-j,-n}\, \xi_{\nu,\mu+\delta_j} = Z_{n,-n}(-g_{-j})\, \xi_{\nu,\mu+\delta_j}$$

and observe that

$$-g_{-j}\, \xi_{\nu,\mu+\delta_j} = m_j\, \xi_{\nu,\mu+\delta_j}.$$

In order to calculate $Z_{n,-n}(m_j)\xi_{\nu,\mu+\delta_j}$ we apply the Lagrange interpolation formula for the polynomial $Z_{n,-n}(u)$ at the interpolation points γ_i with $i = 1, \dots, n$ and then put $u = m_j$. The resulting expression is

$$Z_{n,-n}(m_j)\xi_{\nu,\mu+\delta_j} = \sum_{i=1}^{n} \prod_{a=1,\, a\neq i}^{n} \frac{m_j^2 - \gamma_a^2}{\gamma_i^2 - \gamma_a^2}\, \xi_{\nu+\delta_i,\mu+\delta_j}$$

which also holds for $\nu_j = \mu_j$, as is easily verified.

Next, consider the product $X_{\mu\nu'}\, z_{n-1,j}$ with $j < n-1$. The calculation in this case is similar to the previous one. Assume first that $\nu'_j - \mu_j \geqslant 1$. We have

$$\begin{aligned} X_{\mu\nu'}\, z_{n-1,j} &= X_{\mu+\delta_j,\nu'}\, z_{j,-n+1}\, z_{n-1,j} \\ &= X_{\mu+\delta_j,\nu'}\, z_{n-1,-j}\, z_{-j,-n+1} = X_{\mu+\delta_j,\nu'}\, Z_{n-1,-n+1}(-g_{-j}). \end{aligned}$$

Since

$$-g_{-j}\, \xi_{\nu+\delta_i,\mu+\delta_j} = m_j\, \xi_{\nu+\delta_i,\mu+\delta_j},$$

using the Lagrange interpolation formula for the polynomial $Z_{n-1,-n+1}(u)$ with the interpolation points γ'_r for $r = 1, \dots, n-1$ we get

$$X_{\mu+\delta_j,\nu'} Z_{n-1,-n+1}(m_j) = \sum_{r=1}^{n-1} \prod_{a=1,\, a\neq r}^{n-1} \frac{m_j^2 - {\gamma'_a}^2}{{\gamma'_r}^2 - {\gamma'_a}^2}\, X_{\mu+\delta_j,\nu'+\delta_r}, \tag{9.102}$$

which also holds for $\nu'_j = \mu_j$. If $j = n-1$, then we can write

$$X_{\mu\nu'} = X_{\mu+\delta_{n-1},\nu'}\, Z_{n-1,-n+1}(m_{n-1}),$$

so that (9.102) holds for this case as well.

Thus, writing (9.101) in terms of the normalized vectors

$$\zeta_{\nu\mu\nu'} = \prod_{1\leqslant i<j\leqslant n} (-\gamma_i - \gamma_j)! \prod_{1\leqslant i<j\leqslant n-1} (-\gamma'_i - \gamma'_j)!\, \xi_{\nu\mu\nu'},$$

we finally obtain

$$F_{n-1,-n}\, \zeta_{\nu\mu\nu'} = \sum_{i=1}^{n-1} C_i\, \zeta_{\nu,\mu-\delta_i,\nu'} + \sum_{i=1}^{n} \sum_{j,r=1}^{n-1} D_{ijr}\, \zeta_{\nu+\delta_i,\mu+\delta_j,\nu'+\delta_r},$$

where

$$D_{ijr} = C_j \prod_{a=1,\, a\neq i}^{n} \frac{m_j^2 - \gamma_a^2}{\gamma_a - \gamma_i} \prod_{a=1,\, a\neq r}^{n-1} \frac{m_j^2 - {\gamma'_a}^2}{\gamma'_a - \gamma'_r},$$

completing the proof of the theorem. □

REMARK 9.6.3. Explicit formulas for the action of the simple root generators $F_{k-1,k}$ and $F_{k,k-1}$ with $k = 2,\dots,n$ in the basis $\{\zeta_\Lambda\}$ of $V(\lambda)$ can be found from Theorem 9.6.2 by using the relations $2F_{k-1,k} = [F_{k-1,-k}, F_{-k,k}]$ and $2F_{k,k-1} = [F_{k-1,-k}, F_{-k+1,k-1}]$. However, the coefficients of the basis vectors in the expansions of $F_{k-1,k}\,\zeta_\Lambda$ and $F_{k,k-1}\,\zeta_\Lambda$ are rather complicated, so we will not write them down. □

D **type case.** Define the *D type pattern* Λ associated with λ as an array of the form

$$\begin{array}{ccccccccc}
\lambda_{n1} & & \lambda_{n2} & & \lambda_{n3} & & \cdots & & \lambda_{nn}\\
& \lambda'_{n-1,1} & & \lambda'_{n-1,2} & & \cdots & & \lambda'_{n-1,n-1} & \\
\lambda_{n-1,1} & & \lambda_{n-1,2} & & \cdots & & \lambda_{n-1,n-1} & & \\
& \cdots & & \cdots & & \cdots & & & \\
\lambda_{21} & & \lambda_{22} & & & & & & \\
& \lambda'_{11} & & & & & & & \\
\lambda_{11} & & & & & & & &
\end{array}$$

such that $\lambda_{ni} = \lambda_i$ for $i = 1,\dots,n$, the remaining entries are all simultaneously integers or half-integers together with the λ_i, and the following inequalities hold:

$$-|\lambda_{k1}| \geqslant \lambda'_{k-1,1} \geqslant \lambda_{k2} \geqslant \lambda'_{k-1,2} \geqslant \cdots \geqslant \lambda_{k,k-1} \geqslant \lambda'_{k-1,k-1} \geqslant \lambda_{kk},$$
$$-|\lambda_{k-1,1}| \geqslant \lambda'_{k-1,1} \geqslant \lambda_{k-1,2} \geqslant \lambda'_{k-1,2} \geqslant \cdots \geqslant \lambda_{k-1,k-1} \geqslant \lambda'_{k-1,k-1}$$

for $k = 2,\dots,n$. Set $\lambda'_{k-1,0} = \max\{\lambda_{k1}, \lambda_{k-1,1}\}$. We will use the notation

$$l_{ki} = \lambda_{ki} - i + 3/2, \qquad l'_{ki} = \lambda'_{ki} - i + 1/2. \tag{9.103}$$

COROLLARY 9.6.4. *The vectors*

$$\xi_\Lambda = \prod_{k=2,\dots,n}^{\longrightarrow} \left(\prod_{i=1}^{k-1} z_{ki}^{\lambda'_{k-1,i-1}-\lambda_{k-1,i}} z_{i,-k}^{\lambda'_{k-1,i-1}-\lambda_{ki}} \cdot \prod_{j=l_{kk}}^{l'_{k-1,k-1}-1} Z_{k,-k}(j) \right) \xi$$

parameterized by all patterns Λ associated with λ form a basis of the representation $V(\lambda)$ of $\mathfrak{o}_{2n}$.

PROOF. Arguing by induction on n with the use of Theorem 9.5.2 and the decomposition (9.58), we reduce the proof to the case $n = 2$. Here $\lambda = (\lambda_1, \lambda_2)$ and the corresponding patterns Λ have the form

$$\begin{array}{ccc}
\lambda_{21} & & \lambda_{22}\\
& \lambda'_{11} & \\
\lambda_{11} & &
\end{array}$$

We have $z_{21} = F_{21}$, $z_{1,-2} = F_{1,-2}$ and $Z_{2,-2}(u) = F_{21}F_{1,-2}$. Therefore, the vectors are given by

$$\xi_\Lambda = F_{21}^{\lambda'_{10}-\lambda_{11}} F_{1,-2}^{\lambda'_{10}-\lambda_{21}} (F_{21}F_{1,-2})^{\lambda'_{11}-\lambda_{22}}\,\xi, \tag{9.104}$$

where $\lambda'_{10} = \max\{\lambda_{21}, \lambda_{11}\}$. The Lie algebra $\mathfrak{o}_4$ is isomorphic to the direct sum of two copies of $\mathfrak{sl}_2$ spent respectively by the elements

$$F_{21}, \quad F_{12}, \quad F_{11} - F_{22} \qquad \text{and} \qquad F_{1,-2}, \quad F_{-2,1}, \quad -F_{11} - F_{22}.$$

The representation $V(\lambda)$ is isomorphic to the tensor product of two highest weight $\mathfrak{sl}_2$-modules with the highest weights $\lambda_1 - \lambda_2$ and $-\lambda_1 - \lambda_2$, respectively. So the vectors of the form

$$F_{21}^k\, F_{1,-2}^l\, \xi, \qquad 0 \leqslant k \leqslant \lambda_1 - \lambda_2, \qquad 0 \leqslant l \leqslant -\lambda_1 - \lambda_2 \tag{9.105}$$

form a basis of $V(\lambda)$. However, this set of vectors coincides with the set $\{\xi_\Lambda\}$ parameterized by the patterns Λ. Indeed, a one-to-one correspondence between the parameters is given by

$$\lambda'_{11} = \lambda_2 + \min\{k, l\}, \qquad \lambda_{11} = \lambda_1 - k + l.$$

Hence $\{\xi_\Lambda\}$ is a basis of $V(\lambda)$. □

We will use the normalized basis vectors ζ_Λ of $V(\lambda)$ defined by

$$\zeta_\Lambda = N_\Lambda\, \xi_\Lambda,$$

where

$$N_\Lambda = \prod_{k=2}^{n-1} \prod_{1\leqslant i<j\leqslant k} (-l'_{ki} - l'_{kj})!$$

and the corresponding basis vectors ξ_ν of $V(\lambda)^+_\mu$ defined by

$$\zeta_\nu = \prod_{1\leqslant i<j\leqslant n-1} (-\gamma_i - \gamma_j)!\, \xi_\nu. \tag{9.106}$$

We will be using notation like $\Lambda \pm \delta_{ki}$ or $\Lambda \pm \delta'_{ki}$, which is interpreted in the same way as in the C case. Moreover, ζ_Λ is considered to be equal to zero unless Λ is a pattern.

LEMMA 9.6.5. *The action of the elements F_{11}, F_{21} and $F_{-2,1}$ of the Lie algebra $\mathfrak{o}_{2n}$ in the basis $\{\zeta_\Lambda\}$ of $V(\lambda)$ is given by the relations*

$$F_{11}\, \zeta_\Lambda = \lambda_{11}\, \zeta_\Lambda$$

$$F_{21}\, \zeta_\Lambda = \begin{cases} \zeta_{\Lambda-\delta_{11}} & \text{if } \lambda_{21} \geqslant \lambda_{11} \\ \zeta_{\Lambda+\delta'_{11}-\delta_{11}} & \text{if } \lambda_{21} < \lambda_{11} \end{cases}$$

and

$$F_{-2,1}\, \zeta_\Lambda = \begin{cases} (-\lambda_{11} - \lambda'_{11} + 1)\,(\lambda_{11} + \lambda'_{11} - \lambda_{21} - \lambda_{22})\, \zeta_{\Lambda-\delta_{11}} & \text{if } \lambda_{21} < \lambda_{11} \\ (-\lambda_{21} - \lambda'_{11} + 1)\,(\lambda'_{11} - \lambda_{22})\, \zeta_{\Lambda-\delta'_{11}-\delta_{11}} & \text{if } \lambda_{21} \geqslant \lambda_{11}. \end{cases}$$

PROOF. It is sufficient to verify the formulas in the case $n = 2$. In this case all relations easily follow with the use of the basis (9.105) and then writing the resulting formulas in terms of the entries of Λ. □

The action of the elements F_{22}, $F_{1,-2}$ and F_{12} in the basis $\{\zeta_\Lambda\}$ can also be easily calculated directly by considering the $n = 2$ case. The corresponding formulas will be included in the next lemmas, where they are obtained by ignoring the entries of the patterns Λ with out-of-range indices.

Given a pattern Λ we set for $1 \leqslant i < k \leqslant n$,

$$C_{ki} = \prod_{a=1,\, a\neq i}^{k-1} \frac{1}{(l_{k-1,i} - l_{k-1,a})(l_{k-1,i} + l_{k-1,a} - 1)},$$

$$D_{ki}(x) = \prod_{a=1,\, a\neq i}^{k-1} \frac{x^2 - l'^2_{k-1,a}}{l'_{k-1,a} - l'_{k-1,i}},$$

where x is a variable.

LEMMA 9.6.6. *The action of the elements $F_{k-1,-k}$ with $k = 2, \dots, n$ of the Lie algebra $\mathfrak{o}_{2n}$ in the basis $\{\zeta_\Lambda\}$ of $V(\lambda)$ is given by the relations*

$$F_{k-1,-k}\, \zeta_\Lambda = \sum_{i=1}^{k-1} C_{ki} \Big(\zeta^+_{\Lambda,k,i} - \zeta^-_{\Lambda,k,i} \Big).$$

Here

$$\zeta^+_{\Lambda,k,i} = \sum_{j=1}^{k-1} \sum_{m=1}^{k-2} D_{kj}(l_{k-1,i}) D_{k-1,m}(l_{k-1,i})\, \zeta_{\Lambda + \delta'_{k-1,j} + \delta_{k-1,i} + \delta'_{k-2,m}}$$

for $i = 2, \dots, k-1$ and for $i = 1$ if $\lambda_{k-1,1} < \lambda_{k1}, \lambda_{k-2,1}$. Otherwise,

$$\begin{aligned}
\zeta^+_{\Lambda,k,1} &= \zeta_{\Lambda + \delta_{k-1,1}} && \textit{if} \quad \lambda_{k-1,1} \geqslant \lambda_{k1}, \lambda_{k-2,1}, \\
&= \sum_{j=1}^{k-1} D_{kj}(l_{k-1,1})\, \zeta_{\Lambda + \delta'_{k-1,j} + \delta_{k-1,1}} && \textit{if} \quad \lambda_{k-2,1} \leqslant \lambda_{k-1,1} < \lambda_{k1}, \\
&= \sum_{m=1}^{k-2} D_{k-1,m}(l_{k-1,1})\, \zeta_{\Lambda + \delta_{k-1,1} + \delta'_{k-2,m}} && \textit{if} \quad \lambda_{k1} \leqslant \lambda_{k-1,1} < \lambda_{k-2,1}.
\end{aligned}$$

Furthermore,

$$\zeta^-_{\Lambda,k,i} = \zeta_{\Lambda - \delta_{k-1,i}}$$

for $i = 2, \dots, k-1$ and for $i = 1$ if $\lambda_{k-1,1} \leqslant \lambda_{k1}, \lambda_{k-2,1}$. Otherwise,

$$\begin{aligned}
\zeta^-_{\Lambda,k,1} &= \sum_{j=1}^{k-1} D_{kj}(l_{k-1,1} - 1)\, \zeta_{\Lambda + \delta'_{k-1,j} - \delta_{k-1,1}} && \textit{if} \quad \lambda_{k1} < \lambda_{k-1,1} \leqslant \lambda_{k-2,1}, \\
&= \sum_{m=1}^{k-2} D_{k-1,m}(l_{k-1,1} - 1)\, \zeta_{\Lambda - \delta_{k-1,1} + \delta'_{k-2,m}} && \textit{if} \quad \lambda_{k-2,1} < \lambda_{k-1,1} \leqslant \lambda_{k1}, \\
&= \sum_{j=1}^{k-1} \sum_{m=1}^{k-2} D_{kj}(l_{k-1,1} - 1) D_{k-1,m}(l_{k-1,1} - 1)\, \zeta_{\Lambda + \delta'_{k-1,j} - \delta_{k-1,1} + \delta'_{k-2,m}} \\
& && \textit{if} \quad \lambda_{k-1,1} > \lambda_{k1}, \lambda_{k-2,1}.
\end{aligned}$$

PROOF. As in the proof of Theorem 9.6.2, we may assume that $k = n$ and consider the top part of the pattern Λ of the form

$$\begin{array}{ccccccccc} \lambda_1 & & \lambda_2 & & \lambda_3 & & \cdots & & \lambda_n \\ & \nu_1 & & \nu_2 & & \cdots & & \nu_{n-1} & \\ \mu_1 & & \mu_2 & & \cdots & & \mu_{n-1} & & \\ & \nu'_1 & & \cdots & & \nu'_{n-2} & & & \\ \mu'_1 & & \cdots & & \mu'_{n-2} & & & & \end{array}$$

As in Section 9.5, we will be using the notation $\nu_0 = \max\{\lambda_1, \mu_1\}$. We also set $\nu'_0 = \max\{\mu_1, \mu'_1\}$ and $\gamma'_i = \nu'_i - i + 1/2$ for $i = 1, \dots, n-2$. We will calculate the expansion of $F_{n-1,-n}\, \xi_{\nu\mu\nu'}$, where

$$\xi_{\nu\mu\nu'} = X_{\mu\nu'}\, \xi_{\nu\mu}$$

with

$$X_{\mu\nu'} = \prod_{i=1}^{n-2} z_{n-1,i}^{\nu'_{i-1}-\mu'_i}\, z_{i,-n+1}^{\nu'_{i-1}-\mu_i} \cdot \prod_{r=m_{n-1}}^{\gamma'_{n-1}-1} Z_{n-1,-n+1}(r)$$

and

$$\xi_{\nu\mu} = \prod_{i=1}^{n-1} z_{ni}^{\nu_{i-1}-\mu_i}\, z_{i,-n}^{\nu_{i-1}-\lambda_i} \cdot \prod_{k=l_n}^{\gamma_n-1} Z_{n,-n}(k)\, \xi.$$

Note that $\xi_{\nu\mu} = \xi_\nu \in V(\lambda)^+_\mu$; see (9.87). Arguing as in the proof of Theorem 9.6.2, we find that

$$F_{n-1,-n}\, \xi_{\nu\mu\nu'} = X_{\mu\nu'}\, F_{n-1,-n}\, \xi_{\nu\mu}.$$

Now we apply Corollary 9.4.4 to get

$$F_{n-1,-n}\, \xi_{\nu\mu\nu'} = \sum_{i=1}^{n-1} C_i \Big(X_{\mu\nu'}\, z_{n-1,i}\, z_{i,-n}\, \xi_{\nu\mu} - X_{\mu\nu'}\, z_{i,-n+1}\, z_{ni}\, \xi_{\nu\mu} \Big), \tag{9.107}$$

where

$$C_i = \prod_{a=1,\, a\neq i}^{n-1} \frac{1}{(m_i - m_a)(m_i + m_a - 1)}. \tag{9.108}$$

First we consider the summands in (9.107) corresponding to $i = 1$. If $\lambda_1 \geqslant \mu_1$, then $z_{n1}\, \xi_{\nu\mu} = \xi_{\nu,\mu-\delta_1}$, while for $\lambda_1 < \mu_1$ we have

$$z_{n1}\, \xi_{\nu\mu} = \sum_{i=1}^{n-1} \prod_{a=1,\, a\neq i}^{n-1} \frac{(m_1 - 1)^2 - \gamma_a^2}{\gamma_i^2 - \gamma_a^2}\, \xi_{\nu+\delta_i,\mu-\delta_1}.$$

If $\mu'_1 \geqslant \mu_1$, then $X_{\mu\nu'}\, z_{1,-n+1} = X_{\mu-\delta_1,\nu'}$, while for $\mu'_1 < \mu_1$ we have

$$X_{\mu\nu'}\, z_{1,-n+1} = \sum_{r=1}^{n-2} \prod_{a=1,\, a\neq r}^{n-2} \frac{(m_1 - 1)^2 - \gamma'^2_a}{\gamma'^2_r - \gamma'^2_a}\, X_{\mu-\delta_1,\nu'+\delta_r}.$$

Similarly, if $\lambda_1 \leqslant \mu_1$, then $z_{1,-n}\, \xi_{\nu\mu} = \xi_{\nu,\mu+\delta_1}$; and if $\lambda_1 > \mu_1$, then

$$z_{1,-n}\, \xi_{\nu\mu} = \sum_{i=1}^{n-1} \prod_{a=1,\, a\neq i}^{n-1} \frac{m_1^2 - \gamma_a^2}{\gamma_i^2 - \gamma_a^2}\, \xi_{\nu+\delta_i,\mu+\delta_1}.$$

If $\mu_1' \leqslant \mu_1$, then $X_{\mu\nu'}\, z_{n-1,1} = X_{\mu+\delta_1,\nu'}$; and if $\mu_1' > \mu_1$, then

$$X_{\mu\nu'}\, z_{n-1,1} = \sum_{r=1}^{n-2} \prod_{a=1,\, a\neq r}^{n-2} \frac{m_1^2 - {\gamma'_a}^2}{{\gamma'_r}^2 - {\gamma'_a}^2}\, X_{\mu+\delta_1,\nu'+\delta_r}.$$

The calculation for the summands with $i = 2, \dots, n-1$ in (9.107) is completed in exactly the same way as in the proof of Theorem 9.6.2. Rewriting the result in terms of the normalized basis vectors ζ_Λ gives the desired formulas. □

Unlike the symplectic case, the elements of $\mathfrak{o}_{2n}$ of the form $F_{-k,k}$ are zero. Therefore, the action of the elements $F_{k-1,k}$ of $\mathfrak{o}_{2n}$ in the basis $\{\zeta_\Lambda\}$ has to be calculated by an alternative way; cf. Remark 9.6.3. Namely, we will find the action of certain elements of the universal enveloping algebra $\mathrm{U}(\mathfrak{o}_{2n})$ which will determine the action of all elements of $\mathfrak{o}_{2n}$. For each $k = 2, \dots, n$ we introduce the elements $\Phi_{-k,k}, \Phi_{k,-k} \in \mathrm{U}(\mathfrak{o}_{2n})$ by

$$\Phi_{-k,k} = \sum_{i=1}^{k-1} F_{-k,i} F_{ik}, \qquad \Phi_{k,-k} = \sum_{i=1}^{k-1} F_{ki} F_{i,-k}.$$

LEMMA 9.6.7. *The action of the elements F_{kk}, $\Phi_{k,-k}$ and $\Phi_{-k,k}$ with $k = 2, \dots, n$ in the basis $\{\zeta_\Lambda\}$ of $V(\lambda)$ is given by the relations*

$$F_{kk}\, \zeta_\Lambda = \Big(2\sum_{i=1}^{k} \lambda'_{k-1,i-1} - \sum_{i=1}^{k} \lambda_{ki} - \sum_{i=1}^{k-1} \lambda_{k-1,i} \Big)\, \zeta_\Lambda,$$

$$\Phi_{k,-k}\, \zeta_\Lambda = \sum_{i=1}^{k-1} A_{ki}\, \zeta_{\Lambda+\delta'_{k-1,i}},$$

$$\Phi_{-k,k}\, \zeta_\Lambda = \sum_{i=1}^{k-1} B_{ki}\, \big(F_{kk} - l'_{k-1,i} + 3/2\big)\, \zeta_{\Lambda-\delta'_{k-1,i}},$$

where

$$A_{ki} = \prod_{a=1,\, a\neq i}^{k-1} \frac{1}{l'_{k-1,a} - l'_{k-1,i}},$$

$$B_{ki} = A_{ki} \prod_{a=2}^{k} \big(l_{ka} - l'_{k-1,i}\big) \prod_{a=2}^{k-1} \big(l_{k-1,a} - l'_{k-1,i}\big)$$
$$\times \big(\max\{\lambda_{k1}, \lambda_{k-1,1}\} + l'_{k-1,i} - 1/2\big)\big(\min\{\lambda_{k1}, \lambda_{k-1,1}\} - l'_{k-1,i} + 1/2\big).$$

PROOF. Using the definition of the vectors $\zeta_\nu \in V(\lambda)^+_\mu$ we find that

$$F_{nn}\, \zeta_\nu = \Big(2\sum_{i=0}^{n-1} \nu_i - \sum_{i=1}^{n} \lambda_i - \sum_{i=1}^{n-1} \mu_i \Big)\, \zeta_\nu.$$

This implies the desired formulas for the action of the elements F_{kk}. Now observe that if $a \in \{-n, n\}$, then $2\,\Phi_{a,-a}$ coincides with the element $F^{(2)}_{a,-a}$ defined in (9.30). Hence, $\Phi_{a,-a}$ belongs to the centralizer $\mathrm{U}(\mathfrak{o}_{2n})^{\mathfrak{o}_{2n-2}}$; see the proof of Theorem 9.3.1. Therefore, in order to derive the formulas for the action of the elements $\Phi_{k,-k}$ and $\Phi_{-k,k}$, we need to determine only the action of $\Phi_{a,-a}$ on the basis vectors ζ_ν of $V(\lambda)^+_\mu$. By Corollary 9.3.7, the polynomial $Z_{a,-a}(u)$ has degree $2n-4$. Comparing the expression for $F^{(2)}_{a,-a}$ provided by (9.37) and the formula (9.22) for $Z_{a,-a}(u)$, we conclude that, as an operator on $V(\lambda)^+_\mu$, the coefficient of u^{2n-4} of the polynomial $Z_{a,-a}(u)$ coincides with $\Phi_{a,-a}$. As we observed in the proof of Theorem 9.5.2,

$$Z_{n,-n}(\gamma_i)\,\xi_\nu = \xi_{\nu+\delta_i}, \qquad i=1,\dots,n-1. \tag{9.109}$$

Regarding $Z_{n,-n}(u)$ as a polynomial in u^2 and applying the Lagrange interpolation formula, we derive that

$$\Phi_{n,-n}\,\zeta_\nu = \sum_{i=1}^{n-1} A_i\,\zeta_{\nu+\delta_i}, \qquad A_i = \prod_{a=1,\,a\neq i}^{n-1} \frac{1}{\gamma_a-\gamma_i}.$$

Finally, using (9.90), we come to the following identity of operators in $V(\lambda)^+_\mu$:

$$\Phi_{-n,n} = t^{(1)}_{-n,-n}\,t^{(1)}_{-n,n} - t^{(2)}_{-n,n} + \alpha_0\,t^{(1)}_{-n,n}.$$

Note that the image of the element $s^{(1)}_{nn} \in \mathrm{Y}(\mathfrak{o}_2)$ under the homomorphism (9.29) coincides with F_{nn}; see the proof of Theorem 9.3.1. On the other hand, (4.23) gives another identity of operators in $V(\lambda)^+_\mu$,

$$s^{(1)}_{nn} = t^{(1)}_{nn} - t^{(1)}_{-n,-n} - \alpha_0 - 1/2.$$

Hence, the operator $\Phi_{-n,n}$ can be written as

$$\Phi_{-n,n} = -t^{(2)}_{-n,n} + t^{(1)}_{-n,n}t^{(1)}_{nn} - \big(F_{nn}+3/2\big)\,t^{(1)}_{-n,n}. \tag{9.110}$$

The proof of Theorem 9.5.2 shows that the basis vectors ζ_ν of $V(\lambda)^+_\mu$ are respectively proportional to the basis vectors η_γ for the parameters ν and γ related by (9.88). Furthermore, (9.109) implies

$$Z_{n,-n}(\gamma_i)\,\zeta_\nu = \prod_{j=1,\,j\neq i}^{n-1} (-\gamma_i-\gamma_j)\,\zeta_{\nu+\delta_i}, \qquad i=1,\dots,n-1.$$

Comparing this with (9.91) and using the formulas of Theorem 3.3.8, we conclude that

$$T_{n,-n}(-\gamma_i)\,\zeta_\nu = \frac{1}{\gamma_i-\alpha_0}\,\zeta_{\nu+\delta_i},$$

where the right-hand side is considered to be zero if $\gamma_i - \alpha_0 = 0$; this can occur only if $i=1$ and $|\lambda_1+\mu_1| = 0$ so that $\zeta_{\nu+\delta_1}=0$. This implies

$$T_{-n,n}(-\gamma_i)\,\zeta_\nu = H_i\,\zeta_{\nu-\delta_i}, \qquad i=1,\dots,n-1,$$

where

$$H_i = \prod_{k=0}^{n-1}(\alpha_k-\gamma_i+1)\prod_{k=1}^{n-1}(\beta_k-\gamma_i).$$

Recalling the definition of the parameters α_k and β_k (see Section 9.4), we can also write this as

$$H_i = -\prod_{a=2}^{n}(l_a-\gamma_i)\prod_{a=2}^{n-1}(m_a-\gamma_i) \\ \times\big(\max\{\lambda_1,\mu_1\}+\gamma_i-1/2\big)\big(\min\{\lambda_1,\mu_1\}-\gamma_i+1/2\big).$$

Applying the Lagrange interpolation formula for the polynomial $T_{-n,n}(u)$ we find that

$$T_{-n,n}(u)\,\zeta_\nu = \sum_{i=1}^{n-1} H_i \prod_{a=1,\,a\neq i}^{n-1} \frac{u+\gamma_a}{\gamma_a-\gamma_i}\,\zeta_{\nu-\delta_i}.$$

Taking the coefficients of u^{n-2} and u^{n-3} we get the respective expansions for $t^{(1)}_{-n,n}\,\zeta_\nu$ and $t^{(2)}_{-n,n}\,\zeta_\nu$. Furthermore, by Theorem 3.3.8,

$$T_{nn}(u)\,\zeta_\nu = (u+\gamma_1)\dots(u+\gamma_{n-1})\,\zeta_\nu,$$

and taking the coefficient of u^{n-2} we get

$$t^{(1)}_{nn}\,\zeta_\nu = (\gamma_1+\dots+\gamma_{n-1})\,\zeta_\nu.$$

Thus, using (9.110) we come to

$$\Phi_{-n,n}\,\zeta_\nu = -\sum_{i=1}^{n-1} A_i\,H_i\,\big(F_{nn}-\gamma_i+3/2\big)\,\zeta_{\nu-\delta_i},$$

completing the proof. □

Using the notation of Lemma 9.6.6, introduce a linear operator $\Phi_{k-1,-k}(u)$ on $V(\lambda)$ depending on a complex parameter u by

$$\Phi_{k-1,-k}(u)\,\zeta_\Lambda =$$
$$\sum_{i=1}^{k-1} C_{ki}\left(\frac{1}{u+l_{k-1,i}+F_{kk}-3/2}\,\zeta^{+}_{\Lambda,k,i} - \frac{1}{u-l_{k-1,i}+F_{kk}-1/2}\,\zeta^{-}_{\Lambda,k,i}\right).$$

This is a well-defined element of $V(\lambda)$ for all values of u such that none of the denominators in the formula vanishes when F_{kk} is replaced by its eigenvalue on the vector $\zeta^{+}_{\Lambda,k,i}$ or $\zeta^{-}_{\Lambda,k,i}$, respectively.

THEOREM 9.6.8. *Let the positive integer $n \geqslant 2$ be fixed and let the highest weight λ for $\mathfrak{o}_{2n}$ vary. Then for any $k = 2,\dots,n$ the matrix elements of the operator $F_{k-1,k}$ in the basis $\{\zeta_\Lambda\}$ of $V(\lambda)$ are rational functions in the entries of the pattern Λ associated with λ. These functions can be found from the relation*

$$F_{k-1,k}\,\zeta_\Lambda = \Big(\Phi_{k-1,-k}(2)\,\Phi_{-k,k} - \Phi_{-k,k}\,\Phi_{k-1,-k}(0)\Big)\,\zeta_\Lambda, \tag{9.111}$$

which holds for all patterns Λ such that $\Phi_{k-1,-k}(0)\,\zeta_\Lambda$ is a well-defined element of $V(\lambda)$. Moreover, together with Lemmas 9.6.5, 9.6.6 and 9.6.7 this determines the action of any element of the Lie algebra $\mathfrak{o}_{2n}$ in the basis $\{\zeta_\Lambda\}$ of $V(\lambda)$.

PROOF. Since the elements F_{21}, $F_{-2,1}$ and $F_{k-1,-k}$, $F_{k-1,k}$ with $k=2,\dots,n$ generate $\mathfrak{o}_{2n}$ as a Lie algebra, the last statement of the theorem follows. In order to prove the first part, it is sufficient to consider the case $k=n$ and use the top part of the pattern Λ as defined in the proof of Lemma 9.6.6. Using the notation of that proof, we can write

$$F_{n-1,n}\,\xi_{\nu\mu\nu'} = X_{\mu\nu'}\,F_{n-1,n}\,\xi_{\nu\mu}.$$

Applying Corollary 9.4.4, we get

$$F_{n-1,n}\,\xi_{\nu\mu\nu'} = \sum_{i=1}^{n-1} C_i\,\Big(X_{\mu\nu'}\,z_{n-1,i}\,z_{in}\,\xi_{\nu\mu} + X_{\mu\nu'}\,z_{n-1,-i}\,z_{-i,n}\,\xi_{\nu\mu}\Big), \tag{9.112}$$

where C_i is defined in (9.108). In order to prove the first claim it is sufficient to verify that $z_{\pm i,n}\,\xi_{\nu\mu}$ is a linear combination of the basis vectors of $V(\lambda)^+_{\mu\pm\delta_i}$ whose coefficients are rational functions in the parameters λ_k, ν_k and μ_k. The basis vector $\xi_{\nu\mu}$ coincides with the vector (9.92); see the proof of Theorem 9.5.2. Therefore, if $\gamma_j - \beta_j \geqslant 1$ for some $j \in \{1,\dots,n-1\}$, then using Corollary 9.3.7, we can write $\xi_{\nu\mu}$ as a linear combination of the vectors $z_{nj}\,z_{j,-n}\,\xi_{\nu-\delta_j,\mu}$ whose coefficients are rational functions in the parameters λ_k, ν_k and μ_k. Furthermore, the coefficients in the expansion of $z_{j,-n}\,\xi_{\nu-\delta_j,\mu}$ in terms of the basis vectors of $V(\lambda)^+_{\mu+\delta_j}$ are also rational functions; see the proof of Lemma 9.6.6. So it now suffices to verify that the same property is shared by the coefficients of the expansions of $z_{\pm i,n}\,z_{nj}\,\xi_{\nu,\mu+\delta_j}$. Suppose first that $i=j$. Using (9.22) we obtain

$$z_{in}\,z_{ni}\,\xi_{\nu,\mu+\delta_i} = z_{-n,-i}\,z_{-i,-n}\,\xi_{\nu,\mu+\delta_i} = Z_{-n,-n}(-g_{-i})\,\xi_{\nu,\mu+\delta_i}.$$

Note that

$$-g_{-i}\,\xi_{\nu,\mu+\delta_i} = m_i\,\xi_{\nu,\mu+\delta_i}.$$

In order to calculate $Z_{-n,-n}(m_i)\,\xi_{\nu,\mu+\delta_i}$ we use the relation (9.89). By Theorem 3.3.8 (see also Corollary 3.3.9), the matrix elements of the operators $T_{ab}(u)$ in the basis $\{\xi_{\nu,\mu+\delta_i}\}$ of $V(\lambda)^+_{\mu+\delta_i}$ are rational functions in the λ_k, ν_k and μ_k, thus completing the argument in the case under consideration. In the remaining cases the statement follows by an easy induction with the application of Proposition 9.2.5.

We will now be calculating $F_{n-1,n}\,\xi_{\nu\mu\nu'}$ with the use of (9.112). Let us verify that for any $j \in \{-n+1,\dots,n-1\}$ we have

$$z_{jn}\,\xi_{\nu\mu} = [z_{j,-n},\Phi_{-n,n}]\,\frac{1}{f_j+F_{nn}}\,\xi_{\nu\mu} \tag{9.113}$$

provided that $(f_j+F_{nn})\,\xi_{\nu\mu} \neq 0$. Indeed, calculating in $\mathrm{U}'(\mathfrak{o}_{2n})$ modulo J' we find

$$[F_{j,-n},\Phi_{-n,n}] = \sum_{a<j} F_{ja}F_{an} + F_{jn}(f_j+F_{nn}).$$

As $\Phi_{-n,n}$ commutes with the elements of $\mathfrak{o}_{2n-2}$, using the explicit expressions for $z_{j,-n}$ and z_{jn} provided by (9.13) we derive that

$$[z_{j,-n},\Phi_{-n,n}] = z_{jn}\,(f_j+F_{nn}),$$

proving (9.113). Since $\Phi_{-n,n}F_{nn} = (F_{nn}+2)\,\Phi_{-n,n}$ we come to

$$F_{n-1,n}\,\xi_{\nu\mu\nu'} = X_{\mu\nu'}\left(\widetilde{\Phi}_{n-1,-n}(2)\,\Phi_{-n,n} - \Phi_{-n,n}\widetilde{\Phi}_{n-1,-n}(0)\right)\xi_{\nu\mu},$$

where

$$\widetilde{\Phi}_{n-1,-n}(u) = \sum_{j=-n+1}^{n-1} z_{n-1,j}\,z_{j,-n} \prod_{a=-n+1,\,a\neq\pm j}^{n-1} \frac{1}{f_j-f_a}\cdot\frac{1}{u+f_j+F_{nn}},$$

assuming that $(f_j+F_{nn})\,\xi_{\nu\mu}\neq 0$ for all $j=-n+1,\dots,n-1$. Now the proof of (9.111) is completed by calculating

$$X_{\mu\nu'}\,\widetilde{\Phi}_{n-1,-n}(u)\,\xi_{\nu\mu}$$

as in the proof of Lemma 9.6.6.

Finally, for any fixed $k\in\{2,\dots,n\}$ and any pattern Λ we have

$$F_{k-1,k}\,\zeta_\Lambda = \sum_{\Lambda'} c_{\Lambda\Lambda'}\,\zeta_{\Lambda'},$$

for some coefficients $c_{\Lambda\Lambda'}$. The coefficient $c_{\Lambda\Lambda'}$ is nonzero only if

$$\Lambda' = \Lambda + \sum_{r,i} m_{ri}\, \delta_{ri} + \sum_{r,i} m'_{ri}\, \delta'_{ri}$$

for some integers m_{ri} and m'_{ri}, where the sums are taken over the pairs (i, r) such that $1 \leqslant i \leqslant r \leqslant n-1$. Suppose now that certain integers m_{ri} and m'_{ri} are fixed for each such pair (i, r) while the highest weight λ and corresponding pattern Λ vary. The coefficients $c_{\Lambda\Lambda'}$ are rational functions in the entries of Λ which are determined by (9.111) except for those patterns Λ where at least one of the denominators in the formula defining $\Phi_{k-1,-k}(0)\, \zeta_\Lambda$ is zero. However, the rational function $c_{\Lambda\Lambda'}$ is determined by its values on the infinite set of patterns Λ for which none of the denominators vanishes. □

EXAMPLE 9.6.9. Suppose that $n = 2$. Here $\Phi_{-2,2} = F_{-2,1} F_{12}$ and the calculation of $F_{12}\, \zeta_\Lambda$ used in the proof of Theorem 9.6.8 relies on the identity

$$F_{12}\,(F_{11} + F_{22}) = \big[F_{1,-2}, \Phi_{-2,2}\big].$$

Let us assume that the entries of the pattern Λ satisfy $\lambda_{11} \geqslant \lambda_{21}$. Then Lemmas 9.6.5 and 9.6.7 give

$$(F_{11} + F_{22})\, \zeta_\Lambda = \Big(2\,(\lambda_{11} + \lambda'_{11}) - \lambda_{21} - \lambda_{22}\Big)\, \zeta_\Lambda.$$

Hence, assuming also that $2\,(\lambda_{11} + \lambda'_{11}) - \lambda_{21} - \lambda_{22} \neq 0$ we have

$$F_{12}\, \zeta_\Lambda = \frac{1}{2\,(\lambda_{11} + \lambda'_{11}) - \lambda_{21} - \lambda_{22}} \Big(F_{1,-2}\, \Phi_{-2,2} - \Phi_{-2,2}\, F_{1,-2}\Big)\, \zeta_\Lambda.$$

Now using the formulas for the action of $F_{1,-2}$ and $\Phi_{-2,2}$ provided by Lemmas 9.6.6 and 9.6.7, respectively, we get the formula

$$F_{12}\, \zeta_\Lambda = \big(\lambda'_{11} - \lambda_{21} - 1\big)\big(\lambda_{22} - \lambda'_{11}\big)\, \zeta_{\Lambda - \delta'_{11} + \delta_{11}},$$

which holds for all patterns Λ with $\lambda_{11} \geqslant \lambda_{21}$. A similar calculation leads to a formula for the patterns with $\lambda_{11} < \lambda_{21}$. The same formulas for $F_{12}\, \zeta_\Lambda$ follow by a more direct calculation with the use of (9.104). □

***B* type case.** Define the *B type pattern* Λ associated with λ as an array of the form

$$\begin{array}{ccccccccc}
\sigma_n & & \lambda_{n1} & & \lambda_{n2} & & \cdots & & \lambda_{nn} \\
& \lambda'_{n1} & & \lambda'_{n2} & & \cdots & & \lambda'_{nn} & \\
\sigma_{n-1} & & \lambda_{n-1,1} & & \cdots & & \lambda_{n-1,n-1} & & \\
& \lambda'_{n-1,1} & & \cdots & & \lambda'_{n-1,n-1} & & & \\
\cdots & & \cdots & & \cdots & & & & \\
\sigma_1 & & \lambda_{11} & & & & & & \\
& \lambda'_{11} & & & & & & &
\end{array}$$

such that $\lambda_{ni} = \lambda_i$ for $i = 1, \ldots, n$, each σ_k is 0 or 1, the remaining entries are all nonpositive integers or nonpositive half-integers together with the λ_i, and the following inequalities hold:

$$\lambda'_{k1} \geqslant \lambda_{k1} \geqslant \lambda'_{k2} \geqslant \lambda_{k2} \geqslant \cdots \geqslant \lambda'_{k,k-1} \geqslant \lambda_{k,k-1} \geqslant \lambda'_{kk} \geqslant \lambda_{kk}$$

for $k = 1, \ldots, n$, and

$$\lambda'_{k1} \geqslant \lambda_{k-1,1} \geqslant \lambda'_{k2} \geqslant \lambda_{k-1,2} \geqslant \cdots \geqslant \lambda'_{k,k-1} \geqslant \lambda_{k-1,k-1} \geqslant \lambda'_{kk}$$

for $k = 2, \ldots, n$, together with the condition

$$\lambda'_{k1} = 0 \qquad \text{implies} \qquad \sigma_k = 0$$

for all $k = 1, \ldots, n$. We will use the notation

$$l_{ki} = \lambda_{ki} - i + 1, \qquad l'_{ki} = \lambda'_{ki} - i + 1, \tag{9.114}$$

and set $l_{k0} = 0$ for all k.

COROLLARY 9.6.10. *The vectors*

$$\xi_\Lambda = \prod_{k=1,\ldots,n}^{\longrightarrow} \left(z_{k0}^{\sigma_k} \cdot \prod_{i=1}^{k-1} z_{ki}^{\lambda'_{ki}-\lambda_{k-1,i}} z_{i,-k}^{\lambda'_{ki}-\lambda_{ki}} \cdot \prod_{j=l_{kk}}^{l'_{kk}-1} Z_{k,-k}(j) \right) \xi$$

parameterized by all patterns Λ associated with λ form a basis of the representation $V(\lambda)$ of $\mathfrak{o}_{2n+1}$.

PROOF. Arguing by induction on n with the use of Theorem 9.5.3 and the decomposition (9.58), we reduce the proof to the case $n = 1$. Here $\lambda = (\lambda_1)$ and the corresponding patterns Λ have the form

$$\begin{array}{ccc} \sigma_1 & & \lambda_{11} \\ & \lambda'_{11} & \end{array}$$

We have $z_{10} = F_{10}$, $z_{0,-1} = F_{0,-1}$ and $Z_{1,-1}(u) = F_{10} F_{0,-1}$. Therefore, the vectors are given by

$$\xi_\Lambda = F_{10}^{\sigma_1} (F_{10} F_{0,-1})^{\lambda'_{11}-\lambda_{11}} \xi = (-1)^{\lambda'_{11}-\lambda_{11}} F_{10}^{\sigma_1 + 2(\lambda'_{11}-\lambda_{11})} \xi. \tag{9.115}$$

It is clear that $\{\xi_\Lambda\}$ is a basis of $V(\lambda)$. □

As in the C and D cases, we will use the normalized basis vectors ζ_Λ of $V(\lambda)$ defined by

$$\zeta_\Lambda = N_\Lambda \, \xi_\Lambda,$$

where

$$N_\Lambda = \prod_{k=2}^{n} \prod_{1 \leqslant i < j \leqslant k} (-l'_{ki} - l'_{kj})!$$

and the corresponding basis vectors ξ_ν of $V(\lambda)^+_\mu$ defined by

$$\zeta_\nu = \prod_{1 \leqslant i < j \leqslant n} (-\gamma_i - \gamma_j)! \, \xi_\nu. \tag{9.116}$$

We will keep using notation like $\Lambda \pm \delta_{ki}$ or $\Lambda \pm \delta'_{ki}$, which is interpreted in the same way as in the C and D cases. We consider ζ_Λ to be equal to zero unless Λ is a pattern.

Given a pattern Λ we set for $0 \leqslant i < k \leqslant n$,

$$C_{ki} = \prod_{a=1,\, a \neq i}^{k-1} \frac{1}{l_{k-1,i} - l_{k-1,a}} \cdot \prod_{a=1}^{k-1} \frac{1}{l_{k-1,i} + l_{k-1,a} - 1},$$

$$D_{ki}(x) = \prod_{a=1,\, a \neq i}^{k} \frac{x^2 - l'^2_{ka}}{l'_{ka} - l'_{ki}},$$

where x is a variable. In the following lemma the formulas for the case $k = 1$ are obtained by ignoring the entries of the patterns Λ with out-of-range indices. They also follow easily by using (9.115).

LEMMA 9.6.11. *The action of the elements $F_{k-1,-k}$ with $k = 1, \dots, n$ of the Lie algebra $\mathfrak{o}_{2n+1}$ in the basis $\{\zeta_\Lambda\}$ of $V(\lambda)$ is given by the relations*

$$F_{k-1,-k}\,\zeta_\Lambda = C_{k0}\,\zeta_{\Lambda,k,0} + \sum_{i=1}^{k-1} C_{ki}\left(\frac{1}{l_{k-1,i}}\,\zeta^+_{\Lambda,k,i} - \frac{1}{l_{k-1,i}-1}\,\zeta^-_{\Lambda,k,i}\right),$$

where

$$\zeta^-_{\Lambda,k,i} = \zeta_{\Lambda-\delta_{k-1,i}},$$

$$\zeta^+_{\Lambda,k,i} = \sum_{j=1}^{k}\sum_{m=1}^{k-1} D_{kj}(l_{k-1,i})\,D_{k-1,m}(l_{k-1,i})\,\zeta_{\Lambda+\delta'_{kj}+\delta_{k-1,i}+\delta'_{k-1,m}};$$

and

$$\begin{aligned}
\zeta_{\Lambda,k,0} &= (-1)^k\,\zeta_{\bar\Lambda} && \text{if}\quad \sigma_k = \sigma_{k-1} = 0,\\
&= \sum_{j=1}^{k} D_{kj}(0)\,\zeta_{\bar\Lambda+\delta'_{kj}} && \text{if}\quad \sigma_k = 1,\ \sigma_{k-1} = 0,\\
&= -\sum_{m=1}^{k-1} D_{k-1,m}(0)\,\zeta_{\bar\Lambda+\delta'_{k-1,m}} && \text{if}\quad \sigma_k = 0,\ \sigma_{k-1} = 1,\\
&= (-1)^{k-1}\sum_{j=1}^{k}\sum_{m=1}^{k-1} D_{kj}(0)\,D_{k-1,m}(0)\,\zeta_{\bar\Lambda+\delta'_{kj}+\delta'_{k-1,m}} && \\
& && \text{if}\quad \sigma_k = \sigma_{k-1} = 1;
\end{aligned}$$

the array $\bar\Lambda = \bar\Lambda^{(k)}$ is obtained from Λ by replacing σ_k and σ_{k-1} respectively with $\sigma_k + 1$ and $\sigma_{k-1} + 1$ modulo 2. In the case where $l_{k-1,i} = 0$ the vector $\zeta^+_{\Lambda,k,i}$ is also zero and the corresponding summand in the expansion of $F_{k-1,-k}\,\zeta_\Lambda$ is understood as being equal to zero.

PROOF. Arguing as in the proof of Theorem 9.6.2 and Lemma 9.6.6 we may assume that $k = n$ and consider the top part of the pattern Λ of the form

$$\begin{array}{ccccccccccc}
\sigma & & \lambda_1 & & \lambda_2 & & \lambda_3 & & \cdots & & \lambda_n\\
 & \nu_1 & & \nu_2 & & \nu_3 & & \cdots & & \nu_n & \\
\sigma' & & \mu_1 & & \mu_2 & & \cdots & & \mu_{n-1} & & \\
 & \nu'_1 & & \nu'_2 & & \cdots & & \nu'_{n-1} & & & \\
\sigma'' & & \mu'_1 & & \cdots & & \mu'_{n-2} & & & &
\end{array}$$

We set $\nu' = (\sigma', \nu'_1, \dots, \nu'_{n-1})$ and $\gamma'_i = \nu'_i - i + 1$. We will calculate the expansion of $F_{n-1,-n}\,\xi_{\nu\mu\nu'}$, where

$$\xi_{\nu\mu\nu'} = X_{\mu\nu'}\,\xi_{\nu\mu}$$

with

$$X_{\mu\nu'} = z^{\sigma'}_{n-1,0}\prod_{i=1}^{n-2} z^{\nu'_i-\mu'_i}_{n-1,i}\,z^{\nu'_i-\mu_i}_{i,-n+1}\cdot\prod_{r=m_{n-1}}^{\gamma'_{n-1}-1} Z_{n-1,-n+1}(r)$$

and

$$\xi_{\nu\mu} = z_{n0}^{\sigma} \prod_{i=1}^{n-1} z_{ni}^{\nu_i-\mu_i} z_{i,-n}^{\nu_i-\lambda_i} \cdot \prod_{k=l_n}^{\gamma_n-1} Z_{n,-n}(k)\,\xi.$$

Note that $\xi_{\nu\mu}$ coincides with the basis vector $\xi_\nu \in V(\lambda)^+_\mu$ defined in (9.93). As $F_{n-1,-n}$ is permutable with the elements $z_{n-1,i}$ and $Z_{n-1,-n+1}(u)$, we have

$$F_{n-1,-n}\,\xi_{\nu\mu\nu'} = X_{\mu\nu'}\,F_{n-1,-n}\,\xi_{\nu\mu}.$$

Now we use Corollary 9.4.4. Recalling that $f_0 = -1/2$ and assuming that $\mu_1 \neq 0$, we obtain

$$F_{n-1,-n}\,\xi_{\nu\mu\nu'} = C_0\,X_{\mu\nu'}\,z_{n-1,0}\,z_{0,-n}\,\xi_{\nu\mu} + \sum_{i=1}^{n-1} C_i\Big(\frac{1}{m_i}\,X_{\mu\nu'}\,z_{n-1,i}\,z_{i,-n}\,\xi_{\nu\mu} - \frac{1}{m_i-1}\,X_{\mu\nu'}\,z_{i,-n+1}\,z_{ni}\,\xi_{\nu\mu}\Big),$$

where

$$C_i = \prod_{a=1,\,a\neq i}^{n-1} \frac{1}{m_i-m_a} \cdot \prod_{a=1}^{n-1} \frac{1}{m_i+m_a-1}, \qquad i=0,1,\dots,n-1,$$

and $m_0 = 0$. If $\mu_1 = 0$, then $z_{0,-n}\,\xi_{\nu\mu} = z_{1,-n}\,\xi_{\nu\mu} = 0$ and the corresponding terms in the expression for $F_{n-1,-n}\,\xi_{\nu\mu\nu'}$ are omitted; see Corollary 9.4.4. We proceed in the same way as in the proof of Theorem 9.6.2 to calculate $X_{\mu\nu'}\,z_{n-1,i}\,z_{i,-n}\,\xi_{\nu\mu}$ and $X_{\mu\nu'}\,z_{i,-n+1}\,z_{ni}\,\xi_{\nu\mu}$ for $i \neq 0$. In order to expand $\xi_0 := X_{\mu\nu'}\,z_{n-1,0}\,z_{0,-n}\,\xi_{\nu\mu}$ we need to consider different cases depending on the values of σ and σ'. We will use the notation $\overline{\nu} = (\sigma+1, \nu_1, \dots, \nu_n)$ with addition modulo 2. If $\sigma = 0$, then

$$z_{0,-n}\,\xi_{\nu\mu} = (-1)^n\,z_{n0}\,\xi_{\nu\mu} = (-1)^n\,\xi_{\overline{\nu}\mu}.$$

If $\sigma = 1$, then

$$z_{0,-n}\,\xi_{\nu\mu} = z_{n0}z_{0,-n}\,\xi_{\overline{\nu}\mu},$$

which coincides with $Z_{n,-n}(0)\,\xi_{\overline{\nu}\mu}$, as $g_0 = 0$. Recall that $\xi_{\nu\mu}$ coincides with the vector (9.97). As $Z_{n,-n}(u)$ is a polynomial in u^2 of degree $n-1$, using the Lagrange interpolation formula with the points $\gamma_1, \dots, \gamma_n$ we calculate $Z_{n,-n}(u)\,\xi_{\overline{\nu}\mu}$ and put $u = 0$ to get

$$Z_{n,-n}(0)\,\xi_{\overline{\nu}\mu} = \sum_{j=1}^{n} \prod_{a=1,\,a\neq j}^{n} \frac{-\gamma_a^2}{\gamma_j^2-\gamma_a^2}\,\xi_{\overline{\nu}+\delta_j,\mu}.$$

The product $X_{\mu\nu'}\,z_{n-1,0}$ is calculated in a similar way, and we obtain

$$\begin{aligned}
\xi_0 &= (-1)^n \xi_{\overline{\nu}\mu\overline{\nu}'} && \text{if}\quad \sigma=\sigma'=0,\\
&= \sum_{j=1}^{n} \prod_{a=1,\,a\neq j}^{n} \frac{-\gamma_a^2}{\gamma_j^2-\gamma_a^2}\,\xi_{\overline{\nu}+\delta_j,\mu,\overline{\nu}'} && \text{if}\quad \sigma=1,\ \sigma'=0,\\
&= -\sum_{m=1}^{n-1} \prod_{a=1,\,a\neq m}^{n-1} \frac{-\gamma_a'^2}{\gamma_m'^2-\gamma_a'^2}\,\xi_{\overline{\nu},\mu,\overline{\nu}'+\delta_m} && \text{if}\quad \sigma=0,\ \sigma'=1,\\
&= (-1)^{n-1}\sum_{j=1}^{n}\sum_{m=1}^{n-1} \prod_{a=1,\,a\neq j}^{n} \frac{-\gamma_a^2}{\gamma_j^2-\gamma_a^2} \prod_{a=1,\,a\neq m}^{n-1} \frac{-\gamma_a'^2}{\gamma_m'^2-\gamma_a'^2}\,\xi_{\overline{\nu}+\delta_j,\mu,\overline{\nu}'+\delta_m}\\
& && \text{if}\quad \sigma=\sigma'=1,
\end{aligned}$$

where $\overline{\nu}' = (\sigma' + 1, \nu'_1, \dots, \nu'_{n-1})$ with addition modulo 2. The proof is completed by writing the formulas in terms of the normalized basis vectors ζ_Λ. □

As in the D case, we will find the action of certain elements of the universal enveloping algebra $\mathrm{U}'(\mathfrak{o}_{2n+1})$ in the basis $\{\zeta_\Lambda\}$ of $V(\lambda)$ which will determine the action of all elements of $\mathfrak{o}_{2n+1}$. For each $k = 1, \dots, n$ we introduce the elements $\Phi_{-k,k}, \Phi_{k,-k} \in \mathrm{U}(\mathfrak{o}_{2n+1})$ by

$$\Phi_{-k,k} = \sum_{i=1}^{k-1} F_{-k,i} F_{ik} - \frac{1}{2} F_{0k}^2, \qquad \Phi_{k,-k} = \sum_{i=1}^{k-1} F_{ki} F_{i,-k} - \frac{1}{2} F_{k0}^2.$$

LEMMA 9.6.12. *The action of the elements F_{kk}, $\Phi_{k,-k}$ and $\Phi_{-k,k}$ with $k = 1, \dots, n$ in the basis $\{\zeta_\Lambda\}$ of $V(\lambda)$ is given by the relations*

$$F_{kk}\, \zeta_\Lambda = \Big(\sigma_k + 2 \sum_{i=1}^{k} \lambda'_{ki} - \sum_{i=1}^{k} \lambda_{ki} - \sum_{i=1}^{k-1} \lambda_{k-1,i} \Big) \zeta_\Lambda,$$

$$\Phi_{k,-k}\, \zeta_\Lambda = \sum_{i=1}^{k} A_{ki}\, \zeta_{\Lambda + \delta'_{ki}},$$

$$\Phi_{-k,k}\, \zeta_\Lambda = \sum_{i=1}^{k} B_{ki} \left(F_{kk} - l'_{ki} + 3/2 \right) \zeta_{\Lambda - \delta'_{ki}},$$

where

$$A_{ki} = \prod_{a=1,\, a \neq i}^{k} \frac{1}{l'_{ka} - l'_{ki}},$$

$$B_{ki} = A_{ki} \left(2\, l'_{ki} - 1 \right) \left(1 - \sigma_k - l'_{ki} \right) \prod_{a=1}^{k} \left(l_{ka} - l'_{ki} \right) \prod_{a=1}^{k-1} \left(l_{k-1,a} - l'_{ki} \right).$$

PROOF. By the definition of the vectors $\zeta_\nu \in V(\lambda)^+_\mu$ we have

$$F_{nn}\, \zeta_\nu = \Big(\sigma + 2 \sum_{i=1}^{n} \nu_i - \sum_{i=1}^{n} \lambda_i - \sum_{i=1}^{n-1} \mu_i \Big) \zeta_\nu,$$

proving the formulas for the action of the elements F_{kk}. For any $a \in \{-n, n\}$ we have $2\,\Phi_{a,-a} = F^{(2)}_{a,-a}$; see (9.30). Therefore $\Phi_{a,-a}$ belongs to the centralizer $\mathrm{U}(\mathfrak{o}_{2n+1})^{\mathfrak{o}_{2n-1}}$; see the proof of Theorem 9.3.1. So we need to determine only the action of $\Phi_{a,-a}$ on the basis vectors ζ_ν of $V(\lambda)^+_\mu$. The rest of the proof is not essentially different from the proof of Lemma 9.6.7. In particular, we have

$$\Phi_{n,-n}\, \zeta_\nu = \sum_{i=1}^{n} A_i\, \zeta_{\nu + \delta_i}, \qquad A_i = \prod_{a=1,\, a \neq i}^{n} \frac{1}{\gamma_a - \gamma_i}.$$

Furthermore, we get the following identity of operators in $V(\lambda)^+_\mu$:

$$\Phi_{-n,n} = -t^{(2)}_{-n,n} + t^{(1)}_{-n,n} t^{(1)}_{nn} - \left(F_{nn} + 3/2 \right) t^{(1)}_{-n,n}; \tag{9.117}$$

cf. (9.110). This implies

$$\Phi_{-n,n}\, \zeta_\nu = \sum_{i=1}^{n} A_i\, H_i \left(F_{nn} - \gamma_i + 3/2 \right) \zeta_{\nu - \delta_i},$$

where

$$H_i = (2\gamma_i - 1)(1 - \sigma - \gamma_i) \prod_{a=1}^{n} (l_a - \gamma_i) \prod_{a=1}^{n-1} (m_a - \gamma_i),$$

completing the proof. □

In the notation of Lemma 9.6.11, introduce a linear operator $\Phi_{k-1,-k}(u)$ on $V(\lambda)$ depending on a complex parameter u by

$$\begin{aligned}\Phi_{k-1,-k}(u)\,\zeta_\Lambda = {}& \frac{C_{k0}}{u + F_{kk} - 3/2}\,\zeta_{\Lambda,k,0} \\ &+ \sum_{i=1}^{k-1} C_{ki} \bigg(\frac{1}{l_{k-1,i}\,(u + l_{k-1,i} + F_{kk} - 3/2)}\,\zeta^{+}_{\Lambda,k,i} \\ &\qquad - \frac{1}{(l_{k-1,i} - 1)(u - l_{k-1,i} + F_{kk} - 1/2)}\,\zeta^{-}_{\Lambda,k,i} \bigg).\end{aligned}$$

This is a well-defined element of $V(\lambda)$ for all values of u such that none of the denominators in the formula vanishes when F_{kk} is replaced by its eigenvalue on the vector $\zeta_{\Lambda,k,0}$, $\zeta^{+}_{\Lambda,k,i}$ or $\zeta^{-}_{\Lambda,k,i}$, respectively.

THEOREM 9.6.13. *Let the positive integer $n \geqslant 1$ be fixed and let the highest weight λ for $\mathfrak{o}_{2n+1}$ vary. Then for any $k = 1, \dots, n$ the matrix elements of the operator $F_{k-1,k}$ in the basis $\{\zeta_\Lambda\}$ of $V(\lambda)$ are rational functions in the entries of the pattern Λ associated with λ. These functions can be found from the relation*

$$F_{k-1,k}\,\zeta_\Lambda = \Big(\Phi_{k-1,-k}(2)\,\Phi_{-k,k} - \Phi_{-k,k}\,\Phi_{k-1,-k}(0)\Big)\,\zeta_\Lambda, \tag{9.118}$$

which holds for all patterns Λ such that $\Phi_{k-1,-k}(0)\,\zeta_\Lambda$ is a well-defined element of $V(\lambda)$. Moreover, together with Lemmas 9.6.11 and 9.6.12 this determines the action of any element of the Lie algebra $\mathfrak{o}_{2n+1}$ in the basis $\{\zeta_\Lambda\}$ of $V(\lambda)$.

PROOF. The elements $F_{k-1,-k}$ and $F_{k-1,k}$ with $k = 1, \dots, n$ generate $\mathfrak{o}_{2n+1}$ as a Lie algebra, which implies the last statement of the theorem. For the proof of the first part it is sufficient to consider the case $k = n$ and use the top part of the pattern Λ as in the proof of Lemma 9.6.11. Using the notation of that proof, we can write

$$F_{n-1,n}\,\xi_{\nu\mu\nu'} = X_{\mu\nu'}\,F_{n-1,n}\,\xi_{\nu\mu}$$

so that Corollary 9.4.4 gives

$$\begin{aligned}F_{n-1,n}\,\xi_{\nu\mu\nu'} = {}& C_0\,X_{\mu\nu'}\,z_{n-1,0}\,z_{0,n}\,\xi_{\nu\mu} \\ &+ \sum_{i=1}^{n-1} C_i \Big(\frac{1}{m_i}\,X_{\mu\nu'}\,z_{n-1,i}\,z_{in}\,\xi_{\nu\mu} + \frac{1}{m_i - 1}\,X_{\mu\nu'}\,z_{n-1,-i}\,z_{-i,n}\,\xi_{\nu\mu} \Big).\end{aligned}$$

The argument is now completed in the same way as in the proof of Theorem 9.6.8 with slight modifications in the formulas. In particular, for any value of the index $j \in \{-n+1, \dots, n-1\}$ we have

$$z_{jn}\,\xi_{\nu\mu} = [z_{j,-n}, \Phi_{-n,n}]\,\frac{1}{f_j + F_{nn}}\,\xi_{\nu\mu}$$

provided that $(f_j + F_{nn})\,\xi_{\nu\mu} \neq 0$. This gives

$$F_{n-1,n}\,\xi_{\nu\mu\nu'} = X_{\mu\nu'}\,\Big(\widetilde{\Phi}_{n-1,-n}(2)\,\Phi_{-n,n} - \Phi_{-n,n}\widetilde{\Phi}_{n-1,-n}(0)\Big)\,\xi_{\nu\mu},$$

where

$$\widetilde{\Phi}_{n-1,-n}(u) = \sum_{j=-n+1}^{n-1} z_{n-1,j}\, z_{j,-n} \prod_{a=-n+1,\, a\neq j}^{n-1} \frac{1}{f_j - f_a} \cdot \frac{1}{u + f_j + F_{nn}},$$

assuming that $(f_j + F_{nn})\,\xi_{\nu\mu} \neq 0$ for all $j = -n+1, \dots, n-1$. □

EXAMPLE 9.6.14. In the case $n = 1$ the basis vectors of $V(\lambda)$ are defined in (9.115) and the operator $\Phi_{-1,1}$ is given by

$$\Phi_{-1,1} = -\frac{1}{2}\, F_{01}^2.$$

The calculation of $F_{01}\,\zeta_\Lambda$ used in the proof of Theorem 9.6.13 relies on the identity

$$F_{01}\,(F_{11} - 1/2) = \big[F_{0,-1}, \Phi_{-1,1}\big].$$

Suppose that $\sigma_1 = 1$. If $(F_{11} - 1/2)\,\zeta_\Lambda \neq 0$, then the identity can be used to show that

$$F_{01}\,\zeta_\Lambda = -\lambda'_{11}\,(2\lambda'_{11} - 2\lambda_{11} + 1)\,\zeta_{\bar{\Lambda}}.$$

This holds for all patterns Λ with $\sigma_1 = 1$. A similar calculation leads to a formula for the patterns with $\sigma_1 = 0$. □

9.7. Examples

1. Let $N \geqslant 1$ and $M \geqslant 0$ be integers. Let p denote the extremal projector for the Lie algebra $\mathfrak{gl}_M$. By Theorem 9.1.3, the Mickelsson–Zhelobenko algebra $\mathrm{Z}(\mathfrak{gl}_{M+N}, \mathfrak{gl}_M)$ is generated by the elements

$$E_{ab}, \quad pE_{ia}, \quad pE_{ai}, \qquad i \in \{1, \dots, M\}, \quad a, b \in \{M+1, \dots, M+N\}.$$

Introduce the corresponding normalized generators z_{ia} and z_{ai} by

$$z_{ia} = pE_{ia}\,(h_i - h_{i-1}) \dots (h_i - h_1),$$
$$z_{ai} = pE_{ai}\,(h_i - h_{i+1}) \dots (h_i - h_M),$$

where $h_i = E_{ii} - i + 1$. The normalized generators z_{ia} and z_{ai} can be regarded as elements of the Mickelsson algebra $\mathrm{S}(\mathfrak{gl}_{M+N}, \mathfrak{gl}_M)$; cf. Corollary 9.2.3. Identify the universal enveloping algebra $\mathrm{U}(\mathfrak{gl}_{M+N})$ with $\mathrm{Y}_1(\mathfrak{gl}_{M+N})$, the Yangian of level 1; see Section 5.1. Consider the quantum minors defined in (5.7) and (5.8) as elements of the algebra $\mathrm{U}(\mathfrak{gl}_{M+N}) \otimes \mathbb{C}[u]$. Then the elements z_{ia} and z_{ai} admit the following quantum minor presentation:

$$z_{ia} = (-1)^{i-1}\, T^{1\dots i}_{1\dots i-1,a}(-h_i),$$
$$z_{ai} = T^{i+1\dots M,a}_{i\dots M}(-h_i - i + 1).$$

2. The elements of the Mickelsson algebra $\mathrm{S}(\mathfrak{gl}_{M+N}, \mathfrak{gl}_M)$ can be regarded as operators in the subspace of $\mathfrak{gl}_M$-highest vectors of the representation $L(\lambda)$ of $\mathfrak{gl}_{M+N}$; see Section 8.5. More precisely, for any indices $i \in \{1, \dots, M\}$ and $a \in \{M+1, \dots, M+N\}$ we have

$$z_{ia} : L(\lambda)^+_\mu \to L(\lambda)^+_{\mu+\delta_i}, \qquad z_{ai} : L(\lambda)^+_\mu \to L(\lambda)^+_{\mu-\delta_i};$$

cf. (9.59). Suppose that the components of λ and μ are nonnegative integers and consider the corresponding skew diagram λ/μ (see Corollary 8.5.5). Introduce the row order on the cells of λ/μ which corresponds to reading the diagram by rows from left to right starting from the top row. For a cell $\alpha \in \lambda/\mu$ denote by $r(\alpha)$ the row number of α and by $l(\alpha)$ the (increased) leg length of α which equals 1 plus the number of cells of λ/μ in the column containing α which are below α. Define the vector $\zeta_\mu \in L(\lambda)$ by

$$\zeta_\mu = \prod_{\alpha\in\lambda/\mu,\ r(\alpha)\leqslant M} z_{M+l(\alpha),r(\alpha)}\,\zeta,$$

where ζ is the highest vector of $L(\lambda)$ and the product is taken in the row order. Then $\zeta_\mu \in L(\lambda)^+_\mu$ and ζ_μ is the highest vector of the skew representation $L(\lambda)^+_\mu$ of the Yangian $\mathrm{Y}(\mathfrak{gl}_N)$; cf. Proposition 8.5.2. In particular, for the skew diagram of Example 8.5.6 we have

$$\zeta_\mu = (z_{41})^2\,(z_{31})^2\,z_{52}\,z_{42}\,(z_{32})^3\,\zeta.$$

3. The mapping

$$t_{kl}(u) \mapsto \alpha(u)\Big(\delta_{kl}(u-M) + E_{M+k,M+l}\Big)\prod_{i=1}^{M}(u+h_i)$$
$$-\,\alpha(u)\sum_{i=1}^{M} z_{M+k,i}\,z_{i,M+l}\prod_{j=1,\,j\neq i}^{M}\frac{u+h_j}{h_i-h_j}$$

with $k,l\in\{1,\dots,N\}$ and $\alpha(u)=\big(u(u-1)\dots(u-M)\big)^{-1}$ defines a homomorphism $\mathrm{Y}(\mathfrak{gl}_N)\to \mathrm{Z}(\mathfrak{gl}_{M+N},\mathfrak{gl}_M)$; cf. Theorem 9.3.1.

4. Let $N=2n$ or $N=2n+1$ be a positive integer. Suppose that M is a nonnegative integer such that $N-M$ is even and positive. So, $M=2m$ or $M=2m+1$ for some $m<n$. We will identify the Lie algebra $\mathfrak{g}_M$ with the subalgebra of $\mathfrak{g}_N$ spanned by the elements F_{ij} with the indices satisfying $-m\leqslant i,j\leqslant m$. Denote by $V(\lambda)^+$ the subspace of $\mathfrak{g}_M$-highest vectors in $V(\lambda)$:

$$V(\lambda)^+ = \{\eta\in V(\lambda)\mid F_{ij}\,\eta=0,\qquad -m\leqslant i<j\leqslant m\}.$$

Given a $\mathfrak{g}_M$-highest weight $\mu=(\mu_1,\dots,\mu_m)$ we denote by $V(\lambda)^+_\mu$ the corresponding weight subspace in $V(\lambda)^+$:

$$V(\lambda)^+_\mu = \{\eta\in V(\lambda)^+\mid F_{ii}\,\eta=\mu_i\,\eta,\qquad i=1,\dots,m\}.$$

Note that in the case $m=n-1$ it coincides with the multiplicity space introduced in Section 9.4. Consider the homomorphism $\mathrm{Y}(\mathfrak{g}_{N-M})\to\mathrm{Y}(\mathfrak{g}_N)$ provided by Proposition 4.1.11, and take its composition with the evaluation homomorphism ϱ_N defined in (2.106). By Corollary 4.1.9, the image of this composition is contained in the centralizer $\mathrm{U}(\mathfrak{g}_N)^{\mathfrak{g}_M}$. This gives a homomorphism $\varphi:\mathrm{Y}(\mathfrak{g}_{N-M})\to\mathrm{U}(\mathfrak{g}_N)^{\mathfrak{g}_M}$. The vector space $V(\lambda)^+_\mu$ is a representation of $\mathrm{U}(\mathfrak{g}_N)^{\mathfrak{g}_M}$ and hence $V(\lambda)^+_\mu$ becomes equipped with the $\mathrm{Y}(\mathfrak{g}_{N-M})$-module structure defined via the homomorphism φ. We call this module the *skew representation* of $\mathrm{Y}(\mathfrak{g}_{N-M})$. In the particular case $M=0$ (with even N) the skew representation is just the evaluation module $V(\lambda)$ over $\mathrm{Y}(\mathfrak{g}_N)$ defined via the evaluation homomorphism (2.106). In the case $M=N-2$ the structure of the skew representation is described in Theorems 9.4.11, 9.4.13 and 9.4.15.

5. Suppose that $\mathfrak{g}_N=\mathfrak{sp}_{2n}$. Then the skew representation $V(\lambda)^+_\mu$ of the twisted Yangian $\mathrm{Y}(\mathfrak{sp}_{2n-2m})$ is irreducible. Moreover, the results of Section 9.6 imply that

$V(\lambda)^+_\mu$ admits a basis parameterized by trapezium patterns with the fixed top row λ and the bottom row μ, as illustrated:

$$\begin{array}{ccccccccccc} & \lambda_1 & & \lambda_2 & & \lambda_3 & & \cdots & & \lambda_n & \\ \lambda'_{n1} & & \lambda'_{n2} & & \lambda'_{n3} & & \cdots & & \lambda'_{nn} & & \\ & \cdots & & \cdots & & \cdots & & \cdots & & & \\ \lambda'_{m+1,1} & & \lambda'_{m+1,2} & & & \cdots & & \lambda'_{m+1,m+1} & & & \\ & \mu_1 & & \mu_2 & & \cdots & & \mu_m & & & \end{array}$$

The entries are assumed to satisfy the betweenness conditions for the C type patterns; see Section 9.6. In particular, the space $V(\lambda)^+_\mu$ is nonzero if and only if

$$\mu_i \geqslant \lambda_{i+n-m}, \qquad i = 1, \dots, m,$$

and

$$\lambda_i \geqslant \mu_{i+n-m}, \qquad i = 1, \dots, n,$$

assuming that $\mu_i = -\infty$ for $i > m$ and $\mu_i = 0$ for $i \leqslant 0$.

6. In the notation of the previous example suppose that the space $V(\lambda)^+_\mu$ is nonzero. For any three integers i, j, k we will denote by $\mathrm{mid}\{i, j, k\}$ the one which is between the other two. If one of the indices, say k, is the symbol $-\infty$, then $\mathrm{mid}\{i, j, k\}$ is understood as $\min\{i, j\}$. Consider the trapezium array Λ^0 whose entries are determined by

$$\lambda_{ki} = \mathrm{mid}\{\lambda_i, \mu_{i+k-m}, \mu_{i+m-k}\}$$

and

$$\lambda'_{ki} = \mathrm{mid}\{\lambda_i, \mu_{i+k-m-1}, \mu_{i+m-k}\}$$

for all possible values of i and k. Then Λ^0 is a pattern and the corresponding basis vector is the highest vector of the $\mathrm{Y}(\mathfrak{sp}_{2n-2m})$-module $V(\lambda)^+_\mu$.

7. Let the series $\nu(u)$ be defined by

$$\nu(u) = \prod_{i=1}^{m} \frac{(u + \mu_i - i + 1/2)\,(u - \mu_i + i + 1/2)}{(u - i + 1/2)\,(u + i + 1/2)}.$$

The highest weight $\mu(u) = \big(\mu_{m+1}(u), \dots, \mu_n(u)\big)$ of the $\mathrm{Y}(\mathfrak{sp}_{2n-2m})$-module $V(\lambda)^+_\mu$ is given by the formulas

$$\mu_k(u) = \nu(u) \cdot \prod_{\substack{i=1\\ \lambda_i < \mu_{i+k-m-1}}}^{k-1} \frac{u - \max\{\lambda_i, \mu_{i+k-m}\} + k - m + i - 1/2}{u - \mu_{i+k-m-1} + k - m + i - 1/2}$$

$$\times \prod_{\substack{i=1\\ \lambda_i > \mu_{i+m-k+1}}}^{k-1} \frac{u + \min\{\lambda_i, \mu_{i+m-k}\} + k - m - i - 1/2}{u + \mu_{i+m-k+1} + k - m - i - 1/2}$$

$$\times \frac{u + \min\{\lambda_k, \mu_m\} - m - 1/2}{u - m - 1/2},$$

where $k = m + 1, \dots, n$.

8. The Drinfeld polynomials $P_1(u), \dots, P_{n-m}(u)$ for the $\mathrm{Y}(\mathfrak{sp}_{2n-2m})$-module $V(\lambda)^+_\mu$ can be calculated by the following combinatorial rule. Given any $\mathfrak{sp}_{2n}$-highest weight $\lambda = (\lambda_1, \dots, \lambda_n)$, set $\lambda_{-i} = -\lambda_i$ for $i = 1, \dots, n$. We also assume that $\lambda_0 = 0$ with $\lambda_k = -\infty$ and $\lambda_{-k} = +\infty$ for $k > n$. Introduce the *diagram* $\Gamma(\lambda)$ as an infinite set of unit squares (cells) on the plane whose centers have integer coordinates. The coordinates (i, j) of a cell are interpreted as the row and column number so that i increases from top to bottom and j increases from left to right. With these assumptions,

$$\Gamma(\lambda) = \{(i, j) \in \mathbb{Z}^2 \mid -n \leqslant i \leqslant n+1, \quad \lambda_i \leqslant j < \lambda_{i-1}\}.$$

The diagram has a central symmetry, as illustrated below for $\lambda = (-4, -7)$ and $n = 2$:

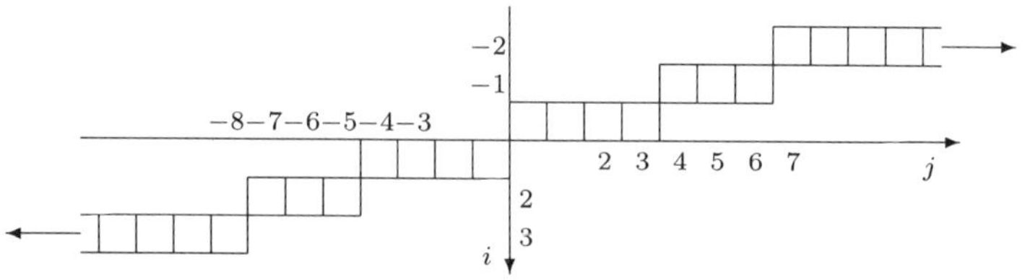

By the *content* of a cell $\alpha = (i, j)$ with coordinates i and j we will mean the number $c(\alpha) = j - i$. For any nonnegative integer p we will denote by $\Gamma(\lambda)^{(p)}$ the image of $\Gamma(\lambda)$ with respect to the shift operator $(i, j) \mapsto (i - p, j)$. In other words, $\Gamma(\lambda)^{(p)}$ is obtained from the diagram $\Gamma(\lambda)$ by lifting each cell p units up. Then for each $k = 1, \dots, n - m$ the Drinfeld polynomial $P_k(u)$ is given by

$$P_k(u) = \prod_{\alpha} \big(u + c(\alpha) + 1/2\big),$$

where α runs over the cells of the intersection $\Gamma(\mu) \cap \Gamma(\lambda)^{(k-1)}$.

9. Let $\lambda = (-2, -8, -10, -13)$ and $\mu = (-4, -7)$ so that $n = 4$ and $m = 2$. The polynomial $P_1(u)$ is calculated from the figure

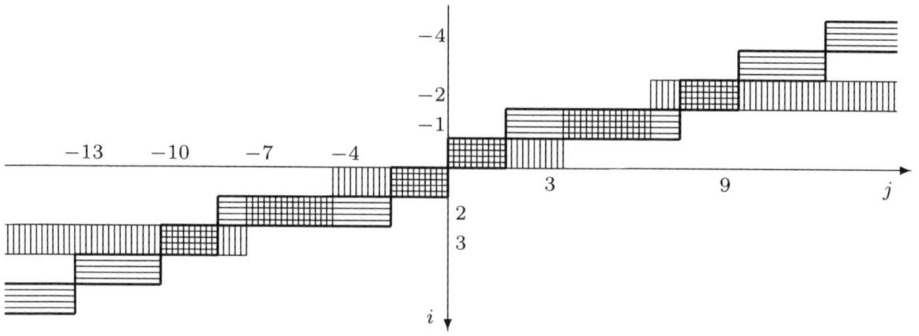

and it is given by

$$\begin{aligned} P_1(u) = {} & (u - 25/2)(u - 23/2)(u - 17/2)(u - 15/2)(u - 13/2)(u - 5/2)(u - 3/2) \\ & (u + 1/2)(u + 3/2)(u + 11/2)(u + 13/2)(u + 15/2)(u + 21/2)(u + 23/2). \end{aligned}$$

Note that the property $P_1(u) = P_1(-u+1)$ is implied by the central symmetry of $\Gamma(\mu) \cap \Gamma(\lambda)$. The horizontal and vertical shadings indicate the diagrams $\Gamma(\lambda)$ and $\Gamma(\mu)$, respectively. The polynomial $P_2(u)$ is calculated from the figure

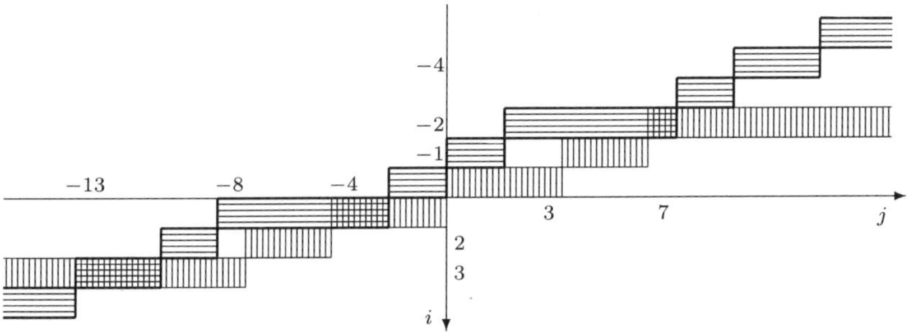

and it is given by

$$P_2(u) = (u - 31/2)(u - 29/2)(u - 27/2)(u - 9/2)(u - 7/2)(u + 19/2).$$

Bibliographical notes

9.1. The extremal projector was discovered by Asherova, Smirnov and Tolstoy [**15, 16, 17**]. The projector turned out to be a powerful instrument in the representation theory of the simple Lie algebras. It plays an essential role in the theory of Mickelsson algebras, which has a wide spectrum of applications from the branching rules and reduction problems to the classification of Harish-Chandra modules; see Zhelobenko's expository paper [**486**] and his book [**487**]. Analogues of the extremal projector for various classes of quantum and superalgebras were produced by Tolstoy [**443, 444**]. A generalized 'relative extremal projector' was constructed by Conley and Sepanski [**79, 80**].

9.2–9.6. The exposition here is based upon the author's papers [**303, 304, 305**]. The branching rules for all classical reductions $\mathfrak{o}_N \downarrow \mathfrak{o}_{N-1}$ and $\mathfrak{sp}_{2n} \downarrow \mathfrak{sp}_{2n-2}$ are due to Zhelobenko [**480**]; see also Hegerfeldt [**172**], King [**218**], Proctor [**395**], Okounkov [**367**], and Goodman and Wallach [**142**]. The lowering operators for the symplectic Lie algebras were first constructed by Mickelsson [**288**]; see also Bincer [**36**]. The explicit relations in the algebra $\mathrm{Z}(\mathfrak{sp}_{2n}, \mathfrak{sp}_{2n-2})$ were calculated by Zhelobenko [**483**]. In some particular cases, bases in $V(\lambda)$ were constructed, e.g., by Wong and Yeh [**469**], and Smirnov and Tolstoy [**421**]. Weight bases for the fundamental representations of $\mathfrak{sp}_{2n}$ and $\mathfrak{o}_{2n+1}$ were constructed by Donnelly [**94, 95**]. He also demonstrated that the bases of his coincide with those given by Corollaries 9.6.1 and 9.6.10, up to a diagonal equivalence. Some combinatorial applications of the properties of the bases in the B case were obtained by Donnelly, Lewis and Pervine [**96**]; see also Alverson *et al.* [**5**]. Harada [**168**] employed the results of [**303**] to construct a new integrable system on the coadjoint orbits of the symplectic groups. This provides an analogue of the Guillemin–Sternberg construction [**164**] for the unitary groups. The weight bases in the orthogonal case were used by Shchepetilov [**415**] in his study of the quantum mechanical two-body problem.

9.7. Examples 1–3 are due to the author [**307**]. The homomorphism from the Yangian to the Mickelsson–Zhelobenko algebra provides an alternative realization of the skew representations of the Yangian; cf. Section 8.5. From a different perspective, connections between the (twisted) Yangians and the Mickelsson algebras have been studied in a series of papers by Khoroshkin and Nazarov [**210, 211, 212**]. In Examples 4–9 the main results of the author's paper [**311**] are sketched. An alternative 'tensor' approach to the skew representations of the twisted Yangians was developed by Nazarov [**344, 345**]. Some irreducibility properties for the restriction of the tensor products of the skew representations of $\mathrm{Y}(\mathfrak{gl}_N)$ to the subalgebra $\mathrm{Y}(\mathfrak{g}_N)$ were established by Mudrov [**325**].

Bibliography

[1] A. Abderrezzak, *Généralisation d'identités de Carlitz, Howard et Lehmer*, Aequat. Math. **49** (1995), 36–46.

[2] C. Ahn and W. M. Koo, *$gl(n|m)$ color Calogero-Sutherland models and super Yangian algebra*, Phys. Lett. B **365** (1996), 105–112.

[3] C. Ahn and S. Nam, *Yangian symmetries in the $SU(N)_1$ WZW model and the Calogero-Sutherland model*, Phys. Lett. B **378** (1996), 107–112.

[4] T. Akasaka and M. Kashiwara, *Finite-dimensional representations of quantum affine algebras*, Publ. Res. Inst. Math. Sci. **33** (1997), 839–867.

[5] L. W. Alverson II, R. G. Donnelly, S. J. Lewis, M. McClard, R. Pervine, R. A. Proctor, and N. J. Wildberger, *Distributive lattices defined for representations of rank two semisimple Lie algebras*, preprint `arXiv:0707.2421`.

[6] T. Arakawa, *Drinfeld functor and finite-dimensional representations of Yangian*, Comm. Math. Phys. **205** (1999), 1–18.

[7] T. Arakawa and T. Suzuki, *Lie algebras and degenerate affine Hecke algebras of type A*, J. Algebra **209** (1998), 288–304.

[8] D. Arnaudon, J. Avan, L. Frappat, E. Ragoucy and M. Rossi, *On the quasi-Hopf structure of deformed double Yangians*, Lett. Math. Phys. **51** (2000), 193–204.

[9] D. Arnaudon, J. Avan, L. Frappat and E. Ragoucy, *Yangian and quantum universal solutions of Gervais–Neveu–Felder equations*, Comm. Math. Phys. **226** (2002), 183–203.

[10] D. Arnaudon, J. Avan, N. Crampé, L. Frappat and E. Ragoucy, *R-matrix presentation for super-Yangians $Y(\mathrm{osp}(m|2n))$*, J. Math. Phys. **44** (2003), 302–308.

[11] D. Arnaudon, J. Avan, N. Crampé, A. Doikou, L. Frappat and E. Ragoucy, *General boundary conditions for the $sl(N)$ and $sl(M|N)$ open spin chains*, J. Stat. Mech. Theory Exp. 2004, no. 8, P08005.

[12] D. Arnaudon, N. Crampé, A. Doikou, L. Frappat and E. Ragoucy, *Analytical Bethe ansatz for closed and open $gl(N)$-spin chains in any representation*, J. Stat. Mech. Theory Exp. 2005, no. 2, P02007.

[13] D. Arnaudon, N. Crampé, A. Doikou, L. Frappat and E. Ragoucy, *Analytical Bethe ansatz for open spin chains with soliton nonpreserving boundary conditions*, Int. J. Mod. Phys. A **21** (2006), 1537–1554.

[14] D. Arnaudon, A. Molev and E. Ragoucy, *On the R-matrix realization of Yangians and their representations*, Annales Henri Poincaré **7** (2006), 1269–1325.

[15] R. M. Asherova, Yu. F. Smirnov and V. N. Tolstoy, *Projection operators for simple Lie groups*, Theor. Math. Phys. **8** (1971), 813–825.

[16] R. M. Asherova, Yu. F. Smirnov and V. N. Tolstoy, *Projection operators for simple Lie groups. II. General scheme for constructing lowering operators. The groups $SU(n)$*, Theor. Math. Phys. **15** (1973), 392–401.

[17] R. M. Asherova, Yu. F. Smirnov and V. N. Tolstoy, *Description of a certain class of projection operators for complex semisimple Lie algebras*, Math. Notes **26** (1979), 499–504.

[18] J. Avan, A. Jevicki and J. Lee, *Yangian-invariant field theory of matrix-vector models*, Nucl. Phys. B **486** (1997), 650–672.

[19] G. E. Baird and L. C. Biedenharn, *On the representations of the semisimple Lie groups. II*, J. Math. Phys. **4** (1963), 1449–1466.

[20] A. O. Barut and R. Rączka, *Theory of group representations and applications*, 2nd edition, World Scientific, Singapore, 1986.

[21] B. Basu-Mallick and A. Kundu, *Multi-parameter deformed and nonstandard* $Y(\mathfrak{gl}_M)$ *Yangian symmetry in a novel class of spin Calogero-Sutherland models*, Nucl. Phys. B **509** (1998), 705–728.

[22] R. J. Baxter, *Exactly solved models in statistical mechanics*, Academic Press, New York, 1982.

[23] V. V. Bazhanov and N. Reshetikhin, *Restricted solid-on-solid models connected with simply laced algebras and conformal field theory*, J. Phys. A **23** (1990), 1477–1492.

[24] J. Beck, *Braid group action and quantum affine algebras*, Comm. Math. Phys. **165** (1994), 555–568.

[25] G. M. Benkart, D. J. Britten, and F. W. Lemire, *Stability in modules for classical Lie algebras – a constructive approach*, Memoirs AMS **85** (1990), no. 430.

[26] G. Benkart and P. Terwillinger, *Irreducible modules for the quantum affine algebra* $U_q(\widehat{\mathfrak{sl}}_2)$ *and its Borel subalgebra* $U_q(\widehat{\mathfrak{sl}}_2)^{\geqslant}$, J. Algebra **282** (2004), 172–194.

[27] A. D. Berenstein and A. V. Zelevinsky, *Involutions on Gel'fand-Tsetlin schemes and multiplicities in skew* GL_n*-modules*, Soviet Math. Dokl. **37** (1988), 799–802.

[28] D. Bernard, *An introduction to Yangian symmetries*, Int. J. Mod. Phys. B **7** (1993), 3517–3530.

[29] D. Bernard and A. LeClair, *Quantum group symmetries and non-local currents in 2D QFT*, Comm. Math. Phys. **142** (1991), 99–138.

[30] D. Bernard, M. Gaudin, F. Haldane and V. Pasquier, *Yang-Baxter equation in long-range interacting systems*, J. Phys. **A 26** (1993), 5219–5236.

[31] D. Bernard, Z. Maassarani and P. Mathieu, *Logarithmic Yangians in WZW models*, Mod. Phys. Lett. A **12** (1997), 535–544.

[32] L. C. Biedenharn and J. D. Louck, *Angular momentum in quantum physics: theory and application*, Addison-Wesley, Reading, Mass., 1981.

[33] L. C. Biedenharn and J. D. Louck, *A new class of symmetric polynomials defined in terms of tableaux*, Advances in Appl. Math. **10** (1989), 396–438.

[34] L. C. Biedenharn and J. D. Louck, *Inhomogeneous basis set of symmetric polynomials defined by tableaux*, Proc. Nat. Acad. Sci. U.S.A. **87** (1990), 1441–1445.

[35] Y. Billig, V. Futorny and A. Molev, *Verma modules for Yangians*, Lett. Math. Phys. **78** (2006), 1–16.

[36] A. Bincer, *Missing label operators in the reduction* $Sp(2n) \downarrow Sp(2n-2)$, J. Math. Phys. **21** (1980), 671–674.

[37] A. Bincer, *Mickelsson lowering operators for the symplectic group*, J. Math. Phys. **23** (1982), 347–349.

[38] P. P. Boalch, *Stokes matrices, Poisson Lie groups and Frobenius manifolds*, Invent. Math. **146** (2001), 479–506.

[39] A. I. Bondal, *A symplectic groupoid of triangular bilinear forms and the braid group*, Izvestiya Mathematics **68** (2004), 659–708.

[40] A. I. Bondal, *Symplectic groupoids related to Poisson–Lie groups*, Proc. Steklov Inst. Math. **246** (2004), 34–53.

[41] P. Bouwknegt, A. W. W. Ludwig and K. Schoutens, *Spinon bases, Yangian symmetry and fermionic representations of Virasoro characters in conformal field theory*, Phys. Lett. B **338** (1994), 448–456.

[42] P. Bouwknegt and K. Schoutens, *The* $SU(n)_1$ *WZW models: spinon decomposition and Yangian structure*, Nucl. Phys. B **482** (1996), 345–372.

[43] P. Bouwknegt and K. Schoutens, *Spinon decomposition and Yangian structure of* $\widehat{\mathfrak{sl}}_n$ *modules*, in: "Geometric Analysis and Lie Theory in Mathematics and Physics", Austral. Math. Soc. Lect. Ser. **11**, Cambridge Univ. Press, Cambridge, 1998, pp. 105–131.

[44] J. Bowman, *Irreducible modules for the quantum affine algebra* $U_q(\mathfrak{g})$ *and its Borel subalgebra* $U_q(\mathfrak{g})^{\geqslant}$, preprint `arXiv:math/0606627`.

[45] S. I. Boyarchenko and S. Z. Levendorskiĭ, *On affine Yangians*, Lett. Math. Phys. **32** (1994), 269–274.

[46] A. J. Bracken and H. S. Green, *Vector operators and a polynomial identity for SO(n)*, J. Math. Phys. **12** (1971), 2099–2106.

[47] C. Briot and E. Ragoucy, *RTT presentation of finite* $\mathcal{W}$*-algebras*, J. Phys. A **34** (2001), 7287–7310.

[48] C. Briot and E. Ragoucy, *Twisted super-Yangians and their representations*, J. Math. Phys. **44** (2003), 1252–1275.

[49] C. Briot and E. Ragoucy, *$\mathcal{W}$-superalgebras as truncations of super-Yangians*, J. Phys. A **36** (2003), 1057–1081.

[50] J. Brown and J. Brundan, *Elementary invariants for centralizers of nilpotent matrices*, preprint arXiv:math/0611024.

[51] J. Brundan and A. Kleshchev, *Parabolic presentations of the Yangian* $\mathrm{Y}(\mathfrak{gl}_n)$, Comm. Math. Phys. **254** (2005), 191–220.

[52] J. Brundan and A. Kleshchev, *Shifted Yangians and finite W-algebras*, Adv. Math. **200** (2006), 136–195.

[53] J. Brundan and A. Kleshchev, *Representations of shifted Yangians and finite W-algebras*, preprint, Memoirs AMS, to appear; arXiv:math/0508003.

[54] E. R. Caianiello, *Proprietà di Pfaffiani e Hafniani*, Ricerca, Napoli **7** (1956), 25–31.

[55] A. Capelli, *Über die Zurückführung der Cayley'schen Operation Ω auf gewöhnliche Polar-Operationen*, Math. Ann. **29** (1887), 331–338.

[56] A. Capelli, *Sur les opérations dans la théorie des formes algébriques*, Math. Ann. **37** (1890), 1–37.

[57] V. Caudrelier and E. Ragoucy, *Lax pair and super-Yangian symmetry of the nonlinear super-Schrödinger equation*, J. Math. Phys. **44** (2003), 5706–5732.

[58] V. Chari, *On the fermionic formula and the Kirillov–Reshetikhin conjecture*, Internat. Math. Res. Notices (2001), 629–654.

[59] V. Chari, *Braid group actions and tensor products*, Int. Math. Res. Not. **7** (2002), 357–382.

[60] V. Chari and S. Ilangovan, *On the Harish-Chandra homomorphism for infinite-dimensional Lie algebras*, J. Algebra **90** (1984), 476–494.

[61] V. Chari and A. A. Moura, *Characters of fundamental representations of quantum affine algebras*, Acta Appl. Math. **90** (2006), 43–63.

[62] V. Chari and A. Pressley, *Yangians and R-matrices*, L'Enseign. Math. **36** (1990), 267–302.

[63] V. Chari and A. Pressley, *Fundamental representations of Yangians and rational R-matrices*, J. Reine Angew. Math. **417** (1991), 87–128.

[64] V. Chari and A. Pressley, *Quantum affine algebras*, Comm. Math. Phys. **142** (1991), 261–283.

[65] V. Chari and A. Pressley, *A guide to quantum groups*, Cambridge University Press, Cambridge, 1994.

[66] V. Chari and A. Pressley, *Yangians: their representations and characters,* Acta Appl. Math. **44** (1996), 39–58.

[67] V. Chari and A. Pressley, *Yangians, integrable quantum systems and Dorey's rule*, Comm. Math. Phys. **181** (1996), 265–302.

[68] V. Chari and N. Xi, *Monomial bases of quantized enveloping algebras*, in: "Recent developments in quantum affine algebras and related topics" (Raleigh, NC, 1998), pp. 69–81, Contemp. Math., 248, Amer. Math. Soc., Providence, RI, 1999.

[69] L. O. Chekhov, *Teichmüller theory of bordered surfaces*, preprint arXiv:math/0610872.

[70] W. Y. C. Chen, B. Li and J. D. Louck, *The flagged double Schur function*, J. Alg. Comb. **15** (2002), 7–26.

[71] I. V. Cherednik, *Factorized particles on the half-line and root systems*, Theor. Math. Phys. **61** (1984), 977–983.

[72] I. V. Cherednik, *On special bases of irreducible finite-dimensional representations of the degenerate affine Hecke algebra*, Funct. Analysis Appl. **20** (1986), 87–89.

[73] I. V. Cherednik, *A new interpretation of Gelfand–Tzetlin bases*, Duke Math. J. **54** (1987), 563–577.

[74] I. V. Cherednik, *Quantum groups as hidden symmetries of classic representation theory*, in: "Differential Geometric Methods in Physics (Chester, 1988)", World Scientific, Teaneck, NJ, 1989, pp. 47–54.

[75] I. Cherednik, *Lectures on Knizhnik-Zamolodchikov equations and Hecke algebras*, Math. Soc. Japan Memoirs **1** (1998), 1–96.

[76] A. Chervov and D. Talalaev, *Quantum spectral curves, quantum integrable systems and the geometric Langlands correspondence*, preprint arXiv:hep-th/0604128.

[77] A. Chervov and D. Talalaev, *KZ equation, G-opers, quantum Drinfeld-Sokolov reduction and quantum Cayley–Hamilton identity*, preprint arXiv:hep-th/0607250.

[78] N. Ciccoli and F. Gavarini, *A quantum duality principle for coisotropic subgroups and Poisson quotients*, Adv. Math. **199** (2006), 104–135.

[79] C. H. Conley and M. R. Sepanski, *Relative extremal projectors*, Adv. Math. **174** (2003), 155–166.

[80] C. H. Conley and M. R. Sepanski, *Infinite commutative product formulas for relative extremal projectors*, Adv. Math. **196** (2005), 52–77.

[81] N. Crampé, *Hopf structure of the Yangian $Y(\mathfrak{sl}_n)$ in the Drinfeld realization*, J. Math. Phys. **45** (2004), 434–447.

[82] T. Curtright and C. Zachos, *Supersymmetry and the nonlocal Yangian deformation symmetry*, Nucl. Phys. B **402** (1993), 604–612.

[83] C. De Concini and D. Kazhdan, *Special bases for S_N and $GL(n)$*, Israel J. Math. **40** (1981), 275–290.

[84] G. W. Delius and N. J. MacKay, *Quantum group symmetry in sine-Gordon and affine Toda field theories on the half-line*, Comm. Math. Phys. **233** (2003), 173–190.

[85] G. W. Delius, N. J. MacKay and B. J. Short, *Boundary remnant of Yangian symmetry and the structure of rational reflection matrices*, Phys. Lett. B **522** (2001), 335–344; Erratum, ibid. **524** (2002), 401.

[86] J. Ding and I. Frenkel, *Isomorphism of two realizations of quantum affine algebra* $\mathrm{U}(\widehat{gl}(n))$, Comm. Math. Phys. **156** (1993), 277–300.

[87] X. M. Ding, Bo-Yu Hou and L. Zhao, *$\hbar$ (Yangian) deformation of the Miura map and Virasoro algebra*, Internat. J. Modern Phys. A **13** (1998), 1129–1144.

[88] X. M. Ding and L. Zhao, *Free boson representation of $DY_\hbar(sl_2)_k$ and the deformation of the Feigin-Fuchs*, Comm. Theor. Phys. **32** (1999), 103–108.

[89] J. Dixmier, *Algèbres Enveloppantes*, Gauthier-Villars, Paris, 1974.

[90] V. K. Dobrev and P. Truini, *Polynomial realization of the $U_q(\mathrm{sl}(3))$ Gelfand-(Weyl)-Zetlin basis*, J. Math. Phys. **38** (1997), 3750–3767.

[91] V. K. Dobrev, A. D. Mitov and P. Truini, *Normalized $U_q(\mathrm{sl}(3))$ Gelfand-(Weyl)-Zetlin basis and new summation formulas for q-hypergeometric functions*, J. Math. Phys. **41** (2000), 7752–7768.

[92] L. Dolan, C. R. Nappi and E. Witten, *Yangian symmetry in $D = 4$ superconformal Yang-Mills theory*, in: "Quantum Theory and Symmetries", World Sci. Publ., Hackensack, NJ, 2004, pp. 300–315.

[93] R. G. Donnelly, *Symplectic analogs of the distributive lattices $L(m,n)$*, J. Combin. Theory Ser. A **88** (1999), 217–234.

[94] R. G. Donnelly, *Explicit constructions of the fundamental representations of the symplectic Lie algebras*, J. Algebra **233** (2000), 37–64.

[95] R. G. Donnelly, *Extremal properties of bases for representations of semisimple Lie algebras*, J. Algebraic Combin. **17** (2003), 255–282.

[96] R. G. Donnelly, S. J. Lewis and R. Pervine, *Constructions of representations of $o(2n+1, C)$ that imply Molev and Reiner-Stanton lattices are strongly Sperner*, Discr. Math. **263** (2003), 61–79.

[97] V. G. Drinfeld, *Hopf algebras and the quantum Yang–Baxter equation*, Soviet Math. Dokl. **32** (1985), 254–258.

[98] V. G. Drinfeld, *Degenerate affine Hecke algebras and Yangians*, Funct. Anal. Appl. **20** (1986), 56–58.

[99] V. G. Drinfeld, *A new realization of Yangians and quantized affine algebras*, Soviet Math. Dokl. **36** (1988), 212–216.

[100] V. G. Drinfeld, *Quantum groups*, in: "International Congress of Mathematicians (Berkeley, 1986)", Amer. Math. Soc., Providence, RI, 1987, pp. 798–820.

[101] Yu. A. Drozd, V. M. Futorny and S. A. Ovsienko, *On Gel'fand-Zetlin modules*, in: "Proceedings of the Winter School on Geometry and Physics" (Srní, 1990), Rend. Circ. Mat. Palermo (2) Suppl. No. 26 (1991), 143–147.

[102] Yu. A. Drozd, V. M. Futorny and S. A. Ovsienko, *Gelfand-Zetlin modules over Lie algebra* sl(3), Contemp. Math. **131** (1992), 23–29.

[103] Yu. A. Drozd, V. M. Futorny and S. A. Ovsienko, *Harish-Chandra subalgebras and Gel'fand-Zetlin modules*, in: "Finite-Dimensional Algebras and Related Topics" (Ottawa, ON, 1992), pp. 79–93, NATO Adv. Sci. Inst. Ser. C Math. Phys. Sci., 424, Kluwer Acad. Publ., Dordrecht, 1994.

[104] J. Du, *Canonical bases for irreducible representations of quantum* GL_n, Bull. London Math. Soc. **24** (1992), 325–334.
[105] J. Du, *Canonical bases for irreducible representations of quantum* GL_n. *II*, J. London Math. Soc. **51** (1995), 461–470.
[106] B. Dubrovin, *Geometry of 2D topological field theory*, in: "Integrable Systems and Quantum Groups" (M. Francaviglia, S. Greco, Eds.), Lect. Notes. Math. **1620**, Springer, 1996, pp. 120–348.
[107] B. Enriquez and G. Felder, *A construction of Hopf algebra cocycles for the Yangian double* $DY(\mathfrak{sl}_2)$, J. Phys. A **31** (1998), 2401–2413.
[108] B. Enriquez and G. Felder, *Coinvariants for Yangian doubles and quantum Knizhnik-Zamolodchikov equations*, Internat. Math. Res. Notices (1999), 81–104.
[109] P. Etingof and V. Retakh, *Quantum determinants and quasideterminants*, Asian J. Math. **3** (1999), 345–352.
[110] L. D. Faddeev, *Integrable models in* $(1+1)$*-dimensional quantum field theory*, in: "Recent Advances in Field Theory and Statistical Mechanics (Les Houches, 1982)", North-Holland, Amsterdam, 1984, pp. 561–608.
[111] L. D. Faddeev, *Algebraic aspects of the Bethe ansatz*, Internat. J. Modern Phys. A **10** (1995), 1845–1878.
[112] L. D. Faddeev and L. A. Takhtajan, *Spectrum and scattering of excitations in the one-dimensional isotropic Heisenberg model*, J. Soviet Math. **24** (1984), 241–267.
[113] B. Feigin, E. Frenkel and N. Reshetikhin, *Gaudin model, Bethe ansatz and critical level*, Comm. Math. Phys. **166** (1994), 27–62.
[114] E. Frenkel and E. Mukhin, *Combinatorics of q-characters of finite-dimensional representations of quantum affine algebras*, Comm. Math. Phys. **216** (2001), 23–57.
[115] E. Frenkel and E. Mukhin, *The Hopf algebra* $\mathrm{Rep}\, U_q\widehat{\mathfrak{gl}}_\infty$, Selecta Math. **8** (2002), 537–635.
[116] E. Frenkel and N. Reshetikhin, *The q-characters of representations of quantum affine algebras and deformations of* $\mathcal{W}$*-algebras*, Contemp. Math. **248** (1999), 163–205.
[117] W. Fulton, *Young tableaux. With applications to representation theory and geometry.* London Mathematical Society Student Texts, 35, Cambridge University Press, Cambridge, 1997.
[118] W. Fulton, *Equivariant cohomology in algebraic geometry*, Eilenberg lectures, Columbia University, Spring 2007; `http://www.math.lsa.umich.edu/~dandersn/eilenberg`
[119] W. Fulton and J. Harris, *Representation theory. A first course*, Graduate Texts in Mathematics, 129, Readings in Mathematics, Springer-Verlag, New York, 1991.
[120] V. Futorny, A. Molev and S. Ovsienko, *Harish-Chandra modules for Yangians*, Represent. Theory **9** (2005), 426–454.
[121] V. Futorny and S. Ovsienko, *Kostant's theorem for special filtered algebras*, Bull. London Math. Soc. **37** (2005), 187–199.
[122] V. Futorny and S. Ovsienko, *Galois algebras I: Structure theory*, preprint `arXiv:math/0610069`.
[123] V. Futorny and S. Ovsienko, *Galois algebras II: Representation theory*, preprint `arXiv:math/0610071`.
[124] F. Gavarini, *Presentation by Borel subalgebras and Chevalley generators for quantum enveloping algebras*, Proc. Edinburgh Math. Soc. **49** (2006), 291–308.
[125] A. M. Gavrilik and N. Z. Iorgov, *q-deformed algebras* $U_q(\mathrm{so}_n)$ *and their representations*, Methods of Funct. Anal. Topology **3** (1997), 51–63.
[126] A. M. Gavrilik and N. Z. Iorgov, *On Casimir elements of q-algebras* $U_q'(\mathrm{so}_n)$ *and their eigenvalues in representations*, in: "Symmetry in Nonlinear Mathematical Physics", Proc. Inst. Mat. Ukr. Nat. Acad. Sci. **30**, Kyiv, 1999, pp. 310–314.
[127] A. M. Gavrilik and A. U. Klimyk, *q-deformed orthogonal and pseudo-orthogonal algebras and their representations*, Lett. Math. Phys. **21** (1991), 215–220.
[128] I. M. Gelfand, *Center of the infinitesimal group ring*, Mat. Sbornik **26** (1950), 103–112 (Russian). English transl. in: I. M. Gelfand, "Collected Papers", Vol. II, Berlin: Springer-Verlag, 1988, pp. 22–30.
[129] I. Gelfand, S. Gelfand, V. Retakh and R. Wilson, *Quasideterminants*, Adv. Math. **193** (2005), 56–141.
[130] I. M. Gelfand and M. I. Graev, *Finite-dimensional irreducible representations of the unitary and the full linear groups, and related special functions*, Izv. AN SSSR, Ser. Mat. **29** (1965),

1329–1356 (Russian). English transl. in: I. M. Gelfand, "Collected papers", Vol. II, Berlin: Springer-Verlag, 1988, pp. 662–692.

[131] I. M. Gelfand, D. Krob, A. Lascoux, B. Leclerc, V. S. Retakh and J.-Y. Thibon, *Noncommutative symmetric functions*, Adv. Math. **112** (1995), 218–348.

[132] I. M. Gelfand and V. S. Retakh, *Determinants of matrices over noncommutative rings*, Funct. Anal. Appl. **25** (1991), 91-102.

[133] I. M. Gelfand and V. S. Retakh, *A theory of noncommutative determinants and characteristic functions of graphs*, Funct. Anal. Appl. **26** (1992), 231–246.

[134] I. M. Gelfand and M. L. Tsetlin, *Finite-dimensional representations of the group of unimodular matrices*, Dokl. Akad. Nauk SSSR **71** (1950), 825–828 (Russian). English transl. in: I. M. Gelfand, "Collected papers", Vol. II, Berlin: Springer-Verlag, 1988, pp. 653–656.

[135] I. M. Gelfand and M. L. Tsetlin, *Finite-dimensional representations of groups of orthogonal matrices*, Dokl. Akad. Nauk SSSR **71** (1950), 1017–1020 (Russian). English transl. in: I. M. Gelfand, "Collected papers", Vol. II, Berlin: Springer-Verlag, 1988, pp. 657–661.

[136] I. M. Gelfand and A. Zelevinsky, *Models of representations of classical groups and their hidden symmetries*, Funct. Anal. Appl. **18** (1984), 183–198.

[137] I. M. Gelfand and A. Zelevinsky, *Multiplicities and proper bases for gl_n*, in: "Group Theoretical Methods in Physics", Vol. II, Yurmala, 1985, pp. 147–159. Utrecht: VNU Sci. Press, 1986.

[138] F. Geoffriau, *Homomorphisme de Harish-Chandra pour les algèbres de Takiff généralisées*, J. Algebra **171** (1995), 444–456.

[139] A. Gerasimov, S. Kharchev and D. Lebedev, *Representation theory and quantum inverse scattering method: the open Toda chain and the hyperbolic Sutherland model*, Internat. Math. Res. Notices (2004), 823–854.

[140] A. Gerasimov, S. Kharchev, D. Lebedev and S. Oblezin, *On a class of representations of the Yangian and moduli space of monopoles*, Comm. Math. Phys. **260** (2005), 511–525.

[141] V. Ginzburg and E. Vasserot, *Langlands reciprocity for affine quantum groups of type A_n*, Internat. Math. Res. Notices (1993), 67–85.

[142] R. Goodman and N. R. Wallach, *Representations and invariants of the classical groups*, Cambridge University Press, 1998.

[143] M. D. Gould, *The characteristic identities and reduced matrix elements of the unitary and orthogonal groups*, J. Austral. Math. Soc. B **20** (1978), 401–433.

[144] M. D. Gould, *On an infinitesimal approach to semisimple Lie groups and raising and lowering operators of $O(n)$ and $U(n)$*, J. Math. Phys. **21** (1980), 444–453.

[145] M. D. Gould, *On the matrix elements of the $U(n)$ generators*, J. Math. Phys. **22** (1981), 15–22.

[146] M. D. Gould, *General $U(N)$ raising and lowering operators*, J. Math. Phys. **22** (1981), 267–270.

[147] M. D. Gould, *Wigner coefficients for a semisimple Lie group and the matrix elements of the $O(n)$ generators*, J. Math. Phys. **22** (1981), 2376–2388.

[148] M. D. Gould, *Characteristic identities for semisimple Lie algebras*, J. Austral. Math. Soc. B **26** (1985), 257–283.

[149] M. D. Gould, *Representation theory of the symplectic groups. I*, J. Math. Phys. **30** (1989), 1205–1218.

[150] M. D. Gould and L. C. Biedenharn, *The pattern calculus for tensor operators in quantum groups*, J. Math. Phys. **33** (1992), 3613–3635.

[151] M. D. Gould and P. D. Jarvis, *Characteristic identities for Kac-Moody algebras*, Lett. Math. Phys. **22** (1991), 91–100.

[152] M. D. Gould and E. G. Kalnins, *A projection-based solution to the $Sp(2n)$ state labeling problem*, J. Math. Phys. **26** (1985), 1446–1457.

[153] M. D. Gould and N. I. Stoilova, *Casimir invariants and characteristic identities for* $\mathrm{gl}(\infty)$, J. Math. Phys. **38** (1997), 4783–4793.

[154] M. D. Gould, R. B. Zhang and A. J. Bracken, *Generalized Gel'fand invariants and characteristic identities for quantum groups*, J. Math. Phys. **32** (1991), 2298–2303.

[155] M. D. Gould and Y.-Z. Zhang, *On super-RS algebra and Drinfeld realization of quantum affine superalgebras*, Lett. Math. Phys. **44** (1998), 291–308.

[156] I. Goulden and C. Greene, *A new tableau representation for supersymmetric Schur functions*, J. Algebra. **170** (1994), 687–703.

[157] L. Gow, *On the Yangian $Y(\mathfrak{gl}_{m|n})$ and its quantum Berezinian*, Czech. J. Phys. **55** (2005), 1415–1420.

[158] L. Gow, *Gauss decomposition of the Yangian* $\mathrm{Y}(\mathfrak{gl}_{m|n})$, Comm. Math. Phys., to appear; `arXiv:math/0605219`.

[159] W. Graham, *Positivity in equivariant Schubert calculus*, Duke Math. J. **109** (2001), 599–614.

[160] H. S. Green, *Characteristic identities for generators of GL(n), O(n) and Sp(n)*, J. Math. Phys. **12** (1971), 2106–2113.

[161] H. S. Green and P. D. Jarvis, *Casimir invariants, characteristic identities, and Young diagrams for color algebras and superalgebras*, J. Math. Phys. **24** (1983), 1681–1687.

[162] J. Grime, *The hook fusion procedure*, Electron. J. Comb. **12** (2005), R26.

[163] J. Grime, *The hook fusion procedure for Hecke algebras*, J. Alg. **309** (2007), 744–759.

[164] V. Guillemin and S. Sternberg, *The Gelfand–Cetlin system and quantization of the complex flag manifolds*, J. Funct. Anal. **52** (1983), 106–128.

[165] V. Guizzi and P. Papi, *A combinatorial approach to the fusion process for the symmetric group*, Europ. J. Comb. **19** (1998), 835–845.

[166] D. I. Gurevich, P. N. Pyatov and P. A. Saponov, *Hecke symmetries and characteristic relations on reflection equation algebras*, Lett. Math. Phys. **41** (1997), 255–264.

[167] F. D. M. Haldane, Z. N. C. Ha, J. C. Talstra, D. Bernard and V. Pasquier, *Yangian symmetry of integrable quantum chains with long-range interactions and a new description of states in conformal field theory*, Phys. Rev. Lett. **69** (1992), 2021–2025.

[168] M. Harada, *The symplectic geometry of the Gel'fand–Cetlin–Molev basis for representations of* $Sp(2n, \mathbb{C})$, J. Symplectic Geom. **4** (2006), 1–41.

[169] G. Hatayama, A. Kuniba, M. Okado, T. Takagi and Y. Yamada, *Remarks on fermionic formula*, Contemp. Math. **248** (1999), 243–291.

[170] T. Hauer, *Systematic proof of the existence of Yangian symmetry in chiral Gross-Neveu models*, Phys. Lett. B **417** (1998), 297–302.

[171] M. Havlíček, A. U. Klimyk and S. Pošta, *Central elements of the algebras $U'_q(\mathrm{so}_m)$ and $U_q(\mathrm{iso}_m)$*, Czech. J. Phys. **50** (2000), 79–84.

[172] G. C. Hegerfeldt, *Branching theorem for the symplectic groups*, J. Math. Phys. **8** (1967), 1195–1196.

[173] D. Hernandez, *Algebraic approach to q,t-characters*, Adv. Math. **187** (2004), 1–52.

[174] M. J. Hopkins and A. I. Molev, *A q-analogue of the centralizer construction and skew representations of the quantum affine algebra*, Symmetry, Integrability and Geometry: Methods and Applications **2** (2006), paper 092, 29 pp.

[175] Hou Pei-yu, *Orthonormal bases and infinitesimal operators of the irreducible representations of group U_n*, Scientia Sinica **15** (1966), 763–772.

[176] R. Howe, *Remarks on classical invariant theory*, Trans. Amer. Math. Soc. **313** (1989), 539–570.

[177] R. Howe, *Perspectives on invariant theory: Schur duality, multiplicity-free actions and beyond*, Israel Math. Conf. Proc. **8** (1995), 1–182.

[178] R. Howe and T. Umeda, *The Capelli identity, the double commutant theorem, and multiplicity-free actions*, Math. Ann. **290** (1991), 569–619.

[179] J. E. Humphreys, *Introduction to Lie algebras and representation theory*, Graduate Texts in Math. **9**, Springer, Berlin, 1972.

[180] K. Iohara, *Bosonic representations of Yangian double $DY_\hbar(\mathfrak{g})$ with $\mathfrak{g} = \mathfrak{gl}_N, \mathfrak{sl}_N$*, J. Phys. A **29** (1996), 4593–4621.

[181] N. Z. Iorgov and A. U. Klimyk, *Classification theorem on irreducible representations of the q-deformed algebra $\mathrm{U}'_q(\mathfrak{so}_n)$*, Int. J. Math. Sci. 2005, no. 2, 225–262.

[182] A. Isaev, *Quantum groups and Yang-Baxter equations*, preprint MPIM2004-132, Max-Planck-Institut für Mathematik, Bonn.

[183] A. Isaev, O. Ogievetsky and P. Pyatov, *On quantum matrix algebras satisfying the Cayley-Hamilton-Newton identities*, J. Phys. A **32** (1999), L115–L121.

[184] M. Itoh, *Explicit Newton's formulas for $\mathfrak{gl}_n$*, J. Algebra **208** (1998), 687–697.

[185] M. Itoh, *Capelli elements for the orthogonal Lie algebras*, J. Lie Theory **10** (2000), 463–489.

[186] M. Itoh, *Cayley–Hamilton theorem for the skew Capelli elements*, J. Algebra **242** (2001), 740–761.

[187] M. Itoh, *Correspondence of the Gelfand invariants in reductive dual pairs*, J. Aust. Math. Soc. **75** (2003), 263–277.

[188] M. Itoh, *Capelli elements for the dual pair* (O_M, Sp_N), Math. Z. **246** (2004), 125–154.
[189] M. Itoh, *Capelli identities for reductive dual pairs*, Adv. Math. **194** (2005), 345–397.
[190] M. Itoh, *Two determinants in the universal enveloping algebras of the orthogonal Lie algebras*, preprint, 2006.
[191] M. Itoh, *Two permanents in the universal enveloping algebras of the symplectic Lie algebras*, preprint, 2006.
[192] M. Itoh and T. Umeda, *On central elements in the universal enveloping algebras of the orthogonal Lie algebras*, Compositio Math. **127** (2001), 333–359.
[193] A. G. Izergin and V. E. Korepin, *A lattice model related to the nonlinear Schrödinger equation*, Sov. Phys. Dokl. **26** (1981), 653–654.
[194] G. James and A. Kerber, *The representation theory of the symmetric group*, Encyclopedia of Mathematics and its Applications, **16**, Reading, MA/London, Addison–Wesley, 1981.
[195] P. D. Jarvis and H. S. Green, *Casimir invariants and characteristic identities for generators of the general linear, special linear and orthosymplectic graded Lie algebras*, J. Math. Phys. **20** (1979), 2115–2122.
[196] P. D. Jarvis and M. K. Murray, *Casimir invariants, characteristic identities, and tensor operators for "strange" superalgebras*, J. Math. Phys. **24** (1983), 1705–1710.
[197] M. Jimbo, *A q-difference analogue of* $\mathrm{U}(\mathfrak{g})$ *and the Yang–Baxter equation*, Lett. Math. Phys. **10** (1985), 63–69.
[198] M. Jimbo, *A q-analogue of* $U_q(\mathfrak{gl}(N+1))$*, Hecke algebra and the Yang–Baxter equation*, Lett. Math. Phys. **11** (1986), 247–252.
[199] M. Jimbo, *Quantum R-matrix for the generalized Toda system*, Comm. Math. Phys. **102** (1986), 537–547.
[200] M. Jimbo, A. Kuniba, T. Miwa and M. Okado, *The* $A_n^{(1)}$ *face models*, Comm. Math. Phys. **119** (1988), 543–565.
[201] A. Joseph, *Quantum groups and their primitive ideals*, Springer-Verlag, Berlin, 1995.
[202] Guo-xing Ju, Shi-kun Wang and Ke Wu, *The algebraic structure of the* $gl(n|m)$ *color Calogero-Sutherland models*, J. Math. Phys. **39** (1998), 2813–2820.
[203] A. Jucys, *On the Young operators of the symmetric group*, Lietuvos Fizikos Rinkinys **6** (1966), 163–180.
[204] A. A. Jucys, *Symmetric polynomials and the center of the symmetric group ring*, Rep. Math. Phys. **5** (1974), 107–112.
[205] V. G. Kac, *Infinite-dimensional Lie algebras*, Cambridge University Press, Cambridge, 1990.
[206] M. Kashiwara, *Crystalizing the q-analogue of universal enveloping algebras*, Comm. Math. Phys. **133** (1990), 249–260.
[207] M. Kashiwara, *On level-zero representation of quantized affine algebras*, Duke Math. J. **112** (2002), 117–195.
[208] S. V. Kerov, A. N. Kirillov and N. Yu. Reshetikhin, *Combinatorics, the Bethe ansatz and representations of the symmetric group*, J. Sov. Math. **41** (1988), 916–924.
[209] S. Khoroshkin, D. Lebedev and S. Pakuliak, *Traces of intertwining operators for the Yangian double*, Lett. Math. Phys. **41** (1997), 31–47.
[210] S. Khoroshkin and M. Nazarov, *Yangians and Mickelsson algebras I*, Transform. Groups **11** (2006), 625–658.
[211] S. Khoroshkin and M. Nazarov, *Yangians and Mickelsson algebras II*, Mosc. Math. J. **6** (2006), 477–504.
[212] S. Khoroshkin and M. Nazarov, *Twisted Yangians and Mickelsson algebras I*, Selecta Math., to appear; `arXiv:math/0703651`.
[213] S. Khoroshkin and O. Ogievetsky, *Mickelsson algebras and Zhelobenko operators*, preprint `arXiv:math/0606259`.
[214] S. M. Khoroshkin, A. A. Stolin and V. N. Tolstoy, *Rational solutions of Yang-Baxter equation and deformation of Yangians*, in: "From Field Theory to Quantum Groups", World Scientific, River Edge, NJ, 1996, pp. 53–75.
[215] S. M. Khoroshkin, A. A. Stolin and V. N. Tolstoy, *Deformation of Yangian* $Y(\mathrm{sl}_2)$, Comm. Algebra **26** (1998), 1041–1055.
[216] S. M. Khoroshkin and V. N. Tolstoy, *Extremal projector and universal R-matrix for quantum contragredient Lie (super)algebras*, in: "Quantum Group and Related Topics" (Wrocław, 1991), 23–32, Math. Phys. Stud., **13**, Kluwer Academic Publishers, 1992.
[217] S. M. Khoroshkin and V. N. Tolstoy, *Yangian double*, Lett. Math. Phys. **36** (1996), 373–402.

[218] R. C. King, *Weight multiplicities for the classical groups*, in: "Group Theoretical Methods in Physics", Fourth Internat. Colloq., Nijmegen, 1975, Lecture Notes in Phys., Vol. 50, pp. 490–499, Berlin: Springer 1976.

[219] A. N. Kirillov, *Combinatorial identities and completeness of states for the Heisenberg magnet*, J. Sov. Math. **30** (1985), 2298–3310.

[220] A. N. Kirillov, *Completeness of states of the generalized Heisenberg magnet*, J. Sov. Math. **36** (1987), 115–128.

[221] A. A. Kirillov, *A remark on the Gelfand-Tsetlin patterns for symplectic groups*, J. Geom. Phys. **5** (1988), 473–482.

[222] A. A. Kirillov, *Family algebras*, Electron. Res. Announc. Amer. Math. Soc. **6** (2000), 7–20.

[223] A. A. Kirillov, *Introduction to family algebras*, Mosc. Math. J. **1** (2001), 49–63.

[224] A. N. Kirillov and A. D. Berenstein, *Groups generated by involutions, Gel'fand-Tsetlin patterns, and combinatorics of Young tableaux*, St. Petersburg Math. J. **7** (1996), 77–127.

[225] A. N. Kirillov, A. Kuniba and T. Nakanishi, *Skew Young diagram method in spectral decomposition of integrable lattice models*, Comm. Math. Phys. **185** (1997), 441–465.

[226] A. N. Kirillov, A. Kuniba and T. Nakanishi, *Skew Young diagram method in spectral decomposition of integrable lattice models. II. Higher levels*, Nuclear Phys. B **529** (1998), 611–638.

[227] A. N. Kirillov and N. Yu. Reshetikhin, *Yangians, Bethe ansatz and combinatorics*, Lett. Math. Phys. **12** (1986) 199–208.

[228] A. N. Kirillov and N. Yu. Reshetikhin, *The Bethe ansatz and the combinatorics of Young tableaux*, J. Soviet Math. **41** (1988), 925–955.

[229] A. N. Kirillov and N. Yu. Reshetikhin, *Representations of Yangians and multiplicities of the inclusion of the irreducible components of the tensor product of representations of simple Lie algebras*, J. Soviet Math. **52** (1990), 3156–3164.

[230] N. Kitanine, J.-M. Maillet and V. Terras, *Form factors of the XXZ Heisenberg spin-$\frac{1}{2}$ finite chain*, Nucl. Phys. B **554** (1999), 647–678.

[231] M. Kleber, *Combinatorial structure of finite-dimensional representations of Yangians: the simply-laced case*, Internat. Math. Res. Notices **7** (1997), 187–201.

[232] A. Klimyk and K. Schmüdgen, *Quantum groups and their representations*, Springer, Berlin, 1997.

[233] H. Knight, *Spectra of tensor products of finite-dimensional representations of Yangians*, J. Algebra **174** (1995), 187–196.

[234] F. Knop, *A Harish-Chandra homomorphism for reductive group actions*, Ann. Math. **140** (1994), 253–288.

[235] F. Knop, *Symmetric and non-symmetric quantum Capelli polynomials*, Comment. Math. Helvet. **72** (1997), 84–100.

[236] A. Knutson and T. Tao, *Puzzles and (equivariant) cohomology of Grassmannians*, Duke Math. J. **119** (2003), 221–260.

[237] K. Koike and I. Terada, *Young-diagrammatic methods for the representation theory of the classical groups of type B_n, C_n, D_n*, J. Algebra **107** (1987), 466–511.

[238] H. Konno, *Free field representation of level-k Yangian double $\mathcal{D}Y(sl_2)_k$ and deformation of Wakimoto*, Lett. Math. Phys. **40** (1997), 321–336.

[239] M. Konvalinka, *Non-communtative Sylverster's determinantal identity*, preprint `arXiv:math/0703213`.

[240] T. H. Koornwinder and V. B. Kuznetsov, *Gauss hypergeometric function and quadratic R-matrix algebras*, St. Petersburg Math. J. **6** (1994), 161–184.

[241] V. E. Korepin, N. M. Bogoliubov and A. G. Izergin, *Quantum inverse scattering method and correlation functions*, Cambridge Monographs on Mathematical Physics, Cambridge University Press, Cambridge, 1993.

[242] D. Korotkin and H. Samtleben, *Yangian symmetry in integrable quantum gravity*, Nucl. Phys. B **527** (1998), 657–689.

[243] B. Kostant and S. Sahi, *The Capelli identity, tube domains and the generalized Laplace transform*, Adv. Math. **87** (1991), 71–92.

[244] B. Kostant and S. Sahi, *Jordan algebras and Capelli identities*, Invent. Math. **112** (1993), 657–664.

[245] V. Kreiman, *Equivariant Littlewood-Richardson tableaux*, preprint `arXiv:0706.3738`.

[246] D. Krob and B. Leclerc, *Minor identities for quasi-determinants and quantum determinants*, Comm. Math. Phys. **169** (1995), 1–23.

[247] P. P. Kulish and N. Yu. Reshetikhin, *Diagonalisation of* GL(N) *invariant transfer matrices and quantum N-wave system (Lee model)*, J. Phys. A **16** (1983), L591–L596.

[248] P. P. Kulish and N. Yu. Reshetikhin, *GL_3-invariant solutions of the Yang-Baxter equation*, J. Soviet Math. **34** (1986), 1948–1971.

[249] P. P. Kulish, N. Yu. Reshetikhin and E. K. Sklyanin, *Yang–Baxter equation and representation theory*, Lett. Math. Phys. **5** (1981), 393–403.

[250] P. P. Kulish and E. K. Sklyanin, *On the solutions of the Yang–Baxter equation*, J. Soviet Math. **19** (1982), 1596–1620.

[251] P. P. Kulish and E. K. Sklyanin, *Quantum spectral transform method: recent developments*, in: "Integrable Quantum Field Theories", Lecture Notes in Phys. **151**, Springer, Berlin, 1982, pp. 61–119.

[252] P. P. Kulish and E. K. Sklyanin, *Algebraic structures related to reflection equations*, J. Phys. A **25** (1992), 5963–5975.

[253] P. P. Kulish, R. Sasaki and G. Schwiebert, *Constant solutions of reflection equations and quantum groups*, J. Math. Phys. **34** (1993), 286–304.

[254] P. P. Kulish and A. A. Stolin, *Deformed Yangians and integrable models*, Czech. J. Phys. **47** (1997), 1207–1212.

[255] A. Kuniba and J. Suzuki, *Analytic Bethe ansatz for fundamental representations of Yangians*, Comm. Math. Phys. **173** (1995), 225–264.

[256] G. Kuperberg, *Symmetries of plane partitions and the permanent-determinant method*, J. Comb. Theory **A 68** (1994), 115–151.

[257] V. B. Kuznetsov, M. F. Jørgensen and P. L. Christiansen, *New boundary conditions for integrable lattices*, J. Phys. A **28** (1995), 4639–4654.

[258] V. Lakshmibai, C. Musili and C. S. Seshadri, *Geometry of G/P. IV. Standard monomial theory for classical types*, Proc. Indian Acad. Sci. Sect. A Math. Sci. **88** (1979), 279–362.

[259] A. Lascoux, *Notes on interpolation in one and several variables*, Lectures at Tianjin University, June 1996; `http://www-igm.univ-mlv.fr/~al/pub_engl.html`

[260] A. Lauve, *Flag varieties for the Yangian* $\mathrm{Y}(\mathfrak{gl}_n)$, preprint `arXiv:math/0601056`.

[261] A. LeClair and F. Smirnov, *Infinite quantum group symmetry of fields in massive 2D quantum field theory*, Int. J. Mod. Phys. A **7** (1992), 2997–3022.

[262] B. Leclerc, *A Littlewood-Richardson rule for evaluation representations of* $U_q(\widehat{\mathfrak{sl}}_n)$, Sém. Lothar. Combin. **50** (2003/04), Art. B50e, 12 pp.

[263] B. Leclerc, M. Nazarov and J.-Y. Thibon, *Induced representations of affine Hecke algebras and the canonical bases for quantum groups*, in: "Studies in Memory of Issai Schur", Progress in Mathematics **210**, Birkhaüser, Basel, 2003, pp. 115–153.

[264] R. Leduc and A. Ram, *A ribbon Hopf algebra approach to the irreducible representations of centralizer algebras: the Brauer, Birman–Wenzl, and type A Iwahori–Hecke algebras*, Adv. Math. **125** (1997), 1–94.

[265] S. T. Lee, K. Nishiyama and A. Wachi, *Intersection of harmonics and Capelli identities for symmetric pairs*, preprint `arXiv:math/0510033`.

[266] F. Lemire and J. Patera, *Formal analytic continuation of Gelfand's finite-dimensional representations of* $gl(n, \mathbb{C})$, J. Math. Phys. **20** (1979), 820–829.

[267] G. Letzter, *Coideal subalgebras and quantum symmetric pairs*, in: "New Directions in Hopf Algebras", Math. Sci. Res. Inst. Publ. **43**, Cambridge Univ. Press, Cambridge, 2002, pp. 117–165.

[268] S. Z. Levendorskiĭ, *On PBW bases for Yangians*, Lett. Math. Phys. **27** (1993), 37–42.

[269] S. Z. Levendorskiĭ, *On generators and defining relations of Yangians*, J. Geom. Phys. **12** (1993), 1–11.

[270] S. Z. Levendorskiĭ and A. Sudbery, *Yangian construction of the Virasoro algebra*, Lett. Math. Phys. **37** (1996), 243–247.

[271] A. Liguori, M. Mintchev and L. Zhao, *Boundary exchange algebras and scattering on the half line*, Comm. Math. Phys. **194** (1998), 569–589.

[272] P. Littelmann, *An algorithm to compute bases and representation matrices for SL_{n+1}-representations*, J. Pure Appl. Algebra **117/118** (1997), 447–468.

[273] P. Littelmann, *Cones, crystals, and patterns*, Transform. Groups **3** (1998), 145–179.

[274] G. Lusztig, *Quantum deformations of certain simple modules over enveloping algeras*, Adv. Math. **70** (1988), 237–249.

[275] G. Lusztig, *Affine Hecke algebras and their graded version*, J. Amer. Math. Soc. **2** (1989), 599–635.

[276] G. Lusztig, *Finite-dimensional Hopf algebras arising from quantized universal enveloping algebras*, J. Amer. Math. Soc. **3** (1990), 257–296.

[277] G. Lusztig, *Canonical bases arising from quantized enveloping algebras*, J. Amer. Math. Soc. **3** (1990), 447–498.

[278] G. Lusztig, *Introduction to quantum groups*, Progress in Mathematics, **110**, Birkhäuser Boston, Boston, 1993.

[279] I. G. Macdonald, *Schur functions: theme and variations*, in: "Actes 28-e Séminaire Lotharingien", pp. 5–39, Publ. I.R.M.A. Strasbourg, 1992, 498/S–27.

[280] I. G. Macdonald, *Symmetric Functions and Hall Polynomials*, Oxford University Press, Oxford, 1995.

[281] J. M. Maillet and J. Sanchez de Santos, *Drinfeld twists and algebraic Bethe ansatz*, in: "L. D. Faddeev's Seminar on Mathematical Physics", Amer. Math. Soc. Transl. **201**, Amer. Math. Soc., Providence, RI, 2000, pp. 137–178.

[282] J. M. Maillet and V. Terras, *On the quantum inverse scattering problem*, Nucl. Phys. B **575** (2000), 627–644.

[283] O. Mathieu, *Bases des représentations des groupes simples complexes (d'après Kashiwara, Lusztig, Ringel et al.)*, Sémin. Bourbaki, Vol. 1990/91, Astérisque no. 201–203, Exp. no. 743 (1992), 421–442.

[284] V. Mazorchuk, *Generalized Verma modules*, Mathematical Studies Monograph Series, 8, VNTL Publishers, L'viv, 2000.

[285] V. Mazorchuk, *On categories of Gelfand-Zetlin modules*, in: "Noncommutative Structures in Mathematics and Physics" (Kiev, 2000), pp. 299–307, NATO Sci. Ser. II Math. Phys. Chem., 22, Kluwer Acad. Publ., Dordrecht, 2001.

[286] V. Mazorchuk, *On Gelfand-Zetlin modules over orthogonal Lie algebras*, Algebra Colloq. **8** (2001), 345–360.

[287] V. Mazorchuk and L. Turowska, *On Gelfand-Zetlin modules over* $U_q(\mathrm{gl}_n)$, in: "Quantum Groups and Integrable Systems" (Prague, 1999), Czech. J. Phys. **50** (2000), 139–144.

[288] J. Mickelsson, *Lowering operators and the symplectic group*, Rep. Math. Phys. **3** (1972), 193–199.

[289] J. Mickelsson, *Step algebras of semisimple subalgebras of Lie algebras*, Rep. Math. Phys. **4** (1973), 307–318.

[290] J. Mickelsson, *Lowering operators for the reduction* $U(n) \downarrow SO(n)$, Rep. Math. Phys. **4** (1973), 319–332.

[291] M. Mintchev, E. Ragoucy, P. Sorba and Ph. Zaugg, *Yangian symmetry in the nonlinear Schrödinger hierarchy*, J. Phys. A **32** (1999), 5885–5900.

[292] M. Mintchev, E. Ragoucy and P. Sorba, *Spontaneous symmetry breaking in the* $gl(N)$*-NLS hierarchy on the half line*, J. Phys. A **34** (2001), 8345–8364.

[293] A. S. Mishchenko and A. T. Fomenko, *Euler equations on finite-dimensional Lie groups*, Izv. AN SSSR Ser. Math. **42** (1978), 396–415.

[294] A. I. Molev, *Representations of twisted Yangians*, Lett. Math. Phys. **26** (1992), 211–218.

[295] A. I. Molev, *Gelfand–Tsetlin basis for representations of Yangians*, Lett. Math. Phys. **30** (1994), 53–60.

[296] A. I. Molev, *Sklyanin determinant, Laplace operators and characteristic identities for classical Lie algebras*, J. Math. Phys. **36** (1995), 923-943.

[297] A. I. Molev, *Noncommutative symmetric functions and Laplace operators for classical Lie algebras*, Lett. Math. Phys. **35** (1995), 135-143.

[298] A. I. Molev, *Yangians and Laplace operators for classical Lie algebras*, in: "Confronting the Infinite. Proceedings of the Conference in Celebration of the 70th Years of H. S. Green and C. A. Hurst", World Scientific, Singapore, 1995, pp. 239–245.

[299] A. I. Molev, *Casimir elements for certain polynomial current Lie algebras*, in: "Group 21, Physical Applications and Mathematical Aspects of Geometry, Groups, and Algebras", Vol. 1 (H.-D. Doebner, W. Scherer, P. Nattermann, Eds.), World Scientific, Singapore, 1997, pp. 172–176.

[300] A. I. Molev, *Stirling partitions of the symmetric group and Laplace operators for the orthogonal Lie algebra*, Discrete Math. **180** (1998), 281–300.

[301] A. Molev, *Factorial supersymmetric Schur functions and super Capelli identities*, in: "Kirillov's Seminar on Representation Theory" (G. I. Olshanski, Ed.), Amer. Math. Soc. Transl. **181**, AMS, Providence, RI, 1998, pp. 109–137.

[302] A. I. Molev, *Finite-dimensional irreducible representations of twisted Yangians*, J. Math. Phys. **39** (1998), 5559–5600.

[303] A. I. Molev, *A basis for representations of symplectic Lie algebras*, Comm. Math. Phys. **201** (1999), 591–618.

[304] A. I. Molev, *A weight basis for representations of even orthogonal Lie algebras*, in: "Combinatorial Methods in Representation Theory (Kyoto, 1998)", Adv. Studies in Pure Math. **28**, Kinokuniya, Tokyo, 2000, pp. 223–242.

[305] A. I. Molev, *Weight bases of Gelfand–Tsetlin type for representations of classical Lie algebras*, J. Phys. A **33** (2000), 4143–4168.

[306] A. I. Molev, *Irreducibility criterion for tensor products of Yangian evaluation modules*, Duke Math. J. **112** (2002), 307–341.

[307] A. I. Molev, *Yangians and transvector algebras*, Discrete Math. **246** (2002), 231–253.

[308] A. I. Molev, *Yangians and their applications*, in: "Handbook of Algebra", Vol. 3 (M. Hazewinkel, Ed.), Elsevier, 2003, pp. 907–959.

[309] A. I. Molev, *A new quantum analog of the Brauer algebra*, Czech. J. Phys. **53** (2003), 1073–1078.

[310] A. I. Molev, *Gelfand-Tsetlin bases for classical Lie algebras*, in: "Handbook of Algebra", Vol. 4 (M. Hazewinkel, Ed.), Elsevier, 2006, pp. 109–170.

[311] A. I. Molev, *Skew representations of twisted Yangians*, Selecta Math. **12** (2006), 1–38.

[312] A. I. Molev, *Representations of the twisted quantized enveloping algebra of type C_n*, Moscow Math. J. **6** (2006), 531–551.

[313] A. I. Molev, *On the matrix units for the symmetric group*, preprint `arXiv:math/0612207`.

[314] A. I. Molev, *Littlewood-Richardson polynomials*, preprint `arXiv:0704.0065`.

[315] A. Molev and M. Nazarov, *Capelli identities for classical Lie algebras*, Math. Ann. **313** (1999), 315–357.

[316] A. Molev, M. Nazarov and G. Olshanski, *Yangians and classical Lie algebras*, Russian Math. Surveys **51** (1996), 205–282.

[317] A. Molev and G. Olshanski, *Centralizer construction for twisted Yangians*, Selecta Math. **6** (2000), 269–317.

[318] A. Molev and G. Olshanski, *Degenerate affine Hecke algebras and centralizer construction for the symmetric groups*, J. Alg. **237** (2001), 302–341.

[319] A. Molev and E. Ragoucy, *Representations of reflection algebras*, Rev. Math. Phys. **14** (2002), 317–342.

[320] A. Molev and E. Ragoucy, *Symmetries and invariants of twisted quantum algebras and associated Poisson algebras*, preprint `arXiv:math/0701902`.

[321] A. Molev, E. Ragoucy and P. Sorba, *Coideal subalgebras in quantum affine algebras*, Rev. Math. Phys. **15** (2003), 789–822.

[322] A. Molev and V. Retakh, *Quasideterminants and Casimir elements for the general linear Lie superalgebra*, Internat. Math. Res. Notices (2004), 611–619.

[323] A. I. Molev and B. E. Sagan, *A Littlewood-Richardson rule for factorial Schur functions*, Trans. Amer. Math. Soc. **351** (1999), 4429–4443.

[324] A. I. Molev, V. N. Tolstoy and R. B. Zhang, *On irreducibility of tensor products of evaluation modules for the quantum affine algebra*, J. Phys. A **37** (2004), 2385–2399.

[325] A. I. Mudrov, *Irreducibility of fusion modules over twisted Yangians at generic point*, preprint `arXiv:math/0612738`.

[326] A. I. Mudrov, *Reflection equation and twisted Yangians*, preprint `arXiv:math/0612737`.

[327] E. Mukhin, V. Tarasov and A. Varchenko, *Bethe eigenvectors of higher transfer matrices*, J. Stat. Mech. Theory Exp. 2006, no. 8, P08002.

[328] E. Mukhin, V. Tarasov and A. Varchenko, *Generating operator of XXX or Gaudin transfer matrices has quasi-exponential kernel*, Symmetry, Integrability and Geometry: Methods and Applications **3** (2007), paper 060, 31 pp.

[329] E. Mukhin, V. Tarasov and A. Varchenko, *A generalization of the Capelli identity*, preprint `arXiv:math/0610799`.

[330] S. Murakami and F. Göhmann, *Yangian symmetry and quantum inverse scattering method for the one-dimensional Hubbard model*, Phys. Lett. A **227** (1997), 216–226.

[331] G. E. Murphy, *A new construction of Young's seminormal representation of the symmetric group*, J. Algebra **69** (1981), 287–291.

[332] G. E. Murphy, *On the representation theory of the symmetric groups and associated Hecke algebras*, J. Algebra **152** (1992), 492–513.

[333] J. G. Nagel and M. Moshinsky, *Operators that lower or raise the irreducible vector spaces of* U_{n-1} *contained in an irreducible vector space of* U_n, J. Math. Phys. **6** (1965), 682–694.

[334] W. Nakai and T. Nakanishi, *Paths, tableaux and q-characters of quantum affine algebras: the* C_n *case*, J. Phys. A **39** (2006), 2083–2115.

[335] H. Nakajima, *Quiver varieties and finite-dimensional representations of quantum affine algebras*, J. Amer. Math. Soc. **14** (2001), 145–238.

[336] H. Nakajima, *Quiver varieties and t-analogs of q-characters of quantum affine algebras*, Ann. of Math. (2) **160** (2004), 1057–1097.

[337] T. Nakanishi, *Fusion, mass, and representation theory of the Yangian algebra*, Nucl. Phys. B **439** (1995), 441–460.

[338] M. L. Nazarov, *Quantum Berezinian and the classical Capelli identity*, Lett. Math. Phys. **21** (1991), 123–131.

[339] M. L. Nazarov, *Yangians of the "strange" Lie superalgebras*, in: "Quantum Groups (Leningrad, 1990)", Lecture Notes in Math. **1510**, Springer, Berlin, 1992, pp. 90–97.

[340] M. Nazarov, *Capelli identities for Lie superalgebras*, Ann. Sci. École Norm. Sup. (4) **30** (1997), 847–872.

[341] M. Nazarov, *Yangians and Capelli identities*, in: "Kirillov's Seminar on Representation Theory" (G. I. Olshanski, Ed.), Amer. Math. Soc. Transl. **181**, Amer. Math. Soc., Providence, RI, 1998, pp. 139–163.

[342] M. Nazarov, *Yangian of the queer Lie superalgebra*, Comm. Math. Phys. **208** (1999), 195–223.

[343] M. Nazarov, *Capelli elements in the classical universal enveloping algebras*, in: "Combinatorial methods in representation theory (Kyoto, 1998)", Adv. Stud. Pure Math. **28**, Kinokuniya, Tokyo, 2000, pp. 261–285.

[344] M. Nazarov, *Representations of Yangians associated with skew Young diagrams*, in: "International Congress of Mathematicians (Beijing, 2002)", Higher Education Press, Beijing, 2002, Vol. II, pp. 643–654.

[345] M. Nazarov, *Representations of twisted Yangians associated with skew Young diagrams*, Selecta Math. (N.S.) **10** (2004), 71-129.

[346] M. Nazarov, *Rational representations of Yangians associated with skew Young diagrams*, Math. Z. **247** (2004), 21–63.

[347] M. Nazarov, *A mixed hook-length formula for affine Hecke algebras*, European J. Combin. **25** (2004), 1345–1376.

[348] M. Nazarov and G. Olshanski, *Bethe subalgebras in twisted Yangians*, Comm. Math. Phys. **178** (1996), 483–506.

[349] M. Nazarov and A. Sergeev, *Centralizer construction of the Yangian of the queer Lie superalgebra*, in: "Studies in Lie Theory", Progr. Math., 243, Birkhaüser Boston, Boston, MA, 2006, pp. 417–441.

[350] M. Nazarov and V. Tarasov, *Yangians and Gelfand–Zetlin bases*, Publ. Res. Inst. Math. Sci. Kyoto Univ. **30** (1994), 459–478.

[351] M. Nazarov and V. Tarasov, *Representations of Yangians with Gelfand–Zetlin bases*, J. Reine Angew. Math. **496** (1998), 181–212.

[352] M. Nazarov and V. Tarasov, *On irreducibility of tensor products of Yangian modules*, Internat. Math. Res. Notices (1998), 125–150.

[353] M. Nazarov and V. Tarasov, *On irreducibility of tensor products of Yangian modules associated with skew Young diagrams*, Duke Math. J. **112** (2002), 343–378.

[354] J. E. Nelson and T. Regge, $2+1$ *quantum gravity*, Phys. Lett. B **272** (1991), 213–216.

[355] J. E. Nelson and T. Regge, $2+1$ *gravity for genus* > 1, Comm. Math. Phys. **141** (1991), 211–223.

[356] J. E. Nelson and T. Regge, *Invariants of* $2+1$ *gravity*, Comm. Math. Phys. **155** (1993), 561–568.

[357] M. Noumi, *Macdonald's symmetric polynomials as zonal spherical functions on quantum homogeneous spaces*, Adv. Math. **123** (1996), 16–77.

[358] M. Noumi and T. Sugitani, *Quantum symmetric spaces and related q-orthogonal polynomials*, in: "Group Theoretical Methods in Physics (Toyonaka, 1994)", World Scientific, River Edge, NJ, 1995, pp. 28–40.

[359] M. Noumi, T. Umeda and M. Wakayama, *A quantum analogue of the Capelli identity and an elementary differential calculus on* $GL_q(n)$, Duke Math. J. **76** (1994), 567–594.

[360] M. Noumi, T. Umeda and M. Wakayama, *A quantum dual pair* $(\mathfrak{sl}_2, \mathfrak{o}_n)$ *and the associated Capelli identity*, Lett. Math. Phys. **34** (1995), 1–8.

[361] M. Noumi, T. Umeda and M. Wakayama, *Dual pairs, spherical harmonics and a Capelli identity in quantum group theory*, Compositio Math. **104** (1996), 227–277.

[362] G. Ochiai, *A Capelli-type identity associated with the dual pair* (O_n, Sp_{2m}), Master's Thesis, Kyoto University, 1996 (Japanese).

[363] O. Ogievetsky and P. Pyatov, *Orthogonal and symplectic quantum matrix algebras and Cayley-Hamilton theorem for them*, preprint `arXiv:math/0511618`.

[364] E. Ogievetsky, N. Reshetikhin and P. Wiegmann, *The principal chiral field in two dimensions on classical Lie algebras*, Nucl. Phys. B **280** (1987), 45–96.

[365] A. Okounkov, *Quantum immanants and higher Capelli identities*, Transform. Groups **1** (1996), 99–126.

[366] A. Okounkov, *Young basis, Wick formula, and higher Capelli identities*, Int. Math. Research Not. (1996), 817–839.

[367] A. Okounkov, *Multiplicities and Newton polytopes*, in: "Kirillov's Seminar on Representation Theory" (G. Olshanski, Ed.), Amer. Math. Soc. Transl. **181**, AMS, Providence, RI, 1998, pp. 231–244.

[368] A. Okounkov and G. Olshanski, *Shifted Schur functions*, St. Petersburg Math. J. **9** (1998), 239–300.

[369] A. Okounkov and G. Olshanski, *Shifted Schur functions II. Binomial formula for characters of classical groups and applications*, in: "Kirillov's Seminar on Representation Theory" (G. Olshanski, Ed.), Amer. Math. Soc. Transl. **181**, AMS, Providence, RI, 1998, pp. 245–271.

[370] A. Okounkov and A. Vershik, *A new approach to representation theory of symmetric groups*, Selecta Math. (N.S.) **2** (1996), 581–605.

[371] G. I. Olshanski, *Description of unitary representations with highest weight for the groups* $U(p,q)^{\sim}$, Funct. Anal. Appl. **14** (1981), 190–200.

[372] G. I. Olshanski, *Extension of the algebra* $U(g)$ *for infinite-dimensional classical Lie algebras* g, *and the Yangians* $Y(gl(m))$, Soviet Math. Dokl. **36** (1988), 569–573.

[373] G. I. Olshanski, *Yangians and universal enveloping algebras*, J. Soviet Math. **47** (1989), 2466–2473.

[374] G. I. Olshanski, *Representations of infinite-dimensional classical groups, limits of enveloping algebras, and Yangians*, in: "Topics in Representation Theory" (A. A. Kirillov, Ed.), Advances in Soviet Math. **2**, AMS, Providence, RI, 1991, pp. 1–66.

[375] G. I. Olshanski, *Irreducible unitary representations of the groups* $U(p,q)$ *sustaining passage to the limit as* $q \to \infty$, Zapiski Nauchn. Semin. LOMI, vol. 172 (1989), 114–120 (Russian); English transl. in: J. Soviet Math. **59** (1992), 1102–1107.

[376] G. Olshanski, *Twisted Yangians and infinite-dimensional classical Lie algebras*, in: "Quantum Groups (Leningrad, 1990)", Lecture Notes in Math. **1510**, Springer, Berlin, 1992, pp. 103–120.

[377] G. Olshanski, *Generalized symmetrization in enveloping algebras*, Transform. Groups **2** (1997), 197–213.

[378] U. Ottoson, *A classification of the unitary irreducible representations of* $\mathrm{SO}_0\,(N,\,1)$, Comm. Math. Phys. **8** (1968), 228–244.

[379] U. Ottoson, *A classification of the unitary irreducible representations of* $\mathrm{SU}(N,\,1)$, Comm. Math. Phys. **10** (1968), 114–131.

[380] S. Ovsienko, *Finiteness statements for Gelfand–Tsetlin modules*, in: "Algebraic Structures and Their Applications", Math. Inst., Kiev, 2002.

[381] T. D. Palev, *Irreducible finite-dimensional representations of the Lie superalgebra* $gl(n|1)$ *in a Gel'fand-Zetlin basis*, J. Math. Phys. **30** (1989), 1433–1442.

[382] T. D. Palev, *Essentially typical representations of the Lie superalgebras $gl(n/m)$ in a Gel'fand-Zetlin basis*, Funct. Anal. Appl. **23** (1989), 141–142.
[383] T. D. Palev, *Highest weight irreducible unitary representations of the Lie algebras of infinite matrices. I. The algebra $gl(\infty)$*, J. Math. Phys. **31** (1990), 579–586.
[384] T. D. Palev, N. I. Stoilova and J. van der Jeugt, *Finite-dimensional representations of the quantum superalgebra $U_q[gl(n/m)]$ and related q-identities*, Comm. Math. Phys. **166** (1994), 367–378.
[385] T. D. Palev and V. N. Tolstoy, *Finite-dimensional irreducible representations of the quantum superalgebra $U_q[\mathrm{gl}(n/1)]$*, Comm. Math. Phys. **141** (1991), 549–558.
[386] S. C. Pang and K. T. Hecht, *Lowering and raising operators for the orthogonal group in the chain $O(n) \supset O(n-1) \supset \cdots$, and their graphs*. J. Math. Phys. **8** (1967), 1233–1251.
[387] A. M. Perelomov and V. S. Popov, *Casimir operators for $U(n)$ and $SU(n)$*, Soviet J. Nucl. Phys. **3** (1966), 676–680.
[388] A. M. Perelomov and V. S. Popov, *Casimir operators for the orthogonal and symplectic groups*, Soviet J. Nucl. Phys. **3** (1966), 819–824.
[389] A. M. Perelomov and V. S. Popov, *Casimir operators for semisimple Lie algebras*, Izv. AN SSSR Ser. Mat. **32** (1968), 1368–1390.
[390] H. Pfeiffer, *Factorizing twists and the universal R-matrix of the Yangian* $\mathrm{Y}(\mathfrak{sl}_2)$, J. Phys. A **33** (2000), 8929–8951.
[391] R. Proctor, *Representations of $\mathfrak{sl}(2, C)$ on posets and the Sperner property*, SIAM J. Algebraic Discrete Methods **3** (1982), 275–280.
[392] R. Proctor, *Bruhat lattices, plane partition generating functions, and minuscule representations*, European J. Combin. **5** (1984), 331–350.
[393] R. Proctor, *Odd symplectic groups*, Invent. Math. **92** (1988), 307–332.
[394] R. Proctor, *Solution of a Sperner conjecture of Stanley with a construction of Gel'fand*, J. Combin. Theory Ser. A **54** (1990), 225–234.
[395] R. Proctor, *Young tableaux, Gelfand patterns, and branching rules for classical groups*, J. Algebra **164** (1994), 299–360.
[396] E. Ragoucy, *Twisted Yangians and folded $\mathcal{W}$-algebras*, Internat. J. Modern Phys. A **16** (2001), 2411–2433.
[397] E. Ragoucy and P. Sorba, *Yangians and finite $\mathcal{W}$-algebras*, Czech. J. Phys. **48** (1998), 1483–1487.
[398] E. Ragoucy and P. Sorba, *Yangian realisations from finite $\mathcal{W}$-algebras*, Comm. Math. Phys. **203** (1999), 551–572.
[399] M. Rais and P. Tauvel, *Indice et polynômes invariants pour certaines algèbres de Lie*, J. Reine Angew. Math. **425** (1992), 123–140.
[400] A. Ram, *Seminormal representations of Weyl groups and Iwahori-Hecke algebras*, Proc. London Math. Soc. **75** (1997), 99–133.
[401] N. Yu. Reshetikhin, *Integrable models of quantum one-dimensional magnets with $O(n)$ and $Sp(2k)$-symmetry*, Theoret. Math. Phys. **63** (1985), 555–569.
[402] N. Yu. Reshetikhin and M. A. Semenov-Tian-Shansky, *Central extensions of quantum current groups*, Lett. Math. Phys. **19** (1990), 133–142.
[403] N. Yu. Reshetikhin, L. A. Takhtajan and L. D. Faddeev, *Quantization of Lie Groups and Lie algebras*, Leningrad Math. J. **1** (1990), 193–225.
[404] V. Retakh and A. Zelevinsky, *Base affine space and canonical basis in irreducible representations of $Sp(4)$*, Dokl. Acad. Nauk USSR **300** (1988), 31–35.
[405] J. Rogawski, *On modules over the Hecke algebra of a p-adic group*, Invent. Math. **79** (1985), 443–465.
[406] M. Rosso, *Finite-dimensional representations of the quantum analog of the enveloping algebra of a complex simple Lie algebra*, Comm. Math. Phys. **177** (1988), 581–593.
[407] N. Rozhkovskaya, *Commutativity of quantum family algebras*, Lett. Math. Phys. **63** (2003), 87–103.
[408] N. Rozhkovskaya, *Braided central elements*, preprint `arXiv:math/0510226`.
[409] L. G. Rybnikov, *The shift of invariants method and the Gaudin model*, Funct. Anal. Appl. **40** (2006), 188–199.
[410] L. G. Rybnikov, *Uniqueness of higher Gaudin hamiltonians*, preprint `arXiv:math/0608588`.
[411] B. E. Sagan, *The symmetric group. Representations, combinatorial algorithms, and symmetric functions*, 2nd edition, Grad. Texts in Math., 203, Springer-Verlag, New York, 2001.

[412] S. Sahi, *The spectrum of certain invariant differential operators associated to a Hermitian symmetric space*, in: "Lie Theory and Geometry" (J.-L. Brylinski, R. Brylinski, V. Guillemin, V. Kac, Eds.), Progress in Math. **123**, Birkhäuser, Boston, 1994, pp. 569–576.

[413] K. Schoutens, *Yangian symmetry in conformal field theory*, Phys. Lett. B **331** (1994), 335–341.

[414] M. A. Semenov-Tian-Shansky, *Quantum and classical integrable systems*, in: "Integrability of Nonlinear Systems" (Pondicherry, 1996), pp. 314–377, Lect. Notes Phys., **495**, Springer, Berlin, 1997.

[415] A. V. Shchepetilov, *Two-body quantum mechanical problem on spheres*, J. Phys. A **39** (2006), 4011–4046.

[416] V. V. Shtepin, *Intermediate Lie algebras and their finite-dimensional representations*, Russian Akad. Sci. Izv. Math. **43** (1994), 559–579.

[417] E. K. Sklyanin, *Quantum version of the method of inverse scattering problem*, J. Soviet. Math. **19** (1982), 1546–1596.

[418] E. K. Sklyanin, *Boundary conditions for integrable quantum systems*, J. Phys. **A21** (1988), 2375–2389.

[419] E. K. Sklyanin, *Quantum inverse scattering method. Selected topics*, in: "Quantum Groups and Quantum Integrable Systems", pp. 63–97, Nankai Lectures Math. Phys., World Sci. Publ., River Edge, NJ, 1992.

[420] E. K. Sklyanin, *Separation of variables in the quantum integrable models related to the Yangian $Y[sl(3)]$*, J. Math. Sci. **80** (1996), 1861–1871.

[421] Yu. F. Smirnov and V. N. Tolstoy, *A new projected basis in the theory of five-dimensional quasi-spin*, Rept. Math. Phys. **4** (1973), 97–111.

[422] Yu. F. Smirnov and V. N. Tolstoy, *Extremal projectors for usual, super and quantum algebras and their use for solving Yang-Baxter problem*, in: "Selected Topics in Mathematical Physics", pp. 347–359, World Scientific, Teaneck, NJ, 1990.

[423] A. Solov'ev, *Cartan-Weyl basis for Yangian double $DY(sl_3)$*, Theoret. Math. Phys. **111** (1997), 731–743.

[424] J. R. Stembridge, *On minuscule representations, plane partitions and involutions in complex Lie groups*, Duke Math. J. **73** (1994), 469–490.

[425] A. Stolin, *On rational solutions of Yang-Baxter equation for $\mathfrak{sl}(n)$*, Math. Scand. **69** (1991), 57–80.

[426] A. Stolin, *On rational solutions of Yang-Baxter equations. Maximal orders in loop algebra*, Comm. Math. Phys. **141** (1991), 533–548.

[427] A. Stolin and P. Kulish, *New rational solutions of Yang-Baxter equation and deformed Yangians*, Czech. J. Phys. **47** (1997), 123–129.

[428] V. A. Stukopin, *Yangians of Lie superalgebras of type $A(m,n)$*, Funct. Anal. Appl. **28** (1994), 217–219.

[429] V. A. Stukopin, *Yangians of classical lie superalgebras: basic constructions, quantum double and universal R-matrix*, Proceedings of the Institute of Mathematics of NAS of Ukraine **50** (2004), 1195–1201.

[430] V. Stukopin, *Yangian of the strange Lie superalgebra of Q_{n-1} type, Drinfeld approach*, Symmetry, Integrability and Geometry: Methods and Applications **3** (2007), paper 069, 12 pp.

[431] T. Suzuki, *Rogawski's conjecture on the Jantzen filtration for the degenerate affine Hecke algebra of type A*, Electronic J. Representation Theory **2** (1998), 393–409.

[432] T. Suzuki, *Representations of degenerate affine Hecke algebra and $\mathfrak{gl}_n$*, Adv. Stud. Pure Math. **28** (2000), 343–372.

[433] K. Takemura, *The Yangian symmetry in the spin Calogero model and its applications*, J. Phys. A **30** (1997), 6185–6204.

[434] K. Takemura and D. Uglov, *The orthogonal eigenbasis and norms of eigenvectors in the Spin Calogero-Sutherland Model*, J. Phys. A **30** (1997), 3685–3717.

[435] L. A. Takhtadzhan and L. D. Faddeev, *Quantum method of the inverse problem and the Heisenberg XYZ-model*, Russian Math. Surv. **34** (1979), no. 5, 11–68.

[436] D. V. Talalaev, *The quantum Gaudin system*, Funct. Anal. Appl. **40** (2006), 73–77.

[437] V. O. Tarasov, *Structure of quantum L-operators for the R-matrix of the XXZ-model*, Theor. Math. Phys. **61** (1984), 1065–1071.

[438] V. O. Tarasov, *Irreducible monodromy matrices for the R-matrix of the XXZ-model and lattice local quantum Hamiltonians*, Theor. Math. Phys. **63** (1985), 440–454.

[439] V. O. Tarasov, *Cyclic monodromy matrices for* $sl(n)$ *trigonometric R-matrices*, Comm. Math. Phys. **158** (1993), 459–484.

[440] V. Tarasov and A. Varchenko, *Difference equations compatible with trigonometric KZ differential equations*, Internat. Math. Res. Notices (2000), 801–829.

[441] V. Tarasov and A. Varchenko, *Duality for Knizhnik-Zamolodchikov and dynamical equations*, Acta Appl. Math. **73** (2002), 141–154.

[442] V. Tarasov and A. Varchenko, *Combinatorial formulae for nested Bethe vectors*, preprint arXiv:math/0702277.

[443] V. N. Tolstoy, *Extremal projectors for contragredient Lie algebras and superalgebras of finite growth*, Russ. Math. Surveys **44** (1989), 257–258.

[444] V. N. Tolstoy, *Extremal projectors for quantized Kac-Moody superalgebras and some of their applications*, in: "Quantum Groups" (Clausthal, 1989), pp. 118–125, Lect. Notes Phys., **370**, Springer, Berlin, 1990.

[445] V. N. Tolstoy, *Connection between Yangians and quantum affine algebras*, in: "New Symmetries in the Theories of Fundamental Interactions" (Karpacz, 1996), PWN, Warsaw, 1997, pp. 99–117.

[446] V. N. Tolstoy, *Drinfeldians*, in: "Lie Theory and Its Applications in Physics II" (H.-D. Doebner, V. K. Dobrev and J. Hilgert, Eds.), World Scientific, Singapore, 1998, pp. 325–337.

[447] V. N. Tolstoy and J. P. Draayer, *New appoach in theory of Clebsch-Gordan coefficients for* $u(n)$ *and* $U_q(u(n))$, Czech. J. Phys. **50** (2000), 1359–1370.

[448] V. N. Tolstoy, I. F. Istomina and Yu. F. Smirnov, *The Gel'fand-Tseĭtlin basis for the Lie superalgebra* $\mathrm{gl}(n/m)$, in: "Group Theoretical Methods in Physics", Vol. I (Yurmala, 1985), VNU Sci. Press, Utrecht, 1986, pp. 337–348.

[449] K. Ueno, T. Takebayashi and Y. Shibukawa, *Gelfand–Zetlin basis for* $U_q(gl(N+1))$*-modules*, Lett. Math. Phys. **18** (1989), 215–221.

[450] M. Ugaglia, *On a Poisson structure on the space of Stokes matrices*, Internat. Math. Res. Notices **6** (1999), 473–493.

[451] D. Uglov, *Symmetric functions and the Yangian decomposition of the Fock and basic modules of the affine Lie algebra* $\widehat{\mathfrak{sl}}_N$, in: "Quantum Many-Body Problems and Representation Theory", Math. Soc. Japan Mem. **1**, Math. Soc. Japan, Tokyo, 1998, pp. 183–241.

[452] D. Uglov, *Yangian Gelfand-Zetlin bases,* $\mathfrak{gl}_N$*-Jack polynomials and computation of dynamical correlation functions in the spin Calogero-Sutherland model*, Comm. Math. Phys. **191** (1998), 663–696.

[453] D. Uglov, *Skew Schur functions and Yangian actions on irreducible integrable modules of* $\widehat{\mathfrak{gl}}_N$, Ann. Comb. **4** (2000), 383–400.

[454] D. B. Uglov and V. E. Korepin, *The Yangian symmetry of the Hubbard model*, Phys. Lett. A **190** (1994), 238–242.

[455] T. Umeda, *Newton's formula for* $\mathfrak{gl}_n$, Proc. Amer. Math. Soc. **126** (1998), 3169–3175.

[456] T. Umeda, *The Capelli identities, a century after*, in: "Selected Papers on Harmonic Analysis, Groups, and Invariants", pp. 51–78, Amer. Math. Soc. Transl. Ser. 2, 183, AMS, Providence, RI, 1998.

[457] T. Umeda and M. Wakayama, *Another look at the differential operators on quantum matrix spaces and its applications*, Comment. Math. Univ. St. Paul. **47** (1998), 53–80.

[458] A. van den Hombergh, *A note on Mickelsson's step algebra*, Indag. Math. **37** (1975), 42–47.

[459] M. Varagnolo, *Quiver varieties and Yangians*, Lett. Math. Phys. **53** (2000), 273–283.

[460] M. Varagnolo and E. Vasserot, *Standard modules of quantum affine algebras*, Duke Math. J. **111** (2002), 509–533.

[461] E. Vasserot, *Affine quantum groups and equivariant K-theory*, Transform. Groups **3** (1998), 269–299.

[462] E. B. Vinberg, *On certain commutative subalgebras of a universal enveloping algebra*, Izv. AN SSSR Ser. Math. **54** (1990), 3–25.

[463] A. Wachi, *Central elements in the universal enveloping algebras for the split realization of the orthogonal Lie algebras*, Lett. Math. Phys. **77** (2006), 155–168.

[464] H. Weyl, *Classical groups, their invariants and representations*, Princeton Univ. Press, Princeton NJ, 1946.

[465] N. J. Wildberger, *A combinatorial construction for simply-laced Lie algebras*, Adv. in Appl. Math. **30** (2003), 385–396.

[466] N. J. Wildberger, *A combinatorial construction of G_2*, J. Lie Theory **13** (2003), 155–165.

[467] N. J. Wildberger, *Minuscule posets from neighbourly graph sequences*, European J. Combin. **24** (2003), no. 6, 741–757.

[468] M. K. F. Wong, *Representations of the orthogonal group. I. Lowering and raising operators of the orthogonal group and matrix elements of the generators.* J. Math. Phys. **8** (1967), 1899–1911.

[469] M. K. F. Wong and H.-Y. Yeh, *The most degenerate irreducible representations of the symplectic group*, J. Math. Phys. **21** (1980), 630–635.

[470] N. Xi, *Special bases of irreducible modules of the quantized universal enveloping algebra* $U_v(\mathrm{gl}(n))$, J. Algebra **154** (1993), 377–386.

[471] Ping Xu, *Dirac submanifolds and Poisson involutions*, Ann. Sci. École Norm. Sup. **36** (2003), 403–430.

[472] A. B. Zamolodchikov and Al. B. Zamolodchikov, *Factorized S-matrices in two dimensions as the exact solutions of certain relativistic quantum field models*, Ann. Phys. **120** (1979), 253–291.

[473] A. Zelevinsky, *The p-adic analogue of the Kazhdan-Lusztig conjecture*, Funct. Anal. Appl. **15** (1981), 83–92.

[474] A. Zelevinsky, *Resolvents, dual pairs, and character formulas*, Funct. Anal. Appl. **21** (1987), 152–154.

[475] R. B. Zhang, *Representations of super Yangian*, J. Math. Phys. **36** (1995), 3854–3865.

[476] R. B. Zhang, *The quantum super-Yangian and Casimir operators of* $U_q(\mathrm{gl}(M|N))$, Lett. Math. Phys. **33** (1995), 263–272.

[477] R. B. Zhang, *The* $\mathrm{gl}(M|N)$ *super Yangian and its finite-dimensional representations*, Lett. Math. Phys. **37** (1996), 419–434.

[478] Y.-Z. Zhang, *Super-Yangian double and its central extension*, Phys. Lett. A **234** (1997), 20–26.

[479] R. B. Zhang, M. D. Gould and A. J. Bracken, *Generalized Gel'fand invariants of quantum groups*, J. Phys. A **24** (1991), 937–943.

[480] D. P. Zhelobenko, *The classical groups. Spectral analysis of their finite-dimensional representations*, Russ. Math. Surv. **17** (1962), 1–94.

[481] D. P. Želobenko, *Compact Lie groups and their representations*, Transl. of Math. Monographs **40**, AMS, Providence, RI, 1973.

[482] D. P. Zhelobenko, *S-algebras and Verma modules over reductive Lie algebras*, Soviet. Math. Dokl. **28** (1983), 696–700.

[483] D. P. Zhelobenko, *Z-algebras over reductive Lie algebras*, Soviet. Math. Dokl. **28** (1983), 777–781.

[484] D. P. Zhelobenko, *On Gelfand–Zetlin bases for classical Lie algebras*, in: "Representations of Lie Groups and Lie Algebras" (A. A. Kirillov, Ed.), pp. 79–106, Budapest: Akademiai Kiado, 1985.

[485] D. P. Zhelobenko, *Extremal projectors and generalized Mickelsson algebras on reductive Lie algebras*, Math. USSR-Izv. **33** (1989), 85–100.

[486] D. P. Zhelobenko, *An introduction to the theory of S-algebras over reductive Lie algebras*, in: "Representations of Lie Groups and Related Topics" (A. M. Vershik and D. P. Zhelobenko, Eds.), Adv. Studies in Contemp. Math. **7**, New York, Gordon and Breach Science Publishers, 1990, pp. 155–221.

[487] D. P. Zhelobenko, *Representations of reductive Lie algebras*, VO "Nauka", Moscow, 1994 (Russian).

Index

Titles in This Series

143 **Alexander Molev,** Yangians and classical Lie algebras, 2007
142 **Joseph A. Wolf,** Harmonic analysis on commutative spaces, 2007
141 **Vladimir Maz′ya and Gunther Schmidt,** Approximate approximations, 2007
140 **Elisabetta Barletta, Sorin Dragomir, and Krishan L. Duggal,** Foliations in Cauchy-Riemann geometry, 2007
139 **Michael Tsfasman, Serge Vlăduţ, and Dmitry Nogin,** Algebraic geometric codes: Basic notions, 2007
138 **Kehe Zhu,** Operator theory in function spaces, 2007
137 **Mikhail G. Katz,** Systolic geometry and topology, 2007
136 **Jean-Michel Coron,** Control and nonlinearity, 2007
135 **Bennett Chow, Sun-Chin Chu, David Glickenstein, Christine Guenther, James Isenberg, Tom Ivey, Dan Knopf, Peng Lu, Feng Luo, and Lei Ni,** The Ricci flow: Techniques and applications, Part I: Geometric aspects, 2007
134 **Dana P. Williams,** Crossed products of C^*-algebras, 2007
133 **Andrew Knightly and Charles Li,** Traces of Hecke operators, 2006
132 **J. P. May and J. Sigurdsson,** Parametrized homotopy theory, 2006
131 **Jin Feng and Thomas G. Kurtz,** Large deviations for stochastic processes, 2006
130 **Qing Han and Jia-Xing Hong,** Isometric embedding of Riemannian manifolds in Euclidean spaces, 2006
129 **William M. Singer,** Steenrod squares in spectral sequences, 2006
128 **Athanassios S. Fokas, Alexander R. Its, Andrei A. Kapaev, and Victor Yu. Novokshenov,** Painlevé transcendents, 2006
127 **Nikolai Chernov and Roberto Markarian,** Chaotic billiards, 2006
126 **Sen-Zhong Huang,** Gradient inequalities, 2006
125 **Joseph A. Cima, Alec L. Matheson, and William T. Ross,** The Cauchy Transform, 2006
124 **Ido Efrat, Editor,** Valuations, orderings, and Milnor K-Theory, 2006
123 **Barbara Fantechi, Lothar Göttsche, Luc Illusie, Steven L. Kleiman, Nitin Nitsure, and Angelo Vistoli,** Fundamental algebraic geometry: Grothendieck's FGA explained, 2005
122 **Antonio Giambruno and Mikhail Zaicev, Editors,** Polynomial identities and asymptotic methods, 2005
121 **Anton Zettl,** Sturm-Liouville theory, 2005
120 **Barry Simon,** Trace ideals and their applications, 2005
119 **Tian Ma and Shouhong Wang,** Geometric theory of incompressible flows with applications to fluid dynamics, 2005
118 **Alexandru Buium,** Arithmetic differential equations, 2005
117 **Volodymyr Nekrashevych,** Self-similar groups, 2005
116 **Alexander Koldobsky,** Fourier analysis in convex geometry, 2005
115 **Carlos Julio Moreno,** Advanced analytic number theory: L-functions, 2005
114 **Gregory F. Lawler,** Conformally invariant processes in the plane, 2005
113 **William G. Dwyer, Philip S. Hirschhorn, Daniel M. Kan, and Jeffrey H. Smith,** Homotopy limit functors on model categories and homotopical categories, 2004
112 **Michael Aschbacher and Stephen D. Smith,** The classification of quasithin groups II. Main theorems: The classification of simple QTKE-groups, 2004
111 **Michael Aschbacher and Stephen D. Smith,** The classification of quasithin groups I. Structure of strongly quasithin K-groups, 2004
110 **Bennett Chow and Dan Knopf,** The Ricci flow: An introduction, 2004

For a complete list of titles in this series, visit the
AMS Bookstore at **www.ams.org/bookstore/**.